D1537930

Prefixes for Powers of 10*

Multiple	Prefix	Abbreviation
10^{24}	yotta	Y
10^{21}	zetta	Z
10^{18}	exa	E
10^{15}	peta	P
10^{12}	tera	T
10^{9}	giga	G
10^{6}	mega	M
10^{3}	kilo	k
10^{2}	hecto	h
10^{1}	deka	da
10^{-1}	deci	d
10^{-2}	centi	c
10^{-3}	milli	m
10^{-6}	micro	μ
10^{-9}	nano	n
10^{-12}	pico	p
10^{-15}	femto	f
10^{-18}	atto	a
10^{-21}	zepto	z
10^{-24}	yocto	y

* Commonly used prefixes are in red. All prefixes are pronounced with the accent on the first syllable.

The Greek Alphabet

alpha	A	α	Nu	N	ν
beta	B	β	Xi	Ξ	ξ
gamma	Γ	γ	Omicron	O	o
delta	Δ	δ	Pi	Π	π
epsilon	E	ϵ, ε	Rho	P	ρ
zeta	Z	ζ	Sigma	Σ	σ
eta	H	η	Tau	T	τ
theta	Θ	θ	Upsilon	Y	υ
iota	I	ι	Phi	Φ	ϕ
kappa	K	κ	Chi	X	χ
lambda	Λ	λ	Psi	Ψ	ψ
mu	M	μ	Omega	Ω	ω

Terrestrial and Astronomical Data*

acceleration of gravity at Earth's surface	g	9.80 m/s^2
radius of Earth R_E	R_E	$6.38 \times 10^6 \text{ m}$
mass of Earth	M_E	$5.98 \times 10^{24} \text{ kg}$
mass of the Sun		$1.99 \times 10^{30} \text{ kg}$
mass of the Moon		$7.35 \times 10^{22} \text{ kg}$
escape speed at Earth's surface		$11.2 \text{ km/s} = 6.96 \text{ mi/s}$
standard temperature and pressure (STP)		$0°C = 273.15 \text{ K}$ $1 \text{ atm} = 101.3 \text{ kPa}$
Earth–Moon distance†		$3.84 \times 10^8 \text{ m} = 2.39 \times 10^5 \text{ mi}$
Earth–Sun distance (mean)†		$1.50 \times 10^{11} \text{ m} = 9.32 \times 10^7 \text{ mi}$
speed of sound in dry air (20°C, 1 atm)		344 m/s
density of dry air (STP)		1.29 kg/m^3
density of dry air (20°C, 1 atm)		1.20 kg/m^3
density of water (4°C, 1 atm)		1000 kg/m^3
latent heat of fusion of water (0°C, 1 atm)	L_F	334 kJ/kg
latent heat of vaporization of water (100°C, 1 atm)	L_V	2260 kJ/kg

* Additional data on the solar system can be found in Appendix B and at http://nssdc.gsfc.nasa.gov/planetary/planetfact.html.
† Center to center.

Mathematical Symbols

$=$	is equal to
$\equiv$	is defined by
$\neq$	is not equal to
$\approx$	is approximately equal to
$\sim$	is of the order of
$\propto$	is proportional to
$>$	is greater than
$\geq$	is greater than or equal to
$>>$	is much greater than
$<$	is less than
$\leq$	is less than or equal to
$<<$	is much less than
Δx	change in x
$\|x\|$	absolute value of x
$n!$	$n(n-1)(n-2)...1$
Σ	sum
$\Delta t \to 0$	Δt approaches zero

Abbreviations for Units

A	ampere	h	hour	N	newton
Å	angstrom (10^{-10} m)	Hz	hertz	nm	nanometer (10^{-9} m)
atm	atmosphere	in.	inch	Pa	pascal
BTU	British thermal unit	J	joule	rad	radians
Bq	becquerel	K	kelvin	rev	revolution
C	coulomb	kg	kilogram	R	roentgen
°C	degree Celsius	km	kilometer	Sv	sievert
cal	calorie	keV	kilo-electron volt	s	second
Ci	curie	lb	pound	T	tesla
cm	centimeter	L	liter	u	unified mass unit
eV	electron volt	m	meter	V	volt
°F	degree Fahrenheit	MeV	mega-electron volt	W	watt
fm	femtometer, fermi (10^{-15} m)	mi	mile	Wb	weber
ft	foot	min	minute	y	year
G	gauss	mm	millimeter	μm	micrometer (10^{-6} m)
Gy	gray	mmHg	millimeters of mercury	μs	microsecond
g	gram	mol	mole	μC	microcoulomb
H	henry	ms	millisecond	Ω	ohm

Some Conversion Factors

Length
1 m = 39.37 in. = 3.281 ft = 1.094 yard
1 m = 10^{15} fm = 10^{10} Å = 10^9 nm
1 km = 0.6214 mi
1 mi = 5280 ft = 1.609 km
1 light-year = 1 $c \cdot y$ = 9.461 × 10^{15} m
1 in. = 2.540 cm

Volume
1 L = 10^3 cm^3 = 10^{-3} m^3 = 1.057 qt

Time
1 h = 3600 s = 3.6 ks
1 y = 365.24 day = 3.156 × 10^7 s

Speed
1 km/h = 0.278 m/s = 0.6214 mi/h
1 ft/s = 0.3048 m/s = 0.6818 mi/h

Angle–angular speed
1 rev = 2π rad = 360°
1 rad = 57.30°
1 rev/min (rpm) = 0.1047 rad/s

Force–pressure
1 N = 10^5 dyn = 0.2248 lb
1 lb = 4.448 N
1 atm = 101.3 kPa = 1.013 bar = 760 mmHg = 14.70 lb/in.^{2}

Mass
1 u = [(10^{-3} mol^{-1})/N_A] kg = 1.661 × 10^{-27} kg
1 tonne = 10^3 kg = 1 Mg
1 kg = 2.205 lb

Energy–power
1 J = 10^7 erg = 0.7376 ft $\cdot$ lb = 9.869 × 10^{-3} L $\cdot$ atm
1 kW $\cdot$ h = 3.6 MJ
1 cal = 4.186 J
1 L $\cdot$ atm = 101.325 J = 24.22 cal
1 eV = 1.602 × 10^{-19} J
1 BTU = 778 ft $\cdot$ lb = 252 cal = 1054 J
1 horsepower = 550 ft $\cdot$ lb/s = 746 W

Thermal conductivity
1 W/(m $\cdot$ K) = 6.938 BTU $\cdot$ in./(h $\cdot$ ft^2 $\cdot$ °F)

Magnetic field
1 T = 10^4 G

Viscosity
1 Pa $\cdot$ s = 10 poise

COLLEGE PHYSICS

COLLEGE PHYSICS

Volume II

Roger A. Freedman
Todd G. Ruskell
Philip R. Kesten
David L. Tauck

W. H. FREEMAN AND COMPANY
New York

Publisher: *Jessica Fiorillo*
Development Editor: *Blythe Robbins*
Acquisitions Editor: *Alicia Brady*
Associate Director of Marketing: *Debbie Clare*
Senior Market Development Manager: *Kirsten Watrud*
Media Acquisitions Editor: *Dave Quinn*
Media Producer: *Jenny Chiu*
Associate Editor: *Heidi Bamatter*
Assistant Editor: *Courtney Lyons*
Editorial Assistant: *Tue Tran*
Marketing Assistant: *Samantha Zimbler*
Senior Project Editor: *Georgia Lee Hadler*
Copy Editor: *Betty Pessagno*
Photo Editor: *Ted Szczepanski*
Photo Researcher: *Christina Micek*
Text and Cover Designer: *Blake Logan*
Illustration Coordinator: *Janice Donnola*
Illustrations: *Precision Graphics*
Production Coordinator: *Julia DeRosa*
Composition: *cMPreparé*
Printing and Binding: *Quad Graphics*

Cover photo by Kim Taylor/Hotspot Media
Barn Owl (Tyto alba) landing. Three images at 20 millisecond intervals.

Library of Congress Control Number: 2013938909

Volume II:
ISBN-10: 1-4641-0201-5
ISBN-13: 978-1-4641-0201-1

© 2014 by W. H. Freeman and Company
All rights reserved

Printed in the United States of America
First printing

MCAT® is a registered trademark of the Association of American Medical Colleges. MCAT material included is printed with permission of the AAMC. Additional duplication is prohibited without written permission of the AAMC. The appearance of MCAT material in this book does not constitute sponsorship or endorsement by the AAMC.

W. H. Freeman and Company
41 Madison Avenue
New York, NY 10010
Houndmills, Basingstoke RG21 6XS, England
www.whfreeman.com

Roger:

To the memory of S/Sgt Ann Kazmierczak Freedman,
WAC and Pvt. Richard Freedman, AUS

Todd:

I dedicate this book to Susan and Allison, whose never-ending patience,
love, and support made it possible. And to my parents, from whom I learned
so much—especially my father, who so effectively demonstrates what it means
to be an effective teacher both in and out of the classroom.

Phil:

To my parents, for instilling in me a love of learning, to my wife for her
unconditional support, and to my children for letting their kooky dad
infuse so much of their lives with science.

Dave:

To my parents, Bill and Jean, for showing me how to lead a wonderful life,
and to my sister and friends, teachers and students for helping me do it.

Contents in Brief

Contents

Biological Applications

Unique and fully integrated physiological and biological applications are found throughout the text and are indicated by an owl icon. Below is a list of select biological applications organized by chapter section for easy reference.

Biological Applications

A UNIQUE AUTHOR TEAM

This ground-breaking text boasts an exceptionally strong writing team that is uniquely qualified to write a college physics textbook. The *College Physics* author team is led by **Roger Freedman**, accomplished textbook author of such bestselling titles as *Universe* (W. H. Freeman), *Investigating Astronomy* (W. H. Freeman), and *University Physics* (Pearson). Dr. Freedman is a Lecturer in Physics at the University of California, Santa Barbara. He was an undergraduate at the University of California campuses in San Diego and Los Angeles, and did his doctoral research in theoretical nuclear physics at Stanford University. He came to UCSB in 1981 after three years of teaching and doing research at the University of Washington. At UCSB, Dr. Freedman has taught in both the Department of Physics and the College of Creative Studies, a branch of the university intended for highly gifted and motivated undergraduates. In recent years, he has helped to develop computer-based tools for learning introductory physics and astronomy and has been a pioneer in the use of classroom response systems and the "flipped" classroom model at UCSB. Roger holds a commercial pilot's license and was an early organizer of the San Diego Comic-Con, now the world's largest popular culture convention.

As a Teaching Professor of Physics at the Colorado School of Mines, **Todd Ruskell** focuses on teaching at the introductory level, and continually develops more effective ways to help students learn. One method used in large enrollment introductory courses is Studio Physics. This collaborative, hands-on environment helps students develop better intuition about, and conceptual models of, physical phenomena through an active learning approach. Dr. Ruskell brings his experience in improving students' conceptual understanding to the text, as well as a strong liberal arts perspective. Dr. Ruskell's love of physics began with a B.A. in physics from Lawrence University in Appleton, Wisconsin. He went on to receive an M.S. and Ph.D. in optical sciences from the University of Arizona. He has received awards for teaching excellence, including Colorado School of Mines' Alumni Teaching Award. Dr. Ruskell currently serves on the physics panel and advisory board for the NANSLO (North American Network of Science Labs Online) project.

Philip Kesten, Associate Professor of Physics and Associate Provost for Residential Learning Communities at Santa Clara University, holds a B.S. in physics from the Massachusetts Institute of Technology and received his Ph.D. in high-energy particle physics from the University of Michigan. Since joining the Santa Clara faculty in 1990, Dr. Kesten has also served as Chair of Physics, Faculty Director of the ATOM and da Vinci Residential Learning Communities, and Director of the Ricard Memorial Observatory. He has received awards for teaching excellence and curriculum innovation, was Santa Clara's Faculty Development Professor for 2004–2005, and was named the California Professor of the Year in 2005 by the Carnegie Foundation for the Advancement of Education. Dr. Kesten is co-founder of Docutek (A SirsiDynix Company), an Internet software company, and has served as the Senior Editor for *Modern Dad*, a newsstand magazine.

Unlike any other physics text on the market, this project includes a physiologist as primary author. **David Tauck**, Associate Professor of Biology, holds both a B.A. in biology and an M.A. in Spanish from Middlebury College. He earned his Ph.D. in physiology at Duke University and completed postdoctoral fellowships at Stanford University and Harvard University in anesthesia and neuroscience, respectively. Since joining the Santa Clara University faculty in 1987, he has served as Chair of the Biology Department, the College Committee on Rank and Tenure, and the Institutional Animal Care and Use Committee; he has also served as president of the local chapter of Phi Beta Kappa. Dr. Tauck currently serves as the Faculty Director in Residence of the da Vinci Residential Learning Community.

Dear Students and Instructors,

Welcome to *College Physics*! We are excited to bring you this innovative text. No other college physics text presents material in quite this way. Our unique author team includes an experienced and highly successful textbook author, physicists who have spent years focusing on how students learn physics best, and even a biology professor who brings his perspective on what makes physics interesting to the students who take this course.

Our innovations help students master concepts and succeed in developing and practicing the problem-solving skills they need to do well in this course. The visual impact of this text is something totally new: Word balloons throughout the text highlight key physics concepts in figures and equations so that students can learn the concepts in easy-to-manage pieces.

In addition, numerous features in the text are also designed to help students succeed in this course. Look for the **Chapter Goals** outlined at the beginning of each chapter, **Watch Out!** boxes that address misconceptions; **Got the Concept?** questions that test students' understanding of the material; **Take-Home Messages** that directly link to the chapter goals and help students focus on the important concepts presented in each chapter section; **Examples** that are broken down into key steps with an easy-to-follow Set-Up, Solve, Reflect structure; **Questions and Problems** at the end of each chapter; and we even have a special **MCAT Appendix** to help pre-med students prepare for this important test.

Our aim is to instill in students a deeper appreciation of physics—by showing them how it connects to their lives and their future careers, but also by helping them succeed in the course. We hope you enjoy exploring and using this text!

Best Regards,

Roger, Todd, Phil, Dave

Preface

TEXT FEATURES

College Physics provides instructors and students with a fresh approach to the algebra-based physics course. With art designed to teach and a focus on real-life biological and medical applications, *College Physics* provides students with a deeper understanding of physics principles and a greater appreciation for why physics is important to their future work in the life sciences.

A Focus on Developing Problem-Solving Skills

Many textbooks encourage memorization of equations over comprehension of concepts, allowing students to look for shortcuts in doing—and therefore learning—physics. To better align physics concepts with problem solving, *College Physics* incorporates student-friendly pedagogy by presenting worked examples in a model that emphasizes reasoning and analysis. Presented in a flexible format that easily demonstrates text, equations, and figures, the worked examples use a common procedure throughout to help build students' confidence when approaching problem solving. This procedure, summarized by the key phrases "Set Up," "Solve," and "Reflect," mirrors the approach scientists take in solving problems:

Set Up. The first step in each problem is to determine an overall approach and to gather the necessary pieces of information needed to solve it. These might include sketches, equations related to the physics, and concepts. The authors rely heavily on diagrams since they can be extraordinarily beneficial in helping students understand how to approach a problem.

Solve. Rather than simply summarizing the mathematical manipulations required to move from first principles to the final answer, the authors show many intermediate steps in working out solutions to the sample problems, highlighting a crucial part of the problem-solving process that is otherwise often overlooked. While many other texts only include figures in the Set Up portion of the worked example, the authors use sketches throughout each section whenever they are helpful in clarifying the physics for students.

Reflect. An important part of the process of solving a problem is to reflect on the meaning, implications, and validity of the answer. Is it physically reasonable? Do the units make sense? Is there a deeper or wider understanding that can be drawn from the result? The authors address these and related questions when appropriate.

Example 6-2 Work Done by Actin

In order to fertilize eggs, the sperm of the horseshoe crab (Figure 6-3) must penetrate two protective layers of the egg with a combined thickness of about $40\ \mu m$. To achieve this, a bundle of the protein actin $60\ \mu m$ in length pushes from the outer surface of the sperm with a constant force of 1.9×10^{-9} N. How much work does the actin bundle do in this process?

Figure 6-3 **A horseshoe crab** Horseshoe crabs (family *Limulidae*) are similar to crustaceans, but are more closely related to spiders and scorpions. Like the cells of other animals and plants, horseshoe crab cells contain an important protein called actin.

Set Up
The force exerted is both constant and in the same direction as motion of the end of the bundle. Hence we can once again use Equation 6-1.

$$W = Fd \qquad (6\text{-}1)$$

Solve
The work done by the actin bundle equals the product of force and distance. The actin bundle has to push through the outer layers of the egg.

$$
\begin{aligned}
W &= Fd \\
&= (1.9 \times 10^{-9}\ \text{N})(40 \times 10^{-6}\ \text{m}) \\
&= 7.6 \times 10^{-14}\ \text{J}
\end{aligned}
$$

Reflect
The amount of work done by the actin bundle seems ridiculously small. But such a bundle is microscopic, with a mass of only about 10^{-16} kg. Hence the actin bundle does about 10^3 J of work per kilogram of mass. If a 70-kg person could do that much work on a per-kilogram basis on the book in Example 6-1, he or she could lift the book 3.5 km (nearly 12,000 ft)! An actin bundle is small, but it packs a big punch.

$$\frac{\text{Work done by actin bundle}}{\text{Mass of actin bundle}} = \frac{7.6 \times 10^{-14}\ \text{J}}{10^{-16}\ \text{kg}} = 10^3\ \text{J/kg}$$

If a 70-kg person could do that much work on a per-kilogram basis,

$$\text{Work done} = W = (10^3\ \text{J/kg})(70\ \text{kg}) = 7.0 \times 10^4\ \text{J}$$

Height d to which this amount of work could raise a book of mass 2.00 kg and weight 19.8 N:

$$W = mgd, \text{ so } d = \frac{W}{mg} = \frac{7.0 \times 10^4\ \text{J}}{19.8\ \text{N}} = 3.5 \times 10^3\ \text{m} = 3.5\ \text{km}$$

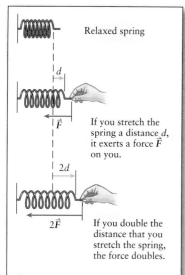

Relaxed spring

d

If you stretch the spring a distance d, it exerts a force $\vec{F}$ on you.

$|\vec{F}|$

$2d$

$2\vec{F}$

If you double the distance that you stretch the spring, the force doubles.

Figure 6-17 **Hooke's law** If you stretch an ideal spring, the force that it exerts on you is directly proportional to its extension.

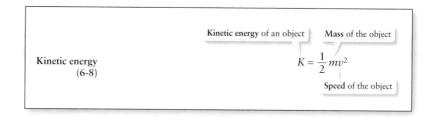

Kinetic energy of an object   Mass of the object

Kinetic energy
(6-8)

$$K = \frac{1}{2}mv^2$$

Speed of the object

Innovative Visual Features

This text incorporates groundbreaking visual features that bridge physics concepts to equations and figures throughout the book. In the form of "balloon art," the authors break down important equations and figures into smaller amounts of information.

The visual word balloons reinforce key ideas from the text and guide students to understanding the physics principles that are important to grasp in each chapter. The figures are simple, colorful, and approachable, inviting students to explore them rather than intimidating students into ignoring them.

A Seamless Blend of Physics and Biology

This book guides students through the fundamentals of introductory physics, weaving physiology and biomedical topics throughout the text and helping life science students discover the reasons why learning physics is important to their own fields of study. The authors' aim is to instill in students a deeper appreciation of physics by showing how it determines many characteristics of living systems. A complete list of biological applications can be found at the beginning of this Preface, and these applications are marked throughout the chapters with a brown owl icon like the one below. We offer an electronic biology appendix that professors can reference to learn more about the biology included in the text.

Figure 6-4 **Getting tired while doing zero work** Weights that you hold stationary in your outstretched arms undergo no displacement, so you do zero work on them. Why, then, do your arms get tired?

Muscles and Doing Work

Pick up a book or other heavy object and hold it in your hand at arm's length (Figure 6-4). After awhile you'll notice your arm getting tired: It feels like you're doing work to hold the object in midair. But Equation 6-1 says that you're doing *no* work on the book, because the book isn't moving (its displacement d is zero). So if you're not doing any work, why does your arm feel tired?

To see the explanation for this seeming paradox, notice that the muscles in your arm (and elsewhere in your body) exert forces on their ends by contracting. Skeletal muscles (those that control the motions of your arms, legs, fingers, toes, and other structures) are made up of bundles of muscle cells (Figure 6-5a). Muscle cells consist of bundles of myofibrils that are segmented into thousands of tiny structures called sarcomeres. Connected to each other, end to end, sarcomeres contain interdigitated filaments of the contractile proteins actin and myosin (Figure 6-5b). As the filaments slide past each other the sarcomeres can shorten, resulting in the muscle contracting; when the muscle relaxes, sarcomeres lengthen.

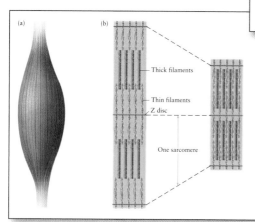

(a)   (b)

Thick filaments

Thin filaments

Z disc

One sarcomere

Figure 6-5 **Muscles and muscle function** (a) The muscles of your body are composed of many individual muscle fibers. (b) The internal structure of a muscle fiber when relaxed (left) and contracted (right).

LEARNING TOOLS

Carefully crafted to support students new to college-level physics, learning tools placed throughout the text guide students in developing the crucial conceptual foundation that they need to be successful in the course.

Outcome-Based Learning Objectives

To help outline assessment criteria and key goals throughout the chapter, the beginning of each chapter highlights outcome-based learning objectives that allow students to judge if they've reached the objective set forth for each chapter section. The listed learning objectives include both short-term (chapter sections) goals and long-term comprehensive (chapter) goals that highlight the important pedagogy that students should focus on throughout the text. The learning objectives of each chapter section are reinforced by Take-Home Messages at the end of each chapter section.

6

Work and Energy

In this chapter, your goals are to:
- (6-1) Explain the relationship between work and energy.
- (6-2) Calculate the work done by a constant force on an object moving in a straight line.
- (6-3) Describe what kinetic energy is and understand the work-energy theorem.
- (6-4) Apply the work-energy theorem to solve problems.
- (6-5) Recognize why the work-energy theorem applies even for curved paths and varying forces like the spring force.
- (6-6) Explain the meaning of potential energy and how conservative forces such as the gravitational force and the spring force give rise to gravitational potential energy and spring potential energy.
- (6-7) Explain the differences between, and when you can apply, the generalized law of conservation of energy and the conservation of total mechanical energy.
- (6-8) Identify which kinds of problems are best solved with energy conservation and the steps to follow in solving these problems.

To master this chapter, you should review:
- (2-5) The equations for constant-acceleration motion in a straight line.
- (4-2 and 4-5) Newton's second and third laws of motion.

What do you think?
Kangaroos use a bounding gait to move rapidly across the countryside. The large muscles in its hind legs help a kangaroo to start bounding. Which anatomical structure is most important for sustaining the kangaroo's bounding movement? (a) The kangaroo's knees; (b) its head; (c) its tendons; (d) its front legs.

Take-Home Messages

This boxed feature appears at the end of each chapter section and summarizes the main physics principles introduced in that section. These **Take-Home Messages** are directly tied to the outcome-based learning objectives listed at the beginning of each chapter, so that students can check for understanding before moving on to the next chapter section.

Special Attention to Common Misconceptions

Having taught physics for many years, the authors know which topics are prone to cause misunderstandings for students. So they created the "**Watch Out!**" boxed feature, which draws attention to these misconceptions and corrects them immediately as students read the text. By addressing these head on, students will have a deeper grasp of the physics presented throughout the text.

Conceptual Problems Built into the Flow of the Text

Health sciences and biological sciences students are used to a conceptual approach to both problem solving and learning in general. For that reason, the authors created a similar approach with the "**Got the Concept?**" questions that appear throughout the text. These questions almost never require numeric calculations but instead invite students to think through the implications or connections of a physics concept.

Take-Home Message for Section 6-2

✔ If a force acts on an object that undergoes a displacement, the force can do work on that object.

✔ For a constant force and straight-line displacement, the amount of work done is given by the displacement multiplied by the component of the the force parallel to that displacement.

✔ Whether the work done is positive, negative, or zero depends on the angle between the direction of the force and the direction of the displacement.

✔ If one object does negative work on a second object, the second object must do an equal amount of positive work on the first object.

! Watch Out! Displacement and distance are not the same thing.

The fundamental difference is that distance tells you "how far," while displacement tells you "how far and in what direction." Displacement is a vector. That's why displacements can be positive or negative, but distances are *always* positive. (If you were asked, "How far is it from New York to Boston?" you wouldn't answer "Negative 300 km.") In Figure 2-2 each swimmer travels the same distance, 50.0 m, but their displacements are different (+50.0 m for swimmer A, −50.0 m for swimmer B) because the *directions* of their motion are different.

? Got the Concept? 2-2 Average Velocity

It takes you 15 min to drive 6.0 mi in a straight line to the local hospital. It takes 10 min to go the last 3.0 mi, 2.0 min to go the last mile, and only 30 s (= 0.50 min) to go the last 0.50 mile. What is your average velocity for the trip? Take the positive x direction to be from your starting point toward the hospital.

End-of-Chapter Summary

The end-of-chapter summary incorporates the key concepts from each chapter into an easy-to-follow table. Each table identifies the important topics covered in the chapter and provides students with a quick summary of each concept, as well as the important equation(s) or figure(s) associated with the topic.

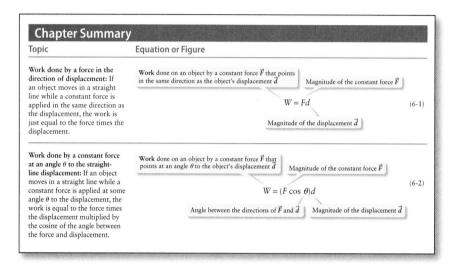

End-of-Chapter Problems

To reinforce the problem-solving skills taught in the chapters, the end-of-chapter problems incorporate conceptual questions and problems in the following format:

Conceptual Problems
Multiple-Choice Problems
Estimation/Numerical Analysis
Problems by Chapter Section
General Problems

Problems include three levels of difficulty: basic or single-concept problems (•), intermediate problems that may require synthesis of concepts and multiple steps (••), and challenging problems (•••). Problems are also designated when they pertain to certain topics, such as Biology, Medical, Sports, and Astronomy.

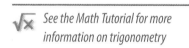

√x̄ *See the Math Tutorial for more information on trigonometry*

Math Tutorial

Students may find that it's been a while since they studied a particular math concept and they need a reminder. Margin notes link concepts in the text to the **Math Appendix** found in the back of the book. The Math Tutorial serves as a refresher and includes worked examples as well as practice problems.

Special MCAT Appendix

Since many students taking this course are preparing for a medical career, there is a special MCAT appendix in the back of the book to help students prepare for the MCAT test. This appendix includes actual MCAT problems provided by the Association of American Medical Colleges (AAMC) for students to practice with before taking the actual test.

MCAT® Appendix

The section that follows includes material from previously administered MCAT® items and is reprinted with permission of the Association of American Medical Colleges (AAMC).

Passage 13 (81–85)

Tennis balls must pass a rebound test before they can be certified for tournament play. To qualify, balls dropped from a given height must rebound within a specified range of heights. Measuring rebound height can be difficult because the ball is at its maximum height for only a brief time. It is possible to perform a simpler indirect measurement to calculate the height of rebound by measuring how long it takes the ball to rebound and hit the floor again. The diagram below illustrates the experimental setup used to make the measurement.

The ball is dropped from a height of 2.0 m, and it hits the floor and then rebounds to a height h. A microphone detects the sound of the ball each time it hits the floor, and a timer connected to the microphone measures the time (t) between the two impacts. The height of the rebound is $h = gt^2/8$ where $g = 9.8$ m/s^2 is the acceleration due to gravity. Care must be taken so that the measured times do not contain systematic error. Both the speed of sound and the time of impact of the ball with the floor must be considered. Four balls were tested using the method. The results are listed in the table below.

Ball	Time (s)
A	1.01
B	1.05
C	0.97
D	1.09

81. If NO air resistance is present, which of the following quantities remains constant while the ball is in the air between the first and second impacts?
A. Kinetic energy of the ball
B. Potential energy of the ball
C. Momentum of the ball
D. Horizontal speed of the ball

STUDENT ANCILLARY SUPPORT

Supplemental learning materials allow students to interact with concepts in a variety of scenarios. By analyzing figures, reinforcing problem-solving methods, reviewing chapter objectives, and reading through worked-out solutions, students obtain a practical understanding of the core concepts. With that in mind, W. H. Freeman has developed the most comprehensive student learning package—both printed and online—available today.

Printed Student Resources

- **Problem-Solving Guide with Solutions** by Timothy A. French, DePaul University
  Volume I (Chapters 1–15) ISBN: 1-4641-0137-X
  Volume II (Chapters 16–28) ISBN: 1-4641-0138-8

The *Problem-Solving Guide with Solutions* takes a unique approach to promoting students' problem-solving skills by providing detailed and annotated solutions to selected problems. Unlike other solutions manuals, this guide follows the "Set Up," "Solve," and "Reflect" format outlined in the Worked Examples in the text for worked-out solutions to selected odd-numbered end-of-chapter problems in the textbook. It also includes integrated media icons which point to selected problem-solving tools that can be accessed.

- **Student Workbook** by Jason Bryslawskyj, CUNY Graduate Center, CUNY Baruch College and Mark Kanner, CUNY Graduate Center, CUNY City College
  Volume I (Chapters 1–15) ISBN: 1-4641-4957-7
  Volume II (Chapters 16–28) ISBN: 1-4641-4958-5

This workbook helps students put concepts from the text into practice before applying them to homework problems. Short exercises highlight the main goals in each chapter section, allowing students to develop the skills they need to apply these concepts when problem solving. By reiterating the key concepts, problem-solving strategies, and learning objectives from the main text, the workbook allows students to build the confidence they need to successfully tackle homework problems.

Premium Media Resources

The *College Physics* Companion Website, www.whfreeman. com/collegephysics1e, provides a range of tools for problem solving and conceptual support. These Premium Media Resources can be purchased for a small fee. Select resources are also embedded in the Multimedia-Enhanced e-Book and WebAssign Premium.

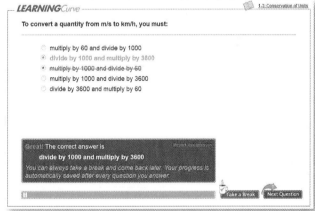

- **LearningCurve** incorporates adaptive question selection, personalized study plans, and state-of-the-art question analysis reports in activities with a game-like feel that keeps students engaged with the material. Integrated e-Book sections give students additional exposure to the course text. An innovative scoring system ensures that students who need more help with the material spend more time quizzing themselves than students who are already proficient. There is a LearningCurve activity available for each chapter in the book.

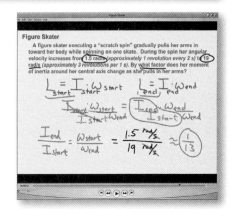

- **P'Casts** are videos that emulate the face-to-face experience of watching an instructor work a problem. Using a virtual whiteboard, the P'Casts' tutors demonstrate the steps involved in solving key worked examples, while explaining concepts along the way. The worked examples were chosen with the input of physics students and instructors across the country. There are 250 P'Casts total, all of which can be viewed online or downloaded to portable media devices.

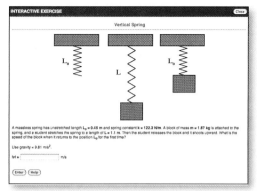

- **Interactive Exercises** are active learning, problem-solving activities. Each Interactive Exercise consists of a parent problem accompanied by a Socratic-dialog "help" sequence designed to encourage critical thinking as users do a guided conceptual analysis before attempting the mathematics. Immediate feedback for both correct and incorrect responses is provided through each problem-solving step. Over 75 Interactive Exercises are available.

- **Balloon Art Concept Checks** guide students through the process of identifying important physics concepts in key figures and equations. Based on the colorful unique "balloon art" featured in *College Physics*, these interactive questions reinforce key ideas from the text, highlight important physics principles they must grasp in each chapter, and invite students to explore physics in an exciting and colorful way.

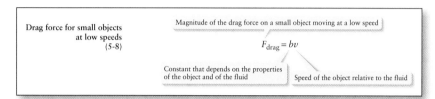

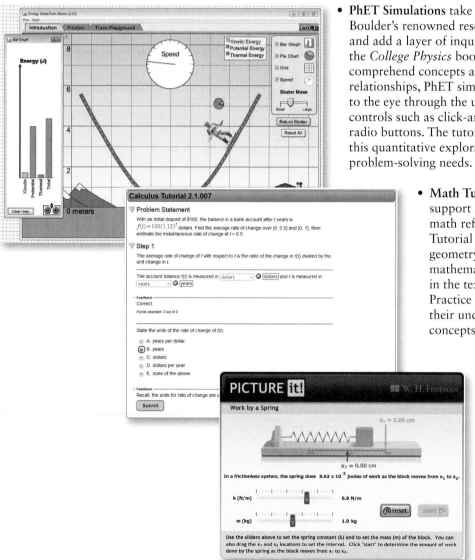

- **PhET Simulations** take the University of Colorado at Boulder's renowned research-based physics simulations and add a layer of inquiry-based questions unique to the *College Physics* book. To help students visually comprehend concepts and illustrate cause-and-effect relationships, PhET simulations animate what is invisible to the eye through the use of graphics and intuitive controls such as click-and-drag manipulation, sliders, and radio buttons. The tutorial questions further encourage this quantitative exploration, while addressing specific problem-solving needs.

- **Math Tutorial** offers improved mathematic support for students who may need a little math refresher. The comprehensive Math Tutorial reviews basic results of algebra, geometry, trigonometry, and calculus; links mathematical concepts to physics concepts in the text; and provides Examples and Practice Problems so students may check their understanding of mathematical concepts.

- **Picture Its** help bring static figures from the text to life. By manipulating variables within each animated figure, students visualize a variety of physics concepts. Approximately 50 activities are available.

- **Pocket Worked Examples** put the problem-solving worked examples in the palm of your hand. All Worked Examples from the *College Physics* text are available as downloadable PDF files, which can be printed or sent to your mobile device.

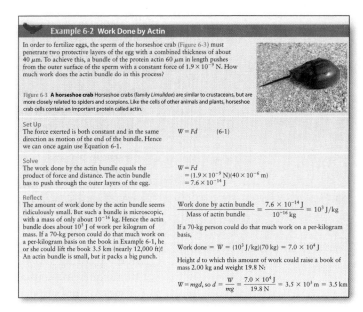

Example 6-2 Work Done by Actin

In order to fertilize eggs, the sperm of the horseshoe crab (Figure 6-3) must penetrate two protective layers of the egg with a combined thickness of about 40 μm. To achieve this, a bundle of the protein actin 60 μm in length pushes from the outer surface of the sperm with a constant force of 1.9×10^{-9} N. How much work does the actin bundle do in this process?

Figure 6-3 **A horseshoe crab** Horseshoe crabs (family *Limulidae*) are similar to crustaceans, but are more closely related to spiders and scorpions. Like the cells of other animals and plants, horseshoe crab cells contain an important protein called actin.

Set Up
The force exerted is both constant and in the same direction as motion of the end of the bundle. Hence we can once again use Equation 6-1.

$$W = Fd \qquad (6-1)$$

Solve
The work done by the actin bundle equals the product of force and distance. The actin bundle has to push through the outer layers of the egg.

$$W = Fd$$
$$= (1.9 \times 10^{-9} \text{ N})(40 \times 10^{-6} \text{ m})$$
$$= 7.6 \times 10^{-14} \text{ J}$$

Reflect
The amount of work done by the actin bundle seems ridiculously small. But such a bundle is microscopic, with a mass of only about 10^{-16} kg. Hence the actin bundle does about 10^3 J of work per kilogram of mass. If a 70-kg person could do that much work on a per-kilogram basis on the book in Example 6-1, he or she could lift the book 3.5 km (nearly 12,000 ft)! An actin bundle is small, but it packs a big punch.

$$\frac{\text{Work done by actin bundle}}{\text{Mass of actin bundle}} = \frac{7.6 \times 10^{-14} \text{ J}}{10^{-16} \text{ kg}} = 10^3 \text{ J/kg}$$

If a 70-kg person could do that much work on a per-kilogram basis,

$$\text{Work done} = W = (10^3 \text{ J/kg})(70 \text{ kg}) = 7.0 \times 10^4 \text{ J}$$

Height d to which this amount of work could raise a book of mass 2.00 kg and weight 19.8 N:

$$W = mgd, \text{ so } d = \frac{W}{mg} = \frac{7.0 \times 10^4 \text{ J}}{19.8 \text{ N}} = 3.5 \times 10^3 \text{ m} = 3.5 \text{ km}$$

- **PhysicsNews** from *Scientific American* provides an up-to-the-minute streaming feed of new physics-related stories direct from *Scientific American* magazine. Stay on top of the latest happenings in physics all in one easy place.

SCIENTIFIC AMERICAN™

ELECTRONIC TEXTBOOKS

For students interested in digital textbooks, W. H. Freeman offers the complete *College Physics* book in two easy-to-use formats.

The Multimedia-Enhanced e-Book

The Multimedia-Enhanced e-Book contains the complete text with a wealth of helpful interactive functions. All student Premium Resources are linked directly from the e-Book pages. Students are thus able to access supporting resources when they need them, taking advantage of the "teachable moment" as they read. Customization functions include instructor and student notes, highlighting, document linking, and editing capabilities. Access to the Multimedia-Enhanced e-Book can be purchased from the Book Companion Website.

The CourseSmart eTextbook

Though it does not include any Premium Resources, the CourseSmart eTextbook does provide the full digital text, along with tools to take notes, search, and highlight passages. A free app allows access to CourseSmart eTextbooks on Android and Apple devices, such as the iPad. They can also be downloaded to your computer and accessed without an Internet connection, removing any limitations for students when it comes to reading digital text. The CourseSmart eTextbook can be purchased at www.coursesmart.com.

ONLINE TUTORING

W. H. Freeman knows that learning physics can be hard and that sometimes students need some extra support. That's why we've partnered with NetTutor to bring students tutoring they can access anytime, anywhere.

NetTutor®

www.nettutor.com

NetTutor® is an online tutoring service that specializes in a customized tutoring experience. Using the Socratic Method, tutors guide students through problems or concepts, but never present answers. This approach develops critical thinking

skills and encourages students to persevere. Powered by an easy-to-use interactive interface, NetTutor® provides the most student-friendly online tutoring environment available, conducted by local tutors who really understand how to help today's students learn.

With NetTutor®, students get:

- Expert, U.S.-based tutors, ready to help your students with problems from your textbook and your assignments.
- Tutors who use questions and familiar concepts to guide students through the learning process.
- A virtual whiteboard, which enables students to interact with tutors as they work with you to solve problems, share information, and ask questions.
- The ability to print or save copies of each tutoring session for reference.
- The option for students to extend access via direct online purchase.

Assistance is always available!

Online tutoring seven days a week.
Offline question-and-answer 24 hours a day.
For more information about packaging tutoring access with your *College Physics* text, please contact your Publisher's Representative.

INSTRUCTOR RESOURCES

For instructors using *College Physics*, W. H. Freeman provides a complete suite of assessment tools and course materials for the taking.

Computerized Test Bank

ISBN: 1-4641-0133-7
The Test Bank offers over 2,000 multiple-choice questions, tackling both core physics concepts and various life-science applications. While the Test Bank is also available in downloadable Word files on the book companion website, the easy-to-use CD includes Windows and Macintosh versions of the widely used Diploma test generation software, allowing instructors to add, edit, and sequence questions to suit their testing needs.

Electronic Instructor Resources

Instructors can access valuable teaching tools through www.whfreeman.com/collegephysics1e. These password-protected resources are designed to enhance lecture presentations, and include Textbook Images (available in .JPEG and PowerPoint format), Clicker Questions, Lecture PowerPoint slides, Art PowerPoint slides, Instructor Solutions, Biology Appendix, and more.

Course Management System Cartridges

W. H. Freeman provides seamless integration of resources in your Course Management Systems. Four cartridges are available (Blackboard, Canvas, Desire2Learn, and Angel), and other select cartridges (Moodle, Sakai, etc.) can be produced upon request.

ONLINE HOMEWORK SYSTEMS

W. H. Freeman offers the widest variety of online homework options on the market.

www.saplinglearning.com

Sapling Learning provides highly effective interactive homework and instruction that improve student learning outcomes for the problem-solving disciplines. They offer an enjoyable teaching and effective learning experience that is distinctive in three important ways:

- **Targeted Instructional Content:** Sapling Learning increases student engagement and comprehension by delivering immediate feedback and targeted instructional content.
- **Performance Tracking:** Sapling Learning grades assignments, tracks student participation and progress, and compiles performance analytics—helping instructors save time and tailor assignments to address student needs.

- **Proven Results:** Independent university studies have shown Sapling Learning improves student performance by three-fourths to a full letter grade.
- **Unsurpassed Service and Support:** Sapling Learning makes teaching more enjoyable by providing a dedicated Master's- and Ph.D.-level colleague who provides software, course development, and consulting support throughout the semester. Our Tech TAs help instructors to:

  - Customize assignments by editing our existing questions or creating new questions that support the course, not just the text.
  - Choose point values and grading policies to help achieve educational goals, analyze class statistics, and apply the results to help students learn more efficiently.
  - Provide instructors with one-on-one training on use of Sapling Learning's online homework system and customization tools.
  - Resolve any issues that may arise with the software or online assignments so that instructors spend less time managing technology and more time with their students.

Sapling Learning is not tied to a specific textbook or edition, giving instructors freedom to customize Sapling to their syllabus and students the flexibility to choose more affordable used or rental textbooks.

www.webassign.com

For instructors interested in online homework management, WebAssign Premium features a time-tested secure online environment already used by millions of students worldwide. Featuring algorithmic problem generation and supported by a wealth of physics-specific learning tools, WebAssign Premium for *College Physics* presents instructors with a powerful assignment manager and student environment. WebAssign Premium provides the following resources:

- **Algorithmically generated problems:** Students receive homework problems containing unique values for computation, encouraging them to work out the problems on their own.
- **Complete access to the Multimedia-Enhanced e-Book** is available from a live table of contents, as well as from relevant problem statements.
- **Links to select Premium Multimedia Resources** (P'Casts, Interactive Exercises, and Picture Its) are provided as hints and feedback to ensure a clearer understanding of the problems and the concepts they reinforce.
- **Personal Study Plan** allowing students to review key prerequisite algebra concepts at their own pace.
- **Additional Resource Question Collections** from expert physics education researchers that have been thoroughly class-tested and researched are also available to add to your WebAssign course at no extra charge.

For budget-conscious students, a lower-priced, homework-only option is also available. This version does not include the e-Book or any of the Premium Multimedia Resources.

ACKNOWLEDGMENTS

Creating a first edition textbook requires the coordinated effort of an enormous number of talented professionals. We are grateful for the dedicated support of our in-house team at W. H. Freeman; thank you for transforming our concept into a beautiful book.

We especially want to thank our development editor, Blythe Robbins, for guiding us through the process of creating a textbook. We would also like to thank Associate Publisher Jessica Fiorillo for encouraging us and leading our editorial team, Media Acquisitions Editor Dave Quinn for producing gorgeous ancillaries and contributing significantly to the design of the book

and cover, and of course, we thank Associate Editor Heidi Bamatter for coordinating reviews, in addition to Assistant Editor Courtney Lyons and Editorial Assistant Tue Tran for their hard work on the project. Randi Rossignol and Susan Brennan made important contributions to the early stages of this project, and they deserve special thanks for motivating many of the unique features of this book. Special thanks also go to our skilled in-house production team, Julia DeRosa, Ted Szczepanski, Bill Page, Janice Donnola, Blake Logan, and Georgia Hadler, for their patience, dedication, and attention to detail. And of course, Senior Market Development Manager Kirsten

Watrud and Senior Marketing Manager Alicia Brady deserve a special thank you for providing us with key market data.

We are particularly grateful for the substantial contributions of Accuracy Checker Valerie Walters, Problems Editor Mark Hollabaugh, Solutions Manual Author Tim French, Workbook Authors Jason Bryslawskyj and Mark Kanner. Francesca Monaco of cMPreparé also deserves a special thank you for coordinating the compositor process.

Friends and Family

One of us (RAF) thanks his wife Caroline Robillard for her patience with the seemingly endless hours that went into preparing this textbook. I also thank my students at the University of California, Santa Barbara, for giving me the opportunity to test and refine new ideas for making physics more accessible.

One of us (TGR) thanks his wife Susan and daughter Allison for their limitless patience and understanding with the countless hours spent working on this book. I also thank my parents who showed me how to live a balanced life.

One of us (PRK) would like to acknowledge valuable and insightful conversations on physics and physics teaching with Richard Barber, John Birmingham, and J. Patrick Dishaw of Santa Clara University, and to offer these colleagues my gratitude. Finally, I offer my gratitude to my wife Kathy and my children Sam and Chloe for their unflagging support during the arduous process that led to the book you hold in your hands.

One of us (DLT) thanks his family and friends for accommodating my tight schedule during the years that writing this book consumed. I especially want to thank my parents, Bill and Jean, for their boundless encouragement and support, and for teaching me everything I've ever really needed to know; they've shown me how to live a good life, be happy, and age gently. I greatly appreciate my sister for encouraging me not to abandon a healthy lifestyle just to write a book. I also want to thank my nonbiological family, Holly and Geoff, for leading me to Sonoma County and for making Sebastopol feel like home.

Accuracy Review

We know that instructors have particular concerns about using first edition texts because they are prone to errors. We have done everything in our power to alleviate this concern in using our text by submitting chapters and end-of-chapter problem sets to several rounds of accuracy reviews and detailed error-checking, by the following:

Wayne R. Anderson
Sacramento City College

Marisa Bauza Roman
Drexel University

Kevin W. Cooper
Ohio University

Mark Hollabaugh
Normandale Community College

Dr. Guy Letteer
Sacred Heart Preparatory and Santa Clara University

Linghong Li
The University of Tennessee at Martin

Alan Meert
University of Pennsylvania

Dr. Valerie A. Walters
V Walters Consulting

Class Testers

We thank the faculty below and their students for class testing the text.

Miah Muhammad Adel, *University of Arkansas, Pine Bluff*
Jeremy Armstrong, *Winona State University*
Harold Bank, *Clayton State University*
Jeff J. Bechtold, *Austin Community College*
Michael Bates, *Moraine Valley Community College*
Bill Brandon, *University of North Carolina, Pembroke*
Allison Bruce, *El Paso Community College*
Andrew Cahoon, *Colby Sawyer College*
Zengjun Chen, *Tuskegee University*
Kathryn Devine, *The College of Idaho*
Carl Drake, *Jackson State University*
Diana Driscoll, *Case Western Reserve University*
Greg Falabella, *Wagner College*
Brian Geislinger, *Gadsden State Community College*
Zvonko Hlousek, *California State University, Long Beach*
Olenka Hubrickyj Cabot, *San Jose State University*
Charles Johnson, *South Georgia College*
Anthony Karmis, *University of California, Santa Barbara*
Elena Kuchina, *Thomas Nelson Community College*
Cecille Labuda, *University of Mississippi*
Shelly R. Lesher, *University of Wisconsin, La Crosse*
Susannah Lomant, *Georgia Perimeter College*
Ron MacTaylor, *Salem State University*
Firas Mansour, *University of Waterloo*
Maxim Marienko, *Hofstra University/Queens College CUNY*
Eric Martell, *Millikin University*
Karie A. Meyers, *Pima Community College*
Tamar More, *University of Portland*
Sandra J. Rhoades, *Kennesaw State University*
Bill Robinson, *North Carolina State University*
Michael Sampogna, *Pima Community College*
Tumer Sayman, *Eastern Michigan University*
Pete Schwart, *California Polytechnic State University*
Malav Shah, *Macon State College*
Peter Sheldon, *Randolph College*
Jason Shulman, *Richard Stockton College of New Jersey*
Glenn Spiczak, *University of Wisconsin, River Falls*
Melody Thomas, *Northwest Arkansas Community College*
Greg Thompson, *Adrian University*
Vijayalakshmi Varadarajan, *Des Moines Area Community College*
Brian Zulkoskey, *University of Saskatchewan, Saskatoon*

Reviewers

We would also like to thank the many colleagues who carefully reviewed chapters for us. Their insightful comments significantly improved our book.

Don Abernathy, *North Central Texas College*
Elise Adamson, *Wayland Baptist University*
Miah Muhammad Adel, *University of Arkansas, Pine Bluff*
Ricardo Alarcon, *Arizona State University*
Z. Altounian, *McGill University*
Abu Amin, *Riverland Community College*
Sanjeev Arora, *Fort Valley State University*
Llani Attygalle, *Bowling Green State University*
Yiyan Bai, *Houston Community College*
Michael Bates, *Moraine Valley Community College*
Luc Beaulieu, *Memorial University*
Jeff J. Bechtold, *Austin Community College*
David Bennum, *University of Nevada*
Satinder Bhagat, *University of Maryland*
Dan Boye, *Davidson University*
Jeff Bronson, *Blinn College*
Douglas Brumm, *Florida State College at Jacksonville*
Mark S. Bruno, *Gateway Community College*
Brian K. Bucklein, *Missouri Western State University*
Michaela Burkardt, *New Mexico State University*
Kris Byboth, *Blinn College*
Joel W. Cannon, *Washington & Jefferson College*
Kapila Clara Castoldi, *Oakland University*
Paola M. Cereghetti, *Lehigh University*
Hong Chen, *University of North Florida*
Zengjun Chen, *Tuskegee University*
Uma Choppali, *Dallas County Community College*
Todd Coleman, *Century College*
José D'Arruda, *University of North Carolina, Pembroke*
Tinanjan Datta, *Georgia Regents University*
Chad L. Davies, *Gordon College*
Brett DePaola, *Kansas State University*
Sandra Doty, *Ohio University*
James Dove, *Metro Community College*
Carl T. Drake, *Jackson State University*
Rodney Dunning, *Longwood University*
Vernessa M. Edwards, *Alabama A & M University*
Davene Eyres, *North Seattle Community College*
Hasan Fakhruddin, *Ball State University*
Paul Fields, *Pima Community College*
Lewis Ford, *Texas A & M University*
J.A. Forrest, *University of Waterloo*
Scott Freedman, *Philadelphia Academy Charter High School*
Tim French, *Harvard University*
James Friedrichsen III, *Austin Community College*
Sambandamurthy Ganapathy, *SUNY, Buffalo*
J. William Gary, *University of California, Riverside*
L. Gasparov, *University of North Florida*
Gasparyan, *California State University, Bakersfield*
Brian Geislinger, *Gadsden State Community College*
Oommen George, *San Jacinto College*
Anindita Ghosh, *Suffolk County Community College*
Alan I. Goldman, *Iowa State University*
Richard Goulding, *Memorial University of Newfoundland*

Morris C. Greenwood, *San Jacinto College Central*
Thomas P. Guella, *Worcester State University*
Alec Habig, *University of Minnesota, Duluth*
Edward Hamilton, *Gonzaga University*
C. A. Haselwandter, *University of Southern California*
Zvonko Hlousek, *California State University, Long Beach*
Micky Holcomb, *West Virginia University*
Kevin M. Hope, *University of Montevallo*
J. Johanna Hopp, *University of Wisconsin, Stout*
Leon Hsu, *University of Minnesota*
Olenka Hubickyj Cabot, *San Jose State University*
Richard Ignace, *East Tennessee State University*
Elizabeth Jeffery, *James Madison University*
Yong Joe, *Ball State University*
Darrin Eric Johnson, *University of Minnesota, Duluth*
David Kardelis, *Utah State University, College of Eastern Utah*
Agnes Kim, *Georgia State College*
Ju H. Kim, *University of North Dakota*
Seth T. King, *University of Wisconsin, La Crosse*
Kathleen Koenig, *University of Cincinnati*
Olga Korotkova, *University of Miami*
Minjoon Kouh, *Drew University*
Tatiana Krivosheev, *Clayton State University*
Michael Kruger, *University of Missouri, Kansas City*
Jessica C. Lair, *Eastern Kentucky University*
Josephine M. Lamela, *Middlesex County College*
Patrick M. Len, *Cuesta College*
Shelly R. Lesher, *University of Wisconsin, La Crosse*
Zhujun Li, *Richland College*
Bruce W. Liby, *Manhattan College*
David M. Lind, *Florida State University*
Jeff Loats, *Metropolitan State College of Denver*
Susannah E. Lomant, *Georgia Perimeter College*
Jia Grace Lu, *University of Southern California*
Mark Lucas, *Ohio University*
Lianxi Ma, *Blinn College*
Aklilu Maasho, *Dyersburg State Community College*
Ron MacTaylor, *Salem State University*
Eric Mandell, *Bowling Green State University*
Maxim Marienko, *Hofstra University*
Mark Matlin, *Bryn Mawr College*
Dan Mattern, *Butler Community College*
Mark E. Mattson, *James Madison University*
Jo Ann Merrell, *Saddleback College*
Michael R. Meyer, *Michigan Technological University*
Karie A. Meyers, *Pima Community College*
Andrew Meyertholen, *University of the Redlands*
John H. Miller, Jr., *University of Houston*
Ronald C. Miller, *University of Central Oklahoma*
Ronald Miller, *Texas State Technical College System*
Hector Mireles, *California State University, Pomona*
Ted Monchesky, *Dalhousie University*
Steven W. Moore, *California State University, Monterey Bay*
Mark Morgan-Tracy, *University of New Mexico*
Dennis Nemeschansky, *University of Southern California*
Terry F. O'Dwyer, *Nassau Community College*
John S. Ochab, *J. Sargeant Reynolds Community College*

Umesh C. Pandey, *Central New Mexico Community College*
Archie Paulson, *Madison Area Tech College*
Christian Poppeliers, *Augusta State University*
James R. Powell, *University of Texas, San Antonio*
Michael Pravica, *University of Nevada, Las Vegas*
Kenneth M. Purcell, *University of Southern Indiana*
Kenneth Ragan, *McGill University*
Milun Rakovic, *Grand Valley State University*
Jyothi Raman, *Oakland University*
Ravi, *Spelman College*
Lou Reinisch, *Jacksonville State University*
Sandra J. Rhoades, *Kennesaw State University*
John Rollino, *Rutgers University, Newark*
Rodney Rossow, *Tarrant County College*
Larry Rowan, *University of North Carolina*
Michael Sampogna, *Pima Community College*
Tumer Sayman, *Eastern Michigan University*
Jim Scheidhauer, *DePaul University*
Paul Schmidt, *Ball State University*
Morton Seitelman, *Farmingdale State College*
Saeed Shadfar, *Oklahoma City University*
Weidian Shen, *Eastern Michigan University*
Jason Shulman, *Richard Stockton College of New Jersey*
Michael J. Shumila, *Mercer County Community College*

R. Seth Smith, *Francis Marion University*
Frank Somer, *Columbia College*
Chad Sosolik, *Clemson University*
Brian Steinkamp, *University of Southern Indiana*
Narasimhan Sujatha, *Wake Tech Community College*
Maxim Sukharev, *Arizona State University*
James H. Taylor, *University of Central Missouri*
Richard Taylor, *University of Oregon*
E. Tetteh-Lartey, *Blinn Community College*
Fiorella Terenzi, *Brevard Community College*
Gregory B. Thompson, *Adrian College*
Marshall Thomsen, *Eastern Michigan University*
Som Tyagi, *Drexel University*
Vijayalakshmi Varadarajan, *Des Moines Area Community College*
John Vasut, *Baylor University*
Dimitrios Vavylonis, *Lehigh University*
Kendra L. Wallis, *Eastfield College*
Laura Weinkauf, *Jacksonville State University*
Heather M. Whitney, *Wheaton College*
Capp Yess, *Morehead State University*
Chadwick Young, *Nicholls State University*
Yifu Zhu, *Florida International University*
Raymond L. Zich, *Illinois State University*

COLLEGE PHYSICS

16

Electrostatics I: Electric Charge, Forces, and Fields

In this chapter, your goals are to:
- (16-1) Explain why studying electric phenomena is important.
- (16-2) Describe how objects acquire a net electric charge.
- (16-3) Recognize the differences between insulators, conductors, and semiconductors.
- (16-4) Use Coulomb's law to quantitatively describe the force that one charged particle exerts on another.
- (16-5) Explain the relationship between electric force and electric field.
- (16-6) Describe the connection between enclosed charge and electric flux described by Gauss's law.
- (16-7) Apply Gauss's law to symmetric situations and to the distribution of excess charge on conductors.

To master this chapter, you should review:
- (3-2, 3-3, and 3-4) How to add vectors and how to do vector calculations using components.
- (4-2) Newton's second law.

> **What do you think?**
> A shark can detect the electric field produced by a small, electrically charged object. If a charged object 2 m from a shark's nose is moved to be only 1 m from the nose, the magnitude of the electric field that the shark detects will (a) decrease; (b) double; (c) increase by a factor of 4; (d) increase by a factor of 8; (e) increase by a factor of 16.

16-1 Electric forces and electric charges are all around you—and within you

You may think of *electricity* as something that gives you a shock on a dry day when you walk across a carpet and then touch a metal doorknob. Or perhaps it's just a commodity that you purchase from the power company. The reality is that electric phenomena are all around you, from the behavior of grains of pollen to the drama of a thunderstorm (Figure 16-1).

Electricity is an essential part of how you learn physics. As you read this sentence, your eye casts an image of the text onto your retina, and this image is transmitted to your brain along the optic nerve as a stream of electrical impulses. Electricity even explains why you don't fall through the floor: The molecules on the surface of the floor exert electric forces that repel the molecules on the underside of your shoes, and these forces are strong enough to balance Earth's downward gravitational force on you.

Electric phenomena are important because *all* ordinary matter is made of electrically charged objects. That's why we'll spend the next several chapters learning about electric charges and their physics. The simplest place to begin our study of electricity is with *electrostatics*, the branch of physics that deals with electric charges at rest and how they interact with each other. In this chapter we'll first learn about the different kinds

Figure 16-1 Electric charges and forces Electrostatics—the physics that describes how opposite charges attract and like charges repel—plays an important role in (a) biology and (b) the weather.

(a)
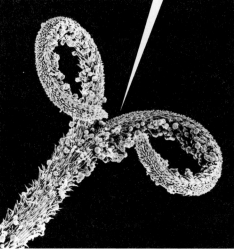
Grains of pollen wafting through the air carry a small amount of electric charge. This attracts them to the oppositely charged stigma of a flower and so aids in pollination.

(b)

The motion of air in a thundercloud causes positive and negative charges to separate and build up within the cloud. The release of these charges causes lightning.

Take-Home Message for Section 16-1

✔ Electric phenomena are important because all ordinary matter is made of electrically charged objects.

✔ Electrostatics is the study of electric charges at rest and the forces that they exert on each other.

of electrical charge in nature, and describe the forces that two electrically charged objects exert on each other. We'll see how these forces manifest themselves in two important classes of materials called conductors and insulators. We'll go on to learn about the concept of an *electric field,* which will give us a powerful way to understand how electric charges can exert forces on each other even over a distance. We'll conclude this chapter with a look at *Gauss's law,* an important tool for understanding electric fields.

16-2 Matter contains positive and negative electric charge

Figure 16-2 shows some simple experiments that demonstrate the nature of electric charge. If you hold two rubber rods next to each other, they neither attract nor repel each other. The same is true for two glass rods held next to each other, or for a rubber rod and a glass rod held next to each other (Figure 16-2a). But if you rub each rubber rod with fur and each glass rod with silk, we find that the rubbed ends of the rubber rods repel each other and the rubbed ends of the glass rods repel each other, but the rubbed ends of the rubber and glass rod attract each other (Figure 16-2b). What's more, the rubbed end of the rubber rod attracts the piece of fur used to rub it, and the rubbed end of the glass rod attracts the piece of silk used to rub it (Figure 16-2c). In each case where there is a repulsion or an attraction, the repulsive or attractive force is stronger the closer the objects are held to each other.

Here's how we explain these experiments and others like them:

(1) *Matter has electric charge.* In addition to having mass, matter also possesses a property called **electric charge**. This charge comes in two forms, positive and negative. An object that has a net electric charge (more positive than negative, or more negative than positive) is **charged**. Most matter, however, contains equal amounts of positive and negative charge and is electrically **neutral**. In Figure 16-2a the rubber and glass rods are electrically neutral, as are the pieces of fur and silk.

(2) *Electric charge is a property of the constituents of atoms.* All ordinary matter is made up of atoms. Atoms are in turn composed of more fundamental particles called electrons, protons, and (except for the most common type of hydrogen atom) neutrons. Electrons carry a negative charge, while protons carry an equally large positive charge; neutrons are neutral. Protons and neutrons have about the same mass as each other and are found in the dense nucleus at the center of an atom. A negatively charged electron has only about 0.05% the mass of a proton,

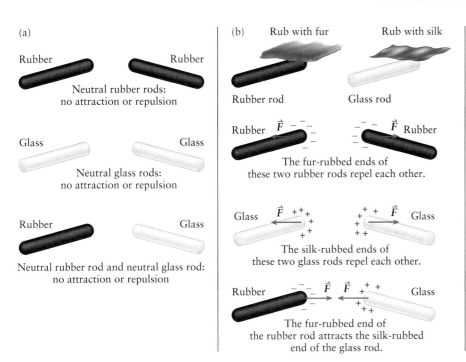

(a)

Rubber   Rubber

Neutral rubber rods:
no attraction or repulsion

Glass   Glass

Neutral glass rods:
no attraction or repulsion

Rubber   Glass

Neutral rubber rod and neutral glass rod:
no attraction or repulsion

(b) Rub with fur   Rub with silk

Rubber rod   Glass rod

Rubber $\vec{F}$ $\vec{F}$ Rubber

The fur-rubbed ends of
these two rubber rods repel each other.

Glass $\vec{F}$ $\vec{F}$ Glass

The silk-rubbed ends of
these two glass rods repel each other.

Rubber $\vec{F}$ $\vec{F}$ Glass

The fur-rubbed end of
the rubber rod attracts the silk-rubbed
end of the glass rod.

(c)
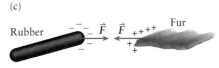
Rubber $\vec{F}$ $\vec{F}$ Fur

The fur-rubbed end of the rubber rod
attracts the fur with which it was rubbed.

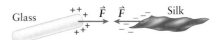
Glass $\vec{F}$ $\vec{F}$ Silk

The silk-rubbed end of the glass rod
attracts the silk with which it was rubbed.

Figure 16-2 **Experiments in electrostatics** Electric charge can be transferred from one object to another by rubbing. Electrically charged objects exert forces on each other.

so the electrons move around the massive, positively charged nucleus (**Figure 16-3**). A normal atom contains as many electrons as it does protons and so is electrically neutral.

(3) *Charge can be transferred between objects.* When dissimilar materials are rubbed together, negatively charged electrons can be transferred from one material to the other. As an example, consider the rubber rod and fur shown in Figure 16-2b. Initially, the rod and fur are electrically neutral, so each has equal amounts of positive and negative charge. But when the rubber rod and fur are rubbed together, electrons are transferred from the fur to the rubber. (This is a consequence of the internal structures of the two materials. It turns out that electrons can lower their electric potential energy by moving from the fur to the rubber, so they move in that direction. In the same way, a ball rolls from the top of a hill to the bottom because that's the direction that lowers its gravitational potential energy. We'll discuss electric potential energy in detail in Chapter 17.) This leaves the fur with more positive charge than negative and the rubber with more negative charge than positive. So the fur ends up with a *net* positive charge, and the rubber rod ends up with a *net* negative charge. By contrast, when the glass rod and silk in Figure 16-2b (both of which are initially neutral) are rubbed together, electrons are transferred from the glass to the silk. So the silk ends up with a net negative charge and the glass rod with a net positive charge. In Figure 16-2b we use plus and minus signs to indicate the net charge on each object (+ for positive, − for negative).

(4) *Electrically charged objects exert forces on each other.* Objects with electric charges of the same sign (both positive or both negative) repel each other, while objects with electric charges of opposite sign (one positive and the other negative) attract each other (Figure 16-2c). The force between charges is called the **electric force**. The magnitude of the force depends on the amount of charge on each object and on the distance between the objects. The force increases for a greater amount of charge, and it increases if the charged objects are brought closer to each other.

(5) *Charge is never created or destroyed.* In the experiments shown in Figure 16-2 charge is *transferred* between objects. Although charge moves between the rubber rod and the fur or between the glass rod and the silk, in each case the *total* charge of the two objects remains the same as before they were rubbed together. That is, charge is *conserved*; it is never created or destroyed. No one has ever observed a process in which charge is not conserved, so to the best of our knowledge the conservation of electric charge is an absolute law of nature.

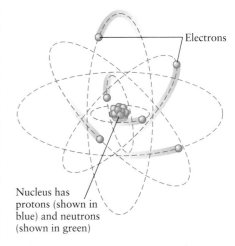

Electrons

Nucleus has protons (shown in blue) and neutrons (shown in green)

Figure 16-3 **A simplified model of an atom** Negatively charged electrons orbit the atom's nucleus, which contains most of the atom's mass. The nucleus contains two types of particles, positively charged protons and uncharged neutrons.

The character of the electric force explains the remarkable experiment shown in Figure 16-4a. A balloon is first rubbed against the girl's sleeve, causing electrons to transfer from her sleeve onto the rubber surface of the balloon (just like the rubber rod and fur in Figure 16-2b). When the negatively charged balloon is held next to the girl's head, electric forces attract the hair to the balloon—even though the hair is *neutral*. What happens is that the negatively charged electrons on the balloon repel the electrons in the atoms that comprise the hair and attract the positively charged nuclei of these atoms. The electrons and nuclei remain within their atoms but end up slightly displaced from each other (Figure 16-4b). This effect is called **polarization**, and we say that the hair becomes *polarized*. As we described above, the electric force is greater if charged objects are close to each other than if they are far apart. Figure 16-4b shows that the nuclei in the polarized hair are slightly closer to the negatively charged balloon than are the electrons in the hair. So the attractive force between the hair nuclei and the balloon is slightly greater than the repulsive force between the hair electrons and the balloon. The result is that the hair feels a net attractive force toward the balloon.

The unit of electric charge is the **coulomb** (C), named after the eighteenth-century French physicist Charles-Augustin de Coulomb who uncovered the fundamental law that governs the interaction of charges. (We'll discuss this law in Section 16-4.) One coulomb turns out to be quite large compared to the amount of charge on a single electron or proton. As we mentioned above, electrons and protons have the same magnitude of electric charge; we call this magnitude e. Precise measurements show that

$$e = 1.60217657 \times 10^{-19}\,\text{C}$$

We use q or Q as the symbol for the charge of an object, so the charge on a proton is (to four significant figures) $q_{\text{proton}} = +e = +1.602 \times 10^{-19}$ C and the charge on an electron is $q_{\text{electron}} = -e = -1.602 \times 10^{-19}$ C. No free particle has ever been detected with a charge smaller in magnitude than e. For that matter, no object has ever been found whose charge was not a multiple of e such as $+2e$, $-3e$, and so on. We summarize this by saying that charge is *quantized*: The charge of an object is always increased or decreased by an amount equal to an integer multiple of the fundamental charge e. It is impossible to add a fraction of a proton or electron to an object. (There is strong evidence that protons and neutrons are composed of more fundamental particles called *quarks* that have charges of $+2e/3$ and $-e/3$. However, no isolated quarks have ever been observed outside of a larger particle such as a proton or neutron. We'll learn more about quarks in Chapter 28.)

Note that while electrons and protons have opposite signs, it's completely arbitrary whether we choose electrons to have negative charge and protons to have positive charge or the other way around. The choice of sign is actually due to the American scientist and statesman Benjamin Franklin, who decided in the mid-eighteenth century

(a)

(b)

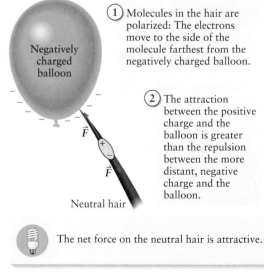

① Molecules in the hair are polarized: The electrons move to the side of the molecule farthest from the negatively charged balloon.

② The attraction between the positive charge and the balloon is greater than the repulsion between the more distant, negative charge and the balloon.

The net force on the neutral hair is attractive.

Figure 16-4 **A hair-raising experiment** (a) Rubbing a balloon on your sleeve gives the balloon an electric charge. A charged balloon can attract your hair and make it stand up, even though your hair has no net charge. (b) The nature of the electric force explains why this attraction happens.

that the sign of charge on a glass rod rubbed with silk (see Figure 16-2b) is positive. The electron and proton were not discovered until much later (1897 and 1919, respectively).

Example 16-1 Electrons in a Raindrop

A water molecule is made up of two hydrogen atoms, each of which has one electron, and an oxygen atom, which has eight electrons. One mole of water has a mass of $18.0 \text{ g} = 18.0 \times 10^{-3}$ kg. How many electrons are there in a single raindrop with a radius of $1.00 \text{ mm} = 1.00 \times 10^{-3}$ m? What is the total charge of all of these electrons? The density of liquid water is $1.00 \times 10^3 \text{ kg/m}^3$.

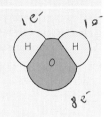

Set Up

Each water molecule has 10 electrons (two from the hydrogen atoms and eight from the oxygen atom). So we need to determine the number of water molecules in the raindrop and multiply by 10 to find the number of electrons. We'll use the definition of density from Section 11-2 and the number of molecules in a mole (Avogadro's number) from Section 14-3. We'll also need the formula for the volume of a sphere of radius r.

Definition of density:

$$\rho = \frac{m}{V} \qquad (11\text{-}1)$$

Number of molecules in one mole:

$N_A = 6.022 \times 10^{23}$ molecules/mol

Charge of an electron:

$q_{\text{electron}} = -e = -1.602 \times 10^{-19}$ C

Volume of a sphere of radius r:

$$V = \frac{4\pi}{3}r^3$$

Solve

First find the mass of the raindrop, then determine the number of moles of water that corresponds to this mass.

Volume of the raindrop of radius $r = 1.00 \times 10^{-3}$ m:

$$V = \frac{4\pi}{3}r^3 = \frac{4\pi}{3}(1.00 \times 10^{-3} \text{ m})^3 = 4.19 \times 10^{-9} \text{ m}^3$$

From Equation 11-1, the mass of such a raindrop is

$$m = \rho V = (1.00 \times 10^3 \text{ kg/m}^3)(4.19 \times 10^{-9} \text{ m}^3) = 4.19 \times 10^{-6} \text{ kg}$$

The number of moles of water in this mass is

$$n = \frac{4.19 \times 10^{-6} \text{ kg}}{18.0 \times 10^{-3} \text{ kg/mol}} = 2.33 \times 10^{-4} \text{ mol}$$

Then determine the number of molecules in the raindrop, and from that determine the number of electrons.

Each mole contains N_A molecules, so the total number of molecules in the raindrop is

$$nN_A = (2.33 \times 10^{-4} \text{ mol})(6.022 \times 10^{23} \text{ molecules/mol})$$
$$= 1.40 \times 10^{20} \text{ molecules}$$

The number of electrons in the raindrop is

$$N = (1.40 \times 10^{20} \text{ molecules})(10 \text{ electrons/molecule})$$
$$= 1.40 \times 10^{21} \text{ electrons}$$

To find the charge of this number of electrons, multiply by the charge per electron.

Total charge of this number of electrons:

$$q = Nq_{\text{electron}} = (1.40 \times 10^{21} \text{ electrons})(-1.602 \times 10^{-19} \text{ C/electron})$$
$$= -224 \text{ C}$$

Reflect

How large a number is 1.40×10^{21}? The human body contains about 10^{14} cells. Even if you include all of the cells in all of the approximately 9 million (9×10^6) people in New York City, that's only about 9×10^{20} cells—still fewer than the number of electrons in a single raindrop.

The charge of -224 C is quite substantial, but remember that for every electron in a water molecule there is also one proton (one in each hydrogen atom and eight in each oxygen atom). Since a proton carries exactly as much positive charge as an electron carries negative charge, our raindrop has zero *net* charge. In Section 16-4 we'll see how difficult it would be to separate the positive and negative charges of this raindrop.

? Got the Concept? 16-1 Transferring Charge

Consider the rubber rod and glass rod in Figure 16-2b. Compared to their masses before they are rubbed with fur and silk, respectively, what are their masses after being rubbed? (a) Both rods have more mass; (b) both rods have less mass; (c) the rubber rod has more mass and the glass rod has less mass; (d) the rubber rod has less mass and the glass rod has more mass; (e) the masses of both rods are unchanged.

Take-Home Message for Section 16-2

✔ Electric charge is quantized. The smallest amount of charge that can be added to or removed from an object is equal to the fundamental charge $e = 1.602 \times 10^{-19}$ C.

✔ All ordinary matter contains positive charge in the form of protons, which have a charge of $+e$, and negative charge in the form of electrons, which have a charge of $-e$. An object is electrically neutral if it contains equal amounts of positive and negative charge. If these amounts are not equal, the object has a net charge.

✔ Charge can neither be created nor destroyed. However, charge can be transferred from one object to another, for example, by moving electrons between objects.

✔ Objects with a net charge exert electric forces on each other. These forces become weaker with increasing distance. If the objects have the same sign of charge (both positive or both negative), the forces are repulsive. If the objects have opposite signs of charge (one positive and one negative), the forces are attractive.

16-3 Charge can flow freely in a conductor but not in an insulator

All substances contain positive and negative charges. But how *mobile* those charges are depends on the specific material. The rubber, glass, silk, and fur shown in Figure 16-2 are all examples of **insulators**, substances in which charges are not able to move freely. All of the electrons in an insulator are bound tightly to the nuclei of atoms, and any excess charge added to an insulator tends to stay wherever it is placed. So when electrons get placed on one end of a rubber rod by rubbing it with fur, the excess electrons stay on that end (Figure 16-5a). The electrons cannot redistribute themselves along the rod, so there is no excess charge on the other, unrubbed end of the rod to attract the positively charged fur. (Due to a polarization effect like that shown in Figure 16-4, there is still a very weak attraction between the positively charged fur and the neutral, unrubbed end of the rubber rod.) Most nonmetals are insulators.

By contrast, a metallic substance such as copper is an example of a **conductor**, a substance in which charges *can* move freely. In copper the outermost, or valence, electron of an atom can easily be dislodged from that atom. (The valence electron is relatively far from the positively charged nucleus, and the many electrons closer to the nucleus tend to shield the valence electron from the charge of the nucleus.) As a result, the valence electrons can move between copper atoms relatively freely. Excess charges

(a)

The fur-rubbed end of the rubber rod attracts the fur with which it was rubbed...

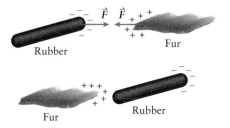

...but the end of the rubber rod that was not rubbed does not feel a strong attraction.

(b)

The nylon-rubbed end of the copper rod attracts the nylon with which it was rubbed...

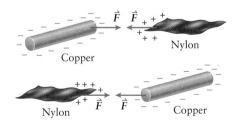

...and the end of the copper rod that was not rubbed also feels a strong attraction.

Figure 16-5 Insulators versus conductors (a) If you place excess charge at one location on an insulator, it remains at that location. (b) If you place excess charge at one location on a conductor, the excess charge can move freely through the conductor.

placed on a conductor can also move freely through the material. This explains what happens when you rub one end of a copper rod with nylon (Figure 16-5b). Electrons are transferred from the nylon to the copper, giving the copper rod a net negative charge and the nylon a net positive charge. But after one end of the copper rod has been rubbed, the nylon attracts *both* ends of the copper rod (rubbed and unrubbed) equally. That's because the excess electrons deposited on one end of the copper rod can easily move within the rod. When the positively charged nylon is brought close to *either* end of the copper rod, the excess electrons on the copper rush to that end. As a result, that end has a net negative charge and is attracted to the nylon.

Conductors and insulators are an essential part of all electric circuits, which are systems in which there is an ongoing current (a flow of charge) around a closed path. Electric circuits are at the heart of any device that uses a battery (such as a mobile phone, a flashlight, or an electric vehicle) or that you plug into a wall socket (such as a desktop computer, a toaster, or an electric fan). An electric circuit uses moving charges (typically electrons) to transfer energy along a conductor from one point in a circuit to another. In a flashlight, for example, electrons flowing through a copper wire carry energy from the battery (which is a repository of *electric potential energy*) to the light bulb, where the energy is converted into visible light. The electrons then return to the battery through a second copper wire to pick up more energy and repeat the process. Insulators play a crucial role in this process: Each conducting wire is clad in a sheath made of an insulator, which helps ensure that the electrons flow only along the length of the wires. (The visible part of an ordinary extension cord or power cord is actually the insulating sheath. The copper wires are contained within the sheath.) We'll explore the idea of electric potential energy in detail in Chapter 17, and we'll devote Chapters 18 and 21 to exploring various important kinds of electric circuits.

Most metals are good electrical conductors *and* also good conductors of heat, with large values of thermal conductivity (see Section 14-7). That's because the flow of electric charge within a material and the flow of heat within that material both require that particles of the material be free to move. In Section 14-3 we learned that the temperature of a material is related to the kinetic energy of that material's particles. For heat to flow from one part of a material to another, the faster-moving particles in the high-temperature part of the material must transfer kinetic energy to the slower-moving particles in the low-temperature part of the object. This can only happen if particles such as electrons are free to move from the high-temperature part of the material to the low-temperature part, where they collide with atoms and transfer their kinetic energy. So the presence of free electrons makes metals good thermal conductors as well as good conductors of electricity. Materials that are electrical insulators, in which the electrons are generally not free to move between atoms, also tend to be thermal insulators with low thermal conductivity.

Not all electric conductors are metals. In biological systems a very important conductor of electricity is water. That's because in nearly all biological systems, electric charges are carried by *ions*—atoms with an excess or deficit of electrons. These ions are suspended in water, that is, in aqueous solution. Water is particularly good at holding ions in solution because the water molecule (H_2O), while electrically neutral, has slightly more negative charge at the oxygen atom of the molecule and slightly more positive charge at the hydrogen atoms. So when an ionic compound such as sodium chloride (NaCl) is dissolved in water, the positive ions (Na^+) are attracted to the negative (oxygen) end of H_2O molecules and the negative ions (Cl^-) are attracted to the positive (hydrogen) end of H_2O molecules. Because the H_2O molecules to which the ions are attached are free to move, the aqueous solutions which make up living things are conductors. It doesn't matter whether charge is carried by an electron or by an ion. As we'll see in Chapter 18, it also doesn't matter whether the moving charges are positive or negative: In either case there is a flow of charge.

? Got the Concept? 16-2   Sodium in Water

A normal atom of sodium (chemical symbol Na) has 11 electrons and is electrically neutral. But in the compound sodium chloride (NaCl) the sodium atom is actually an ion of charge $+e$. How does a sodium atom acquire this charge? (a) by annihilating one of its electrons; (b) by creating a new electron; (c) by transferring an electron to another atom or molecule; (d) by acquiring an electron from another atom or molecule; (e) by acquiring a proton from another atom or molecule.

There's an important third class of substances called **semiconductors**. These substances have electrical properties that are intermediate between those of insulators and conductors. A common example of a semiconductor is silicon. A silicon atom has four outer electrons as compared to just one for copper, which might suggest that it would be a good conductor. However, in pure silicon each of those electrons is part of a chemical bond with a neighboring silicon atom, so the outer electrons have limited mobility. As a result, pure silicon conducts electricity far worse than a conductor such as copper, though still far better than an insulator such as rubber.

Semiconductors turn out to be of tremendous practical use in electric circuits. That's because it's possible to adjust their electrical properties by *doping*—that is, by adding a small amount of a second substance. Here's an example: If we take pieces of silicon that have been doped in different ways and put them in contact, we can arrange it so that charges will flow through the combination in one direction but will *not* flow in the opposite direction! Such a combination, called a *diode*, plays the same role in an electric circuit as the valves in the human heart, which allow blood to flow through the heart in one direction only. In Chapter 21 we'll look at some of the applications of semiconductors to modern technologies such as light-emitting diodes (LEDs), solar cells, and integrated circuits.

Take-Home Message for Section 16-3

✔ Charges are free to move within a conductor but can move very little in an insulator.

✔ In a metal conductor such as copper the moving charges are typically electrons. In biological systems water acts as a conductor, with ions as the moving charges.

✔ Semiconductors have electrical properties intermediate between those of conductors and insulators.

16-4 Coulomb's law describes the force between charged objects

We learned in Section 16-2 that electrically charged objects, or *charges* for short, exert forces on each other (see Figure 16-2). Careful measurements reveal that the force between two charges is directly proportional to the amount, or magnitude, of each of the interacting charges. The greater the amount of charge on the objects, the larger the force that acts on each charge. The force also depends on the distance between the charged objects; the closer the charges, the larger the force.

Coulomb's law is a summary in mathematical form of the results of these measurements. Specifically, this law tells us about the force between two **point charges**, which are very small charged objects whose size is much smaller than the separation between the charges (Figure 16-6a).

A point charge is an idealization, just like a massless rope or a frictionless incline, but it's a good description in many situations where charged objects interact with each other. Coulomb's law tells us the magnitude of the electric force that two point charges q_1 and q_2 separated by a distance r exert on each other:

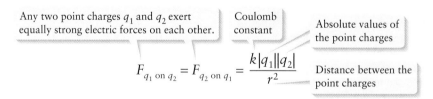

Any two point charges q_1 and q_2 exert equally strong electric forces on each other.

Coulomb constant

Absolute values of the point charges

Distance between the point charges

$$F_{q_1 \text{ on } q_2} = F_{q_2 \text{ on } q_1} = \frac{k|q_1||q_2|}{r^2}$$

Coulomb's law
(16-1)

The value of the **Coulomb constant** k in Equation 16-1 is to three significant figures:

$$k = 8.99 \times 10^9 \text{ N} \cdot \text{m}^2/\text{C}^2$$

Note that Equation 16-1 just tells you the *magnitude* of the electric force between two point charges. As Figure 16-6b shows, the *direction* of the electric force is such that charges of the same sign (both positive or both negative) repel each other, while charges of opposite sign (one positive and one negative) attract each other. Figure 16-7 shows an application of this principle.

(a)

Point charges: Their size is actually much smaller than their separation r.

q_1 ⚪   ⚪ q_2

r

(b)

$\vec{F}$ ⊕   ⊕ $\vec{F}$

Charges of the same sign repel each other...

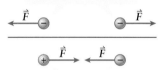

$\vec{F}$ ⊖   ⊖ $\vec{F}$

⊕ $\vec{F}$ $\vec{F}$ ⊖

...and charges of opposite signs attract each other.

⊖ $\vec{F}$ $\vec{F}$ ⊕

Figure 16-6 Coulomb's law
(a) Coulomb's law describes the electric force that two point charges (not drawn to scale) exert on each other. (b) The direction of the electric force between point charges depends on the signs (positive or negative) of the charges.

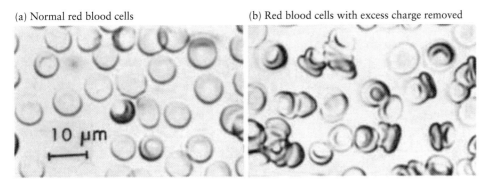

(a) Normal red blood cells   (b) Red blood cells with excess charge removed

10 μm

Figure 16-7 **Coulomb's law and red blood cells** Red blood cells carry oxygen from your lungs to other parts of your body through the circulatory system. (a) Each red blood cell has a slight excess of electrons that gives it a net negative charge. Since all cells are negatively charged, they exert repulsive electric forces on each other that help keep the cells apart. (b) The red blood cells in this photo have been treated with an enzyme that removes the excess electrons. Without electric forces to keep them apart, the red blood cells tend to clump. If red blood cells in your body behaved like this, their flow through your circulatory system would be impeded and your body would be starved for oxygen.

You should notice the similarity between Coulomb's law and Newton's law of universal gravitation:

Gravitational constant (same for any two objects) | Masses of the two objects

Any two objects (1 and 2) exert equally strong gravitational forces on each other.

$$F_{1\,on\,2} = F_{2\,on\,1} = \frac{Gm_1m_2}{r^2}$$

Center-to-center distance between the two objects

Newton's law of universal gravitation (10-2)

The gravitational forces are attractive: $\vec{F}_{1\,on\,2}$ pulls object 2 toward object 1 and $\vec{F}_{2\,on\,1}$ pulls object 1 toward object 2.

In both Coulomb's law (Equation 16-1) and the law of universal gravitation (Equation 10-2), the force that one object exerts on the other is inversely proportional to the square of the distance between them. In addition, the gravitational force is proportional to the product of the masses of the two particles, while the electric force is proportional to the product of the charges of the two particles. As remarkable as these similarities are, there is one essential difference between the two laws: Newton's law of universal gravitation tells us that two objects with mass always attract each other, but Coulomb's law states that two charged objects can attract or repel each other, depending on the sign of the charges they carry.

The electric force also shares some properties with every other force we've discussed. The electric force is a vector and has a direction. The net electric force on an object is the vector sum of every separate electric force that acts on it. Furthermore, the electric forces that two charged objects exert on each other obey Newton's third law. When point objects with charges q_1 and q_2 interact, the force that object 1 exerts on object 2 is equal in magnitude to the force that object 2 exerts on object 1, but the two forces are in opposite directions (see Figure 16-6b).

Example 16-2 Calculating Electric Force

(a) What is the electric force (magnitude and direction) between two electrons separated by a distance of 10.0 cm = 0.100 m? (b) Suppose you could remove all the electrons from a drop of water 1.00 mm in radius (see Example 16-1 in Section 16-2) and clump them into a ball 1.00 mm in radius. If this ball of electrons is 10.0 cm from the drop of water from which they were removed, what is the magnitude of the electric force between the drop of water and the ball of electrons?

Set Up

The two electrons repel because both have a negative charge $q = -e = -1.602 \times 10^{-19}$ C. From Example 16-1, the combined charge of all of the electrons in a water drop of this size is -224 C; the water drop was initially neutral, so the charge of the water drop after all of the electrons have been removed is $+224$ C. Since the ball of electrons and the water drop (with electrons removed) have opposite signs of charge, they attract each other.

We can use Coulomb's law (Equation 16-1) in both parts of this problem. That's because in both cases the charged objects are much smaller than the distance that separates them, so we can treat them as point charges. (Indeed, electrons are very small even compared to the dimensions of an atom.)

Coulomb's law:

$$F_{q_1 \text{ on } q_2} = F_{q_2 \text{ on } q_1} = \frac{k|q_1||q_2|}{r^2} \quad (16\text{-}1)$$

(a)

(b)

Solve

(a) Find the magnitude of the force that each electron exerts on the other.

For the two electrons, the charges are

$$q_1 = q_2 = -e = -1.602 \times 10^{-19} \text{ C}$$

The distance between the electrons is $r = 0.100$ m. From Equation 16-1, the magnitude of the force that each electron exerts on each other is

$$F = \frac{(8.99 \times 10^9 \text{ N} \cdot \text{m}^2/\text{C}^2)|-1.602 \times 10^{-19} \text{ C}||-1.602 \times 10^{-19} \text{ C}|}{(0.100 \text{ m})^2}$$
$$= 2.31 \times 10^{-26} \text{ N}$$

(b) Find the magnitude of the force between the ball of electrons and the water drop from which they were extracted.

For the ball of electrons and the water drop with electrons removed, the charges are

$$q_1 = -224 \text{ C}, q_2 = +224 \text{ C}$$

The distance between the two objects is $r = 0.100$ m. From Equation 16-1, the two objects exert forces on each other of magnitude

$$F = \frac{(8.99 \times 10^9 \text{ N} \cdot \text{m}^2/\text{C}^2)|-224 \text{ C}||+224 \text{ C}|}{(0.100 \text{ m})^2}$$
$$= 4.51 \times 10^{16} \text{ N}$$

Reflect

The repulsive force between the two electrons is tiny because each particle carries only a tiny amount of charge. By contrast, the attractive force between the ball of electrons and the electron-free water drop is immense. To put this force into perspective, a solid cube of lead with a weight of 4.51×10^{16} N would be 7.4 *kilometers* on a side! This is the magnitude of force that you would have to exert to keep the electrons from flying back into the water drop. There is no known way to produce a force of this magnitude, which is why you'll never see an object with all of its electrons removed. It's relatively easy to remove a small fraction of an object's electrons, as for the fur and the glass rod in Figure 16-2b. But removing *all* of the electrons from a piece of fur or a glass rod is not a practical thing to do.

Example 16-3 Three Charges in a Line

A particle with negative charge q is placed halfway between two identical particles, each of which carries the same positive charge: $Q_1 = Q_2 = +Q$. The distance between adjacent charges is d. If each of the three particles experiences a net electric force of zero, what is the magnitude of charge q in terms of Q?

Set Up

We want the net force on each point charge—that is, the *vector* sum of the forces on that charge due to the other two charges—to be equal to zero. We'll use Coulomb's law, Equation 16-1, to solve for the magnitude of the force of one charge on another. We'll also use the idea that charges of the same signs repel while charges of opposite signs attract.

Coulomb's law:

$$F_{q_1 \text{ on } q_2} = F_{q_2 \text{ on } q_1} = \frac{k|q_1||q_2|}{r^2}$$

(16-1)

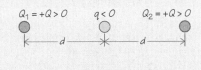

Solve

✓ Let's start by considering the forces on positive charge Q_2. The other positive charge, Q_1, exerts a repulsive force $\vec{F}_{Q_1 \text{ on } Q_2}$ that pushes Q_2 away from Q_1, that is, to the right. The negative charge q exerts an attractive force $\vec{F}_{q \text{ on } Q_2}$ that pulls Q_2 toward q, that is, to the left. In order for the net force on Q_2 to be zero, these two forces must have the same magnitude.

The net force on charge Q_2 must be zero:

$$\vec{F}_{Q_1 \text{ on } Q_2} + \vec{F}_{q \text{ on } Q_2} = 0$$

For this to be true, $\vec{F}_{Q_1 \text{ on } Q_2}$ and $\vec{F}_{q \text{ on } Q_2}$ must have the same magnitude:

$$F_{Q_1 \text{ on } Q_2} = F_{q \text{ on } Q_2}$$

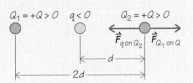

✓ Use Coulomb's law (Equation 16-1) to find the magnitudes $F_{Q_1 \text{ on } Q_2}$ and $F_{q \text{ on } Q_2}$. Set these equal to each other and solve for the magnitude (absolute value) of q.

The distance between Q_1 and Q_2, each of which has a charge of magnitude Q, is $2d$. From Equation 16-1, the force of Q_1 on Q_2 has magnitude

$$F_{Q_1 \text{ on } Q_2} = \frac{k|Q_1||Q_2|}{(2d)^2} = \frac{kQ^2}{4d^2}$$

d is the distance between Q_1 and q, and also between Q_2 and q. Notice that Q_1 and Q_2 are separated by distance $2d$. Remember that q is negative, so we need to keep its absolute value.

The distance between q and Q_2 is d, so the magnitude of the force of q on Q_2 is

$$F_{q \text{ on } Q_2} = \frac{k|q||Q_2|}{d^2} = \frac{k|q|Q}{d^2}$$

In order for the forces to have the same magnitude,

$$\frac{kQ^2}{4d^2} = \frac{k|q|Q}{d^2}$$

Solve for the absolute value of q:

$$|q| = \frac{Q}{4}$$

Reflect

The value $|q| = Q/4$ satisfies the condition that there is zero net force on Q_2. Because Q_1 is twice as far from Q_2 as q, and because the electric force is inversely proportional to the square of the distance, the charge Q_1 must be $(2)^2 = 4$ times greater than the magnitude of q in order for the forces these charges exert on Q_2 to have the same magnitude.

You can see that since Q_1 and Q_2 have the same charge, the forces on Q_1 are the mirror images of those on Q_2 (a repulsive force to the *left* exerted by Q_2, which is a distance $2d$ from Q_1, and an attractive force to the *right* exerted by q, which is a distance d from Q_1). So the net force on Q_1 will be zero, too.

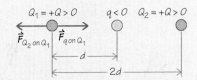

You can see that the net force on the negative charge q is also guaranteed to be zero. This charge is the same distance d from the two equal positive charges Q_1 and Q_2, so the force from Q_1 that pulls q to the left is just as great as the force from Q_2 that pulls q to the right.

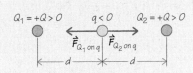

Example 16-4 Three Charges in a Plane

Charges Q_1 and Q_3 are both positive and equal to 1.50×10^{-6} C; charge Q_2 is negative and equal to -1.50×10^{-6} C. Charges Q_1 and Q_2 are placed at a fixed position a distance $D = 6.00$ cm apart, and Q_3 is placed at a fixed position a distance $H = 4.00$ cm above the midpoint of the line that connects Q_1 and Q_2. Calculate the magnitude and direction of the force on Q_3 due to the other two charges.

Set Up

The net force on positive charge Q_3 is the vector sum of $\vec{F}_{1\,\text{on}\,3}$, the *repulsive* force that the positive charge Q_1 exerts on Q_3, and $\vec{F}_{2\,\text{on}\,3}$, the *attractive* force that the negative charge Q_2 exerts on Q_3. We'll use Coulomb's law to find the magnitude of each of these forces. We'll then add the two force vectors using components.

Coulomb's law:

$$F_{q_1 \text{ on } q_2} = F_{q_2 \text{ on } q_1} = \frac{k|q_1||q_2|}{r^2} \quad (16\text{-}1)$$

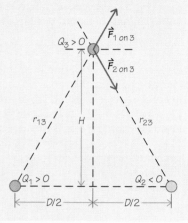

Solve

Use Equation 16-1 to find the magnitudes of the forces $\vec{F}_{1\,\text{on}\,3}$ and $\vec{F}_{2\,\text{on}\,3}$.

The figure above shows that the distance from Q_1 to Q_3 is the same as the distance from Q_2 to Q_3:

$$r_{13} = r_{23} = \sqrt{\left(\frac{D}{2}\right)^2 + H^2} = \sqrt{\left(\frac{6.00 \text{ cm}}{2}\right)^2 + (4.00 \text{ cm})^2}$$
$$= 5.00 \text{ cm} = 5.00 \times 10^{-2} \text{ m}$$

All three charges have the same magnitude:

$$|Q_1| = |Q_2| = |Q_3| = 1.50 \times 10^{-6} \text{ C}$$

So from Equation 16-1, $\vec{F}_{1\,\text{on}\,3}$ and $\vec{F}_{2\,\text{on}\,3}$ have the same magnitude:

$$F_{1 \text{ on } 3} = \frac{k|Q_1||Q_3|}{r_{13}^2}$$
$$= \frac{(8.99 \times 10^9 \text{ N} \cdot \text{m}^2/\text{C}^2)(1.50 \times 10^{-6} \text{ C})^2}{(5.00 \times 10^{-2} \text{ m})^2}$$
$$= 8.09 \text{ N}$$

$$F_{2 \text{ on } 3} = \frac{k|Q_2||Q_3|}{r_{23}^2} = F_{1 \text{ on } 3} = 8.09 \text{ N}$$

Choose the positive x direction to be to the right and the positive y direction to be upward. Then find the x and y components of $\vec{F}_{1\,\text{on}\,3}$ and $\vec{F}_{2\,\text{on}\,3}$, and use these to calculate the components of the net force on Q_3.

The components of $\vec{F}_{1\,\text{on}\,3}$ and $\vec{F}_{2\,\text{on}\,3}$ are

$$F_{1 \text{ on } 3,x} = F_{1 \text{ on } 3} \cos\theta$$
$$F_{1 \text{ on } 3,y} = F_{1 \text{ on } 3} \sin\theta$$
$$F_{2 \text{ on } 3,x} = F_{2 \text{ on } 3} \cos\theta$$
$$F_{2 \text{ on } 3,y} = -F_{2 \text{ on } 3} \sin\theta$$

Since the magnitudes $F_{1\,\text{on}\,3}$ and $F_{2\,\text{on}\,3}$ are equal, the components of the net force on Q_3 are

$$F_{\text{net on } 3,x} = F_{1 \text{ on } 3,x} + F_{2 \text{ on } 3,x}$$
$$= F_{1 \text{ on } 3} \cos\theta + F_{2 \text{ on } 3} \cos\theta$$
$$= 2F_{1 \text{ on } 3} \cos\theta$$

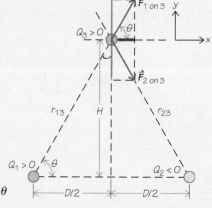

$$F_{\text{net on } 3,y} = F_{1 \text{ on } 3} \sin \theta + (-F_{2 \text{ on } 3} \sin \theta)$$
$$= F_{1 \text{ on } 3} \sin \theta - F_{1 \text{ on } 3} \sin \theta = 0$$

From the figure,

$$\cos \theta = \frac{(D/2)}{r_{13}} = \frac{3.00 \text{ cm}}{5.00 \text{ cm}} = 0.600$$

So

$$F_{\text{net on } 3,x} = 2(8.09 \text{ N})(0.600) = 9.71 \text{ N}$$
$$F_{\text{net on } 3,y} = 0$$

So the net force on Q_3 is to the right and has magnitude 9.71 N.

Reflect
The two individual forces on Q_3 add to a net force that is neither directly away from Q_1 nor directly toward Q_2.

? Got the Concept? 16-3 Electric Force

In part (b) of Example 16-2 we imagined removing all of the electrons from a drop of water and moving them to a given distance from the water drop. We then calculated the electric force between the drop and the electrons. Suppose instead we removed only one-half of the electrons from the water drop, then moved them to the same distance as in Example 16-2. Compared to the force we calculated in part (b) of Example 16-2, what would be the force between the electrons and the water drop in this case? (a) the same; (b) $1/\sqrt{2}$ as great; (c) 1/2 as great; (d) 1/4 as great; (e) 1/16 as great.

Take-Home Message for Section 16-4

✔ Coulomb's law tells us the magnitude of the electric force that two point charges exert on each other. This magnitude is proportional to the product of the magnitudes of the two charges and inversely proportional to the square of the distance between them.

✔ The electric force between two point charges is repulsive if the two charges have the same sign (both positive or both negative) and attractive if the two charges have opposite signs (one positive and one negative).

16-5 The concept of electric field helps us visualize how charges exert forces at a distance

Most forces in our daily experience arise only when one object is in direct contact with another object. Two examples are the normal force that acts on your body when you sit in a chair or when you push directly on an object. Nevertheless, some forces, such as the gravitational force and the Coulomb force, appear to act even between two objects separated by a distance. The fundamental theory of interactions between particles suggests that objects exert forces on each other by exchanging small units of energy or matter. For our purposes, however, a more useful approach employs the concept of the *field*. In this view, every charged object modifies all of space by producing an electric field, which is strongest closest to the object but extends infinitely far away. A second charged object senses this change in space, interacting with the electric field and experiencing an electric force (Figure 16-8).

If we place a particle carrying charge q in an electric field $\vec{E}$, the force experienced by the particle is

If a particle with charge q is placed at a position where the electric field due to other charges is $\vec{E}$...

$$\vec{F} = q\vec{E}$$

...then the electric force on the particle is $\vec{F} = q\vec{E}$.

Electric field and electric force
(16-2)

Figure 16-8 Electric field and electric force The electric field concept helps us visualize how a charge Q exerts an electric force on a second charge q.

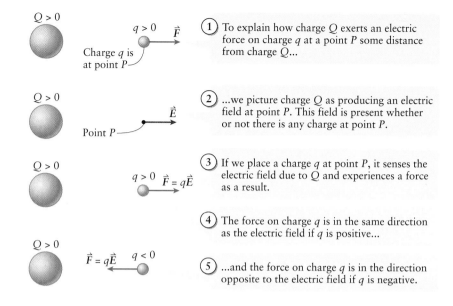

An equivalent way to write Equation 16-2 is

$$\vec{E} = \frac{\vec{F}}{q}$$

(16-3)

Equation 16-3 tells us that we can interpret the electric field $\vec{E}$ at a certain point as the *electric force per charge* that acts on a charged object placed at that point. If an object with double the charge is placed at that point, it will experience double the force.

! Watch Out! The direction of *force* versus the direction of *field*.

Although we can think of the electric field as the electric force per unit charge, you must be careful about the direction of the force. In particular, the direction of the electric force that a charge experiences depends on the *sign* of the charge as well as the *direction* of the electric field. Equation 16-2 tells us that a positive charge ($q > 0$) experiences a force in the same direction as the electric field $\vec{E}$, but a negative charge ($q < 0$) experiences a force in the direction opposite to $\vec{E}$ (Figure 16-8). So the electric field at a certain point is in the direction of the electric force that would be exerted on a *positive* charge placed at that point.

From Equation 16-3 we see that the SI units of electric field are newtons per coulomb, or N/C. If a 1-C charge experiences a 1-N force at a certain point in space, then at that point in space there is an electric field with a magnitude of 1 N/C.

Note that Equations 16-2 and 16-3 are strictly valid *only* if the particle of charge q is a *point* particle (that is, one with a very small size). That's because the value of $\vec{E}$ can be different at different places. If the charge q occupied a large volume (say, spread over a sphere 1 m in diameter), the value of $\vec{E}$ could be different at different points within that volume. In that case it wouldn't be clear which value of $\vec{E}$ to use in Equation 16-2 or 16-3. But if the charge q is within a point particle, it's clear what value of $\vec{E}$ to use: The value at the point where that particle is located.

Although this is the first time we've introduced the idea of a field, we've actually used the concept before. In Chapter 4 we found that we could express the gravitational force $\vec{w}$ on an object of mass m as $\vec{w} = m\vec{g}$, where $\vec{g}$ is the acceleration due to gravity. But we can also think of $\vec{g}$ as the *gravitational field* of Earth. In this picture, Earth sets up a field in the space around it, and $\vec{g}$ represents the value of that field vector (magnitude and direction) at a given point. An object of mass m placed at that point then experiences a gravitational force $\vec{w} = m\vec{g}$ (compare to Equation 16-2). Just as the concept of an electric field gives us a way to visualize how two charges can interact over a distance, the concept of a gravitational field helps us visualize how two objects with mass (Earth and the object of mass m) can interact without touching each other.

Go to Interactive Exercise 16-1 for more practice dealing with electric fields

Example 16-5 Determining Charge-to-Mass Ratio

When released from rest in a uniform electric field of magnitude 1.00×10^4 N/C, a certain charged particle travels 2.00 cm in 2.88×10^{-7} s in the direction of the field. You can ignore any nonelectric forces acting on the particle. (a) What is the *charge-to-mass ratio* of this particle (that is, the ratio of its charge q to its mass m)? (b) Other experiments show that the mass of this particle is 6.64×10^{-27} kg. What is the charge of this particle?

Set Up

The only force that acts on the particle is the electric force given by Equation 16-2. Since the particle accelerates in the direction of the electric field $\vec{E}$, the force on the particle must also be in the direction of $\vec{E}$. So the charge on the particle must be positive. Since $\vec{E}$ is uniform (it has the same value at all points), the force on the particle will be constant. Its acceleration a_x will be constant as well, so we can use one of the constant-acceleration equations from Chapter 2 to determine a_x. We'll use this with Newton's second law and Equation 16-2 to learn what we can about this particle.

Electric field and electric force:
$$\vec{F} = q\vec{E} \qquad (16\text{-}2)$$

Straight-line motion with constant acceleration:
$$x = x_0 + v_{0x}t + \frac{1}{2}a_x t^2 \qquad (2\text{-}9)$$

Newton's second law:
$$\sum \vec{F}_{\text{ext}} = m\vec{a} \qquad (4\text{-}2)$$

Solve

(a) We are given that the particle travels $2.00 \text{ cm} = 2.00 \times 10^{-2}$ m in 2.88×10^{-7} s. Use this to determine the particle's constant acceleration.

Take the positive x direction to be the direction in which the particle moves. The particle begins at rest, so $v_{0x} = 0$. If we take the initial position of the particle to be $x_0 = 0$, then Equation 2-9 becomes
$$x = \frac{1}{2}a_x t^2$$

Solve for the acceleration:
$$a_x = \frac{2x}{t^2} = \frac{2(2.00 \times 10^{-2} \text{ m})}{(2.88 \times 10^{-7} \text{ s})^2} = 4.82 \times 10^{11} \text{ m/s}^2$$

Relate the acceleration to the net external (electric) force on the particle and solve for the charge-to-mass ratio.

The net force on the particle of charge q is the electric force in the x direction, which from Newton's second law is equal to the mass m of the particle multiplied by the acceleration a_x:
$$qE_x = ma_x$$

This says that the acceleration of a particle in an electric field depends on the particle's charge-to-mass ratio:
$$a_x = \frac{q}{m}E_x$$

In this example we know both a_x and E_x, so the charge-to-mass ratio is
$$\frac{q}{m} = \frac{a_x}{E_x} = \frac{4.82 \times 10^{11} \text{ m/s}^2}{1.00 \times 10^4 \text{ N/C}}$$

Since $1 \text{ N} = 1 \text{ kg} \cdot \text{m/s}^2$, this is
$$\frac{q}{m} = 4.82 \times 10^7 \text{ C/kg}$$

(b) Given the charge-to-mass ratio and the mass of the particle, determine the charge q.

The charge of the particle is
$$q = m\left(\frac{q}{m}\right) = (6.64 \times 10^{-27} \text{ kg})(4.82 \times 10^7 \text{ C/kg})$$
$$= 3.20 \times 10^{-19} \text{ C}$$

The charge on a proton is $e = 1.60 \times 10^{-19}$ C; the charge on this particle is $2e$.

Reflect

The particle in this example has about four times the mass of a proton but only double the charge of a proton. For historical reasons it's known as an *alpha particle*; in fact, it's the nucleus of a helium atom, which contains two protons (each with charge e) and two neutrons (each with nearly the same mass as a proton but with zero charge).

Figure 16-9 Paper electrophoresis
This analytical technique used by chemists makes use of the electric field concept.

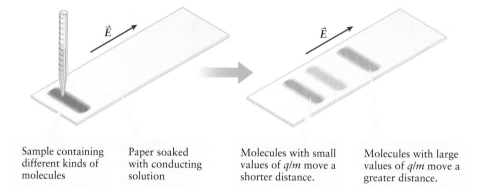

Sample containing different kinds of molecules

Paper soaked with conducting solution

Molecules with small values of q/m move a shorter distance.

Molecules with large values of q/m move a greater distance.

An important application of the force produced by an electric field is *electrophoresis*. Chemists use this technique to separate molecules of different kinds according to their charge and mass. In the simplest kind of electrophoresis, a small amount of a sample containing molecules of different kinds is placed on a strip of filter paper, and the paper is soaked with a solution that conducts electricity (Figure 16-9). An electric field of magnitude E is then applied along the length of the strip. If the molecules could move in the solution without fluid resistance, they would behave like the particle in Example 16-5: Each molecule would accelerate in response to the field, and that acceleration would be proportional to the charge-to-mass ratio q/m of the molecule. As a result, molecules with different values of q/m would move to different locations on the paper. In fact, there *is* fluid resistance on the molecules, but the net result is the same: Molecules with larger values of q/m move farther along the paper than do those with small values of q/m. The many applications of this specific technique, called *paper electrophoresis*, include analyzing currency to determine whether it is counterfeit (a forger's ink may have a different chemical composition than the ink used in legal currency) and looking for the presence of cancer antibodies or human immunodeficiency virus (HIV) in blood.

A different sort of electrophoresis is used for DNA profiling (also called genetic fingerprinting), which is an essential part of modern forensic science. A sample of human DNA is treated with an enzyme that breaks the long DNA strand into shorter segments. The sizes of these segments are characteristic of the person's genetic code, so measuring the segment sizes is a powerful technique for forensic identification. Unfortunately, the ratio of q/m for a segment of DNA is nearly the same for segments of any size, so paper electrophoresis isn't useful. Instead a sample containing the DNA segments is placed in a special gel that is permeated by many microscopic pores. When an electric field is applied, all of the segments move in response, but the smaller segments move through the gel pores more easily than large ones. The result is that the DNA segments are spread out according to their size, allowing a genetic "fingerprint" to be made. Similar techniques are used in medical research for studying both DNA and proteins.

Electric Field of a Point Charge

Equation 16-2 tells us how a charge q responds to a given electric field $\vec{E}$. It also tells us how to determine the value of $\vec{E}$ at any point. As an example, Figure 16-10 shows how we might determine the electric field around a positive point charge Q. To do this,

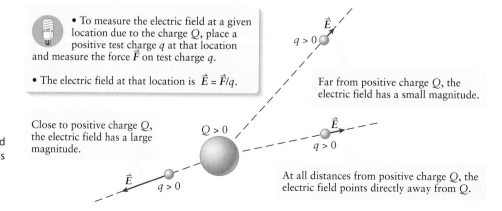

• To measure the electric field at a given location due to the charge Q, place a positive test charge q at that location and measure the force $\vec{F}$ on test charge q.

• The electric field at that location is $\vec{E} = \vec{F}/q$.

$\vec{E}$
$q > 0$

Far from positive charge Q, the electric field has a small magnitude.

Close to positive charge Q, the electric field has a large magnitude.

$Q > 0$

$\vec{E}$
$q > 0$

Figure 16-10 Mapping the electric field We can map out the electric field surrounding a point charge Q—in this case a positive charge—by placing a positive test charge $+q$ at various locations around Q.

$\vec{E}$
$q > 0$

At all distances from positive charge Q, the electric field points directly away from Q.

(a) Positive point charge: electric field vectors

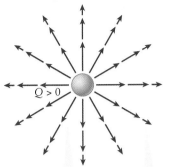
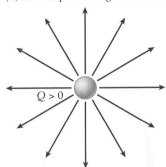

(b) Positive point charge: electric field lines

(c) Negative point charge: electric field lines

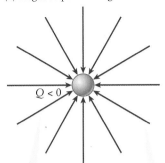

Field due to a positive charge points away from the charge.	Field strength decreases with increasing distance from the charge.	Field due to a positive charge points away from the charge.	Farther from the charge, where the field strength is weaker, field lines are farther apart.	Field lines due to a negative charge point toward the charge.	Farther from the charge, where the field strength is weaker, field lines are farther apart.

Figure 16-11 Electric field and electric field lines The electric field around a charged object can be represented by (a) electric field vectors or (b), (c) electric field lines.

we place a small positive charge q (which we call a *test charge*) at various locations around the charge Q and measure the force $\vec{F}$ on that small charge. The electric field at each location is given by Equation 16-3, $\vec{E} = \vec{F}/q$; since q is positive, the electric field is in the same direction as the force on the test charge. The Coulomb force exerted by Q repels a positive test charge q at any location; thus, the direction of force at each location—and so the direction of the electric field at each location, shown by the blue vectors—is radially away from Q. As the lengths of the vectors show, the magnitude of the electric field decreases with increasing distance from Q. That's because the force between Q and q decreases with increasing distance in accordance with Coulomb's law (Section 16-4).

Figure 16-11 shows two ways we can represent the entire electric field around a positive charge Q. In Figure 16-11a we draw vectors to represent the electric field at a large number of points around Q. Figure 16-11b shows a simpler approach that's easier to draw: We connect adjacent vectors to form lines, called **electric field lines**. The direction of the field line passing through any point represents the direction of the field at that point. The magnitude of the electric field is shown by the *density* of the field lines—that is, how close they are to each other. In regions where the field lines are close together, such as close to Q, the field has a large magnitude. Where the field lines are far apart, such as far from Q, the magnitude of the electric field is smaller. If the charge Q is negative, a positive test charge q would experience a force $\vec{F}$ toward Q, and so the electric field $\vec{E} = \vec{F}/q$ due to charge Q is directed *toward* that charge (Figure 16-11c).

❗ Watch Out! Electric fields are three-dimensional.

Figure 16-11 may give you the incorrect impression that the electric field and electric field lines of a point charge Q lie only on the plane of the page. Not so! The electric field completely surrounds the charge; it is three-dimensional. The field lines are arranged around the charge rather like the spines of a sea urchin (Figure 16-12).

Figure 16-12 An electric field analogy The electric field lines of a point charge are arranged radially around the charge in three dimensions, much like the spines on a sea urchin.

We can use Coulomb's law, Equation 16-1, to write an expression for the *magnitude* of the electric field due to a point charge Q. If we place a test charge q a distance r from charge Q, Equation 16-1 tells us that the magnitude of the Coulomb force on q is $F = k|q||Q|/r^2$. From Equation 16-2, we can also write this force magnitude as

$F = |q|E$, where E is the magnitude of the electric field due to charge Q at the position of test charge q. Setting these two expressions for F equal to each other, we get

$$|q|E = \frac{k|q||Q|}{r^2}$$

or

Magnitude of the electric field due to a point charge (16-4)

Magnitude of the electric field due to a point charge Q ⎯ Coulomb constant ⎯ Absolute value of charge Q

$$E = \frac{k|Q|}{r^2}$$

Distance between point charge Q and the location where the field is measured

The *magnitude* of the electric field due to a point charge Q decreases with increasing distance r from the charge. As Figure 16-11 shows, the *direction* of the electric field depends on the sign of the charge Q. The field points radially outward from a positive point charge and radially inward toward a negative point charge.

Electric Field of an Arrangement of Charges

Go to Picture It 16-1 for more practice dealing with electric fields

The results in Figure 16-11 and Equation 16-4 describe the electric field due to a single point charge (positive or negative). If several point charges $Q_1, Q_2, Q_3, \ldots$ are present, experiment shows that the *net* electric force $\vec{F}$ that these charges exert on a test charge q at any location is just the vector sum of the forces $\vec{F}_1, \vec{F}_2, \vec{F}_3, \ldots$ that these charges *individually* exert on q. If we use the symbol $\vec{E}$ for the net electric field produced at a given location by $Q_1, Q_2, Q_3, \ldots$ together, and the symbols $\vec{E}_1, \vec{E}_2, \vec{E}_3, \ldots$ for the electric fields that these charges produce individually, we can use Equation 16-2 to express this experimental result as

$$\vec{F} = \vec{F}_1 + \vec{F}_2 + \vec{F}_3 + \cdots \quad \text{or} \quad q\vec{E} = q\vec{E}_1 + q\vec{E}_2 + q\vec{E}_3 + \cdots$$

If we divide both sides of the second of these equations by q, we get

(16-5)

$$\vec{E} = \vec{E}_1 + \vec{E}_2 + \vec{E}_3 + \cdots$$

In other words, *when there are two or more point charges present, the electric field at any point in space is the vector sum of the fields due to each charge separately.* The following examples illustrate how to use this principle.

Example 16-6   Where *Is* the Electric Field Zero?

A point charge $Q_1 = +4.00$ nC (1 nC = 1 nanocoulomb = 10^{-9} C) is placed 0.500 m to the left of a point charge $Q_2 = +9.00$ nC. Find the position between the two point charges where the net electric field is zero.

Set Up

At any point between the two charges, the electric field $\vec{E}_1$ due to Q_1 points to the right (away from this positive charge) and the electric field $\vec{E}_2$ due to Q_2 points to the left (away from this positive charge). We want to find the point P where the total field $\vec{E} = \vec{E}_1 + \vec{E}_2$ equals zero. We'll use the symbol D for the 0.500-m distance between the two charges and x for the distance from Q_1 to point P: Then the distance from Q_2 to point P is $D - x$. Our goal is to find the value of x for which $\vec{E} = 0$.

Magnitude of the electric field due to a point charge Q:

$$E = \frac{k|Q|}{r^2} \quad \text{(16-4)}$$

Total electric field:

$$\vec{E} = \vec{E}_1 + \vec{E}_2 \quad \text{(16-5)}$$

$Q_1 = +4.00$ nC $\vec{E}_2$ P $\vec{E}_1$ $Q_2 = +9.00$ nC

x
$D = 0.500$ m

Solve

If the net electric field at P is zero, then $\vec{E}_1$ (to the right) and $\vec{E}_2$ (to the left) must have equal magnitudes so that these two vectors cancel. Use Equation 16-4 to write this statement in equation form.

In order for the net electric field at P to be zero,

$$E_1 = E_2$$

The distance from Q_1 to P is x, and the distance from Q_2 to P is $D - x$, so from Equation 16-4

$$E_1 = \frac{k|Q_1|}{x^2} \quad \text{and} \quad E_2 = \frac{k|Q_2|}{(D-x)^2}$$

If these are equal to each other,

$$\frac{k|Q_1|}{x^2} = \frac{k|Q_2|}{(D-x)^2}$$

Solve this equation for x.

The factors of k cancel in the above equation, so

$$\frac{|Q_1|}{x^2} = \frac{|Q_2|}{(D-x)^2} \quad \text{or}$$

$$|Q_1|(D-x)^2 = |Q_2|x^2$$

Multiply out the quantity $(D-x)^2$ and rearrange:

$$|Q_1|(D^2 - 2Dx + x^2) = |Q_2|x^2$$

$$(|Q_2| - |Q_1|)x^2 + (2D|Q_1|)x + (-D^2|Q_1|) = 0$$

We can simplify this if we divide through by $|Q_1|$:

$$\left(\frac{|Q_2|}{|Q_1|} - 1\right)x^2 + 2Dx + (-D^2) = 0$$

This is a quadratic equation of the form $ax^2 + bx + c = 0$, with $a = |Q_2|/|Q_1| - 1 = |9.00\text{ nC}|/|4.00\text{ nC}| - 1 = 1.25$, $b = 2D = 2(0.500\text{ m}) = 1.00\text{ m}$, and $c = -D^2 = -(0.500\text{ m})^2 = -0.250\text{ m}^2$. The solutions are

$$x = \frac{-b \pm \sqrt{b^2 - 4ac}}{2a}$$

$$= \frac{-(1.00\text{ m}) \pm \sqrt{(1.00\text{ m})^2 - 4(1.25)(-0.250\text{ m}^2)}}{2(1.25)}$$

$$= \frac{-(1.00\text{ m}) \pm \sqrt{2.25\text{ m}^2}}{2.50} = \frac{-(1.00\text{ m}) \pm (1.50\text{ m})}{2.50}$$

$$= +0.200\text{ m or} -1.00\text{ m}$$

We want a positive value of x to correspond to a point to the right of Q_1, so the solution we want is $x = +0.200$ m. We conclude that point P is a distance $x = 0.200$ m to the right of charge Q_1 and a distance $D - x = 0.500\text{ m} - 0.200\text{ m} = 0.300$ m to the left of charge Q_2.

Ask prof how you got this equation from first [handwritten annotation]

Reflect

Because charge Q_1 is smaller than Q_2, the location of the point where the net electric field is zero must be closer to Q_1 than to Q_2. That's just what we found. You can check the result $x = 0.200$ m by substituting this value into the above expressions for E_1 and E_2 and confirming that $E_1 = E_2$ for this value of x.

But what's the significance of the second solution, $x = -1.00$ m? This refers to a point 1.00 m to the *left* of charge Q_1 and 1.00 m + 0.500 m = 1.50 m to the left of charge Q_2. Our calculation shows that this point E_1 is equal to E_2. However, at this point $\vec{E}_1$ and $\vec{E}_2$ *both* point to the *left* (both fields point away from the positive charges that produce them). So at this point the electric fields $\vec{E}_1$ and $\vec{E}_2$ do *not* cancel, and the total field is not zero.

[Figure: $\vec{E}_1$ and $\vec{E}_2$ arrows pointing left from point P; $Q_1 = +4.00$ nC, $Q_2 = +9.00$ nC; distances marked 1.00 m and $D = 0.500$ m]

Example 16-7 Field of an Electric Dipole

A combination of two point charges of the same magnitude but opposite signs is called an **electric dipole**. Figure 16-13 shows an electric dipole made up of a point charge $+q$ and a point charge $-q$ separated by a distance $2d$. Derive expressions for the magnitude and direction of the net electric field due to these two charges at a point P a distance y along the midline of the dipole.

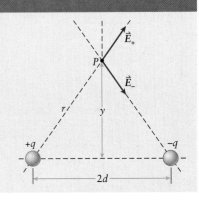

Figure 16-13 **An electric dipole** The field produced by an electric dipole at any point is the vector sum of the fields $\vec{E}_+$ and $\vec{E}_-$ caused by the positive charge $+q$ and the negative charge $-q$, respectively.

Set Up

The net field is the vector sum of the field $\vec{E}_+$ due to the charge $+q$ (which points away from $+q$) and the field $\vec{E}_-$ due to the charge $-q$ (which points toward $-q$). Note that this problem is very similar to Example 16-4 in Section 16-4, in which we used vector addition to find the net electric *force* exerted by two charges (one positive and one negative) on a third charge; here we use vector addition to find the net electric *field* due to the positive and negative charge. As in Example 16-4, we'll choose the positive x direction to be to the right and the positive y axis to be upward and add the two vectors using components.

Magnitude of the electric field due to a point charge Q:

$$E = \frac{k|Q|}{r^2} \tag{16-4}$$

Total electric field:

$$\vec{E} = \vec{E}_1 + \vec{E}_2 \tag{16-5}$$

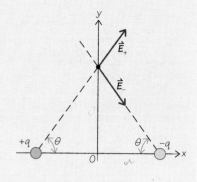

Solve

Use Equation 16-4 to find the magnitudes of the fields $\vec{E}_+$ and $\vec{E}_-$ at P.

The distance from $+q$ to point P is the same as the distance from $-q$ to P. Call this distance r:

$$r = \sqrt{y^2 + d^2}$$

Since $+q$ and $-q$ have the same magnitude (q, which is positive) and are the same distance from P, Equation 16-4 tells us that the fields that the two charges produce at P have the same magnitude:

$$E_+ = E_- = \frac{kq}{r^2} = \frac{kq}{y^2 + d^2}$$

Find the x and y components of $\vec{E}_+$ and $\vec{E}_-$, and use these to calculate the components of the net field at P.

The components of $\vec{E}_+$ and $\vec{E}_-$ are

$$E_{+,x} = E_+ \cos\theta$$
$$E_{+,y} = E_+ \sin\theta$$
$$E_{-,x} = E_- \cos\theta$$
$$E_{-,y} = -E_- \sin\theta$$

Since the magnitudes E_+ and E_- are equal, the components of the net field at P are

$$E_x = E_{+,x} + E_{-,x} = E_+ \cos\theta + E_- \cos\theta$$
$$= \frac{2kq}{y^2 + d^2} \cos\theta$$

$$E_y = E_{+,y} + E_{-,y} = E_+ \sin\theta + (-E_- \sin\theta)$$
$$= \frac{kq}{y^2 + d^2} \sin\theta - \frac{kq}{y^2 + d^2} \sin\theta = 0$$

From the figure,

$$\cos\theta = \frac{d}{r} = \frac{d}{\sqrt{y^2 + d^2}}$$

So the components of the net electric field are

$$E_x = \frac{2kq}{(y^2 + d^2)} \frac{d}{\sqrt{y^2 + d^2}} = \frac{2kqd}{(y^2 + d^2)^{3/2}}$$
$$E_y = 0$$

The net electric field at point P is to the right and has magnitude $E = 2kqd/(y^2 + d^2)^{3/2}$.

Reflect

We can check our result by substituting $y = 0$, so that the point P is directly between the two charges and a distance d from each charge. Then $\vec{E}_+$ and $\vec{E}_-$ both point to the right, and the magnitude of the net electric field should be equal to the sum of the magnitudes of $\vec{E}_+$ and $\vec{E}_-$.

Note that at points very far from the dipole, so that y is much greater than d, the magnitude of the field is inversely proportional to the *cube* of y: At double the distance, the field of a dipole is $(1/2)^3 = 1/8$ as great. This is a much more rapid decrease with distance than the field of a single point charge, for which E is inversely proportional to the *square* of the distance: At double the distance, the field of a point charge is $(1/2)^2 = 1/4$ as great. The dipole field decreases much more rapidly because the fields of $+q$ and $-q$ partially cancel each other.

At $y = 0$, the net electric field has magnitude

$$E = \frac{2kqd}{(0 + d^2)^{3/2}} = \frac{2kqd}{d^3} = \frac{2kq}{d^2} = 2\left(\frac{kq}{d^2}\right)$$

This is just twice the magnitude of the field due to each individual charge:

$$E_+ = E_- = \frac{kq}{d^2}$$

If y is much greater than d, $y^2 + d^2$ is approximately equal to y^2. Then the magnitude of the net electric field due to the dipole is approximately

$$E_{\text{net}} = \frac{2kqd}{(y^2)^{3/2}} = \frac{2kqd}{y^3}$$

By using techniques like the ones we employed in Example 16-7, it's possible to calculate and map out the electric field at *all* points around an electric dipole. **Figure 16-14** shows the field lines. Note that as you move away from the dipole along its midline, the field lines become farther apart. This is a graphical way of showing that the magnitude of the field decreases with increasing distance, just as we found in Example 16-7.

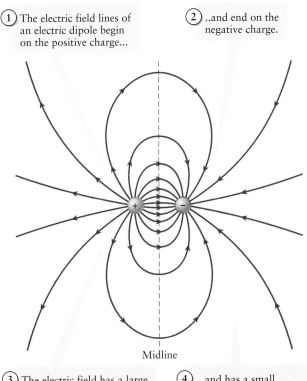

(1) The electric field lines of an electric dipole begin on the positive charge...

(2) ..and end on the negative charge.

(3) The electric field has a large magnitude where the field lines are close together...

(4) ...and has a small magnitude where the field lines are far apart.

Midline

Figure 16-14 Field lines of an electric dipole At any point, the electric field due to an electric dipole is the vector sum of the field due to the positive charge and the field due to the negative charge.

❓ Got the Concept? 16-4 Electric Field I

The positive charge in the dipole shown in Figure 16-14 is attracted to the negative charge. To find the force of this attraction, the electric field $\vec{E}$ to use in Equation 16-2 is (a) the field due to the positive charge; (b) the field due to the negative charge; (c) the net field due to both charges; (d) none of these.

❓ Got the Concept? 16-5 Electric Field II

Suppose both of the charges in Figure 16-13 were negative and had the same magnitude. At point P in that figure, the net electric field due to these charges would (a) point to the left; (b) point to the right; (c) point straight up; (d) point straight down; (e) be zero.

Take-Home Message for Section 16-5

✔ Any charged object produces an electric field in the space around it. A second charged object responds to this electric field; this is the origin of the electric force that the first object exerts on the second.

✔ The electric field of a positive point charge points directly away from that charge; the electric field of a negative point charge points directly toward that charge.

✔ The net electric field due to two or more charges is the vector sum of the electric fields due to the individual charges.

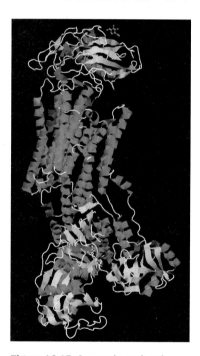

Figure 16-15 A protein molecule This illustration shows the structure of a particular protein molecule, a long chain of amino acids that coils on itself in a complex way. The coiling takes place because different parts of the chain carry different amounts of electric charge. The electric forces between the parts pull the chain into the complex shape shown here. Understanding the details of protein folding requires knowing the electric field at any point due to the arrangement of charges along the chain. This very complicated problem can only be solved using a computer.

16-6 Gauss's law gives us more insight into the electric field

Example 16-7 in the previous section shows how we can find the net electric field due to two point charges. In principle, we can extend this approach to calculate the net electric field due to a collection of any number of point charges. If there are many such charges, however, the calculations can become very complex (Figure 16-15).

Happily, there's an alternative and a much easier approach that we can use to find the electric field if the charges are arranged in a very *symmetrical* fashion—for example, uniformly distributed over a spherical volume. This approach uses a principle called *Gauss's law*, which is an alternative way to express Coulomb's law (Equation 16-1). We'll develop Gauss's law in this section and apply it to a variety of physical situations in the following section.

To understand Gauss's law, we first need to define a new quantity called *electric flux*. We'll do this by making an analogy between electric fields and the flow of water.

Water Flux and Electric Flux

Figure 16-16 shows water flowing through a pipe. The vectors labeled $\vec{v}$ in Figure 16-16a represent the velocity of the water at each point. For simplicity we've assumed that the water velocity is the same everywhere. Now imagine that we place a rectangular wire frame of area A in the flow. We define the *flux* of water through this wire frame as the product of (i) the area A of the frame and (ii) the component of flow velocity $\vec{v}$ that's perpendicular to the plane of the frame. If we orient the frame so that it's face-on to the

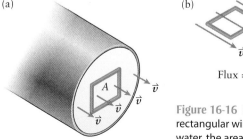

(a)

Flux = Av

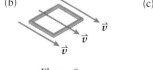

(b)

Flux = 0

(c)

Flux = $Av \cos\theta$

Figure 16-16 Flux of water The flux of water through a rectangular wire frame depends on the velocity $\vec{v}$ of the water, the area A of the wire frame, and the angle between the direction of $\vec{v}$ and the perpendicular to the plane of the frame.

water flow as in Figure 16-16a, then $\vec{v}$ is perpendicular to the plane of the frame and the perpendicular component is just v. The flux is then equal to Av. If instead we orient the frame so that it is edge-on to the flow as in Figure 16-16b, the flow velocity $\vec{v}$ has no component perpendicular to the plane of the frame. In this case the flux is zero. If we orient the frame so that a line perpendicular to the frame is at an angle θ to the direction of $\vec{v}$ as in Figure 16-16c, the perpendicular component of $\vec{v}$ is $v \cos \theta$ and the flux is $A(v \cos \theta) = Av \cos \theta$.

Note that we've actually encountered the concept of flux before. In Section 11-9 we learned that if an incompressible fluid is in steady flow through a pipe of varying cross-sectional area A, the product of the area and the flow speed v has the same value at any two points 1 and 2 along the pipe: $A_1 v_1 = A_2 v_2$ (Equation 11-19). In the language we've just introduced, this says that the *flux* of the fluid through the entire pipe maintains the same value even if the cross-sectional area of the pipe changes.

Figure 16-17 shows how we extend the idea of flux to the electric field. Instead of a pipe carrying a fluid, let's look at a region of space where there is an electric field $\vec{E}$. We saw in Section 16-5 that the value of $\vec{E}$ can vary from point to point, so we consider a small enough region that we can treat $\vec{E}$ as having essentially the same value over that region. We then imagine a small rectangular area A that we can orient however we like. By analogy to the flux of water in Figure 16-16, we define the **electric flux** Φ (the upper-case Greek letter phi) through the area A as follows:

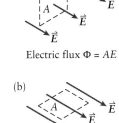

Electric flux $\Phi = AE$

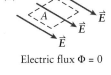

Electric flux $\Phi = 0$

Electric flux $\Phi = AE \cos \theta$

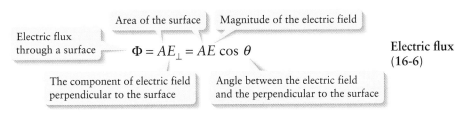

Electric flux through a surface — Area of the surface — Magnitude of the electric field

$$\Phi = AE_\perp = AE \cos \theta$$

The component of electric field perpendicular to the surface — Angle between the electric field and the perpendicular to the surface

Electric flux (16-6)

In Figure 16-17a we orient the area so that it is face-on to the direction of the electric field $\vec{E}$. Then the electric field is perpendicular to the surface and the angle $\theta = 0$. Therefore $E_\perp$ is equal to the field magnitude E and the electric flux through the surface is $\Phi = AE_\perp = AE$; alternatively, since $\theta = 0$, $\Phi = AE \cos 0 = AE(1) = AE$. In Figure 16-17b the area is edge-on to the electric field, so the electric field has zero component perpendicular to the area A. Then $E_\perp = 0$ and so $\Phi = AE_\perp = 0$. Alternatively, with this orientation $\theta = 90°$ and so $\Phi = AE \cos 90° = AE (0) = 0$. Finally, in Figure 16-17c the area is oriented at an angle θ between 0 and 90°, so the component of the electric field perpendicular to the area is $E_\perp = E \cos \theta$. Then the electric flux $\Phi = AE \cos \theta$ has a value between AE and zero.

Note that unlike the flow of water in Figure 16-16, there's nothing "flowing" through the area A in Figure 16-17. Unlike velocity $\vec{v}$, the electric field $\vec{E}$ does *not* represent motion of any kind. But we can still use the analogy between $\vec{v}$ and $\vec{E}$ as expressed by Figures 16-16 and 16-17 to define electric flux Φ in the manner given by Equation 16-6.

Figure 16-17 Electric flux The electric flux through a small rectangular surface depends on the electric field $\vec{E}$, the area A of the surface, and the angle between the direction of $\vec{E}$ and the perpendicular to the plane of the surface. (Compare Figure 16-16.)

 Watch Out! In calculating electric flux, the area is "imaginary."

In Figure 16-16 we measured the flux of water through an area outlined by a real, physical wire frame. But we really didn't need the frame: It was simply there to help us visualize the area in question. In defining electric flux as in Figure 16-17, we've done away with the wire frame entirely. If we actually used a frame made of a conducting wire such as copper, the mobile charges in the wire would move in response to the electric field, and the field of these charges themselves could modify the field that we're trying to analyze. You can think of the area A in Figure 16-17 and Equation 16-6 as "imaginary" in the sense that there's no physical object outlining the area.

Electric Flux through a Closed Surface: Gauss's Law

Why is the idea of electric flux through a surface a useful one? To see the answer, let's consider a *closed* surface—that is, one that encloses a volume. For example, in Figure 16-18a we've drawn a spherical surface of radius r that is centered on and encloses a positive point charge q. To find the electric flux through this surface, first note

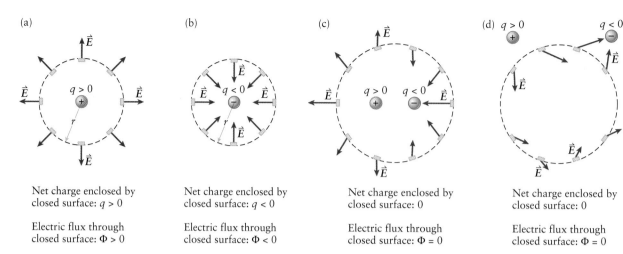

Figure 16-18 **Electric flux through a closed surface** The net electric flux through a closed spherical surface is (a) positive if a positive charge is enclosed by the surface, (b) negative if a negative charge is enclosed by the surface, (c) zero if equal amounts of positive and negative charge are enclosed by the surface, and (d) zero if no charge at all is enclosed by the surface.

from Equation 16-4 that the electric field $\vec{E}$ due to the point charge has the same magnitude $E = kq/r^2$ at every point on the surface, because every point is the same distance r from the point charge. However, $\vec{E}$ points in different directions at different points on the spherical surface: straight upward at the top of the surface, to the left at the leftmost point on the surface, and so on. However, if we look at a very small rectangular portion of the surface, then $\vec{E}$ points in essentially the same direction at every point on that rectangle. In fact, $\vec{E}$ is perpendicular to the rectangle of area ΔA, so $\theta = 0$ in Equation 16-6 and the flux through that rectangle is $(\Delta A)E = (\Delta A)(kq/r^2)$. If we now imagine that the entire spherical surface is made up of a very large number of such rectangles of area $\Delta A_1, \Delta A_2, \Delta A_3, \ldots, \Delta A_N$, the *total* electric flux through the spherical surface as a whole is just the sum of the fluxes through the individual rectangles:

$$\Phi = (\Delta A_1)\left(\frac{kq}{r^2}\right) + (\Delta A_2)\left(\frac{kq}{r^2}\right) + (\Delta A_3)\left(\frac{kq}{r^2}\right) + \cdots + (\Delta A_N)\left(\frac{kq}{r^2}\right)$$

(16-7)

$$= (\Delta A_1 + \Delta A_2 + \Delta A_3 + \cdots + \Delta A_N)\left(\frac{kq}{r^2}\right)$$

In Equation 16-7 the quantity $\Delta A_1 + \Delta A_2 + \Delta A_3 + \cdots + \Delta A_N$ is the sum of the areas of all of the individual rectangles that make up the spherical surface, and so is equal to the total area A of the surface. From the Math Tutorial the surface area of a sphere of radius r is $A = 4\pi r^2$, so Equation 16-7 becomes

(16-8)

$$\Phi = (4\pi r^2)\left(\frac{kq}{r^2}\right) = 4\pi kq$$

Notice that the radius of the spherical surface surrounding the point charge q cancels out in this equation. That's because the magnitude of the electric field decreases as the square of the sphere's radius r and the surface area increases at the same rate. So their product—the electric flux Φ—does not depend on the radius and is directly proportional to the charge q enclosed within the spherical surface. This result is a direct consequence of Coulomb's law (Equation 16-1) and the definition of electric field (Equation 16-4), which describe the $1/r^2$ behavior of electric force and electric field for a point charge.

Equation 16-8 also holds true if the charge q is negative (Figure 16-18b). The equation then tells us that the electric flux through the closed surface is *negative* in this case. The interpretation is that the flux is positive if the electric field points out of the closed surface as in Figure 16-18a, but negative if the electric field points into the closed surface as in Figure 16-18b.

What happens if we replace the single charge q with an electric dipole like that shown in Figure 16-14, with both a positive charge and a negative charge of equal

? Got the Concept? 16-6
Electric Flux I

When a certain charged particle is placed at the center of a sphere, the net electric flux through the surface of the sphere is Φ_0. If the radius of the sphere is doubled, what would be the new flux through the sphere? (a) $\Phi_0/4$; (b) $\Phi_0/2$; (c) Φ_0; (d) $2\Phi_0$; (e) $4\Phi_0$.

magnitude? In Figure 16-18c the spherical surface is centered on the dipole and encloses both the positive and negative charges, so the *net* enclosed charge is zero. The figure shows that for each small rectangle on the surface where the electric field $\vec{E}$ points outward, giving a positive contribution to the flux, there is another small rectangle elsewhere on the surface where $\vec{E}$ points inward, giving an equally large negative contribution to the flux. These positive and negative contributions to the flux cancel, so the net electric flux through this surface—which encloses zero net charge—is zero. The same cancellation also happens for the spherical surface in Figure 16-18d, which does not enclose either charge. In this case as well, the net enclosed charge is zero and the net electric flux through the surface is zero.

Here's what we've concluded so far from Figure 16-18:

- Closed surfaces that enclose a point charge q, as in Figure 16-18a and Figure 16-18b, have a nonzero electric flux through them. This flux is proportional to the enclosed charge q, as in Equation 16-8.

- Closed surfaces that enclose zero net charge—in Figure 16-18c because the surface encloses equal amounts of positive and negative charge, in Figure 16-18d because the surface encloses no charge at all—have zero net electric flux through them. This is consistent with Equation 16-8, but with $q = 0$.

We were able to draw these conclusions because the closed surfaces in the figure are spherical in shape and placed symmetrically with respect to the charges inside them. But it can be shown that the same conclusions hold true for a closed surface of *any* shape or placement. We can summarize these conclusions by rewriting Equation 16-8 with q replaced by q_{encl}, which represents the *net* charge enclosed by the closed surface:

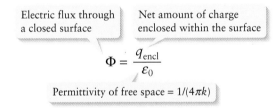

Electric flux through a closed surface

Net amount of charge enclosed within the surface

$$\Phi = \frac{q_{encl}}{\varepsilon_0}$$

Permittivity of free space $= 1/(4\pi k)$

Gauss's law
(16-9)

This relationship is called **Gauss's law** after the scientist who first deduced it, the nineteenth-century German mathematician and physicist Carl Friedrich Gauss. In honor of Gauss, a closed surface used to enclose charge in order to apply Gauss's law is referred to as a **Gaussian surface**. Equation 16-9 holds true for *any* Gaussian surface (Figure 16-19), no matter what its shape or size and no matter where the charges are located inside the surface. As we mentioned previously, such surfaces are imaginary: The surface does not need to be made of any physical substance.

① The net charge enclosed by this spherical Gaussian surface is q, so the net electric flux through this surface is $\Phi = q/\varepsilon_0$.

② The net charge enclosed by this irregular Gaussian surface is also q, so the net electric flux through this surface is also $\Phi = q/\varepsilon_0$.

 The net electric flux through a Gaussian surface depends only on the net charge that it encloses, not the shape or size of the surface.

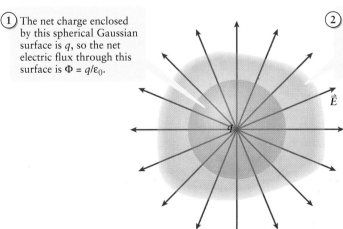

$\vec{E}$

Figure 16-19 **Gauss's law** Two different Gaussian surfaces that enclose the same net charge.

? Got the Concept? 16-7 Electric Flux II

When a certain charged particle is placed at the center of a sphere, the net electric flux through the surface of the sphere is Φ_0. If the sphere were elongated into an oblate spheroid (such as a rugby ball or an American football) but still enclosed the same charge, what would be the new electric flux through the surface? (a) Less than Φ_0; (b) Φ_0; (c) more than Φ_0; (d) not enough information given to decide.

In Equation 16-9 we have replaced the combination $4\pi k$ that appears in Equation 16-8 with $1/\varepsilon_0$. Here ε_0, called the **permittivity of free space** for historic reasons, is equal to $1/(4\pi k) = 8.85 \times 10^{-12}\,\mathrm{C}^2/(\mathrm{N}\cdot\mathrm{m}^2)$ to three significant figures. So we can state Gauss's law as

The net electric flux through any closed surface equals the net charge enclosed by that surface divided by the permittivity ε_0. Charges outside the surface have no effect on the net electric flux through the surface.

This law holds true because of the $1/r^2$ character of the electric field due to a point charge (see the discussion following Equation 16-8).

We'll see in the following section how we can use Gauss's law to determine the electric field due to certain charge distributions, and what Gauss's law tells us about how charges distribute themselves in a conductor.

? Got the Concept? 16-8 Electric Flux III

When a certain charged particle is placed at the center of a sphere, the net electric flux through the surface of the sphere is Φ_0. If a second, identical charged particle is placed outside the sphere, what would be the new electric flux through the surface? (a) Less than Φ_0; (b) Φ_0; (c) more than Φ_0; (d) not enough information given to decide.

Take-Home Message for Section 16-6

✔   The electric flux through a surface is analogous to the flux of water through a wire frame. Electric flux depends on the magnitude of the electric field, the area of the surface, and the relative orientation of the field and the surface.

✔   For a closed surface (one that encloses a volume), an electric field $\vec{E}$ that points out of the closed surface makes a positive contribution to the electric flux. An electric field that points into the closed surface makes a negative contribution to the electric flux.

✔   Gauss's law states that the net electric flux through a closed surface is proportional to the amount of charge enclosed by the surface. If the net enclosed charge is zero, the net electric flux through the surface is zero.

16-7 In certain situations, Gauss's law helps us to calculate the electric field and to determine how charge is distributed

Gauss's law by itself doesn't tell us the value of the electric field $\vec{E}$ at any one point. Rather, it tells us something about the values of $\vec{E}$ at *every* point on a closed Gaussian surface. For this reason, Gauss's law is useless for finding the value of the electric field at points on the closed surfaces shown in Figures 16-18c and 16-18d: While we know the value of the net electric flux through these surfaces, that isn't enough to tell us the value of $\vec{E}$ at any one point on the surface. Yet we *can* use Gauss's law to determine $\vec{E}$ in cases where the charge that produces the field is distributed in a particularly symmetric way. Let's look at a couple of examples.

Electric Field of a Spherical Charge Distribution

In Figure 16-20 charge Q is distributed uniformly throughout a spherical volume of radius R. (This could be a model of how electric charge is distributed over the volume of an atomic nucleus.) This distribution is *spherically symmetric*: You can rotate the sphere through any angle around its center and it looks exactly the same. Hence the electric field $\vec{E}$ caused by the charge distribution must also be spherically symmetric. This implies that the direction of the electric field must be either radially inward or radially outward, like the sea urchin spines shown in Figure 16-12. (If the field pointed in any other direction, the field lines would look different after rotating the sphere through some angle. The spherical symmetry says that's impossible.) So at any point $\vec{E}$ can have only a radial component E_r, which points either directly away from the center of the sphere ($E_r > 0$) or directly toward the center of the sphere ($E_r < 0$). That's why we've drawn the electric field in Figure 16-20 as pointing in a purely radial direction. This must be true for points inside the sphere as well as outside the sphere.

The spherical symmetry of the charge distribution in Figure 16-20 tells us something more: The value of E_r at a given point can depend only on the radial distance r from the center of the sphere. It can't depend on where around the sphere the point is. (If it did, that would mean that the electric field would look different after rotating the sphere through some angle. Spherical symmetry says that, too, is impossible.)

Given what spherical symmetry tells us about the electric field for the situation in Figure 16-20, what more can we learn by using Gauss's law? In Figure 16-21a we've drawn a spherical Gaussian surface that's centered on the sphere and that encloses the entire sphere. This Gaussian surface has radius $r > R$ and surface area $A = 4\pi r^2$. Just like the spherical surface in Figure 16-18a or Figure 16-18b, the electric field is perpendicular to the surface at every point and has the same magnitude at every point. Using the same reasoning that we used for Equation 16-8, we can say that the net electric flux through this surface is just the area of the surface multiplied by the radial electric field:

$$\Phi = AE_r = 4\pi r^2 E_r \qquad (16\text{-}10)$$

This flux is positive if the electric field points outward (so $E_r > 0$) and negative if the electric field points inward ($E_r < 0$). Since the entire charged sphere is enclosed within the

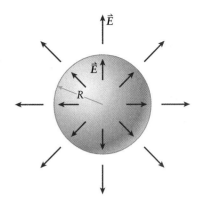

Charge Q is uniformly distributed throughout the volume of this sphere of radius R.

Figure 16-20 A uniformly charged sphere The spherical symmetry of this charge distribution tells us that the electric field $\vec{E}$ that it produces must be radial and that the magnitude of $\vec{E}$ can depend only on the distance from the center of the sphere.

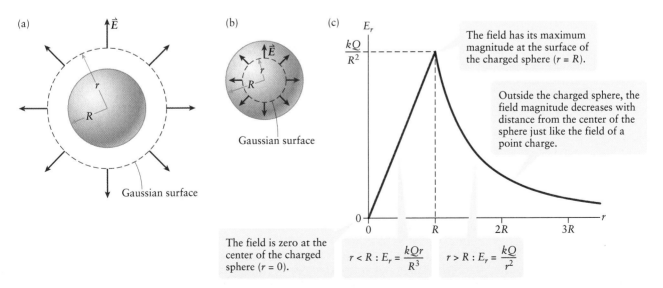

(a)

Gaussian surface

(b)

Gaussian surface

(c)

The field has its maximum magnitude at the surface of the charged sphere ($r = R$).

Outside the charged sphere, the field magnitude decreases with distance from the center of the sphere just like the field of a point charge.

The field is zero at the center of the charged sphere ($r = 0$).

$r < R : E_r = \dfrac{kQr}{R^3}$ $r > R : E_r = \dfrac{kQ}{r^2}$

Figure 16-21 Finding the electric field of a uniformly charged sphere To find how $\vec{E}$ depends on distance from the center of the sphere of charge Q, we use (a) one Gaussian surface that encloses the entire sphere and (b) a second Gaussian surface that encloses only part of the sphere. (c) The radial electric field E_r as a function of the distance r from the center, graphed for the case $Q > 0$.

Gaussian surface, the enclosed charge is $q_{encl} = Q$. Gauss's law, Equation 16-9, tell us that the flux is also equal to q_{encl}/ε_0. Equating Equations 16-9 and 16-10 yields

$$(16\text{-}11) \qquad 4\pi r^2 E_r = \frac{Q}{\varepsilon_0} \quad \text{and} \quad E_r = \frac{Q}{4\pi\varepsilon_0 r^2} = \frac{kQ}{r^2} \quad (r > R)$$

Equation 16-11 says that at points outside the sphere, the electric field is *exactly the same* as the field due to a point charge Q (Equation 16-4). The field is purely radial, is inversely proportional to the square of the distance r, and is proportional to the charge Q on the sphere. The field points radially outward ($E_r > 0$) if the charge Q is positive, and points radially inward ($E_r < 0$) if the charge Q is negative. (Compare Figure 16-11.)

We can also use Gauss's law to tell us about the electric field at points *inside* the sphere. In Figure 16-21b we've drawn another spherical Gaussian surface of radius r and surface area $A = 4\pi r^2$, but with a radius r that's less than the radius R of the charged sphere. Again the net electric flux through this Gaussian surface is given by Equation 16-10. But now the enclosed charge is less than the total charge Q on the sphere. Since the charge is distributed uniformly, the fraction of charge that's enclosed by the Gaussian surface is just equal to the ratio of the volume within the Gaussian surface (a sphere of radius r, with volume $4\pi r^3/3$) to the total volume occupied by the charge Q (a sphere of radius R, with volume $4\pi R^3/3$). So Gauss's law applied to the Gaussian surface in Figure 16-21b tells us

$$4\pi r^2 E_r = \frac{q_{encl}}{\varepsilon_0} = \frac{1}{\varepsilon_0}\left[Q\left(\frac{4\pi r^3/3}{4\pi R^3/3}\right)\right] = \frac{Qr^3}{\varepsilon_0 R^3}$$

If we divide through by $4\pi r^2$, we get

$$(16\text{-}12) \qquad E_r = \frac{Qr^3}{4\pi\varepsilon_0 r^2 R^3} = \frac{Qr}{4\pi\varepsilon_0 R^3} = \frac{kQr}{R^3} \quad (r < R)$$

Equation 16-12 tells us that inside the charged sphere, the direction of the electric field is the same as outside the sphere: $E_r > 0$ (an outward field) if Q is positive and $E_r < 0$ (an inward field) if Q is negative. It also says that inside the sphere, the magnitude of the field *increases* with increasing distance r from the center of the sphere. At the surface of the charged sphere, $r = R$, both Equation 16-11 and Equation 16-12 give the same result: $E_r = kQ/R^2$. At the very center of the sphere ($r = 0$) the electric field is zero. Figure 16-21c shows a graph of E_r as a function of r for the case of a positively charged sphere ($Q > 0$).

These conclusions about the electric field of a charged sphere would have been *very* difficult to obtain without using Gauss's law. (The alternative approach is to divide the charged sphere into a very large number of small segments, treating each segment as an individual point charge, and using vector addition to add the individual electric fields produced by all of the segments. That would take a lot of strenuous mathematics.) The relative ease with which we came to these conclusions shows the power of Gauss's law.

Electric Field of a Large, Flat, Charged Disk

Figure 16-22a shows a charge distribution with a different kind of symmetry: a uniformly charged plate or disk. (We'll see later that charged disks of this kind are found in an important device called a *capacitor*, used in many electric circuits.) This charge distribution has *rotational symmetry*, which means that it looks the same if you rotate it through any angle around an axis that passes vertically through the center of the disk. However, the electric field $\vec{E}$ due to this charge distribution can (and does) vary in a complicated way as you move from the center of the disk toward the edges.

To simplify the problem, let's just consider what the electric field is like at points around the disk near the disk center. We imagine that the edges of the disk are so far away that we can regard them as being infinitely distant. Then our problem is that of finding the electric field due to an *infinite sheet* of charge (Figure 16-22b).

We'll use the symbol σ (the lowercase Greek letter sigma) for the amount of charge per unit area on the sheet, also called the **surface charge density**. The sheet is uniformly

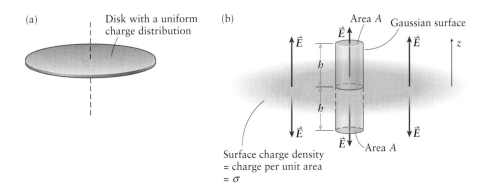

(a) Disk with a uniform charge distribution

(b) Area A Gaussian surface

Surface charge density = charge per unit area = σ

Figure 16-22 **Finding the electric field of a uniformly charged disk** (a) A uniformly charged disk; (b) to find the electric field of the disk at points close to the center of the disk, we can replace the disk by an infinitely large charged sheet.

charged, so σ has the same value everywhere on the sheet. The units of σ are coulombs per square meter (C/m^2).

Our infinite sheet of charge has *translational* symmetry: No matter which way or how far you move parallel to the disk, the charge distribution looks exactly the same. The same must therefore be true of the electric field produced by the charge distribution. So if the disk lies in the xy plane, the electric field $\vec{E}$ at any point cannot depend on the x or y coordinate of that point. The field can depend only on the z coordinate, which is the coordinate measured perpendicular to the plane in Figure 16-22.

Translational symmetry also tells us that $\vec{E}$ must be *perpendicular* to the plane of the sheet. This means that a point charge q placed in that field will feel a force $\vec{F} = q\vec{E}$ either directly toward or directly away from the sheet, depending on the sign of q. (If there were a component of $\vec{E}$ in some direction parallel to the plane of the sheet, a positive point charge would be pushed in that direction. This would only be the case if the point charge were repelled or attracted by one part of the sheet. But since the charge distribution is the same everywhere on the sheet, no part attracts or repels the positive charge more than any other part. So there can't be any component of $\vec{E}$ parallel to the sheet.) Thus $\vec{E}$ can depend only on the z coordinate and can have only a z component.

The infinite sheet also has *reflection* symmetry: It looks the same if we flip the sheet upside down. The same must be true of the electric field. So if the field points upward above the sheet, it must point downward below the sheet, as we've indicated in Figure 16-22b.

To find the electric field at a certain distance h from the plane of the sheet, we'll use a cylindrical Gaussian surface as in Figure 16-22b. The top and bottom faces of the cylinder each have area A, and each is the same distance h from the sheet. The symmetries we have described tell us that the electric field $\vec{E}$ is perpendicular to the top and bottom faces and has the same value everywhere on each of these faces. Because $\vec{E}$ is parallel to the sides of the cylinder, there is *zero* electric flux through the sides: $\Phi_{\text{sides}} = 0$. The flux through the top and bottom faces of the cylinder, however, is not zero. If E_z is the z component of electric field a distance h above the infinite sheet, the flux through the top face of the cylinder is $\Phi_{\text{top}} = AE_z$. Note that this flux is positive if E_z is positive, which means that the electric field above the sheet is upward and points out of the Gaussian surface; this flux is negative if E_z is negative, which means that the electric field above the sheet is downward and points into the Gaussian surface. The reflection symmetry tells us that the flux through the bottom face is the same, so $\Phi_{\text{top}} + \Phi_{\text{bottom}} = 2\Phi_{\text{top}} = 2AE_z$. The net electric flux through the Gaussian surface in Figure 16-22b is therefore

$$\Phi = \Phi_{\text{sides}} + \Phi_{\text{top}} + \Phi_{\text{bottom}} = 0 + 2AE_z = 2AE_z \tag{16-13}$$

The area of the sheet enclosed within the Gaussian surface is A, so the amount of charge enclosed within the surface is the charge per unit area σ multiplied by the area A: $q_{\text{encl}} = \sigma A$. Using this and Equation 16-13 in Gauss's law, Equation 16-9, gives us an expression for the electric field component E_z a distance h above the sheet:

$$2AE_z = \frac{\sigma A}{\varepsilon_0} \quad \text{or} \quad E_z = \frac{\sigma}{2\varepsilon_0} \tag{16-14}$$

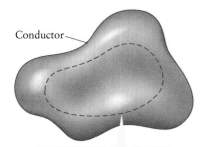

There can be no electric field inside the conductor, so the flux through this Gaussian surface must be zero.

Figure 16-23 A charged conductor Gauss's law tells us that any excess charge on a conductor must reside on the surface of the conductor.

Equation 16-14 tells us that E_z has the same sign as the surface charge density σ. So the electric field $\vec{E}$ above the sheet points upward (away from the sheet) if the surface charge density is positive ($\sigma > 0$); $\vec{E}$ points downward (toward the sheet) if the surface charge density is negative ($\sigma < 0$). This agrees with the idea that electric fields point away from positive charges and toward negative charges.

Equation 16-14 also tells us that the value of E_z at a point a distance h above the sheet does *not* depend on h. (Note that h doesn't appear anywhere in this equation.) So for an infinite sheet of charge, the electric field is the same at all distances from the sheet.

These results are only approximate, since there's no such thing as a truly infinite sheet of charge. But they are valid for a charged disk at points that are relatively close to the disk, so the height h above the disk is small compared to the radius of the disk. We'll make use of Equation 16-14 in later chapters.

Excess Charge on Conductors

Gauss's law leads to a remarkable conclusion about conducting materials to which excess charge is added so that the conductor has a net charge. Charges are free to move in a conductor, so if we add excess charges they will move in the conductor until they come to rest in equilibrium. In this situation the net force on each added charge is zero. Because the electric force on a charge q is directly proportional to the electric field $\vec{E}$ (Equation 16-2), $\vec{E}$ inside the conductor must be *zero*. If it were not, excess charges inside the conductor would experience a force and be pushed to some new location. Note that this statement only applies *inside* the volume of the conductor. Outside the charged block of conductor the electric field need not be zero.

Let's see what Gauss's law tells us in this situation. Imagine a Gaussian surface that lies completely inside the volume of a conductor that carries excess charge (Figure 16-23). Since $\vec{E} = 0$ everywhere inside the conductor, the net electric flux Φ through this surface is zero. From Gauss's law, Equation 16-9, Φ is equal to the net charge q_{encl} enclosed by the surface divided by ε_0. So the net charge inside the Gaussian surface is zero. This holds true for any Gaussian surface, no matter how small, that lies entirely within the conductor. So there can be *no* excess charge within the volume of the conductor. Instead, *all of the excess charge on a conductor in equilibrium must reside on the surface.* That's why we depicted the charged copper rod in Figure 16-5b with its excess charge spread over the surface of the rod, not its interior.

Our conclusion that $\vec{E} = 0$ inside a conductor holds true *only* if all of the charges are at rest. We will see in later chapters that if an electric current is present inside a conductor, there must be a nonzero electric field to sustain the flow of electric charge.

Example 16-8 Gauss's Law and Surface Charge

A conducting sphere of radius R_1 carries excess charge +7 C. This sphere is enclosed within a concentric spherical conducting shell of inner radius R_2 and outer radius R_3. The conducting shell carries a net charge +2 C. How much charge is on the inner surface of the conducting shell? How much is on the outer surface?

Set Up

In equilibrium the electric field inside the volume of the conducting shell must be zero. So if we imagine a spherical Gaussian surface that lies inside the shell, so that its radius r is intermediate between R_2 and R_3, the net electric flux through that surface will be zero. From Gauss's law, that means that the net charge enclosed by the Gaussian surface must be zero. The +2-C charge on the shell will arrange itself on the surfaces of the shell in order to make that happen.

Gauss's law:

$$\Phi = \frac{q_{encl}}{\varepsilon_0} \qquad (16\text{-}9)$$

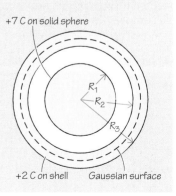

+7 C on solid sphere

+2 C on shell   Gaussian surface

Solve

The charge enclosed within our Gaussian surface includes the charge of the central sphere ($q_{sphere} = +7$ C) and whatever charge q_{inner} is present on the inner surface of the shell. These charges must add to zero.

For the Gaussian surface,

$$q_{encl} = q_{sphere} + q_{inner} = 0$$

So the charge on the inner surface of the shell equals the negative of the charge on the central sphere:

$$q_{inner} = -q_{sphere} = -7 \text{ C}$$

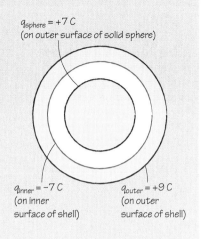

$q_{sphere} = +7$ C
(on outer surface of solid sphere)

$q_{inner} = -7$ C
(on inner surface of shell)

$q_{outer} = +9$ C
(on outer surface of shell)

The total charge on the shell is $q_{shell} = +2$ C. Because all of the excess charge on a conductor resides on its surfaces, q_{shell} is the sum of the charge on the inner and outer surfaces.

$$q_{shell} = q_{inner} + q_{outer}$$

So

$$\begin{aligned} q_{outer} &= q_{shell} - q_{inner} \\ &= (+2 \text{ C}) - (-7 \text{ C}) \\ &= +9 \text{ C} \end{aligned}$$

Reflect

It may seem odd that while the shell as a whole carries a *positive* excess charge of +2 C, the inner surface of the shell has a *negative* charge. What's happened is that when the positively charged sphere is placed inside the shell, some of the free electrons in the shell are drawn to the shell's inner surface, giving it a negative charge of −7 C. This leaves a deficit of electrons—a positive charge of +7 C—on the outside surface. This in turn adds to the +2 C of excess charge to give the outer surface of the shell a charge of +9 C.

? Got the Concept? 16-9   Size of a Gaussian Surface

In our derivation of Equation 16-13 for the electric field due to an infinite charged sheet, we used a Gaussian surface in the shape of a cylinder whose top and bottom faces had area A. If we used a larger cylinder with top and bottom faces of area $2A$, how would this change our result for the electric field? (a) E_z would be four times larger; (b) E_z would be twice as large; (c) E_z would be the same; (d) E_z would be one-half as great; (e) E_z would be one-fourth as great.

Take-Home Message for Section 16-7

✔ Gauss's law can be used to calculate the electric field in certain cases where the charge distribution has a high degree of symmetry. These include a spherical distribution of charge and a very large, uniformly charged sheet.

✔ Gauss's law tells us that when excess charge added to a conductor is allowed to come to rest in equilibrium, the excess charge resides only on the surfaces of the conductor.

Key Terms

charged
conductors
coulomb
Coulomb's constant
Coulomb's law
electric charge
electric dipole

electric field lines
electric flux
electric force
insulators
Gauss's law
Gaussian surface
neutral

permittivity of free space
point charges
polarization
semiconductors
surface charge density

Chapter Summary

Topic	Equation or Figure

Electric charge: Ordinary matter contains equal amounts of positive and negative charge. Charge can be transferred from one object to another, for example, by rubbing. In this and all other processes, charge is conserved: It can be moved from place to place but can neither be created nor destroyed.

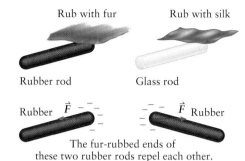

(Figure 16-2b)

Rub with fur   Rub with silk

Rubber rod   Glass rod

The fur-rubbed ends of these two rubber rods repel each other.

The silk-rubbed ends of these two glass rods repel each other.

The fur-rubbed end of the rubber rod attracts the silk-rubbed end of the glass rod.

Conductors, insulators, and semiconductors: In a conductor, charges are free to move with ease; in an insulator, charges can move very little. Semiconductors have properties intermediate between those of conductors and insulators.

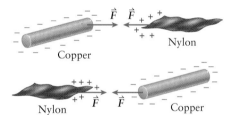

The nylon-rubbed end of the copper rod attracts the nylon with which it was rubbed...

...and the end of the copper rod that was not rubbed also feels a strong attraction.

(Figure 16-5b)

Coulomb's law: The electric forces that two point charges exert on each other are proportional to the magnitudes of the charges and inversely proportional to the square of the distance between the charges. These forces are attractive if the two charges have opposite signs, and repulsive if the two charges have the same sign.

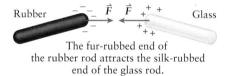

Any two point charges q_1 and q_2 exert equally strong electric forces on each other.

Coulomb constant

Absolute values of the point charges

Distance between the point charges

$$F_{q_1 \text{ on } q_2} = F_{q_2 \text{ on } q_1} = \frac{k|q_1||q_2|}{r^2}$$

(16-1)

Electric field: We can regard the interaction between charges as a two-step process: One charge sets up an electric field, and the other charge responds to that field. The electric field points away from

If a particle with charge q is placed at a position where the electric field due to other charges is $\vec{E}$...

$$\vec{F} = q\vec{E}$$

(16-2)

...then the electric force on the particle is $\vec{F} = q\vec{E}$.

a positive charge and toward a negative charge. The electric field of a single point charge is given by a simple equation; the field due to a combination of charges is the vector sum of the fields due to the individual charges.

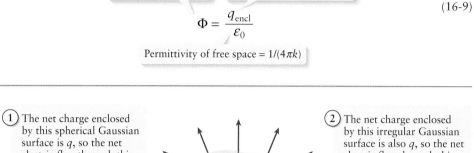

Coulomb constant   Absolute value of charge Q

Magnitude of the electric field due to a point charge Q

$$E = \frac{k|Q|}{r^2} \qquad (16\text{-}4)$$

Distance between point charge Q and the location where the field is measured

Gauss's Law: The electric flux through a surface equals the area of that surface multiplied by the component of electric field perpendicular to the surface. Gauss's law states that the net electric flux through a closed surface (one that encloses a volume) is proportional to the net charge enclosed within that volume. Gauss's law can be used to determine the electric field in situations where the charge distribution is highly symmetric. It also tells us that any excess charge on a conductor resides on the surfaces of the conductor.

Electric flux through a closed surface   Net amount of charge enclosed within the surface

$$\Phi = \frac{q_{\text{encl}}}{\varepsilon_0} \qquad (16\text{-}9)$$

Permittivity of free space = $1/(4\pi k)$

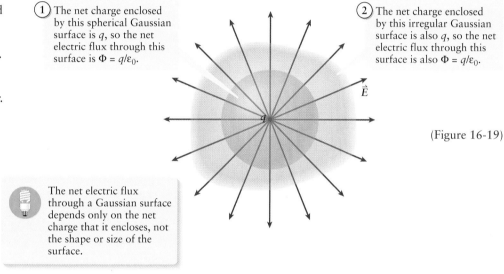

① The net charge enclosed by this spherical Gaussian surface is q, so the net electric flux through this surface is $\Phi = q/\varepsilon_0$.

② The net charge enclosed by this irregular Gaussian surface is also q, so the net electric flux through this surface is also $\Phi = q/\varepsilon_0$.

$\vec{E}$

q

(Figure 16-19)

The net electric flux through a Gaussian surface depends only on the net charge that it encloses, not the shape or size of the surface.

Answer to What do you think? Question

(c) The magnitude of the electric field of a charged object measured at a distance r from the object is proportional to $1/r^2$ (see Section 16-5). If r is made one-half as great (decreased from $r = 2$ m to $r = 1$ m), the electric field magnitude changes by a factor of $1/(1/2)^2 = 1/(1/4) = 4$.

Answers to Got the Concept? Questions

16-1 (c) When the rubber rod is rubbed with fur, electrons are transferred to the rod from the fur. The electrons have mass, so the rod gains a little mass and the fur loses an equal amount of mass. By contrast, electrons are transferred from the glass rod to the silk when these two objects are rubbed together. As a result, the glass rod loses a little mass and the silk gains an equal amount of mass. The amount of mass involved is very small but is not zero.

16-2 (c) A sodium atom is neutral because it has 11 protons in its nucleus (each of charge $+e$) and 11 electrons (each of charge $-e$). If a sodium atom loses one of its 11 electrons by transferring it to another atom, it is left with 11 protons with a combined charge of $+11e$ and 10 electrons with a combined charge of $-10e$, and so it has a net positive charge of $+e$. In NaCl, the electron is transferred to the chlorine atom, which then has a net negative charge of $-e$.

16-3   (d) Equation 16-1 tells us that the magnitude of the electric force between two charged objects is directly proportional to the product of the two charges. If we remove only half of the electrons from the neutral water drop, the ball of electrons has half the negative charge as before and the water drop has half the positive charge as before. So the electric force between the water drop and the ball of electrons will be $\left(\frac{1}{2}\right) \times \left(\frac{1}{2}\right) = \frac{1}{4}$ as great as before.

16-4   (b) Equation 16-2, $\vec{F} = q\vec{E}$, tells us the force $\vec{F}$ that acts on a charge q due to the electric field $\vec{E}$ produced by *other* charges. So to find the force that acts on the positive charge in Figure 16-14 due to the negative charge, in Equation 16-2 we let q be the positive charge (the charge that experiences the force) and let $\vec{E}$ be the electric field produced at the position of the positive charge by the negative charge (the other charge that exerts the force).

16-5   (d) If both charges were negative, the electric field at point P due to each charge would point directly toward that charge. So the electric field due to the right-hand charge would point down and to the right, while the electric field due to the left-hand charge would point down and to the left. Since point P is the same distance from each charge, and each charge has the same magnitude, the electric fields due to the two charges would have the same magnitude. The horizontal components of the two fields cancel (one points left and the other points right), so what remains is a net electric field that points straight down.

16-6   (c) The flux remains the same. According to Gauss's law, the net electric flux due to an enclosed charge does not depend on the size of the surface that encloses the charge.

16-7   (b) The flux remains the same. According to Gauss's law, the net electric flux due to an enclosed charge does not depend on the shape of the surface that encloses the charge.

16-8   (b) The flux remains the same. According to Gauss's law, the electric flux through a closed surface depends only on the charge enclosed *within* the surface. A charge outside the surface will cause positive flux on one part of the surface (where its electric field points out of the surface) and negative flux on another part of the surface (where its electric field points into the surface), but will have no net effect on the flux through the surface as a whole.

16-9   (c) The result does not depend on the size of the cylinder we choose. If we double the area, the flux through the top and bottom faces would double, but the amount of enclosed charge would double as well. In the derivation of Equation 16-14 these factors would cancel, so our result for E_z would be the same.

Questions and Problems

In a few problems, you are given more data than you actually need; in a few other problems, you are required to supply data from your general knowledge, outside sources, or informed estimate.

Interpret as significant all digits in numerical values that have trailing zeros and no decimal points.

For all problems, use $g = 9.80 \text{ m/s}^2$ for the free-fall acceleration due to gravity. Neglect friction and air resistance unless instructed to do otherwise.

• Basic, single-concept problem

•• Intermediate-level problem, may require synthesis of concepts and multiple steps

••• Challenging problem

SSM Solution is in Student Solutions Manual

Conceptual Questions

1. •Discuss the similarities and differences between the gravitational and electric forces.

2. •Why is the gravitational force usually ignored in problems on the scale of particles such as electrons and protons?

3. •How, if at all, would the physical universe be different if the proton were negatively charged and the electron were positively charged?

4. •How, if at all, would the physical universe be different if the proton's charge was very slightly larger in magnitude than the electron's charge?

5. •When an initially electrically neutral object acquires a net positive charge, does its mass increase or decrease? Why? SSM

6. •When you remove socks from a hot dryer, they tend to cling to everything. Two identical socks, however, usually repel. Why?

7. •Describe a set of experiments that might be used to determine if you have discovered a third type of charge other than positive and negative.

8. •How does a person become "charged" as he or she shuffles across a carpet, wearing cloth slippers, on a dry winter day?

9. •After combing your hair with a plastic comb, you find that when you bring the comb near a small bit of paper, the bit of paper moves toward the comb. Then, shortly after the paper touches the comb, it moves away from the comb. Explain these observations. SSM

10. •After combing your hair with a plastic comb, you find that when you bring the comb near an empty aluminum soft-drink can that is lying on its side on a nonconducting table-top, the can rolls toward the comb. After being touched by the comb, the can is still attracted by the comb. Explain these observations.

11. •(a) A positively charged glass rod attracts a lighter object suspended by a thread. Does it follow that the object is negatively charged? (b) If, instead, the rod repels it, does it follow that the suspended object is positively charged?

12. •Some days it can be frustrating to attempt to demonstrate electrostatic phenomena for a physics class. An experiment that works beautifully one day may fail the next day if the weather has changed. Air-conditioning helps a lot while demonstrating the phenomena during the summer. Why?

13. •(a) What are the advantages of thinking of the force on a charge at a point P as being exerted by an electric field at P, rather than by other charges at other locations? (b) Is the convenience of the field as a calculation device worth inventing a new physical quantity? Or is there more to the field concept than that? SSM

14. •Do electric field lines point along the trajectory of positively charged particles? Why or why not?

15. •An electron and a proton are released in a region of space where the electric field is vertically downward. How do the electric forces acting on the electron and proton compare?

16. •Inside a uniform spherical charge distribution, why is it that as one moves out from the center, the electric field increases as r rather than decreases as $1/r^2$?

17. •Is the electric field $\vec{E}$ in Gauss's law only the electric field due to the charge inside the Gaussian surface, or is it the total electric field due to all charges both inside and outside the surface? Explain your answer. SSM

18. •If the net electric flux out of a closed surface is zero, does that mean the charge density must be zero everywhere inside the surface? Explain your answer.

Multiple-Choice Questions

19. •Electric charges of the opposite sign
   A. exert no force on each other.
   B. attract each other.
   C. repel each other.
   D. repel and attract each other.
   E. repel and attract each other depending on the magnitude of the charges.

20. •If two uncharged objects are rubbed together and one of them acquires a negative charge, then the other one
   A. remains uncharged.
   B. also acquires a negative charge.
   C. acquires a positive charge.
   D. acquires a positive charge equal to twice the negative charge.
   E. acquires a positive charge equal to half the negative charge.

21. •Metal sphere A has a charge of $-Q$. An identical metal sphere B has a charge of $+2Q$. The magnitude of the electric force on B due to A is F. The magnitude of the electric force on A due to B is
   A. $F/4$.
   B. $F/2$.
   C. F.
   D. $2F$.
   E. $4F$. SSM

22. •A balloon can be charged by rubbing it with your sleeve while holding it in your hand. You can conclude from this that the balloon is a(n)
   A. conductor.
   B. insulator.
   C. neutral object.
   D. Gaussian surface.
   E. semiconductor.

23. •A positively charged rod is brought near one end of an uncharged metal bar. The end of the metal bar farthest from the charged rod will be charged

   A. positively.
   B. negatively.
   C. neutral.
   D. twice as much as the end nearest the rod.
   E. none of the above ways. SSM

24. •A free positive charge released in an electric field will
   A. remain at rest.
   B. accelerate in the direction opposite to the electric field.
   C. accelerate in the direction perpendicular to the electric field.
   D. accelerate in the same direction as the electric field.
   E. accelerate in a circular path.

25. •Consider a point charge $+Q$ located outside a closed surface such as a sphere bound by the black circle in Figure 16-24. What is the net electric flux through the closed surface?
   A. $\dfrac{+Q}{\varepsilon_0}$
   B. $\dfrac{-Q}{\varepsilon_0}$
   C. 0
   D. $\dfrac{+2Q}{\varepsilon_0}$
   E. $\dfrac{-2Q}{\varepsilon_0}$

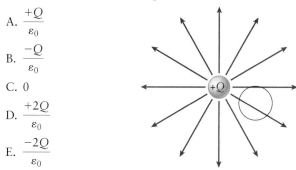

Figure 16-24 Problem 25

26. •If a charge is located at the center of a spherical volume and the electric flux through the surface of the sphere is Φ, what would be the flux through the surface if the radius of the sphere were tripled?
   A. 3Φ
   B. 9Φ
   C. Φ
   D. $\Phi/3$
   E. $\Phi/9$

27. •A point charge $+Q$ is at the center of a spherical conducting shell of inner radius R_1 and outer radius R_2, as shown in Figure 16-25. The charge on the inner surface of the shell is
   A. $+Q$.
   B. $-Q$.
   C. 0.
   D. $+Q/2$.
   E. $-Q/2$. SSM

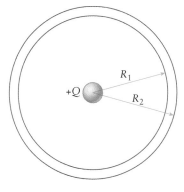

Figure 16-25 Problem 27

Estimation/Numerical Analysis

28. ••(a) Estimate the positive charge in you due to the protons in the molecules present in your body. (b) What is the net charge in your body?

29. • Estimate the amount of charge needed on both a comb and on a bit of tissue paper in order to generate an electric force of sufficient magnitude to support the weight of the paper.

30. •(a) Give a rough value for the force between two charges that possess microcoulombs of charge. (b) What if the charges are in the nanocoulomb range? Assume distances of separation in the centimeter range.

31. •For 10-cm distances of separation, what approximate charges will lead to electrostatic forces of ~10 N? SSM

32. •Estimate the electric field required to levitate a 1-g plastic sphere carrying a net charge of 10 μC near Earth's surface.

33. •Estimate the electric flux passing through the passenger window of your car when you become charged up after sliding into your vehicle.

34. •The magnitude of the repulsive force (F) between two +2.5-μC charges as a function of the distance of separation (r) is listed in the following table. Empirically derive a relationship between F and r by doing a curve fit to find the power n of r in the following formula:

$$F = \frac{kq_1q_2}{r^n}$$

Use a graphing calculator or spreadsheet.

r (m)	F (N)	r (m)	F (N)
0.003	5500	0.080	6
0.004	3000	0.100	5
0.005	2000	0.200	1
0.010	600	0.300	0.5
0.020	175	0.400	0.35
0.040	30	0.500	0.25
0.050	10	0.600	0.15

Problems

16-1 Electric forces and electric charges are all around you— and within you

16-2 Matter contains positive and negative electric charge

35. •The nucleus of a copper atom has 29 protons and 35 neutrons. What is the total charge of the nucleus?

36. •Five electrons are added to 1.00 C of positive charge. What is the net charge of the system?

37. ••How many coulombs of negative charge are there in 0.500 kg of water? SSM

38. •An ion has 17 protons, 18 neutrons, and 18 electrons. What is the net charge of the ion?

39. ••The charge per unit length on a glass rod is 0.00500 C/m. If the rod is 1 mm long, how many electrons have been removed from the glass rod?

40. •How many electrons must be transferred from an object to produce a charge of 1.60 C?

16-3 Charge can flow freely in a conductor but not in an insulator

41. •Suppose 2.00 C of positive charge is distributed evenly throughout a sphere of 1.27-cm radius. (a) What is the charge per unit volume for this situation? (b) Is the sphere insulating or conducting? How do you know?

42. •The maximum amount of charge that can be collected on a Van de Graaff generator's conducting sphere (30-cm diameter) is about 30 μC. Calculate the surface charge density, σ, of the sphere in C/m^2.

43. •**Biology** Most workers in nanotechnology are actively monitored for excess static charge buildup. The human body acts like an insulator as one walks across a carpet, collecting -50 nC per step. (a) What charge buildup will a worker in a manufacturing plant accumulate if she walks 25 steps? (b) How many electrons are present in that amount of charge? (c) If a delicate manufacturing process can be damaged by an electrical discharge of greater than 10^{12} electrons, what is the maximum number of steps that any worker should be allowed to take before touching the components? SSM

16-4 Coulomb's law describes the force between charged objects

44. •Two point charges are separated by a distance of 20.0 cm. The numerical value of one charge is twice that of the other. If each charge exerts a force of magnitude 45.0 N on the other, find the magnitude of the charges.

45. •The mass of an electron is 9.11×10^{-31} kg. How far apart would two electrons have to be in order for the electric force exerted by each on the other to be equal to the weight of an electron? SSM

46. •Charge A, +5.00 μC, is positioned at the origin of a coordinate system. Charge B, -3.00 μC, is fixed on the x axis at $x = 3.00$ m. (a) Determine the magnitude and direction of the force that charge B exerts on charge A. (b) What is the magnitude and direction of the force that charge A exerts on charge B?

47. ••Point charge A with charge $q_A = +3.00$ μC is located at the origin. Point charge B with charge $q_B = -4.00$ μC is on the x axis at $x = 3.00$ m. Point charge C with charge $q_B = -2.00$ μC is on the x axis at $x = 6.00$ m. And point charge D with charge $q_D = +6.00$ μC is on the x axis at $x = 8.00$ m. What is the net electric force on point charge A due to the other three charges?

48. ••A charge of +3.00 μC is located at the origin, and a second charge of -2.00 μC is located on the $x-y$ plane at the point (30.0 cm, 20.0 cm). Determine the electric force exerted by the -2.00 μC charge on the 3.00 μC charge.

49. ••A point charge with charge $q_1 = +5.00$ nC is fixed at the origin. A second point charge with charge $q_2 = -7.00$ nC is located on the x axis at $x = 5.00$ m. Where along the x axis will a third point charge of $q = +2.00$ nC charge need to be for the net electric force on it due to the two fixed charges to be equal to zero? SSM

50. ••Two charges lie on the x axis, a -2.00-μC charge at the origin and a +3.00-μC charge at $x = 0.100$ m. At what position along the x axis, if any, should a third +4.00-μC charge be placed so that the net force on the third charge is equal to zero?

51. ••Point charge A with a charge of +3.00 μC is located at the origin. Point charge B with a charge of +6.00 μC is located on the x axis at $x = 7.00$ cm. And point charge C with a charge of +2.00 μC is located on the y axis at $y = 6.00$ cm. What is the net force (magnitude and direction) exerted on each charge by the others? SSM

52. •A charge q_1 equal to 0.600 μC is at the origin, and a second charge q_2 equal to 0.800 μC is on the x axis at 5.00 cm. (a) Find the force (magnitude and direction) that each charge exerts on the other. (b) How would your answer change if q_2 were -0.800 μC?

16-5 The concept of electric field helps us visualize how charges exert forces at a distance

53. •At point P in Figure 16-26, the electric field is zero. (a) What are the signs of q_1 and q_2? (b) Describe their magnitudes.

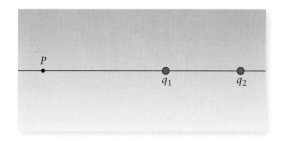

Figure 16-26 Problem 53

54. •Near the surface of Earth, an electric field points radially downward and has a magnitude of approximately 100 N/C. What charge (magnitude and sign) would have to be placed on a penny that has a mass of 3.11 g to cause it to rise into the air with an upward acceleration of 0.190 m/s²?

55. ••Two charges are placed on the x axis, $+5.00$ μC at the origin and -10.0 μC at $x = 10.0$ cm. (a) Find the electric field on the x axis at $x = 6.00$ cm. (b) At what point(s) on the x axis is the electric field zero?

56. ••In Figure 16-27, the electric field at the origin is zero. If q_1 is 1.00×10^{-7} C, what is q_2?

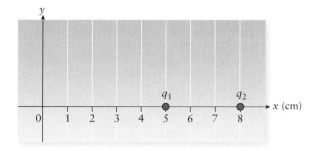

Figure 16-27 Problems 56 and 57

57. •In Figure 16-27, if $q_1 = 1.00 \times 10^{-7}$ C and $q_2 = 2.00 \times 10^{-7}$ C, (a) what is the electric field $\vec{E}$ at the point $(x, y) = (0.00 \text{ cm}, 3.00 \text{ cm})$? (b) What is the force $\vec{F}$ acting on an electron at that position?

58. •In Figure 16-27, if $q_1 = 1.00 \times 10^{-7}$ C and $q_2 = 2.00 \times 10^{-7}$ C, (a) what is the electric field $\vec{E}$ at the point $(x, y) = (6.00 \text{ m}, 3.00 \text{ m})$? (b) What is the force $\vec{F}$ acting on a proton at that position?

59. ••In the Bohr model, the hydrogen atom consists of an electron in a circular orbit of radius $a_0 = 5.29 \times 10^{-11}$ m around the nucleus. Using this model, and ignoring relativistic effects, what is the speed of the electron?

16-6 Gauss's law gives us more insight into the electric field

60. •A rectangular area is rotated in a uniform electric field, from a position where the maximum electric flux goes through it to an orientation where only half the maximum flux goes through it. What is the angle of rotation?

61. •A point charge of 4.00×10^{-12} C is located at the center of a cubical Gaussian surface. What is the electric flux through each face of the cube?

62. •The net electric flux through a cubic box with sides that are 20.0 cm long is 4.80×10^3 N·m²/C. What charge is enclosed by the box?

63. ••A 10.0-cm-long uniformly charged plastic rod is sealed inside a plastic bag. The net electric flux through the bag is 7.50×10^5 N·m²/C. What is the linear charge density (charge per unit length) on the rod? SSM

64. ••Figure 16-28 shows a prism-shaped object that is 40.0 cm high, 30.0 cm deep, and 80.0 cm long. The prism is immersed in a uniform electric field of 500 N/C directed parallel to the x axis. (a) Calculate the electric flux out of each of its five faces and (b) the net electric flux out of the entire closed surface. (c) If in addition to the given electric field the prism also enclosed a point charge of -2.00 μC, qualitatively how would your answers above change, if at all?

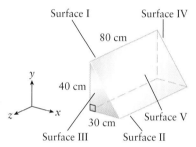

Figure 16-28 Problem 64

16-7 In certain situations, Gauss's law helps us to calculate the electric field and to determine how charge is distributed

65. ••Use Gauss's law to find an expression for the electric field just outside the surface of a sphere carrying a uniform surface charge density σ (charge per unit area). SSM

66. •Determine the charge density for each of the following cases (assume that all densities are uniform): (a) a solid cylinder that has a length L, has a radius R, and carries a charge Q throughout its volume; (b) a flat plate (very thin) that has a width W, has a length L, and carries a charge Q on its surface area; (c) a solid sphere of radius R carrying a charge Q throughout its volume; and (d) a hollow sphere of radius R carrying a charge Q over its surface area.

67. ••An electric field of magnitude 400 N/C exists at all points just outside the surface of a 2.00-cm-diameter steel ball bearing. Assuming the ball bearing is in electrostatic equilibrium, (a) what is the total charge on the ball? (b) What is the surface charge density on the ball?

68. ••Consider an infinite plane with a uniform charge distribution σ. (a) Use Gauss's law to find an expression for the electric field due to the plane. (b) What field would be created by two equal but oppositely charged parallel planes? Consider the region between the planes as well as the two regions outside the planes. The simplest way to express your answers is in terms of the surface charge density σ (the charge per unit area) on the plane.

69. ••A -3.20-μC charge sits in static equilibrium in the center of a conducting spherical shell that has an inner radius of

2.50 cm and an outer radius of 3.50 cm. The shell has a net charge of $-5.80\ \mu\text{C}$. Determine the charge on each surface of the shell and the electric field just outside the shell.

General Problems

70. ••A spherical party balloon that is 25 cm in diameter contains helium at room temperature (20°C) and at a pressure of 1.3 atm. If one electron could be stripped from every helium atom in the balloon and removed to a satellite orbiting Earth 22,000 mi (32,187 km) above the planet, with what force would the balloon and the satellite attract each other?

71. ••A plutonium-242 atom has a nucleus of 94 protons and 148 neutrons and has 94 electrons. The diameter of its nucleus is approximately 15×10^{-15} m. (a) Make a reasonable physical argument as to why we can treat the nucleus as a point charge for points outside of it. (b) Plutonium decays radioactively by emitting an *alpha particle* from its nucleus. The mass of the alpha particle is 6.6×10^{-27} kg, and the particle has two protons and two neutrons. If the alpha particle comes from the surface of the ^{242}Pu nucleus, what is its greatest acceleration?

72. ••When a test charge of $+5.00$ nC is placed at a certain point, the force that acts on it has a magnitude of 0.0800 N and is directed northeast. (a) If the test charge were -2.00 nC instead, what force would act on it? (b) What is the electric field at the point in question?

73. ••**Biology** A red blood cell may carry an excess charge of about -2.5×10^{-12} C distributed uniformly over its surface. The cells, modeled as spheres, are approximately 7.5 μm in diameter and have a mass of 9.0×10^{-14} kg. (a) How many excess electrons does a typical red blood cell carry? (b) Does the mass of the extra electrons appreciably affect the mass of the cell? To find out, calculate the ratio of the mass of the extra electrons to the mass of the cell without the excess charge. (c) What is the surface charge density σ on the red blood cell? Express your answer in C/m^2 and in electrons/m^2. SSM

74. ••Three point charges are placed on the $x-y$ plane: a $+50.0$ nC charge at the origin, a -50.0 nC charge on the x axis at 10.0 cm, and a $+150$ nC charge at the point (10.0 cm, 8.00 cm). (a) Find the total electric force on the $+150$ nC charge due to the other two. (b) What is the electric field at the location of the $+150$ nC charge due to the presence of the other two charges?

75. ••Two small spheres each have a mass m of 0.100 g and are suspended as pendulums by light insulating strings from a common point, as shown in **Figure 16-29**. The spheres are given the same electric charge, and the two come to equilibrium when each string is at an angle of $\theta = 3.00°$ with the vertical. If each string is 1.00 m long, what is the magnitude of the charge on each sphere?

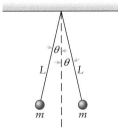

Figure 16-29 Problem 75

76. ••A small 1.00-g plastic ball that has a charge q of 1.00 C is suspended by a string that has a length L of 1.00 m in a uniform electric field, as shown in **Figure 16-30**. If the ball is in equilibrium when the string makes a 9.80° angle with the vertical as indicated by θ, what is the electric field strength?

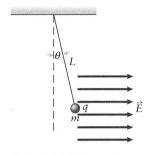

Figure 16-30 Problem 76

77. ••**Biology** The 9-inch-long elephant nose fish in the Congo River generates a weak electric field around its body using an organ in its tail. When small prey (or even potential mates) swim within a few feet of the fish, they perturb the electric field. The change in the field is picked up by electric sensor cells in the skin of the elephant nose. These remarkable fish can detect changes in the electric field as small as $3.0\ \mu\text{N/C}$. (a) How much charge (modeled as a point charge) in the fish would be needed to produce such a change in the electric field at a distance of 75 cm? (b) How many electrons would be required to create the charge? SSM

78. •••Three charges (q_A, q_B, and q_C) are placed at the vertices of the equilateral triangle that has sides of length s in **Figure 16-31**. Derive expressions for the electric field at (a) X (at the center of the triangle), (b) Y (at the midpoint of the side between q_B and q_C), and (c) Z (at the midpoint of the side between q_A and q_C). (d) Now use the following numerical values and calculate the electric field at those same points: $s = 10.0$ cm, $q_A = +20.0$ nC, $q_B = -8.00$ nC, and $q_C = -10.0$ nC.

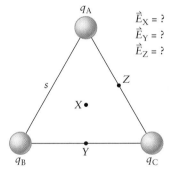

Figure 16-31 Problem 78

79. ••Calculate the electric field at the center of the hexagon shown in **Figure 16-32**. Assume the sides of the hexagon are all 5.00 cm long. SSM

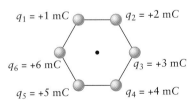

Figure 16-32 Problem 79

80. ••Electric fields up to 2.00×10^5 N/C have been measured inside of clouds during electrical storms. Neglect the drag force due to the air in the cloud and any collisions with air molecules. (a) What acceleration does the maximum electric field produce

for protons in the cloud? Express your answer in SI units and as a fraction of g. (b) If the electric field remains constant, how far will the proton have to travel to reach 10% of the speed of light $(3.00 \times 10^8 \, \text{m/s})$ if it started with negligible speed? (c) Can you neglect the effects of gravity? Explain your answer.

81. ••An electron with an initial speed of $5.00 \times 10^5 \, \text{m/s}$ enters a region in which there is an electric field directed along its direction of motion. If the electron travels 5.00 cm in the field before being stopped, what are the magnitude and direction of the electric field?

82. ••An electron, released in a region where the electric field is uniform, is observed to have an acceleration of $3.00 \times 10^{14} \, \text{m/s}^2$ in the positive x direction. (a) Determine the electric field producing the acceleration. (b) Assuming the electron is released from rest, determine the time required for it to reach a speed of 11,200 m/s, the escape speed from Earth's surface.

83. ••**Chemistry** The iron atom (Fe) has 26 protons, 30 neutrons, and 26 electrons. The diameter of the atom is approximately 1.0×10^{-10} m, while the diameter of its nucleus is about 9.2×10^{-15} m. (You can reasonably model the nucleus as a uniform sphere of charge.) What are the magnitude and direction of the electric field that the nucleus produces (a) just outside the surface of the nucleus and (b) at the distance of the outermost electron? (c) What would be the magnitude and direction of the acceleration of the outermost electron due only to the nucleus, neglecting any force due to the other electrons? SSM

84. •••An electron with kinetic energy K is traveling along the $+x$ axis, which is along the axis of a cathode-ray tube as shown in Figure 16-33. There is an electric field $E = 12.00 \times 10^4 \, \text{N/C}$ pointed in the $+y$ direction between the deflection plates, which are 0.0600 m long and are separated by 0.0200 m. Determine the minimum initial kinetic energy the electron can have and still avoid colliding with one of the plates.

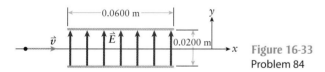
Figure 16-33 Problem 84

85. ••In the famous Millikan oil-drop experiment, tiny spherical droplets of oil are sprayed into a uniform vertical electric field. The drops get a very small charge (just a few electrons) due to friction with the atomizer as they are sprayed. The field is adjusted until the drop (which is viewed through a small telescope) is just balanced against gravity and therefore remains stationary. Using the measured value of the electric field, we can calculate the charge on the drop and from this calculate the charge e of the electron. In one apparatus, the drops are 1.10 μm in diameter and the oil has a density of 0.850 g/cm³. (a) If the drops are negatively charged, which way should the electric field point to hold them stationary (up or down)? (b) Why? (c) If a certain drop contains four excess electrons, what magnitude electric field is needed to hold it stationary? (d) You measure a balancing field of 5183 N/C for another drop. How many excess electrons are on this drop?

86. •••The electric field is zero everywhere except in the region $0 \leq x \leq 3.00$ cm, where there is a uniform electric field of 100 N/C in the $+y$ direction. A proton is moving in the $+x$ direction with a speed of $v = 1.00 \times 10^6$ m/s. When the proton passes through the region $0 \leq x \leq 3.00$ cm, the electric field exerts a force on it. (a) When the x coordinate of the proton's position is 3.00 cm, what is its velocity and what is the y coordinate of its position? (b) When the x coordinate of its position equals 10.0 cm, what is its velocity and what is the y coordinate of its position?

87. •••Two hollow, concentric, spherical shells are covered with charge (Figure 16-34). The inner sphere has a radius R_i and a surface charge density of $+\sigma_i$, while the outer sphere has a radius R_o and a surface charge density of $-\sigma_o$. Derive an expression for the electric field in the following three radial regions: (a) $r < R_i$, (b) $R_i < r < R_o$, and (c) $r > R_o$. SSM

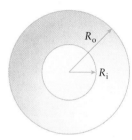
Figure 16-34 Problem 87

88. ••A charge $q_1 = +2q$ is at the origin, and a charge $q_2 = -q$ is on the x axis at $x = a$. Find expressions for the total electric field on the x axis in each of the regions (a) $x < 0$; (b) $0 < x < a$; and (c) $a < x$. (d) Determine all points on the x axis where the electric field is zero. (e) Use a graphing calculator or spreadsheet to make a plot of E_x versus x for all points on the x axis, and (f) qualitatively discuss what happens for $-\infty < x < \infty$.

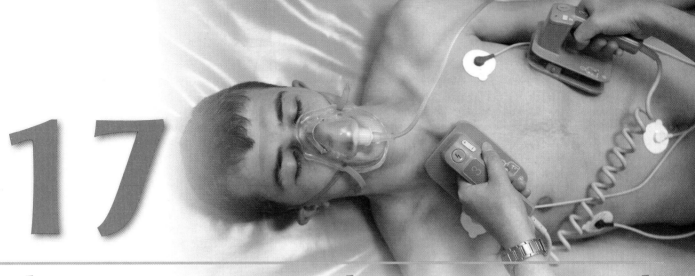

17

Electrostatics II: Electric Potential Energy and Electric Potential

What do you think?

A cardiac defibrillator works by delivering an electric shock to a malfunctioning heart to jump-start it back into its normal rhythm. Electric energy is stored for this purpose in the defibrillator by separating positive charge $+q$ from an equal amount of negative charge $-q$. To release this energy to the heart, conducting paddles are placed on the patient's chest, and charge $-q$ in the form of electrons is allowed to flow through the paddles and patient until it reaches the stored charge $+q$ within the defibrillator. If the value q of the charge stored in a defibrillator is doubled, by what factor will the stored energy increase? (a) $\sqrt{2}$; (b) 2; (c) 4; (d) 8; (e) 16.

In this chapter, your goals are to:

- (17-1) Explain the significance of energy in electrostatics.
- (17-2) Discuss how the work done on a charged particle by the electric field relates to changes in electric potential energy.
- (17-3) Explain the difference between electric potential and electric potential energy.
- (17-4) Recognize why equipotential surfaces are perpendicular to electric field lines.
- (17-5) Explain what is meant by capacitance, and describe how the capacitance of a parallel-plate capacitor depends on the size of the plates and their separation.
- (17-6) Calculate the electric energy stored in a capacitor.
- (17-7) Explain how to treat capacitors attached in series and in parallel as a single equivalent capacitance.
- (17-8) Describe how the capacitance of a capacitor increases when an insulating material other than a vacuum is placed between the plates of the capacitor.

To master this chapter, you should review:

- (6-2)   The work done by a force on a moving object.
- (6-6)   The relationship between a conservative force and the potential energy associated with that force.
- (10-3)   The generalized expression for gravitational potential energy.
- (16-5, 16-7) The electric field due to a point charge and to a sheet of charge.

17-1 Electric energy is important in nature, technology, and biological systems

Everyone is familiar with the idea that *electricity* implies *energy*. The energy to run your computer, lighting, and television is delivered to your home in the form of electricity. A bolt of lightning releases so much energy that it heats air to a temperature of 30,000°C, causing the air to glow with the characteristic light that we call a lightning flash (see Figure 16-1b). A defibrillator saves lives by delivering a sharp punch of electric energy to a malfunctioning heart (see the photo that opens this chapter). When you purchase an ordinary electric battery, you are really paying for the electric energy that the battery can deliver to whatever electric circuit you plug it into (Figure 17-1a). Some specialized species of fish such as electric rays (Figure 17-1b) are equipped with organic batteries; they can deliver an intense burst of electric energy to stun their prey or ward off predators.

In this chapter we'll make clear what is meant by electric potential energy. We'll gain insight into this new kind of energy by considering the similarities to and differ-

(a)

An ordinary battery is a source of electric energy. Devices that use multiple batteries have greater energy requirements.

(b)

An electric ray (order *Torpediniformes*) is equipped with a large number of organic batteries that work together to deliver intense electric shocks.

Figure 17-1 **Electric energy** Sources of electric energy in (a) technology and (b) nature.

ences from gravitational potential energy. Just as there is a change in gravitational potential energy when a massive object moves up or down in the presence of Earth's gravitational field, there is a change in electric potential energy when a charged object moves along with or opposite to an electric field. We'll go on to introduce the useful concepts of electric potential, or electric potential energy per unit charge, and voltage, which is the difference in the value of electric potential at two positions. We'll also learn about equipotential surfaces, which are surfaces on which the electric potential has the same value at every point.

An important device for storing electric energy is a capacitor. In its simplest form a capacitor is just two pieces of metal, called capacitor plates, placed close to each other but not in contact. We'll examine the key properties of capacitors, including how to combine two or more of them for even greater energy storage. We'll conclude the chapter by seeing how capacitors can be made even more effective by inserting an insulating material between the plates. In Chapter 18 we'll learn how batteries like those shown in Figure 17-1a provide electric energy to the components of an electric circuit.

Take-Home Message for Section 17-1

✔ Electric potential energy associated with a point charge is analogous to the gravitational potential energy associated with a massive object.

✔ Electric potential is electric potential energy per charge.

✔ A capacitor is a device for storing electric energy.

17-2 Electric potential energy changes when a charge moves in an electric field

We learned in Section 6-6 that Earth's gravitational force is a *conservative* force: As an object moves from one position to another in the presence of the gravitational force, the work done on the object by the gravitational force depends only on where the object starts and where it ends up, not on how it gets there. This means that we can express the work done by gravity W_{grav} in terms of a change in the gravitational potential energy U_{grav} of the system of Earth and object:

$$W_{grav} = -\Delta U_{grav}$$

(6-16)

The minus sign means that if the object moves downward so that gravity does positive work (the gravitational force and the object's displacement are in the same direction), there is a negative change in gravitational potential energy (U_{grav} decreases). If the object moves upward so that gravity does negative work (force and displacement are in opposite directions), there is a positive change in gravitational potential energy (U_{grav} increases). Note that what matters is the *change* in gravitational potential energy as the object moves from one place to another, not the value of U_{grav} at either place. Once you've chosen the value of U_{grav} at a certain position, you can use Equation 6-16 to tell you the value at any other position.

To be specific, suppose an object of mass m moves in a region where the acceleration $\vec{g}$ due to Earth's gravity is the same at all positions, so the gravitational force $\vec{F} = m\vec{g}$ on the object is uniform. The object's displacement is $\vec{d}$, and that displacement is at an angle θ to the direction of the force $\vec{F} = m\vec{g}$ and so at the same angle θ to the direction of $\vec{g}$ (Figure 17-2a). Then the work done by gravity is $W = Fd \cos \theta$ (Equation 6-2), and the change in gravitational potential energy is

$$\Delta U_{grav} = -W_{grav} = -mgd \cos \theta$$

(17-1)

√x⎯ *See the Math Tutorial for more information on trigonometry*

Figure 17-2 **Change in gravitational potential energy** When an object of mass m moves in the presence of gravity, the gravitational force can do work on it and there can be a change in gravitational potential energy.

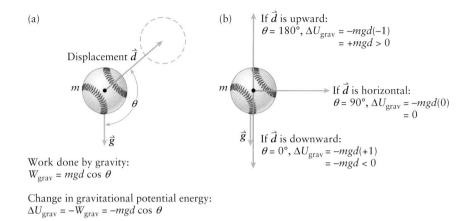

(a) Displacement $\vec{d}$

m

θ

$\vec{g}$

Work done by gravity:
$W_{grav} = mgd \cos \theta$

Change in gravitational potential energy:
$\Delta U_{grav} = -W_{grav} = -mgd \cos \theta$

(b) If $\vec{d}$ is upward:
$\theta = 180°, \Delta U_{grav} = -mgd(-1)$
$= +mgd > 0$

m

If $\vec{d}$ is horizontal:
$\theta = 90°, \Delta U_{grav} = -mgd(0)$
$= 0$

$\vec{g}$ If $\vec{d}$ is downward:
$\theta = 0°, \Delta U_{grav} = -mgd(+1)$
$= -mgd < 0$

Equation 17-1 looks a bit different from the way that we wrote the change in gravitational potential energy in Chapter 6, but we can verify that it makes sense (Figure 17-2b). If the object moves a distance d straight upward, opposite to the direction of the gravitational force, $\theta = 180°$ and $\cos \theta = \cos 180° = -1$. In this case gravity does negative work on the object, and the gravitational potential energy increases: $\Delta U_{grav} = -mgd(-1) = +mgd$. If the object moves a distance d straight downward, in the direction of the gravitational force, $\theta = 0$ and $\cos \theta = \cos 0° = 1$. Then gravity does positive work on the object and the gravitational potential energy decreases: $\Delta U_{grav} = -mgd(+1) = -mgd$. Gravity does zero work, and there is no change in gravitational potential energy, if the object moves horizontally: Then $\theta = 90°$ and $\Delta U_{grav} = -mgd \cos 90° = 0$. These results are the same ones that we found in Section 6-6.

We can apply the same concepts to the idea of *electric potential* energy. That's because like gravity, the electric force due to the interaction between charged objects is a conservative force. You can see this easily for the special case in which an object moves in a region of space where there is a *uniform* electric field $\vec{E}$—that is, where the value of $\vec{E}$ is the same at all positions (Figure 17-3). The electric force on an object of charge q moving in this region is constant and equal to $\vec{F} = q\vec{E}$. That's exactly like the situation in Figure 17-2, which shows an object of mass m moving in a region where $\vec{g}$ is the same at all positions, so the gravitational force $\vec{F} = m\vec{g}$ on the object is uniform. We know that this uniform gravitational force is conservative, so the force $\vec{F} = q\vec{E}$ due to a uniform electric field must be conservative as well. Hence we can express the work done by the electric force in terms of a change in *electric* potential energy.

Electric Potential Energy in a Uniform Field

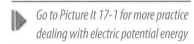

Go to Picture It 17-1 for more practice dealing with electric potential energy

This analogy between gravitational force and electric force tells us how to write the change in electric potential energy for an object of charge q that undergoes a displacement $\vec{d}$ in the presence of a uniform electric field $\vec{E}$ (Figure 17-3a). Following the same steps that we used to find Equation 17-1 above, you can see that if θ is the angle between the directions of $\vec{d}$ and $\vec{E}$, then the work done on the charge by the electric force $\vec{F} = q\vec{E}$ is $W_{electric} = qEd \cos \theta$. The change in **electric potential energy** equals the negative of the electric work done on the charge:

Electric potential energy for a
charge in a uniform electric field
(17-2)

The change in electric potential energy for an object of charge q that moves in a uniform electric field $\vec{E}$...

...equals the negative of the work done on the object by the electric force.

$$\Delta U_{electric} = -W_{electric}$$
$$= -qEd \cos \theta$$

Angle between the displacement and the direction of the electric field

Charge of the object

Magnitude of the electric field

Straight-line displacement of the object

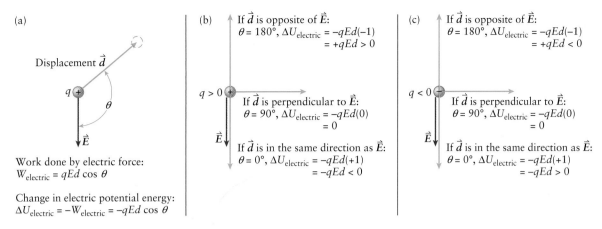

(a)

Displacement $\vec{d}$

q +

θ

$\vec{E}$

Work done by electric force:
$W_{electric} = qEd \cos \theta$

Change in electric potential energy:
$\Delta U_{electric} = -W_{electric} = -qEd \cos \theta$

(b)

$q > 0$ +

$\vec{E}$

If $\vec{d}$ is opposite of $\vec{E}$:
$\theta = 180°$, $\Delta U_{electric} = -qEd(-1)$
$= +qEd > 0$

If $\vec{d}$ is perpendicular to $\vec{E}$:
$\theta = 90°$, $\Delta U_{electric} = -qEd(0)$
$= 0$

If $\vec{d}$ is in the same direction as $\vec{E}$:
$\theta = 0°$, $\Delta U_{electric} = -qEd(+1)$
$= -qEd < 0$

(c)

$q < 0$ −

$\vec{E}$

If $\vec{d}$ is opposite of $\vec{E}$:
$\theta = 180°$, $\Delta U_{electric} = -qEd(-1)$
$= +qEd < 0$

If $\vec{d}$ is perpendicular to $\vec{E}$:
$\theta = 90°$, $\Delta U_{electric} = -qEd(0)$
$= 0$

If $\vec{d}$ is in the same direction as $\vec{E}$:
$\theta = 0°$, $\Delta U_{electric} = -qEd(+1)$
$= -qEd > 0$

Figure 17-3 **Change in electric potential energy** When an object of charge q moves in the presence of a uniform electric field $\vec{E}$, the electric force can do work on it and there can be a change in electric potential energy (compare Figure 17-2).

Just as for the gravitational case, Equation 17-2 tells us how much the electric potential energy *changes* when the charge is displaced as shown in Figure 17-3a. The value of the initial electric potential energy $U_{electric}$ can have any value you choose; once you've chosen that value, the value of $U_{electric}$ at the final position equals the initial value plus $(-qEd \cos \theta)$. Note also that the electric potential energy is a shared property of the charge q and the other charges that produce the electric field $\vec{E}$.

Let's examine Equation 17-2 more closely (Figure 17-3b). Suppose q is positive, so that the force $\vec{F} = q\vec{E}$ on the charged object is in the *same* direction as the uniform electric field $\vec{E}$. If this positive charge moves a distance d opposite to $\vec{E}$, then $\theta = 180°$ and $\cos \theta = -1$. In this case the electric force does negative work on the charge, $\Delta U_{electric} = (-qEd)(-1) = +qEd$ (the negative of the work done by the electric force) is positive, and the electric potential energy of the system increases. That's exactly what happens to *gravitational* potential energy when a massive object moves upward from Earth's surface, in the direction opposite to the gravitational force (Figure 17-2b). If the positive charge q instead moves in the same direction as the electric field, and so in the same direction as the electric force, then $\theta = 0$ and $\cos \theta = +1$. Then the electric force does positive work on the object, so $\Delta U_{electric} = (-qEd)(+1) = -qEd$ is negative and the electric potential energy of the system decreases. That's the same thing that happens when a massive object falls toward Earth's surface, in the same direction as the gravitational force on the object: Gravitational potential energy decreases (Figure 17-2b).

Equation 17-2 also applies to the case of an object with *negative* charge ($q < 0$), so that the force $\vec{F} = q\vec{E}$ on this object is in the direction *opposite* to the electric field $\vec{E}$ (Figure 17-3c). If such a negative charge moves a distance d opposite the direction of $\vec{E}$ so that $\theta = 180°$ and $\cos \theta = -1$ in Equation 17-2, the electric force does *positive* work on the charge because the force and displacement are in the same direction. Then, because $q < 0$, $\Delta U_{electric} = +qEd$ is *negative* because q is negative and the electric potential energy of the system decreases. That's exactly the opposite of what happens for a positive charge that moves a distance d in the direction opposite to the uniform electric field. In the same way, if a negative charge moves a distance d in the direction of $\vec{E}$ so that $\theta = 0$ and $\cos \theta = +1$ in Equation 17-2, the electric force does *negative* work on the object (the electric force is opposite to $\vec{E}$ and so opposite the displacement). The electric potential energy change in this case is $\Delta U_{electric} = (-qEd)(+1) = -qEd$, which is positive because q is negative. So the electric potential energy of the system increases. Again, that's exactly opposite to what happens for a positive charge that moves a distance d in the direction of the electric field.

Example 17-1 Electric Potential Energy Difference in a Uniform Field

An electron (charge $q = -e = -1.60 \times 10^{-19}$ C, mass $m = 9.11 \times 10^{-31}$ kg) is released from rest in a uniform electric field. The field points in the positive x direction and has magnitude 2.00×10^2 N/C. Find the speed of the electron after it has moved 0.300 m.

Set Up

Because the electron has a negative charge, the force $\vec{F} = q\vec{E}$ on the electron is directed opposite to the electric field. So the displacement of the electron will be at an angle $\theta = 180°$ to the direction of the electric field.

If the conservative electric force is the only force acting on the electron, then mechanical energy is conserved. We'll find the change in electric potential energy using Equation 17-2, and from this find the change in kinetic energy of the electron. The electron's initial kinetic energy is zero because it starts at rest; once we know the final kinetic energy, we can determine the electron's final speed.

Electric potential energy for a charge in a uniform electric field:

$$\Delta U_{\text{electric}} = -qEd \cos \theta \quad (17\text{-}2)$$

Conservation of mechanical energy:

$$K_i + U_{\text{electric,i}} = K_f + U_{\text{electric,f}} \quad (6\text{-}23)$$

Kinetic energy:

$$K = \frac{1}{2}mv^2 \quad (6\text{-}8)$$

$E = 2.00 \times 10^2$ N/C

$d = 0.300$ m $\longleftarrow$ o-e

$\longrightarrow +x$

Solve

Use Equation 17-2 to solve for the change in electric potential energy.

The change in electric potential energy equals the difference between the final and initial electric potential energy:

$$\Delta U_{\text{electric}} = U_{\text{electric,f}} - U_{\text{electric,i}}$$

From Equation 17-2,

$$\begin{aligned}
\Delta U_{\text{electric}} &= U_{\text{electric,f}} - U_{\text{electric,i}} \\
&= -qEd \cos \theta \\
&= -(-1.60 \times 10^{-19}\,\text{C})(2.00 \times 10^2\,\text{N/C})(0.300\,\text{m}) \cos 180° \\
&= -(-9.60 \times 10^{-18}\,\text{N} \cdot \text{m})(-1) \\
&= -9.60 \times 10^{-18}\,\text{N} \cdot \text{m} = -9.60 \times 10^{-18}\,\text{J}
\end{aligned}$$

Find the final kinetic energy and the final speed of the electron.

The initial kinetic energy of the electron is $K_i = 0$. Solve the energy conservation equation for the final kinetic energy of the electron:

$$K_f = K_i + U_{\text{electric,i}} - U_{\text{electric,f}} = 0 - (U_{\text{electric,f}} - U_{\text{electric,i}})$$
$$= 0 - (-9.60 \times 10^{-18}\,\text{J}) = 9.60 \times 10^{-18}\,\text{J}$$

Finally, solve for the final speed of the electron:

$$K_f = \frac{1}{2}mv_f^2 \text{ so}$$

$$v_f = \sqrt{\frac{2K_f}{m}} = \sqrt{\frac{2(9.60 \times 10^{-18}\,\text{J})}{9.11 \times 10^{-31}\,\text{kg}}} = 4.59 \times 10^6\,\text{m/s}$$

Reflect

The electric potential energy decreases when we release the electron in the electric field, in the same way that the gravitational potential energy decreases when you release a ball in the presence of Earth's gravity. In both situations, the kinetic energy increases by the same amount that the potential energy decreases.

You should verify that the force on the electron is very small, only 3.2×10^{-17} N in magnitude. The electron has only a tiny mass, however, and our results show that it acquires a ferocious speed (about 1.5% of the speed of light) after moving just 0.300 m.

Electric Potential Energy of Point Charges

Electric potential energy is particularly important for the case of two point charges interacting with each other. For example, many key quantities in chemistry are impossible to understand without first grasping the concept of electric potential energy. An example is the dissociation energy of an ionic compound such as sodium chloride (NaCl)—that is, the energy that must be given to the molecule to break it into its component atoms. The dissociation energy is determined in large part by the change in electric potential energy required to separate the positive sodium ion (Na^+) and the negative chloride ion (Cl^-), both of which behave much like point charges, as well as by the change in electric potential energy required to move an electron from the Cl^- ion to the Na^+ ion to make the atoms neutral.

Just as for a charge moving in a uniform electric field, we can use our knowledge of gravitation to find an expression for the electric potential energy of two point charges q_1 and q_2 separated by a distance r. In Section 10-2 we saw the following expression for the attractive *gravitational* force between two objects with *masses* m_1 and m_2 separated by a distance r:

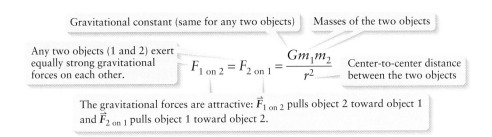

Gravitational constant (same for any two objects)   Masses of the two objects

Any two objects (1 and 2) exert equally strong gravitational forces on each other.

$$F_{1 \text{ on } 2} = F_{2 \text{ on } 1} = \frac{Gm_1m_2}{r^2}$$

Center-to-center distance between the two objects

The gravitational forces are attractive: $\vec{F}_{1 \text{ on } 2}$ pulls object 2 toward object 1 and $\vec{F}_{2 \text{ on } 1}$ pulls object 1 toward object 2.

Newton's law of universal gravitation
(10-2)

The gravitational potential energy of these two objects is given by Equation 10-4 in Section 10-3:

Gravitational constant (same for any two objects)   Masses of the two objects

Gravitational potential energy of a system of two objects (1 and 2)

$$U_{\text{grav}} = -\frac{Gm_1m_2}{r}$$

Center-to-center distance between the two objects

The gravitational potential energy is zero when the two objects are infinitely far apart. If the objects are brought closer together (so r is made smaller), U_{grav} decreases (it becomes more negative).

Gravitational potential energy
(10-4)

Figure 17-4a graphs the gravitational potential energy given by Equation 10-4. We choose the point at which potential energy U_{grav} is zero to be where the two objects are infinitely far apart, so $r \to \infty$. The gravitational potential energy is negative for any finite value of r and increases—that is, becomes less negative—as the objects move farther apart. That's because the work done by the gravitational force and the change in gravitational potential energy are negatives of each other: $\Delta U_{\text{grav}} = -W_{\text{grav}}$ (see Equation 6-16 at the beginning of this chapter section). If we hold object m_1 stationary and move object m_2 farther away, increasing the distance r, the attractive gravitational force does negative work. Then $W_{\text{grav}} < 0$, so $\Delta U_{\text{grav}} > 0$ and the gravitational potential energy increases, as Figure 17-4a shows.

(a) Gravitational potential energy

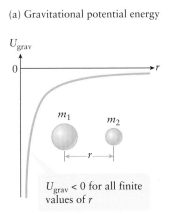

$U_{\text{grav}} < 0$ for all finite values of r

(b) Electric potential energy, charges of opposite sign

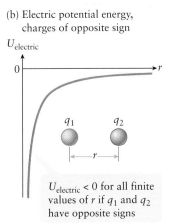

$U_{\text{electric}} < 0$ for all finite values of r if q_1 and q_2 have opposite signs

(c) Electric potential energy, charges of the same sign

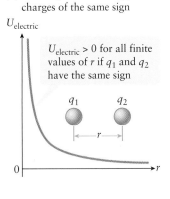

$U_{\text{electric}} > 0$ for all finite values of r if q_1 and q_2 have the same sign

Figure 17-4 **Potential energy for two masses and for two charges** The electric potential energy U_{electric} for two point charges is similar to the gravitational potential energy for two masses. The sign of U_{electric} depends on the signs of the two charges.

Now compare Equation 10-2 to Coulomb's law for the electric force between two point charges, Equation 16-1:

Coulomb's law
(16-1)

Any two point charges q_1 and q_2 exert equally strong electric forces on each other.

Coulomb constant

Absolute values of the point charges

$$F_{q_1 \text{ on } q_2} = F_{q_2 \text{ on } q_1} = \frac{k|q_1||q_2|}{r^2}$$

Distance between the point charges

If we replace m and G in Equation 10-2 with q and k, we see that Equation 16-1 is identical to Equation 10-2 for the gravitational force, but with an important difference: The electric force is *attractive* (like the gravitational force) if the two charges q_1 and q_2 have opposite signs (one negative and the other positive), but *repulsive* if q_1 and q_2 have the same sign (either both positive or both negative). Here's an expression for the electric potential energy of two point charges that accounts for both of these possibilities:

Electric potential energy
of two point charges
(17-3)

Coulomb constant   Values of the two charges

Electric potential energy of two point charges

$$U_{\text{electric}} = \frac{kq_1q_2}{r}$$

Distance between the point charges

This expression is very similar to Equation 10-4 for gravitational potential energy. Equation 17-3 shows that U_{electric} is inversely proportional to the distance r, so the electric potential energy is zero when the two charges are infinitely far apart ($r \to \infty$). But unlike gravitational potential energy, U_{electric} can be either negative or positive depending on the signs of the two charges. If q_1 and q_2 have different signs (one positive and one negative) so that the two charges attract each other, then $U_{\text{electric}} < 0$ for any finite distance r between the charges (Figure 17-4b). This makes sense: If we hold charge q_1 at rest and move charge q_2 farther away, the attractive electric force on q_2 does negative work (the force on q_2 is opposite to the displacement of q_2). In this case the change in electric potential energy is positive and the electric potential energy increases (becomes less negative) with increasing distance r, just like the gravitational potential energy of two masses. But if q_1 and q_2 have the same sign (both positive or both negative) so that the two charges repel each other, then $U_{\text{electric}} > 0$ for any finite distance r between the charges (Figure 17-4c). In this case when q_2 moves away from q_1 the repulsive electric force on q_2 does positive work (the force on q_2 is in the same direction as the displacement of q_2). Then the change in electric potential energy is negative and the electric potential energy *decreases* (becomes less positive) with increasing distance r.

Here's a useful way to interpret U_{electric} as given by Equation 17-3:

The electric potential energy of a pair of point charges equals the amount of work you would have to do to bring the charges to their current positions from infinitely far away.

If the two charges have opposite signs as in Figure 17-4b, you would have to do *negative* work (you would push the charges away from each other as they move toward each other) to oppose the attractive electric force that pulls the two charges together. (If you didn't exert a force to do this work, the electric attraction would make the charges crash into each other instead of stopping a distance r apart.) So in this case the electric potential energy of the two charges is negative, as Figure 17-4b shows. By contrast, if the two charges have the same sign as in Figure 17-4c, you would have to do positive work to push the two charges together (you would push the two charges toward each other as they move toward each other) to overcome the repulsive electric force that pushes the two charges apart. So the electric potential energy of these two charges is positive, as depicted in Figure 17-4c.

The same idea holds for an assemblage of three or more point charges. To put charges q_1, q_2, and q_3 into proximity to each other, you would first have to do work to bring q_1 and q_2 from infinity to a distance r_{12} from each other. If you then brought in the third charge q_3 while keeping the other two charges stationary, you would have to do additional work against the electric force that q_1 exerts on q_3 *and* against the electric force that q_2 exerts on q_3. If q_3 ends up a distance r_{13} from q_1 and a distance r_{23} from q_2, the electric potential energy of the assemblage is the total amount of work that you did:

$$U_{\text{electric}} = \frac{kq_1q_2}{r_{12}} + \frac{kq_1q_3}{r_{13}} + \frac{kq_2q_3}{r_{23}}$$ (17-4)

(electric potential energy of three charges)

The total electric potential energy of a system of three charges is the sum of three terms, one for each pair of charges in the system (q_1 and q_2, q_1 and q_3, and q_2 and q_3). Each such term is the same as Equation 17-3. You can easily extend this idea to an assemblage made up of any number of point charges.

The following examples show how to do calculations using Equations 17-3 and 17-4.

 Go to Interactive Exercise 17-1 for more practice dealing with electric potential energy

---

Example 17-2 Electric Potential Energy and Nuclear Fission

When a nucleus of uranium-235 (92 protons and 143 neutrons) absorbs an additional neutron, it undergoes a process called *nuclear fission* in which it breaks into two smaller nuclei. One possible fission is for the uranium nucleus to divide into two palladium nuclei, each of which has 46 protons and is 5.9×10^{-15} m in radius. The palladium nuclei then fly apart due to their electric repulsion. If we assume that the two palladium nuclei begin at rest and are just touching each other, what is their combined kinetic energy when they are very far apart?

Set Up

We can treat the two spherical nuclei as though they were point charges of $q = +46e$ located at the centers of the two nuclei. (To motivate this, recall from Section 16-7 that the electric field outside a sphere of charge Q is the same as that of a point charge Q located at the sphere's center.) Equation 17-3 then tells us the electric potential energy of the two palladium nuclei when they begin at rest. We'll then use energy conservation to find the combined kinetic energy when the palladium nuclei are very far apart.

Electric potential energy of two point charges:

$$U_{\text{electric}} = \frac{kq_1q_2}{r}$$ (17-3)

Conservation of mechanical energy:

$$K_i + U_{\text{electric,i}} = K_f + U_{\text{electric,f}}$$ (6-23)

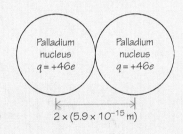

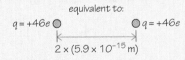

Solve

Use Equation 17-3 to solve for the initial electric potential energy when the two nuclei are just touching.

Each nucleus has charge $+46e$:

$$q_1 = q_2 = +46(1.60 \times 10^{-19} \text{ C}) = +7.36 \times 10^{-18} \text{ C}$$

The separation between the charges is twice the radius of either nucleus:

$$r = 2(5.9 \times 10^{-15} \text{ m})$$

The electric potential energy is

$$U_{\text{electric,i}} = \frac{kq_1q_2}{r} = \frac{(8.99 \times 10^9 \text{ N} \cdot \text{m}^2/\text{C}^2)(7.36 \times 10^{-18} \text{ C})^2}{2(5.9 \times 10^{-15} \text{ m})}$$

$$= 4.1 \times 10^{-11} \text{ N} \cdot \text{m} = 4.1 \times 10^{-11} \text{ J}$$

Use energy conservation to find the combined final kinetic energy of the two palladium nuclei.

The palladium nuclei begin at rest, so the initial kinetic energy is zero:

$$K_i = 0$$

The palladium nuclei end up very far apart, so their separation is essentially infinite ($r \to \infty$). The final electric potential energy is therefore zero:

$$U_{\text{electric,f}} = 0$$

From the energy conservation equation, the final kinetic energy is

$$K_f = K_i + U_{electric,i} - U_{electric,f}$$
$$= 0 + 4.1 \times 10^{-11} \text{ J} - 0 = 4.1 \times 10^{-11} \text{ J}$$

Reflect

All of the initial electric potential energy is converted into kinetic energy of the palladium nuclei. The energy released by the fission of a single uranium-235 nucleus, 4.1×10^{-11} J, is very small. But imagine that you could get 1.0 kg of uranium-235, which contains about 2.6×10^{24} uranium atoms, to undergo fission at once. (This is the principle of a *fission bomb*.) The released energy would be $(2.6 \times 10^{24}) \times (4.1 \times 10^{-11} \text{ J}) = 1.1 \times 10^{14}$ J, equivalent to the energy given off by exploding 26,000 tons of TNT. This gives an inkling of the terrifying amount of energy that can be released by a fission weapon.

Example 17-3 Electric Potential Energy of Three Charges

A particle with charge $q_1 = +4.30 \ \mu C$ is located at $x = 0$, $y = 0$. A second particle with charge $q_2 = -9.80 \ \mu C$ is located at $x = 0$, $y = 4.00$ cm, and a third particle with charge $q_3 = +5.00 \ \mu C$ is located at position $x = 3.00$ cm, $y = 0$. (Note that $1 \ \mu C = 1$ microcoulomb $= 10^{-6}$ C.) What is the total electric potential energy of these three charges?

Set Up

We'll use Equation 17-4 to find the value of $U_{electric}$. We're given the values of the three charges; we'll use the positions of the charges to find the distances r_{12}, r_{13}, and r_{23} between them.

Electric potential energy of three charges:

$$U_{electric} = \frac{kq_1q_2}{r_{12}} + \frac{kq_1q_3}{r_{13}} + \frac{kq_2q_3}{r_{23}}$$

(17-4)

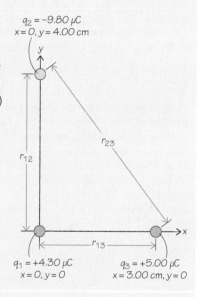

Solve

The figure above shows that charges 1 and 2 are $r_{12} = 4.00$ cm apart, and that charges 1 and 3 are $r_{13} = 3.00$ cm apart. The distance between charges 2 and 3 is the hypotenuse of a right triangle of sides r_{12} and r_{13}.

From the figure above,

$$r_{12} = 4.00 \text{ cm} = 4.00 \times 10^{-2} \text{ m}$$
$$r_{13} = 3.00 \text{ cm} = 3.00 \times 10^{-2} \text{ m}$$

From the Pythagorean theorem,

$$r_{23} = \sqrt{r_{12}^2 + r_{13}^2}$$
$$= \sqrt{(4.00 \times 10^{-2} \text{ m})^2 + (3.00 \times 10^{-2} \text{ m})^2}$$
$$= 5.00 \times 10^{-2} \text{ m}$$

Calculate each term in the expression for electric potential energy, Equation 17-4.

We can write Equation 17-4 as

$$U_{electric} = U_{12} + U_{13} + U_{23}$$

Each term on the right-hand side of this equation represents the contribution to the electric potential energy due to a specific pair of point charges. The contribution due to the interaction between charges q_1 and q_2 is

$$U_{12} = \frac{kq_1q_2}{r_{12}}$$

$$= \frac{(8.99 \times 10^9 \, \text{N} \cdot \text{m}^2/\text{C}^2)(+4.30 \times 10^{-6} \, \text{C})(-9.80 \times 10^{-6} \, \text{C})}{4.00 \times 10^{-2} \, \text{m}}$$

$$= -9.47 \, \text{J}$$

The contribution due to the interaction between charges q_1 and q_3 is

$$U_{13} = \frac{kq_1q_3}{r_{13}}$$

$$= \frac{(8.99 \times 10^9 \, \text{N} \cdot \text{m}^2/\text{C}^2)(+4.30 \times 10^{-6} \, \text{C})(+5.00 \times 10^{-6} \, \text{C})}{3.00 \times 10^{-2} \, \text{m}}$$

$$= +6.44 \, \text{J}$$

The contribution due to the interaction between charges q_2 and q_3 is

$$U_{23} = \frac{kq_2q_3}{r_{23}}$$

$$= \frac{(8.99 \times 10^9 \, \text{N} \cdot \text{m}^2/\text{C}^2)(-9.80 \times 10^{-6} \, \text{C})(+5.00 \times 10^{-6} \, \text{C})}{5.00 \times 10^{-2} \, \text{m}}$$

$$= -8.81 \, \text{J}$$

Finally, calculate the total electric potential energy.

The total electric potential energy is the sum of the three terms calculated above:

$$U_{\text{electric}} = U_{12} + U_{13} + U_{23}$$

$$= (-9.47 \, \text{J}) + (+6.44 \, \text{J}) + (-8.81 \, \text{J})$$

$$= -11.82 \, \text{J}$$

Reflect

The contributions U_{12} and U_{23} are negative because these pairs of charges (q_1 and q_2 for U_{12}, q_2 and q_3 for U_{23}) have opposite signs. The members of these pairs attract each other, so you must do negative work against each attractive force to move the charges from infinity to the positions shown in the figure. The contribution U_{13}, however, is positive because charges q_1 and q_3 have the same sign (both positive). These two charges repel, so you must do positive work against the repulsion to move these charges from infinity to their positions. The total potential energy is negative, which shows that the total amount of work you would do to move all three charges from infinity is negative.

? Got the Concept? 17-1   Electric Potential Energy

Suppose you reversed the signs of the three point charges in Example 17-3 so that $q_1 = -4.30 \, \mu\text{C}$, $q_2 = +9.80 \, \mu\text{C}$, and $q_3 = -5.00 \, \mu\text{C}$. If the positions of the charges remain unchanged, what would be the total electric potential energy of this assemblage? (a) More negative than -11.82 J; (b) -11.82 J; (c) between -11.82 J and $+11.82$ J; (d) $+11.82$ J; (e) more positive than $+11.82$ J.

Take-Home Message for Section 17-2

✔ The electric potential energy associated with a point charge can change if the charge changes position in an electric field.

✔ The electric potential energy of a pair of point charges equals the amount of work you would have to do to move those charges from infinity to their present positions.

✔ The electric potential energy of an assemblage of three or more point charges is the sum of the electric potential energies for each pair of charges in the assemblage.

17-3 Electric potential equals electric potential energy per charge

Our discussion in Section 17-2 shows that if a point charge q changes position, the potential energy change $\Delta U_{electric}$ depends on both the magnitude and the sign (positive or negative) of q (see Figure 17-3). We can simplify things by considering the potential energy *per charge*—that is, the electric potential energy for a charge at a given position divided by the value of that charge. We call this quantity the **electric potential** and denote it by the symbol V:

Electric potential related to electric potential energy (17-5)

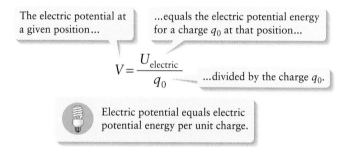

The electric potential at a given position...

...equals the electric potential energy for a charge q_0 at that position...

$$V = \frac{U_{electric}}{q_0}$$

...divided by the charge q_0.

Electric potential equals electric potential energy per unit charge.

We call the charge q_0 the **test charge**: Its charge has such a small magnitude that it doesn't affect the other charges that create the electric field in which q_0 moves. Because we divide out the value of q_0, the value of the potential V at a given position does *not* depend on the value of the point charge q_0 that we place there. Instead, V is determined by the other charges that produce the electric field at the position where we place the test charge.

Note that electric potential V has the same relationship to electric potential energy $U_{electric}$ as electric field $\vec{E}$ has to electric force $\vec{F}$: V is electric potential energy per charge, just as $\vec{E}$ is electric force per charge. Like electric potential energy, but unlike electric field, potential V is a *scalar* quantity.

For electric potential energy, the value of $U_{electric}$ for a charge at a given position is not as important as the potential energy *difference* $\Delta U_{electric}$ when the charge moves from a point a to a different point b. The same is true for electric potential. From Equation 17-5, the **electric potential difference** ΔV between two points is

Electric potential difference related to electric potential energy difference (17-6)

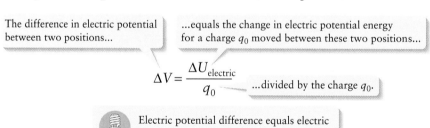

The difference in electric potential between two positions...

...equals the change in electric potential energy for a charge q_0 moved between these two positions...

$$\Delta V = \frac{\Delta U_{electric}}{q_0}$$

...divided by the charge q_0.

Electric potential difference equals electric potential energy difference per unit charge.

The SI unit of electric potential and electric potential difference is the **volt** (V), named after the Italian scientist Alessandro Volta. For example, a common AA or AAA flashlight battery has "1.5 volts" written on its side. This means that the electric potential at the positive terminal of the battery (labeled +) is 1.5 V greater than the electric potential at the negative terminal of the battery (labeled −). In other words, 1.5 V is the electric potential *difference* between the terminals of the battery. In Chapter 18 we'll see how this electric potential difference causes electric charge to flow when a battery is included in an electric circuit. (Note that Volta invented the electric battery in 1800.) Equations 17-5 and 17-6 show that 1 volt is equal to 1 joule per coulomb, and we know that 1 joule is equal to 1 newton multiplied by 1 meter. So

$$1\ V = 1\ \frac{J}{C} = 1\ \frac{N \cdot m}{C}$$

This means that if the electric potential at point b is 1 V higher than at point a, it takes 1 J of work to move a charge of +1 C from a to b. If a charge of +1 C at point b is released from rest and moves to point a, the electric potential that the charge experiences decreases by 1 V, the electric potential energy associated with the charge decreases by 1 J, and the charge acquires 1 J of kinetic energy.

We saw in Chapter 16 that the SI units of electric field are newtons per coulomb, or N/C. Because $1\,V = 1\,N\cdot m/C$, it follows that $1\,V/m = 1\,N/C$. So an equally good set of units for electric field is V/m (volts per meter).

We'll often abbreviate the terms *electric potential* and *electric potential difference* as simply *potential* and *potential difference*, respectively. Since the unit of electric potential is the volt, it's also common to refer to electric potential difference as **voltage**.

 Watch Out! Don't confuse the symbol *V* for electric potential difference and the abbreviation V for volts.

In this chapter and others you'll see mathematical statements such as "$V = 10$ V." Such statements are perfectly legal because *V* and V are different symbols. *V* (italicized) represents the electric potential at a point in space. V (not italicized) is an abbreviation for the unit of electric potential, the volt. In this textbook, you can tell the difference by noticing which font is used and by paying close attention to the context in which *V* and V are used. But it's pretty hard to handwrite something in italics, so make sure you know the distinction between *V* and V in your own work on your homework assignments and exams.

 Watch Out! Electric potential and electric potential energy are not the same.

Remember that electric potential *V* and electric potential energy $U_{electric}$ are related but different quantities. Electric potential energy $U_{electric}$ refers to the energy associated with a particular amount of charge at a given position. Electric potential *V* is the energy associated with a *unit* charge at that position. Note that while *V* and $U_{electric}$ can both be positive or negative, they do not have a direction associated with them. So electric potential and electric potential energy are both *scalar* quantities, not vector quantities.

Electric Potential in a Uniform Electric Field

A simple application of Equation 17-6 is finding the electric potential difference between two points a and b in a uniform electric field (Figure 17-5). From Equation 17-2, the change or difference in electric potential energy when a charge q_0 is moved from a to b is $\Delta U_{electric} = -q_0 Ed\cos\theta$. Equation 17-6 tells us that the electric potential difference between a and b is $\Delta U_{electric}$ divided by q_0, or the electric potential energy change per unit charge:

$$\Delta V = \frac{\Delta U_{electric}}{q_0} = \frac{-q_0 Ed\cos\theta}{q_0}$$
$$= -Ed\cos\theta \qquad (17\text{-}7)$$

Equation 17-7 shows that ΔV is positive if $\cos\theta$ is negative, which is the case if the angle θ is greater than 90° as depicted in Figure 17-5. In other words, *if you move in a direction opposite to the electric field, the electric potential increases*. If the angle θ is less than 90°, $\cos\theta$ is positive and ΔV is negative; *if you move in the direction of the electric field, the electric potential decreases*.

We know that the electric force on a positive charge is in the direction of the electric field $\vec{E}$, and the electric force on a negative charge is in the direction opposite to $\vec{E}$. So our observations about how electric potential changes with position tell us that

If an object has positive charge, the electric force on that object pushes it toward a region of lower electric potential. If an object has negative charge, the electric force on that object pushes it toward a region of higher electric potential.

As we'll see below, these observations hold true whether or not the electric field is uniform.

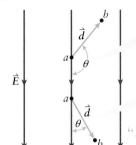

If you move in a direction opposite to the electric field ($\theta > 90°$), the electric potential increases ($\Delta V > 0$).

If you move in the direction of the electric field ($\theta < 90°$), the electric potential decreases ($\Delta V < 0$).

 If you move from point a to point b in a uniform electric field $\vec{E}$, the change in electric potential is $\Delta V = -Ed\cos\theta$.

Figure 17-5 **Change in electric potential** Calculating the difference in electric potential between two points in a uniform electric field $\vec{E}$.

Comparing Equations 17-2 and 17-7 shows that the right-hand side of Equation 17-7 equals the negative of the work done by the electric field ($q_0 Ed \cos \theta$) divided by the charge q_0. So another way to think of the potential difference between two points is as the negative of the work done by the electric field per unit charge when a charged object is moved between those points. Alternatively, the potential difference between two points equals the work that *you* must do per unit charge against the electric force to move a charged object between those points. So if you take a charge of +1 C that is at rest at point *a* and move it to point *b* where it is again at rest, and the potential at *b* is 1 V higher than the potential at *a*, the electric field does −1 J of work on the charge and you do +1 J of work on that charge.

According to Equation 17-7, for a uniform electric field the electric potential difference between two points depends only on the electric field $\vec{E}$ and the displacement $\vec{d}$ between those two points. So knowing how the electric potential varies over a region of space enables us to determine how any charged object will move in that region, regardless of the magnitude of its charge.

▷ *Go to Picture It 17-2 for more practice dealing with electric potential*

Example 17-4   Electric Potential Difference in a Uniform Field I

A uniform electric field points in the positive *x* direction and has magnitude 2.00×10^2 V/m. Points *a* and *b* are both in this field: Point *b* is a distance 0.300 m from *a* in the negative *x* direction. Determine the electric potential difference $V_b - V_a$ between points *a* and *b*.

Set Up

We apply Equation 17-7 to the path shown in the figure. Note that the magnitude of the displacement is $d = 0.300$ m, and the angle between the electric field and the displacement is $\theta = 180°$. Note also that the potential difference ΔV equals the potential at the *end* of the displacement $\vec{d}$ (that is, at point *b*) minus the potential at the *beginning* of the displacement (that is, at point *a*).

Potential difference between two points in a uniform electric field:

$$\Delta V = V_b - V_a = -Ed \cos \theta \quad (17\text{-}7)$$

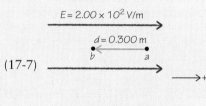

Solve

Use Equation 17-7 to solve for the potential difference.

Calculate the potential difference from the electric field magnitude E, the displacement d, and the angle θ:

$$\begin{aligned} \Delta V = V_b - V_a &= -Ed \cos \theta \\ &= -(2.00 \times 10^2 \text{ V/m})(0.300 \text{ m}) \cos 180° \\ &= -(60.0 \text{ V})(-1) \\ &= +60.0 \text{ V} \end{aligned}$$

Reflect

We can check our result by comparing with Example 17-1 in Section 17-2, where we considered the change in electric potential *energy* $\Delta U_{\text{electric}}$ for an electron that undergoes the same displacement in this same electric field. Using Equation 17-6, we find the same value of $\Delta U_{\text{electric}}$ as in Example 17-1.

The positive value of $\Delta V = V_b - V_a$ means that point *b* is at a higher potential than point *a*. This agrees with our observation above that if you travel opposite to the direction of the electric field, the electric potential increases. The value $E = 2.00 \times 10^2$ V/m means that the electric potential increases by 2.00×10^2 V for every meter that you travel opposite to the direction of $\vec{E}$.

If a charge $q_0 = -1.60 \times 10^{-19}$ C travels from *a* to *b*, the charge in electric potential energy is given by Equation 17-6:

$$\Delta V = \frac{\Delta U_{\text{electric}}}{q_0}$$

so

$$\begin{aligned} \Delta U_{\text{electric}} = q_0 \Delta V &= (-1.60 \times 10^{-19} \text{ C})(+60.0 \text{ V}) \\ &= -9.60 \times 10^{-18} \text{ V} \cdot \text{C} = -9.60 \times 10^{-18} \text{ J} \end{aligned}$$

(Recall from above that 1 V = 1 J/C.)

Example 17-5 Electric Potential Difference in a Uniform Field II

Determine the electric potential difference $\Delta V = V_c - V_a$ between points a and c in the uniform electric field of Example 17-4. Point c is a distance 0.500 m from point a, and the straight-line path from a to c makes an angle of 126.9° with respect to the electric field.

Set Up

Again we'll use Equation 17-7 to calculate the potential difference between the two points. Since the displacement from point a to point c points generally opposite to the direction of the electric field, we expect that $\Delta V = V_c - V_a$ will be positive, just like $\Delta V = V_b - V_a$ in Example 17-4.

Potential difference between two points in a uniform electric field:

$$\Delta V = V_c - V_a = -Ed \cos \theta \qquad (17\text{-}7)$$

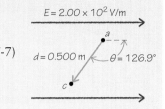

Solve

Use Equation 17-7 to solve for the potential difference.

Calculate the potential difference from the electric field magnitude E, the displacement d, and the angle θ:

$$\begin{aligned} \Delta V = V_c - V_a &= -Ed \cos \theta \\ &= -(2.00 \times 10^2 \text{ V/m})(0.500 \text{ m}) \cos 126.9° \\ &= -(2.00 \times 10^2 \text{ V/m})(0.500 \text{ m})(-0.600) \\ &= +60.0 \text{ V} \end{aligned}$$

Reflect

The potential difference between points a and c is the *same* as the potential difference between points a and b in Example 17-4. Equation 17-7 tells us why this should be: The potential difference $\Delta V = -Ed \cos \theta$ involves the magnitude E of the electric field multiplied by $d \cos \theta$, which is the component of the displacement $\vec{d}$ in the direction of $\vec{E}$. In both examples the electric field magnitude has the same value (2.00×10^2 V/m), as does the component of displacement in the direction of the electric field (-0.300 m).

Another way to come to this same conclusion is to recognize that the displacement from a to c can be broken down into two parts: a displacement in the negative x direction from a to b, followed by a displacement from b to c in the y direction. The potential difference for the displacement from a to b is $V_b - V_a = +60.0$ V as we calculated in Example 17-4; the potential difference for the displacement from b to c is $V_c - V_b = -Ed \cos 90° = 0$ since that displacement is perpendicular to the electric field. So the net potential difference between a and c is $V_c - V_a = (V_c - V_b) + (V_b - V_a) = +60.0$ V $+ 0$ V $= +60.0$ V.

Example 17-6 Transmission Electron Microscope

A *transmission electron microscope* forms an image by sending a beam of fast-moving electrons rather than a beam of light through a thin sample. As we will see in Chapter 26, such fast-moving electrons behave very much like a light wave. If the electrons have sufficiently high energy, the image that they form can show much finer detail than even the best optical microscope. The electrons are emitted from a heated metal filament and are then accelerated toward a second piece of metal called the *anode* that is at a potential 2.50 kV (1 kV = 1 kilovolt = 10^3 V) higher than that of the filament. If the electrons leave the filament initially at rest, how fast are the electrons traveling when they pass the anode?

Set Up

As the electrons move through the potential difference between the filament and the anode, the electric potential energy will change in accordance with Equation 17-6. Each electron has a negative charge $q_0 = -e$, so an *increase* in electric potential ($\Delta V > 0$) means a *decrease* in electric potential energy ($\Delta U_{electric} < 0$). The total mechanical energy (kinetic energy plus potential energy) is conserved as the electron moves because there are no forces acting on it other than the conservative electric force, so the electron kinetic energy will increase as the potential energy decreases.

Electric potential difference related to electric potential energy difference:

$$\Delta V = \frac{\Delta U_{electric}}{q_0} \qquad (17\text{-}6)$$

Mechanical energy is conserved:

$$E = K + U_{electric} = \text{constant}$$

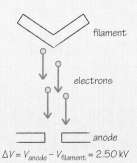

Solve

Solve Equation 17-6 for the change in electric potential energy as the electron moves from the filament to the anode, which is at a potential 2.50 kV higher than the filament.

From Equation 17-6,

$$\Delta U_{electric} = q_0 \Delta V$$

The electron has change $q_0 = -e$, so

$$\Delta U_{electric} = -e \, \Delta V$$
$$= -(1.60 \times 10^{-19} \text{ C})(+2.50 \times 10^3 \text{ V})$$
$$= -4.00 \times 10^{-16} \text{ J}$$

(Recall that 1 V = 1 J/C, so 1 J = 1 C·V.)

The conservation of mechanical energy tells us that the change in the kinetic energy of the electron is equal to the negative of the change in the electric potential energy.

Mechanical energy is conserved:

$$E = K + U_{electric} = \text{constant}$$

So the *change* in mechanical energy is zero:

$$\Delta E = \Delta K + \Delta U_{electric} = 0$$

The change in the kinetic energy of the electron is

$$\Delta K = -\Delta U_{electric} = -(-4.00 \times 10^{-16} \text{ J})$$
$$= +4.00 \times 10^{16} \text{ J}$$

Each electron begins with zero kinetic energy, so the change in its kinetic energy is equal to its final kinetic energy as it reaches the anode.

Since the electron begins with zero kinetic energy at the filament, the change in its kinetic energy is

$$\Delta K = +4.00 \times 10^{-16} \text{ J} = K_{anode} - K_{filament} = K_{anode}$$

Use this to find the speed of the electron at the anode:

$$K_{anode} = \frac{1}{2} m_{electron} v_{anode}^2 \text{ so}$$

$$v_{anode} = \sqrt{\frac{2K_{anode}}{m_{electron}}} = \sqrt{\frac{2(4.00 \times 10^{-16} \text{ J})}{9.11 \times 10^{-31} \text{ kg}}}$$
$$= 2.96 \times 10^7 \text{ m/s}$$

Reflect

The electrons are accelerated to nearly one-tenth of the speed of light ($c = 3.00 \times 10^8$ m/s).

Example 17-6 is just one of many situations in which an object with a charge of $-e$ (such as an electron) or $+e$ (such as a proton) moves through a potential difference. A common unit for the potential energy change in such situations is the **electron volt** (eV), which is equal to the magnitude e of the charge on the electron multiplied by 1 volt. Since $e = 1.60 \times 10^{-19}$ C, it follows that

$$1 \text{ eV} = (1.60 \times 10^{-19} \text{ C})(1 \text{ V}) = 1.60 \times 10^{-19} \text{ C·V} = 1.60 \times 10^{-19} \text{ J}$$

In Example 17-6, the electron moves through a potential difference of $+2.50 \times 10^3$ V, so the change in electric potential energy is -2.50×10^3 eV and the kinetic energy that the electron acquires is $+2.50 \times 10^3$ eV. We also use the abbreviations 1 keV = 10^3 eV, 1 MeV = 10^6 eV, and 1 GeV = 10^9 eV. (The largest particle accelerator in the world, the Large Hadron Collider at the European Organization for Nuclear Research CERN near Geneva, Switzerland, accelerates protons to a kinetic energy of 7×10^{12} eV = 7 TeV. This is equivalent to making the protons pass through a potential difference of 7×10^{12}, or 7 trillion, volts.) In later chapters we'll see that the electron volt is a useful unit for expressing energies on the atomic or nuclear scale.

Electric Potential Due to a Point Charge

Note that Equation 17-7 is useful only in the case of a uniform electric field. Another important case is the electric potential due to a point charge Q. We know from Equation

17-3 that if we place a test charge q_0 a distance r from a point charge Q, the electric potential energy of the system of two charges is

$$U_{\text{electric}} = \frac{kq_0Q}{r}$$

Equation 17-5 tells us that to find the electric potential due to the point charge Q, we must divide U_{electric} by the value of the test charge q_0:

Electric potential due to a point charge Q Coulomb constant

$$V = \frac{kQ}{r}$$

Value of the point charge

Distance from the point charge Q to the location where the potential is measured

Electric potential due to a
point charge
(17-8)

Equation 17-8 says that all points that are the same distance r from a point charge have the same electric potential due to that charge. It also says that if $Q > 0$, the electric potential due to the charge is positive and decreases (becomes less positive) as you move farther away from the charge so that r increases. If $Q < 0$, the electric potential due to the charge is negative and decreases (becomes more negative) as you move closer to the charge. For either sign of Q, the electric potential goes to zero at an infinite distance from the point charge.

These observations about Equation 17-8 are consistent with our previous statements about electric potential and electric force. A positive test charge q_0 placed near a positive charge Q feels an electric force that pushes it farther away from Q, toward regions where the potential V due to Q is lower (less positive). If instead that positive test charge is placed near a negative charge Q, the test charge feels an electric force that pulls it toward Q—again, toward regions where the potential V is lower (in this case, more negative).

! Watch Out! Don't confuse the formulas for electric potential and electric field due to a point charge.

Be sure that you recognize the differences between Equation 17-8, $V = kQ/r$, and Equation 16-4 for the magnitude of the electric field due to a point charge Q, $E = k|Q|/r^2$. Equation 17-8 says that the potential V due to a point charge is inversely proportional to r and can be positive or negative, depending on the sign of Q. By contrast, Equation 16-4 tells us that the magnitude E of the field due to a point charge is inversely proportional to the *square* of r. Furthermore, because E is the magnitude of a vector, it is always positive (it is proportional to the absolute value of Q).

If there is not a single point charge but an assemblage of charges, the total electric potential at a given position due to these charges is the sum of the individual potentials. For example, for the case of three charges Q_1, Q_2, and Q_3, the potential at a point that is a distance r_1 from the first charge, a distance r_2 from the second charge, and a distance r_3 from the third charge is

$$V = \frac{kQ_1}{r_1} + \frac{kQ_2}{r_2} + \frac{kQ_3}{r_3}$$

(electric potential of three charges)

(17-9)

? Got the Concept? 17-2 Electric Potential Difference

Four point charges are each moved from one position to another position where the electric potential has a different value. Rank the four charges in order of the electric potential energy change that takes place, from most positive to most negative. (a) A +0.0010 C charge that moves through a potential increase of 5.0 V. (b) A +0.0020 C charge that moves through a potential decrease of 2.0 V. (c) A −0.0015 C charge that moves through a potential decrease of 4.0 V. (d) A −0.0010 C charge that moves through a potential increase of 2.5 V.

Take-Home Message for Section 17-3

✔   The electric potential at a certain position equals the electric potential energy per unit charge for a test charge at that position. Electric potential is a scalar, not a vector, quantity. Electric potential decreases as you move in the direction of the electric field.

✔   The electric potential due to a point charge is positive if the charge is positive, and negative if the charge is negative.

As you move farther from an isolated point charge, the electric potential becomes closer to zero.

✔   The electric potential due to a collection of charges is the sum of the potentials due to the individual charges.

✔   The electric potential difference between two points equals the difference in electric potential energy per unit charge between those points.

17-4 The electric potential has the same value everywhere on an equipotential surface

In Examples 17-4 and 17-5 in the previous section, we looked at the potential differences between two pairs of points in an electric field, a and b in Example 17-4 and a and c in Example 17-5. Although the electric field is the same in both examples, the distance from a to c in Example 17-5 is clearly longer than the distance from a to b in Example 17-4. So you might have expected that the potential difference $V_c - V_a$ in Example 17-5 would be greater in magnitude than the potential difference $V_b - V_a$ in Example 17-4. In fact we found that the potential differences were equal, which tells us that the potential is the *same* at points b and c. You can see why from Equation 17-7, $\Delta V = -Ed \cos \theta$. A displacement $\vec{d}$ from point b to point c is perpendicular to the electric field $\vec{E}$, so the angle in Equation 17-7 is $\theta = 90°$ and $\cos \theta = \cos 90° = 0$. Thus the potential difference ΔV between points b and c must be zero.

In general, the electric potential will be the same at any two points that lie along a curve perpendicular to electric field lines. Such a curve is called an *equipotential curve* or simply an **equipotential**. You can see that for the case of a uniform electric field, the electric potential will have the same value anywhere on a plane that's perpendicular to the electric field. Such a plane is an example of an **equipotential surface**, one on which the electric potential has the same value at all points (Figure 17-6). No work is required to move a charge from one point to another along any path on an equipotential surface. As we described in Section 17-3, the electric potential decreases as you move in the direction of the electric field $\vec{E}$, so the value of the potential V is lower for equipotential surfaces that are "downstream" in the electric field than on surfaces that are "upstream."

Figure 17-7 shows both electric field lines and equipotentials for the case of a *nonuniform* electric field. Note that the equipotential surfaces are perpendicular at *all* points to the field lines, just as for the case of a uniform field in Figure 17-6. But since

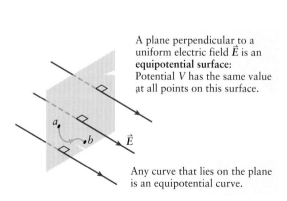

A plane perpendicular to a uniform electric field $\vec{E}$ is an **equipotential surface:** Potential V has the same value at all points on this surface.

$\vec{E}$

Any curve that lies on the plane is an equipotential curve.

Figure 17-6 Equipotential surfaces I In a uniform electric field, equipotential surfaces are planes perpendicular to the electric field lines.

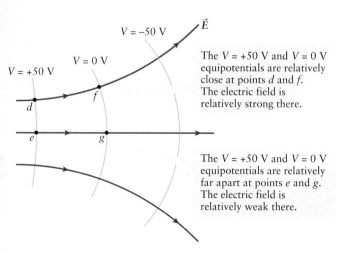

$V = -50$ V

$\vec{E}$

$V = 0$ V

$V = +50$ V

d

f

e

g

The $V = +50$ V and $V = 0$ V equipotentials are relatively close at points d and f. The electric field is relatively strong there.

The $V = +50$ V and $V = 0$ V equipotentials are relatively far apart at points e and g. The electric field is relatively weak there.

Figure 17-7 Equipotential surfaces II In a nonuniform electric field, equipotential surfaces are curved surfaces (seen here from the side) that are everywhere perpendicular to the electric field lines.

the electric field lines are not parallel lines, the equipotential surfaces are not flat planes. In general, *any* surface that is everywhere perpendicular to the electric field is an equipotential surface. Whether the field is uniform or not, the value of the potential V on an equipotential surface is lower the farther "downstream" in the electric field that surface is.

In Figure 17-7 points d and e are both on the equipotential for which $V = +50$ V, and points f and g are both on the equipotential for which $V = 0$ V. So the potential difference between points d and f is the same as the potential difference between points e and g:

$$V_d - V_f = V_e - V_g = (+50 \text{ V}) - (0 \text{ V}) = +50 \text{ V}$$

However, the distance between points d and f is less than that between points e and g. Equation 17-7 tells us that the potential difference between two points is proportional to the distance d between the points and the electric field magnitude E. (This equation is strictly valid only for a uniform field, but is still approximately true for a nonuniform field.) So the electric field must be greater between the points d and f that are closer together, and less between the points e and g that are farther apart. This is an example of a general rule:

Where two adjacent equipotential surfaces are close together, the electric field is relatively strong. Where these surfaces are far apart, the electric field is relatively weak.

Figure 17-8 shows an application of this idea. For a positive point charge (Figure 17-8a) or a negative point charge (Figure 17-8b), the electric field points radially outward or inward. The equipotential surfaces are everywhere perpendicular to the field

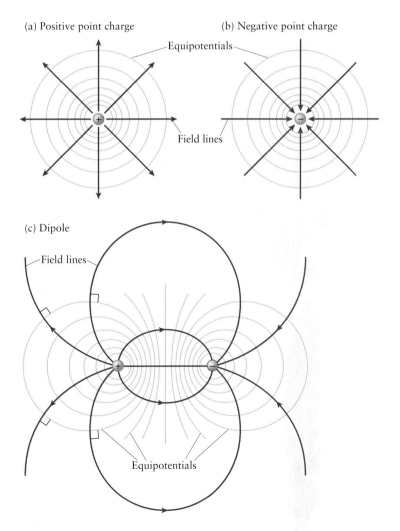

(a) Positive point charge

(b) Negative point charge

Equipotentials

Field lines

(c) Dipole

Field lines

Equipotentials

Figure 17-8 **Equipotential surfaces**
III The equipotential surfaces for a point charge are spheres centered on the point charge.

lines, so they are spheres centered on the point charge. (You could also conclude this from Equation 17-8 for the potential due to a point charge, $V = kQ/r$. This says that V has the same value at all points that are the same distance r from the point charge Q—that is, at all points on a sphere of radius r centered on the point charge.) The radial distance from one spherical equipotential surface to the next is the same no matter where you are around the sphere, so the electric field magnitude is the same at all points a given distance from the point charge. This agrees with Equation 16-4 for the field magnitude E due to a point charge, $E = k|Q|/r^2$; the value of E depends only on the distance from the charge, not on where you are around the charge. For an electric dipole, however, the situation is different (Figure 17-8c). The field lines arc from the positive charge to the negative one, and the equipotential surfaces are neither spherical nor centered on the point charges. (They are, however, everywhere perpendicular to the field lines.) Adjacent equipotential surfaces are close together between the two charges because the electric field is strong there. To the left of the left-hand charge or to the right of the right-hand charge, the electric field is relatively weak and adjacent equipotential surfaces are farther apart.

The equipotential concept is helpful for understanding the electric field around a charged conductor. We learned in Section 16-7 that if we put excess charge on a conductor and allow the individual excess charges to move to their equilibrium positions, all of the excess charge will end up on the conductor's surface and the electric field $\vec{E}$ will be zero everywhere inside the conductor. Since $\vec{E}$ is uniform inside the conductor (it has the same value at all points), we can use Equation 17-7 to calculate the potential difference between two points inside the conductor separated by a distance d: $\Delta V = -Ed \cos \theta = 0$ because $E = 0$. In other words, *the electric potential has the same value everywhere inside a conductor in equilibrium*. We say that a conductor in equilibrium is an **equipotential volume**. That's why we can make statements like "This conductor is at a potential of $+20$ V"—the value of potential is the same everywhere throughout the volume of the conductor. When we say that each of the batteries in Figure 17-1a has a voltage of 1.5 V, we mean that the potential of the conductor at the positive end of the battery is 1.5 V higher than the potential of the conductor at the negative end. The potential difference is the same between any point on the positive terminal and any point on the negative terminal.

The electric field *outside* a charged conductor is *not* zero. If the excess charge on the conductor is positive, the electric field will point away from the surface of the conductor; if the excess charge is negative, the electric field will point toward the surface. Because the surface of the conductor is part of the equipotential volume, it is itself an equipotential surface. Because field lines and equipotential surfaces are always perpendicular, we conclude that *the electric field just outside a conductor in equilibrium must be perpendicular to the surface of the conductor*. Since the electric field always points toward lower electric potential, we can also conclude that *a positively charged conductor is at a higher electric potential than an adjacent negatively charged conductor*. We'll use these ideas about conductors in the following section to help us understand an important device called a *capacitor*.

? Got the Concept? 17-3 Equipotentials and Electric Field

The orange vertical lines in Figure 17-9 represent equipotential curves in some region of space. Which statement is correct about the electric field $\vec{E}$ at point A compared to the electric field at point B? (a) At A the field $\vec{E}$ points to the right and has a greater magnitude than at B; (b) at A the field $\vec{E}$ points to the right and has a smaller magnitude than at B; (c) at A the field $\vec{E}$ points to the left and has a greater magnitude than at B; (d) at A the field $\vec{E}$ points to the left and has a smaller magnitude than at B; (e) none of these.

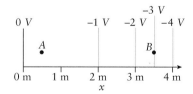

Figure 17-9 What equipotentials tell you about $\vec{E}$ What can you infer from this figure about the electric field at A and B?

Take-Home Message for Section 17-4

✔ The potential is the same everywhere along an equipotential curve or equipotential surface.

✔ Any curve or surface that is perpendicular to the electric field lines is an equipotential.

✔ Where equipotential surfaces are close, the electric field is relatively strong.

✔ A conductor in equilibrium has the same potential throughout its volume. The electric field at the surface of a conductor is perpendicular to the surface.

17-5 A capacitor stores equal amounts of positive and negative charge

The surface of every cell in your body is a *membrane* composed of a phospholipid bilayer that separates the intracellular fluid (inside the cell) and the extracellular fluid (outside the cell). Negative charge accumulates on the membrane's intracellular (interior) surface, and this attracts positive charge onto the extracellular (exterior) surface. The result is a potential difference between the inner and outer surfaces of the membrane, and an electric field within the membrane that points from the outside in. This field and associated potential help drive essential ions through apertures in the membrane. They are also the source of the electrical signal used by the specialized cells called neurons to code, process, and transmit information.

A system or device that can store positive and negative charge like a cell membrane is called a **capacitor**. In technological applications, a capacitor does not use a membrane per se but instead uses two pieces of metal called **plates**. One plate holds a positive charge q, and the other carries a negative charge $-q$. (Note that the *net* charge on the capacitor is zero.) It takes work to separate the positive and negative charges against the electric forces that attract them to each other, and this work goes into increasing the electric potential energy of the system of charge. So a capacitor is a device for storing electric potential energy. As we'll see in the following section, this stored energy is analogous to the elastic potential energy stored in a stretched or compressed spring. We often need to draw capacitors in diagrams that represent electric circuits. The standard symbol for a capacitor in the diagram of a circuit is

The two closely spaced parallel lines represent the two plates of the capacitor, and the straight horizontal lines represent wires that connect the capacitor to the rest of the circuit.

The simplest geometry for a capacitor is two large parallel plates, one with charge $+q$ and the other with charge $-q$. Figure 17-10a shows these two plates separated from each other. We saw in Section 16-7 that close to a large charged plate and far from its edges the electric field due to that plate is uniform, is perpendicular to the plane of the plate, and has magnitude $E = \sigma/(2\varepsilon_0)$ (from Equation 16-14). In this expression σ is the charge per unit area, equal to the charge q divided by the area A of the plate: $\sigma = q/A$. Since each plate has the same magnitude of charge, the field $\vec{E}_+$ due to the charge on the positive plate has the same magnitude as the field $\vec{E}_-$ due to the charge on the negative plate.

For a capacitor the *net* electric field is the vector sum of $\vec{E}_+$ and $\vec{E}_-$: $\vec{E} = \vec{E}_+ + \vec{E}_-$. Figure 17-10b shows the two plates moved into position to form a **parallel-plate capacitor**. Because the magnitude of each field does not depend on the distance from the plate that generates the field, $\vec{E}_+$ cancels $\vec{E}_-$ in the region above the upper plate and below the lower plate. Between the plates the fields $\vec{E}_+$ and $\vec{E}_-$ have the same magnitude and point in the same direction, so the net field between the plates has twice the magnitude of the field due to either plate by itself:

$$E = 2\left(\frac{\sigma}{2\varepsilon_0}\right) = \frac{\sigma}{\varepsilon_0} = \frac{q}{\varepsilon_0 A} \qquad (17\text{-}10)$$

(electric field in a parallel-plate capacitor)

(a) Two large plates (viewed from the side) carry charge $+q$ and $-q$. Close to the plates the electric fields are nearly constant, so we represent them by straight, parallel field lines.

(b) Because the fields are nearly constant, the magnitude of the field due to each plate is the same at any point near the plates...

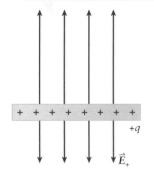

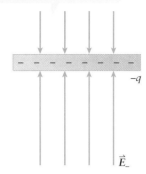

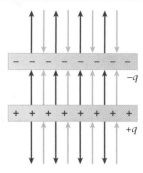

(c) ... so when the plates are placed close together, the fields cancel in the region outside the plates. The fields add in between the plates.

Figure 17-10 **Electric field of a parallel-plate capacitor** The field due to two oppositely charged plates is the vector sum of the fields due to each plate.

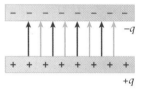

Note that the electric field points from the positive plate to the negative plate, just as we described in our discussion of conductors in Section 17-4.

(Equation 17-10 and the claim that $\vec{E}_+$ and $\vec{E}_-$ cancel outside the capacitor are strictly true only if the plates are large and close together, and when we consider points in space far from the edges of the capacitor. This idealization is physically reasonable for the kinds of problems we will encounter.)

We can substitute Equation 17-10 into Equation 17-7 to calculate the *potential difference* between the plates of a parallel-plate capacitor. The electric field points from the positive plate to the negative plate. If we let d be the distance between the plates, and we travel from the negative plate to the positive plate, the angle between the electric field and the displacement is $\theta = 180°$. The potential difference is therefore

$$V = V_+ - V_- = -Ed \cos\theta = -Ed(-1)$$

$$= Ed = \left(\frac{q}{\varepsilon_0 A}\right)d$$

(17-11)

$$V = \frac{qd}{\varepsilon_0 A}$$

⚠ Watch Out! The symbol V sometimes stands for potential, sometimes for potential difference.

Previously in this chapter we've used V to denote the electric potential at a given point and ΔV to denote the difference in electric potential between two points. In Equation 17-11, and throughout our discussion of capacitors, we'll follow common practice and use V as the symbol for the potential difference, or voltage, between the two capacitor plates. Just remember "for capacitors, V means voltage." We'll frequently refer to V as the "voltage *across* the capacitor."

For a given capacitor of fixed plate area A and plate separation d, the potential difference V between the plates is proportional to the magnitude of the charge q on each plate.

We can rewrite Equation 17-11 as an expression for the amount of charge on each plate. To charge an initially uncharged capacitor, we apply a voltage V between its plates as in **Figure 17-11**. Electrons are driven to one of the plates, giving it a charge $-q$; the plate from which the electrons were taken is left with a charge $+q$. (As we mentioned previously, the net charge on the capacitor remains zero.) Equation 17-11 tells us that the magnitude q of the charge that each plate acquires is proportional to the applied voltage V:

$$q = \frac{\varepsilon_0 A}{d} V \qquad (17\text{-}12)$$

(charge on a parallel-plate capacitor)

The quantity $\varepsilon_0 A/d$ in Equation 17-12 is called the **capacitance** of the capacitor. We use the symbol C for capacitance:

Capacitance of a parallel-plate capacitor   Permittivity of free space = $1/(4\pi k)$

$$C = \frac{\varepsilon_0 A}{d}$$

Area of each capacitor plate

Distance between the capacitor plates

 do not depend on charge or potential difference; only on geometry.

Capacitance of a parallel-plate capacitor
(17-13)

For a parallel-plate capacitor, C depends only on the area A of the plates, the distance d between them, and the material between them. We've assumed that the plates are separated by vacuum; in Section 17-8 we'll explore what happens if the space between the capacitor plates is filled with a different material. Capacitance is therefore a compact way to summarize the electrical properties of a capacitor. In particular, capacitance tells us the amount of charge that can be stored on a capacitor held at a given voltage. From Equations 17-12 and 17-13,

▶ *Go to Picture It 17-3 for more practice dealing with capacitance*

A capacitor carries a charge $+q$ on its positive plate and a charge $-q$ on its negative plate.

The magnitude of q is directly proportional to V, the voltage (potential difference) between the plates.

$$q = CV$$

The constant of proportionality between charge q and voltage V is the capacitance C of the capacitor.

Charge, voltage, and capacitance for a capacitor
(17-14)

This equation says that the magnitude of the charge on the plates of a capacitor is directly proportional to the voltage between the plates. The greater the capacitance, the more charge is present for a given voltage. Note that while Equations 17-12 and 17-13 are valid for a parallel-plate capacitor only, Equation 17-14 is valid for capacitors of *any* geometry. In the problems at the end of this chapter you'll analyze some other simple types of capacitor.

Equation 17-14 shows that the unit of capacitance is the coulomb per volt, also known as the **farad** (symbol F): $1\ \mathrm{F} = 1\ \mathrm{C/V}$. The farad is named for the nineteenth-century English physicist Michael Faraday. The capacitors you'll find in your neighborhood electronics supply store are typically in the range of 10 pF (10×10^{-12} F) to 1000 μF (1000×10^{-6} F); 1 F is an extremely large capacitance. Using $1\ \mathrm{F} = 1\ \mathrm{C/V}$, $1\ \mathrm{V} = 1\ \mathrm{J/C}$, and $1\ \mathrm{J} = 1\ \mathrm{N \cdot m}$, we see that $1\ \mathrm{F} = (1\ \mathrm{C})/(1\ \mathrm{J/C}) = 1\ \mathrm{C^2/J} = 1\ \mathrm{C^2/(N \cdot m)}$. We can express the value of the constant ε_0 (the permittivity of free space) as

When this battery is connected to the plates of a capacitor, it creates a potential difference V between the plates...

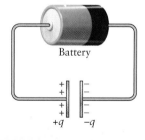

Battery

$+q$ $-q$

...which drives electrons from one plate, through the battery, and onto the other plate. The plate that lost electrons ends up with charge $+q$, and the other plate ends up with charge $-q$.

$$\varepsilon_0 = 8.85 \times 10^{-12}\ \frac{\mathrm{C^2}}{\mathrm{N \cdot m^2}} = 8.85 \times 10^{-12}\ \frac{\mathrm{F}}{\mathrm{m}}$$

Figure 17-11 Charging a capacitor
The charges that appear on the two capacitor plates have the same magnitude but opposite signs.

These units are useful in calculations of capacitance, as we'll see below.

Figure 17-12 Capacitive touchscreens Mobile devices such as these use the physics of capacitors to determine where your finger touches the screen.

An everyday application of Equation 17-13 is the touchscreen on a mobile device such as a smartphone or tablet (Figure 17-12). Behind the device's glass screen is a layer of a special transparent conductor called indium tin oxide (ITO), which is actually a solid mixture of indium oxide, In_2O_3, and tin oxide, SnO_2. When you touch your finger to the screen, the conducting ITO layer acts as one plate of a capacitor and your finger—which is also a conductor—acts as the other plate. Sensor circuits in the mobile device detect where on the screen a capacitance appears due to this capacitor, which is how the device "knows" where on the screen it has been touched. (To verify that the object touching the screen has to be a conductor, try using the rubber eraser on a pencil to touch the screen of a mobile device. Because rubber is an insulator, not a conductor, the screen won't respond.) The sensor circuits are adjusted so that they will only register if the capacitance C is above a certain minimum value. That explains why you have to physically touch the screen: According to Equation 17-13, C increases as the distance d between the plates decreases, so the capacitance is largest when your finger is touching the screen and so closest to the ITO layer. Your finger still acts as a capacitor plate if you hold it a slight distance away from the screen, but the capacitance is now too low to trigger the sensor circuits.

Example 17-7 A Parallel-Plate Capacitor

Two square, parallel conducting plates each have dimensions 5.00 cm by 5.00 cm and are placed 0.100 mm apart. Determine the capacitance of this configuration.

Set Up

We'll use Equation 17-13 to determine the value of C for this capacitor.

Capacitance of a parallel-plate capacitor:

$$C = \frac{\varepsilon_0 A}{d} \qquad (17\text{-}13)$$

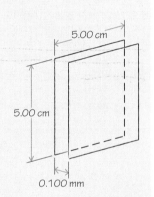

Solve

The area of each plate is the product of the length of the two sides, and the plate separation is given. We need to convert all dimensions into SI units.

Area $A = (5.00 \times 10^{-2}\,\text{m})(5.00 \times 10^{-2}\,\text{m}) = 2.50 \times 10^{-3}\,\text{m}^2$

Plate separation $d = 1.00 \times 10^{-4}\,\text{m}$

Using Equation 17-13,

$$C = \frac{(8.85 \times 10^{-12}\,\text{F/m})(2.50 \times 10^{-3}\,\text{m}^2)}{1.00 \times 10^{-4}\,\text{m}} = 2.21 \times 10^{-10}\,\text{F}$$
$$= 221\,\text{pF}$$

(Recall that the prefix "p" or pico– represents 10^{-12}.)

Reflect

Capacitors with capacitances of a few hundred picofarads are well within the range available from your local electronics supply store and are likely found in calculators, cell phones, and portable music players.

To make a 1-F capacitor with the same separation $d = 0.100$ mm between plates, we would have to increase the plate area by a factor of $(1\,\text{F})/(2.21 \times 10^{-10}\,\text{F}) = 4.52 \times 10^9$. The length of each side would have to be increased by a factor of the square root of 4.52×10^9, or 6.73×10^4. The sides of the plates would be $(6.73 \times 10^4)(5.00 \times 10^{-2}\,\text{m}) = 3.36 \times 10^3$ m, or 3.36 *kilometers* (about 2 miles). One farad is a *very* large capacitance.

Example 17-8 Insulin Release

The hormone insulin minimizes variations in blood glucose levels. Pancreatic beta cells ("β-cells") synthesize insulin and store it in *vesicles*, bubble-like organelles approximately 150 nm in radius within the cytoplasm of the cells. To release insulin, vesicles fuse with the membrane of the β-cell. This increases the surface area of the β-cell by the surface area of the fused vesicles (Figure 17-13). The thickness of the cell membrane does not change, so the increase in surface area increases the capacitance of the β-cell membrane. Experiment shows that the membrane capacitance increases at 1.6×10^{-13} F/s during insulin release. If the membrane capacitance is approximately 1 μF per square centimeter of surface area, estimate the number of vesicles that fuse with the cell membrane per second during insulin release.

An insulin-producing beta cell from the pancreas

Red circles outline vesicles containing insulin.

Cell membrane

4. The vesicle has reformed and is ready to be filled with insulin.

3. The vesicle is reforming.

1. One vesicle is fusing with the cell membrane.

2. Now the vesicle has released its insulin to the outside of the cell. The vesicle membrane is part of the cell membrane.

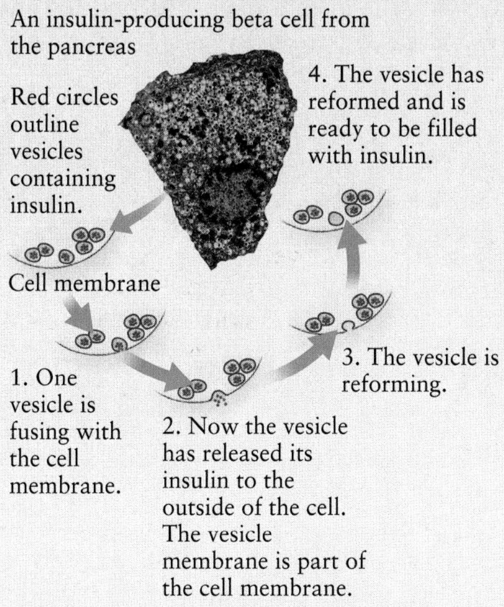

Figure 17-13 **Changing membrane, changing capacitance** To study this important process in a pancreatic β-cell, scientists monitor changes in the capacitance of the cell membrane.

Set Up

Although the cell membrane is not flat like the parallel-plate capacitor shown in Figure 17-10, we can treat it as such because the size of the cell is large compared to the thickness of the cell membrane. (It's like being above Earth's surface at a height that's much smaller than Earth's radius. While Earth is approximately spherical, when seen from such a short distance it appears to be flat.) We'll use Equation 17-13 to calculate the rate at which the area A of the cell membrane must increase to cause the measured rate of capacitance increase. This increase in area comes from fusing vesicles.

Capacitance of a parallel-plate capacitor:

$$C = \frac{\varepsilon_0 A}{d} \qquad (17\text{-}13)$$

Surface area of a sphere of radius r:

$$A_{sphere} = 4\pi r^2$$

Solve

Find the rate at which the membrane area must increase to cause this rate of capacitance increase.

Rate of change of the capacitance of the membrane:

$$\frac{\Delta C_m}{\Delta t} = 1.6 \times 10^{-13} \text{ F/s}$$

Equation 17-13 tells us that the capacitance C_m is directly proportional to the membrane surface area A_m. We are told that the capacitance per unit area is approximately 1 μF/cm^2, so the change in capacitance ΔC_m that corresponds to a given change in area ΔA_m is

$$\Delta C_m = \left(1 \, \frac{\mu F}{cm^2}\right) \Delta A_m$$

Since the capacitance increases by 1.6×10^{-13} F in 1 s, the increase in area of the membrane in 1 s must be

$$\Delta A_m = \frac{\Delta C_m}{1 \, \mu F/cm^2}$$

$$= (1.6 \times 10^{-13} \text{ F})\left(1 \, \frac{cm^2}{\mu F}\right)\left(\frac{1 \, \mu F}{10^{-6} \text{ F}}\right)\left(\frac{1 \text{ m}}{100 \text{ cm}}\right)^2$$

$$= 1.6 \times 10^{-11} \text{ m}^2$$

(We used 1 μF = 1 microfarad = 10^{-6} F and 1 m = 100 cm.)

Now we can calculate how many vesicles must fuse with the membrane per second to cause this increase in area.

The surface area of a single vesicle is

$$A_{\text{vesicle}} = 4\pi r_{\text{vesicle}}^2 = 4\pi(150\text{ nm})^2 = 4\pi(150 \times 10^{-9}\text{ m})^2$$
$$= 2.8 \times 10^{-13}\text{ m}^2$$

The number of vesicles that must add their area to the membrane in 1 s to cause an area increase $\Delta A_m = 1.6 \times 10^{-11}\text{ m}^2$ is

$$\frac{\Delta A_m}{A_{\text{vesicle}}} = \frac{1.6 \times 10^{-11}\text{ m}^2}{2.8 \times 10^{-13}\text{ m}^2/\text{vesicle}} = 57\text{ vesicles}$$

Our final answer should have just one significant figure, since we were given the capacitance per unit area $1\ \mu\text{F}/\text{cm}^2$ to just one significant figure. So our final answer is that about 60 vesicles fuse with the membrane wall per second.

Reflect

The ability to detect small changes in capacitance makes it possible to study the molecular mechanisms of hormone release from single cells and to verify previous biochemical measurements of insulin release from β-cells. After insulin release, new vesicles form by pinching off from the cell membrane and are refilled with insulin. By recycling the vesicle membrane, β-cells maintain their size over the long term.

? Got the Concept? 17-4 Capacitors

A parallel-plate capacitor has a potential difference V between its plates. If the potential difference is increased to $2V$, what effect does this have on the capacitance C? (a) C increases by a factor of 4; (b) C increases by a factor of 2; (c) C becomes $\frac{1}{2}$ as great; (d) C becomes $\frac{1}{4}$ as great; (e) C is unchanged.

Take-Home Message for Section 17-5

✔ A capacitor has two plates, one with charge $+q$ and the other with charge $-q$.

✔ For a given capacitor the potential difference between the plates is proportional to the magnitude of the charge q on each plate. The proportionality constant, called the capacitance, depends on the geometry of the plates and on the material in the space between the plates.

17-6 A capacitor is a storehouse of electric potential energy

In electric circuits, capacitors are most useful because of their ability to store electric potential energy. An applied voltage V such as that supplied by the battery in Figure 17-11 charges a capacitor by effectively pulling negative charges from one of the plates and depositing them on the other. To move the negative charges away from the positive charges requires work, and it is this work that results in electric potential energy being stored in the capacitor. At a later time, the potential energy can be transferred by charge leaving the capacitor and passed on to other parts of the circuit. (That's what happens in the electronic flash unit in a camera or mobile phone. The device's battery charges a capacitor, and the energy stored in the charged capacitor is used to produce a short, intense burst of light.)

Let's see how to calculate the amount of electric potential energy stored in a capacitor that has charge $+q$ and $-q$ on its positive and negative plates, respectively. This is equal to the amount of electric potential energy that's added to the capacitor if we start with both plates uncharged (which we can regard as a state of zero potential energy) and move charge $+q$ from the first plate to the second one, leaving charge $-q$ on the first plate. If the potential difference between the plates had a constant value ΔV,

Equation 17-6 tells us that the change in electric potential energy $\Delta U_{electric}$ in this process would be

$$\Delta V = \frac{\Delta U_{electric}}{q} \quad \text{so} \quad \Delta U_{electric} = q\,\Delta V$$

However, the potential difference between the plates (that is, the voltage across the capacitor) does *not* stay constant as we transfer charge from one plate to the other! Equation 17-14 tells us that the potential difference between the plates is proportional to the amount of charge on the positive plate. So as we transfer more charge from one plate to the other, the potential difference across which the charge must move increases from zero (its starting value) to a final value V given by Equation 17-14: $q = CV$, so $V = q/C$. To correctly calculate the amount of potential energy stored in the capacitor when it is charged, we have to replace ΔV in Equation 17-6 by the *average* value of the potential difference during the charging process. Because potential difference increases in direct proportion to the charge, this average value is just the average of the starting potential difference and the final potential difference V:

$$\Delta U_{electric} = q\,\Delta V_{average}$$

$$= q\left[\frac{(\text{starting potential difference}) + (\text{final potential difference})}{2} \right]$$

$$= q\left(\frac{0 + V}{2} \right) = \frac{1}{2}qV \tag{17-15}$$

Equation 17-14 tells us that $q = CV$ and $V = q/C$, so we can also write Equation 17-15 as

$$\Delta U_{electric} = \frac{1}{2}(CV)V = \frac{1}{2}CV^2 \quad \text{or} \quad \Delta U_{electric} = \frac{1}{2}q\left(\frac{q}{C} \right) = \frac{q^2}{2C} \tag{17-16}$$

If we say that the electric potential energy of the initial uncharged capacitor was zero, the final potential energy $U_{electric}$ is just equal to the increase in potential energy $\Delta U_{electric}$ given by Equation 17-15 or Equation 17-16. So we can write the potential energy stored in the capacitor in three ways:

The electric potential energy stored in a charged capacitor...

$$U_{electric} = \frac{1}{2}qV = \frac{1}{2}CV^2 = \frac{q^2}{2C}$$

...can be expressed in three ways in terms of the charge q, potential difference V, and capacitance C.

Electric potential energy stored in a capacitor
(17-17)

Equation 17-17 says that the energy stored in a charged capacitor is proportional to the *square* of the charge q or, equivalently, to the square of the potential difference (voltage) V across the capacitor. These results are true for capacitors of all kinds, not just the simple parallel-plate capacitor that we discussed in Section 17-5.

The last of the three expressions for electric potential energy in Equation 17-17, $U_{electric} = q^2/(2C)$, is very similar to the equation for *spring* potential energy that we learned in Section 6-6:

Spring potential energy of a stretched or compressed spring   Spring constant of the spring

$$U_{spring} = \frac{1}{2}kx^2$$

Spring potential energy
(6-19)

Extension of the spring ($x > 0$ if spring is stretched, $x < 0$ if spring is compressed)

The only differences between $U_{electric} = q^2/(2C)$ from Equation 17-17 and the expression in Equation 6-19 is that the spring displacement x is replaced by the charge

q and the spring constant k is replaced by the reciprocal of the capacitance, $1/C$. This similarity isn't surprising. To add potential energy to a spring by stretching it, you have to pull against the force of magnitude $F = kx$ that the spring exerts on you. The greater the distance that the spring is already stretched, the more force it exerts and the harder it is to stretch it further. In exactly the same way, to add potential energy to a capacitor by increasing the magnitude of charge on the two plates, you have to transfer charge against a voltage $V = (1/C)q$. The greater the charge that is already on the plates, the greater the voltage and the harder it is to increase q. In later chapters we'll make more use of this analogy between displacement x and force $F = kx$ for a spring and charge q and voltage $V = (1/C)q$ for a capacitor.

Example 17-9 A Defibrillator Capacitor

A defibrillator, like the one shown in the photograph that opens this chapter, is essentially a capacitor that is charged by a high-voltage source and then delivers the stored energy to a patient's heart. (a) How much charge does the 80.0-μF capacitor in a certain defibrillator store when it is fully charged by applying 2.50 kV? (b) How much energy can this defibrillator deliver?

Set Up

We're given the capacitance $C = 80.0$ μF (recall 1 μF = 1 microfarad = 10^{-6} F) and the potential difference $V = 2.50$ kV (recall 1 kV = 1 kilovolt = 10^3 V) between the capacitor plates. We'll use Equation 17-14 to determine the magnitude q of the charge on each capacitor plate, and Equation 17-17 to find the potential energy $U_{electric}$ stored in the charged capacitor. If we assume that no energy is lost in the process of being transferred to the patient, this is equal to the energy that the defibrillator delivers.

Charge, voltage, and capacitance for a capacitor:

$$q = CV \qquad (17\text{-}14)$$

Electric potential energy stored in a capacitor:

$$U_{electric} = \frac{1}{2}qV = \frac{1}{2}CV^2 = \frac{q^2}{2C} \qquad (17\text{-}17)$$

Solve

(a) Substitute the given values of C and V into Equation 17-14 to solve for the charge q.

Charge on the capacitor plates:

$$q = CV = (80.0\ \mu F)(2.50\ kV)$$
$$= (80.0 \times 10^{-6}\ F)(2.50 \times 10^3\ V)$$
$$= 0.200\ F \cdot V = 0.200\ C$$

(Recall that 1 F = 1 C/V, so 1 F$\cdot$V = 1 C.)

(b) Since the values of C and V are given, let's use the second of the three relationships in Equation 17-17 to find the stored electric potential energy.

Energy stored in the capacitor:

$$U_{electric} = \frac{1}{2}CV^2$$

$$= \frac{1}{2}(80.0 \times 10^{-6}\ F)(2.50 \times 10^3\ V)^2$$
$$= 2.50 \times 10^2\ F \cdot V^2$$
$$= 2.50 \times 10^2\ J$$

(Recall that 1 V = 1 J/C and 1 F = 1 C/V, so 1 F$\cdot$V^2 = 1 (C/V)V^2 = 1 C$\cdot$V = 1 C$\cdot$(J/C) = 1 J.)

Reflect

The American Heart Association recommends that a defibrillator shock should deliver between 40 and 360 J in order to be effective, so our numerical result is in the recommended range.

    We can double-check our answers by using the other two relationships in Equation 17-17. Happily, by using the value of q that we

One alternative way to calculate the energy stored in the capacitor:

$$U_{electric} = \frac{1}{2}qV$$

$$= \frac{1}{2}(0.200\ C)(2.50 \times 10^3\ V)$$
$$= 2.50 \times 10^2\ C \cdot V$$
$$= 2.50 \times 10^2\ J$$

calculated in part (a) we get the same result for $U_{electric}$ in part (b) as we found above.

Remember that in solving any problem, you should *always* take advantage of alternative ways to find the answer in order to check your results.

Another alternative way to calculate the energy stored in the capacitor:

$$U_{electric} = \frac{q^2}{2C}$$
$$= \frac{(0.200\ C)^2}{2(80.0 \times 10^{-6}\ F)}$$
$$= 2.50 \times 10^2 \frac{C^2}{F}$$
$$= 2.50 \times 10^2\ J$$

(Recall that $1\ V = 1\ J/C$ and $1\ F = 1\ C/V$. You should verify that $1\ C^2/F = 1\ J$.)

? Got the Concept? 17-5 Spreading the Plates of a Capacitor I

Suppose you increase the distance between the plates of a charged parallel-plate capacitor without changing the amount of charge stored on the plates. What will happen to the energy stored in the capacitor? (a) It will decrease; (b) it will remain the same; (c) it will increase; (d) not enough information given to decide.

? Got the Concept? 17-6 Spreading the Plates of a Capacitor II

Suppose you increase the distance between the plates of a parallel-plate capacitor while holding the potential difference between the plates constant. (You could do this by keeping the plates connected to a battery, as in Figure 17-11.) What will happen to the energy stored in the capacitor? (a) It will decrease; (b) it will remain the same; (c) it will increase; (d) not enough information given to decide.

Take-Home Message for Section 17-6

✔ A charged capacitor stores electric potential energy. The amount of energy stored is proportional to the square of the magnitude of the charge on each plate, and also proportional to the square of the potential difference between the plates.

17-7 Capacitors can be combined in series or in parallel

In both biological systems and electric circuits, it is not uncommon for more than one capacitor to be connected together in some way. The net result is that the capacitor combination behaves as though it were a *single* capacitor, with an **equivalent capacitance** that depends on the properties of the individual capacitors actually present. (An analogy is lifting a heavy conference table to move it across the room. Four people of normal strength could do the job, or it could be done by a single weightlifter. We would say that the weightlifter has an "equivalent strength" equal to that of the combination of four normal people.) Our goal in this section is to find the equivalent capacitance in different situations.

Capacitors in Series

Figure 17-14 shows three initially uncharged capacitors with capacitances C_1, C_2, and C_3 that are connected end to end. The capacitors become charged when the combination is connected to a battery of voltage V as shown in the figure. Note that the negative plate of one capacitor is connected to the positive plate of the next capacitor in the combination. Capacitors connected in this way are said to be in **series**.

What is the equivalent capacitance of this series combination? To answer this question, let's first determine the charges q_1, q_2, and q_3 on the individual capacitors and the

Figure 17-14 **Capacitors in series**
What is the equivalent capacitance of this series combination?

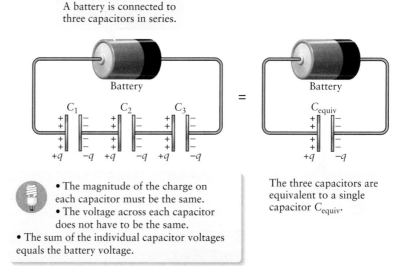

A battery is connected to three capacitors in series.

• The magnitude of the charge on each capacitor must be the same.
• The voltage across each capacitor does not have to be the same.
• The sum of the individual capacitor voltages equals the battery voltage.

The three capacitors are equivalent to a single capacitor C_{equiv}.

voltages V_1, V_2, and V_3 across the individual capacitors. For each capacitor, the magnitudes of the charge on the positive and negative plates must be equal. So as the battery draws negative charge from the left-hand plate of C_1, whatever positive charge $+q$ that plate acquires must be balanced by charge $-q$ on the right plate. The charging of the right plate of C_1 occurs as negative charge is drawn from the left plate of C_2, and because this whole section is initially uncharged, the left plate of C_2 acquires charge $+q$. If we apply the same reasoning to C_3, we see that all three capacitors acquire the *same* charge:

(17-18)
$$q_1 = q_2 = q_3 = q$$
(capacitors in series)

Thus the series combination of three capacitors is equivalent to a single capacitor with charge q. You can think of the charge on the negative plate of C_1 as canceling the charge on the positive plate of C_2, and likewise for the charges on the negative plate of C_2 and the positive plate of C_3. So the three capacitors are equivalent to a single capacitor with charge q.

The charges and voltages for each capacitor are also given by Equation 17-14, $q = CV$. If we substitute this into Equation 17-18, we get

$$q = C_1 V_1 = C_2 V_2 = C_3 V_3$$
(capacitors in series)

which we can rewrite as expressions for the voltages across the individual capacitors:

(17-19)
$$V_1 = \frac{q}{C_1} \quad V_2 = \frac{q}{C_2} \quad V_3 = \frac{q}{C_3}$$
(capacitors in series)

Now, the voltage V of the battery equals the voltage across the combination of three capacitors. This is just the sum of the voltages across the individual capacitors:

(17-20)
$$V = V_1 + V_2 + V_3$$
(capacitors in series)

If we substitute Equations 17-19 into Equation 17-20, we get a relationship between the charge q on each capacitor and the voltage V across the combination—that is, between the charge q on the equivalent capacitor and the voltage V across the equivalent capacitor:

(17-21)
$$V = \frac{q}{C_1} + \frac{q}{C_2} + \frac{q}{C_3} = q\left(\frac{1}{C_1} + \frac{1}{C_2} + \frac{1}{C_3}\right)$$
(capacitors in series)

For the equivalent capacitor alone, of capacitance C_{equiv}, Equation 17-14 says that $q = C_{equiv}V$ or $V = q/C_{equiv}$. If we compare this to Equation 17-21, we see that

 If capacitors are in series, the reciprocal of the equivalent capacitance is the sum of the reciprocals of the individual capacitances.

Equivalent capacitance of capacitors in series (17-22)

$$\frac{1}{C_{equiv}} = \frac{1}{C_1} + \frac{1}{C_2} + \frac{1}{C_3}$$

Equivalent capacitance of capacitors in series

Capacitances of the individual capacitors

If there are more than three capacitors in series, the same rule given in Equation 17-22 applies.

Here's how to follow the prescription of Equation 17-22: (1) Take the reciprocal of each individual capacitance in the series combination, (2) add these reciprocals to find $1/C_{equiv}$, then (3) take the reciprocal of that result to find C_{equiv}. The equivalent capacitance of a series combination is always *less* than the smallest capacitance of any of the individual capacitors.

Capacitors in Parallel

Figure 17-15 shows an alternative way to connect the three initially uncharged capacitors C_1, C_2, and C_3 to a battery with voltage V. Capacitors connected in this way are said to be in **parallel**. Notice the difference in the arrangement of the capacitors in parallel (Figure 17-15) compared to capacitors in series (Figure 17-14). In a series arrangement, the right-hand plate of one capacitor is connected to the left-hand plate of the capacitor to its right. In a parallel arrangement, all of the right-hand plates are connected, and all of the left-hand plates are connected.

To find the equivalent capacitance of capacitors in parallel, first note that all of the right-hand capacitor plates are connected to one terminal of the battery and all of the left-hand plates are connected to the other terminal. So each of the voltages V_1, V_2, and V_3 across the individual capacitors is equal to the voltage V across the battery:

$$V = V_1 = V_2 = V_3 \qquad (17\text{-}23)$$
(capacitors in parallel)

A battery is connected to three capacitors in parallel.

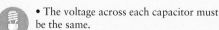

The three capacitors are equivalent to a single capacitor C_{equiv}.

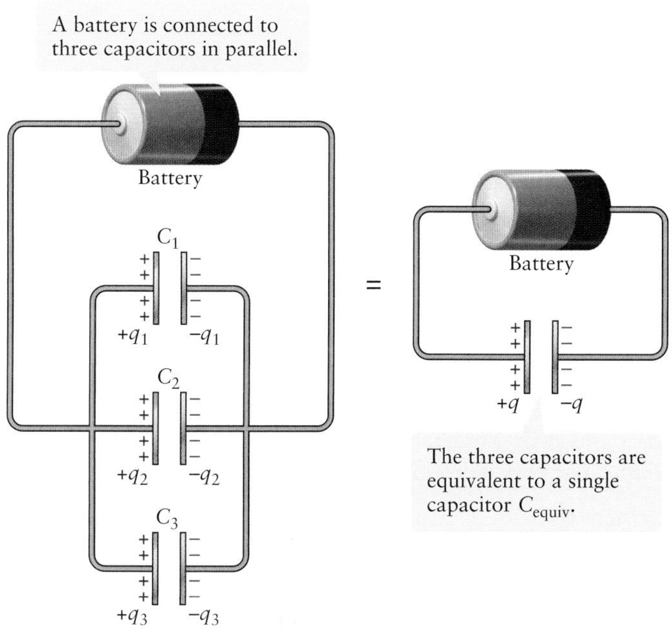

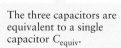 • The voltage across each capacitor must be the same.
• The charges on each capacitor do not have to be the same.
• The sum of the individual capacitor charges equals the charge q on the equivalent capacitor.

Figure 17-15 **Capacitors in parallel**
What is the equivalent capacitance of this parallel combination?

Equation 17-23 coupled with Equation 17-14, $q = CV$, then tells us the charges q_1, q_2, and q_3 on the individual capacitors:

(17-24)

$$q_1 = C_1 V \quad q_2 = C_2 V \quad q_3 = C_3 V$$
$$\text{(capacitors in parallel)}$$

The *total* charge acquired by all three capacitors is the sum of the charges q_1, q_2, and q_3. From Equation 17-24,

$$q = q_1 + q_2 + q_3 = C_1 V + C_2 V + C_3 V$$

(17-25)

$$= (C_1 + C_2 + C_3) V$$
$$\text{(capacitors in parallel)}$$

For the equivalent capacitor of capacitance C_{equiv} that corresponds to the three capacitors in parallel, Equation 17-14 says that $q = C_{\text{equiv}} V$. Comparing this to Equation 17-25 shows that

Equivalent capacitance of capacitors in parallel
(17-26)

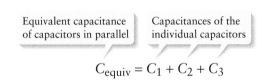

Equivalent capacitance of capacitors in parallel   Capacitances of the individual capacitors

$$C_{\text{equiv}} = C_1 + C_2 + C_3$$

 If capacitors are in parallel, the equivalent capacitance is the sum of the individual capacitances.

> **Watch Out!**
> **Capacitors in series and parallel have different properties.**
>
> Here's a summary of the differences between series and parallel combinations of capacitors. In a *series* combination, each capacitor has the same charge, but there are different voltages across capacitors with different capacitances. In a *parallel* combination, there is the same voltage across each capacitor, but there are different charges on capacitors with different capacitances.

If there are more than three capacitors in parallel, the same rule given in Equation 17-26 applies. The equivalent capacitance of a parallel combination is always *greater* than the smallest capacitance of any of the individual capacitors.

Example 17-10 shows how to do calculations with capacitors in series and in parallel. (We'll see a biological application of these calculations in the following section.) Many real networks of capacitors are more complex: They have a mixture of series and parallel combinations. To find the equivalent capacitance of such a network, we identify any small grouping of capacitors that are either entirely in series or entirely in parallel, find the equivalent capacitance of each group, and then combine them in larger and larger groupings, using the series and parallel rules, until we have accounted for all of the capacitors in the network. Example 17-11 illustrates how to do this.

Example 17-10 Two Capacitors in Series or in Parallel

(a) If two capacitors are connected in series, find their equivalent capacitance for the case when the capacitors have different capacitances $C_1 = 2.00\ \mu\text{F}$ and $C_2 = 4.00\ \mu\text{F}$ and for the case where both have the same capacitance $C_1 = C_2 = 2.00\ \mu\text{F}$. (b) Repeat part (a) for the two capacitors connected in parallel.

Set Up

We'll use Equations 17-22 and 17-26 to find the equivalent capacitance C_{equiv} in each case. Since there are only two capacitors in each combination, we drop the C_3 term from these equations.

Two capacitors in series:

$$\frac{1}{C_{\text{equiv}}} = \frac{1}{C_1} + \frac{1}{C_2} \qquad \text{(17-22)}$$

Two capacitors in parallel:

$$C_{\text{equiv}} = C_1 + C_2 \qquad \text{(17-26)}$$

series

parallel

Solve

(a) We first apply Equation 17-22 to the case where $C_1 = 2.00\ \mu F$ and $C_2 = 4.00\ \mu F$.

With capacitors $C_1 = 2.00\ \mu F$ and $C_2 = 4.00\ \mu F$ in series, Equation 17-22 becomes

$$\frac{1}{C_{equiv}} = \frac{1}{2.00\ \mu F} + \frac{1}{4.00\ \mu F}$$

$$= 0.500\ \mu F^{-1} + 0.250\ \mu F^{-1} = 0.750\ \mu F^{-1}$$

To find C_{equiv}, take the reciprocal of $1/C_{equiv}$:

$$C_{equiv} = \frac{1}{0.750\ \mu F^{-1}} = 1.33\ \mu F$$

Note that C_{equiv} is less than either C_1 or C_2.

Now repeat the calculation for the case where $C_1 = C_2 = 2.00\ \mu F$.

Follow the same steps, but with both capacitances equal to $2.00\ \mu F$:

$$\frac{1}{C_{equiv}} = \frac{1}{2.00\ \mu F} + \frac{1}{2.00\ \mu F}$$

$$= 0.500\ \mu F^{-1} + 0.500\ \mu F^{-1} = 1.00\ \mu F^{-1}$$

$$C_{equiv} = \frac{1}{1.00\ \mu F^{-1}} = 1.00\ \mu F$$

For this case of identical capacitors in series, C_{equiv} is exactly one-half of $C_1 = C_2 = 2.00\ \mu F$.

(b) Apply Equation 17-26 to the case where $C_1 = 2.00\ \mu F$ and $C_2 = 4.00\ \mu F$.

With capacitors $C_1 = 2.00\ \mu F$ and $C_2 = 4.00\ \mu F$ in parallel, Equation 17-26 becomes

$$C_{equiv} = 2.00\ \mu F + 4.00\ \mu F = 6.00\ \mu F$$

Note that C_{equiv} is greater than either C_1 or C_2.

Now repeat the calculation for the case where $C_1 = C_2 = 2.00\ \mu F$.

Follow the same steps, but with both capacitances equal to $2.00\ \mu F$:

$$C_{equiv} = 2.00\ \mu F + 2.00\ \mu F = 4.00\ \mu F$$

For this case of identical capacitors in parallel, C_{equiv} is exactly twice as great as $C_1 = C_2 = 2.00\ \mu F$.

Reflect

Our results illustrate the following general results: Connecting capacitors in series reduces the capacitance, while connecting them in parallel increases the capacitance.

Example 17-11 Multiple Capacitors

Find the equivalent capacitance of the three capacitors shown in Figure 17-16. The individual capacitances are $C_1 = 1.00\ \mu F$, $C_2 = 2.00\ \mu F$, and $C_3 = 6.00\ \mu F$.

Figure 17-16 A capacitor network These three capacitors are neither all in series nor all in parallel.

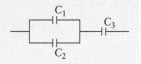

Set Up

Whenever capacitors are combined in ways other than purely in series or purely in parallel, we look for groupings of capacitors that *are* either in parallel or in series, and then we combine the groups one at a time.

Two capacitors in series:

$$\frac{1}{C_{equiv}} = \frac{1}{C_1} + \frac{1}{C_2} \qquad (17\text{-}22)$$

Replace the parallel capacitors C_1 and C_2 by their equivalent capacitor C_{12}:

Then find the equivalent capacitance of C_{12} and C_3 in series.

Notice in Figure 17-16 that C_1 and C_2 are in parallel because their two right plates are directly connected, as are their two left plates. We can therefore use Equation 17-26 to find C_{12} (the equivalent capacitance of the combination of C_1 and C_2) by using our relationship for capacitors in parallel.

Capacitor C_3 is in series with C_{12}, so we can find their combined capacitance by using Equation 17-22. This result is C_{123}, the equivalent capacitance of all three capacitors.

Two capacitors in parallel:

$$C_{equiv} = C_1 + C_2 \qquad (17\text{-}26)$$

Solve

First find the equivalent capacitance of the parallel capacitors C_1 and C_2.

For $C_1 = 1.00 \ \mu F$ and $C_2 = 2.00 \ \mu F$ in parallel, the equivalent capacitance C_{12} is given by Equation 17-26:

$$C_{12} = C_1 + C_2 = 1.00 \ \mu F + 2.00 \ \mu F = 3.00 \ \mu F$$

Then find C_{123}, the equivalent capacitance of the series capacitors C_{12} and C_3. This is the equivalent capacitance of the entire network of C_1, C_2, and C_3.

Since $C_{12} = 3.00 \ \mu F$ and $C_3 = 6.00 \ \mu F$ are in series, their equivalent capacitance C_{123} is given by Equation 17-22:

$$\frac{1}{C_{123}} = \frac{1}{C_{12}} + \frac{1}{C_3} = \frac{1}{3.00 \ \mu F} + \frac{1}{6.00 \ \mu F}$$

$$= 0.333 \ \mu F^{-1} + 0.167 \ \mu F^{-1} = 0.500 \ \mu F^{-1}$$

Take the reciprocal of this to find C_{123}:

$$C_{123} = \frac{1}{0.500 \ \mu F^{-1}} = 2.00 \ \mu F$$

Reflect

As we described previously, when capacitors are connected in parallel, the equivalent capacitance is always greater than the greatest individual capacitor. That's why C_{12} is greater than either C_1 or C_2. When capacitors are connected in series, the equivalent capacitance is always less than the least individual capacitor, which is why C_{123} is less than either C_{12} or C_3.

? Got the Concept? 17-7 Energy in a Capacitor Combination

Capacitor 1 has capacitance $C_1 = 1.0 \ \mu F$, and capacitor 2 has capacitance $C_2 = 2.0 \ \mu F$. You connect the initially uncharged capacitors to each other, then connect the capacitor combination to a battery. Which capacitor stores the greater amount of electric potential energy if the two capacitors are connected in series? If they are connected in parallel?

(a) Capacitor 1 for both the series and parallel cases; (b) capacitor 2 for both the series and parallel cases; (c) capacitor 1 for the series case, capacitor 2 for the parallel case; (d) capacitor 2 for the series case, capacitor 1 for the parallel case; (e) not enough information given to decide.

Take-Home Message for Section 17-7

✔ Whenever two or more capacitors are connected, the equivalent capacitance is the capacitance of a single capacitor that is the equivalent of the combined capacitors.

✔ Capacitors are connected in series when they are connected one after another. In a series combination, all capacitors have the same charge, and the equivalent capacitance is less than that of any of the individual capacitors.

✔ Capacitors are connected in parallel when all of their right-hand plates are directly connected to each other, and all of their left-hand plates are directly connected to each other. In a parallel combination, the voltage is the same across all capacitors, and the equivalent capacitance is greater than that of any of the individual capacitors.

17-8 Placing a dielectric between the plates of a capacitor increases the capacitance

So far in our discussion of capacitors we've assumed that there is only vacuum between the capacitor plates. In most situations, however, the two plates are separated by a layer of a **dielectric**, a material that is both an insulator and *polarizable*—that is, in which there is a separation of positive and negative charge within the material when it's exposed to an electric field. Dielectrics used in commercial capacitors include glass, ceramics, and plastics such as polystyrene. These dielectrics not only help to keep the positive and negative plates from touching each other, but also increase the capacitance of the capacitor. In this section we'll see how dielectrics make this possible.

Let's consider an isolated parallel-plate capacitor with charges $+q$ and $-q$ on its plates and with vacuum in the space between its plates (Figure 17-17a). The charge creates a uniform electric field $\vec{E}_0$ between the plates. We now insert a dielectric material that fills the space between the plates of this capacitor. What happens then depends on the kind of molecules that make up the dielectric. If the molecules are *polar*—that is, if one end of the molecule has a positive charge and the other end has a negative charge of the same magnitude—the molecules will orient themselves so that their positive ends are pointed toward the negatively charged plate and their negative ends toward the positively charged plate, as in Figure 17-17b. (The most common molecule in your body, the water molecule H_2O, is a polar molecule. Others include ammonia, NH_3, and sucrose, $C_{12}H_{22}O_{11}$.) If the molecules are not polar, so there normally is no separation of charge within the molecule, the electric field between the plates of the capacitor will *induce* a slight separation of the positive and negative charges in each molecule. (We described this process in Section 16-2.) The molecules will then orient

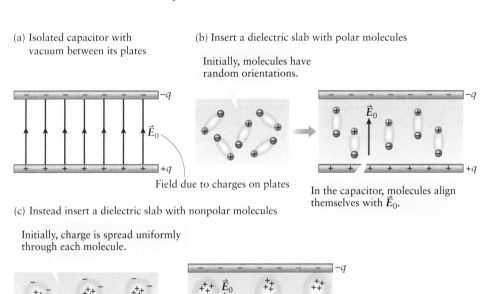

(a) Isolated capacitor with vacuum between its plates

(b) Insert a dielectric slab with polar molecules

Initially, molecules have random orientations.

Field due to charges on plates

In the capacitor, molecules align themselves with $\vec{E}_0$.

(c) Instead insert a dielectric slab with nonpolar molecules

Initially, charge is spread uniformly through each molecule.

In the capacitor, molecules are polarized by $\vec{E}_0$ and then align themselves with $\vec{E}_0$.

(d) With either type of dielectric, the result is layers of positive and negative charge on the dielectric surfaces. These produce a field $\vec{E}_{dielectric}$.

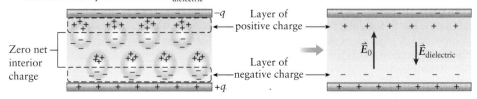

Zero net interior charge

Layer of positive charge

Layer of negative charge

Figure 17-17 A dielectric in a parallel-plate capacitor A dielectric slab inserted between the plates of an isolated capacitor reduces the electric field between the plates.

themselves in the same manner as polar molecules would (Figure 17-17c). In either case we say that the dielectric becomes *polarized* once it has been inserted between the plates of the capacitor.

In the interior of the polarized dielectric in Figure 17-17b or 17-17c, there are as many positive ends of molecules as there are negative ends, so the interior of the dielectric is neutral. But there *is* a net charge at the surfaces of the dielectric: There is a layer of positive charge on the surface next to the negatively charged plate, and a layer of negative charge on the surface next to the positively charged plate. (The dielectric as a whole is neutral, so these two layers have the same magnitude of charge.) These two layers of charge create an electric field $\vec{E}_{\text{dielectric}}$ that points opposite to the electric field $\vec{E}_0$ created by the charges on the plates of the capacitor (Figure 17-17d). Hence $\vec{E}_{\text{dielectric}}$ partially cancels $\vec{E}_0$, and the *net* electric field $\vec{E} = \vec{E}_0 + \vec{E}_{\text{dielectric}}$ between the plates is less than $\vec{E}_0$. So the effect of the dielectric is to *reduce* the electric field between the plates. We can express this as

Electric field in an isolated parallel-plate capacitor with a dielectric (17-27)

Electric field in an isolated parallel-plate capacitor with dielectric between the plates

Electric field in that same capacitor with no dielectric between the plates

$$E = \frac{E_0}{\kappa}$$

Dielectric constant: note that $\kappa \geq 1$

 For an isolated charged capacitor, a dielectric reduces the electric field between the plates.

Table 17-1

Dielectric Constants (at 20°C and 1 atm)

Material	κ
vacuum	1
air	1.00058
lipid	2.2
paraffin	2.2
paper	2.7
ceramic (porcelain)	5.8
water	80

The quantity κ (the Greek letter kappa) is the **dielectric constant** of the material. The greater the value of κ, the more the dielectric is polarized when it is placed between the plates of a charged capacitor, so the greater the magnitude of the field $\vec{E}_{\text{dielectric}}$ due to the dielectric and the smaller the magnitude E of the net field. Table 17-1 lists values of the dielectric constant for a variety of materials.

What does Equation 17-27 tell us about how the dielectric affects the capacitance of the capacitor? We saw in Section 17-5 that the potential difference V between the positive and negative plates is proportional to the magnitude E of the electric field between the plates (see Equation 17-11, which states that $V = Ed$). Equation 17-27 says that the dielectric reduces the value of the electric field by a factor $1/\kappa$, so V is also reduced by a factor $1/\kappa$. Now, Equation 17-14 tells us that the capacitance C is related to potential difference V and the magnitude q of the charges on the plates by $q = CV$, or $C = q/V$. While the dielectric reduces V by a factor $1/\kappa$, it has no effect on the value of q. (Since we specified that the capacitor is isolated, no charge can leave either plate.) Because the capacitance $C = q/V$ is inversely proportional to the potential difference V, which is reduced by a factor $1/\kappa$, it follows that C *increases* by a factor of $1/(1/\kappa) = \kappa$. If C_0 is the capacitance without the dielectric, the capacitance with the dielectric is

Capacitance of a parallel-plate capacitor with a dielectric (17-28)

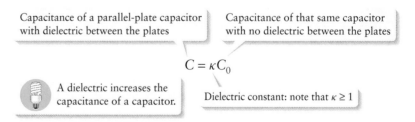

Capacitance of a parallel-plate capacitor with dielectric between the plates

Capacitance of that same capacitor with no dielectric between the plates

$$C = \kappa C_0$$

A dielectric increases the capacitance of a capacitor.

Dielectric constant: note that $\kappa \geq 1$

As an example, Table 17-1 tells us that the dielectric constant of porcelain is $\kappa = 5.8$. From Equation 17-28, the capacitance of a capacitor with porcelain filling the space between its plates is 5.8 times greater than an identical capacitor with vacuum between its plates. That's fundamentally because the electric field in the porcelain-filled capacitor is only $1/5.8 = 0.17$ as great as for an identical vacuum capacitor carrying the same charge (Equation 17-27).

Example 17-12 shows how to apply these ideas to the membrane of a cell, which we can regard as a capacitor with multiple layers of dielectric between its plates.

Example 17-12 Cell Membrane Capacitance

In all cells, it is easier for positive potassium ions (K^+) to flow out of the cell than it is for negative ions. As a result, there is negative charge on the inside of the cell membrane and positive charge on the outside of the membrane, much like a capacitor. Consider a typical cell with a membrane of thickness $7.60 \text{ nm} = 7.60 \times 10^{-9}$ m. (a) Find the capacitance of a square patch of membrane $1.00 \ \mu\text{m} = 1.00 \times 10^{-6}$ m on a side, assuming that the membrane has dielectric constant $\kappa = 1$. (b) The actual structure of the membrane is a layer of lipid surrounded by layers of a polarized aqueous solution and layers of water (Figure 17-18). Taking account of this structure, find the capacitance of a square patch of membrane $1.00 \ \mu\text{m}$ on a side.

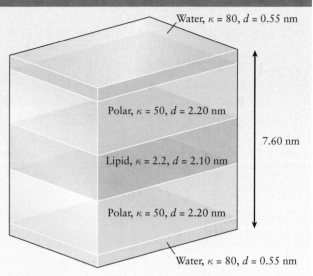

Figure 17-18 **Membrane capacitance** A membrane acts as a capacitor filled with a dielectric. What is the capacitance of a membrane that consists of a layer of lipid surrounded by layers of a polarized aqueous solution and water?

Set Up

A dielectric constant $\kappa = 1$ corresponds to vacuum, so in part (a) we can find the capacitance C_0 of the membrane under the assumption $\kappa = 1$ by using Equation 17-13 from Section 17-5.

In part (b) we must deal with the more complicated situation shown in Figure 17-18. We can't simply use Equation 17-28 to find the capacitance, because there's not a single value of κ for the arrangement of layers that make up the membrane. Instead we'll imagine that there's a very thin conducting sheet separating each layer from the next. If there are charges $+q$ and $-q$ on the outer and inner surfaces of the multilayer membrane, there will also be charges $+q$ and $-q$ on the surfaces of these conducting sheets. This is exactly like the situation with capacitors in series (Section 17-7). So we'll treat each of the five layers in Figure 17-18 as an individual capacitor, then use Equation 17-22 to find the capacitance of the combination.

Parallel-plate capacitor with vacuum between the plates:

$$C_0 = \frac{\varepsilon_0 A}{d} \qquad (17\text{-}13)$$

Parallel-plate capacitor with dielectric:

$$C = \kappa C_0 \qquad (17\text{-}28)$$

Five capacitors in series:

$$\frac{1}{C_{\text{equiv}}} = \frac{1}{C_1} + \frac{1}{C_2} + \frac{1}{C_3} + \frac{1}{C_4} + \frac{1}{C_5} \qquad (17\text{-}22)$$

Solve

(a) First use Equation 17-13 to find the capacitance assuming that the membrane has $\kappa = 1$ (equivalent to vacuum).

The area A for the capacitor of interest is a square $1.00 \ \mu\text{m}$ on a side:

$$A = (1.00 \ \mu\text{m})^2 = (1.00 \times 10^{-6} \text{ m})^2$$
$$= 1.00 \times 10^{-12} \text{ m}^2$$

A parallel-plate capacitor with this area and with $d = 7.60 \times 10^{-9}$ m has capacitance

$$C_0 = \frac{\varepsilon_0 A}{d} = \frac{(8.85 \times 10^{-12} \text{ F/m})(1.00 \times 10^{-12} \text{ m}^2)}{7.60 \times 10^{-9} \text{ m}}$$
$$= 1.16 \times 10^{-15} \text{ F}$$

(b) Calculate the capacitance of each of the five layers separately using Equations 17-13 and 17-28.

The capacitance of the first layer of water is $\kappa_1 = 80$ times the capacitance of a vacuum capacitor of thickness $d_1 = 0.55$ nm:

$$C_1 = \kappa_1 \left(\frac{\varepsilon_0 A}{d_1} \right) = (80)\frac{(8.85 \times 10^{-12} \text{ F/m})(1.00 \times 10^{-12} \text{ m}^2)}{0.55 \times 10^{-9} \text{ m}}$$
$$= 1.3 \times 10^{-12} \text{ F}$$

Similarly, for the second layer of polarized aqueous solution with $\kappa_2 = 50$ and thickness $d_2 = 2.20$ nm the capacitance is

$$C_2 = \kappa_2 \left(\frac{\varepsilon_0 A}{d_2} \right)$$

$$= (50) \frac{(8.85 \times 10^{-12} \text{ F/m})(1.00 \times 10^{-12} \text{ m}^2)}{2.20 \times 10^{-9} \text{ m}}$$

$$= 2.0 \times 10^{-13} \text{ F}$$

The capacitance of the third lipid layer with $\kappa_3 = 2.2$ and thickness $d_3 = 2.10$ nm is

$$C_3 = \kappa_3 \left(\frac{\varepsilon_0 A}{d_3} \right)$$

$$= (2.2) \frac{(8.85 \times 10^{-12} \text{ F/m})(1.00 \times 10^{-12} \text{ m}^2)}{2.10 \times 10^{-9} \text{ m}}$$

$$= 9.3 \times 10^{-15} \text{ F}$$

The fourth layer of polarized aqueous solution is identical to the second layer, and the fifth water layer is identical to the first layer. So

$$C_4 = C_2 = 2.0 \times 10^{-13} \text{ F}$$
$$C_5 = C_1 = 1.3 \times 10^{-12} \text{ F}$$

⌐ Calculate the net capacitance of the five layers in series using Equation 17-22.

The equivalent capacitance C_{equiv} of the five-layer stack is given by

$$\frac{1}{C_{equiv}} = \frac{1}{C_1} + \frac{1}{C_2} + \frac{1}{C_3} + \frac{1}{C_4} + \frac{1}{C_5}$$

Since $C_4 = C_2$ and $C_5 = C_1$, this is

$$\frac{1}{C_{equiv}} = \frac{1}{C_1} + \frac{1}{C_2} + \frac{1}{C_3} + \frac{1}{C_2} + \frac{1}{C_1} = \frac{2}{C_1} + \frac{2}{C_2} + \frac{1}{C_3}$$

$$= \frac{2}{1.3 \times 10^{-12} \text{ F}} + \frac{2}{2.0 \times 10^{-13} \text{ F}} + \frac{1}{9.3 \times 10^{-15} \text{ F}}$$

$$= 1.2 \times 10^{14} \text{ F}^{-1}$$

The reciprocal of this is C_{equiv}:

$$C_{equiv} = \frac{1}{1.2 \times 10^{14} \text{ F}^{-1}} = 8.4 \times 10^{-15} \text{ F}$$

Reflect

The equivalent capacitance of the stack of five layers is less than that of any individual layer. This is just what we saw in Section 17-7 for capacitors in series.

The capacitance $C_{equiv} = 8.4 \times 10^{-15}$ F is greater than the capacitance $C_0 = 1.16 \times 10^{-15}$ F calculated assuming $\kappa = 1$ by a factor $C_{equiv}/C_0 = (8.4 \times 10^{-15} \text{ F})/(1.16 \times 10^{-15} \text{ F}) = 7.2$. So the effective dielectric constant of the membrane is 7.2, intermediate between the largest ($\kappa_1 = 80$) and smallest ($\kappa_3 = 2.2$) values of dielectric constant for the individual layers.

? Got the Concept? 17-8 Inserting a Dielectric into a Capacitor

An isolated parallel-plate capacitor (one that is not connected to anything else) has vacuum between its plates and has charges $+q$ and $-q$ on its plates. If you insert a slab of dielectric with dielectric constant κ that fills the space between the plates, the capacitance increases. What happens to the energy stored in the capacitor as you insert the dielectric slab? Will you have to push the slab into the capacitor, or will you feel the capacitor pulling the slab in? (a) Stored energy increases, you will have to push the slab in; (b) stored energy increases, the slab will be pulled in; (c) stored energy decreases, you will have to push the slab in; (d) stored energy decreases, the slab will be pulled in; (e) not enough information given to decide.

Take-Home Message for Section 17-8

✔ When a dielectric material is placed in an electric field $\vec{E}_0$, it becomes polarized. This produces an additional electric field that partially cancels $\vec{E}_0$ and so reduces the net field in the dielectric. The dielectric constant is a measure of how much the field is reduced.

✔ If a dielectric material fills the space between the plates of a capacitor, the capacitance is greater than if there is vacuum between the plates.

Key Terms

capacitance	electric potential energy	parallel (capacitors)
capacitor	electron volt	parallel-plate capacitor
dielectric	equipotential	plates
dielectric constant	equipotential surface	series (capacitors)
electric potential	equipotential volume	test charge
electric potential	equivalent capacitance	volt
difference	farad	voltage

Chapter Summary

Topic	Equation or Figure

Electric potential energy: Like the gravitational force, the electric force is a conservative force and has an associated potential energy. The change in electric potential energy when a charged object moves from one point to another depends on the sign of the object's charge. The sign of the electric potential energy of two point charges depends on whether the two charges have the same sign or different signs.

The change in electric potential energy for an object of charge q that moves in a uniform electric field $\vec{E}$...

...equals the negative of the work done on the object by the electric force.

$$\Delta U_{\text{electric}} = -W_{\text{electric}}$$

Charge of the object

$$= -qEd\cos\theta$$

Angle between the displacement and the direction of the electric field

Magnitude of the electric field

Straight-line displacement of the object

(17-2)

Coulomb constant   Values of the two charges

Electric potential energy of two point charges

$$U_{\text{electric}} = \frac{kq_1q_2}{r}$$

Distance between the point charges

(17-3)

Electric potential: The electric potential at a given position equals the electric potential energy for a point charge q_0 at that position, divided by the value of q_0. If you move in a direction opposite to the electric field, the electric potential increases; if you move in the direction of the electric field, the electric potential decreases. The electric potential due to a point charge is inversely proportional to the distance from the charge. Voltage is the

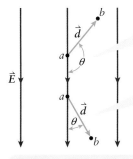

If you move in a direction opposite to the electric field ($\theta > 90°$), the electric potential increases ($\Delta V > 0$).

If you move in the direction of the electric field ($\theta < 90°$), the electric potential decreases ($\Delta V < 0$).

(Figure 17-5)

 If you move from point a to point b in a uniform electric field $\vec{E}$, the change in electric potential is $\Delta V = -Ed\cos\theta$.

difference in electric potential between two locations.

Electric potential due to a point charge Q

Coulomb constant

$$V = \frac{kQ}{r}$$ Value of the point charge   (17-8)

Distance from the point charge Q to the location where the potential is measured

---

Equipotentials: The electric potential has the same value everywhere on an equipotential surface. An equipotential surface is everywhere perpendicular to the electric field. The electric potential has the same value throughout the volume of a conductor in equilibrium.

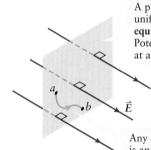

A plane perpendicular to a uniform electric field $\vec{E}$ is an **equipotential surface:** Potential V has the same value at all points on this surface.

(Figure 17-6)

Any curve that lies on the plane is an equipotential curve.

Capacitors: A capacitor has two conducting plates that store charges $+q$ and $-q$, of the same magnitude but opposite sign. The potential difference V between the plates is proportional to the charge q; the proportionality constant, the capacitance C, depends only on the geometry of the capacitor and the substance between the plates.

A capacitor carries a charge $+q$ on its positive plate and a charge $-q$ on its negative plate.

The magnitude of q is directly proportional to V, the voltage (potential difference) between the plates.

$$q = CV$$ (17-14)

The constant of proportionality between charge q and voltage V is the capacitance C of the capacitor.

Energy stored in a capacitor: A charged capacitor stores electric potential energy. The stored energy can be expressed by two of the three quantities: capacitance C, charge q, and potential difference V.

The electric potential energy stored in a charged capacitor...

$$U_{\text{electric}} = \frac{1}{2}qV = \frac{1}{2}CV^2 = \frac{q^2}{2C}$$ (17-17)

...can be expressed in three ways in terms of the charge q, potential difference V, and capacitance C.

Capacitors in series and parallel: A collection of capacitors in a circuit behaves as though it were a single capacitor, with an equivalent capacitance C_{equiv}. The equivalent capacitance is different for capacitors in series (with the positive plate of one capacitor connected to the negative plate of another) than for capacitors in parallel (with positive plates connected to positive plates and negative plates to negative plates).

 If capacitors are in series, the reciprocal of the equivalent capacitance is the sum of the reciprocals of the individual capacitances.

$$\frac{1}{C_{\text{equiv}}} = \frac{1}{C_1} + \frac{1}{C_2} + \frac{1}{C_3}$$ (17-22)

Equivalent capacitance of capacitors in series

Capacitances of the individual capacitors

Equivalent capacitance of capacitors in parallel

Capacitances of the individual capacitors

$$C_{\text{equiv}} = C_1 + C_2 + C_3$$ (17-26)

 If capacitors are in parallel, the equivalent capacitance is the sum of the individual capacitances.

Dielectrics: A dielectric is an insulator that becomes polarized when placed in an electric field. When the space between the plates of a capacitor is filled with a dielectric, the electric field between the plates decreases (for a given charge q on the plates) and the capacitance increases.

Capacitance of a parallel-plate capacitor with dielectric between the plates

Capacitance of that same capacitor with no dielectric between the plates

$$C = \kappa C_0$$

A dielectric increases the capacitance of a capacitor.

Dielectric constant: note that $\kappa \geq 1$

(17-28)

$$C_0 = \frac{\varepsilon_0 A}{d}$$

Answer to What do you think? Question

(c) The energy stored by separating charge $+q$ from charge $-q$ in a capacitor is proportional to q^2 (see Section 17-6, especially Equation 17-17). So doubling the value of q increases the stored energy by a factor of $2^2 = 4$.

Answers to Got the Concept? Questions

17-1 (b) Although q_1, q_2, and q_3 have all reversed sign, the signs of the *products* q_1q_2, q_1q_3, and q_2q_3 in Equation 17-4 remain the same. So if you change all of the positive charges in an assemblage to negative and vice versa, there is no effect on the electric potential energy of the assemblage.

17-2 (c), (a), (d), (b) From Equation 17-6, the potential energy change $\Delta U_{\text{electric}}$ for a point charge q_0 is related to the potential change ΔV by $\Delta V = \Delta U_{\text{electric}}/q_0$, or $\Delta U_{\text{electric}} = q_0 \, \Delta V$. For the four cases we have (a) $\Delta U_{\text{electric}} = (+0.0010 \text{ C})(5.0 \text{ V}) = +0.0050 \text{ C} \cdot \text{V} = +0.0050 \text{ J}$; (b) $\Delta U_{\text{electric}} = (+0.0020 \text{ C})(-2.0 \text{ V}) = -0.0040 \text{ J}$; (c) $\Delta U_{\text{electric}} = (-0.0015 \text{ C})(-4.0 \text{ V}) = +0.0060 \text{ J}$; (d) $\Delta U_{\text{electric}} = (-0.0010 \text{ C})(+2.5 \text{ V}) = -0.0025 \text{ J}$.

17-3 (b) In the vicinity of A, the potential decreases by 1 V (from $V = 0$ V to $V = -1$ V) in the 2-m distance between $x = 0$ and $x = 2$ m. The electric field $\vec{E}$ points in the direction of decreasing potential, which is to the right (the positive x direction) at both points. In the vicinity of B, the equipotentials are much closer together; the potential decreases by 1 V (from $V = -2$ V to $V = -3$ V) in the 0.5-m distance from $x = 3$ m to $x = 3.5$ m. Where equipotentials are farther apart, as at A, the electric field has a smaller magnitude.

17-4 (e) Equation 17-13 shows that the capacitance of a parallel-plate capacitor depends on its geometry only. It does not depend on the amount of charge on the plates or on the potential difference between the plates.

17-5 (c) The stored energy increases. The positive and negative charges on the two plates attract each other, so to pull the plates apart you must do positive work on the system. The work you do increases the electric potential energy stored in the capacitor. To verify this conclusion, note from Equation 17-13 that if you increase the spacing d between the plates, the capacitance $C = \varepsilon_0 A/d$ will decrease. Equation 17-17, $U_{\text{electric}} = q^2/2C$, then tells us that if the charge q remains constant and the capacitance C decreases, the stored energy U_{electric} will increase.

17-6 (a) The stored energy decreases. Equation 17-13 tells us that as you increase the spacing d between the plates, the capacitance $C = \varepsilon_0 A/d$ decreases. Equation 17-17, $U_{\text{electric}} =$

$(1/2)CV^2$, then tells us that if the potential difference V remains constant and C decreases, the stored energy must decrease. Note that in the situation of Got the Concept? 17-5, the stored energy *increases* when the plates are pulled apart. Why is the answer different here? The explanation is that in Got the Concept? 17-5 the charge q remained constant. In the current situation, however, the charge q has to change in accordance with Equation 17-14, $q = CV$: The charge q decreases as the capacitance C decreases. (Electrons flow from the negative plate through the battery and onto the positive plate, so the magnitude of charge on both plates decreases.) Equation 17-17 also tells us that we can express the stored energy as $U_{\text{electric}} = (1/2)qV$, so the stored energy will decrease if q decreases while V remains the same.

17-7 (c) Equation 17-17 gives us two useful expressions for the electric potential energy stored in a capacitor with capacitance C: $U_{\text{electric}} = (1/2)CV^2$ (which says that for a given voltage V, the stored energy is proportional to C) and $U_{\text{electric}} = q^2/2C$ (which says that for a given charge q, the stored energy is inversely proportional to C). In a series combination, both capacitors carry the same charge q, so $U_{\text{electric}} = q^2/2C$ tells us that the capacitor with the smaller capacitance ($C_1 = 1.0 \ \mu\text{F}$) stores the greater amount of energy. In a parallel combination, the voltage V is the same for both capacitors, so $U_{\text{electric}} = (1/2)CV^2$ tells us that the capacitor with the greater capacitance ($C_2 = 2.0 \ \mu\text{F}$) stores the greater amount of energy. We can draw these conclusions without knowing the specific values of q for the series case or V for the parallel case.

17-8 (d) Because the capacitor is isolated, the charge q does not change when the dielectric slab is inserted. So among the expressions for stored energy given in Equation 17-17, the one to use is $U_{\text{electric}} = q^2/2C$. Since q remains constant but capacitance C increases as you insert the slab, it follows that U_{electric} decreases. Just as a ball is pulled downward by the gravitational force, in the direction that decreases gravitational potential energy, the slab must be pulled into the capacitor by an electric force because motion in that direction decreases electric potential energy U_{electric}.

Questions and Problems

In a few problems, you are given more data than you actually need; in a few other problems, you are required to supply data from your general knowledge, outside sources, or informed estimate.

Interpret as significant all digits in numerical values that have trailing zeros and no decimal points. For all problems, use $g = 9.80$ m/s^2 for the free-fall acceleration due to gravity. Neglect friction and air resistance unless instructed to do otherwise.

• Basic, single-concept problem
•• Intermediate-level problem, may require synthesis of concepts and multiple steps
••• Challenging problem
SSM Solution is in Student Solutions Manual

Conceptual Questions

1. •What is the difference between electric potential and electric field?

2. •What is the difference between electric potential and electric potential energy?

3. •Explain why electric potential requires the existence of only one charge, but a finite electric potential energy requires the existence of two charges.

4. •An electron is released from rest in an electric field. Will it accelerate in the direction of increasing or decreasing potential? Why?

5. •Does it make sense to say that the voltage at some point in space is 10.3 V? Explain your answer. SSM

6. ••Explain why an electron will accelerate toward a region of lower electric potential energy but higher electric potential.

7. •Discuss how a topographical map showing various elevations around a mountain is analogous to the equipotential lines surrounding a charged object.

8. •How much work is required to move a charge from one end of an equipotential path to the other? Explain your answer.

9. •(a) If the electric potential throughout some region of space is zero, does it necessarily follow that the electric field is zero? (b) If the electric field throughout a region is zero, does it necessarily follow that the electric potential is zero? SSM

10. •Explain why capacitance depends neither on the stored charge Q nor on the potential difference V between the plates of a capacitor.

11. •Describe three methods by which you might increase the capacitance of a parallel-plate capacitor.

12. •If the voltage across a capacitor is doubled, by how much does the stored energy change?

13. ••The capacitance of several capacitors in series is less than any of the individual capacitances. What, then, is the advantage of having several capacitors in series?

14. •What is the advantage to arranging several capacitors in parallel?

15. ••You charge a capacitor and then remove it from the battery. The capacitor consists of large movable plates with air between them. You pull the plates a bit farther apart. What happens to the stored energy? SSM

16. •Which way of connecting (series or parallel) three identical capacitors to a battery would store more energy?

17. •Qualitatively explain why the equivalent capacitance of a parallel combination of identical capacitors is larger than the individual capacitances.

18. •Does inserting a dielectric into a capacitor increase or decrease the energy stored in the capacitor? Explain your answer.

19. •What are the benefits, if any, of filling a capacitor with a dielectric other than air?

20. ••Capacitors A and B are identical except that the region between the plates of capacitor A is filled with a dielectric. As shown in Figure 17-19, the plates of these capacitors are maintained at the same potential difference by a battery. Is the electric field magnitude in the region between the plates of capacitor A smaller, the same, or larger than the field in the region between the plates of capacitor B? Explain your answer.

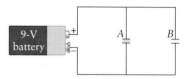

Figure 17-19 Problem 20

Multiple-Choice Questions

21. ••For a positive charge moving in the direction of the electric field,

   A. its potential energy increases and its electric potential increases.

   B. its potential energy increases and its electric potential decreases.

   C. its potential energy decreases and its electric potential increases.

   D. its potential energy decreases and its electric potential decreases.

   E. its potential energy and its electric potential remain constant. SSM

22. ••If a negative charge is released in a uniform electric field, it will move

   A. in the direction of the electric field.

   B. from high potential to low potential.

   C. from low potential to high potential.

   D. in a direction perpendicular to the electric field.

   E. in circular motion.

23. •An equipotential surface must be

   A. parallel to the electric field at every point.

   B. equal to the electric field at every point.

   C. perpendicular to the electric field at every point.

   D. tangent to the electric field at every point.

   E. equal to the inverse of the electric field at every point.

24. •A positive charge is moved from one point to another point along an equipotential surface. The work required to move the charge

   A. is positive.

   B. is negative.

   C. is zero.

   D. depends on the sign of the potential.

   E. depends on the magnitude of the potential.

25. •The electric potential measured at a point equidistant from two particles that have charges equal in magnitude but of opposite sign is
  A. larger than zero.
  B. smaller than zero.
  C. equal to zero.
  D. equal to the average of the two distances times the charges.
  E. equal to the net electric field.

26. •Four point charges of equal magnitude but differing signs are arranged at the corners of a square (Figure 17-20). The electric field E and the potential V at the center of the square are
  A. $E = 0; V \neq 0.$
  B. $E = 0; V = 0.$
  C. $E \neq 0; V \neq 0.$
  D. $E \neq 0; V = 0.$
  E. $E = 2V^{1/2}.$

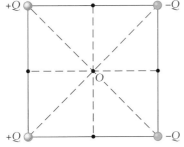

Figure 17-20 Problem 26

27. ••A capacitor consists of a set of two parallel plates of area A separated by a distance d. This capacitor carries a charge Q. If the separation between the plates is doubled, the electrical energy stored in the capacitor will be
  A. halved.
  B. doubled.
  C. unchanged.
  D. quadrupled.
  E. quartered. SSM

28. ••A capacitor consists of a set of two parallel plates of area A separated by a distance d. This capacitor is connected to a battery that maintains a constant potential difference across the plates. If the separation between the plates is doubled, the electrical energy stored in the capacitor will be
  A. halved.
  B. doubled.
  C. unchanged.
  D. quadrupled.
  E. quartered.

29. •When capacitors are connected in series, they have the same
  A. charge.
  B. voltage.
  C. dielectric.
  D. surface area.
  E. separation.

30. •When capacitors are connected in parallel, they have the same
  A. charge.
  B. voltage.
  C. dielectric.
  D. surface area.
  E. separation.

Estimation/Numerical Analysis

31. •If a source charge is in the microcoulomb range, how far from the charge should you be to have an electric potential in the millivolt range as compared to the potential a long way from the charge?

32. •Estimate the number of 100 μF capacitors, connected in parallel, that would provide enough energy to get an electric car moving at 20 m/s (the mass of the car equals 1000 kg, the mass of a 1 μF capacitor equals 0.005 g). Assume the voltage across the capacitors is 12 V.

33. •Estimate the electric potential at the location of an electron bound to the nucleus of an atom. SSM

Problems

17-1 Electric energy is important in nature, technology, and biological systems

17-2 Electric potential energy changes when a charge moves in an electric field

34. ••A point charge q_0 that has a charge of 0.500 μC is at the origin. (a) A second particle q that has a charge of 1.00 μC and a mass of 0.0800 g is placed at $x = 0.800$ m. What is the potential energy of this system of charges? (b) If the particle with charge q is released from rest, what will its speed be when it reaches $x = 2.00$ m?

35. ••A uniform electric field of 2.00 kN/C points in the $+x$ direction. (a) What is the change in potential energy of a $+2.00$ nC test charge, $U_{electric,b} - U_{electric,a}$, as it is moved from point a at $x = -30.0$ cm to point b at $x = +50.0$ cm? (b) The same test charge is released from rest at point a. What is its kinetic energy when it passes through point b? (c) If a negative charge instead of a positive charge were used in this problem, qualitatively how would your answers change? SSM

36. ••**Biology** Two red blood cells each have a mass of 9.0×10^{-14} kg and carry a negative charge spread uniformly over their surfaces. The repulsion arising from the excess charge prevents the cells from clumping together. One cell carries -2.50 pC of charge and the other -3.10 pC, and each cell can be modeled as a sphere 7.5 μm in diameter. (a) What speed would they need when very far away from each other to get close enough to just touch? Assume that there is no viscous drag from any of the surrounding liquid. (b) What is the maximum acceleration of the cells in part (a)?

17-3 Electric potential equals electric potential energy per charge

37. •A uniform electric field of magnitude 28 V/m makes an angle of 30° with a displacement of length 10 m. What is the potential difference of the final position relative to the initial position of this displacement?

38. •At a certain point in space, there is a potential of 800 V relative to zero. What is the potential energy of the system when a $+1.0$ μC charge is placed at that point in space?

39. •How much work is required to move a 2.0-C positive charge from the negative terminal of a 9.0-V battery to the positive terminal? SSM

40. •**Biology** A potential difference exists between the inner and outer surfaces of the membrane of a cell. The inner surface is negative relative to the outer surface. If 1.5×10^{-20} J of work is required to eject a positive sodium ion (Na$^+$) from the interior of the cell, what is the potential difference between the inner and outer surfaces of the cell?

41. •**Chemistry** What is the electric potential due to the nucleus of hydrogen at a distance of 5.00×10^{-11} m? Assume the potential is equal to zero as $r \to \infty$.

42. •(a) What is the electric potential due to a point charge of $+2.00$ μC at a distance of 0.500 cm? (b) How will the answer change if the charge has a value of -2.00 μC? Assume the potential is equal to zero as $r \to \infty$.

43. •The electric potential has a value of -200 V at a distance of 1.25 m from a point charge. What is the value of that charge? Assume the potential is equal to zero as $r \to \infty$.

44. •At point P in Figure 17-21 the electric potential is zero. (As usual, we take the potential to be zero at infinite distance.) (a) What can you say about the two charges? (b) Are there any other points of zero potential on the line connecting P and the two charges?

P

q_1 q_2

Figure 17-21 Problem 44

45. •Two point charges are placed on the x axis: $+0.500$ μC at $x = 0$ and -0.200 μC at $x = 10.0$ cm. At what point(s), if any, on the x axis is the electric potential equal to zero? SSM

46. •A charge of $+2.00$ μC is at the origin and a charge of -3.00 μC is on the y axis at $y = 40.0$ cm. (a) What is the potential at point a, which is on the x axis at $x = 40.0$ cm? (b) What is the potential difference $V_b - V_a$ when point b is at (40.0 cm, 30.0 cm)? (c) How much work is required to move an electron at rest from point a to rest at point b?

47. ••Calculate the electric potential at the origin O due to the point charges in Figure 17-22.

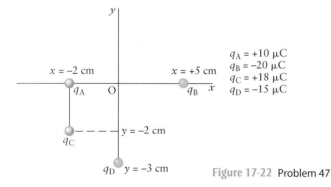

$q_A = +10$ μC
$q_B = -20$ μC
$q_C = +18$ μC
$q_D = -15$ μC

$x = -2$ cm $x = +5$ cm

q_A O q_B x

$y = -2$ cm

q_C

q_D $y = -3$ cm

Figure 17-22 Problem 47

17-4 The electric potential has the same value everywhere on an equipotential surface

48. •Electric field lines for a system of two point charges are shown in Figure 17-23. Reproduce the figure and draw on it some equipotential lines for the system.

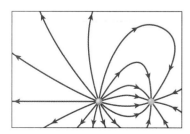

Figure 17-23 Problem 48

49. •In Figure 17-24, equipotential lines are shown at 1-m intervals. What is the electric field at (a) point A and (b) point B?

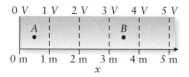

Figure 17-24 Problem 49

50. •Equipotential lines for some region of space are shown in Figure 17-25. What is the approximate electric field at (a) point A and (b) point B?

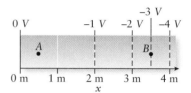

Figure 17-25 Problem 50

51. ••Draw the (a) equipotential lines and (b) electric field lines surrounding pairs of charges with the same sign charge (Figure 17-26). SSM

Figure 17-26 Problem 51

52. ••Draw the (a) equipotential lines and (b) electric field lines surrounding a dipole ($+q$ is a distance L from $-q$) (Figure 17-27). All charges are "points" of very small (negligible) dimensions.

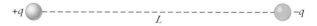

Figure 17-27 Problem 52

53. ••Draw the electric field lines and the electric equipotential lines for the following charge distribution (Figure 17-28).

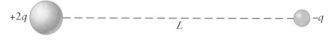

Figure 17-28 Problem 53

17-5 A capacitor stores equal amounts of positive and negative charge

54. •Using a single 10.0-V battery, what capacitance do you need to store 10.0 μC of charge?

55. •A 2.00-μF capacitor is connected to a 12.0-V battery. What is the magnitude of the charge on each plate of the capacitor?

56. •A parallel-plate capacitor has a plate separation of 1.00 mm. If the material between the plates is air, what plate area is required to provide a capacitance of 2.00 pF?

57. •A parallel-plate capacitor has square plates that have edge lengths equal to 1.00×10^2 cm and are separated by 1.00 mm. What is the capacitance of this device?

58. •An air-filled parallel-plate capacitor has plates measuring 10.0 cm × 10.0 cm and a plate separation of 1.00 mm. If you want to construct a parallel-plate capacitor of the same capacitance but with plates measuring 5.00 cm × 5.00 cm, what plate separation do you need?

59. •A parallel-plate capacitor has square plates that have edge length equal to 1.00 m. If the material between the plates is air, what separation distance is required to provide a capacitance of 8850 pF? SSM

17-6 A capacitor is a storehouse of electric potential energy

60. •Using a single 10.0-V battery, what capacitance do you need to store 1.00×10^{-4} J of electric potential energy?

61. •A parallel-plate capacitor has square plates that have edge length equal to 1.00×10^2 cm and are separated by 1.00 mm. It is connected to a battery and is charged to 12.0 V. How much energy is stored in the capacitor?

62. •You charge a 2.00-μF capacitor to 50.0 V. How much additional energy must you add to charge it to 100 V?

63. •A capacitor has a capacitance of 80.0 μF. If you want to store 160 J of electric energy in this capacitor, what potential difference do you need to apply to the plates?

64. ••(a) You want to store 1.00×10^{-5} C of charge on a capacitor, but you only have a 100-V voltage source with which to charge it. What must be the value of the capacitance? (b) You want to store 1.00×10^{-3} J of energy on a capacitor, and you only have a 100-V voltage source with which to charge it. What must be the value of the capacitance?

65. •**Medical** A defibrillator containing a 20.0-μF capacitor is used to shock the heart of a patient by holding it to the patient's chest. Just prior to discharging, the capacitor has a voltage of 10.0 kV across its plates. How much energy is released into the patient, assuming no energy losses? SSM

17-7 Capacitors can be combined in series or in parallel

66. ••How should four 1.0-pF capacitors be connected to have a total capacitance of 0.75 pF?

67. •Three capacitors have capacitances 10.0 μF, 15.0 μF, and 30.0 μF. What is their effective capacitance if the three are connected (a) in parallel and (b) in series?

68. ••A series circuit consists of a 0.50-μF capacitor, a 0.10-μF capacitor, and a 220-V battery. Determine the charge on each of the capacitors.

69. ••Two capacitors provide an equivalent capacitance of 8.00 μF when connected in parallel and 2.00 μF when connected in series. What is the capacitance of each capacitor? SSM

70. ••A 2.00-μF capacitor is first charged by being connected across a 6.00-V battery. It is then disconnected from the battery and connected across an uncharged 4.00-μF capacitor. Calculate the final charge on each of the capacitors.

71. •A 0.0500-μF capacitor and a 0.100-μF capacitor are connected in parallel across a 220-V battery. Determine the charge on each of the capacitors.

72. •What is the equivalent capacitance of the network of three capacitors shown in **Figure 17-29**?

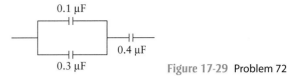

Figure 17-29   Problem 72

73. •Calculate the equivalent capacitance between *a* and *b* for the combination of capacitors shown in **Figure 17-30**.

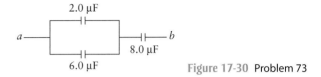

Figure 17-30   Problem 73

74. ••A 10.0-μF capacitor, a 40.0-μF capacitor, and a 100.0-μF capacitor are connected in parallel across a 12.0-V battery. (a) What is the equivalent capacitance of the combination? (b) What is the charge on each capacitor? (c) What is the potential difference across each capacitor?

75. ••A 10.0-μF capacitor, a 40.0-μF capacitor, and a 100.0-μF capacitor are connected in series across a 12.0-V battery. (a) What is the equivalent capacitance of the combination? (b) What is the charge on each capacitor? (c) What is the potential difference across each capacitor? SSM

76. ••For the capacitor network shown in **Figure 17-31**, the potential difference across *ab* is 75.0 V. How much charge and how much energy are stored in this system?

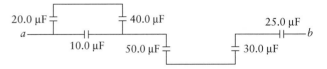

Figure 17-31   Problem 76

17-8 Placing a dielectric between the plates of a capacitor increases the capacitance

77. •What is the dielectric constant of the material that fills the gap between a parallel-plate capacitor with plate area of 20.0 cm² and plate separation of 1.00 cm if the capacitance is measured to be 0.0142 μF?

78. •A parallel-plate capacitor has plates of 1.00 cm by 2.00 cm. The plates are separated by a 1.00-mm-thick piece of paper. What is the capacitance of this capacitor? The dielectric constant for paper is 2.7.

79. ••A 2800-pF air-filled capacitor is connected to a 16-V battery. If you now insert a ceramic dielectric material ($k = 5.8$) that fills the space between the plates, how much charge will flow from the battery? SSM

80. ••A parallel-plate capacitor has square plates that have edge lengths equal to 1.00×10^2 cm and are separated by 1.00 mm. It is connected to a battery and charged to 12.0 V. How much energy would be stored in the capacitor if a ceramic dielectric material (k is 5.8) fills the space between the plates?

81. •••(a) Determine the capacitance of the parallel-plate capacitor shown in Figure 17-32. The dielectric with constant κ_1 fills up one-quarter of the area, but the full separation of the plates. The materials with constants κ_2 and κ_3 fill the other three-quarters of the area, and divide the separation of the plates in half. (b) What happens to the total capacitance if the material with dielectric constant κ_3 is replaced by air?

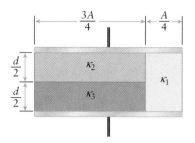

Figure 17-32 Problem 81

82. ••A parallel-plate capacitor that has a plate separation of 0.50 cm is filled halfway with a slab of dielectric material (k is 5.0) (Figure 17-33). If the plates are 1.25 cm by 1.25 cm in area, what is the capacitance of this capacitor?

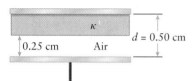

Figure 17-33 Problem 82

General Problems

83. ••A parallel-plate capacitor has a plate separation of 1.5 mm and is charged to 600 V. If an electron leaves the negative plate, starting from rest, how fast is it going when it hits the positive plate?

84. ••Consider a 1.00-m³ cube that has +2.00-μC charges located at seven of its corners as shown in Figure 17-34. (a) Find the potential at the vacant corner. (b) How much work by an external agent is required to bring an additional +2.00-μC charge from rest at infinity to rest at the vacant corner? (c) How does the problem change, if at all, if a -2.00-μC charge is brought in to fill the empty corner rather than a positive charge?

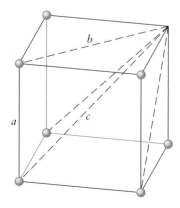

Figure 17-34 Problem 84

85. •••As shown in Figure 17-35, three particles, each with charge q, are at different corners of a rhombus with sides

of length a and with one diagonal of length a and the other of length b. (a) What is the electric potential energy of the charge distribution? (b) How much work by an external agent is required to bring a fourth particle, also of charge q, from rest at infinity to rest at the vacant corner of the rhombus? (c) What is the total electric potential energy of the four charges? SSM

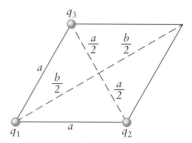

Figure 17-35 Problem 85

86. ••A lightning bolt transfers 20 C of charge to Earth through an average potential difference of 30 MV. (a) How much energy is dissipated in the bolt? (b) How much water (at a temperature of 100°C) can be turned into steam with this energy?

87. ••In 2004, physicists at the Stanford Linear Accelerator Center (SLAC) in California fired electrons toward each other at very high speeds so that they came within 1.0×10^{-15} m of each other (approximately the diameter of a proton). (a) What was the electrical force on each electron at closest approach? (b) Would this force be large enough to move you? (c) How much energy did it take to get the electrons that close together? (d) Ignoring relativistic effects, how fast must the electrons have been traveling initially, when far from each other, to get that close? (e) Do you notice anything suspicious about your answer?

88. ••**Biology** Under certain circumstances, potassium ions (K^+) move across the 8.0-nm-thick cell membrane from the inside to the outside. The potential inside the cell is -70.0 mV, and the potential outside is zero. (a) What is the change in the electrical potential energy of the potassium ions as they move across the membrane? Does their potential energy increase or decrease as they move from the inside to the outside?

89. ••Calculate the equivalent capacitance of the combination in Figure 17-36. SSM

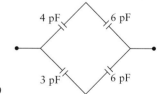

Figure 17-36 Problem 89

90. ••Suppose you are supplied with five identical capacitors (each with a capacitance of 10.0 μF). Determine all of the unique combinations that use all five capacitors and the equivalent capacitance of each combination.

91. ••Determine the equivalent capacitance of the combination in Figure 17-37.

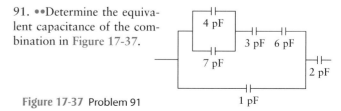

Figure 17-37 Problem 91

92. ••The following arrangement of four capacitors has an equivalent capacitance of 8.00 μF (**Figure 17-38**). Calculate the value of C_x.

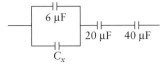

Figure 17-38 Problem 92

93. ••Calculate the charge stored on each capacitor in the circuit shown in **Figure 17-39**.

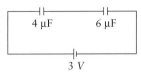

Figure 17-39 Problem 93

94. ••A parallel-plate capacitor is made by sandwiching 0.100-mm sheets of paper (dielectric constant 2.7) between three sheets of aluminum foil (A, B, and C in **Figure 17-40**) and rolling the layers into a cylinder. A capacitor that has an area of 10 m^2 is fabricated this way. (To be practical, it would then have to be folded up so as to fit in a small package.) What is the capacitance of this capacitor?

Figure 17-40 Problem 94

95. ••A parallel-plate capacitor with a capacitance of 5.00 μF is charged with a 12.0-V battery. After fully charging the plates, the battery is removed. How much work is required to increase the separation between the plates by a factor of 3?

96. ••An air-filled, parallel-plate capacitor with area A and gap width d is connected to a battery that maintains the plates at potential difference V. (a) The plates are pulled apart, doubling the gap width, while they remain in electrical contact with the battery terminals. By what factor does the potential energy of the capacitor change? (b) If the capacitor is removed from the battery and thus isolated, what happens to the stored potential energy when the gap width is doubled? Explain your answer.

97. ••An air-filled parallel-plate capacitor is attached to a battery with a voltage V. While attached to the battery, the area of the plates is doubled and the separation of the plates is halved. During this process, what happens to (a) the capacitance, (b) the charged stored on the positive plate of the capacitor, (c) the potential across the plates of the capacitor, and (d) the potential energy stored in the capacitor, as compared to the original configuration? (e) How would your answers change if, once the capacitor was charged by the battery, it was disconnected from the battery while the area and separation were changed as above?

98. •••You have a bucketful of capacitors, each with a capacitance of 1.00 μF and a maximum voltage rating of 250 V. You are to come up with a combination that has a capacitance of 0.75 μF and a maximum voltage rating of 1000 V. What is the minimum number of capacitors you need?

99. •••A parallel-plate capacitor has area A and separation d. (a) How is its capacitance affected if a *conducting* slab of thickness $d' < d$ is inserted between, and parallel to, the plates as shown in **Figure 17-41**? (b) Does your answer depend on where the slab is positioned vertically between the plates? SSM

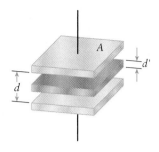

Figure 17-41 Problem 99

100. •••Three 0.18-μF capacitors are connected in parallel across a 12-V battery, as shown in **Figure 17-42**. The battery is then disconnected. Next, one capacitor is carefully disconnected so that it doesn't lose any charge and is reconnected backward, that is, with its positively charged side and its negatively charged side reversed. (a) What is the potential difference across the capacitors now? (b) By how much has the stored energy of the combination of capacitors changed in the process?

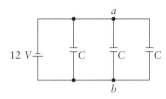

Figure 17-42 Problem 100

101. •(a) Calculate the charge and energy stored on the 25-μF capacitor when the switch S is placed at position A in **Figure 17-43**. (b) Repeat for both the 25-μF and the 20-μF capacitors after the switch is then placed at position B.

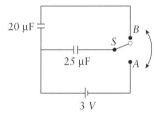

Figure 17-43 Problem 101

102. ••**Figure 17-44** shows equipotential curves for 30 V, 10 V, and −10 V. A proton's speed as it passes point A is 80 km/s. It follows the path shown in the figure. What is the proton's speed at point B?

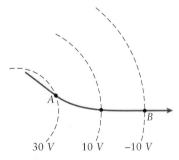

Figure 17-44 Problem 102

103. ••A parallel-plate, air-filled capacitor has a charge of 20.0 μC and a gap width of 0.100 mm. The potential difference between the plates is 200 V. (a) What is the electric field in the region between the plates? (b) What is the surface charge density on the positive plate? (c) If the plates of the capacitor are moved closer together while the charge remains constant, how are the electric field, surface charge density, and potential difference going to change, if at all? Explain your answers.

18

Electric Charges in Motion

What do you think?

The simplest of all electric circuits is a battery connected to a light bulb, such as the one you find inside an ordinary flashlight. The battery causes electrons to move through the circuit. As electrons pass through the light bulb, what is the principal kind of energy that they transfer to the light bulb to make it shine? (a) Kinetic energy; (b) electric potential energy; (c) both kinetic energy and electric potential energy; (d) neither kinetic energy nor electric potential energy; (e) answer depends on the details of how the light bulb is constructed.

In this chapter, your goals are to:

- (18-1) Recognize why moving electric charges are important.
- (18-2) Explain the meaning of current and drift speed, and the difference between direct and alternating current.
- (18-3) Describe the relationships among voltage, current, resistance, and resistivity for charges moving in a wire.
- (18-4) Calculate the resistance of a resistor and the current that a given voltage produces in that resistor.
- (18-5) Discuss Kirchhoff's rules and how to apply them to single-loop and multiloop circuits.
- (18-6) Calculate the power into or out of a circuit element.
- (18-7) Explain what happens when a capacitor in series with a resistor is charged or discharged.

To master this chapter, you should review:

- (5-5)   How fluids exert drag forces on objects moving through them.
- (12-3) How to describe the position of an object oscillating with simple harmonic motion.
- (13-9) How power is the rate at which energy is transferred.
- (16-3) How the loosely bound electrons in metals allow them to conduct electricity.
- (16-7) Why the electric field outside a uniformly charged sphere is the same as if all the charge were located at the center of the sphere.
- (17-3) How electric field and potential difference are related, and how potential difference and electric potential energy difference are related.
- (17-5, The relationship among charge, voltage, and capacitance; the energy stored
  17-6, in a capacitor; and the equivalent capacitance of capacitors attached in
  17-7) series and parallel.

18-1 Life on Earth and our technological society are only possible because of charges in motion

In the previous two chapters, we've examined a variety of topics in the general category of electrostatics. For example, we determined the electric field that a point charge or combination of charges generates at various points in space. In doing so, we considered the charges to be fixed in place. But in many important situations in nature and technology electric charges are in *motion*. You are able to read these words thanks to electric charges that travel along the optic nerve from your eye to your brain, transmitting the

(a)
Wires and cables are made of conductors that allow moving charges to flow along their length.

(b)
The current in a mobile device is provided by a battery (a source of emf).

(c)
A camera's electronic flash uses energy stored in a capacitor to produce a burst of light.

Figure 18-1 **Charges in motion** (a) An essential part of any electric circuit. (b) and (c) Two common devices that use currents (moving electric charges).

image of those words in the form of a coded electrical signal. If you are reading these words after sunset, you are either looking at a printed page illuminated by a light bulb that's powered by moving electric charges, or else reading them on a tablet or other electronic device that operates using complex electric circuitry.

In this chapter we'll look at the basic physics of electric charges in motion. We'll introduce the idea of *current*, which measures the rate at which charges move through a conductor (Figure 18-1a). The value of current will prove to be related to the speed at which moving charges drift through the conductor. Ordinary conductors have *resistance* to the flow of charge, so it's necessary to set up a voltage between the ends of a conductor in order to produce a current. This is the role of a *source of emf*, the most common example of which is an ordinary battery (Figure 18-1b). We'll analyze simple circuits that include a source of emf and one or more *resistors* (circuit elements that have resistance), and see how to treat combinations of resistors.

Fundamentally, an electric circuit is used to transfer energy from one place to another (for instance, from a battery to a light bulb, as in the photograph that opens this chapter). We'll see how to describe the *power*, or rate of energy transfer, associated with any circuit element. Finally, we'll study circuits that include a capacitor, which make it possible to deliver energy in a quick burst (Figure 18-1c).

Take-Home Message for Section 18-1

✔ Technological devices and biological systems depend on electric charges in motion.

✔ An electric circuit is fundamentally a means of transferring energy.

18-2 Electric current equals the rate at which charge flows

Figure 18-2 shows a common situation in which electric charges are in motion. What sets charges into motion is the **battery**, also known as an *electrochemical cell*. Inside the cell are two different substances that, due to their different chemical properties, undergo a chemical reaction so that each substance ends up with an excess or deficit of electrons. In an ordinary alkaline battery, like an AA or D cell, the two substances are zinc and manganese dioxide. The zinc ends up with an electron excess, while the manganese dioxide ends up with an electron deficit.

The two terminals of the battery are each connected to one of these substances. Electrons can flow between each substance and the metal terminal to which it is attached. As a result, the terminal attached to the substance with an electron excess also has an electron excess and is called the *negative* terminal. The other terminal, attached to the substance with an electron deficit, also has an electron deficit and is called the *positive* terminal. These terminals are marked on the battery by a minus sign and a plus sign, respectively. Because of this charge imbalance, the positive terminal is at a higher electric potential than the negative terminal. The potential difference, or voltage, between the terminals depends on the two substances within the battery. For zinc and manganese dioxide, the voltage is 1.5 V, a number that you will see written on the case of any alkaline battery.

Figure 18-2 Current in a circuit
Charge flows in a wire loop due to the potential difference (voltage) supplied by a battery.

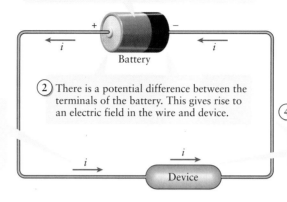

① A simple electric circuit is made up of a battery connected to a wire and a device (such as a light bulb), forming a closed circuit.

② There is a potential difference between the terminals of the battery. This gives rise to an electric field in the wire and device.

③ The electric field causes mobile charges in the wire to flow. The current—the rate at which charge flows past a given point in the wire—is *i*.

④ Charge does not pile up at any point in the circuit. So the current has the same value *i* at all points in the circuit, including inside the battery, in the wire, and inside the device.

We know from Section 17-3 that charges tend to flow between two points when there is a potential difference between those points: The electric force pushes positive charges from high to low potential, and negative charges from low to high potential. If a battery is isolated and not connected to anything, there is no way for charge to travel outside the battery from one terminal to another. So in this case there is no charge flow. But charge *can* flow if a metallic wire is used to connect the two terminals to each other as in Figure 18-2. (As we saw in Section 16-3, metals are especially good electrical conductors because one or more of the electrons associated with a metal atom are only weakly bound to that atom. As a result, these loosely bound electrons can move with relative freedom within the metal.) So the battery and wire form a complete loop or **circuit**, and charge flows continuously through the circuit from one terminal of the battery to the other. Note that the circuit *must* be complete in order for charges to move. If the wire in Figure 18-2 is broken or disconnected from the battery, the flow of charge stops.

Here's another way to see why charges move in the wire. The potential difference between the battery terminals means that there is an electric *field* $\vec{E}$ within the wire. This field points along the length of the wire in the direction from the positive terminal to the negative terminal. A mobile charge with charge q within the wire will feel a force $\vec{F} = q\vec{E}$ from the electric field, and this force is what pushes the charges through the wire. This force is necessary because mobile charges within the wire collide very frequently with the atoms that make up the wire. The net effect of the collisions is to slow or retard the motion of the mobile charges through the wire, rather like how a ball falling through the air or an algal spore moving through water is retarded by fluid resistance (see Section 5-5). The speed at which charge flows is determined by the balance between the retarding force due to collisions and the forward electric force $\vec{F} = q\vec{E}$.

Later in this chapter we'll see how the flow of charge delivers energy to a device (like a light bulb) that's part of the circuit. For now, however, let's concentrate on the properties of the charge flow itself.

Current

$\sqrt{x}$ *See the Math Tutorial for more information on direct and inverse proportions*

The **current** in a circuit like that in Figure 18-2 equals the rate at which charge flows past any point in the circuit. In particular, if an amount of charge Δq moves past a certain point in a time Δt, the current *i* is

Definition of current
(18-1)

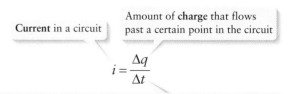

Current in a circuit

Amount of **charge** that flows past a certain point in the circuit

$$i = \frac{\Delta q}{\Delta t}$$

Time required for that amount of charge to flow past that point

The SI unit of current is the **ampere**, named after the French scientist and mathematician André Ampère. One ampere (abbreviated amp or A) is equivalent to one coulomb of charge passing a given point per second: 1 A = 1 C/s.

In much the same way as we treated fluid flow in Chapter 11, we'll begin by limiting our discussion of currents to a *steady* flow of charge. Then the current i has the same value at all *times*. In addition, the current has the same value at all *points* in a simple circuit like that in Figure 18-2 in which charges move around a single loop. The moving charges cannot "pile up" or accumulate at any point in the circuit (if they did, their mutual electrostatic repulsion would make them spread apart again). There is also no way for more moving charges to join the flow or for charges to leave the flow. So the value of the current is the same everywhere in the circuit.

Note that current is *not* a vector quantity. For example, in Figure 18-2 the current is to the left in the upper part of the circuit, downward in the left-hand part of the circuit, to the right in the lower part of the circuit, and upward in the right-hand part of the circuit. So there is no single vector that describes the direction of current in every part of the circuit. Instead, we simply say that the current in Figure 18-2 is counterclockwise around the circuit. If we reversed the battery so that the positive terminal was on the right rather than on the left, the current would be clockwise.

! Watch Out! The direction of current is chosen to be the direction in which *positive* charges would flow.

In ordinary wires and in common electric devices such as light bulbs and toasters, the charges that are free to move are negatively charged electrons. In an electric circuit such as that shown in Figure 18-2, electrons tend to flow through the wire from low potential to high potential, and so from the negative terminal of the battery toward the positive terminal. But in Figure 18-2 we show the current flowing through the wire from the battery's *positive* terminal to its *negative* terminal. That's because the convention is to take current to flow in the direction that positively charged objects would move, regardless of whether the moving objects are positively or negatively charged.

We'll use this convention throughout this book. (This convention is often attributed to the eighteenth-century American scientist and statesman Benjamin Franklin. The discovery that the moving charges in wires are negatively charged came decades after Franklin's death.) Note that in most biological systems such as neurons and muscle cells, the moving charges are in fact positive (they are positive ions, atoms that have lost one or more electrons each). But no matter what the nature of the moving charges, the principles governing current are the same.

Because an electron has a very small charge ($q = -e = -1.60 \times 10^{-19}$ C), typical currents involve the flow of a very large number of electrons. For example, suppose the current i in the circuit shown in Figure 18-2 is 1.00 A or 1.00 C/s. (That's roughly the current in a large flashlight.) Then the rate at which electrons move past any specific point on the wire is

$$1.00 \text{ A} = 1.00 \frac{C}{s} = \left(1.00 \frac{C}{s}\right)\left(\frac{1 \text{ electron}}{1.60 \times 10^{-19} \text{ C}}\right) = 6.25 \times 10^{18} \frac{\text{electron}}{s}$$

(Here we're considering the *magnitude* of the current, so we're ignoring the negative sign on the charge of the electron.) This result says that if we were to pass an imaginary plane through the cross section of the wire, we would find about 6×10^{18} electrons crossing that plane per second.

Example 18-1 Charging a Sphere

A large, hollow metal sphere is electrically isolated from its surroundings, except for a wire that can carry a current to charge the sphere. Initially, the sphere is uncharged and is at electrical potential zero. If the sphere has a radius of 0.150 m and the current is a steady 5.00 μA = 5.00×10^{-6} A, how long does it take for the sphere to attain a potential of 4.00×10^5 V?

Set Up

The electric potential at the surface of a charged sphere of radius R depends on the amount of excess charge Q on the sphere. In Section 16-7 we learned that the electric *field* outside a uniformly charged sphere is identical to that of a particle with the same charge, located at the center of the sphere. So the electric *potential* at a point just outside the sphere must be given by Equation 17-8, the equation for the potential of a charged particle with total charge Q located at the center of the sphere. We'll rearrange this equation to solve for the amount of charge required to generate the given electric potential. Then we'll find the time required for this charge to reach the sphere.

Electric potential of a charged sphere:

$$V = \frac{kQ}{R} \qquad (17\text{-}8)$$

Definition of current:

$$i = \frac{\Delta q}{\Delta t} \qquad (18\text{-}1)$$

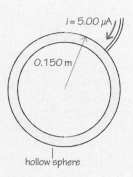

$i = 5.00\ \mu\text{A}$

0.150 m

hollow sphere

Solve

Use Equation 17-8 to solve for the final charge on the sphere.

From Equation 17-8,

$$Q = \frac{VR}{k} = \frac{(4.00 \times 10^5\ \text{V})(0.150\ \text{m})}{(8.99 \times 10^9\ \text{N}\cdot\text{m}^2/\text{C}^2)}$$

$$= 6.67 \times 10^{-6} \frac{\text{V}\cdot\text{C}^2}{\text{N}\cdot\text{m}}$$

Since $1\ \text{V} = 1\ \text{J/C} = 1\ \text{N}\cdot\text{m/C}$, this becomes

$$Q = \left(6.67 \times 10^{-6} \frac{\text{V}\cdot\text{C}^2}{\text{N}\cdot\text{m}}\right)\left(\frac{1\ \text{N}\cdot\text{m/C}}{1\ \text{V}}\right)$$

$$= 6.67 \times 10^{-6}\ \text{C} = 6.67\ \mu\text{C}$$

Our result $Q = 6.67\ \mu\text{C}$ is the amount of charge Δq that must flow onto the sphere in a time Δt. Use Equation 18-1 to solve for Δt. Recall that $1\ \text{A} = 1\ \text{C/s}$.

From Equation 18-1,

$$\Delta t = \frac{\Delta q}{i} = \frac{6.67 \times 10^{-6}\ \text{C}}{5.00 \times 10^{-6}\ \text{A}} = \frac{6.67 \times 10^{-6}\ \text{C}}{5.00 \times 10^{-6}\ \text{C/s}}$$

$$= 1.33\ \text{s}$$

Reflect

A charged sphere like this is used in a Van de Graaff generator, which you may have seen demonstrated in your physics class. If you have, you know that when the generator is turned on to charge the sphere, the sphere can begin to throw off sparks within a second or so. So a result for Δt on the order of 1 s is reasonable.

Drift Speed

The value of the current i tells us what quantity of charge flows past a given point in a circuit per second. This is related to the **drift speed** v_{drift}, which is the average speed at which charges move ("drift") through the circuit. Let's see how to relate these two quantities.

Figure 18-3 shows a wire that has a cross-sectional area A and carries a current i. Charges are moving ("drifting") through the green region at an average speed of v_{drift}. At any time, the total moving charge in that region is Δq. This charge can be expressed in terms of n (the number of moving charges per volume), the volume $A\,\Delta x$ of the green region (a cylinder of area A and length Δx), and the amount of charge e on each moving charge:

(18-2)

$$\Delta q = n(\text{Volume})e = n(A\Delta x)e$$

If the moving charges are electrons, each has a charge $-e$ rather than e. But since we take the direction of current to be the direction in which positive charges would flow, we'll use the convenient fiction that the charge is positive. The time required for this volume of charge to drift the distance Δx along the length of the wire is $\Delta t = \Delta x / v_{\text{drift}}$

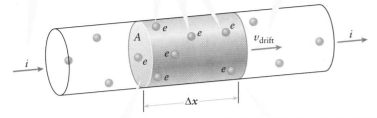

① Current i is present in the wire of cross-sectional area A.

② Within the wire there are n mobile charges per unit volume. Each has a charge e.

③ The mobile charges drift through the wire with speed v_{drift}, and travel a distance Δx in a time $\Delta t = \Delta x / v_{\text{drift}}$.

Figure 18-3 **Drift speed** The current i in a wire is proportional to the speed v_{drift} at which moving charges drift through the wire.

④ A length Δx of the wire (shown in green) has volume $A\Delta x$. This length contains $n(A\Delta x)$ mobile charges, with total charge $\Delta q = n(A\Delta x)e$.

⑤ The current in the wire equals the amount of charge Δq that passes this point divided by the time Δt it takes them to pass this point:

$$i = \frac{\Delta q}{\Delta t} = \frac{n(A\Delta x)e}{\Delta x/v_{\text{drift}}} = nAev_{\text{drift}}$$

(time equals distance divided by speed). So from Equations 18-1 and 18-2, the current in the wire is

$$i = \frac{\Delta q}{\Delta t} = \frac{n(A\Delta x)e}{\Delta x/v_{\text{drift}}}$$

or, simplifying,

Current in a wire   Number of moving charges per unit volume

$$i = nAev_{\text{drift}}$$

Speed at which charges drift through the wire

Cross-sectional area of the wire   Magnitude of the charge on each moving charge

Current and drift speed
(18-3)

The number of moving charges per volume (n) is different for different materials. Equation 18-3 tells us that for a wire made of a given material and with a given cross-sectional area A, the current i is directly proportional to the drift speed.

❗ Watch Out! The drift speed is not the same as the speed at which individual charges move.

Even if there is no current through a wire, the mobile charges within that wire are still in motion. However, their motions are in random directions, so that there is no *net* motion of charge. It's rather like a swarm of bees flying around a hive: Individual bees move in different directions, but the swarm as a whole stays in the same place. If the swarm moves to a different hive, the motions of individual bees will be different but the swarm as a whole will "drift" together to the new hive. In the same way, if a current is present in the wire in Figure 18-3, the individual charges within the green volume can be moving at different speeds and in different directions, but this "swarm" of charge drifts along the wire at speed v_{drift}.

Example 18-2 Electron Drift Speed in a Flashlight

In a large flashlight, the distance from the on-off switch to the light bulb is 10.0 cm. How long does it takes electrons to drift this distance if the flashlight wires are made of copper, are 0.512 mm in radius, and carry a current of 1.00 A? There are 8.49×10^{28} atoms in 1 m^3 of copper, and one electron per copper atom is able to move freely through the metal.

Set Up

We'll use Equation 18-3 to determine the drift speed from the information given about the number of mobile electrons, the radius of the wire, and the current. From this we'll be able to find the drift time by using the familiar relationship between speed, distance, and time.

Current and drift speed:

$$i = nAev_{drift} \qquad (18\text{-}3)$$

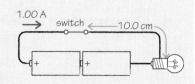

Solve

Calculate the drift speed of the electrons in the wire.

Solve Equation 18-3 for the drift speed:

$$v_{drift} = \frac{i}{nAe}$$

There is one mobile electron per atom and 8.49×10^{28} atoms/m³, so the value of n is 8.49×10^{28} electrons/m³. The wire has a circular cross section with radius $r = 0.512$ mm $= 0.512 \times 10^{-3}$ m, so its cross-sectional area is

$$A = \pi r^2 = \pi(0.512 \times 10^{-3} \text{ m})^2 = 8.24 \times 10^{-7} \text{ m}^2$$

The magnitude of the charge per electron is $e = 1.60 \times 10^{-19}$ C/electron, and the current is $i = 1.00$ A $= 1.00$ C/s. So the drift speed is

$$v_{drift} = \frac{i}{nAe}$$

$$= \frac{1.00 \text{ C/s}}{(8.49 \times 10^{28} \text{ electrons/m}^3)(8.24 \times 10^{-7} \text{ m}^2)(1.60 \times 10^{-19} \text{ C/electron})}$$

$$= 8.94 \times 10^{-5} \text{ m/s} = 0.0894 \text{ mm/s}$$

Then calculate how long it takes electrons to drift the distance from switch to light bulb.

At this drift speed, the time it takes electrons to travel a distance $d = 10.0$ cm $= 0.100$ m from the flashlight on-off switch to the light bulb is

$$t = \frac{d}{v_{drift}} = \frac{0.100 \text{ m}}{8.94 \times 10^{-5} \text{ m/s}} = 1.12 \times 10^3 \text{ s}$$

$$= (1.12 \times 10^3 \text{ s})\left(\frac{1 \text{ min}}{60 \text{ s}}\right) = 18.6 \text{ min}$$

Reflect

The phrase "a snail's pace" refers to something that moves very slowly. But an ordinary snail moves at about *twice* the drift speed of electrons in this wire. The drift speed is so slow because electrons in the wire are continually colliding with the copper atoms, which slows their progress tremendously. (More sophisticated physics shows that an electron in copper moves in *random* motion at an average speed of about 10^6 m/s, about 10^{10} times faster than the drift speed. In the analogy we made earlier between electrons and a swarm of bees, you should think of the electrons in this wire as *very* fast-moving bees within a swarm that's drifting very, very slowly.)

At their slower-than-a-snail's pace, it takes electrons more than a quarter of an hour to travel from the on-off switch to the light bulb. Why, then, does the flashlight turn on immediately when you move the switch to the "on" position?

The explanation is that the wire is full of mobile electrons, and these electrons drift in response to the electric field in the wire. With the switch in the off position, there is no electric field and so no drift. An electric field is only set up when the switch is put in the on position, making a complete circuit like that shown in Figure 18-2. Changes in the electric field propagate through the wire at close to the speed of light (3×10^8 m/s), so the field is set up throughout the circuit, and the electrons begin to drift, in a tiny fraction of a second. That's why the light comes on nearly instantaneously.

Direct Current and Alternating Current

In the circuit shown in Figure 18-2 the current always flows around the circuit in the same direction, as shown by the arrows labeled i. Current of this kind is called **direct current** or **DC** for short. You'll find direct current in any device that's powered by a battery, such as a flashlight, a television remote control, or a mobile phone. That's because the potential difference between the two terminals of the battery that powers the circuit always has the same sign: The positive terminal is always at a higher potential than the negative terminal.

Something very different happens in a toaster or a table lamp that you plug into a wall socket. The potential difference between the two terminals in a wall socket is not constant, but instead oscillates or *alternates*. At one instant the left-hand terminal is at a higher potential than the right-hand terminal; a short time later the left-hand terminal is the one at the lower potential, and a short time after that the left-hand terminal is again at the higher potential. As a result, the current in a device plugged into a wall socket alternates direction (**Figure 18-4**). This is called **alternating current** or **AC** for short. (The third terminal found in most wall sockets, called the *ground*, remains at a constant potential and is not directly involved in producing the current.)

The potential difference or voltage between the two terminals of a wall socket varies with time in a sinusoidal way, just as does the position of an object in simple harmonic motion (Section 12-3). We can write this voltage as

$$V(t) = V_0 \sin \omega t = V_0 \sin 2\pi f t \qquad (18\text{-}4)$$

This expression says that the potential difference between the left-hand and right-hand terminals in a wall socket varies between $+V_0$ and $-V_0$. In Equation 18-4 f is the frequency at which the voltage oscillates, and $\omega = 2\pi f$ is the corresponding angular frequency. In North America, Central America, and much of South America, the frequency used is $f = 60$ Hz; in most of the rest of the world, $f = 50$ Hz is used. The current in a circuit driven by such a wall socket oscillates with the same frequency. So in an AC circuit, electrons just oscillate back and forth around an equilibrium position. That's very different from a DC circuit, in which electrons plod slowly in the same direction around the circuit.

If $f = 60$ Hz, the *period* of oscillation of the current is $T = 1/f = 1/(60\ \text{Hz}) = 1/60$ s. This means that the current in the circuit shown in Figure 18-4 moves in one direction for $(1/2) \times (1/60)$ s $= 1/120$ s, then moves in the opposite direction for $1/120$ s, and so on. Since the drift speed of electrons in a wire is typically very slow (see Example 18-2), electrons can move only a very short distance (typically around 10^{-6} m or less) in $1/120$ s.

In Chapter 21 we'll study alternating current in detail, learn how an alternating voltage is generated, and see why it's used instead of direct current (DC) for wall sockets. For the remainder of this chapter we'll concentrate exclusively on DC circuits.

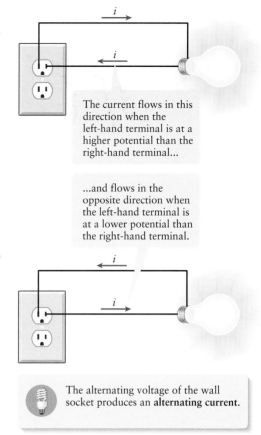

The current flows in this direction when the left-hand terminal is at a higher potential than the right-hand terminal...

...and flows in the opposite direction when the left-hand terminal is at a lower potential than the right-hand terminal.

The alternating voltage of the wall socket produces an **alternating current**.

Figure 18-4 Alternating current The potential difference between the terminals of a wall socket varies sinusoidally. As a result, the current in a circuit that contains this socket alternates direction.

? Got the Concept? 18-1   Current in a Wire

Suppose that part of the wire in Example 18-2 were only half the thickness of the rest of the wire, with a radius of 0.256 mm instead of 0.512 mm. Compared to the rest of the wire, would the current in the thinner part be (a) four times greater, (b) twice as great, (c) the same, (d) one-half as great, or (e) one-fourth as great?

? Got the Concept? 18-2   Drift Speed in a Wire

Suppose that part of the wire in Example 18-2 were only half the thickness of the rest of the wire, with a radius of 0.256 mm instead of 0.512 mm. Compared to the rest of the wire, would the drift speed in the thinner part be (a) four times faster, (b) twice as fast, (c) the same, (d) one-half as fast, or (e) one-fourth as fast?

Take-Home Message for Section 18-2

✔ An electric field applied in a wire results in an electric current (a net flow of charge in the wire).

✔ The SI unit of current is the ampere (A): 1 A = 1 C/s.

✔ Current has a magnitude and a direction, but it is not a vector.

✔ By convention, the direction assigned to a current is the direction in which positive charge carriers would move. In typical metals like those used in wires, it is the negative electrons that actually move and carry current.

✔ Direct current (DC) always travels the same direction around a circuit. Alternating current (AC) continually changes direction.

current stays same regaldless of A change, but Vdrift compensates so that this is held true.

18-3 The resistance to current through an object depends on the object's resistivity and dimensions

nice summary

We saw in Section 18-2 that a current can flow in a wire only if there is a potential difference between the ends of the wire. This gives rise to an electric field inside the wire, and this field exerts a force that causes mobile charges to move. However, the motion of these mobile charges is slowed due to collisions between the charges and the atoms of the wire. The net result is that charges drift through the wire at a relatively slow speed. To better understand current, we need to answer three questions:

(1) How is the electric field in a current-carrying wire related to the potential difference between the ends of the wire?
(2) How is the resulting current related to the electric field in the wire?
(3) How large is the current in the wire as a consequence of the potential difference between the ends of the wire?

The first of these questions is easily answered for a straight wire of uniform cross-sectional area A and length L (Figure 18-5). If there is a potential difference V between the ends of the wire, the electric field $\vec{E}$ points along the length of the wire from the high-potential end toward the low-potential end. Since the wire is uniform, we expect that the magnitude E of the field will be uniform as well. From Chapter 17, the potential difference V is just equal to the field magnitude multiplied by the length of the wire:

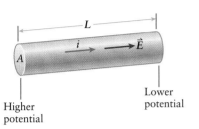

L

i ⟶ $\vec{E}$

A

Higher potential

Lower potential

Figure 18-5 Inside a current-carrying wire The electric field in a current-carrying wire points from the higher-potential end toward the lower-potential end. The current is in the same direction.

(18-5)
$$V = EL \quad \text{or} \quad E = \frac{V}{L}$$

This is the same as Equation 17-7, $\Delta V = -Ed \cos\theta$. We've changed the symbol for potential difference from ΔV to V, replaced the distance d by L, and used $\theta = 180°$ for the angle between the direction of $\vec{E}$ and the direction that we imagine traveling along the length of the wire to measure the potential difference. With this choice the potential difference V is positive. Remember that potential always increases as we travel opposite to $\vec{E}$ and decreases as we travel in the direction of $\vec{E}$. Remember also that the direction of current, which we choose to be the direction in which positive charges would move, is in the direction of $\vec{E}$ and so from high potential to low potential. For example, if there is a 1.5-V potential difference ($V = 1.5$ V) between the ends of a wire 10 cm in length ($L = 10$ cm $= 0.10$ m), Equation 18-5 tells us that the electric field inside the wire has magnitude $E = V/L = (1.5 \text{ V})/(0.10 \text{ m}) = 15$ V/m.

For the answer to the second question, we must turn to experiments on current-carrying wires. For many materials (including common conductors such as copper), experiments show that the current i that arises in a wire when an electric field is present inside the wire is directly proportional to both the electric field magnitude E and the cross-sectional area A of the wire. We can write this relationship as

(18-6)
$$i = \frac{EA}{\rho}$$

In Equation 18-6 the quantity ρ (the Greek letter rho) is called the **resistivity**. The value of ρ depends on the material of which the wire is made and tells us how well or poorly this material inhibits the flow of electric charge. Equation 18-6 says that for a given cross-sectional area A and a given electric field magnitude E, a *greater* value of the resistivity means a *smaller* amount of current i. (It's unfortunate that the same Greek letter that we use as the symbol for density is also used as the symbol for resistivity. We promise never to use density and resistivity in the same equation!)

The units of resistivity are voltmeters per ampere, or V·m/A. For reasons that we will see below, the unit V/A (volts per ampere) appears very often and is given its own name, the **ohm** (symbol Ω, the uppercase Greek letter omega): $1\ \Omega = 1$ V/A. In terms of this, the units of resistivity can be written as ohmmeters, or $\Omega \cdot$ m.

For copper, which is a good conductor of electricity, the resistivity is very low: $\rho = 1.725 \times 10^{-8}$ $\Omega \cdot$ m at 21°C. For hard rubber, which is a very poor conductor (and a good insulator), $\rho = 10^{13}$ $\Omega \cdot$ m at 21°C. The explanation for the huge ratio between these two values of resistivity is that copper has many mobile electrons per cubic meter, while hard rubber has hardly any. The value of resistivity also depends on temperature. In metals, atoms move more rapidly and are likely to be less well organized at higher temperatures compared to lower temperatures. As a result, moving charges suffer more collisions with the atoms in a material when the temperature is higher. Consequently, the resistivity of most metals increases as temperature increases. In other materials, such as ceramics, resistivity *decreases* with increasing temperature. Table 18-1 lists the values of resistivity for a variety of substances at 21°C.

In practice, it's more useful to know the answer to the third question: how the current i in a straight wire is related to the potential difference V between the ends of the wire, rather than to the electric field E inside the wire. (That's because electrical meters measure potential difference directly, not electric field.) To obtain such a relationship, substitute the expression $E = V/L$ from Equation 18-5 into Equation 18-6:

$$i = \frac{EA}{\rho} = \left(\frac{V}{L}\right)\left(\frac{A}{\rho}\right) = V\left(\frac{A}{\rho L}\right) \quad \text{or} \quad V = i\left(\frac{\rho L}{A}\right) \qquad (18\text{-}7)$$

We define the **resistance** R of a wire as

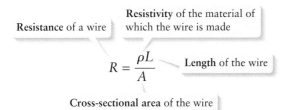

Resistance of a wire
Resistivity of the material of which the wire is made

$$R = \frac{\rho L}{A}$$

Length of the wire
Cross-sectional area of the wire

In terms of this new quantity, we can rewrite Equation 18-7 as

Potential difference between the ends of a wire

$$V = iR$$

Current in the wire   Resistance of the wire

Equation 18-9 says the potential difference, or voltage, V required to produce a current i in a wire is proportional to i and to the resistance of the wire. The greater the resistance, the greater the voltage required for a given current. Equation 18-8 further tells us that the resistance of a wire depends on the resistivity of the material of which the wire is made, the length of the wire, and the cross-sectional area of the wire. For a given material, the resistance is greater for a wire that is long (large L) and thin (small A) than for a wire that is short (small L) and thick (large A). You can see an example by looking inside the slots of an ordinary kitchen toaster. The wire that makes up the toaster's heating coils is very thin, and if uncoiled the wire would be very long. This tells us that the coils have a high resistance. The resistance is made even higher by making the wire out of nichrome, which has a resistivity about 10^2 times greater than that of copper (see Table 18-1).

Since resistivity ρ has units of ohmmeters ($\Omega \cdot$ m), length L has units of meters, and area A has units of meters squared, Equation 18-8 says that the units of resistance R are ohms. This agrees with Equation 18-9, since 1 Ω = 1 V/A: To produce a 1-A current in a wire with a resistance of 1 Ω = 1 V/A, a voltage of 1 V is required.

An ohm is a relatively small resistance. The heating coils of a toaster have a resistance R of about 10 to 20 Ω. A typical **resistor**—a circuit component intended to add resistance to the flow of current—such as you will find for sale in an electronics store will likely have a resistance in the range from 1 kΩ (1 kΩ = 1 kilohm = 10^3 Ω) to 10 MΩ

Table 18-1 Resistivity of Some Conductors and Insulators at 21°C

Conductor	ρ ($\Omega \cdot$ m)
aluminum	2.733×10^{-8}
copper	1.725×10^{-8}
gold	2.271×10^{-8}
iron	9.98×10^{-8}
nichrome	150×10^{-8}
nickel	7.2×10^{-8}
silver	1.629×10^{-8}
titanium	43.1×10^{-8}
tungsten	5.4×10^{-8}

Insulator	ρ ($\Omega \cdot$ m)
glass	10^{12}
hard rubber	10^{13}
fused quartz	7.5×10^{17}

Definition of resistance
(18-8)

Relationship among potential difference, current, and resistance
(18-9)

$(1 \text{ M}\Omega = 1 \text{ megohm} = 10^6 \ \Omega)$. The resistance of a human body measured on the skin can be more than 0.5 MΩ. The standard symbol for a resistor is a jagged line:

Equation 18-9 is often referred to as "Ohm's law," after the nineteenth-century Bavarian physicist Georg Ohm whose pioneering experiments increased our understanding of electric current (and for whom the ohm, or volt per ampere, is named). By itself it suggests that voltage and current are directly proportional to each other. This proportionality holds true if the resistance R remains constant as the voltage is changed, and this is in fact the case for many conducting materials over a wide range of voltages. Materials that have this property are referred to as *ohmic*. However, for many materials (including those used in the filament of an incandescent light bulb as well as in a variety of electronic devices), the value of the resistance changes as the potential difference across the material changes. We can still use Equation 18-9 for such *nonohmic* materials, provided we keep in mind that R is not a constant.

Example 18-3 Stretching a Wire

A 10.0-m-long wire has a radius of 2.00 mm and a resistance of 50.0 Ω. If the wire is stretched to 10.0 times its original length, what will be its new resistance?

Set Up

Equation 18-8 tells us that resistance of the wire depends on its resistivity (which is a property of the material of which the wire is made, and doesn't change if the dimensions change). It also depends on the length and cross-sectional area, both of which change when the wire is stretched. To find the new cross-sectional area, we'll use the idea that the volume of the wire (the product of its length and cross-sectional area) does not change as it's stretched.

Definition of resistance:

$$R = \frac{\rho L}{A} \qquad (18\text{-}8)$$

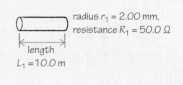

Solve

Find the cross-sectional area of the wire after it has been stretched.

The original cross-sectional area A_1 of the wire of radius $r_1 = 2.00$ mm $= 2.00 \times 10^{-3}$ m is

$$A_1 = \pi r_1^2 = \pi (2.00 \times 10^{-3} \text{ m})^2 = 1.26 \times 10^{-5} \text{ m}^2$$

The volume of the wire is $A_1 L_1$, where $L_1 = 10.0$ m is the wire's initial length. If we stretch the wire to a new length $L_2 = 10.0 L_1 = 1.00 \times 10^2$ m, the cross-sectional area will change to a new value A_2, but the volume will be the same:

$$A_2 L_2 = A_1 L_1$$

$$A_2 = \frac{A_1 L_1}{L_2} = \frac{(1.26 \times 10^{-5} \text{ m}^2)(10.0 \text{ m})}{1.00 \times 10^2 \text{ m}} = 1.26 \times 10^{-6} \text{ m}^2$$

The new cross-sectional area is 1/10.0 of the initial cross-sectional area.

Calculate the new resistance of the wire.

The initial resistance of the wire (before being stretched) is

$$R_1 = \frac{\rho L_1}{A_1} = 50.0 \ \Omega$$

The resistance of the wire after being stretched is

$$R_2 = \frac{\rho L_2}{A_2}$$

If we take the ratio of the two resistances, the unknown value of resistivity will cancel out:

$$\frac{R_2}{R_1} = \frac{\rho L_2/A_2}{\rho L_1/A_1} = \frac{L_2 A_1}{L_1 A_2}$$

$$= \left(\frac{1.00 \times 10^2 \text{ m}}{10.0 \text{ m}}\right)\left(\frac{1.26 \times 10^{-5} \text{ m}^2}{1.26 \times 10^{-6} \text{ m}^2}\right)$$

$$= (10.0)(10.0) = 1.00 \times 10^2$$

The length has increased by a factor of 10.0 and the cross-sectional area has decreased by a factor of 10.0, so the new value of resistivity is greater than the old value by a factor of $(10.0)^2 = 1.00 \times 10^2$:

$$R_2 = 1.00 \times 10^2 \, R_1 = (1.00 \times 10^2)(50.0 \, \Omega)$$

$$= 5.00 \times 10^3 \, \Omega = 5.00 \text{ k}\Omega$$

Reflect

Stretching the wire makes it longer and thinner. A longer wire has more resistance than a shorter one when both are made from the same material, and a thinner wire has more resistance than a thicker one. So, we should expect the stretched wire in this problem to have a much higher resistance than the wire had initially.

Example 18-4 Calculating Current

If a 12.0-V potential difference is set up between the ends of each wire in Example 18-3, how much current will flow in each wire?

Set Up

We know the resistance of each wire from Example 18-3, and so we can use Equation 18-9 to solve for the current i in each wire.

Relationship between potential difference, current, and resistance:

$$V = iR \qquad (18\text{-}9)$$

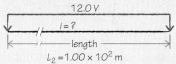

Solve

Rewrite Equation 18-9 as an expression for the current in terms of the voltage and resistance. Then solve for the current in each case.

From Equation 18-9,

$$i = \frac{V}{R}$$

For the wire before it is stretched, $R = R_1 = 50.0 \, \Omega$. The current that results from a 12.0-V potential difference between the ends of this wire is

$$i_1 = \frac{V}{R_1} = \frac{12.0 \text{ V}}{50.0 \, \Omega} = 0.240 \frac{V}{\Omega}$$

Recall that $1 \, \Omega = 1 \text{ V/A}$, so $1 \text{ A} = 1 \text{ V}/\Omega$. So the current in the wire before it is stretched is

$$i_1 = 0.240 \text{ A}$$

After the wire is stretched, the resistance is $R = R_2 = 5.00 \times 10^3 \, \Omega$. A 12.0-V potential difference between the ends of this wire produces a current

$$i_2 = \frac{V}{R_2} = \frac{12.0 \text{ V}}{5.00 \times 10^3 \, \Omega} = 2.40 \times 10^{-3} \text{ A} = 2.40 \text{ mA}$$

Reflect

The stretched wire has a greater resistance than the original wire, so the same potential difference produces a smaller current. Note that the milliampere (1 mA = 10^{-3} A) is a commonly used unit of current, as are the microampere (1 μA = 10^{-6} A), the nanoampere (1 nA = 10^{-9} A), and the picoampere (1 pA = 10^{-12} A).

? Got the Concept? 18-3 Is It Ohmic?

A wire is made of a certain conducting material. You apply different potential differences V across the wire and measure the current i that results. Your results are $V = 2.0$ V, $i = 0.15$ A; $V = 4.0$ V, $i = 0.28$ A; $V = 8.0$ V, $i = 0.50$ A. Is the material of which the wire is made (a) ohmic or (b) nonohmic?

Take-Home Message for Section 18-3

✔ Resistivity ρ is a measure of how well or poorly a particular material inhibits an electric current. The SI units of resistivity are voltmeters per ampere or ohmmeters.

✔ The value of resistivity differs from material to material. The resistivity is low for conductors and high for insulators.

✔ The resistance R of an object is a measure of the current through the object for a given potential difference between its ends. The value of R depends on the object's shape and on the resistivity of the material from which it is made. A long, narrow wire has a much higher resistance than a short, thick one.

The circuit elements with colored bands are resistors. The particular colors on each resistor indicate the value of its resistance.

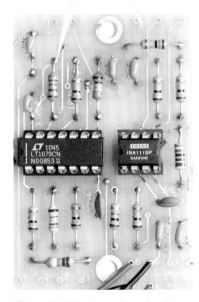

Figure 18-6 Resistors in a circuit
A typical electronic circuit is likely to contain many resistors.

18-4 Resistance is important in both technology and physiology

Electrical resistance can be found in every technological device, from the wires in an automobile ignition system to the resistive elements (resistors) in the circuits of a computer or mobile phone. In electronic devices, resistors are often small and cylindrical with colored bands (Figure 18-6). Other resistors look more like tiny, black cubes with the number of ohms etched on one face.

What purpose does a resistor serve? As we have seen, a potential difference V between two points in a conducting material causes a current i. But according to Equation 18-9, $V = iR$, the amount of current that the potential difference produces depends on the resistance of the conducting material. The greater the resistance R, the smaller the current i. Stated another way, the resistance allows us to control the current due to any particular applied voltage.

An important application of this idea takes place in every one of the millions of cells in your body. In order for a cell to live, there must be a higher concentration of positively charged potassium ions (K^+) inside the cell than in the surrounding fluid. This difference in concentration means that K^+ ions tend to leak out of the cell through pathways called *potassium channels* (Figure 18-7). The flow of K^+ ions constitutes a current, and each potassium channel acts like a resistor. For a given potential difference between the interior and exterior of the cell, the amount of current that flows through these channels depends on their resistance. So the flow of potassium through the membranes of your cells is determined by the electrical resistance of the membrane channels.

Why doesn't the flow of K^+ continue until there is equal concentration of these ions inside and outside the cell? Figure 18-7 shows the reason: As positive ions leave the cell, the cell interior is left with a net negative charge and a lower potential than the outside of the cell. This potential difference, called the *membrane potential*, gives rise to an electric field across the membrane that points from the outside to the inside of the cell. This field opposes additional flow of K^+ ions out of the cell. There is still some random flow of K^+ ions in and out of the cell, but the *net* flow stops when the membrane potential reaches an equilibrium value.

As the following example shows, we can use Equation 18-9 to determine the resistance of the membrane to potassium ion flow.

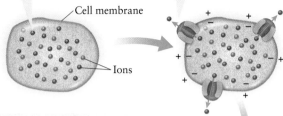

① Potassium ions (shown as green circles) cannot diffuse through a membrane.

② Potassium channels embedded in the membrane allow only potassium ions to diffuse out of the cell.

Cell membrane

Ions

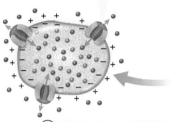

⑤ At equilibrium, the membrane potential prevents any additional net movement of positively charged potassium ions out of the cell.

③ Positively charged potassium ions leave the cell but negative charges (orange circles) are unable to follow.

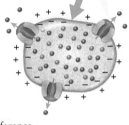

④ The result is a potential difference (called the membrane potential) across the membrane.

electrochemical gradient

Figure 18-7 **Membrane potential** The concentration of positively changed potassium ions (K^+) is always higher inside cells than outside, which causes K^+ ions to leak out. This flow of K^+ ions is a current; the channels that allow K^+ ions to leak out of the cell behave like resistors.

Example 18-5 Resistance of a Potassium Channel

Using an instrument called a patch clamp, scientists are able to control the electrical potential difference across a tiny patch of cell membrane and measure the current through individual potassium channels. In one experiment a voltage of 0.120 V was applied across a patch of membrane, and as a result K^+ ions carried 6.60 pA of current (1 pA = 1 picoampere = 10^{-12} A). What is the resistance of this K^+ channel?

Set Up

Whether the mobile charges are negative electrons or positive ions, we can use Equation 18-9 to relate potential difference (voltage), current, and resistance.

Relationship among potential difference, current, and resistance:

$$V = iR \qquad (18\text{-}9)$$

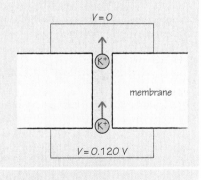

Solve

Rewrite Equation 18-9 to solve for the resistance.

From Equation 18-9,

$$R = \frac{V}{i}$$

The current is $i = 6.60$ pA $= 6.60 \times 10^{-12}$ A, so

$$R = \frac{0.120 \text{ V}}{6.60 \times 10^{-12} \text{ A}} = 1.82 \times 10^{10} \ \Omega$$

Reflect

This is an immensely high resistance compared to the values found in circuits like that shown in Figure 18-6. Yet it is very characteristic of the channels in cell membranes, which typically have resistances in the range from 10^9 Ω to 10^{11} Ω. A potassium channel is only about 10^{-9} m wide—so small that K^+ ions must pass through it in single file—so it's not surprising that the resistance is so high.

Although the current through a single potassium channel is very small, the electric field required to produce that current is tremendous. The thickness of the cell membrane is only about 7.5 nm = 7.5×10^{-9} m. Can you use Equation 18-5 to show that a potential difference of 0.120 V across this distance corresponds to an electric field magnitude of 1.6×10^7 V/m? (By comparison, the electric field inside the wires of a flashlight is only about 10 V/m.)

An important application of the membrane potential is in the nervous system. When a nerve signal propagates along an axon—a long, cable-like fiber that extends from a nerve cell—it does so in the form of a *variation* in the membrane potential of the axon. This variation, which travels along the length of the axon, is called an *action potential*. At rest, the baseline membrane potential is lower on the inside compared to the outside surface of a cell. But if the potential on the interior of the membrane becomes just a little less negative, sodium ion (Na^+) channels in the membrane open. Because the concentration of Na^+ is always higher outside of cells compared to inside, Na^+ ions rush into the axon. This flow is so great that the interior surface of the axon membrane ends up with a positive charge and the exterior with a negative charge, just the reverse of what's shown in Figure 18-7. This inversion of charge is the action potential. The Na^+ channels then start closing as potassium (K^+) channels begin to open, allowing K^+ ions to flow out of the axon and restore the resting polarity of the cell membrane. You can think of an action potential as a momentary voltage "flip" that propagates along the length of the axon like a wave pulse along a stretched string.

The speed with which an action potential can propagate along an axon depends in part on the resistance of the axon to current flowing along its length. As Equation 18-8 tells us, this longitudinal resistance depends on both the length and diameter of the axon. The larger the diameter of the axon, for example, the lower the longitudinal resistance and the faster the electrical signal propagates. (In mammals, the signal propagates at about 10 m/s in axons of 3 μm diameter, but at about 50 m/s in axons of 10 μm diameter.) Some axons, such as the giant ones that mediate the escape reflex in squid, have exceptionally large diameters.

(handwritten margin note:) Neurobiology connection. Resistivity and action potential speed.

Example 18-6 Giant Axons in Squid

Running along each side of the back of a squid is a tube-like structure that can be as large as 1.50 mm in diameter in some species. Originally thought to be blood vessels, these structures are actually giant axons that are part of the squid's nervous system. Although the vast majority of axons in the squid range in diameter from about 10.0 μm to 50.0 μm, as squid develop, axons from about 30,000 neurons fuse together to form these giant axons. Compare the resistance of a giant axon to current along its length to the resistance of an axon with the same length but a more typical diameter of 15.0 μm.

Set Up

Equation 18-8 tells us the resistance of an axon in terms of its resistivity, length, and cross-sectional area. We'll use this to express the ratio of the resistance of a giant axon to that of an ordinary axon of the same composition (and hence same resistivity) and length but of smaller cross-sectional area.

Definition of resistance:

$$R = \frac{\rho L}{A} \qquad (18\text{-}8)$$

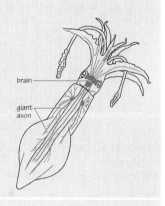

brain

giant axon

Solve

Rewrite Equation 18-8 in terms of the diameter of an axon.

An axon has a circular cross section of radius r, so its cross-sectional area is

$$A = \pi r^2$$

The radius is equal to one-half the diameter d, so

$$A = \pi\left(\frac{d}{2}\right)^2 = \frac{\pi d^2}{4}$$

If we substitute this into Equation 18-8, we get an expression for the resistance of an axon of length L and diameter d:

$$R = \frac{\rho L}{(\pi d^2/4)} = \frac{4\rho L}{\pi d^2}$$

Write the ratio of the resistance of a giant axon to that of a typical axon.

Let d_{giant} be the diameter of a giant axon and $d_{typical}$ be the diameter of a typical axon. If the two axons have the same length L and contain the same material of resistivity ρ,

$$R_{giant} = \frac{4\rho L}{\pi d_{giant}^2}$$

$$R_{typical} = \frac{4\rho L}{\pi d_{typical}^2}$$

The ratio of these two resistances is

$$\frac{R_{giant}}{R_{typical}} = R_{giant} \times \frac{1}{R_{typical}} = \frac{4\rho L}{\pi d_{giant}^2} \times \frac{\pi d_{typical}^2}{4\rho L}$$

The factors of 4, ρ, L, and π cancel, so

$$\frac{R_{giant}}{R_{typical}} = \frac{d_{typical}^2}{d_{giant}^2} = \left(\frac{d_{typical}}{d_{giant}}\right)^2$$

Substitute the numerical values of the diameters of the two axons.

We are given $d_{giant} = 1.50$ mm $= 1.50 \times 10^{-3}$ m and $d_{typical} = 15.0\ \mu$m $= 15.0 \times 10^{-6}$ m $= 1.50 \times 10^{-5}$ m. So the ratio of resistances of the two axons is

$$\frac{R_{giant}}{R_{typical}} = \left(\frac{1.50 \times 10^{-5}\,\text{m}}{1.50 \times 10^{-3}\,\text{m}}\right)^2 = (1.00 \times 10^{-2})^2$$

$$= 1.00 \times 10^{-4} = \frac{1}{1.00 \times 10^4}$$

The resistance of the giant axon is one ten-thousandth as great as that of a typical axon.

Reflect

When a squid recognizes danger, it sends electrical signals (action potentials) along the giant axons to trigger muscle contraction. Because the longitudinal resistance along an axon determines the speed at which nerve signals propagate, signals get to the muscle as much as 10,000 times more quickly through the giant axon than they would if the squid's nervous system was comprised entirely of ordinary nerve cells with small-diameter axons.

❓ Got the Concept? 18-4   Resistance and Diameter

Two copper wires have the same length, but one has four times the diameter of the other. Compared to the wire that has the smaller diameter, the wire that has the larger diameter has a resistance that is (a) 16 times greater; (b) 4 times greater; (c) the same; (d) 1/4 as great; (e) 1/16 as great.

Take-Home Message for Section 18-4

✔ Electrical resistance is important in technology and in the physiology of cells.

✔ For a given applied voltage, the greater the resistance in an electrical system, the smaller the current.

18-5 Kirchhoff's rules help us to analyze simple electric circuits

Many simple electric circuits are made up of a source of electric potential difference and one or more resistors. An example is a light like that shown in the photograph that opens this chapter: The batteries provide the potential difference, and the filament of the light bulb acts as a resistor. We'll refer to batteries and resistors collectively as **circuit elements**. (Later in this chapter we'll consider circuits that also include capacitors as circuit elements.)

In this section we'll see how to analyze circuits made up of a battery and one or more resistors. In particular, we'll see how to determine the voltage across each circuit element as well as the current through each circuit element.

A Single-Loop Circuit and Kirchhoff's Loop Rule

Figure 18-8a shows a battery connected to a single resistor of resistance R. In fact, there are *two* resistors in this circuit; the other one, which we label r, is the **internal resistance** of the battery itself. This reflects the resistance that mobile charges encounter as they pass through the battery. So we can think of the battery as having two components, its **emf** (pronounced "ee-em-eff")—the aspect of the battery that causes charges in the circuit to move—and its internal resistance. Although we draw these as separate entities in Figure 18-8a, they cannot in fact be separated. When we say that a D or AA cell is a "1.5-volt battery" or that the battery in an automobile is "a 12-volt battery," we're actually stating the value of the battery's emf. We use the symbol ε, an uppercase script "e," for emf. So $\varepsilon = 1.5$ V for a D or AA cell and $\varepsilon = 12$ V for a standard automotive battery. (Note that the term "emf" comes from the older term "electromotive force." Although an emf is what pushes mobile charges through the circuit, it has units of volts, not newtons. So it's not accurate to call an emf a "force.") We'll often refer to a battery as a **source of emf**. In later chapters we'll encounter other sources of emf.

The symbol in Figure 18-8a for a source of constant emf is similar to the symbol for a capacitor (see Section 17-5), but with two parallel lines of unequal length:

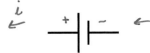

The longer of the two parallel lines represents the positive terminal of the source. Since we regard current as flowing in the direction that positive charges would move, such a source of emf causes current to flow out of its positive terminal and into its negative terminal (see Figure 18-2).

We call the circuit in Figure 18-8a a **single-loop circuit** because there's only a single path that moving charges can follow around the circuit. Think about what happens to such a charge as it travels through this circuit in the direction of the current i. (We'll continue to use the convenient fiction that these are positive charges.) As the charge passes through the source of emf (the battery) from the negative terminal to the positive terminal, it experiences an *increase* in electric potential (a *voltage rise*) of ε. When it moves through the internal resistance, however, the charge experiences a *decrease* in electric potential (a *voltage drop*) of ir. (Remember that the current moves in the direction of the electric field, and the electric field points in the direction from high potential to low potential. So the potential decreases as you move with the current through a resistor.) The charge experiences an additional decrease in potential (voltage drop) of iR when it moves through the resistor of resistance R.

There's also resistance in the wires that connect the battery and the resistor. However, the resistance of the connecting wires is generally quite small compared to the values of r and R. So we'll ignore the wire resistance and

(a)

A battery has an emf ε...   ...and an internal resistance r.   The battery is connected to a resistor of resistance R.

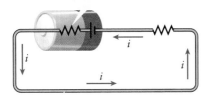

The connecting wires have negligible resistance.

 The sum of the potential changes around the closed loop of this circuit must be zero.

(b)

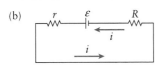

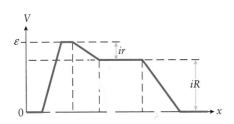

The potential increases by ε from the negative terminal to the positive terminal of the emf.   The potential drops by ir across the internal resistance and drops by iR across the resistor.

Figure 18-8 Kirchhoff's loop rule (a) A circuit made up of a battery (which has an emf ε and an internal resistance r) and a resistor R. (b) The changes in electric potential around the circuit.

assume that a moving charge experiences no change in potential as it traverses the wires.

The current in Figure 18-8a is steady and does not change with time. So when the charge finishes a complete trip around the circuit and returns to where it started, the value of the potential it experiences at the starting point must have the same value as when the charge left that point. In other words, the *net change* in potential for a round trip around the loop must be *zero* (Figure 18-8b). This idea was first proposed by the Prussian physicist Gustav Kirchhoff in the mid-nineteenth century and is often referred to as **Kirchhoff's loop rule**:

The sum of the changes in potential around a closed loop in a circuit must equal zero.

In equation form, we can write Kirchhoff's loop rule for the circuit in Figure 18-8a as

$$\varepsilon - ir - iR = 0 \qquad\qquad (18\text{-}10)$$

We can rearrange Equation 18-10 to solve for the current in the circuit:

$$\varepsilon = ir + iR = i(r + R)$$
$$i = \frac{\varepsilon}{r + R}$$

To give a specific numerical example, suppose $\varepsilon = 12.0$ V, $r = 1.00\ \Omega$, and $R = 19.0\ \Omega$. Then the current is

$$i = \frac{12.0\ \text{V}}{1.00\ \Omega + 19.0\ \Omega} = \frac{12.0\ \text{V}}{20.0\ \Omega} = 0.600\ \text{A}$$

(Recall that 1 A = 1 V/Ω.) You can see that if the battery had a greater internal resistance r, the current would be smaller. As an example, the internal resistance r of a disposable battery in a flashlight or television remote control increases as the battery is used. Eventually r becomes so great that the current becomes too small to make the device work, which means that it's time to replace the battery. The emf ε of the used battery is almost the same as when it was new; it's the internal resistance that makes the battery no longer useful.

so current i gets too small.

Note that for this numerical example, the voltage drop across the internal resistance is $ir = (0.600\ \text{A})(1.00\ \Omega) = 0.600$ V and the voltage drop across the 19.0-Ω resistor is $iR = (0.600\ \text{A})(19.0\ \Omega) = 11.4$ V. The sum of the potential changes is zero, just as Kirchhoff's loop rule says it must be:

$$\varepsilon - ir - iR = 12.0\ \text{V} - 0.600\ \text{V} - 11.4\ \text{V} = 0$$

For this single-loop circuit, the *current* $i = 0.600$ A is the same through the internal resistance r as through the resistor R. That's because no charges can appear or disappear at any place in the circuit (see Section 18-2). However, the *voltage drops* are different for r and R because the values of resistance are different.

! Watch Out! **The voltage across a battery in a circuit is less than the emf.**

The example we have just given shows that when a battery is in a circuit, the voltage V across the battery (that is, the potential difference between its terminals) is *not* equal to the emf ε. Rather, the voltage across the battery is equal to the emf minus the potential drop ir across the internal resistance of the battery. For the battery with emf $\varepsilon = 12.0$ V described above, the voltage is $V = \varepsilon - ir = 12.0\ \text{V} - (0.600\ \text{A})(1.00\ \Omega) =$ 12.0 V $-$ 0.600 V $=$ 11.4 V. The only time when the voltage $V = \varepsilon - ir$ across a battery is equal to the emf is when $i = 0$: that is, when the battery is disconnected from the circuit, so there is no current flowing through the battery. So a 1.5-V battery has a 1.5-V potential difference between its terminals only when the battery isn't connected to anything!

There is always resistance to take into account

Resistors in Series

An important application of Kirchhoff's loop rule is to resistors in **series**, so that the resistors are connected end to end. (We used the same nomenclature for capacitors in

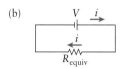

(a)

The current through each resistor in series is the same.

(b)

• The current i through R_{equiv} is the same as through each of the resistors in series.
• The voltage drop V across R_{equiv} is the same as across the combination of resistors in series.

Figure 18-9 Resistors in series (a) A circuit contains three resistors connected in series to a source of emf. (b) The three resistors have been replaced by a single, equivalent resistor.

series in Section 17-7.) Figure 18-9a shows a circuit that has three resistors R_1, R_2, and R_3 in series connected to a source of emf such as a battery with voltage V. This voltage includes the emf and internal resistance of the battery. Imagine we replace the three separate resistors with an **equivalent resistance**—that is, a single resistor of a resistance R_{equiv} that gives the same current as the series combination, as shown in Figure 18-9b. What is the equivalent resistance in terms of R_1, R_2, and R_3?

For R_{equiv} to have the same effect in the circuit as R_1, R_2, and R_3 together, it must give rise to the same current i and the same voltage drop. As we discussed above, the voltage drop across a resistance R that carries current i is iR. The same current is present through each of the three resistors shown in Figure 18-9a, and the total voltage drop across the three resistors in series is the sum of the individual voltage drops: $iR_1 + iR_2 + iR_3$. The voltage drop through the equivalent resistance is iR_{equiv}, so

$$iR_{equiv} = iR_1 + iR_2 + iR_3 = i(R_1 + R_2 + R_3)$$

So the equivalent resistance of the three resistors in series is

$$R_{equiv} = R_1 + R_2 + R_3$$

For example, if the three resistors have resistances $R_1 = 25.0\ \Omega$, $R_2 = 12.0\ \Omega$, and $R_3 = 36.0\ \Omega$, they are equivalent to a single resistor with resistance $R_{equiv} = 25.0\ \Omega + 12.0\ \Omega + 36.0\ \Omega = 73.0\ \Omega$. This means that if we replace the three resistors by a single 73.0-Ω resistor, the current in the circuit will be exactly the same. The net voltage drop is also the same. If the current in our example is $i = 1.00$ A, the voltage drops across the individual resistors are $V_1 = iR_1 = (1.00\ \text{A})(25.0\ \Omega) = 25.0$ V, $V_2 = iR_2 = (1.00\ \text{A})(12.0\ \Omega) = 12.0$ V, and $V_3 = iR_3 = (1.00\ \text{A})(36.0\ \Omega) = 36.0$ V, and the net voltage drop is 25.0 V $+ 12.0$ V $+ 36.0$ V $= 73.0$ V. The voltage drop across the equivalent resistance is $iR_{equiv} = (1.00\ \text{A})(73.0\ \Omega) = 73.0$ V, the same as for the three resistors in series.

In general, if there are N resistors arranged in series, the equivalent resistance is

Equivalent resistance of resistors in series (18-11)

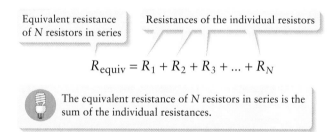

Equivalent resistance of N resistors in series   Resistances of the individual resistors

$$R_{equiv} = R_1 + R_2 + R_3 + ... + R_N$$

The equivalent resistance of N resistors in series is the sum of the individual resistances.

Equation 18-11 tells us that by combining resistors in series, we create a circuit with a higher equivalent resistance than that of any of the individual resistors. A special case is when we put two identical resistors R in series. The equivalent resistance is

$$R_{equiv} = R + R = 2R$$

The equivalent resistance of two identical resistors in series is twice that of each individual resistor.

Our analysis tells us that for resistors in series, the *current* is the same through each resistor, but the *voltage drop* is different for different resistors. As we will see, this is not the case if the resistors are in an arrangement other than series. To demonstrate this, we'll need to introduce the second of Kirchhoff's laws.

Kirchhoff's Junction Rule and Resistors in Parallel

Figure 18-10 shows a circuit with two resistors R_1 and R_2 connected in **parallel** to a battery. (As in Figure 18-9, the potential V includes both the emf and the internal resistance of the battery.) Unlike the single-loop circuits shown in Figures 18-8 and 18-9, this is a **multiloop circuit**: There is more than one pathway that a moving charge can

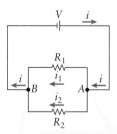

① At junction A, current i splits into current i_1 and current i_2.

② At junction B, current i_1 and current i_2 recombine to current i.

Figure 18-10 Kirchhoff's junction rule At the circuit junctions A and B, the net current into the junction must equal the net current out of that junction.

take from the positive terminal of the battery through the circuit to the negative terminal. In particular, the circuit in Figure 18-10 has two **junctions** at A and B where the current either breaks into two currents (as at A) or comes together (as at B). What is the relationship among the current i that passes through the battery, the current i_1 that passes through resistor R_1, and the current i_2 that passes through resistor R_2?

One condition on i, i_1, and i_2 is that charge can neither be created nor destroyed, nor can it pile up anywhere in the circuit. This means that the rate at which charge *arrives* at a junction must be equal to the rate at which charge *leaves* that junction. This is our second rule of electric circuits. Like the loop rule, this was also proposed by Kirchhoff and is often referred to as **Kirchhoff's junction rule:**

The sum of the currents flowing into a junction equals the sum of the currents flowing out of it.

Let's apply this rule to the junctions shown in Figure 18-10. Current i flows into junction A, and currents i_1 and i_2 flow out of it. So the junction rule tells us that i (the sole current flowing into junction A) must equal $i_1 + i_2$ (the sum of the currents flowing out of A). At junction B, the sum of the currents flowing in is $i_1 + i_2$ and the sole current flowing out is i. So by analyzing either junction we can conclude that

$$i = i_1 + i_2 \qquad (18\text{-}12)$$

This says that the current divides itself into i_1 and i_2 when it reaches junction A, with no extra current being added and no current being lost. These currents rejoin at junction B.

Equation 18-12 tells us that the current splits up when it reaches junction A, but by itself it doesn't tell us how much of current i takes the branch through resistor R_1 (as current i_1) and how much takes the branch through resistor R_2 (as current i_2). To determine this, let's apply the *loop* rule to two different loops through the circuit in Figure 18-10. First consider a loop that starts at the negative terminal of the battery, passes through the battery to the positive terminal (voltage rise V), then follows the path of current i_1 through resistor R_1 (voltage drop i_1R_1), and returns to the negative terminal of the battery. (As before, we'll ignore the resistance of the wires, so there is no voltage drop as a charge traverses the wires.) From the loop rule, the net change in electric potential for this loop is zero:

$$V - i_1R_1 = 0 \qquad (18\text{-}13)$$

The second loop we'll consider also starts at the negative terminal of the battery and passes through the battery to the positive terminal (voltage rise V), but then follows the path of current i_2 through resistor R_2 (voltage drop i_2R_2) before returning to the battery's negative terminal. The loop rule says that the net change in electric potential is also zero for this loop:

$$V - i_2R_2 = 0 \qquad (18\text{-}14)$$

If you compare Equations 18-13 and 18-14, you'll see that these equations can both be true only if

$$i_1R_1 = i_2R_2 \qquad (18\text{-}15)$$

In other words, the voltage drop must be the *same* for each of the resistors in parallel. If the resistances R_1 and R_2 are different, the currents i_1 and i_2 will be different; the current will be greater in the resistor with the smaller resistance and smaller in the resistor with the greater resistance. So for resistors in parallel, the *voltage drop* is the same for both resistors, but the *currents* are different for different resistors. (Compare this to resistors in series, for which the current is the same but the voltage drops are different for different resistors.)

As an illustration, suppose $V = 12.0$ V, $R_1 = 3.00\ \Omega$, and $R_2 = 6.00\ \Omega$. From Equation 18-13, the current i_1 through resistor R_1 is given by

$$V - i_1R_1 = 0 \quad \text{so} \quad i_1R_1 = V \quad \text{and} \quad i_1 = \frac{V}{R_1} = \frac{12.0\ \text{V}}{3.00\ \Omega} = 4.00\ \text{A}$$

(a)

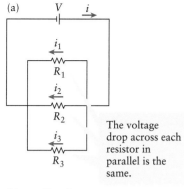

The voltage drop across each resistor in parallel is the same.

(b)

• The current i through R_{equiv} is the same as through the combination of resistors in parallel.
• The voltage drop V across R_{equiv} is the same as across each of the resistors in parallel.

Figure 18-11 Resistors in parallel (a) A circuit contains three resistors connected in parallel to a source of emf. (b) The three resistors have been replaced by a single, equivalent resistor.

 Go to Picture It 18-1 for more practice dealing with resistors in combination

We can then find the current i_2 using Equation 18-15:

$$i_1 R_1 = i_2 R_2 \quad \text{and} \quad i_2 = \frac{i_1 R_1}{R_2} = \frac{(4.00\text{ A})(3.00\text{ }\Omega)}{6.00\text{ }\Omega} = 2.00\text{ A}$$

Current $i_1 = 4.00$ A is twice as great as current $i_2 = 2.00$ A because resistance $R_1 = 3.00$ Ω is half as great as resistance $R_2 = 6.00$ Ω. Note that the total current i that passes through the battery is $i = i_1 + i_2 = 4.00$ A $+ 2.00$ A $= 6.00$ A.

We now have the tools that we need to find the equivalent resistance of a set of resistors in parallel. Figure 18-11a shows three resistors R_1, R_2, and R_3 connected in parallel to a battery with voltage V (including the emf and internal resistance). The currents through these three resistors are i_1, i_2, and i_3, respectively. Figure 18-11b shows the three resistors replaced by an equivalent resistance R_{equiv}. As we did for the case of three resistors in series, we'll determine R_{equiv} by demanding that the net current i and the voltage drop be the same for the actual set of resistors in Figure 18-11a and the equivalent resistance in Figure 18-11b. From our discussion above, we see that the voltage drop through each of the resistors in Figure 18-11a is equal to V:

$$V = i_1 R_1, \quad V = i_2 R_2, \quad V = i_3 R_3$$

If we divide each of these equations by R_1, R_2, and R_3 respectively, we get expressions for the current through each resistor:

(18-16) $$i_1 = \frac{V}{R_1}, \quad i_2 = \frac{V}{R_2}, \quad i_3 = \frac{V}{R_3}$$

The junction rule tells us that the total current that passes through the battery is the sum of the currents through the individual resistors: $i = i_1 + i_2 + i_3$. From Equations 18-16, we can write this as

(18-17) $$i = \frac{V}{R_1} + \frac{V}{R_2} + \frac{V}{R_3} = V\left(\frac{1}{R_1} + \frac{1}{R_2} + \frac{1}{R_3}\right)$$

The total current must be the same for the circuit in Figure 18-11b with the equivalent resistance. If we apply the loop theorem to this circuit, we get $V - iR_{\text{equiv}} = 0$, so $iR_{\text{equiv}} = V$ or

(18-18) $$i = \frac{V}{R_{\text{equiv}}} = V\left(\frac{1}{R_{\text{equiv}}}\right)$$

Since Equations 18-17 and 18-18 are both expressions for the total current i, they can both be valid only if the right-hand sides of these equations are equal to each other. This gives us an equation for the equivalent resistance that corresponds to the three resistors in Figure 18-11a:

(18-19) $$\frac{1}{R_{\text{equiv}}} = \frac{1}{R_1} + \frac{1}{R_2} + \frac{1}{R_3}$$

As an illustration, suppose the three resistors in Figure 18-11a are $R_1 = 25.0$ Ω, $R_2 = 12.0$ Ω, and $R_3 = 36.0$ Ω. According to Equation 18-19, they are equivalent to a single resistor with resistance R_{equiv} given by

$$\frac{1}{R_{\text{equiv}}} = \frac{1}{25.0\text{ }\Omega} + \frac{1}{12.0\text{ }\Omega} + \frac{1}{36.0\text{ }\Omega} = 0.151\text{ }\Omega^{-1}$$

$$R_{\text{equiv}} = \frac{1}{0.151\text{ }\Omega^{-1}} = 6.62\text{ }\Omega$$

This is *less* than the resistance of any one of the individual resistors. If the net current through the battery is 1.00 A, the voltage drop across the equivalent resistance is $iR_{\text{equiv}} = (1.00\text{ A})(6.62\text{ }\Omega) = 6.62$ V. The voltage drop across each of the individual resistors in parallel must be the same, so 6.62 V $= i_1 R_1 = i_2 R_2 = i_3 R_3$. You can use these relationships to show that $i_1 = 0.265$ A, $i_2 = 0.551$ A, and $i_3 = 0.184$ A (note that the current is greatest through resistor R_2, which has the smallest of the three resistances). The net current through the battery is $i = i_1 + i_2 + i_3 = 1.00$ A, just as for the battery connected to the equivalent resistance.

We can generalize Equation 18-19 to the case where we have N resistors arranged in parallel:

> The reciprocal of the equivalent resistance of N resistors in parallel is the sum of the reciprocals of the individual resistances.

$$\frac{1}{R_{equiv}} = \frac{1}{R_1} + \frac{1}{R_2} + \frac{1}{R_3} + \ldots + \frac{1}{R_N}$$

Equivalent resistance of resistors in parallel (18-20)

Equivalent resistance of N resistors in parallel

Resistances of the individual resistors

Equation 18-20 tells us that by combining resistors in parallel, we create a circuit with a smaller equivalent resistance than any of the individual resistors. For the special case of two identical resistors R in parallel, the equivalent resistance is given by

$$\frac{1}{R_{equiv}} = \frac{1}{R} + \frac{1}{R} = \frac{2}{R} \quad \text{so} \quad R_{equiv} = \frac{R}{2}$$

The equivalent resistance of two identical resistors in parallel is one-half that of each individual resistor.

 Watch Out! Resistors do not combine in the same way as capacitors.

Be careful to distinguish between the rules for equivalent *resistance* that we've developed in this section and the rules for equivalent *capacitance* that we found in Section 17-7. For resistors in series, we find the equivalent resistance by adding the individual resistances ($R_{equiv} = R_1 + R_2 + \ldots$); for resistors in parallel, we find the reciprocal of the equivalent resistance by adding the reciprocals of the individual resistances ($1/R_{equiv} = 1/R_1 + 1/R_2 + \ldots$). These rules are reversed for capacitors. For capacitors in series, we find the reciprocal of the equivalent capacitance by adding the reciprocals of the individual capacitances ($1/C_{equiv} = 1/C_1 + 1/C_2 + \ldots$); for capacitors in parallel, we find the equivalent capacitance by adding the individual capacitances ($C_{equiv} = C_1 + C_2 + \ldots$).

The following examples illustrate some applications of the relationships we've developed in this section. In the second example, we'll see how to analyze resistors arranged in a combination that is neither purely series nor purely parallel.

Example 18-7 Giant Axons in Squid Revisited

As we saw in Example 18-6 (Section 18-4), in a squid axons of approximately 30,000 nerve cells fuse together to form each giant axon. A typical axon has a diameter of 15.0 μm, is 10.0 cm long, and has a resistivity of about 3100 $\Omega \cdot$m. Find the resistance of a giant squid axon by considering it as 30,000 separate axons in parallel.

Set Up
We can find the resistance of a typical axon by using Equation 18-8. Each of these individual axons acts as a separate conducting path that mobile charges can follow between a point of high potential and low potential. That's just like the three resistors in parallel shown in Figure 18-11a, so we can use Equation 18-20 to determine the equivalent resistance of the giant axon as a whole in terms of the resistances of the individual axons.

Definition of resistance:

$$R = \frac{\rho L}{A} \qquad (18\text{-}8)$$

Equivalent resistance of resistors in parallel:

$$\frac{1}{R_{equiv}} = \frac{1}{R_1} + \frac{1}{R_2} + \frac{1}{R_3} + \ldots + \frac{1}{R_N} \qquad (18\text{-}20)$$

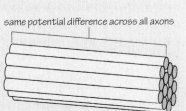

same potential difference across all axons

Solve

Calculate the resistance of an individual axon.

Each individual axon has resistivity $\rho = 3100 \; \Omega \cdot$ m, length $L = 10.0$ cm $= 0.100$ m, and diameter $15.0 \; \mu$m $= 15.0 \times 10^{-6}$ m. The radius r is one-half of the diameter:

$$r = \frac{1}{2}(15.0 \times 10^{-6} \text{ m}) = 7.50 \times 10^{-6} \text{ m}$$

If the axon has a circular cross section, its cross-sectional area is $A = \pi r^2$. So from Equation 18-8 the resistance is

$$R = \frac{\rho L}{A} = \frac{\rho L}{\pi r^2} = \frac{(3100 \; \Omega \cdot \text{m})(0.100 \text{ m})}{\pi (7.50 \times 10^{-6} \text{ m})^2} = 1.75 \times 10^{12} \; \Omega$$

This resistance is very large because the axon is long and thin and the axon resistivity is much higher than that of metallic conductors (see Table 18-1).

Calculate the resistance of the giant axon, as if it were 30,000 individual axons in parallel.

If N axons are in parallel and each has the same resistance R, there are N identical terms on the right-hand side of Equation 18-20. So

$$\frac{1}{R_{\text{equiv}}} = \frac{1}{R_1} + \frac{1}{R_2} + \frac{1}{R_3} + \cdots + \frac{1}{R_N} = \frac{N}{R}$$

Take the reciprocal of both sides:

$$R_{\text{equiv}} = \frac{R}{N}$$

There are $N = 30,000$ individual axons, each of which has resistance $R = 1.75 \times 10^{12} \; \Omega$, so the equivalent resistance of the giant axon is

$$R_{\text{equiv}} = \frac{1.75 \times 10^{12} \; \Omega}{30,000} = 5.85 \times 10^7 \; \Omega$$

Reflect

In Example 18-6, we found that the resistance of a giant squid axon is about 10,000 times smaller than that of a normal axon. But in this example we've found that the equivalent resistance of a giant axon is 30,000 times smaller than that of an individual axon. Why do our answers differ by a factor of 3?

The explanation is that in this example, we made the implicit assumption that the total cross-sectional area of the giant axon is 30,000 times the cross-sectional area of a small axon. However, in an actual squid the cross-sectional area of the giant axon with a diameter of 1.50 mm is smaller than the total cross-sectional area of 30,000 small axons each of diameter 15.0 μm. Although about 30,000 cells contribute to the formation of the giant axon, the process by which the giant axon forms is *not* an actual fusing together of 30,000 fully formed smaller axons. Nevertheless, it's encouraging that our calculations in Example 18-6 and this problem are pretty close.

Example 18-8 Resistors in Combination

Figure 18-12 shows two different combinations of three identical resistors, each with resistance R. Find the equivalent resistance of the combination in (a) Figure 18-12a and (b) Figure 18-12b.

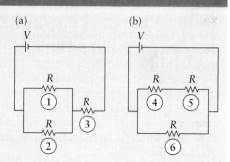

Figure 18-12 Two combinations of three identical resistors What is the equivalent resistance of each combination?

Set Up

Neither combination in Figure 18-12 is a simple series or parallel arrangement of resistors. But in Figure 18-12a resistors 1 and 2 are in parallel with each other, and that combination is in series with resistor 3. Similarly, in Figure 18-12b resistors 4 and 5 are in series with each other, and that combination is in parallel with resistor 6. So we can use Equations 18-11 and 18-20 together to find the equivalent resistance of both arrangements of resistors.

Equivalent resistance of resistors in series:

$$R_{equiv} = R_1 + R_2 + R_3 + \ldots + R_N \qquad (18\text{-}11)$$

Equivalent resistance of resistors in parallel:

$$\frac{1}{R_{equiv}} = \frac{1}{R_1} + \frac{1}{R_2} + \frac{1}{R_3} + \ldots + \frac{1}{R_N} \qquad (18\text{-}20)$$

resistors in series

$R_1 \quad R_2$

resistors in parallel

R_1

R_2

Solve

(a) For the arrangement in Figure 18-12a, first find the equivalent resistance of resistors 1 and 2.

Resistors 1 and 2 in Figure 18-12a are in parallel, so their equivalent resistance R_{12} is given by Equation 18-20:

$$\frac{1}{R_{12}} = \frac{1}{R} + \frac{1}{R} = \frac{2}{R}$$
$$R_{12} = \frac{R}{2}$$

The combination of resistors 1 and 2 is in series with resistor 3. This tells us the overall equivalent resistance.

Equivalent resistor R_{12} is in series with resistor 3. The equivalent resistance R_{123} of the entire combination is given by Equation 18-11:

$$R_{123} = R_{12} + R = \frac{R}{2} + R = \frac{3R}{2}$$

(b) For the arrangement in Figure 18-12b, first find the equivalent resistance of resistors 4 and 5.

Resistors 4 and 5 in Figure 18-12b are in series, so their equivalent resistance R_{45} is given by Equation 18-11:

$$R_{45} = R + R = 2R$$

The combination of resistors 4 and 5 is in parallel with resistor 6. This tells us the overall equivalent resistance.

Equivalent resistor R_{45} is in parallel with resistor 6. The equivalent resistance R_{456} of the entire combination is given by Equation 18-20:

$$\frac{1}{R_{456}} = \frac{1}{R_{45}} + \frac{1}{R} = \frac{1}{2R} + \frac{1}{R} = \frac{3}{2R}$$
$$R_{456} = \frac{2R}{3}$$

Reflect

If we had more than three resistors, or if the resistors had different values, we could create a large number of combinations and equivalent resistances.

parallel decreases equivalent resistance as opposed to series.

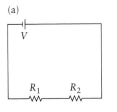
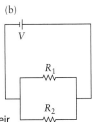
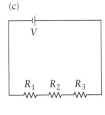
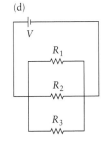

❓ Got the Concept? 18-5

Combinations of Resistors I

Rank the four circuits shown in Figure 18-13 in order of their equivalent resistance, from highest to lowest.

Figure 18-13 Rank the circuits How do these four circuits compare in their equivalent resistances? In the net current through their sources of emf?

❓ Got the Concept? 18-6 Combinations of Resistors II

Rank the four circuits shown in Figure 18-13 in order of the current through the source of emf V, from highest to lowest.

Take-Home Message for Section 18-5

✔ In traversing a battery from its negative terminal to its positive terminal, the potential increases (a voltage rise) by an amount that depends on the battery's emf, its internal resistance, and the current i. In traversing a resistor of resistance R in the direction of the current, the potential drops by iR (a voltage drop).

✔ Kirchhoff's loop rule says that the sum of potential changes around a closed loop in a circuit is zero.

✔ Kirchhoff's junction rule says that the net current into a circuit junction equals the net current out of that junction.

✔ If resistors are connected in series, the current is the same in each resistor but the voltage drops are different across different resistors. The equivalent resistance equals the sum of the individual resistances.

✔ If resistors are connected in parallel, the voltage drop is the same for each resistor but the currents are different for different resistors. The reciprocal of the equivalent resistance equals the sum of the reciprocals of the individual resistances.

18-6 The rate at which energy is produced or taken in by a circuit element depends on current and voltage

The aspects of electric circuits that we've concentrated on so far are voltage, current, and resistance. But an electric circuit is fundamentally a way to transfer *energy* from one place to another, such as from a battery to a flashlight bulb (where the energy is converted into visible light) or from a wall socket to a toaster (where the energy is used to heat your morning bread or bagel). In most applications what's of interest is the *rate* at which energy is transferred into or out of a circuit element. For example, in order for a toaster to be useful, it must heat the bread rapidly enough that it becomes toast in a minute or so, not an hour.

As we learned in Section 13-9, power is the rate at which energy is transferred into or out of an object. The unit of power is the joule per second, or watt (abbreviated W): 1 W = 1 J/s. You can see the importance of power in electric circuits from the numbers that are used to describe various electric devices: An amplifier for a home audio system is rated by its power output (perhaps 75 to 100 W), and any light bulb is stamped with the power that must be supplied to it for normal operation (say, 13 or 60 W).

In this section we'll see how to calculate the power *output* of a source of emf such as a battery, which is fundamentally a source of electric potential energy. We'll also see how to calculate the power *input* of a resistor, which absorbs electric potential energy and converts it into other forms of energy.

Power in a Circuit Element

The key to understanding energy and power in electric circuits is that there is a potential difference across each circuit element, which means that a change takes place in electric potential energy when a moving charge traverses a circuit element. Remember from Section 17-3 the relationship between electric potential difference and electric potential energy difference:

Electric potential difference related to electric potential energy difference (17-6)

The difference in electric potential between two positions...

...equals the change in electric potential energy for a charge q_0 moved between these two positions...

$$\Delta V = \frac{\Delta U_{electric}}{q_0}$$

...divided by the charge q_0.

Electric potential difference equals electric potential energy difference per unit charge.

Let's rewrite this equation for the case in which a small quantity Δq of moving charge moves from one end of a circuit element to the other. If the potential difference between the ends of the element is ΔV, then Equation 17-6 becomes

(18-21)

$$\Delta V = \frac{\Delta U_{electric}}{\Delta q} \quad \text{or} \quad \Delta U_{electric} = (\Delta q)\,\Delta V$$

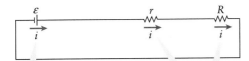

Charges moving through the emf undergo a potential change of $+\varepsilon$, so electric potential energy increases; energy is extracted from the emf.

Charges undergo a potential change of $-ir$ moving through the internal resistance and $-iR$ moving through the resistor. Electric potential energy decreases; energy goes into the resistances.

Figure 18-14 **Energy and power in a single-loop circuit** Energy flows from the emf into the moving charges; energy flows from the moving charges into the internal resistance and the resistor.

We'll continue to use the idea that current is in the direction in which positive charges would flow, so the moving charge is positive: $\Delta q > 0$. Then Equation 18-21 tells us that there is an *increase* in electric potential energy ($\Delta U_{electric} > 0$) if the charge Δq traverses a circuit element from low potential to high potential, so $\Delta V > 0$. That's the case for a charge that travels through the source of emf in Figure 18-14 from the negative terminal to the positive terminal. Since energy is conserved, it must be that the amount of energy extracted from the source is equal to the increase in electric potential energy.

There is a *decrease* in electric potential energy ($\Delta U_{electric} < 0$) if the charge Δq traverses a circuit element from high potential to low potential, so $\Delta V < 0$. That's what happens when charge Δq travels through the resistor R in Figure 18-14 in the direction of the current. The lost electric potential energy is deposited into the resistor, which causes an increase in the resistor's temperature. If the resistor is the filament of a conventional flashlight bulb, the increased temperature causes the filament to emit radiation, some of which is in the form of visible light (see Section 14-7).

> **Watch Out!** Potential energy may change as charges move around a circuit, but current does not.
>
> It's a common misconception that current is "used up" when it passes through a resistor, and that current is "added to" when it passes through a source of emf. In fact, there is *no* change in the current as it passes through either a resistor or a source: Moving charges leave any circuit element at the same rate as they enter the element. All that changes is the electric potential energy associated with the charges.

With a current i through the circuit, the power P for each circuit element is just the rate at which electric potential energy changes in that element. This equals the change in electric potential energy $\Delta U_{electric}$ for a charge Δq that enters the element, as given by Equation 18-21, divided by the time Δt that it takes each new bit of charge Δq to enter the element:

$$P = \frac{\Delta U_{electric}}{\Delta t} = \frac{\Delta q \, \Delta V}{\Delta t} = \left(\frac{\Delta q}{\Delta t}\right)\Delta V = (i)\Delta V \qquad (18\text{-}22)$$

In the last part of Equation 18-22 we've used the idea that the ratio $\Delta q/\Delta t$ equals the rate at which charge enters the circuit element—in other words, this ratio equals the current i through the element (see Equation 18-1).

Equation 18-22 states that the power can be positive or negative depending on the sign of the potential difference ΔV between the ends of the circuit element. Instead of worrying about these signs, we'll replace ΔV in Equation 18-22 with the voltage V across the circuit element, which we'll regard as the absolute value of the potential difference between the ends of the element. Then P is always positive, and we can write

Power produced by or transferred into a circuit element

$$P = iV$$

Current through the circuit element

Voltage (absolute value) across the circuit element

Power for a circuit element
(18-23)

 Source of emf: Power flows out of the source and into the moving charges.
Resistor: Power flows out of the charges and into the resistor.

Equation 18-23 applies to *any* circuit element: The amount of power that flows into or out of a circuit element is equal to the product of the current through the element multiplied by the voltage across the element. For a given voltage V, each small amount of charge Δq that traverses the circuit element transfers the same amount of energy into or out of the element; the more charge that traverses the element per unit time and so the greater the current i, the greater the *rate* of energy transfer P.

! Watch Out! **The units of power take time into account.**

The units of power sometimes cause confusion because it seems that watts require an additional time measurement. For example, to find the power in watts that goes into a flashlight bulb, a student might ask "Find the power for what amount of time?" That's not a sensible question: Time is already included in the units of power, since 1 watt is equal to 1 joule *per second*. When a light bulb is rated at 120 W, that means it requires 120 J of energy every second to operate. Time is already taken into account.

Note that the power company charges its customers based not on how much *power* they use at a given time, but on the total amount of *energy* that they use in one billing period. Since power is energy per time, the units of energy are the units of power multiplied by the units of time. That's why the power company bills on the number of kilowatt-hours (kWh) that a customer uses: 1 kWh = 1000 watt-hours, or the amount of energy it takes to run a device that uses 1000 W of power (typical for a microwave oven) for one hour. Since 1 h = 3600 s, you can see that 1 kWh = (1000 W)(3600 s) = 3.6×10^6 W·s = 3.6×10^6 J.

For the special case where the circuit element is a resistor with resistance R, Equation 18-9 tells us that the voltage V across the resistor is equal to the product of the current and the resistance: $V = iR$, or equivalently $i = V/R$. We can use these two expressions to rewrite Equation 18-23 in two equivalent forms for the special case of a resistor:

$$P = i(iR) \quad \text{or} \quad P = \left(\frac{V}{R}\right)V$$

We can simplify these to

Power for a resistor
(18-24)

Power into a resistor   Current through the resistor

$$P = i^2R = \frac{V^2}{R}$$ Voltage across the resistor

Resistance of the resistor

The expression $P = i^2R$ is useful if we know the current through a resistor of known resistance, while $P = V^2/R$ is useful if we know the voltage across that resistor. The following examples show how to use Equations 18-23 and 18-24.

Example 18-9 Power in a Single-Loop Circuit

A battery with emf 12.0 V and internal resistance 1.00 Ω is connected to a resistor with resistance 19.0 Ω. Find (a) the rate at which energy is supplied by the emf, (b) the rate at which energy flows into the internal resistance, and (c) the rate at which energy flows into the resistor.

Set Up

We can find the current in the circuit using Kirchhoff's loop rule (the sum of the changes in potential around a closed loop in a circuit

Relationship among potential difference, current, and resistance:

$$V = iR \qquad (18-9)$$

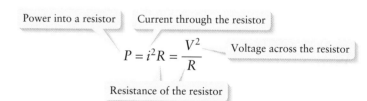

must equal zero). Equation 18-9 then tells us the voltage across either resistance. We'll use Equation 18-23 to determine the power for each element of the circuit, and check our results for the two resistances using Equations 18-24.

Power for a circuit element:

$$P = iV \qquad (18\text{-}23)$$

Power for a resistor:

$$P = i^2 R = \frac{V^2}{R} \qquad (18\text{-}24)$$

Solve

(a) First apply Kirchhoff's loop rule to determine the current i. Start at the point a in the circuit and go around in the direction of the current.

There is a voltage rise of $\varepsilon = 12.0$ V going through the emf, a voltage drop of ir going through the internal resistance $r = 1.00\ \Omega$, and a voltage drop of iR going the resistor with resistance $R = 19.0\ \Omega$. The sum of the potential changes around the circuit is zero:

$$+\varepsilon + (-ir) + (-iR) = 0$$

Rearrange this to solve for the current i:

$$\varepsilon - i(r + R) = 0, \text{ so } i(r + R) = \varepsilon \text{ and}$$

$$i = \frac{\varepsilon}{r + R} = \frac{12.0\ \text{V}}{1.00\ \Omega + 19.0\ \Omega} = \frac{12.0\ \text{V}}{20.0\ \Omega} = 0.600\ \text{A}$$

(Recall that $1\ \text{A} = 1\ \text{V}/\Omega$.)

Use Equation 18-23 to find the power extracted from the emf.

The voltage across the emf itself is $\varepsilon = 12.0$ V. The rate at which energy is extracted from the emf is

$$P_{\text{emf}} = i\varepsilon = (0.600\ \text{A})(12.0\ \text{V}) = 7.20\ \text{A} \cdot \text{V}$$

Since $1\ \text{A} = 1\ \text{C}/\text{s}$, $1\ \text{V} = 1\ \text{J}/\text{C}$, and $1\ \text{W} = 1\ \text{J}/\text{s}$,

$$P_{\text{emf}} = 7.20\frac{\text{C}}{\text{s}} \cdot \frac{\text{J}}{\text{C}} = 7.20\frac{\text{J}}{\text{s}} = 7.20\ \text{W}$$

(b) First use Equation 18-9 to find the potential difference across the internal resistance $r = 1.00\ \Omega$. Then find the power in the internal resistance using Equation 18-23.

From Equation 18-9, the voltage across the internal resistance is

$$V_r = ir = (0.600\ \text{A})(1.00\ \Omega) = 0.600\ \text{V}$$

From Equation 18-23, the rate at which energy flows into the internal resistance is

$$P_r = iV_r = (0.600\ \text{A})(0.600\ \text{V}) = 0.360\ \text{W}$$

This power into the internal resistance goes into heating the battery.

(c) Repeat part (b) for the resistor of resistance $R = 19.0\ \Omega$.

The voltage across the resistor is

$$V_R = iR = (0.600\ \text{A})(19.0\ \Omega) = 11.4\ \text{V}$$

The rate at which energy flows into the resistor is

$$P_R = iV_R = (0.600\ \text{A})(11.4\ \text{V}) = 6.84\ \text{W}$$

Reflect

Note that the rate at which energy is extracted from the source of emf is equal to the *net* rate at which energy flows into the internal resistance and the resistor. This is equivalent to saying that energy is conserved in the circuit.

The net rate of energy flow into the two resistances is

$$P_r + P_R = 0.360\ \text{W} + 6.84\ \text{W} = 7.20\ \text{W}$$

This is the same as the rate at which energy flows out of the source of emf, $P_{\text{emf}} = 7.20$ W.

We can check our result $P_R = 6.84$ W for the power into the resistor by showing that we get the same results using Equations 18-24. Can you use the same approach to check the result $P_r = 0.360$ W for the internal resistance?

From the first of Equations 18-24, the power into the 19.0-Ω resistor is

$$P_R = i^2 R = (0.600 \text{ A})^2 (19.0 \ \Omega) = 6.84 \text{ W}$$

(Note that $1 \text{ A}^2 \cdot \Omega = 1 \text{ A} \cdot \text{V} = 1 \text{ W}$.)

To use the second of Equations 18-24, use the voltage $V_R = 11.4$ V across the resistor:

$$P_R = \frac{V_R^2}{R} = \frac{(11.4 \text{ V})^2}{19.0 \ \Omega} = 6.84 \text{ W}$$

(Note that $1 \text{ V}^2 / \Omega = 1 \text{ V} \cdot (\text{V}/\Omega) = 1 \text{ V} \cdot \text{A} = 1 \text{ W}$.)

This agrees with the result for P_R found above.

Example 18-10 Power in Series and Parallel Circuits

Two resistors, one with $R_1 = 2.00 \ \Omega$ and one with $R_2 = 3.00 \ \Omega$, are both connected to a battery with emf $\varepsilon = 12.0$ V and negligible internal resistance. Find the power delivered by the battery and the power absorbed by each resistor if the resistors are connected to the battery (a) in series and (b) in parallel.

Set Up

We'll use the same tools as in the previous example, plus Equations 18-11 and 18-20 for the equivalent resistance of resistors in series and in parallel. In each case we'll use the equivalent resistance to find the current through the battery, then use Equation 18-23 to determine the power provided by the battery. If the resistors are in series, the current through each is the same as the current through the battery, so we'll use the first of Equations 18-24 ($P = i^2/R$) to determine the power into each resistor. For resistors in parallel the voltage is the same across each resistor, so in that case we'll find the power into each resistor using the second of Equations 18-24 ($P = V^2/R$).

Relationship among potential difference, current, and resistance:

$$V = iR \qquad (18\text{-}9)$$

Equivalent resistance of two resistors in series:

$$R_{\text{equiv}} = R_1 + R_2 \qquad (18\text{-}11)$$

Equivalent resistance of two resistors in parallel:

$$\frac{1}{R_{\text{equiv}}} = \frac{1}{R_1} + \frac{1}{R_2} \qquad (18\text{-}20)$$

Power for a circuit element:

$$P = iV \qquad (18\text{-}23)$$

Power for a resistor:

$$P = i^2 R = \frac{V^2}{R} \qquad (18\text{-}24)$$

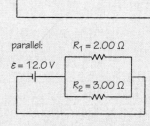

Solve

(a) The two resistors in series are equivalent to a single resistor R_{equiv}. This is connected directly to the terminals of the battery, across which the voltage is ε (we're told to ignore the internal resistance). So the voltage across R_{equiv} is also equal to ε, which tells us the current through the circuit. Equation 18-23 then tells us the power delivered by the battery.

From Equation 18-11, the equivalent resistance of the two resistors in series is

$$R_{\text{equiv}} = R_1 + R_2 = 2.00 \ \Omega + 3.00 \ \Omega = 5.00 \ \Omega$$

The voltage drop across this equivalent resistance is $V = iR_{\text{equiv}}$ from Equation 18-9, which is also equal to the emf $\varepsilon = 12.0$ V of the battery. So the current through the equivalent resistance is given by

$$\varepsilon = iR_{\text{equiv}} \quad \text{or} \quad i = \frac{\varepsilon}{R_{\text{equiv}}} = \frac{12.0 \text{ V}}{5.00 \ \Omega} = 2.40 \text{ A}$$

This is also the current through the battery. Since the voltage across the battery is equal to $\varepsilon = 12.0$ V, the power delivered by the battery is

$$P_{\text{battery}} = i\varepsilon = (2.40 \text{ A})(12.0 \text{ V}) = 28.8 \text{ W}$$

This must be equal to the power that goes into the two resistors.

The current $i = 2.40$ A is the same through both resistors in this series circuit. Use this to calculate the power that goes into each resistor.

Using the first of Equations 18-24, the power into resistor $R_1 = 2.00$ Ω is

$$P_1 = i^2 R_1 = (2.40 \text{ A})^2 (2.00 \text{ } \Omega) = 11.5 \text{ W}$$

The power into resistor $R_2 = 3.00$ Ω is

$$P_2 = i^2 R_2 = (2.40 \text{ A})^2 (3.00 \text{ } \Omega) = 17.3 \text{ W}$$

The net power into the two resistors is equal to the power supplied by the battery, as it should be:

$$P_1 + P_2 = 11.5 \text{ W} + 17.3 \text{ W} = 28.8 \text{ W} = P_{\text{battery}}$$

(b) Follow the same steps as in part (a) to find the equivalent resistance of the two resistors in parallel, the current through the battery connected to that parallel arrangement, and the power delivered by the battery in this situation.

From Equation 18-20, the equivalent resistance of the two resistors in parallel is given by

$$\frac{1}{R_{\text{equiv}}} = \frac{1}{R_1} + \frac{1}{R_2} = \frac{1}{2.00 \text{ } \Omega} + \frac{1}{3.00 \text{ } \Omega} = 0.833 \text{ } \Omega^{-1}$$

$$R_{\text{equiv}} = \frac{1}{0.833 \text{ } \Omega^{-1}} = 1.20 \text{ } \Omega$$

Equation 18-9 tells us that the voltage drop across this equivalent resistance is $V = iR_{\text{equiv}}$; since this equivalent resistance is connected to the terminals of the battery, the voltage drop is also equal to the battery emf $\varepsilon = 12.0$ V. So the current through the equivalent resistance and through the battery is given by

$$\varepsilon = iR_{\text{equiv}} \text{ or } i = \frac{\varepsilon}{R_{\text{equiv}}} = \frac{12.0 \text{ V}}{1.20 \text{ } \Omega} = 10.0 \text{ A}$$

From Equation 18-23, the power delivered by the battery is

$$P_{\text{battery}} = i\varepsilon = (10.0 \text{ A})(12.0 \text{ V}) = 1.20 \times 10^2 \text{ W}$$

Note that this is more than four times as much power as the same battery delivers to the resistors in series.

The voltage is the same across resistors in parallel. Each resistor is effectively connected directly to the terminals of the battery, so the voltage across each resistor is $V = \varepsilon = 12.0$ V. Use this to calculate the power that goes into each resistor.

Using the second of Equations 18-24, we find that the power into resistor $R_1 = 2.00$ Ω is

$$P_1 = \frac{V^2}{R_1} = \frac{\varepsilon^2}{R_1} = \frac{(12.0 \text{ V})^2}{2.00 \text{ } \Omega} = 72.0 \text{ W}$$

The power into resistor $R_2 = 3.00$ Ω is

$$P_2 = \frac{V^2}{R_2} = \frac{\varepsilon^2}{R_2} = \frac{(12.0 \text{ V})^2}{3.00 \text{ } \Omega} = 48.0 \text{ W}$$

As for the series case, the net power into the two resistors is equal to the power supplied by the battery:

$$P_1 + P_2 = 72.0 \text{ W} + 48.0 \text{ W} = 1.20 \times 10^2 \text{ W} = P_{\text{battery}}$$

Reflect

Although the same battery and same resistors are used in both circuits, the power provided by the battery is *much* different in the two circuits. That's because the current through the battery is different for the two circuits: $i = 2.40$ A for the series circuit, $i = 10.0$ A for the parallel circuit.

What's more, the power into each resistor is very different in the two circuits, because the voltage and current for each resistor are greater for the parallel circuit than for the series circuit. If the resistors are light bulbs that take in power and use it to produce light, the lights

Use Equation 18-9 to find the voltage across each resistor in the series circuit:

$$V_1 = iR_1 = (2.40 \text{ A})(2.00 \text{ } \Omega) = 4.80 \text{ V}$$
$$V_2 = iR_2 = (2.40 \text{ A})(3.00 \text{ } \Omega) = 7.20 \text{ V}$$

(compared to $V = \varepsilon = 12.0$ V for each resistor in the parallel circuit) will glow brighter in the parallel circuit than in the series circuit.

Use Equation 18-9 to find the current through each resistor in the parallel circuit:

$$i_1 = \frac{\varepsilon}{R_1} = \frac{12.0 \text{ V}}{2.00 \text{ }\Omega} = 6.00 \text{ A}$$

$$i_1 = \frac{\varepsilon}{R_2} = \frac{12.0 \text{ V}}{3.00 \text{ }\Omega} = 4.00 \text{ A}$$

(compared to $i = 2.40$ A for each resistor in the series circuit)

Note that the current is the same for both resistors in the series circuit, so more power goes into the resistor with the greater resistance ($P_2 = 17.3$ W into $R_2 = 3.00$ Ω versus $P_1 = 11.5$ W into $R_1 = 2.00$ Ω). That follows from the relationship $P = i^2R$ for resistors. By contrast, the voltage is the same for the two resistors in the parallel circuit, so more power goes into the resistor with the smaller resistance ($P_1 = 72.0$ W into $R_1 = 2.00$ Ω versus $P_2 = 48.0$ W into $R_2 = 3.00$ Ω). That agrees with the relationship $P = V^2/R$ for resistors. So if the resistors are light bulbs, the light bulb with $R_2 = 3.00$ Ω is the brighter one in the series circuit, but the bulb with $R_1 = 2.00$ Ω is the brighter one in the parallel circuit! When comparing the power in resistors, choose wisely among the relationships $P = iV$, $P = i^2R$, and $P = V^2/R$.

We've seen how to apply Equation 18-23, $P = iV$, and Equations 18-24, $P = i^2R = V^2/R$, to DC circuits in which there is a steady current that does not vary with time. But these same equations also apply to circuits in which the current is *not* constant and varies. In the following section, we'll use Equations 18-23 and 18-24 to help us understand the time-varying current in a circuit that includes both a resistor and a capacitor.

? Got the Concept? 18-7 Batteries in Series

Consider a resistor R connected to a single source of emf ε and negligible internal resistance (Figure 18-15a). If the same resistor is instead connected to two sources of emf ε in series, each with negligible internal resistance (Figure 18-15b), by what factor does the power into the resistor increase? (a) $\sqrt{2}$; (b) 2; (c) 4; (d) 8; (e) 16.

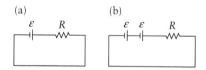

Figure 18-15 **One emf or two** How does adding a second source of emf affect the power delivered to the resistor?

? Got the Concept? 18-8 Ranking the Power

Five identical resistors, A, B, C, D, and E, are connected to a source of emf as shown in Figure 18-16. The source has negligible internal resistance. Rank the five resistors in order of the amount of power that goes into each resistor, from greatest to smallest. If the power in two resistors is the same, indicate as such.

Figure 18-16 **Rank the resistors** How do these five identical resistors compare in the power that they absorb?

Take-Home Message for Section 18-6

✔ Power is the rate at which energy is transferred into or out of a system. The unit of power is the watt (1 W = 1 J/s).

✔ The power into or out of any circuit element equals the current through the element multiplied by the voltage across the element.

✔ The power into a resistor is proportional to the square of the current through the resistor or, equivalently, the square of the voltage across the resistor.

18-7 A circuit containing a resistor and capacitor has a current that varies with time

In an ordinary DC circuit, the battery provides a steady current and delivers energy to the other circuit elements at a steady rate. In some circuits, however, what's required is a short burst of energy. That's the case for the electronic flash in a camera or mobile phone (Figure 18-1c): Energy has to be delivered to the flash lamp in a very brief time interval to produce a short-duration, high-intensity flash.

The simplest way to deliver a short burst of electric potential energy to a circuit element is by using a *capacitor* as an energy source. (This would be a good time to review our discussion of capacitors in Sections 17-5 and 17-6.) We'll begin by examining a circuit in which a battery is used to charge a capacitor. Once we've done this, we'll see what happens when the charged capacitor is discharged and its stored energy is delivered to a resistor.

A Series *RC* Circuit: Charging the Capacitor

Figure 18-17a shows a circuit with a resistor of resistance R and a capacitor of capacitance C connected in series to a battery of emf ε. (We will assume that the internal resistance of the battery is so small compared to R that it can be ignored.) Such a circuit that contains both a resistor and capacitor in series is called a **series *RC*** circuit.

Initially, the capacitor in Figure 18-17a is uncharged. If the switch in this circuit is moved to the up position (Figure 18-17b), the circuit is completed and the battery will cause charge to begin to flow. We'll call $t = 0$ the time when the switch is thrown. As time passes, positive charge $+q$ builds up on the upper capacitor plate, and an equal amount of negative charge $-q$ builds up on the lower capacitor plate. (The two charges must be of equal magnitude since the current is the same throughout the circuit. Hence positive charge leaves the initially uncharged lower plate at the same rate that it arrives at the initially uncharged upper plate, leaving as much negative charge on the lower plate as there is positive charge on the upper plate.) So the circuit in Figure 18-17b is a *charging* series *RC* circuit.

Let's apply Kirchhoff's loop rule to the circuit shown in Figure 18-17b. In Section 17-5 we found that the magnitude of the charge q on the capacitor plates is proportional to the voltage V across the capacitor:

A capacitor carries a charge $+q$ on its positive plate and a charge $-q$ on its negative plate.

The magnitude of q is directly proportional to V, the voltage (potential difference) between the plates.

$$q = CV$$

The constant of proportionality between charge q and voltage V is the capacitance C of the capacitor.

We can rewrite Equation 17-14 as $V = q/C$. If we start at point p in Figure 18-17b and move in a clockwise round trip through the circuit, we encounter a voltage rise of $+\varepsilon$ as we pass through the source of emf, a voltage drop $-iR$ as we pass through the resistor, and a voltage drop $-q/C$ as we pass through the capacitor from the positive plate (which is at higher potential) to the negative plate (which is at lower potential). Kirchhoff's loop rule tells us that the sum of these voltages must be zero:

$$\varepsilon + (-iR) + \left(-\frac{q}{C}\right) = 0 \qquad (18\text{-}25)$$

If we solve Equation 18-25 for the current i in the circuit, we get

$$i = \frac{\varepsilon}{R} - \frac{q}{RC} \qquad (18\text{-}26)$$

Equation 18-26 tells us that the current in the circuit of Figure 18-17b *cannot* be constant. As charge builds up on the capacitor, the value of q increases. As a result, the right-hand side of Equation 18-26 must decrease with time, so the current i must decrease with time as well. Physically what's happening is that the voltage $V = q/C$ across the capacitor increases as the charge increases. This voltage is opposed to that of the source of emf, so the current decreases. Since the current is what's causing the charge to increase, this means that q increases at an ever-slower rate as time goes by.

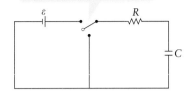

(a) The position of this switch determines whether or not the source of emf ε is part of the circuit.

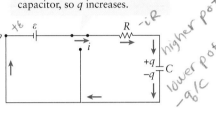

(b) With the switch in the up position, the emf causes charge to flow. This current charges the capacitor, so q increases.

Figure 18-17 Charging a series *RC* circuit (a) In this circuit a resistor and an initially uncharged capacitor are connected in series to a source of emf. (b) The switch is moved to the up position at $t = 0$, beginning the charging process.

Charge, voltage, and capacitance for a capacitor
(17-14)

how does current and charge on capacitor change with time?

We'd like to know the capacitor charge *q* and the current *i* as functions of time. Solving Equation 18-26 for these functions is a problem in calculus that's beyond our scope. Instead, we'll present the solutions and see that they make sense:

Capacitor charge and current in a charging series *RC* circuit (18-27)

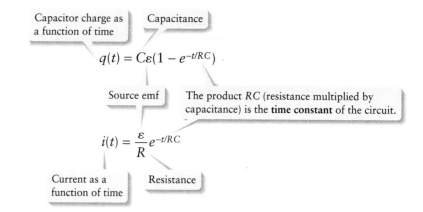

Capacitor charge as a function of time

Capacitance

$$q(t) = C\varepsilon(1 - e^{-t/RC})$$

Source emf

The product *RC* (resistance multiplied by capacitance) is the **time constant** of the circuit.

$$i(t) = \frac{\varepsilon}{R}e^{-t/RC}$$

Current as a function of time

Resistance

In these equations, *t* = 0 is when the switch in Figure 18-17b is moved to the up position.

√x̄ *See the Math Tutorial for more information on exponents and logarithms*

Figure 18-18 graphs the charge *q* and current *i* as given by Equations 18-27. Both of these equations involve the **exponential function**, the irrational number *e* = 2.71828... raised to a power. (Note that *e* is *not* the same as the magnitude of the charge on the electron, for which we unfortunately use the same symbol.) The number *e* has special properties: If a population has N_0 members and grows at a rate *r* per year (for example, if the population grows by 2% per year, *r* = 0.02), then after *t* years the population will be $N(t) = N_0 e^{rt}$. In both of Equations 18-27 the exponential function has a *negative* exponent, which means that this function decreases with time. In particular, $e^{-t/RC}$ decreases by a factor of $1/e = 0.36787...$ every time the quantity *t/RC* increases by 1—that is, whenever time *t* increases by *RC*. At *t* = 0 (when the switch in Figure 18-17

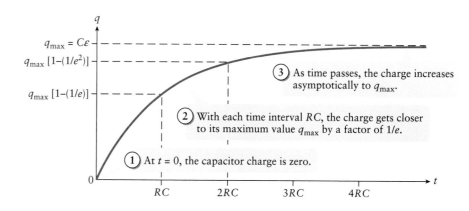

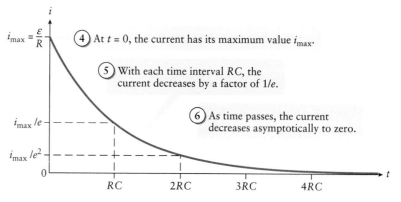

Figure 18-18 Charge and current in a charging series *RC* circuit Capacitor charge *q* starts at zero and approaches a maximum value; current *i* starts at a maximum value and approaches zero.

is closed and charge begins to flow), $e^{-t/RC} = e^0 = 1$; at $t = RC$, $e^{-t/RC} = e^{-1} = 1/e = 0.368$ to three significant figures; at $t = 2RC$, $e^{-t/RC} = e^{-2} = 1/e^2 = 0.135$; and so on.

The quantity RC is called the **time constant** of a series RC circuit. (Note that the product RC has units of ohms times farads. Since $1\ \Omega = 1\ \text{V/A}$, $1\ \text{F} = 1\ \text{C/V}$, and $1\ \text{A} = 1\ \text{C/s}$, it follows that $1\ \Omega \cdot \text{F} = 1\ (\text{V/A})(\text{C/V}) = 1\ \text{C/A} = 1\ \text{s}$. So the quantity RC does indeed have units of time.) The smaller the time constant, the more rapidly the charge q approaches its maximum value q_{max} and the more rapidly the current decreases to its final value of zero. This makes sense: A smaller resistance R (and so a smaller value of RC) allows a greater current, which means that the capacitor can accumulate charge at a faster rate. A smaller capacitance C also decreases the time constant because the capacitor can store less charge for a given voltage and so can be charged more rapidly.

No matter what the value of RC, at $t = 0$ the capacitor is uncharged, so there is zero voltage across the capacitor. As a result, at $t = 0$ the voltage iR across the resistor is equal to the voltage ε across the source, so $\varepsilon = iR$ and $i = i_{max} = \varepsilon/R$. After a very long time the current has dropped to zero, so the voltage iR across the resistor is zero. As a result, the voltage q/C across the capacitor is equal to the voltage ε across the source, so $\varepsilon = q/C$ and $q = q_{max} = C\varepsilon$.

A Series RC Circuit: Discharging the Capacitor

Suppose a long time has elapsed since the switch in Figure 18-17b was thrown to the up position, so the capacitor is charged to a charge q_{max}. (This may be less than $C\varepsilon$, depending on how long the source has had to charge the capacitor.) We now throw the switch to the down position as in **Figure 18-19** and restart our clock so that $t = 0$ is the time when the switch is moved to the new position. Now the emf is no longer part of the circuit, so the positive charge on the upper plate of the capacitor is free to move counterclockwise around the circuit to cancel the negative charge on the lower plate. So the value of the capacitor charge q decreases, and the electric potential energy stored in the capacitor is transferred into the resistor. We call this a *discharging* series RC circuit.

Kirchhoff's loop rule for the discharging circuit is now the same as Equation 18-25, but with the emf ε removed:

$$(-iR) + \left(-\frac{q}{C}\right) = 0 \tag{18-28}$$

Solving Equation 18-28 for the current i in the circuit gives

$$i = -\frac{q}{RC} \tag{18-29}$$

The minus sign in Equation 18-29 means that the current flows in the direction opposite to that in the charging RC circuit of Figure 18-17b. (You can see this in Figure 18-19.) Equation 18-29 says that as the capacitor discharges and the charge q decreases, the current i will decrease in magnitude. The charge and current as functions of time are

> Capacitor charge as a function of time

$$q(t) = q_{max}\, e^{-t/RC}$$

> Capacitor charge at $t = 0$

> The product RC (resistance multiplied by capacitance) is the **time constant** of the circuit.

$$i(t) = -\frac{q_{max}}{RC} e^{-t/RC}$$

> Current as a function of time

With the switch in the down position, the voltage across the capacitor causes a current. This current discharges the capacitor, so q decreases.

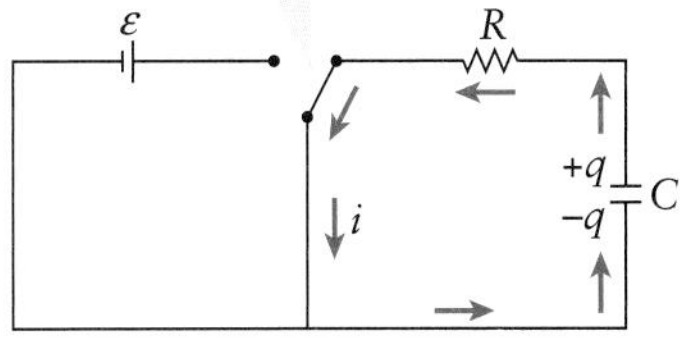

Figure 18-19 **Discharging a series *RC* circuit** When the switch in the circuit of Figure 18-17 is moved to the down position, the discharging process begins.

Capacitor charge and current in a discharging series RC circuit (18-30)

 In these equations, $t = 0$ is when the switch in Figure 18-19 is moved to the down position.

Figure 18-20 Charge and current in a discharging series *RC* circuit
Capacitor charge q starts at a maximum value and approaches zero; current i starts at a maximum (negative) value and approaches zero.

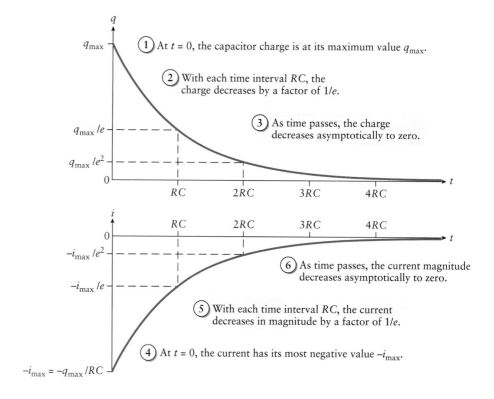

① At $t = 0$, the capacitor charge is at its maximum value q_{max}.

② With each time interval RC, the charge decreases by a factor of $1/e$.

③ As time passes, the charge decreases asymptotically to zero.

⑥ As time passes, the current magnitude decreases asymptotically to zero.

⑤ With each time interval RC, the current decreases in magnitude by a factor of $1/e$.

④ At $t = 0$, the current has its most negative value $-i_{max}$.

$-i_{max} = -q_{max}/RC$

 Go to Interactive Exercise 18-1 and 18-2 for more practice dealing with RC circuits

 Go to Interactive Exercise 18-3 for more practice dealing with time constants

The graphs in Figure 18-20 show the charge q and current i as given by Equations 18-30. Both functions are proportional to $e^{-t/RC}$, so both decrease by a factor of $1/e$ whenever time t increases by one time constant RC. So the value of RC determines both how rapidly the capacitor charges and how rapidly it discharges.

Example 18-11 A Charging Series *RC* Circuit

A 10.0-MΩ resistor is connected in series with a 5.00-μF capacitor. When a switch is thrown, these circuit elements are connected to a 24.0-V battery of negligible internal resistance. The capacitor is initially uncharged. (a) What is the current in the circuit immediately after the switch is moved so that charging begins? (b) What is the charge on the capacitor once it is fully charged? (c) Find the capacitor charge, current, power provided by the battery, power taken in by the resistor, and power taken in by the capacitor at $t = 50.0$ s. (d) When the capacitor is fully charged, find the total energy that has been delivered by the battery and the total energy that has been delivered to the capacitor.

Set Up

We are given $R = 10.0$ MΩ $= 10.0 \times 10^6$ Ω, $C = 5.00$ μF $= 5.00 \times 10^{-6}$ F, and $\varepsilon = 24.0$ V. Equations 18-27 tell us the capacitor charge and current at any time, including at $t = 0$ (when the switch is first closed) and $t \to \infty$ (long after the switch is closed, so the capacitor is fully charged). We'll use Equations 18-23 and 18-24 to find the power out of the battery and into the resistor and capacitor. (Equation 17-14 will help us in this.) In order to charge the capacitor to its maximum charge q_{max}, the total charge that must pass through the battery is q_{max}; we'll use this and Equation 17-6 to find the total energy delivered by the battery. Equation 17-17 tells us the total energy that is stored in the charged capacitor.

Capacitor charge and current in a charging series *RC* circuit:

$$q(t) = C\varepsilon(1 - e^{-t/RC})$$
$$i(t) = \frac{\varepsilon}{R}e^{-t/RC} \qquad (18\text{-}27)$$

Power for a circuit element:

$$P = iV \qquad (18\text{-}23)$$

Power for a resistor:

$$P = i^2 R = \frac{V^2}{R} \qquad (18\text{-}24)$$

Charge, voltage, and capacitance for a capacitor:

$$q = CV \qquad (17\text{-}14)$$

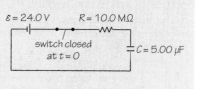

Electric potential difference related to electric potential energy difference:

$$\Delta V = \frac{\Delta U_{electric}}{q_0} \tag{17-6}$$

Electric potential energy stored in a capacitor:

$$U_{electric} = \frac{1}{2}qV = \frac{1}{2}CV^2 = \frac{q^2}{2C} \tag{17-17}$$

Solve

(a) Find the current at $t = 0$.

From the second of Equations 18-27, the current when the switch is first closed at $t = 0$ is

$$i(0) = \frac{\varepsilon}{R}e^{-(0)/RC} = \frac{\varepsilon}{R}e^0$$

Since any number raised to the power 0 equals 1, we have $e^0 = 1$ and

$$i(0) = i_{max} = \frac{\varepsilon}{R} = \frac{24.0\text{ V}}{10.0 \times 10^6\ \Omega}$$
$$= 2.40 \times 10^{-6}\text{ A} = 2.40\ \mu\text{A}$$

(b) Find the capacitor charge long after the switch is closed ($t \to \infty$).

The first of Equations 18-27 tells us the capacitor charge $q(t)$. As $t \to \infty$, the exponent $-t/RC \to -\infty$. Any number raised to the power $-\infty$ is zero, so

$$e^{-t/RC} \to 0$$

$$q(t) = C\varepsilon(1 - e^{-t/RC}) \to q_{max} = C\varepsilon(1 - 0) = C\varepsilon$$
$$= (5.00 \times 10^{-6}\text{ F})(24.0\text{ V})$$
$$= 1.20 \times 10^{-4}\text{ C} = 0.120\text{ mC}$$

(c) The time constant for this circuit is $RC = 50.0$ s, so we are actually being asked about the behavior of the circuit one time constant after the switch is closed. Use this to find charge q, current i, and the power out of or into each circuit element.

The time constant for this circuit is

$$RC = (10.0 \times 10^6\ \Omega)(5.00 \times 10^{-6}\text{ F}) = 50.0\text{ s}$$

so at $t = 50.0$ s, $t/RC = (50.0\text{ s})/(50.0\text{ s}) = 1.00$

From Equations 18-27,

$$q = C\varepsilon(1 - e^{-1.00}) = (5.00 \times 10^{-6}\text{ F})(24.0\text{ V})(1 - 0.368)$$
$$= 7.58 \times 10^{-5}\text{ C} = 0.632 q_{max}$$

$$i = \frac{\varepsilon}{R}e^{-1.00} = \frac{24.0\text{ V}}{10.0 \times 10^6\ \Omega}(0.368)$$
$$= 8.83 \times 10^{-7}\text{ A} = 0.368 i_{max}$$

The voltage across the battery is $\varepsilon = 24.0$ V, so from Equation 18-23 the power out of the battery is

$$P_{battery} = i\varepsilon = (8.83 \times 10^{-7}\text{ A})(24.0\text{ V}) \quad at\ t=50.0s$$
$$= 2.12 \times 10^{-5}\text{ W} = 21.2\ \mu\text{W}$$

From the first of Equations 18-24, the power into the resistor is

$$P_R = i^2R = (8.83 \times 10^{-7}\text{ A})^2(10.0 \times 10^6\ \Omega)$$
$$= 7.80 \times 10^{-6}\text{ W} = 7.80\ \mu\text{W}$$

Equation 17-14, $q = CV$, tells us that the voltage across the capacitor is $V = q/C$. Combining this with Equation 18-23 gives the power into the capacitor:

$$P_C = i\left(\frac{q}{C}\right) = (8.83 \times 10^{-7}\text{A})\left(\frac{7.58 \times 10^{-5}\text{ C}}{5.00 \times 10^{-6}\text{ F}}\right) \quad at\ t = 50.0s$$
$$= 1.34 \times 10^{-5}\text{ W} = 13.4\ \mu\text{W}$$

(d) Use the maximum charge stored by the capacitor, which is the total charge moved across the battery, and Equation 7-6 to calculate the change in electric potential energy imparted by the battery. Use Equations 17-7 to calculate the electric potential energy stored in the capacitor.

Note that the net power into the resistor and capacitor combined equals the power out of the battery:

$$P_R + P_C = 7.80 \ \mu\text{W} + 13.4 \ \mu\text{W} = 21.2 \ \mu\text{W} = P_{\text{battery}}$$

Long after the switch is closed, the total amount of charge that has passed through the battery and to the positive capacitor plate is $q_{\text{max}} = 1.20 \times 10^{-4}$ C. From Equation 17-6, the potential energy change that was imparted by moving this charge across the 24.0-V emf of the battery is

$$\Delta U_{\text{battery}} = q_{\text{max}} \ \varepsilon = (1.20 \times 10^{-4} \text{ C})(24.0 \text{ V})$$
$$= 2.88 \times 10^{-3} \text{ J} = 2.88 \text{ mJ}$$

The last of Equations 17-17 tells us the amount of energy that went into the capacitor to store charge q_{max} there:

$$U_C = \frac{q^2_{\text{max}}}{2C} = \frac{(1.20 \times 10^{-4} \text{ C})^2}{2(5.00 \times 10^{-6} \text{ F})}$$
$$= 1.44 \times 10^{-3} \text{ J} = 1.44 \text{ mJ}$$

So exactly one-half of the energy taken from the battery goes into the capacitor: $U_C = (1/2)\Delta U_{\text{battery}}$.

Reflect

Our results for charge q and current i in part (c) agree with Figure 18-18: After one time constant the capacitor charge has reached $[1 - (1/e)] = 0.632 = 63.2\%$ of its fully charged value q_{max}, and the current has decreased to $1/e = 0.368 = 36.8\%$ of its initial value i_{max}. Can you show that after 5 time constants ($t = 5RC = 250$ s) the charge will have reached 99.3% of q_{max} and the current will have decreased to just 0.674% of i_{max}?

The power calculations in part (c) show that all of the power extracted from the battery is accounted for: Part of the energy extracted from the battery goes into the resistor, and the rest goes into adding to the electric potential energy stored in the capacitor. We've shown this for a specific instant, but it's true at *all* times during the charging process.

These results from part (c) also help us understand our calculations in part (d): Only one-half of the energy extracted from the battery goes into the capacitor, so the other half must have gone into the resistor. This is a general result for charging *any* series RC circuit.

Axon (stained white) seen end-on.

Myelin (stained-blue) seen end-on.

Figure 18-21 A myelinated axon In vertebrates, special cells wrap themselves around some axons forming a thick insulating layer called myelin. This decreases the effective capacitance of the axon and increases the speed at which nerve signals can propagate along the length of the axon.

The physics of capacitors in circuits helps explain what happens in the propagation of action potentials along axons, which we discussed in Section 18-4. All biological membranes, including the axonal membrane, act like a capacitor. Under resting conditions, the inner surface of the membrane is negatively charged and the outer surface is positively charged (see Figure 18-7). As we described in Section 18-4, as an action potential propagates along an axon the resting charge distribution reverses: At the peak of an action potential, the inner surface of the membrane is positively charged and the outer surface is negatively charged. When the membrane returns to its initial charge state, it's analogous to a capacitor discharging as in the circuit of Figure 18-19. The value of C is the capacitance of a segment of axon membrane, and the value of R is the resistance of that segment to charge flowing through channels in the membrane. (This is different from the resistance we discussed in Section 18-4, especially Example 18-6. There we considered the *longitudinal* resistance for current flowing along the length of the axon; the value of R we're considering here is the *transverse* resistance to charges flowing across the axon membrane, perpendicular to the axon's length.)

The value of the time constant RC determines the rate of discharge: The smaller the value of RC, the faster the capacitance of the membrane discharges and the more rapidly the signal propagates along the axon. We saw in Section 18-4 that some invertebrates such as squid have large-diameter axons to allow action potentials to propagate quickly from the brain to the muscles that mediate escape reflexes. Another way to make signals propagate more rapidly along an axon is to decrease the capacitance C of the membrane, thus reducing the value of RC. In vertebrates, special cells wrap themselves around some axons, forming a thick insulating layer called *myelin* (Figure 18-21). The myelin acts as a capacitor between the surface of the axon and its environment. This additional capacitance is in series with the capacitance of the axon's cell membrane and so decreases the effective capacitance of the axon. (Recall from Section 17-7 that when capacitors are placed in series, the effective capacitance is less than that of any individual capacitor.)

Only certain sections of the axon are coated with myelin. The action potential in the nonmyelinated gaps, or *nodes*, produces a signal that propagates rapidly along the myelinated regions of the axon. The signal weakens in strength in the myelinated region where no action potential is generated, but is reinvigorated when it reaches the next node.

? Got the Concept? 18-9   Discharging an *RC* Circuit   ~brightness relates to power.~

The switch in the circuit of Figure 18-19 is moved to the down position, causing the capacitor to discharge. Suppose that the resistor is a light bulb and that its brightness is proportional to the power into the resistor. Compared to the brightness of the bulb when the switch is moved to the down position at $t = 0$, the brightness of the bulb at $t = RC$ is (a) the same; (b) $1 - (1/e^2) = 0.865$ as great; (c) $1 - 1/e = 0.632$ as great; (d) $1/e = 0.368$ as great; (e) $1/e^2 = 0.135$ as great.

~The i at t=RC is 0.368 but power is 0.135~

Take-Home Message for Section 18-7

✔ A circuit containing a resistor and a capacitor connected in series is a series *RC* circuit.

✔ Connecting the resistor and an uncharged capacitor to a source of emf causes a current that charges the capacitor. Switching the circuit to remove the source of emf causes the capacitor to discharge through the resistor. In either case, an exponential function describes how the current and the capacitor charge vary with time.

✔ The rate of charge or discharge in a series *RC* circuit depends on the time constant, equal to the product *RC*. In a time interval *RC*, the current decreases by a factor of $1/e$.

Key Terms

AC (alternating current)	emf	power
alternating current (AC)	equivalent resistance	resistance
ampere	exponential function	resistivity
battery	internal resistance	resistor
circuit	junction	series (resistors)
circuit element	Kirchhoff's junction rule	series *RC* circuit
current	Kirchhoff's loop rule	single-loop circuit
DC (direct current)	multiloop circuit	source of emf
direct current (DC)	ohm	time constant
drift speed	parallel (resistors)	

Chapter Summary

Topic	Equation or Figure	
Current: Electric charge flows around a circuit in response to an electric potential difference, such as that provided by a battery (a source of emf). Current is the rate of charge flow, and is related to the speed at which charges drift through the circuit. In ordinary metals the moving charges are negatively charged electrons, but it's conventional to take the direction of the current to be the direction in which positive charges would flow.	Current in a circuit · Amount of **charge** that flows past a certain point in the circuit $$i = \frac{\Delta q}{\Delta t}$$ **Time** required for that amount of charge to flow past that point	(18-1)
	Current in a wire · Number of moving charges per unit volume $$i = nAev_{\text{drift}}$$ Speed at which charges drift through the wire · Cross-sectional area of the wire · Magnitude of the charge on each moving charge	(18-3)

Resistance: Resistivity is a measure of how difficult it is for charges to flow through a material. The resistance of a wire or circuit element depends on the dimensions of the object as well as the resistivity of the material of which the object is made. Resistance and resistivity are not constants, but vary with temperature. A resistor is a circuit element whose most important property is its resistance.

Resistance of a wire | Resistivity of the material of which the wire is made

$$R = \frac{\rho L}{A}$$ Length of the wire

Cross-sectional area of the wire

(18-8)

Voltage, current, and resistance: In order for current to exist in a conductor with resistance, there must be a potential difference (voltage) between the ends of the conductor. For a given resistance, a greater voltage produces a greater current.

Potential difference between the ends of a wire

$$V = iR$$

Current in the wire   Resistance of the wire

(18-9)

Rules for circuits: Kirchhoff's loop rule states that the sum of the changes in electric potential around a closed loop in a circuit must equal zero. Kirchhoff's junction rule states that sum of the currents into a junction equals the sum of the currents out of it. These rules allow us to analyze the currents and voltages in electric circuits.

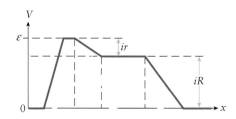

(Figure 18-8b)

The potential increases by ε from the negative terminal to the positive terminal of the emf.

The potential drops by ir across the internal resistance and drops by iR across the resistor.

Resistors in series and parallel: A collection of resistors in a circuit behaves as though it were a single resistor, with an equivalent resistance R_{equiv}. The equivalent resistance is different for resistors in series (all of which carry the same current) than for

Equivalent resistance of N resistors in series | Resistances of the individual resistors

same(i)

$$R_{\text{equiv}} = R_1 + R_2 + R_3 + ... + R_N$$

(18-11)

The equivalent resistance of N resistors in series is the sum of the individual resistances.

resistors in parallel (all of which have the same voltage).

The reciprocal of the equivalent resistance of N resistors in parallel is the sum of the reciprocals of the individual resistances.

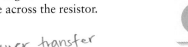

$$\frac{1}{R_{\text{equiv}}} = \frac{1}{R_1} + \frac{1}{R_2} + \frac{1}{R_3} + \dots + \frac{1}{R_N} \tag{18-20}$$

Equivalent resistance of N resistors in parallel

Resistances of the individual resistors

Power in circuits: Power is the rate of energy transfer. In an electric circuit, energy flows from a source of emf into the other circuit elements. The power into a resistor can be expressed in terms of the resistance and either the current through the resistor or the voltage across the resistor.

Power produced by or transferred into a circuit element

$$P = iV \tag{18-23}$$

Current through the circuit element

Voltage (absolute value) across the circuit element

Source of emf: Power flows out of the source and into the moving charges.
Resistor: Power flows out of the charges and into the resistor.

Power into a resistor   Current through the resistor

$$P = i^2 R = \frac{V^2}{R} \tag{18-24}$$

Voltage across the resistor

Resistance of the resistor

Series RC circuits: If a resistor R and capacitor C are connected in series to a source of emf, the capacitor charge increases toward a maximum value. The current in the circuit (which carries the charge to the capacitor) begins with a large value and gradually decreases to zero. If the source is taken out of the circuit, the capacitor discharges. The current in the circuit now carries the charge away from the capacitor, and again gradually decreases to zero. The rate at which the current decreases depends on the time constant of the circuit, equal to the product RC.

With the switch in the up position, the emf causes charge to flow. This current charges the capacitor, so *q increases.*

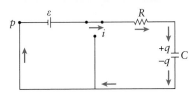

(Figure 18-17b)

With the switch in the down position, the voltage across the capacitor causes a current. This current discharges the capacitor, so *q decreases.*

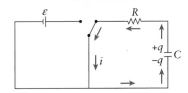

(Figure 18-19)

Answer to What do you think? Question

(b) When electrons pass through the battery, they gain electric potential energy. They lose most of this potential energy as they pass through the filament of the light bulb, and this lost energy goes into heating the filament and causing it to glow. The electrons lose the rest of the potential energy as they pass through the wires that connect the battery and light bulb. (This is a relatively small amount of energy because the connecting wires have low resistance.) If electrons lost *kinetic* energy at a certain point in the circuit, they would slow down and create an electron "traffic jam." The electric repulsion between electrons prevents such "jams" from happening, so the electrons must maintain the same average speed and so the same average kinetic energy as they travel around the circuit.

Answers to Got the Concept? Questions

18-1 (c) The current i in a simple circuit (in which the moving charges travel around a single loop) has the same value at all points in the circuit. This is true whether the thickness of the wire varies or is the same at all points.

18-2 (a) As we saw in the answer to the previous Got the Concept? question, the current i has the same value at all points in the circuit. Equation 18-3 shows that for this to be true, the product Av_{drift} (the cross-sectional area of the wire multiplied by the drift speed) must also have the same value at all points. (The quantity n, the number of mobile charges per unit volume, depends only on what the wire is made of, not on its dimensions.) The thinner wire has a radius r that is half as great and a cross-sectional area $A = \pi r^2$ that is one-quarter as great as the rest of the wire. So the drift speed of the electrons must be four times greater in the thinner part of the wire.

18-3 (b) For an ohmic material, the current and voltage are directly proportional, so doubling the voltage should cause a doubling of current. That is *not* the case for this material: Doubling the voltage from 2.0 V to 4.0 V increases the current by a factor of only $(0.28\,\text{A})/(0.15\,\text{A}) = 1.9$, and doubling the voltage again from 4.0 V to 8.0 V increases the current by a factor of only $(0.50\,\text{A})/(0.28\,\text{A}) = 1.8$. So this material is nonohmic.

18-4 (e) As in Example 18-6, the resistance of wire of resistivity ρ, length L, and diameter d is $R = 4\rho L/\pi d^2$. This says that the resistance is inversely proportional to the square of the diameter. So if the diameter is increased by a factor of 4 while leaving the resistivity and length the same, the resistance will change by a factor of $1/4^2 = 1/16$.

18-5 (c), (a), (b), (d) Combining resistors in series results in a circuit with larger equivalent resistance; combining them in parallel results in a smaller equivalent resistance. Therefore circuits (a) and (c) have higher resistance than circuits (b) and (d). For the same reason, the equivalent resistance of (c) with three resistors in series is higher than that of circuit (a) with two resistors in series, and the equivalent resistance of circuit (b) with two resistors in parallel is higher than that of circuit (d) with three resistors in parallel. So in order of equivalent resistance from highest to lowest, the circuits are (c), (a), (b), and (d). This conclusion doesn't depend on how the value of R_3 compares to the values of R_1 or R_2: Adding any resistance in series to a group of resistors increases the equivalent resistance, and adding any resistance in parallel to a group of resistors decreases the equivalent resistance.

18-6 (d), (b), (a), (c) The larger the equivalent resistance, the smaller the current for a fixed voltage. So a ranking of the circuits in order of decreasing current is just the opposite of a ranking in order of decreasing equivalent resistance. So the ranking in this question is the opposite of the ranking in Got the Concept? 18-5.

18-7 (c) Placing two sources of emf in series is equivalent to increasing the emf from ε to 2ε. This doubles the voltage V across the resistor, which from $V = iR$ (Equation 18-9) means that the current i through the resistor doubles as well. From Equations 18-24, $P = i^2R = V^2/R$, the power into the resistor is proportional to either the square of the current i or the square of the resistor voltage V. So the power into the resistor increases by a factor of $2^2 = 4$. A device such as a flashlight or television remote control that uses two batteries uses four times as much power as one with a single battery; a device with four batteries provides $4^2 = 16$ times as much power.

18-8 $P_A = P_E > P_D > P_B = P_C$ The full current that passes through the source also passes through resistors A and E. This current divides up in order to pass through two branches—one with resistors B and C, one with resistor D—so only a fraction of the full current passes through resistors B, C, and D. Since the power into a resistor is given by $P = i^2R$ (the first of Equations 18-24), it follows that the same power goes into A and E ($P_A = P_E$), and that this is greater than the power into any of the other three resistors. The voltage across the branch with resistors B and C is the same as that across the branch with resistor D (the branches are in parallel). All of this voltage is across resistor D, but only half of this voltage is across resistor B and half across resistor C. From the second of Equations 18-24, $P = V^2/R$, this means that there is more power into D than into B or C, but equal amounts into B and C (so $P_D > P_B = P_C$). So $P_A = P_E > P_D > P_B = P_C$. If the resistors are light bulbs, their ranking by brightness will be the same as this ranking by power.

18-9 (e) The second of Equations 18-30 shows that the current in a discharging series RC circuit is proportional to $e^{-t/RC}$, so at $t = RC$ the current has decreased by a factor of $e^{-1} = 1/e = 0.368$ compared to its value at $t = 0$ (when $e^{-t/RC} = e^0 = 1$). The power into a resistor is $P = i^2R$ (Equations 18-24). Since the power is proportional to the square of i, and i is $1/e$ as great as at $t = 0$, the power is $1/e^2 = (0.368)^2 = 0.135$ as great as at $t = 0$.

Questions and Problems

In a few problems, you are given more data than you actually need; in a few other problems, you are required to supply data from your general knowledge, outside sources, or informed estimate.

Interpret as significant all digits in numerical values that have trailing zeros and no decimal points.

For all problems, use $g = 9.80\,\text{m/s}^2$ for the free-fall acceleration due to gravity. Neglect friction and air resistance unless instructed to do otherwise.

• Basic, single-concept problem

•• Intermediate-level problem, may require synthesis of concepts and multiple steps

••• Challenging problem

SSM *Solution is in Student Solutions Manual*

Conceptual Questions

1. •We distinguish the direction of current in a circuit. Why don't we consider it a vector quantity?

2. •Is current dissipated when it passes through a resistor? Explain your answer.

3. •We justified a number of electrostatic phenomena by the argument that there can be no electric field in a conductor. Now we say that the current in a conductor is driven by a potential difference and thus there is an electric field in the conductor. Is this statement a contradiction? SSM

4. •Two wires, A and B, have the same physical dimensions but are made of different materials. If A has twice the resistance of B, how do their resistivities compare?

5. •The average drift velocity of electrons in a wire carrying a steady current is constant even though the electric field within the wire is doing work on the electrons. What happens to this energy?

6. •Under ordinary conditions, the drift speed of electrons in a metal is around 10^{-4} m/s or less. Why doesn't it take a long time for a light bulb to come on when you flip the wall switch that is several meters away?

7. •For a given source of constant voltage, will more heat develop in a large external resistance connected across it or a small one? SSM

8. •An ammeter measures the current through a particular circuit element. (a) How should it be connected with that element, in parallel or in series? (b) Should an ammeter have a very large or a very small resistance? Why?

9. •**Biology** When a bird lands on a high voltage wire and grabs the wire with its feet, will the bird be electrocuted? Explain your answer.

10. •If the only voltage source you have is 36 V, how could you light some 6-V light bulbs without burning them out?

11. ••Many ordinary strings of Christmas-tree lights contain about 50 bulbs connected in parallel across a 110 V line. Sixty years ago most strings contained 50 bulbs connected in series across the line. What would happen if you could put one of the old-style bulbs into a modern Christmas-tree light set? (The light sockets are made differently to prevent this.) SSM

12. •**Biology** Explain how an action potential is generated.

13. •Give a simple physical explanation for why the charge on a capacitor in an *RC* circuit can't be changed instantaneously.

14. •(a) Does the time required to fully charge a capacitor through a given resistor with a battery depend on the voltage of the battery? (b) Does it depend on the total amount of charge to be placed on the capacitor? Explain your answers.

Multiple-Choice Questions

15. •If a current-carrying wire has a cross-sectional area that gradually becomes smaller along the length of the wire, the drift velocity
  A. increases along the length of the wire.
  B. decreases along the length of the wire.
  C. remains the same along the length of the wire.
  D. increases along the length of the wire if the resistance increases too.
  E. decreases along the length of the wire if the resistance decreases too. SSM

16. •What causes an electric shock?
  A. current
  B. voltage
  C. both current and voltage
  D. resistance and current
  E. resistance and voltage

17. •Two copper wires have the same length, but one has twice the diameter of the other. Compared to the one that has the smaller diameter, the one that has the larger diameter has a resistance that is
  A. larger by a factor of 2.
  B. larger by a factor of 4.
  C. the same.
  D. smaller by a factor of 1/2.
  E. smaller by a factor of 1/4.

18. ••A wire has a length L and a resistance R. It is stretched uniformly to a length of $2L$. The resistance of the wire after it has been stretched is
  A. $(1/4)\, R$.
  B. $(1/2)\, R$.
  C. R.
  D. $2\, R$.
  E. $4\, R$.

19. •When a thin wire is connected across a voltage of 1 V, the current is 1 A. If we connect the same wire across a voltage of 2 V, the current is
  A. $(1/4)$ A.
  B. $(1/2)$ A.
  C. 1 A.
  D. 2 A.
  E. 4 A. SSM

20. •When a wire that has a large diameter and a length L is connected across the terminals of an automobile battery, the current is 40 A. If we cut the wire to half of its original length and connect one piece that has a length $L/2$ across the terminals of the same battery, the current will be
  A. 10 A.
  B. 20 A.
  C. 40 A.
  D. 80 A.
  E. 160 A.

21. •A charge flows from the positive terminal of a 6-V battery, through a light bulb, and through the battery back to the positive terminal. The total voltage drop experienced by the charge is
  A. 1 V.
  B. 6 V.
  C. 0 V.
  D. −1 V.
  E. −6 V.

22. •When a second bulb is added in series to a circuit with a single bulb, the resistance of the circuit
  A. increases.
  B. decreases.
  C. remains the same.
  D. doubles.
  E. triples.

23. •When a bulb is added in parallel to a circuit with a single bulb, the resistance of the circuit
  A. increases.
  B. decreases.
  C. remains the same.
  D. doubles.
  E. triples. SSM

24. •If we use a 2-V battery instead of a 1-V battery to charge the capacitor shown in Figure 18-22, the time constant will
    A. be four times greater.
    B. double.
    C. remain the same.
    D. be half as much.
    E. be four times less.

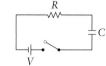

Figure 18-22 Problem 24

Estimation/Numerical Analysis

25. •Choose three appliances that you use on a daily basis and estimate the amount of electric current that is drawn by each.

26. •Estimate the current in a high voltage transmission line such as you might see in rural areas, sometimes close to remote interstate highways.

27. •Estimate the resistance of the electric coils in your toaster.

28. ••Estimate the electric energy usage of a citizen of a developed country such as Canada or the United States compared to a citizen of a developing nation.

29. •Estimate the current provided by the battery in a cellular phone.

30. •Estimate the electric current of a typical lightning bolt.

31. •Estimate the value of resistors and capacitors that are used in a common flash attachment of a disposable camera. SSM

Problems

18-1 Life on Earth and our technological society are only possible because of charges in motion

18-2 Electric current equals the rate at which charge flows

32. •A steady current of 35 mA exists in a wire. How many electrons pass any given point in the wire per second?

33. •A light bulb requires a current of 0.50 A to emit a normal amount of light. If the light is left on for 1 h, how many electrons pass through the bulb?

34. •During a thunderstorm, a lightning bolt carries current between a cloud and the ground below. If a lightning bolt transports a total charge of 80 C in 0.001 s, what is the magnitude of the average current?

35. •A synchrotron radiation facility consists of a circular ring that has a 40.0-m radius and creates an electron beam with a current of 487 mA when the electrons have a speed approximately equal to the speed of light. How many electrons pass a given point in the accelerator per hour? SSM

36. ••A copper wire that has a diameter of 2.00 mm carries a current of 10.0 A. Assuming that each copper atom contributes one free electron to the metal, find the drift velocity of the electrons in the wire. The molar mass of copper is 63.5 g/mol and the density of copper is 8.95 g/cm^3.

37. •Biology 2.00 × 10^{-3} mol of potassium ions pass through a cell membrane in 4.00 × 10^{-2} s. (a) Calculate the electric current and (b) describe its direction relative to the motion of the potassium ions.

18-3 The resistance to current through an object depends on the object's resistivity and dimensions

38. •If a copper wire has a resistance of 2.00 Ω and a diameter of 1.00 mm, how long is it?

39. •Calculate the resistance of a piece of copper wire that is 1.00 m long and has a diameter of 1.00 mm.

40. ••The resistance ratio of two conductors that have equal cross-sectional areas and equal lengths is 1:3. What is the ratio of the resistivities of the materials from which they are made?

41. ••An 8.00-m-long length of wire has a resistance of 4.00 Ω. The wire is uniformly stretched to a length of 16.0 m. Find the resistance of the wire after it has been stretched. SSM

42. ••When 120 V is applied to the filament of a 75-W light bulb, the current drawn is 0.63 A. When a potential difference of 3.0 V is applied to the same filament, the current is 0.086 A. Is the filament made of an ohmic material? Explain your answer.

43. •A power transmission line is made of copper that is 1.80 cm in diameter. What is the resistance of 1 mi of the line?

18-4 Resistance is important in both technology and physiology

44. •A common resistor used in a class lab has a color code shown in Figure 18-23. Write the numerical value of the resistor and give its *tolerance* (the uncertainty in the stated value). You may need to look up the resistor code using your favorite online search engine.

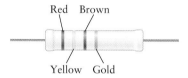

Red Brown

Yellow Gold

Figure 18-23 Problem 44

45. ••There is a current of 112 pA when a certain potential is applied across a certain resistor. When that same potential is applied across a resistor made of the identical material but 25 times longer, the current is 0.044 pA. Compare the effective diameters of the two resistors.

46. •What are the first three color codes depicting a resistor with a value of 20 GΩ? You may need to look up the resistor code using your favorite online search engine.

47. ••Electric cables often consist of a large number of strands of conducting wire. Due to variances in manufacturing, cables from a certain company can have between 850 and 950 separate strands, each with a mean diameter of 0.72 ± 0.07 mm. Determine the ratio of the (a) lowest and (b) highest possible resistance of one of these cables to the average resistance of many individual strands of conducting wire. SSM

48. ••A certain flexible conducting wire changes shape as environmental variables, such as temperature, change. If the diameter of the wire increases by 25% while the length decreases by 12%, by what factor does the resistance of the wire change?

49. ••Biology Cell membranes contain channels that allow ions to cross the phospholipid bilayer. A particular K$^+$ channel carries a current of 1.9 pA. How many K$^+$ ions pass through it in 1.0 ms?

50. ••Biology Cell membranes contain channels that allow K$^+$ ions to leak out. Consider a channel that has a diameter of 1.0 nm and a length of 10 nm. If the channel has a resistance of 18 GΩ, what is the resistivity of the solution in the channel?

18-5 Kirchhoff's rules help us to analyze simple electric circuits

51. •If a light bulb draws a current of 1.0 A when connected to a 12-V circuit, what is the resistance of its filament?

52. •If a flashlight bulb has a resistance of 12.0 Ω, how much current will the bulb draw when it is connected to a 6.0-V circuit?

53. •A light bulb has a resistance of 8.0 Ω and a current of 0.5 A when it is connected to a voltage. At what voltage is it operating?

54. •Calculate the voltage difference $V_A - V_B$ in each of the situations shown in **Figure 18-24**.

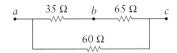

(a)

(b)

(c)

(d)

Figure 18-24 Problem 54

55. •An 18.0-Ω resistor and a 6.00-Ω resistor are connected in series. What is the equivalent resistance of the resistors?

56. •An 18.0-Ω resistor and a 6.00-Ω resistor are connected in parallel. What is the equivalent resistance of the resistors?

57. ••A 9.00-Ω resistor and a 3.00-Ω resistor are connected in series across a 9.00-V battery. Find (a) the current through each resistor and (b) the voltage drop across each resistor. SSM

58. ••A 9.00-Ω resistor and a 3.00-Ω resistor are connected in parallel across a 9.00-V battery. Find (a) the current through each resistor and (b) the voltage drop across each resistor.

59. ••A potential difference of 3.6 V is applied between points a and b in **Figure 18-25**. Find (a) the current in each of the resistors and (b) the total current the three resistors draw from the power source.

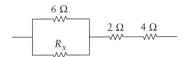

Figure 18-25 Problem 59

60. ••The four resistors in **Figure 18-26** have an equivalent resistance of 8 Ω. Calculate the value of R_x.

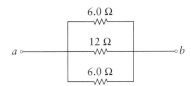

Figure 18-26 Problem 60

61. ••Find the equivalent resistance of the combination of resistors shown in **Figure 18-27**. SSM

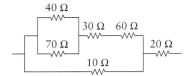

Figure 18-27 Problem 61

62. ••A metal wire of resistance 48-Ω is cut into four equal pieces that are then connected side by side to form a new wire which is one-quarter of the original length. What is the resistance of the new wire?

63. ••Two resistors A and B are connected in series to a 6.0-V battery; the voltage across resistor A is 4.0 V. When A and B are connected in parallel across a 6.0-V battery, the current through B is 2.0 A. What are the resistances of A and B? The batteries have negligible internal resistance.

64. ••A potential difference of 7.50 V is applied between points a and c in **Figure 18-28**. (a) Find the difference in potential between points b and c. (b) Is the current through the 60-Ω resistor larger or smaller than that through the 35-Ω resistor? Why?

a ———35 Ω——— b ———65 Ω——— c

60 Ω

Figure 18-28 Problem 64

18-6 The rate at which energy is produced or taken in by a circuit element depends on current and voltage

65. •A heater is rated at 1500 W. How much current does it draw when it is connected to a 120-V voltage source? SSM

66. •A 4.0-Ω resistor is connected to a 12-V voltage source. What is the power dissipated by the resistor?

67. •A coffeemaker has a resistance of 12 Ω and draws a current of 15 A. How much power does it use?

68. ••When connected in parallel across a 120-V source, two light bulbs consume 60 W and 120 W, respectively. What powers do the light bulbs consume if instead they are connected in series across the same source? Assume the resistance of each light bulb is constant.

69. •A transmission line carries 1200 A and has a resistance of 40 Ω. Calculate the energy lost per second due to *joule heating*, the transformation of electrical energy to thermal energy when current through wires encounters resistance.

70. •A stereo speaker has a resistance of 8 Ω. If the power output is 40 W, calculate the current passing through the speaker wires.

71. •If your local power company charges $0.11 per kW·h, what would it cost to run a 1500-W heater continuously during an 8-h night? SSM

72. ••A house is heated by a 24-kW electric furnace using resistance heating, and the local power company charges $0.10 per kW·h. The heating bill for January is $218. How long must the furnace have been running on an average January day?

18-7 A circuit containing a resistor and capacitor has a current that varies with time

73. •A 4.00-MΩ resistor and a 3.00-μF capacitor are connected in series with a power supply. What is the time constant for the circuit? SSM

74. •A capacitor of 20.0 μF and a resistor of 100 Ω are quickly connected in series to a battery of 6.00 V. What is the charge on the capacitor 0.00100 s after the connection is made?

75. •A 10.0-μF capacitor has an initial charge of 100.0 μC. If a resistance of 20.0 Ω is connected across it, what is the initial current through the resistor?

76. ••A 10.0-μF capacitor carries an initial charge of 80.0 μC. (a) If a resistance of 25.0-Ω is connected across it, what is the

initial current in the resistor? (b) What is the time constant of the circuit?

77. ••A 12.5-μF capacitor is charged to a potential of 50.0 V and then discharged through a 75.0-Ω resistor. (a) How long after discharge begins does it take for the capacitor to lose 90.0% of its initial (i) charge and (ii) energy? (b) What is the current through the resistor at both times in part (a)?

78. ••**Biology** A single ion channel is selectively permeable to K+ and has a resistance of 1.0 GΩ. During an experiment the channel is open for approximately 1.0 ms while the voltage across the channel is maintained at +80 mV with a patch clamp. How many K+ ions travel through the channel?

General Problems

79. ••How much power is dissipated in each resistor shown in Figure 18-29?

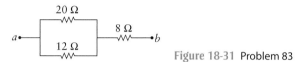

Figure 18-29   Problem 79

80. •Determine the current through each resistor in Figure 18-30.

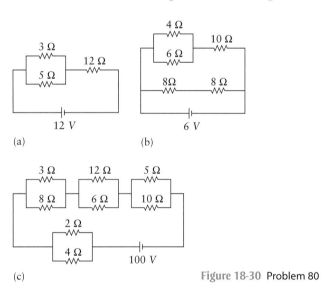

(a)   (b)

(c)   **Figure 18-30**   Problem 80

81. •••An electric heater consists of a single resistor connected across 110 V. It is used to heat 200.0 g of water in a coffee cup from 20°C to 90°C in 2.70 min. (a) Assuming that 90% of the energy drawn from the power source goes into heating the water, what is the resistance of the heater? (b) Assuming no other heat losses, how much longer will it take to heat your water if you had to power the water heater with your 12.0-V car battery?

82. •••Measurements made during severe electrical storms reveal that lightning bolts can carry as much as 30 C of charge and can travel between a cloud and the ground in time intervals of around 100 μs (and sometimes even shorter). Potential differences have been measured as high as 400 million volts. (a) What is the current in such a lightning strike? How does it compare with typical household currents? (b) What is the resistance of the air during such a strike? (c) How much energy is transferred during a severe strike? (d) What mass of water at 100°C could the lightning bolt in part (c) evaporate?

83. ••In the circuit shown in Figure 18-31, a potential difference of 5.00 V is applied between points a and b. Determine (a) the equivalent total resistance, (b) the current in each resistor, and (c) the power dissipated in each resistor. (d) How does the amount of power dissipated in the 20-Ω resistor compare to power dissipated in the other two?

Figure 18-31   Problem 83

84. ••**Biology** Giant electric eels can deliver a voltage shock of 500 V and up to 1.0 A of current for a brief time. For a snorkeler swimming in salt water, the resistance of her skin is negligible, so her body resistance is about 600 Ω. A current of about 500 mA can cause heart fibrillation and death if it lasts too long. (a) What is the maximum power a giant electric eel can deliver to its prey? (b) If the snorkeler is unfortunate enough to be struck by the eel, what current will pass through her body? Is this large enough to be dangerous? (c) What power does the snorkeler receive from the eel?

85. ••**Biology** The resistance of the body is made up of two basic parts: the resistance of the skin (see the following problem) and the resistance of the interior of the body. The internal resistance varies somewhat, but about 500 Ω measured between the two hands is typical. We can model this part of the body as a cylinder 1.6 m long and 14 cm in diameter. (a) What is the average internal resistivity of the body? (b) How much power will be delivered to the body of a person who accidentally grabs the ends of wires connected across a 110-V power source?

86. ••**Biology** Most of the resistance of the human body comes from the skin, as the interior of the body contains aqueous solutions that are good electrical conductors. For dry skin, the resistance between a person's hands is measured at typically 500 kΩ. The skin varies in thickness, but on the average it is about 2.0 mm thick. We can model the body between the hands as a cylinder 1.6 m long and 14 cm in diameter with the skin wrapped around it. (a) What is the resistivity of the skin? (b) Compare your answer with that found in part (a) of the previous problem for the internal resistivity of the body. How do you explain the fact that the resistance of the skin is about 1000 times greater than the internal resistance of the body, even though the resistivity of the skin is only about 60 times that of the interior of the body?

87. ••The capacitor in the flash of a disposable camera has a value of 160 μF. What is the resistance of the filament in the bulb if it takes 10 s to charge the capacitor to 80% of its maximum charge?

88. ••A discharging capacitor starts with a maximum charge and exponentially decays to zero. (a) How much time (in terms of the time constant RC) will it take before a discharging capacitor holds only 50% of the maximum charge? (b) What percentage of the original charge will the capacitor hold at a time that is 3 times the time constant?

89. ••A 3.0-μF capacitor is put across a 12-V battery. After a long time (~20 time constants), the capacitor is disconnected and placed in series through an open switch, with a 200-Ω resistor. (a) Determine the charge on the capacitor before it is discharged. (b) What is the initial current through the resistor

when the switch is closed? (c) At what time will the current reach 37% of its initial value?

90. ••You are working late on a project and find that you need a 75-Ω resistor, but unfortunately you only have a box of 50-Ω resistors on hand. All the electronics shops are closed, so you cannot buy the needed resistor. (a) How can you make a 75-Ω resistor using the resistors you have on hand? (b) How could you use your 50-Ω resistors to make a 60-Ω resistor? (c) Suppose you have only 50-μF capacitors. How could you use them to make a 75-μF capacitor and (d) to make a 60-μF capacitor?

91. ••A circuit probe is any device, such as a meter or an oscilloscope, that gives information about the circuit. Oscilloscopes and voltmeters are designed to measure the voltage across a circuit element, so they are connected across that element. Ideally, the probe should not disturb the circuit in any way by introducing resistance or capacitance, but in reality we cannot avoid some disturbance. A probe should disturb the circuit as little as possible. (a) When it is connected across a circuit element, is the probe in series or in parallel with that element? (b) Should a probe connected in this way have (i) a very large or a very small resistance, (ii) a very large or a very small capacitance? Explain your reasoning and use the properties of series and parallel connections to justify your answer.

92. ••For most purposes, we store electrical energy in batteries. But there are drawbacks to batteries: They release their energy rather slowly and are very damaging environmentally. Capacitors would be much cleaner for the environment and can be quickly recharged. Unfortunately they don't store much energy. (a) A new 1.5-V AAA battery has a "capacity" (*not* capacitance) of 1250 mA · h. What does this "capacity" actually represent? Express it in standard SI units. (b) How many joules of energy can be stored in the AAA battery? (c) At a steady current of 400 mA, how many hours will the AAA battery last? (d) How much energy can be stored in a typical 10-μF capacitor charged to a potential of 1.5 V? How does that compare to the energy stored in the AAA battery?

93. ••An *ultracapacitor* is a very high-capacitance device capable of storing much more energy than an ordinary capacitor. It is designed so that the spacing of the plates is around 1000 times smaller than in ordinary capacitors. Furthermore the plates contain millions of microscopic nanotubes, which increase their effective area 100,000 times. (a) If the plate separation and effective area of a 10-μF capacitor are changed as described above to make an ultracapacitor, what is its new capacitance? (b) How much energy does the ultracapacitor store if charged to 1.5 V? Compare this to the energy stored in an ordinary 10-μF capacitor at that potential. (c) Compare the energy stored in the ultracapacitor to that of the AAA battery in the previous problem. (d) If the ultracapacitor is to take 1.0 min to decrease to $1/e$ of its initial maximum charge, what resistance must it discharge through?

19

Magnetism

What do you think?

The aurora borealis ("northern lights") is caused by fast-moving, electrically charged subatomic particles ejected from the Sun. Earth's magnetic field exerts a magnetic force on these particles that steers them toward our planet's north magnetic pole. When these particles enter Earth's upper atmosphere, they collide with the atoms there and cause the atoms to emit an eerie glow. (A similar effect in the southern hemisphere is called the aurora australis.) In what direction are these subatomic particles moving when the magnetic force on them is strongest? (a) In the same direction as the magnetic field; (b) opposite to the magnetic field; (c) perpendicular to the magnetic field; (d) either (a) or (b); (e) the magnetic force doesn't depend on the direction of motion.

In this chapter, your goals are to:

- (19-1) Recognize that magnetic forces can act over large distances.
- (19-2) Recognize that magnetism is fundamentally an interaction between moving electric charges.
- (19-3) Calculate the magnitude and direction of the magnetic force on a charged particle.
- (19-4) Describe how a mass spectrometer uses magnetic fields to sort atoms according to their mass.
- (19-5) Calculate the magnetic force on a current-carrying wire.
- (19-6) Explain why a current loop in a uniform magnetic field experiences a net torque but zero net force.
- (19-7) Describe the principle of Ampère's law and how to use it.
- (19-8) Calculate the magnetic force that parallel current-carrying wires exert on each other.

To master this chapter, you should review:

- (5-6) Uniform circular motion.
- (8-6) The torque generated by a force.
- (16-5) The properties of electric field lines and the nature of electric dipoles.
- (18-2) The relationship between the current in a wire and the drift speed of charges in the wire.
- (18-6) Power into a current-carrying resistor.

19-1 Magnetic forces are interactions between two magnets

If you rub a balloon on your head, then move the balloon a few centimeters away from your head, the hairs on your head will stand up (see Figure 16-4). This phenomenon is a result of *electric* forces. During the rubbing, electrons are transferred between your hair and the balloon, giving one a net negative charge and the other a net positive charge; these attract each other, even over a distance, and this attraction makes your hair stand up. The effect is rather feeble, however, and only the finest hairs on your head will respond noticeably to these electric forces.

Figure 19-1a shows another force that acts at a distance like the electric force, but can be *much* stronger than that between the balloon and your hair. Certain objects called **magnets**, made of one of a handful of special materials such as iron, cobalt, and nickel, can exert strong **magnetic forces** on other magnets. (The name *magnet* comes

(a)
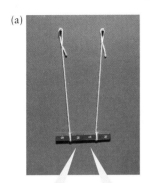

The north pole of the left-hand bar magnet...   ...attracts the south pole of the right-hand bar magnet.

(b)

The magnetized compass needle always points toward Earth's magnetic north pole.

(c)

The magnetic field required to pick up pieces of iron at a scrap yard comes from an electromagnet—a device that generates a magnetic field when there is current through its wires.

Figure 19-1 **Magnetic forces** (a) The force between two bar magnets. (b) A compass needle interacting with Earth's magnet. (c) A large electromagnet.

from the region of ancient Greece known as Magnesia, where these objects were discovered more than 2500 years ago.) Just as we use the umbrella term *electricity* to refer to interactions between electrically charged objects, we use the term **magnetism** to describe the interactions between magnets. One application of the magnetic force is the compass (**Figure 19-1b**). Earth's core acts like a giant magnet. The magnetic compass needle interacts with Earth's magnet in such a way that the compass needle always points toward the north pole. We will see that moving charges such as those in the large electromagnet in **Figure 19-1c** create magnetic fields.

Our goal in this chapter is to understand magnetism. We begin in the next section by investigating the properties of magnets, and realizing that they are created by moving charges. From there, we develop a full understanding of magnetic forces, which requires us to understand two things: (1) the nature of the magnetic field that moving charges *produce* and (2) how moving charges *respond* to the magnetic fields produced by other moving charges. It turns out to be easiest to look at the second of these first. We'll analyze the magnetic force that acts on a moving charge placed in a magnetic field, as well as the magnetic force on a wire that carries a current (a collection of charges moving in the wire) and is placed in a magnetic field. Later in the chapter we'll see how to calculate the magnetic field produced by a given collection of charges in motion.

Take-Home Message for Section 19-1

✔ Magnetic forces can act over large distances, much like electric forces.

19-2 Magnetism is an interaction between moving charges

Like the electric force, magnetic forces become stronger as the objects are moved closer together. However, the magnetic force is not simply a form of the electric force, because the attracting magnets in Figure 19-1a are *not* electrically charged: Their atoms are made of positively charged nuclei and negatively charged electrons, but their net charges are zero. Any electric forces between these magnets are very weak and are not responsible for the strong attraction shown in Figure 19-1a.

As we did for the electric force, we can explain how the magnetic force acts at a distance by invoking the idea of a field. Just as an electrically charged object sets up an electric field in the space around it, a magnet sets up a **magnetic field** in the space around it (**Figure 19-2**). A second magnet placed in this field experiences a magnetic force that depends on the magnitude and direction of the field. We use the symbol $\vec{B}$ for magnetic field. Unlike the electric field $\vec{E}$ of a point charge, which points either directly away from or directly toward the charge, the magnetic field $\vec{B}$ of a magnet points away from one end of a magnet toward the other end. These two ends are called the **magnetic poles** of the magnet, one of which is called the *north pole* and the other the *south pole*. (The meaning of the names is that Earth itself acts like a giant magnet. If allowed to swing freely, a magnet will orient itself so that its north

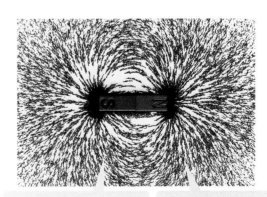

Bar magnets interact because each magnet produces a magnetic field, to which the other magnet responds.   Iron filings align with the magnetic field, making it visible.

Figure 19-2 The magnetic field of a bar magnet.

(a) These magnets attract.

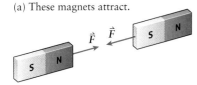

(b) These magnets repel.

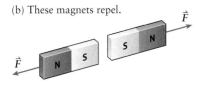

(c) If we cut a magnet in half, each half has a north and south pole.

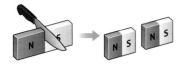

Figure 19-3 Bar magnets and magnetic poles (a) The opposite poles of these magnets attract. (b) Like poles of these magnets repel. (c) Every magnet has both a north and south pole; they cannot be separated.

pole is toward the north and its south pole is toward the south, as in Figure 19-1b.) By convention we choose the direction of the magnetic field $\vec{B}$ produced by a magnet to be such that $\vec{B}$ points away from the magnet's north pole and toward the magnet's south pole. This should remind you of an electric dipole (see Section 16-5) made up of a positive charge $+q$ and a negative charge $-q$, for which the electric field $\vec{E}$ points away from the positively charged end and toward the negatively charged end. Indeed, a magnet like that shown in Figure 19-2 is often called a **magnetic dipole**.

The magnetic forces between magnets can be either attractive or repulsive. If the north pole of one magnet is close to the south pole of a second magnet, the force attracts the two magnets toward each other (Figure 19-3a); if the north poles are close together or the south poles are close together, the force makes the two magnets repel each other (Figure 19-3b). This is very different from the behavior of electrically charged objects, but analogous to the way in which electric dipoles interact. However, there is an important distinction between electric and magnetic dipoles. You can take an electric dipole apart by separating its component positive and negative charges, but you *cannot* separate the north and south poles of a magnet. If you cut a magnet in half, each half has a north pole and a south pole (Figure 19-3c). The same is true no matter how many small pieces you cut a magnet into; a very small piece produces only a weak magnetic field, but is still a magnetic dipole with both a north pole and a south pole. (Physicists have speculated about the existence of *magnetic monopoles*, particles that have the properties of an isolated north or south magnetic pole. Many experiments have been performed to look for evidence of a magnetic monopole. None has been found.)

What is it about magnets that causes them to produce and respond to magnetic fields? And why is it impossible to separate their north and south poles? The answers to these questions were revealed by a set of crucial experiments in the nineteenth century. Figure 19-4a shows a version of one of these experiments. A coil of copper wire loops through a flat piece of clear plastic, and iron filings are spread over the plastic. Since copper is not magnetic, the filings lie wherever they were placed. But when the coil is connected to a source of emf that establishes a current through the coil, the filings line up just as they do around the magnet in Figure 19-2. This demonstrates that the moving charges that make up the current in the coil *produce* a magnetic field. The field points out of the coil at one end, which is the "north pole" of the coil, and points into the other end, which is the coil's "south pole."

We can verify that the coil produces a magnetic field by carrying out the experiment shown in Figure 19-4b, in which we've replaced one of the two magnets from Figure 19-3a with a coil. When the current in the coil is turned on, the coil and the magnet are attracted to each other. This can only happen if the current-carrying coil is indeed producing a magnetic field, and the magnet is responding to that field. If we flip the coil over as shown in Figure 19-4c, the coil and the magnet now repel each other.

(a) The magnetic field created by current in a coil.

(b) This coil and magnet attract.

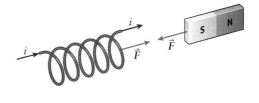

(c) This coil and magnet repel.

(d) These coils attract, just like two magnets.

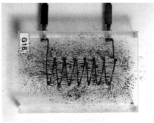

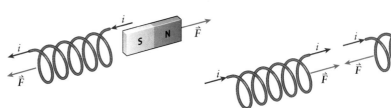

Figure 19-4 A current-carrying coil acts like a magnet (a) The field of a current-carrying coil is very similar to that of a bar magnet (Figure 19-1a). (b), (c) Such a coil interacts with a bar magnet just as though it were a magnet itself. (d) Two current-carrying coils interact like two bar magnets.

This is just what would happen if the coil were an iron magnet that we flipped over to interchange the north and south poles (see Figures 19-3a and 19-3b).

In order for the coil in Figures 19-4b and 19-4c to be attracted to or repelled from the magnet, it must also be true that the moving charges that make up the current in the coil respond to the magnetic field of the magnet. In other words, a current-carrying coil and a magnet are fundamentally the same, in that both objects *produce* as well as *respond to* magnetic fields. Figure 19-4d shows an experiment that verifies this: Two current-carrying coils with the same orientation attract each other, just as do two magnets with the same orientation.

The series of experiments shown in Figure 19-4 suggests that *magnetism is an interaction between charges in motion.* In order for two objects to exert magnetic forces on each other, there must be moving charges in both objects. We use the umbrella term **electromagnetism** to include both electricity and magnetism, since both involve interactions between charges. The distinction between electricity and magnetism is that two charges exert *electric* forces on each other whether or not the charges are moving, while two charges exert *magnetic* forces on each other only if both charges are in motion. (The objects in the experiments shown in Figure 19-4 are all electrically neutral. So there are no electric forces in those experiments, only magnetic forces.)

You may be wondering how magnetism can involve charges in motion, since the ordinary magnets shown in Figures 19-1, 19-2, and 19-3 are not attached to sources of emf and so do not carry currents. The answer is that there *are* charges in motion inside an ordinary iron magnet: The moving charges are the electrons within the atoms of the magnet, and their motions within the atom are like that of electrons in the circular coils of Figure 19-4. Each individual atom in a magnet produces only a tiny magnetic field. But because a large fraction of the atoms that make up the magnet are oriented so that their electron motions are in the same direction as in the surrounding atoms, their magnetic fields add together to make a substantial total field. This explains why cutting a magnet into pieces doesn't leave you with a separate north pole and south pole: Each piece is simply a smaller version of the original magnet.

? Got the Concept? 19-1 Electric and Magnetic Forces

Two identical objects each have a positive charge Q. One of the objects is held in place, while the other object is in motion a short distance from the first object. What kinds of forces do the two objects exert on each other? (a) Both electric and magnetic forces; (b) electric forces but not magnetic forces; (c) magnetic forces but not electric forces; (d) neither electric forces nor magnetic forces; (e) answer depends on how the second object is moving.

Take-Home Message for Section 19-2

✔ Magnetism is an interaction between moving charged particles. A moving charged particle produces a magnetic field, and a moving charged particle in a magnetic field experiences a force. A stationary charged particle neither produces nor responds to a magnetic field.

✔ In a magnet, charges are in motion at the atomic level to produce a magnetic field.

19-3 A moving point charge can experience a magnetic force

We'll begin our study of magnetic forces by considering the force on a single charged particle moving in a magnetic field $\vec{B}$. For example, this could be an electron moving in a current-carrying wire in the vicinity of a magnet, or a moving proton in Earth's upper atmosphere that is acted on by our planet's magnetic field.

Here's what experiments tell us about the magnetic force on such a moving charged particle:

(1) A charged particle can experience a force when placed in a magnetic field, but *only when it is moving.*

(2) The magnetic force depends on the direction of the charged particle's velocity relative to the direction of the magnetic field. The charged particle does not experience a magnetic force when its velocity is parallel to or opposite to the magnetic field.

(3) The magnitude of the force is proportional to the charge on the particle, to the magnitude of the magnetic field, and to the speed of the particle.

(4) The direction of the force is perpendicular to both the direction of motion of the charged particle and the direction of the magnetic field.

(5) The direction of the force depends on the sign of the charge.

Figure 19-5a shows the magnetic force on a particle with positive charge q moving with velocity $\vec{v}$ in the presence of a magnetic field $\vec{B}$. If θ is the angle between the velocity of the charged particle and the magnetic field, the magnitude of the magnetic force $\vec{F}$ is

Magnitude of magnetic force on a moving charged particle (19-1)

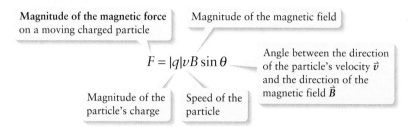

Magnitude of the magnetic force on a moving charged particle

Magnitude of the magnetic field

$$F = |q|vB\sin\theta$$

Angle between the direction of the particle's velocity $\vec{v}$ and the direction of the magnetic field $\vec{B}$

Magnitude of the particle's charge

Speed of the particle

The SI units of magnetic field are **tesla** (T). In Equation 19-1, if charge q is in coulombs (C), speed v is in meters per second (m/s), and magnetic field magnitude B is in tesla (T), the force F is in newtons (N). So 1 T is the same as $1\,\mathrm{N\cdot s/(C\cdot m)}$ or $1\,\mathrm{N/(A\cdot m)}$. (Recall that one ampere equals one coulomb per second: $1\,\mathrm{A} = 1\,\mathrm{C/s}$.) The strongest magnetic field most of us will experience directly is the roughly 2-T field inside a magnetic resonance imaging (MRI) scanner. Other magnetic fields in our environment are much weaker: The strength of Earth's magnetic field is about 5×10^{-5} T, the electrical impulses that drive the contraction and relaxation of your heart produce

(a) Magnetic force on a positive charge

Your right hand helps determine the direction of the magnetic force.

$q > 0$ $\vec{F}$ $\vec{B}$ θ $\vec{v}$

The force $\vec{F}$ is perpendicular to both the velocity $\vec{v}$ and the magnetic field $\vec{B}$.

(b) Magnetic force on a negative charge

If $q < 0$, the magnetic force is opposite to the direction given by your right hand.

$q < 0$ $\vec{B}$ θ $\vec{v}$ $\vec{F}$

Again, the force $\vec{F}$ is perpendicular to both the velocity $\vec{v}$ and the magnetic field $\vec{B}$.

(c) Magnetic force on a positive charge, $\theta = 90°$

For a given speed v, the magnetic force is greatest if the velocity $\vec{v}$ is perpendicular to the magnetic field $\vec{B}$.

$\vec{F}$ $\vec{B}$ $90°$ $\vec{v}$

(d) Magnetic force on a positive charge, $\theta = 0$ or $180°$

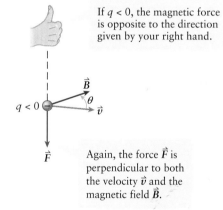

$\vec{v}$ $\vec{B}$ $\theta = 0$

$\vec{B}$ $\theta = 180°$ $\vec{v}$

In these cases, the magnetic force is zero.

Figure 19-5 Magnetic force on a moving charged particle The direction of the magnetic force on a particle with charge q depends on whether (a) q is positive or (b) q is negative. The magnitude of the force $F = |q|vB\sin\theta$ goes from (c) a maximum value if $\theta = 90°$ to (d) zero if $\theta = 0$ or $\theta = 180°$.

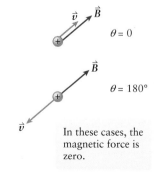

magnetic fields of about 5×10^{-11} T, and the magnetic fields in your brain are on the order of 10^{-13} T. An alternative unit for magnetic field is the *gauss* (G): 1 G = 10^{-4} T.

The direction of the force on the charge is given by a **right-hand rule** (Figure 19-5a). First extend the fingers of your right hand along the direction of the velocity $\vec{v}$, and orient your hand so that the palm is facing the magnetic field vector $\vec{B}$. With your right thumb extended as though you were giving the "thumbs-up" signal, swivel your hand as though you were slapping the magnetic field vector with your open palm. If the particle has a positive charge, so $q > 0$, your thumb points in the direction of the force $\vec{F}$. If the particle has a negative charge, so $q < 0$ (Figure 19-5b), the force $\vec{F}$ points in the direction opposite to that given by the right-hand rule. In either case, the force $\vec{F}$ is perpendicular to the velocity $\vec{v}$ *and* perpendicular to the magnetic field $\vec{B}$. If we draw the vectors $\vec{v}$ and $\vec{B}$ with their tails together, they form a plane; the force $\vec{F}$ is perpendicular to that plane, with a direction given by the right-hand rule. (We will encounter a number of other right-hand rules for magnetism in this chapter.)

Note that Equation 19-1 involves the magnitude or absolute value of the charge, $|q|$, which is always positive. So the magnitude F of the magnetic force does *not* depend on whether the charge is positive or negative.

 Watch Out! The magnetic force on a moving charged particle is never in the same direction as the magnetic field.

From Chapter 16, you're used to the idea that the electric force on a particle with charge q placed in an electric field $\vec{E}$ is $\vec{F}_{electric} = q\vec{E}$. This force is in the same direction as $\vec{E}$ if q is positive; if q is negative, it is in the direction opposite to $\vec{E}$.

Figures 19-5a and 19-5b show that the magnetic force on a moving charged particle is very different: This force is *perpendicular* to the direction of the magnetic field. Electric and magnetic forces are quite different from each other!

one is parallel and antiparallel and the other is perpendicular

Equation 19-1 tells us that the magnitude of the magnetic force on a moving charged particle depends on the angle θ between the directions of $\vec{v}$ and $\vec{B}$. For a given particle speed v, the magnetic force has its greatest magnitude if the velocity $\vec{v}$ is perpendicular to the magnetic field $\vec{B}$; then $\theta = 90°$, $\sin \theta = 1$, and $F = |q|vB$ (Figure 19-5c). If $\theta = 0$ or $\theta = 180°$, so the particle is moving either in the same direction as $\vec{B}$ or in the direction opposite to $\vec{B}$, $\sin \theta = 0$ and the force on the particle is *zero*. Thus a charged particle moving in the direction of the magnetic field, or in the direction opposite to the magnetic field, experiences no magnetic force (Figure 19-5d).

We've shown the vectors $\vec{v}$, $\vec{B}$, and $\vec{F}$ in perspective in Figure 19-5. Drawing in perspective isn't easy to do, so we'll often use a simple convention for drawing vectors that are pointed either into or out of the pages of this book (Figure 19-6). Think of a vector as an arrow, with a sharp point at its head and feathers at its tail. If a vector is directed perpendicular to the plane of the page and pointed toward you—that is, *out of* the page—we'll depict it with the symbol ⊙. The dot in the center of this symbol represents the sharp point of the arrowhead. If a vector is directed perpendicular to the plane of the page and pointed away from you—that is, *into* the page—we'll depict it with the symbol ⊗. The "X" in the center of this symbol represents the feathers on the tail of the arrow.

Go to Interactive Exercise 19-1 and 19-2 for more practice dealing with magnetic forces

(a)
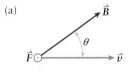

If $q > 0$, the force vector in this situation points out of the page (as indicated by the circle with a dot).

(b)
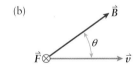

If $q < 0$, the force vector in this situation points into the page (as indicated by the circle with an X).

Figure 19-6 **Vectors out of the page and into the page** We use the symbols ⊙ and ⊗ to denote vectors that point out of the page and into the page, respectively.

Example 19-1 Magnetic Forces on a Proton and an Electron

At a location near our planet's equator, the direction of Earth's magnetic field is horizontal (that is, parallel to the ground) and due north, and the magnitude of the field is 2.5×10^{-5} T. Find the direction and magnitude of the magnetic force on a particle moving at 1.0×10^4 m/s if the particle is (a) a proton moving horizontally and due east, (b) an electron moving horizontally and due east, and (c) a proton moving horizontally in a direction 25° east of north. Recall that a proton has charge $e = 1.60 \times 10^{-19}$ C and an electron has charge $-e$.

Set Up

In each case we'll use Equation 19-1 to find the magnitude of the force on the moving proton or electron. The right-hand rule (Figure 19-5) will tell us the direction of the magnetic force.

Magnetic force on a moving charged particle:

$$F = |q|vB \sin \theta \qquad (19\text{-}1)$$

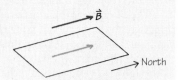

Solve

(a) The velocity vector $\vec{v}$ of the proton is perpendicular to the magnetic field vector $\vec{B}$, so in Equation 19-1 the angle $\theta = 90°$ and $\sin \theta = 1$. The vectors $\vec{v}$ and $\vec{B}$ both lie in a horizontal plane; the magnetic force vector $\vec{F}$ must be perpendicular to this plane, so $\vec{F}$ is vertical. The right-hand rule tells us that the force points vertically upward.

Charge on the proton:

$$q = e = 1.60 \times 10^{-19} \text{ C}$$

The proton velocity vector is perpendicular to the magnetic field vector ($\theta = 90°$), so the magnitude of the magnetic force on the proton is

$$F = |q|vB \sin \theta = evB \sin \theta$$
$$= (1.60 \times 10^{-19} \text{ C})(1.0 \times 10^4 \text{ m/s})(2.5 \times 10^{-5} \text{ T})\sin 90°$$
$$= 4.0 \times 10^{-20} \text{ N}$$

(b) The electron has the same magnitude of charge as the proton and the same velocity, so it experiences the same magnitude of magnetic force. But its charge is negative, so the direction of the magnetic force $\vec{F}$ on the electron is *opposite* to the direction given by the right-hand rule. Hence the force $\vec{F}$ points vertically downward.

Charge on the electron:

$$q = -e = -1.60 \times 10^{-19} \text{ C}$$

The electron velocity vector is perpendicular to the magnetic field vector ($\theta = 90°$), so the magnitude of the magnetic force on the electron is

$$F = |q|vB \sin \theta = evB \sin \theta$$
$$= (1.60 \times 10^{-19} \text{ C})(1.0 \times 10^4 \text{ m/s})(2.5 \times 10^{-5} \text{ T})\sin 90°$$
$$= 4.0 \times 10^{-20} \text{ N}$$

(c) The proton velocity $\vec{v}$ and the magnetic field $\vec{B}$ again both lie in a horizontal plane, but now the angle between these vectors is $\theta = 25°$. So the magnitude of the magnetic force is less than in part (a). The direction of the force is the same as in part (a), however.

The proton velocity vector is at an angle $\theta = 25°$ to the magnetic field vector, so the magnitude of the magnetic force on the proton is

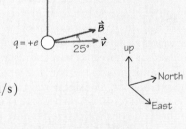

$$F = |q|vB \sin \theta = evB \sin \theta$$
$$= (1.60 \times 10^{-19} \text{ C})(1.0 \times 10^4 \text{ m/s})$$
$$\times (2.5 \times 10^{-5} \text{ T}) \sin 25°$$
$$= 1.7 \times 10^{-20} \text{ N}$$

This is smaller than in part (a) because $\sin 25° = 0.42$ compared to $\sin 90° = 1$.

Reflect

This example illustrates how the magnetic force on a moving charged particle depends on the direction in which the particle is moving. Note that the force magnitudes in parts (a), (b), and (c) are very small because a single electron or proton carries very little charge. These particles also have very small mass, however (1.67×10^{-27} kg for the proton and 9.11×10^{-31} kg for the electron), so the resulting *accelerations* are tremendous. Can you use Newton's second law to show that the proton in part (a) and the electron in part (b) have accelerations of 2.4×10^7 m/s^2 and 4.4×10^{10} m/s^2 respectively?

? Got the Concept? 19-2   A Proton in a Magnetic Field

A proton is fired into a region of uniform magnetic field pointing into the page (Figure 19-7). The proton's initial velocity is shown by the blue vector, which lies in the plane of the page. In which direction does the proton's trajectory bend? (a) Toward the top of the figure; (b) toward the bottom of the figure; (c) out of the figure; (d) into the figure; (e) answer depends on the speed of the proton.

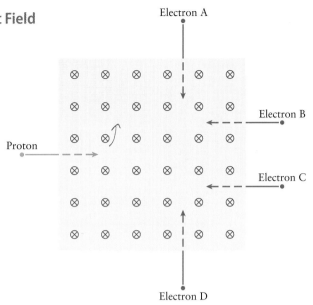

Bending trajectory.

Figure 19-7 Charged particles in a magnetic field How will each particle be deflected when it enters this region of uniform magnetic field?

? Got the Concept? 19-3   Electrons in a Magnetic Field

Four electrons, A, B, C, and D, are fired into a region of uniform magnetic field pointing into the page (Figure 19-7). The initial velocities of the electrons are shown by the red vectors, which lie in the plane of the page. Each electron will follow a trajectory that is bent by the magnetic force. For which electron could that trajectory lead to the point on the left-hand side where the proton enters the field region? (There may be more than one correct answer.) (a) Electron A; (b) electron B; (c) electron C; (d) electron D; (e) none of the electrons.

Take-Home Message for Section 19-3

✔ A charged particle moving in a magnetic field can experience a magnetic force. The magnitude of this force depends on the angle θ between the particle's velocity $\vec{v}$ and the direction of the magnetic field $\vec{B}$. The force is maximum if $\theta = 90°$ ($\vec{v}$ and $\vec{B}$ are perpendicular) and zero if $\theta = 0$ or $180°$ ($\vec{v}$ and $\vec{B}$ are either in the same direction or opposite directions).

✔ The magnitude of the magnetic force on a charged particle is proportional to the amount of charge, to the speed of the charged particle, and to the magnitude of the field.

✔ The direction of the magnetic force is perpendicular to both the particle's velocity $\vec{v}$ and the direction of the magnetic field $\vec{B}$, and depends on the sign of the charge. A right-hand rule helps tell us the force direction.

✔ The SI unit of magnetic field is the tesla (T).

19-4 A mass spectrometer uses magnetic forces to differentiate atoms of different masses

An important application of the magnetic force on a moving charged particle is the **mass spectrometer**, a device used by chemists to determine the masses of individual atoms and molecules. Figure 19-8 shows a simplified version of one type of mass spectrometer. A sample to be analyzed is first vaporized and its atoms and molecules are ionized by removing one or more electrons, so all of the ions have a positive charge. A beam of the moving ions then passes through a *velocity selector*, a region where there is *both* an electric field $\vec{E}$ and a magnetic field $\vec{B}$. The fields $\vec{E}$ and $\vec{B}$ are oriented perpendicular to the ion beam as well as perpendicular to each other. You should be able to show in Figure 19-8 that in this region the electric force on a positive ion acts to the left while the magnetic force acts to the right. An ion will pass through this region without deflection only if these two forces cancel so that the *net* force on the ion is zero. For an ion of positive charge q, the electric force has magnitude $F_{electric} = qE$ and the magnetic force has magnitude $F_{magnetic} = qvB \sin 90° = qvB$ (the ion velocity and the magnetic field are perpendicular, so $\theta = 90°$ in Equation 19-1). So the condition for an ion to continue through this region without being deflected is

(19-2)
$$qE = qvB \quad \text{or} \quad v = \frac{E}{B}$$

(speed of ions emerging from a velocity selector)

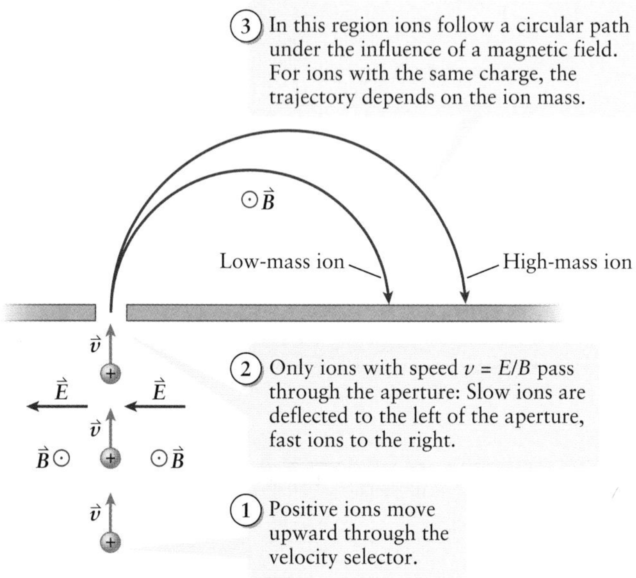

③ In this region ions follow a circular path under the influence of a magnetic field. For ions with the same charge, the trajectory depends on the ion mass.

$\odot \vec{B}$

Low-mass ion   High-mass ion

$\vec{v}$

$\vec{E}$ $\vec{E}$

$\vec{v}$

$\vec{B} \odot$ $\odot \vec{B}$

$\vec{v}$

② Only ions with speed $v = E/B$ pass through the aperture: Slow ions are deflected to the left of the aperture, fast ions to the right.

① Positive ions move upward through the velocity selector.

Figure 19-8 **A mass spectrometer** Electric and magnetic forces are used in this device to separate different ions according to their mass.

If an ion has a speed different from $v = E/B$, it will be deflected out of the beam. This is why we call this arrangement of fields a velocity selector: After passing through the $\vec{E}$ and $\vec{B}$ fields, the only ions that remain in the beam are those with a speed given by Equation 19-2.

As Figure 19-8 shows, the ion beam then passes into a second region with a magnetic field as before but no electric field. The beam deflects to the right as a result of the magnetic force and continues to deflect. That's because as the direction of the velocity vector changes, the direction of the magnetic force also changes in order to remain perpendicular to the ion velocity $\vec{v}$. Because $\vec{v}$ and $\vec{B}$ remain perpendicular, the angle θ in Equation 19-1 remains equal to 90° as the ions deflect and the magnitude of the magnetic force remains the same, $F_{magnetic} = qvB \sin 90° = qvB$. You probably recognize that what's going on here is exactly the situation that we described in Section 5-6. Because an ion is acted on by a force of constant magnitude that points perpendicular to the ion's velocity, the ion moves at a constant speed in a circular path—that is, in uniform circular motion. We can use Newton's second law to find the radius r of the circular path followed by an ion of charge q and mass m that moves with speed v. The acceleration in uniform circular motion has magnitude $a = v^2/r$, and the net force on the ion has magnitude qvB, so

(19-3)
$$qvB = \frac{mv^2}{r} \quad \text{or} \quad r = \frac{mv^2}{qvB} = \frac{mv}{qB}$$

(radius of circular path followed by charged particles in a uniform magnetic field)

Equation 19-3 is the key to understanding how the mass spectrometer works. All of the ions are moving at the same speed v (thanks to the velocity selector), and all are exposed to the same magnitude B of the magnetic field. But ions of different mass m (and the same charge q) will follow paths with different radii and so will land at different locations on the detector in Figure 19-8. By counting how many ions land at each location, we can learn about the composition of the sample being analyzed. Example 19-2 illustrates how this works.

Example 19-2 Measuring Isotopes with a Mass Spectrometer

Most oxygen atoms are the isotope ^{16}O ("oxygen-16"), which contains eight electrons, eight protons, and eight neutrons. The mass of this atom is 16.0 u, where 1 u = 1 atomic mass unit = 1.66×10^{-27} kg. The second most common isotope is ^{18}O ("oxygen-18"), which has two additional neutrons and so is more massive than an atom of ^{16}O: Each atom has a mass of 18.0 u. To determine the relative abundances of ^{16}O and ^{18}O, you send a beam of singly ionized oxygen atoms ($^{16}O^+$ and $^{18}O^+$, each with a charge $+e = +1.60 \times 10^{-19}$ C) through a mass spectrometer like that shown in Figure 19-8. The magnetic field strength in both parts of the spectrometer is 0.0800 T, and the electric field strength in the velocity selector is 4.00×10^3 V/m. (a) What is the speed of the beam that emerges from the velocity selector? (b) How far apart are the points where the ^{16}O and ^{18}O ions land?

Set Up

We use Equation 19-2 to find the speed v of ions that pass undeflected through the velocity selector, and Equation 19-3 to find the radius of the circular path that each isotope follows. The distance from where each ion enters the region of uniform magnetic field to where it strikes the detector is the diameter (twice the radius) of the circular path.

Speed of ions emerging from a velocity selector:

$$v = \frac{E}{B} \qquad (19\text{-}2)$$

Radius of circular path followed by charged particles in a uniform magnetic field:

$$r = \frac{mv}{qB} \qquad (19\text{-}3)$$

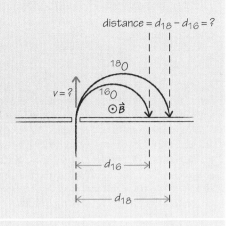

Solve

(a) Calculate the speed of ions that emerge from the velocity selector. Note that this speed does not depend on the charge or mass of the ion, so the $^{16}O^+$ and $^{18}O^+$ ions both emerge with this speed.

From Equation 19-2,

$$v = \frac{E}{B} = \frac{4.00 \times 10^3 \text{ V/m}}{0.0800 \text{ T}} = 5.00 \times 10^4 \frac{\text{V}}{\text{T} \cdot \text{m}}$$

From above, $1 \text{ T} = 1 \text{ N} \cdot \text{s}/(\text{C} \cdot \text{m})$, so $1 \text{ T} \cdot \text{m} = 1 \text{ N} \cdot \text{s}/\text{C}$. We also recall from Chapter 17 that $1 \text{ V} = 1 \text{ J/C} = 1 \text{ N} \cdot \text{m}/\text{C}$. So we can write the speed as

$$v = 5.00 \times 10^4 \frac{\text{V}}{\text{T} \cdot \text{m}} = 5.00 \times 10^4 \left(\frac{\text{N} \cdot \text{m}}{\text{C}} \right) \left(\frac{\text{C}}{\text{N} \cdot \text{s}} \right)$$

$$= 5.00 \times 10^4 \text{ m/s}$$

(b) Calculate the radius r and diameter d of the circular path followed by each isotope.

The masses of the two atoms are

For ^{16}O: $m_{16} = (16.0 \text{ u})(1.66 \times 10^{-27} \text{ kg/u}) = 2.66 \times 10^{-26} \text{ kg}$

For ^{18}O: $m_{18} = (18.0 \text{ u})(1.66 \times 10^{-27} \text{ kg/u}) = 2.99 \times 10^{-26} \text{ kg}$

The mass of each positive ion ($^{16}O^+$ and $^{18}O^+$) is slightly less than the mass of the neutral atom; the difference is the mass of one electron, which is $9.11 \times 10^{-31} \text{ kg} = 0.0000911 \times 10^{-26} \text{ kg}$. This difference is so small that we can ignore it.

For ^{16}O,

$$r_{16} = \frac{m_{16}v}{qB} = \frac{(2.66 \times 10^{-26} \text{ kg})(5.00 \times 10^4 \text{ m/s})}{(1.60 \times 10^{-19} \text{ C})(0.0800 \text{ T})}$$

$$= 0.104 \frac{\text{kg} \cdot \text{m}}{\text{T} \cdot \text{C} \cdot \text{s}}$$

Since $1 \text{ T} = 1 \text{ N} \cdot \text{s}/(\text{C} \cdot \text{m})$ and $1 \text{ N} = 1 \text{ kg} \cdot \text{m/s}^2$, it follows that $1 \text{ T} = \text{kg}/(\text{C} \cdot \text{s})$ and $1 \text{ T} \cdot \text{C} \cdot \text{s} = 1 \text{ kg}$. So for ^{16}O we have

$$r_{16} = 0.104 \frac{\text{kg} \cdot \text{m}}{\text{kg}} = 0.104 \text{ m}$$

$$d_{16} = 2r_{16} = 0.208 \text{ m} = 20.8 \text{ cm}$$

For ^{18}O,

$$r_{18} = \frac{m_{18}v}{qB} = \frac{(2.99 \times 10^{-26} \text{ kg})(5.00 \times 10^4 \text{ m/s})}{(1.60 \times 10^{-19} \text{ C})(0.0800 \text{ T})}$$

$$= 0.117 \text{ m}$$

$$d_{18} = 2r_{18} = 0.234 \text{ m} = 23.4 \text{ cm}$$

The distance between the positions where the ^{16}O and ^{18}O ions land is the difference between the two diameters.

The distance between where the ^{16}O and ^{18}O ions land is

$d_{18} - d_{16} = 23.4 \text{ cm} - 20.8 \text{ cm} = 2.6 \text{ cm}$

This is a substantial distance, so the mass spectrometer does a good job of separating the two isotopes.

Reflect

Experiments like these show that, on average, 99.8% of the oxygen atoms are ^{16}O and 0.2% are ^{18}O. (Making up a small fraction of a percent is a third isotope, ^{17}O.)

Measurements of the ratio of ^{18}O to ^{16}O are important to the science of *paleoclimatology*, the study of Earth's ancient climate. One way to determine the average temperature of the planet in the distant past is to examine ancient ice deposits in Greenland and Antarctica. These deposits endure for hundreds of thousands of years; deposits near the surface are more recent, while deeper deposits are older. The ice comes from ocean water that evaporated closer to the equator and then fell as snow in the far north or far south. Each molecule of water is made up of two hydrogen atoms and one oxygen atom, which could be ^{16}O or ^{18}O. A water molecule can more easily evaporate if it contains lighter ^{16}O than if it contains heavier ^{18}O, so the water that evaporated and fell on Greenland and Antarctica as snow contains an even smaller percentage of ^{18}O than ocean water. This deficiency becomes even more pronounced for colder climates. It has been shown that a decrease of one part per million of ^{18}O in ice indicates a 1.5°C drop in sea-level air temperature at the time it originally evaporated from the oceans.

Using mass spectrometers to analyze ancient ice from Greenland and Antarctica, paleoclimatologists have been able to determine the variation in Earth's average temperature over the past 160,000 years. They have also analyzed the amount of atmospheric carbon dioxide (CO_2) that was trapped in the ice as it froze. An important result of these studies is that higher levels of atmospheric CO_2 have gone hand-in-hand with elevated temperatures for the last 160,000 years, which is just what we would expect from our discussion of global warming in Section 14-7. In the same way, the tremendous increase in CO_2 levels in the past century due to burning fossil fuels has gone hand-in-hand with recent dramatic increases in our planet's average temperature.

? Got the Concept? 19-4 Ions in a Mass Spectrometer

When passed through the mass spectrometer of Example 19-2, which of the following ions would follow nearly the same path as a ^{16}O$^+$ ion? Assume that all ions are moving at the speed given by Equation 19-2. (a) A doubly charged oxygen-16 ion (^{16}O^{2+}); (b) a singly charged sulfur-32 ion (^{32}S$^+$); (c) a doubly charged sulfur-32 ion (^{32}S^{2+}); (d) more than one of these; (e) none of these.

Take-Home Message for Section 19-4

✔ In a mass spectrometer, a magnetic field causes ions of the same speed and charge but different masses to follow different paths.

✔ In a velocity selector, the only charged objects that pass through undeflected are those traveling at the speed for which the electric and magnetic forces have equal magnitudes but opposite directions.

19-5 Magnetic fields exert forces on current-carrying wires

The magnetic force that we've described may seem to apply only in certain very special circumstances. But in fact you use magnetic forces whenever you listen to recorded music through earbuds (**Figure 19-9a**). Within each earbud is a small but powerful magnet adjacent to a flexible plastic cone with a coil of wire attached to it (**Figure 19-9b**). This coil is connected through the earbud wires to your MP3 player, which sends the musical signal to the coil in the form of a varying electric current. Charges within the coil are thus set into motion, and these moving charges experience magnetic forces exerted by the magnet. These forces pull on the coil that contains the charges and on the plastic cone to which the coil is attached. We learned in Section 19-3 that the direction of magnetic force depends on the direction in which charges move. So the force on the charges,

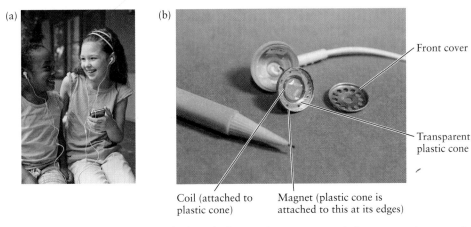

(a)   (b)

Front cover

Transparent
plastic cone

Coil (attached to   Magnet (plastic cone is
plastic cone)   attached to this at its edges)

Figure 19-9 **Earbud physics** (a) Earbuds and other speakers use magnetic forces to produce sound.
(b) Internal construction of an earbud.

coil, and plastic cone reverses whenever the current in the coil changes direction. The
result is that the plastic cone oscillates back and forth in response to the signal coming
from your MP3 player. This oscillation pushes on the surrounding air, producing a
sound wave—the sound of music—that travels to your eardrum.

Let's see how to find the magnitude of the magnetic force on a current-carrying
wire, such as a segment of the coil in Figure 19-9b. Figure 19-10 shows a straight wire
of length ℓ that carries a current i. A magnetic field $\vec{B}$ points at an angle θ to the direction of the current, and has the same value along the entire length of the wire. We
learned in Section 18-2 that the current in the wire is related to the speed at which individual charges drift through the wire:

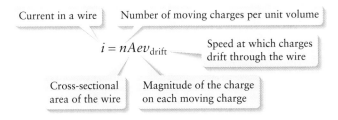

Current in a wire   Number of moving charges per unit volume

$$i = nAev_{drift}$$

Speed at which charges
drift through the wire

Cross-sectional   Magnitude of the charge
area of the wire   on each moving charge

From Equation 19-1, the magnetic force on an individual charge e moving through
the wire at speed v_{drift} has magnitude $ev_{drift}B \sin \theta$. Each such moving charge feels a
force of the same magnitude and in the same direction. The total number of such moving charges in the wire shown in Figure 19-10 is n (the number of moving charges per
unit volume) multiplied by the volume V of the wire, where $V = A\ell$ (the cross-sectional
area of the wire multiplied by its length). So the magnitude of the *net* magnetic force on
all of the moving charges in the wire is

$$F = (nA\ell)(ev_{drift}B \sin \theta)$$

If we rearrange the terms in this equation, we get

$$F = (nAev_{drift})\ell B \sin \theta \qquad (19\text{-}4)$$

The quantity in parentheses in Equation 19-4 is just the current i as given by Equation 18-3. So the magnitude of the magnetic force on a current-carrying wire is

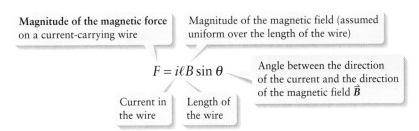

Magnitude of the magnetic force   Magnitude of the magnetic field (assumed
on a current-carrying wire   uniform over the length of the wire)

$$F = i\ell B \sin \theta$$

Angle between the direction
of the current and the direction
of the magnetic field $\vec{B}$

Current in   Length of
the wire   the wire

The force $\vec{F}$ on a current-carrying
wire in a magnetic field $\vec{B}$ is
perpendicular to both $\vec{B}$ and the
length of the wire. The direction
is given by a right-hand rule.

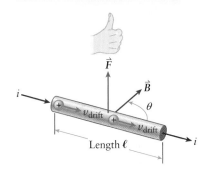

Figure 19-10 **Magnetic force on a
current-carrying wire** The force that
a magnetic field exerts on this wire is
the sum of the forces on all the moving
charges within the wire.

Current and drift speed
(18-3)

Magnitude of magnetic force on
a current-carrying wire
(19-5)

Just as for the magnetic force on a moving charged particle, a right-hand rule tells you the direction of the magnetic force on a current-carrying wire. Extend the fingers of your right hand in the direction of the current, and orient your hand so that the palm is facing the magnetic field vector $\vec{B}$. With your right thumb extended, swivel your hand as though you were slapping the magnetic field vector with your open palm. Your outstretched thumb points in the direction of the magnetic field exerted on the wire (Figure 19-10).

Equation 19-5 says that there is *no* magnetic force on the wire if the axis of the wire lies along the direction of the magnetic field $\vec{B}$: that is, if $\vec{B}$ points either in the same direction as the current ($\theta = 0$) or in the direction opposite to the current ($\theta = 180°$). In either case, $\sin \theta = 0$ and $F = 0$. For a given magnetic field of magnitude B, the force is greatest if $\vec{B}$ points perpendicular to the current so $\theta = 90°$ and $\sin \theta = \sin 90° = 1$. This idea is used in the design of the earbuds shown in Figure 19-9. The wires in each earbud are curved into a circular coil, not straight as in Figure 19-10. But we can treat the coil as being made up of many short segments, each of which is effectively a straight piece of wire. As Figure 19-11a shows, the magnetic field from the magnet in each earbud is perpendicular to each such segment. The forces on different segments are in different directions, but the vector sum of these forces is toward the magnet. If the current is in the reverse direction as in Figure 19-11b, all of the individual forces also reverse direction and the vector sum of the forces is away from the magnet. The current from the MP3 player continually reverses direction, so the coil and attached diaphragm oscillate back and forth, producing a sound wave.

In the following section we'll see how to use Equation 19-5 to help us understand how electric motors work. For now, let's see how to use a magnetic field to levitate a current-carrying wire.

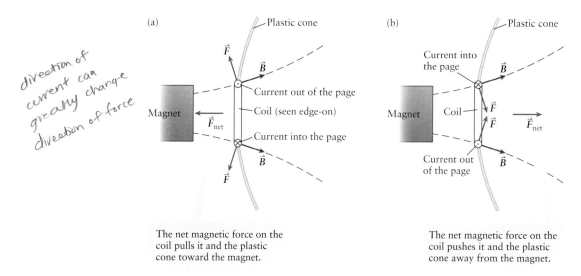

direction of current can greatly change direction of force

The net magnetic force on the coil pulls it and the plastic cone toward the magnet.

The net magnetic force on the coil pushes it and the plastic cone away from the magnet.

Figure 19-11 **Magnetic forces in an earbud** (a) The magnetic field produced by an earbud magnet is not uniform, so the field exerts a net force on the coil and the plastic cone to which it is attached. (b) Reversing the direction of the current reverses the direction of the net force.

Example 19-3 Magnetic Levitation

You set up a uniform horizontal magnetic field that points from south to north and has magnitude 2.00×10^{-2} T. (This is about 400 times stronger than Earth's magnetic field, but easily achievable with common magnets.) You want to place a straight copper wire of diameter 0.812 mm in this field, then run enough current through the wire so that the magnetic force will make the wire "float" in midair. This is called *magnetic levitation*. What minimum current is required to make this happen? The density of copper is 8.96×10^3 kg/m^3.

Set Up

In order to make the wire "float," there must be an upward magnetic force on the wire that just balances the downward gravitational force. We know from Equation 19-5 that to maximize the magnetic force, the current direction should be perpendicular to the magnetic field $\vec{B}$. The right-hand rule then shows that the current should flow from west to east so that the magnetic force is directed upward. We're not given the mass of the wire, but we can express the mass (and hence the gravitational force on the wire) in terms of its density. We're also not given the length of the wire; as we'll see, this will cancel out of the calculation.

Magnetic force on a current-carrying wire:

$$F = i\ell B \sin\theta \qquad (19\text{-}5)$$

Definition of density:

$$\rho = \frac{m}{V} \qquad (11\text{-}1)$$

Volume of a cylindrical wire of cross-sectional area A and length ℓ:

$$V = A\ell$$

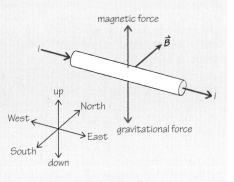

Solve

Write expressions for the two forces (magnetic and gravitational) that act on the wire.

Assume the wire has length ℓ. Since the current flows in a direction perpendicular to the magnetic field, the angle θ in Equation 19-5 is 90°. So the magnitude of the upward magnetic force is

$$F = i\ell B \sin 90° = i\ell B$$

The magnitude of the downward gravitational force on the wire of mass m is

$$w = mg$$

From Equation 11-1, the mass equals the density of copper multiplied by the volume of the wire:

$$m = \rho V = \rho A\ell$$

The cross-sectional area A of the wire is that of a circle of radius r:

$$A = \pi r^2 \quad \text{so} \quad m = \rho A\ell = \rho(\pi r^2)\ell$$

So the magnitude of the gravitational force is

$$w = mg = \rho(\pi r^2)\ell g$$

If the wire is floating in equilibrium, the upward magnetic force must just balance the downward gravitational force. Use this to solve for the required current i.

In equilibrium the net vertical force on the wire is zero:

$$F - w = 0, \text{ so } F = w \text{ and } i\ell B = \rho(\pi r^2)\ell g$$

The length ℓ of the wire cancels out of this equation (both the magnetic force and the gravitational force are proportional to ℓ):

$$iB = \rho(\pi r^2)g$$

Solve for the current i:

$$
\begin{aligned}
i &= \frac{\rho(\pi r^2)g}{B} \\[2mm]
&= \frac{(8.96 \times 10^3 \text{ kg/m}^3)(\pi)(0.406 \times 10^{-3} \text{ m})^2(9.80 \text{ m/s}^2)}{2.00 \times 10^{-2} \text{ T}} \\[2mm]
&= 2.27 \frac{\text{kg}}{\text{T} \cdot \text{s}^2} = 2.27 \text{ A}
\end{aligned}
$$

[Check on units: We know from Section 19-3 that $1\text{ T} = 1\text{ N}/(\text{A} \cdot \text{m})$, and we also know that $1\text{ N} = 1\text{ kg} \cdot \text{m/s}^2$. Therefore $1\text{ T} = 1\text{ kg}/(\text{A} \cdot \text{s}^2)$, and so $1\text{ kg}/(\text{T} \cdot \text{s}^2) = 1\text{ A}.$]

Reflect

A current of 2.27 A is relatively small, so this experiment in magnetic levitation is not too difficult to perform. Note that the required current i is inversely proportional to the magnitude B of the magnetic field. You can see that if you tried to make a wire "float" using Earth's magnetic field, which is about 1/400 as strong as the field used here, you would need to use an immense current of 400×2.27 A = 909 A. That's not practical because a current of that magnitude would cause the wire in this example to melt! (Recall from Equation 18-24 in Section 18-6 that the power into a resistor with resistance R that carries current i is $P = i^2 R$. The wire in this example has a small cross-sectional area, so its resistance R will be fairly large. The power delivered to the wire by a 909-A current will quickly increase its temperature to above the melting point of copper, 1085°C.) So you needn't worry about any of your electrical devices floating in midair when you turn on the current.

A practical application of magnetic levitation is train design. By using magnetic forces to make a train float just above the track, the rolling friction between the wheels and the track can be completely eliminated and very high speeds achieved. (Magnetic forces are also used to propel the train forward.) A train of this type in commercial operation in Shanghai, China reaches a top speed of 431 km/h (268 mi/h). Such train lines require special magnets and wires capable of sustaining very high currents.

? Got the Concept? 19-5 Direction of Magnetic Force on a Wire

A long, straight wire carrying a current can be placed in various orientations with respect to a constant magnetic field as shown in Figure 19-12a, b, c, d, e, and f. What is the direction of the force in each case? Give your answers in terms of the positive and negative x, y, and z directions. If the force is zero, say so. (In each case the positive z direction is out of the plane of the figures.)

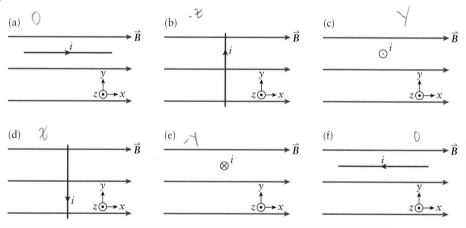

Figure 19-12 **Which way is the force?** What is the direction of the magnetic force on the wire in each case?

Take-Home Message for Section 19-5

✔ The current-carrying wire in a magnetic field experiences a magnetic force that is perpendicular to both the current direction and the magnetic field direction.

✔ The magnitude of this magnetic force is maximum if the current and magnetic field directions are perpendicular, and zero if the wire axis is along the magnetic field.

19-6 A magnetic field can exert a torque on a current loop

A common everyday application of magnetic forces on current-carrying wires is an electric motor. An electric motor is at the heart of a kitchen blender, a vacuum cleaner, the starter for an internal-combustion automobile, and many other devices (Figure 19-13a). Inside any electric motor you'll find a rotating portion (called the *rotor*) that's wrapped with coils of wire (Figure 19-13b), as well as magnets in the stationary part of the motor. When you turn the motor on, causing a current through the coils, the magnets exert forces on the current-carrying wire of the coils. The net effect is that there is a magnetic *torque* that makes the rotor spin. Let's look at a simplified version of this process to see how such a magnetic torque arises.

Figure 19-14a shows a straight wire bent into a single rectangular loop of wire with sides of length L and W. The loop carries a current i provided by a source of emf (not shown), so we call it a **current loop**. This loop is immersed in a uniform magnetic field of magnitude B and is free to rotate around an axis (shown in green). You should apply the right-hand rule for magnetic forces on a current-carrying wire (see Section 19-5) to each side of this loop. You'll see that the left-hand side of the loop experiences a force to the left, the right-hand side feels a force to the right, the top of the loop feels an upward force, and the bottom of the loop feels a downward force, as Figure 19-14a shows. Each segment of the loop carries the same current and is in the same magnetic field. Since the left- and right-hand sides are the same length and at the same angle to the magnetic field, the forces on these two sides have the same magnitude but opposite directions. Thus these forces cancel, and there is zero net force to the left or the right. The forces on the top and bottom segments of the wire also cancel for the same reason. So the net *force* on this rectangular loop is zero. (You can convince yourself that the same would be true if we reversed the direction of the current around the loop or the direction of the magnetic field, or changed the angle between the direction of the field and the plane of the loop.) This result is not just true for rectangular loops like the one shown in Figure 19-14a. It turns out that there is zero net magnetic force on *any* closed current-carrying loop in a uniform magnetic field, no matter what the shape of the loop.

Although the net magnetic force on the current loop in Figure 19-14a is zero, the net magnetic *torque* around the axis is not. Figure 19-14b is a side view of the current

Figure 19-13 Electric motors and magnetic torque (a) This fan uses an electric motor to rotate the fan blades. (b) The key components of any electric motor are current-carrying coils and a source of magnetic field. The magnetic field exerts forces on the current-carrying wires of the coils, and these produce a torque that makes the rotating part of the motor spin.

! Watch Out!

A current-carrying coil can feel a net magnetic force if the $\vec{B}$ field is not uniform.

In Figure 19-14a we assumed that the magnetic field is uniform (its magnitude and direction are the same at all points), and found that the net magnetic force on the current loop is zero. But if the magnetic field magnitude were greater at the left-hand side of the loop than at the right-hand side, then the left-hand side would experience a greater magnetic force. This would result in a net force on the loop to the left. In this chapter we'll restrict our discussion to the simple case in which the field is uniform.

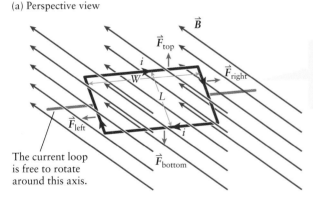

(a) Perspective view

There is zero net magnetic force on a current loop in a uniform magnetic field:
$$\vec{F}_{top} + \vec{F}_{bottom} + \vec{F}_{left} + \vec{F}_{right} = 0$$

The current loop is free to rotate around this axis.

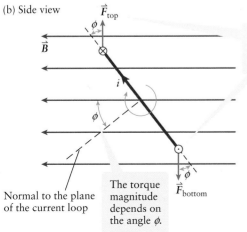

(b) Side view

There can be a nonzero net magnetic torque on a current loop in a uniform magnetic field.

Normal to the plane of the current loop

The torque magnitude depends on the angle ϕ.

Figure 19-14 A current loop in a uniform magnetic field (a) The magnetic forces on the sides of the current loop. (b) These forces can give rise to a net magnetic torque.

loop; the axis shown in green in Figure 19-14a is perpendicular to the plane of this figure and passes through the center of the loop. The forces on the near and far sides of the loop, which point out of and into the page, respectively, are directed parallel to this axis and so produce no torque. However, the forces that act on the top and bottom sides of the loop both give rise to a torque. For the situation shown in Figure 19-14b, both of these forces tend to make the loop rotate clockwise around the axis, so the net torque is clockwise.

Let's calculate the magnitude of the magnetic torque on the loop in Figure 19-14b. First note that the magnetic field is perpendicular to the current in both the top side of the loop and the bottom side of the loop. So in Equation 19-5 for the magnetic force on a current-carrying wire the angle θ equals 90°. Both the top and bottom sides of the loop have length W (see Figure 19-14a), so the magnitude of the magnetic force on each of these sides is

(19-6)

$$F = iWB \sin 90° = iWB$$

 See the Math Tutorial for more information on trigonometry

To find the torque around the axis that each of these forces produces, we use Equation 8-18:

Magnitude of torque **(8-18)**

Magnitude of the **torque** produced by a force acting on an object

Magnitude of the force

$$\tau = rF \sin \phi$$

Distance from the rotation axis of the object to where the force is applied

Angle between the vector $\vec{r}$ (from the rotation axis to where the force is applied) and the force vector $\vec{F}$

In Figure 19-14b the distance r is one-half of the length L of the near or far side of the loop, and ϕ is the angle between the direction of the near or far side of the current loop and the force exerted on either the top or bottom of the loop. This angle phi is also equal to the angle between the direction of the magnetic field and an imaginary line that's perpendicular to the plane of the current loop. We call this imaginary line the **normal** to the plane of the loop. From Equations 8-18 and 19-6, the magnitude of the torque due to the force on either the top or bottom side of the loop is

(19-7)

$$\tau_{\text{one side}} = \left(\frac{L}{2}\right) F \sin \phi = \left(\frac{L}{2}\right)(iWB) \sin \phi = \frac{1}{2}i(LW)B \sin \phi$$

The product LW (length times width) in Equation 19-7 is just the area of the current loop: $A = LW$. As we mentioned above, the torque from the top of the loop and the torque from the bottom of the loop both act in the same direction, so the *net* torque on the loop is just double the torque on one side given by Equation 19-7. Using $A = LW$, we have

$$\tau = 2\tau_{\text{one side}} = 2\left(\frac{1}{2}\right)i(LW)B \sin \phi = 2\left(\frac{1}{2}\right)iAB \sin \phi$$

or

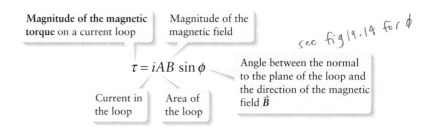

Magnitude of magnetic torque on a current loop **(19-8)**

Magnitude of the **magnetic torque** on a current loop

Magnitude of the magnetic field

see fig 19.14 for ϕ

$$\tau = iAB \sin \phi$$

Angle between the normal to the plane of the loop and the direction of the magnetic field $\vec{B}$

Current in the loop

Area of the loop

Although we've assumed a rectangular loop, Equation 19-8 applies to a current loop of area A of *any* shape.

Typically, the rotating coil in an electric motor is not just a single loop of wire, but has many *turns* (equivalent to many single coils stacked on top of each other). If a coil has N turns, the magnetic torque on it is N times greater than the value given by Equation 19-8.

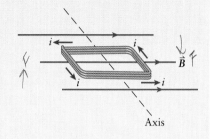

N turns

N × equation 19-8
for total magnetic torque on coil

Example 19-4 Angular Acceleration of a Current Loop

A length of copper wire is formed into a square loop with 50 turns. The loop is free to turn about a frictionless axis that lies in the plane of the loop and passes through its center. Each side of the loop is 2.00 cm long, and the moment of inertia of the loop about the axis of rotation is 4.00×10^{-6} kg·m². The loop lies in a region where there is a uniform magnetic field of magnitude 1.50×10^{-2} T that is perpendicular to the rotation axis of the loop. The current in the loop is 0.500 A. Find the angular acceleration of the loop (magnitude and direction) (a) when the loop is released from rest from the orientation shown in Figure 19-15, (b) after the loop has rotated 90°, and (c) after the loop has rotated 180°.

Figure 19-15 A square current loop in a magnetic field When this square current loop is released from rest, how will it begin to rotate?

Set Up

For each orientation we'll use Equation 19-8 to find the magnitude of the magnetic torque on the loop. There is no other torque on the loop (there is no friction, and gravity exerts zero torque since the center of mass of the loop lies on the rotation axis). So the magnetic torque is also the net torque, and we can use Equation 8-20 to find the magnitude of the angular acceleration. We'll find the direction of the angular acceleration by looking at the directions of the magnetic forces on the individual sides of the loop. Figure 19-15 shows that forces on the sides parallel to the rotation axis tend to affect rotation. The forces on the other sides cause no torque (compare Figure 19-14a).

Magnitude of magnetic torque on a current loop:

$$\tau = iAB \sin \phi \qquad (19\text{-}8)$$

Newton's second law for rotational motion:

$$\sum \tau_z = I\alpha_z \qquad (8\text{-}20)$$

Solve

(a) First find the direction of the magnetic torque, and hence the direction of the angular acceleration.

The right-hand rule for the magnetic force on a current-carrying wire shows that the forces on the wire tend to cause a clockwise rotation. So the angular acceleration is clockwise, and when the loop is released it will begin to rotate in the clockwise direction.

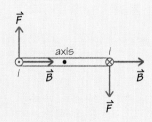

Find the magnitude of the angular acceleration.

The normal to the loop is perpendicular to the magnetic field, so in Equation 19-8 $\sin \phi = \sin 90° = 1$. The loop is square, so the cross-sectional area of the loop is just the square of the length of one side (2.00 cm = 0.0200 m). Since there are 50 turns, the total torque is N = 50 times greater than that given by Equation 19-8:

$$\tau = NiAB \sin \phi$$
$$= (50)(0.500\ \text{A})(0.0200\ \text{m})^2(1.50 \times 10^{-2}\ \text{T})(1)$$
$$= 1.50 \times 10^{-4}\ \text{T·A·m}^2 = 1.50 \times 10^{-4}\ \text{N·m}$$

(Recall that 1 T = 1 N/(A·m), so 1 T·A·m² = 1 N·m.)

This is the net torque on the loop, so from Equation 8-20 the magnitude of the loop's angular acceleration is

$$\alpha = \frac{\tau}{I} = \frac{1.50 \times 10^{-4}\,\text{N}\cdot\text{m}}{4.00 \times 10^{-6}\,\text{kg}\cdot\text{m}^2}$$

$$= 37.5\,\text{rad/s}^2$$

(If the torque is in N·m and the moment of inertia is in kg·m², the angular acceleration in Equation 8-20 is in rad/s².)
When released, the loop will begin to rotate in the clockwise direction as shown above. Note that as the loop rotates, the angle ϕ and the value of sin ϕ will decrease, so the torque and angular acceleration will decrease: This is *not* a situation with constant angular acceleration.

(b) Find the angular acceleration when the loop has rotated 90° from its initial orientation.

When the loop has rotated 90°, the normal to the loop is in the same direction as the magnetic field so $\phi = 0$ and sin $\phi = 0$. From Equation 19-8 it follows that there is zero torque on the loop at this point in its rotation. We can also see this using the right-hand rule for the magnetic forces on the wires of the loop: These forces do not exert any torque around the axis. Therefore this orientation represents an equilibrium position for the loop, and the angular acceleration is zero. Note that the loop will be in motion as it passes through this position (there has been an angular acceleration ever since the loop was released). As a result, the loop doesn't stop at this position but keeps on rotating.

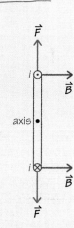

(c) Find the angular acceleration when the loop has rotated by 180° from its initial orientation.

At the 180° position, the normal to the loop is again perpendicular to the direction of the magnetic field, just as in part (a). So again the angle $\phi = 90°$, and again the magnitude of the angular acceleration is $\alpha = 37.5$ rad/s². However, since the loop has been flipped over relative to its original orientation, the directions of the forces are reversed. So the torque and angular acceleration are now *counterclockwise*. In fact, the angular acceleration has been increasingly counterclockwise ever since the loop moved past the position in (b). Since passing that position, the loop has been rotating in a clockwise direction but has been slowing down due to the counterclockwise angular acceleration.

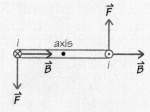

Reflect

The motion of the loop in this example should remind you of the motion of a pendulum (Section 12-5). When displaced from equilibrium and released, the pendulum will swing toward its equilibrium orientation (hanging straight down), but overshoot that equilibrium and swing to the other side of equilibrium. If there is no friction, the pendulum will keep swinging back and forth indefinitely. The same is true for this current loop: If the axis on which it rotates is frictionless, it will oscillate back and forth between the orientation in part (a) and the orientation in part (c).

In an electric motor we want the coil to continue rotating in the same direction, not oscillate back and forth like the coil in Example 19-4. To make this happen, the connection between the coil and the source of emf is arranged so that when the coil is at its equilibrium position (as in part (b) of Example 19-4), the direction of the current *reverses*. As a result, when the coil moves past this equilibrium position, the torque is now in the same direction as the rotation and the rotation continues to speed up. The current reverses again after another half-rotation, so the torque is always in the same direction.

With this arrangement, the coil would continue to gain rotational speed without limit if there were no other torques acting on it. In practice there are other torques that oppose the rotation, and the rotational speed reaches an upper limit. That's the case for the electric fan in Figure 19-13a: Air resistance on the fan blades increases as the fan turns faster, and the fan speed stabilizes when the torque due to air resistance just balances the torque of the electric motor.

? Got the Concept? 19-6 Magnetic Torque

Suppose the number of turns in the current loop of Example 19-4 were increased from 50 to 100. Would this make the maximum angular acceleration of the loop (a) Four times greater, (b) twice as great, (c) 1/2 as great, (d) 1/4 as great, or (e) none of these?

Take-Home Message for Section 19-6

✔ There is zero net force on a current loop in a uniform magnetic field.

✔ A uniform magnetic field can exert a torque on a current loop. The torque is maximum if the normal to the plane of the loop is perpendicular to the magnetic field direction.

19-7 Ampère's law describes the magnetic field created by current-carrying wires

So far in our discussion of magnetism we've looked at the forces and torques that a magnetic field exerts on a current-carrying wire. But current-carrying wires also *produce* magnetic fields (see Figure 19-1c, and Figure 19-4). As we described in Section 19-2, all magnetic fields are produced by electric charges in motion, whether it's a current in the coils of the electromagnet shown in Figure 19-1c, electrons in motion within the atoms of an iron bar magnet, or electric currents in the human brain as shown in Figure 19-16. To complete our understanding of magnetic forces, we need to be able to calculate the magnetic field produced by an arrangement of charges in motion. This is much like Chapter 16, where we needed to learn how to calculate the electric field due to an arrangement of charges in order to complete our understanding of electric forces.

We saw in Section 16-5 that the electric field that is due to a charge distribution is just the vector sum of the electric fields due to all of the charges in the distribution. These calculations can be rather challenging unless the charge distribution is very simple. When we look at the analogous problem of finding the *magnetic* field due to a distribution of moving charges, we find that the problem is even more complicated. That's because the magnetic field due to even a single moving charged particle is itself rather complex: The field does not point directly away from or toward the moving charge, but in a direction determined in part by the velocity (and perpendicular to the velocity vector). Rather than looking at how to do calculations of this kind, we'll just look at the result for one important situation, the magnetic field due to a long, straight, current-carrying wire. We'll then see an alternative approach for calculating a magnetic field using *Ampère's law*. This law is a very powerful one, but like Gauss's law for electric fields (Sections 16-6 and 16-7) it can be used for field calculations only in certain simple situations. We'll conclude the section with a look at the magnetic field produced by a current loop, as well as some of the applications of this field.

Electric currents in the human brain generate weak magnetic fields. The colors in this magnetoencephalogram represent the strength of the magnetic field produced in different regions.

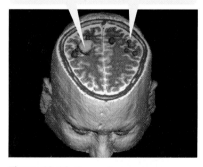

Figure 19-16 **Currents as sources as magnetic field** Electric currents in biological systems produce magnetic fields.

A Long, Straight Wire and Ampère's Law

Consider a very long, straight wire that carries a constant current i. Experiment and calculation both show that the magnetic field due to this current has the properties shown in Figure 19-17. Note that the wire and the magnetic field that it produces have *cylindrical symmetry*: The wire and the field pattern look exactly the same if you rotate the wire around its length. The magnitude of the field is given by

Magnitude of magnetic field due to a long, straight wire
(19-9)

Only if wire is infinitely long

| **Magnitude of the magnetic field** due to a long, straight wire | Permeability of free space |

$$B = \frac{\mu_0 i}{2\pi r}$$ Current in the wire

Distance from the wire to the location where the field is measured

Equation 19-9 and the field properties shown in Figure 19-17 are strictly correct only if the wire is infinitely long. But they are very good approximations if the distance r is small compared to the length of the wire.

The constant μ_0 in Equation 19-9, called the **permeability of free space** for historic reasons, plays a role in magnetism that's comparable to the role of the constant ε_0, the *permittivity* of free space, in electricity (see Section 16-6). Its value is *exactly* $\mu_0 = 4\pi \times 10^{-7}\,\text{T}\cdot\text{m/A}$. The value is known exactly because of the way that the tesla is defined: At a distance of 1 m from a long, straight wire carrying a current of 1 A, the magnetic field strength is exactly

$$B = \frac{\mu_0 i}{2\pi r} = \frac{(4\pi \times 10^{-7}\,\text{T}\cdot\text{m/A})(1\,\text{A})}{2\pi(1\,\text{m})} = 2 \times 10^{-7}\,\text{T}$$

(**1**) The magnetic field lines due to a long, straight, current-carrying wire are circles that lie in planes perpendicular to the axis of the wire.

(**2**) The magnetic field vectors are tangential to the circular magnetic field lines, and perpendicular to the axis of the wire.

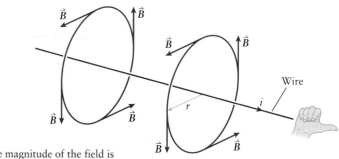

(**3**) The magnitude of the field is proportional to the current i and inversely proportional to the distance r from the axis of the wire.

(**4**) Right-hand rule for the field direction: If you point your right thumb in the direction of the current and curl your fingers, the magnetic field curls around the field lines in the direction of the curled fingers of your right hand.

Figure 19-17 **Magnetic field of a long, straight, current-carrying wire** The magnitude of this field is given by Equation 19-9.

! **Watch Out! Remember that vectors are straight, never curved.**

When drawing the magnetic field around a long, straight wire, it may be tempting to draw curved arrows to represent how the magnetic field lines curl around the wire. But a vector always denotes a *single* direction and so *cannot* be curved.

At any point along a magnetic field line, the direction of the field is always along the *tangent* to the field line as shown in Figure 19-17.

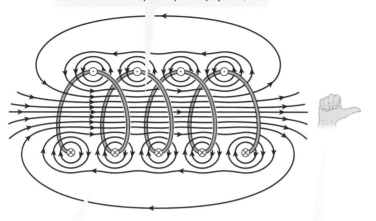

① The magnetic field in the interior of a long, straight solenoid is essentially uniform (the field lines are very nearly evenly spaced).

② The magnetic field outside the solenoid is very weak (the field lines are far apart).

③ Right-hand rule for the field direction: If you curl the fingers of your right hand around the solenoid in the direction of the current, the magnetic field inside the solenoid points in the direction of your right thumb.

Figure 19-18 **Magnetic field of a solenoid** Compare this illustration to the photograph of a solenoid and its field in Figure 19-4a.

The magnetic field around current-carrying wires with other geometries is much more complicated. As an example, Figure 19-18 shows some of the magnetic field lines for a straight helical coil of wire. Such a coil is called a **solenoid**. Close to an individual wire, the field lines resemble those around the long, straight wire shown in Figure 19-17. In the space outside the solenoid the magnetic field is very weak, as you can see from the large spacing between adjacent field lines. (Recall from Section 16-5 that the same is true for electric fields: Where field lines are far apart, the field magnitude is small.) But in the interior of the solenoid the magnetic field lines are close together and nearly evenly spaced, indicating that the magnetic field there is strong and nearly uniform.

This property of solenoids explains why a conventional magnetic resonance imaging (MRI) scanner is in the form of a long tube inside which the patient lies (Figure 19-19). This tube is actually the interior of a solenoid like that shown in Figure 19-18, so the patient is bathed in a strong, uniform magnetic field—which is just what MRI requires. (In Chapter 27 we'll learn more about the physics of MRI.)

Can we use Equation 19-9 to calculate the field inside a solenoid? Not directly, no, because the wires that make up the solenoid are not straight. But we can use Equation 19-9 to illustrate a useful principle about magnetic fields and their sources, and then use that principle to determine the solenoid field. Imagine that we draw a circle of radius r around a long, straight wire as shown in Figure 19-20. Imagine further that we break the circle into a number of segments of length $\Delta \ell$. If $\Delta \ell$ is sufficiently small, we can treat each segment as being straight. Then for each segment, take the component $B_\parallel$ of magnetic field parallel to that segment and multiply it by the segment length $\Delta \ell$. If we add up the values of these products for every segment in the circle, the result is a quantity called the **circulation** of the magnetic field around the circle:

$$\text{Circulation} = \sum B_\parallel \Delta \ell \qquad (19\text{-}10)$$

For the circle shown in Figure 19-20, Equation 19-9 tells us that $B_\parallel$ is equal to $\mu_0 i / (2\pi r)$ at every point around the circle. That's because the magnetic field is everywhere tangent to the circle and so parallel to a short segment of length on the circle. Therefore the circulation of the magnetic field as defined by Equation 19-10 is

$$\text{Circulation} = \sum \left(\frac{\mu_0 i}{2\pi r} \right) \Delta \ell = \frac{\mu_0 i}{2\pi r} \sum \Delta \ell \qquad (19\text{-}11)$$

In Equation 19-11 we've taken the quantity $\mu_0 i / (2\pi r)$ outside the sum because it has the same value everywhere around the circle and so in all terms of the sum.

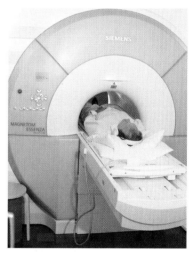

Figure 19-19 **Magnetic resonance imaging (MRI)** The medical imaging technique known as MRI requires that the patient be immersed in a strong, uniform magnetic field. In this MRI device this is done by having the patient lie inside a solenoid.

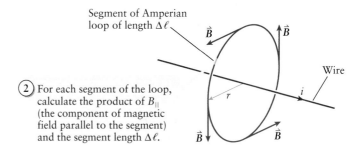

① We draw an Amperian loop that encircles the current-carrying wire. (This particular loop is a circle, and so coincides with a field line.)

Segment of Amperian loop of length $\Delta \ell$

Wire

② For each segment of the loop, calculate the product of $B_\parallel$ (the component of magnetic field parallel to the segment) and the segment length $\Delta \ell$.

③ The circulation of magnetic field around the loop is the sum of all these products for the entire loop:

$$\sum B_\parallel \Delta \ell$$

The quantity $\sum \Delta \ell$ is the sum of the lengths of all of the segments that make up the circle—that is, the circumference of the circle, which is equal to $2\pi r$. So the circulation in Equation 19-11 is equal to

(19-12)

$$\text{Circulation} = \frac{\mu_0 i}{2\pi r}(2\pi r) = \mu_0 i$$

The circulation as given by Equation 19-12 does *not* depend on the radius r of the circle. (If the circle is larger, the magnetic field has a smaller magnitude, but it takes more segments of length $\Delta \ell$ to go all the way around the circle.) Remarkably, the same result holds true even if we draw a noncircular path around the wire: The circulation around *any* path that encloses the wire is equal to $\mu_0 i$. This is an example of **Ampère's law**, which was discovered by the French physicist André-Marie Ampère (pronounced "ahm-pair") in 1826:

Circulation of magnetic field around an Amperian loop

Permeability of free space

Ampère's law
(19-13)

$$\sum B_\parallel \Delta \ell = \mu_0 i_{\text{through}}$$

Current through the Amperian loop

An **Amperian loop** is simply a closed path in space; the circle in Figure 19-20 is an example. The subscript "through" reminds us that the right-hand side of Equation 19-13 should only include current that passes through the interior of the Amperian loop.

It can be shown that Ampère's law is true for *any* magnetic field and *any* Amperian loop, no matter what the geometry of the current that produces the magnetic field. (The proof is beyond our scope.) However, it's only in certain very symmetric situations that Ampère's law helps us calculate the value of magnetic field. (This is much like Gauss's law for electric fields. We saw in Sections 16-6 and 16-7 that Gauss's law is always true but is only useful for electric field calculations in certain situations.) Happily, one such situation is the field of a solenoid.

Using Ampère's Law: Magnetic Field of a Solenoid

As Figure 19-18 shows, inside the solenoid the fields from each loop add to create a total field that runs parallel to the central axis; this is especially noticeable close to the axis and far from the ends. Outside the solenoid the fields nearly cancel, especially close to the solenoid and far from the ends. The fields outside the solenoid cancel more completely when the length of the solenoid is large compared to its diameter and when the coils of the solenoid are tightly wound. If there are N turns or *windings* of wire in the solenoid, the *winding density* is $n = N/L$. (The units of n are windings per meter.) Let's see how to use Ampère's law to find the magnetic field of a long, narrow, tightly wound (ideal) solenoid of length L, diameter D, and winding density n carrying a current i. In particular, we'll apply Ampère's law to find the field inside the coil and far from the ends—that is, where we expect the field to be relatively uniform.

To use Ampère's law effectively, we must first select an Amperian loop *through* which current flows and *around* which we can calculate the magnetic field. The point at which we want to determine the field—in this case, a point inside the solenoid—must lie on the Amperian loop. Figure 19-21 shows our choice of Amperian loop on a cut-away view of a section of the solenoid. This view is similar to Figure 19-18, in which the current is shown coming out of the page along the top part of the windings and going into the page along the bottom. We picked a rectangle as the Amperian loop and positioned it so that the bottom of the loop is parallel to one of the field lines of the uniform field inside the solenoid. The left-hand side of Ampère's law, Equation 19-13, can then be written as

$$\sum B_{\parallel} \Delta \ell = B_{\parallel 1} \ell_1 + B_{\parallel 2} \ell_2 + B_{\parallel 3} \ell_3 + B_{\parallel 4} \ell_4 \qquad (19\text{-}14)$$

Side 4 is in the region of zero magnetic field outside the solenoid, so $B_{\parallel 4} = 0$ and $B_{\parallel 4} \ell_4 = 0$. Side 1 runs partially through this region of zero field, so $B_{\parallel 1} = 0$ there; inside the solenoid the field is nonzero but is perpendicular to side 1, so $B_{\parallel 1} = 0$ there as well. The same is true for side 3, so in Equation 19-14 $B_{\parallel 1} \ell_1 = B_{\parallel 3} \ell_3 = 0$. The only side of the Amperian loop that makes a nonzero contribution to the path integral is side 2. Here the magnetic field of magnitude B is parallel to the side of length ℓ_2, so $B_{\parallel 2} = B$ and Equation 19-14 becomes

$$\sum B_{\parallel} \Delta \ell = B \ell_2 \qquad (19\text{-}15)$$

On the right-hand side of the Ampère's law equation, we need to evaluate i_{through}, the current that passes through the loop. This quantity is *not* just the current in the solenoid i, because every winding of the coil that passes through the loop brings a contribution i to i_{through}. The length of solenoid enclosed by the loop is ℓ_2 and there are n windings per meter, so there are $n\ell_2$ windings that pass through the

(1) These are the windings of the solenoid. The current points out of the page along the top and into the page along the bottom.

(2) We use a rectangular Amperian loop. Sides 1, 3, and 4 make zero contribution to the circulation around this loop (the field is either zero or perpendicular to the loop on these sides).

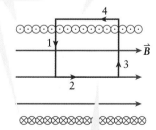

(3) The only contribution to the circulation is from side 2, of length ℓ_2:

$$\sum B_{\parallel} \Delta \ell = B \ell_2$$

Figure 19-21 **Ampère's law and the field of a solenoid** The Amperian loop shown here helps us to calculate the magnetic field inside a long, straight solenoid.

loop. Each winding carries current i, so the total current through the loop is $n\ell_2$ multiplied by i:

$$i_{\text{through}} = n\ell_2 i \qquad (19\text{-}16)$$

We can now substitute Equations 19-15 and 19-16 into the two sides of Equation 19-13 for Ampère's law:

$$B\ell_2 = \mu_0(n\ell_2 i)$$

The length ℓ_2 of the Amperian loop cancels, and we are left with

Magnetic field inside a long, straight solenoid (19-17)

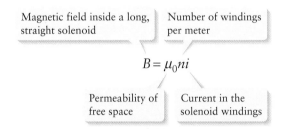

Magnetic field inside a long, straight solenoid | Number of windings per meter

$$B = \mu_0 n i$$

Permeability of free space | Current in the solenoid windings

In deriving Equation 19-17, we've assumed that the field inside the solenoid is perfectly uniform and parallel to the solenoid axis and that the field outside the solenoid is exactly zero. These assumptions are strictly valid only for an infinitely long solenoid. But Equation 19-17 is a good approximation at points near the middle of any solenoid, especially one whose length is large compared to its diameter.

Note that Ampère's law also tells us the *direction* of the magnetic field inside the solenoid. Just as for the case of a long, straight wire (Figure 19-17), point your right thumb in the direction of the current through the Amperian loop in Figure 19-21. This current points out of the page, so you should point your thumb in that direction. If you curl the fingers of your right hand, they will curl in a counterclockwise direction: from left to right below the enclosed windings, and from right to left above the enclosed windings. That's consistent with the direction of the magnetic field inside the solenoid, which is from left to right. (The field has zero magnitude above the windings and outside the solenoid, so the direction is meaningless there.)

Equation 19-17 tells us that the magnitude of the uniform field created by a long solenoid is proportional to n, the number of windings per length. The more windings that can be packed into a length of the solenoid, the greater the field magnitude. A typical solenoid used for electronic applications is therefore more likely to look like the one in Figure 19-22, with many layers of tightly packed windings, than the one shown in Figure 19-18.

In the following example, we use Ampère's law to find the magnetic field due to a rather different distribution of current.

Figure 19-22 A real-life solenoid When the current to this solenoid is turned on, the iron rod that sticks out from its end experiences a strong magnetic force and is pulled into the solenoid. Such a device can be used to unlock a security door, as well as many other applications.

Example 19-5 Magnetic Field Due to a Coaxial Cable

A coaxial cable consists of a solid conductor of radius R_1 surrounded by insulation, which in turn is surrounded by a thin conducting shell of radius R_2 made of either fine wire mesh or a thin metallic foil (Figure 19-23). The combination is enclosed in an outer layer of insulation. The inner conductor carries current in one direction, and the outer conductor carries it back in the opposite direction. For a coaxial cable that carries a constant current i, find expressions for the magnetic field (a) inside the inner conductor at a distance $r < R_1$ from its central axis, (b) in the space between the two conductors, and (c) outside the coaxial cable. Assume that the moving charge in the inner conductor is distributed uniformly over the volume of the conductor.

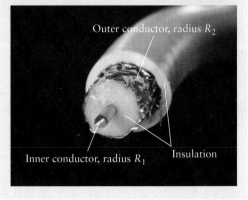

Outer conductor, radius R_2

Inner conductor, radius R_1 Insulation

Figure 19-23 A coaxial cable Equal amounts of current flow in opposite directions in the inner and outer conductors of this cable.

Set Up

Both the inner conductor separately and the coaxial cable as a whole have the same cylindrical symmetry as a long, straight wire (Figure 19-17). So we expect that the field lines are circles concentric with the axis of the cable. Just as for the long, straight wire, this means that it's natural to choose circular paths concentric with the cable axis as the Amperian loops. To find the field in the three regions, we'll choose the radius r of the Amperian loop to be less than R_1 in part (a), between R_1 and R_2 in part (b), and greater than R_2 in part (c).

Ampère's law:

$$\sum B_{\parallel}\Delta\ell = \mu_0 i_{through} \quad (19\text{-}13)$$

Dashed circles labeled I, II, and III: Amperian loops for parts (a), (b), and (c) respectively

Solve

(a) Find the field inside the inner conductor by using an Amperian loop of radius $r < R_1$.

Inside the inner conductor, the magnetic field has magnitude B_{inner} and points tangent to the Amperian loop, so $B_{\parallel} = B_{inner}$. The left-hand side of the Ampère's law equation is

$$\sum B_{\parallel}\Delta\ell = B_{inner}\sum\Delta\ell = B_{inner}(2\pi r)$$

The Amperian loop encloses area πr^2, which is less than the cross-sectional area πR_1^2 of the inner conductor. The current through the loop is therefore a fraction $(\pi r^2)/(\pi R_1^2)$ of the total current i in the inner conductor:

$$i_{through} = i\left(\frac{\pi r^2}{\pi R_1^2}\right) = i\frac{r^2}{R_1^2}$$

Insert these into Equation 19-13 and solve for B_{inner}:

$$B_{inner}(2\pi r) = \mu_0 i\frac{r^2}{R_1^2}$$

$$B_{inner} = \mu_0 i\frac{r^2}{2\pi r R_1^2} = \frac{\mu_0 i r}{2\pi R_1^2}$$

(b) Find the field between the conductors by using an Amperian loop of radius r, where $R_1 < r < R_2$.

Between the two conductors the magnetic field of magnitude $B_{between}$ also points tangent to the Amperian loop, so as in part (a) the left-hand side of Equation 19-13 is

$$\sum B_{\parallel}\Delta\ell = B_{between}(2\pi r)$$

The Amperian loop encloses the entire inner conductor, so $i_{through} = i$. Insert these into Equation 19-13 and solve for $B_{between}$:

$$B_{between}(2\pi r) = \mu_0 i$$

$$B_{between} = \frac{\mu_0 i}{2\pi r}$$

(c) Find the field outside the cable by using an Amperian loop of radius $r > R_2$.

Just as in parts (a) and (b), outside the outer conductor the magnetic field of magnitude B_{outer} points tangent to the Amperian loop, so

$$\sum B_{\parallel}\Delta\ell = B_{outer}(2\pi r)$$

The Amperian loop encloses both conductors, each of which carries current i. Since the currents flow in opposite directions, the *net* current through the loop is $i_{through} = 0$. So Equation 19-13 tells us

$$B_{outer}(2\pi r) = \mu_0(0)$$

$$B_{outer} = 0$$

Reflect

Our result from (a) says that the magnetic field is zero at the center of the inner conductor ($r = 0$), then increases in direct proportion to r with increasing distance from the center. The field reaches its maximum value at the outer surface of the inner conductor ($r = R_1$). Between the conductors the field is inversely proportional to r, so the magnitude decreases with increasing distance from the center of the cable. Outside the outer conductor there is *zero* magnetic field.

Coaxial cables are often referred to as "shielded" cables. The arrangement of the two conductors eliminates the presence of stray magnetic fields outside the cable. The shielding also serves to isolate the inner conductor from external electromagnetic signals. You'll find a coaxial cable connected to the back of most television sets (it's the "cable" in the term "cable TV"); the signal carried by this cable involves an alternating current and hence a varying magnetic field rather than a steady one, but the shielding principle is the same.

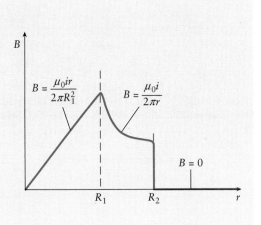

Magnetic Field of a Current Loop

An important special case for which Ampère's law is *not* helpful is the magnetic field produced by a current loop (a current-carrying wire bent into a circle). **Figure 19-24** shows some of the magnetic field lines for such a loop.

Unlike the case for a long, straight wire (Figure 19-17) or for a coaxial cable (Example 19-5), the magnetic field does *not* have the same magnitude at all points on a field line: The magnitude B is greater where the lines are closer together. So choosing an Amperian loop that coincides with a field line will not give us a simple equation for the magnitude B, as was the case in Example 19-5. To calculate the magnetic field at a given point in this situation, it's necessary to find the contribution to the field due to each short segment of the loop, then add those contributions using vector arithmetic.

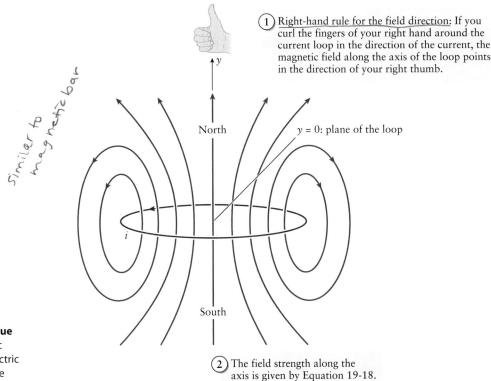

Figure 19-24 Magnetic field due to a current loop This magnetic field pattern is similar to the electric field pattern of an electric dipole (Figure 16-14).

Such a calculation is beyond our scope. Here's the result for the magnitude of the magnetic field along the axis of the current loop, labeled y in Figure 19-24:

$$\boxed{B} = \frac{\mu_0 i}{2} \frac{R^2}{(R^2 + y^2)^{3/2}} \tag{19-18}$$

In Equation 19-18 i is the current in the loop, R is the radius of the loop, and y is the coordinate along the y axis, where $y = 0$ represents the plane of the loop. If we substitute $y = 0$ into Equation 19-18, we get the field magnitude at the very center of the loop. At points far from the loop, so y is much greater than R, we can replace $R^2 + y^2$ with y^2 to good approximation. Equation 19-18 then becomes

$$B = \frac{\mu_0 i}{2} \frac{R^2}{(y^2)^{3/2}} = \frac{\mu_0 i}{2} \frac{R^2}{|y|^3} \tag{19-19}$$

(We've added the absolute value signs because y can be positive or negative, but the field magnitude B must be positive.) So at large distances from a current loop, the magnetic field is inversely proportional to the *cube* of the distance from the loop.

This result is reminiscent of the *electric* field of an electric dipole (a combination of a positive charge q and a negative charge $-q$): We found in Example 16-7 (Section 16-5) that at large distances from an electric dipole, the electric field due to that dipole is inversely proportional to the cube of the distance. The overall magnetic field pattern of a current loop also has some similarities to the electric field pattern of an electric dipole. (Compare Figure 19-24 and Figure 16-14; if you rotate Figure 16-14 clockwise 90°, the similarities will be more evident.)

As a result of these similarities, we use the term *magnetic dipole* to refer to a current loop. The two "poles" of a magnetic dipole are the points just above and just below the loop, as Figure 19-24 shows. We call these poles north and south by analogy to the poles of a bar magnet: The magnetic field points away from the current loop at its north pole and points toward the current loop at its south pole, just as for a bar magnet (see Section 19-2).

Note that unlike the two charges that make up an electric dipole, the north and south poles of a current loop can never be separated: The current loop must always have two sides! As we discussed in Section 19-2, a permanent magnet such as a bar magnet acts as a magnetic dipole. That's because a permanent magnet is really just a collection of *atomic* current loops, each the result of electron motions within the atom. Their combined effect is the same as electrons moving around the circular loop of wire in Figure 19-24. The poles of such a magnet can no more be separated than can the two sides of a current loop.

Earth's magnetic field is nearly that of a dipole, with the axis of the field tilted about 11° from Earth's rotation axis. (As a result, the magnetic poles—the place on Earth's surface where the field lines point straight and vertical to the surface—are displaced by several hundreds of kilometers from Earth's geographic poles.) The magnetic field is produced because molten material in the outer regions of Earth's core is in a state of continuous motion, and this motion gives rise to electric currents that generate the field. The photo that opens this chapter illustrates one dynamic consequence of our planet having a magnetic field.

Anyone who uses a compass to navigate makes use of Earth's magnetic field, which points generally from south to north. Other living organisms also take advantage of Earth's field to guide them from location to location. Pigeons, honeybees, and sea turtles among others rely to some extent on an internal magnetic compass to navigate. Sea turtles, for example, have been observed to travel hundreds of kilometers and still find their way back to their nesting sites along relatively direct paths. Yet when the turtles are transported away from their nests after a magnet has been attached to their heads, they take wildly circuitous routes back to the nesting site. The field of the attached magnet clearly disrupts the turtles' ability to determine their position using Earth's magnetic field.

Magnetic Materials

While there are circulating electrons within every kind of atom, not all materials have the same magnetic properties. In some materials there is a net rotation of electron charge within the atom, so each atom behaves like a current loop. In most cases these atomic current loops are randomly oriented, so their effects cancel out. But if the material is placed in a strong magnetic field, the atomic current loops experience a torque and align themselves with the magnetic field (see Figure 19-14b). As a result, the material behaves like a much larger current loop. If the magnetic field is turned off, random thermal motion will cause the atomic current loops to return to their original, nonaligned orientations. Materials that display this behavior are called **paramagnetic**. Everyday paramagnetic materials include aluminum and sodium. The net magnetic effect in a paramagnetic material is generally quite small; while an empty can made of (paramagnetic) aluminum acts like a current loop when brought next to a magnet, the magnetic force on the aluminum can is so small that a magnet can't pick it up.

In a handful of materials, the interactions between adjacent atomic current loops are very strong. As a result, once the material is placed in a magnetic field, the atomic current loops not only align with the field but can *remain* aligned after the field is turned off, leaving the material permanently magnetized. Iron is the most common of these materials, which are called **ferromagnetic** ("ferro" derives from the Latin word for iron). Any permanent magnet, such as one you might use to attach notes to a refrigerator door, is made of a ferromagnetic material. A permanent magnet can pick up objects made of a ferromagnetic material, such as a steel paper clip. The field of the permanent magnet causes the atomic current loops in the paper clip to align, making the paper clip a magnet itself. The magnetized paper clip is then attracted to the permanent magnet.

! Watch Out! Earth is not a permanent magnet.

Our planet's core is made primarily of iron and nickel, both of which are ferromagnetic materials. It's common to conclude from this that the core is magnetized like a permanent magnet and that this gives rise to our planet's magnetic field. However, this cannot be true. Any ferromagnetic material loses its magnetism if it is heated above a certain temperature specific to that material: This critical temperature is 773°C for iron and 354°C for nickel. The temperature in Earth's core is in excess of 4400°C, so the iron and nickel in the core do *not* act like ferromagnetic materials. Instead, Earth's magnetic field is caused by electric currents in the molten material that makes up the outer regions of the core.

Most materials are neither paramagnetic nor ferromagnetic because their atoms have zero net electron current. When placed in a magnetic field, a small amount of atomic current appears, but the current loops end up aligned in the direction *opposite* to what happens for paramagnetic or ferromagnetic materials. (This is a consequence of *electromagnetic induction*, which we'll discuss in Chapter 20.) As a result, these materials, called **diamagnetic**, are slightly repelled by magnets rather than being attracted. In most cases, however, the repulsion is very weak.

? Got the Concept? 19-7 Ampère's Law

The field lines in Figure 19-24 are closed curves. If the distance around one such curve is L and the current in the loop is i, what is the average value around the closed curve of the component of magnetic field $B_{\parallel}$ parallel to the curve? (a) $\mu_0 i$; (b) $\mu_0 i/L$; (c) $2\mu_0 i$; (d) $2\mu_0 i/L$; (e) not enough information given to decide.

Take-Home Message for Section 19-7

✔ Ampère's law relates the current through a wire to the magnetic field it generates.

✔ Current through a long, straight wire produces a magnetic field with circular field lines centered on the wire.

✔ Current through a solenoid (a straight, helical coil of wire) produces a relatively uniform magnetic field along its axis that is proportional to the winding density as well as the current.

✔ A current loop produces a more complicated magnetic field. A magnetic material can be thought of as a collection of atomic current loops.

19-8 Two current-carrying wires exert magnetic forces on each other

We've seen that a current-carrying wire experiences a force when placed in a magnetic field, and also that a current-carrying wire generates a magnetic field. Let's put these ideas together and look at the magnetic interaction between *two* current-carrying wires. (Note that these two wires do not exert *electric* forces on each other. That's because each wire has as much positive charge as negative charge, and so is electrically neutral.)

André-Marie Ampère was the first to study the magnetic interaction between current-carrying wires theoretically and experimentally. He was also the first to show that two parallel, straight wires carrying current in the same direction attract each other, and that two parallel, straight wires carrying current in opposite directions repel. Let's see why this is the case.

Figure 19-25 shows the situation. Wires 1 and 2 are long, straight, and parallel to each other and separated by a distance d. The current i_1 in wire 1 sets up a magnetic field $\vec{B}_1$ at the position of wire 2, which carries current i_2. The magnitude B_1 of this field is given by Equation 19-20:

$$B_1 = \frac{\mu_0 i_1}{2\pi d} \tag{19-20}$$

The right-hand rule for the field produced by a long, straight wire (Section 19-7) tells us that at the position of wire 2, $\vec{B}_1$ points upward and perpendicular to wire 2. To find the direction of the force $\vec{F}_{1 \text{ on } 2}$ that this field exerts on wire 2, use the right-hand rule for the direction of the magnetic force on a current-carrying wire (Section 19-5): This tells us that $\vec{F}_{1 \text{ on } 2}$ points toward wire 1, so the force attracts wire 2 to wire 1. The magnitude of the force on wire 2, of length ℓ_2, is given by Equation 19-5 with $\theta = 90°$ (since the direction of the current in wire 2 is perpendicular to the direction of $\vec{B}_1$):

$$F_{1 \text{ on } 2} = i_2 \ell_2 B_1 \sin 90° = i_2 \ell_2 B_1 \tag{19-21}$$

If we substitute B_1 from Equation 19-20 into Equation 19-21, we get

$$F_{1 \text{ on } 2} = i_2 \ell_2 \left(\frac{\mu_0 i_1}{2\pi d} \right) = \frac{\mu_0 i_1 i_2 \ell_2}{2\pi d} \tag{19-22}$$

The force per unit length on wire 2 is $F_{1 \text{ on } 2}$ (given by Equation 19-22) divided by the length ℓ_2 of wire 2:

$$\text{Magnetic force per unit length exerted by wire 1 on wire 2} = F_{1 \text{ on } 2}/\ell_2 = \frac{\mu_0 i_1 i_2}{2\pi d} \tag{19-23}$$

We can use the same procedure to find the magnetic force per unit length that wire 2 exerts on wire 1. The field $\vec{B}_2$ that the current i_2 in wire 2 produces at the position of wire 1 has magnitude $B_2 = \mu_0 i_2 / (2\pi d)$ and points *downward* in Figure 19-25. From the right-hand rule for the force on a current-carrying wire, the force $\vec{F}_{2 \text{ on } 1}$ on wire 1 points toward wire 2 (the force is attractive); its magnitude is $F_{2 \text{ on } 1} = i_1 \ell_1 B_2 = i_1 \ell_1 [\mu_0 i_2 / (2\pi d)] = \mu_0 i_1 i_2 \ell_1 / (2\pi d)$, where ℓ_1 is the length of wire 1. The force per unit length on wire 1 is then $F_{1 \text{ on } 2}$ divided by ℓ_1:

$$\text{Magnetic force per unit length exerted by wire 2 on wire 1} = F_{2 \text{ on } 1}/\ell_1 = \frac{\mu_0 i_1 i_2}{2\pi d} \tag{19-24}$$

The force magnitudes per unit length in Equations 19-23 and 19-24 are equal, and the forces $\vec{F}_{1 \text{ on } 2}$ and $\vec{F}_{2 \text{ on } 1}$ are opposite in direction. That's just what we would expect from Newton's third law.

What changes if we reverse the direction of the current i_1 in wire 1? The force *magnitudes* given by Equations 19-23 and 19-24 won't be affected, but the force *directions* will be. This will reverse the direction of the magnetic field $\vec{B}_1$ that wire 1 produces at the position of wire 2, and so will reverse the direction of the force $\vec{F}_{1 \text{ on } 2}$ on wire 2. So in this case wire 2 will be pushed away from wire 1 (it will be repelled). Reversing the direction

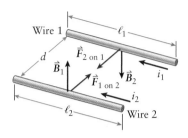

$\vec{B}_1$ = Magnetic field due to wire 1 at the position of wire 2

$\vec{B}_2$ = Magnetic field due to wire 2 at the position of wire 1

✓Figure 19-25 **Magnetic forces between two current-carrying wires** These two parallel wires carry current in the same direction and exert attractive magnetic forces on each other. If we reverse the direction of one of the currents, the forces become repulsive.

of i_1 will also reverse the direction of the force $\vec{F}_{2 \text{ on } 1}$ that wire 2 exerts on wire 1, so this force will push wire 1 away from 2 (again, it will be repelled). So we conclude that

> *Two parallel current-carrying wires attract each other if they carry current in the same direction, and repel each other if they carry current in opposite directions.*

Example 19-6　Wires in a Computer

The two long, straight wires that run along the back of a computer case to power the cooling fan carry 0.110 A in opposite directions. The wires are separated by 5.00 mm. (a) Find the force per unit length (magnitude and direction) that these wires exert on each other. (b) The mass per unit length of the wire is 5.00×10^{-3} kg/m. What acceleration does one of the wires experience due to this force?

Set Up

We'll use Equation 19-23 to find the force per unit length that one wire exerts on the other. (As we saw with Equation 19-24, the force per unit length has the same magnitude for either wire.) If we assume that this force equals the net external force on the wire, we can use Newton's second law to calculate the acceleration of the wire.

Magnetic force per unit length exerted by wire 1 on wire 2:

$$F_{1 \text{ on } 2} / \ell_2 = \frac{\mu_0 i_1 i_2}{2\pi d} \qquad (19\text{-}23)$$

Newton's second law:

$$\sum \vec{F}_{\text{ext}} = m\vec{a} \qquad (4\text{-}2)$$

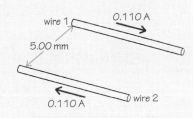

Solve

(a) The currents are in opposite directions, so the force is repulsive (it pushes the two wires apart). Use Equation 19-23 to find the magnitude of the force per unit length.

The two wires are separated by $d = 5.00$ mm $= 5.00 \times 10^{-3}$ m and carry currents with the same magnitude: $i_1 = i_2 = 0.110$ A. The force per unit length on either wire is

$$= \frac{(4\pi \times 10^{-7} \, \text{T} \cdot \text{m/A})(0.110 \, \text{A})(0.110 \, \text{A})}{2\pi(5.00 \times 10^{-3} \, \text{m})}$$
$$= 4.84 \times 10^{-7} \, \text{T} \cdot \text{A} = 4.84 \times 10^{-7} \, \text{N/m}$$

(Recall that $1 \, \text{T} = 1 \, \text{N}/(\text{A} \cdot \text{m})$.)

(b) Use Newton's second law to find the acceleration of the wire.

Our result for part (a) says that a 1-m length of wire would experience a force of magnitude 4.84×10^{-7} N. Since this wire has mass per unit length 5.00×10^{-3} kg/m, a 1-m length would have mass 5.00×10^{-3} kg. The acceleration is

$$a = |\vec{a}| = \frac{|\sum \vec{F}_{\text{ext}}|}{m} = \frac{4.84 \times 10^{-7} \, \text{N}}{5.00 \times 10^{-3} \, \text{kg}}$$
$$= 9.68 \times 10^{-5} \, \text{m/s}^2$$

Reflect

The force and acceleration are both very gentle, so the effect on these wires will be almost imperceptible. In applications with very large currents, however, the magnetic forces between conductors can be substantial.

❓ Got the Concept? 19-8　Forces on a Current-carrying Coil

A very flexible helical coil is suspended as shown in Figure 19-26. What will happen when a sizable current i is sent through the coil? (a) The coils will be pulled together; (b) the coils will be pushed apart; (c) some of the coils will be pulled together, while others will be pulled apart; (d) there will be no net effect on the coils.

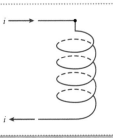

Figure 19-26 **A hanging coil** What happens to this coil when the current is turned on?

Take-Home Message for Section 19-8

✔ Two wires attract each other when carrying current in the same direction and repel each other when carrying currents in opposite directions.

✔ The forces per unit length on the wires are equal in magnitude and opposite in direction, exactly as required by Newton's third law.

Key Terms

Ampère's law
Amperian loop
circulation
current loop
diamagnetic
electromagnetism
ferromagnetic

magnet
magnetic dipole
magnetic field
magnetic force
magnetic poles
magnetism
mass spectrometer

normal
paramagnetic
permeability of free space
right-hand rule
solenoid
tesla

Chapter Summary

Topic	Equation or Figure
Magnetism and magnetic forces: Magnetic forces are present whenever moving charged objects interact with each other. (By comparison, electric forces are present whenever charged objects interact with each other, whether moving or not.) A magnet is an object that contains charges in continuous motion. A magnet sets up a magnetic field in the space around it; a second magnet responds to that field and can be attracted or repelled, depending on its orientation.	(a) These magnets attract. (b) These magnets repel. (Figure 19-3 a/b)

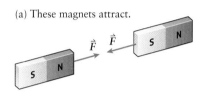

(a) These magnets attract.

(b) These magnets repel.

(Figure 19-3 a/b)

Magnetic force on a moving charged particle: A single charged particle can experience a magnetic force when moving in a magnetic field. The magnitude of the force depends on both the speed of the particle and the direction of the particle relative to the magnetic field. The direction of the force is perpendicular to both the velocity $\vec{v}$ and the magnetic field $\vec{B}$, and is given by a right-hand rule.

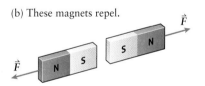

Magnitude of the magnetic force on a moving charged particle

Magnitude of the magnetic field

$$F = |q|vB\sin\theta$$

(19-1)

Magnitude of the particle's charge

Speed of the particle

Angle between the direction of the particle's velocity $\vec{v}$ and the direction of the magnetic field $\vec{B}$

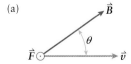

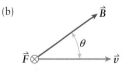

(Figure 19-6)

(a)

If $q > 0$, the force vector in this situation points out of the page (as indicated by the circle with a dot).

(b)

If $q < 0$, the force vector in this situation points into the page (as indicated by the circle with an X).

Particle trajectories in a magnetic field: A charged particle moving in a magnetic field and subject to no other forces can move in a circular trajectory whose radius depends on its speed, mass, and charge as well as the magnetic field magnitude. This is the principle of the mass spectrometer.

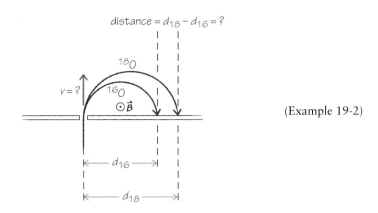

(Example 19-2)

Magnetic forces on current-carrying wires: If a wire carries a current and is placed in a magnetic field, it experiences a magnetic force. This force is the sum of the magnetic forces acting on the individual moving charges within the wire. The force magnitude depends on the amount of current and the orientation of the wire relative to the magnetic field; its direction is given by a right-hand rule.

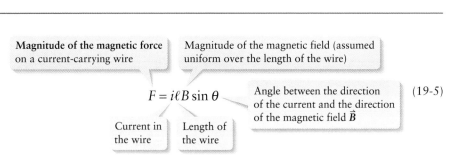

Magnitude of the magnetic force on a current-carrying wire

Magnitude of the magnetic field (assumed uniform over the length of the wire)

$$F = i\ell B \sin\theta$$

Angle between the direction of the current and the direction of the magnetic field $\vec{B}$

Current in the wire

Length of the wire

(19-5)

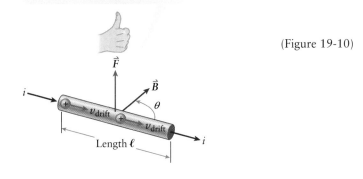

The force $\vec{F}$ on a current-carrying wire in a magnetic field $\vec{B}$ is perpendicular to both $\vec{B}$ and the length of the wire. The direction is given by a right-hand rule.

(Figure 19-10)

Magnetic torque on a current loop: A current-carrying loop of wire can experience a torque when placed in a magnetic field. The magnitude and direction of the torque depend on how the current loop is oriented relative to the direction of the magnetic field. Electric motors make use of this principle.

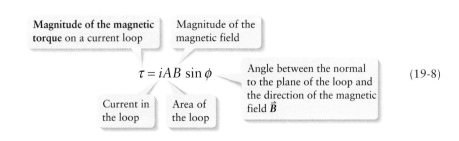

Magnitude of the magnetic torque on a current loop

Magnitude of the magnetic field

$$\tau = iAB \sin\phi$$

Angle between the normal to the plane of the loop and the direction of the magnetic field $\vec{B}$

Current in the loop

Area of the loop

(19-8)

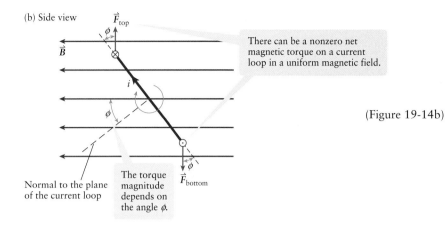

(b) Side view

There can be a nonzero net magnetic torque on a current loop in a uniform magnetic field.

Normal to the plane of the current loop

The torque magnitude depends on the angle ϕ.

(Figure 19-14b)

Ampère's law and the field produced by moving electric currents: A long, straight, current-carrying wire produces a relatively simple magnetic field in the space around it. Ampère's law—which relates the circulation of magnetic field around a closed loop to the amount of current through that loop—can be used to find the magnetic field produced by currents with other simple geometries.

Magnitude of the magnetic field due to a long, straight wire

Permeability of free space

$$B = \frac{\mu_0 i}{2\pi r}$$

Current in the wire

Distance from the wire to the location where the field is measured

(19-9)

Circulation of magnetic field around an Amperian loop

Permeability of free space

$$\sum B_\parallel \Delta \ell = \mu_0 i_{\text{through}}$$

Current through the Amperian loop

(19-13)

(1) We draw an Amperian loop that encircles the current-carrying wire. (This particular loop is a circle, and so coincides with a field line.)

Segment of Amperian loop of length $\Delta \ell$

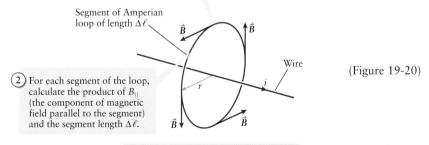

Wire

(2) For each segment of the loop, calculate the product of $B_\parallel$ (the component of magnetic field parallel to the segment) and the segment length $\Delta \ell$.

(Figure 19-20)

(3) The circulation of magnetic field around the loop is the sum of all these products for the entire loop:

$$\sum B_\parallel \Delta \ell$$

Current loops and magnetic materials: A current loop is called a magnetic dipole because the magnetic field that it produces is similar to the electric field produced by an electric dipole. A permanent magnet is a material in which the atoms behave like individual current loops, many of which are oriented in the same direction so that their individual magnetic fields add to make a strong field.

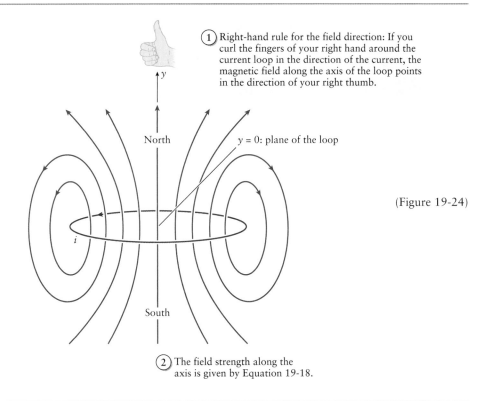

1 Right-hand rule for the field direction: If you curl the fingers of your right hand around the current loop in the direction of the current, the magnetic field along the axis of the loop points in the direction of your right thumb.

North

$y = 0$: plane of the loop

South

2 The field strength along the axis is given by Equation 19-18.

(Figure 19-24)

Force between current-carrying wires: Two parallel current-carrying wires exert magnetic forces on each other: The current in one wire produces a magnetic field, and the current in the other wire responds to that field. The two wires attract if the currents are in the same direction and repel if the currents are in opposite directions.

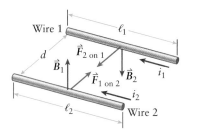

$\vec{B}_1$ = Magnetic field due to wire 1 at the position of wire 2

$\vec{B}_2$ = Magnetic field due to wire 2 at the position of wire 1

(Figure 19-25)

Answer to What do you think? Question

(c) Equation 19-1 (Section 19-3) gives the magnitude of the magnetic force on a charged particle moving in a magnetic field. The magnitude is proportional to sin θ, where θ is the angle between the magnetic field and the velocity of the particle. The sine function is greatest when $\theta = 90°$, in which case the particle is moving perpendicular to the field. This is in stark contrast to the *electric* force on a charged particle, which does not depend on the direction or magnitude of the particle's velocity.

Answers to Got the Concept? Questions

19-1 (b) Two charged objects exert *electric* forces on each other whether or not the objects are moving. But in order for these objects to exert *magnetic* forces on each other, *both* objects must be in motion. (One must be in motion to produce a magnetic field, and the other must be in motion to experience a force due to that field.) Since only one object is in motion, there is no magnetic force between the two objects.

19-2 (a) To apply the right-hand rule to the proton in Figure 19-7, start by pointing the fingers of your right hand in the direction of the proton's initial velocity. Orient your palm so that when you make a "slapping" motion with your hand, your palm points in the direction of the magnetic field (into the page). If you then stick your right thumb out straight, it points toward the top of the page. This is the direction of the force on the proton. This force makes the proton's trajectory bend upward in the figure.

19-3 (a), (c) Electrons have a negative charge, so the direction of the magnetic force each electron experiences is opposite to the direction your thumb points when applying the right-hand rule. For electron A, point the fingers of your right hand in the direction of the electron's initial velocity, then orient your palm so that when you make a "slapping" motion with your hand, your palm points in the direction of the magnetic field (into the page). If you then stick your right thumb out straight, it points to the right, which means the electron feels a force to the left. So electron A feels a force toward the point at which the proton enters the field. For electron B, the right-hand rule has your right thumb pointing toward the bottom of the page, so the force on the electron is toward the top of the page. It does not bend toward the proton's entry point. The right-hand rule predicts the same direction for the magnetic force on electron C, but in this case a force toward the top of the page does bend the electron's trajectory toward the point where the proton enters the field. Finally, when you apply the right-hand rule to electron D, your right thumb sticks out to the left. For the negatively charged electron the magnetic force is therefore to the right, so the trajectory of electron D does not bend toward the proton's entry point into the field.

19-4 (c) Equation 19-3 shows that for a given speed v, the radius of the circular path followed by an ion in a mass spectrometer is $r = mv/(qB)$. This is directly proportional to the ratio of the ion mass m to its charge q. Compared to the value of the ratio m/q for a $^{16}O^+$ ion ($m = 16$ u, $q = e$), the value for a $^{16}O^{2+}$ ion is $1/2$ as great ($m = 16$ u, $q = 2e$), the value for a $^{32}S^+$ ion is twice as great ($m = 32$ u, $q = e$), and the value for a $^{32}S^{2+}$ ion is the same ($m = 32$ u, $q = 2e$).

19-5 (a) Force is zero; (b) $-z$ direction; (c) $+y$ direction; (d) $+z$ direction; (e) $-y$ direction; (f) force is zero. The force on a current-carrying wire is zero if the wire axis lies along the direction of the magnetic field, as in cases (a) and (f). In cases (b), (c), (d), and (e), we use the right-hand rule for the magnetic force on a current-carrying wire to find the direction of the force: Swing the extended fingers of your right hand, palm first, from the direction of the current to the direction of $\vec{B}$. Your extended right thumb then points in the direction of the force on the wire.

19-6 (e) Doubling the number of turns of wire will double the magnetic torque on the current loop. But this also doubles the mass and the moment of inertia of the current loop, so the angular acceleration—equal to the torque divided by the moment of inertia—will be unaffected. For a real electric motor, however, the moment of inertia is due partially to the mass of the coil and partially to the mass of what the motor is turning, such as the fan blades in Figure 19-13a. As a result, doubling the number of turns will increase the maximum torque by a factor of two but increase the moment of inertia by less than a factor of two, and the maximum angular acceleration will in fact increase.

19-7 (b) The net current through the closed curve is i, since the current loop passes once through the plane of the curve. From Ampère's law the circulation of the magnetic field around the closed curve is $\sum B_\parallel \Delta \ell = \mu_0 i_{\text{through}} = \mu_0 i$. The left-hand side of this equation is the average value of $B_\parallel$ multiplied by the total distance around the closed curve, so $(B_\parallel)_{\text{average}} L = \mu_0 i$ and $(B_\parallel)_{\text{average}} = \mu_0 i / L$.

19-8 (a) Each segment of the coil is a piece of wire and is attracted to the piece of wire in the coils directly above and below it (in each of which the current flows in the same direction). Each piece of wire is also repelled by the pieces of wire on the opposite side of the coil above it and the opposite side of the coil below it. But these pieces are at a greater distance d, so these repulsive forces are smaller than the attractive forces (see Equation 19-23). The net result is that the coils attract each other and so pull together.

Questions and Problems

In a few problems, you are given more data than you actually need; in a few other problems, you are required to supply data from your general knowledge, outside sources, or informed estimate.
Interpret as significant all digits in numerical values that have trailing zeros and no decimal points.
For all problems, use $g = 9.80 \text{ m/s}^2$ for the free-fall acceleration due to gravity. Neglect friction and air resistance unless instructed to do otherwise.
• Basic, single-concept problem
•• Intermediate-level problem, may require synthesis of concepts and multiple steps
••• Challenging problem
SSM *Solution is in Student Solutions Manual*

Conceptual Questions

1. •You are given three iron rods. Two of them are magnets but the third one is not. How could you use the two magnets to find that the third rod is not magnetized? SSM

2. •If a magnetic field exerts a force on moving charged particles, is it capable of doing work on the particles? Explain your answer.

3. •A current-carrying wire is in a region where there is a magnetic field, but there is no magnetic force acting on the wire. How can this be?

4. •How is it possible for an object that experiences no net magnetic force to experience a net magnetic torque?

5. ••A velocity selector consists of crossed electric and magnetic fields, with the magnetic field directed toward the top of the page. A beam of positively charged particles passing through the velocity selector from left to right is undeflected by the fields. (a) In what direction is the electric field? (left, right, toward the top of the page, toward the bottom of the page, into the page, out of the page) (b) The direction of the particle beam is reversed so that it travels from right to left. Is it deflected? If so, in what direction? (c) A beam of electrons (negatively charged) moving with the same speed is passed through from left to right. Is it deflected? If so, in what direction? SSM

6. •Physicists refer to crossed electric and magnetic fields as a *velocity selector*. In the same sense, the deflection of charged particles in a strong magnetic field perpendicular to their motion can be thought of as a *momentum selector*. Why?

7. •A long, straight current-carrying wire is placed in a cubic region that has a uniform magnetic field as shown in **Figure 19-27**. Does the force on the wire depend on the width of the magnetic field? Explain your answer.

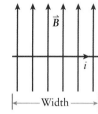

Figure 19-27 Problem 7

8. •In a lightning strike, there is a negative charge moving rapidly from a cloud to the ground. In what direction is a lightning strike deflected by Earth's magnetic field?

9. •In telephone lines, two wires carrying currents in opposite directions are twisted together. How does this reduce the magnetic fields surrounding the wires? SSM

10. •A power cord for an electronic device consists of two parallel straight wires carrying currents in opposite directions. Is there any force between them? Explain your answer.

11. •Parallel wires exert magnetic forces on each other. What about perpendicular wires? Explain your answer.

Multiple-Choice Questions

12. •The magnetic force on a charged moving particle
   A. depends on the sign of the charge on the particle.
   B. depends on the magnetic field at the particle's instantaneous position.
   C. is in the direction which is mutually perpendicular to the direction of motion of the charge and the direction of the magnetic field.
   D. is proportional both to the charge and to the magnitude of the magnetic field.
   E. is described by all of the above options, A through D.

13. •A proton traveling to the right enters a region of uniform magnetic field that points into the page. When the proton enters this region, it will be
   A. deflected out of the plane of page.
   B. deflected into the plane of page.
   C. deflected toward the top of the page.
   D. deflected toward the bottom of the page.
   E. unaffected in its direction of motion. SSM

14. •An electron is moving northward in a magnetic field. The magnetic force on the electron is toward the northeast. What is the direction of the magnetic field?
   A. up
   B. down
   C. west
   D. south
   E. This situation cannot exist because of the orientation of the velocity and force vectors.

15. •A proton with a velocity along the $+x$ axis enters a region where there is a uniform magnetic field $\vec{B}$ in the $+y$ direction. You want to balance the magnetic force with an electric field so that the proton will continue along a straight line. The electric field should be in the
   A. $+x$ direction.
   B. $-x$ direction.

   C. $+z$ direction.
   D. $-z$ direction.
   E. $-y$ direction.

16. •A circular flat coil that has N turns, encloses an area A, and carries a current i has its central axis parallel to a uniform magnetic field $\vec{B}$ in which it is immersed. The net force on the coil is
   A. zero.
   B. $NiAB$.
   C. NiB.
   D. iBA.
   E. NiA.

17. •A circular flat coil that has N turns, encloses an area A, and carries a current i, has its central axis parallel to a uniform magnetic field $\vec{B}$ in which it is immersed. The net torque on the coil is
   A. zero.
   B. $NiAB$.
   C. NiB.
   D. iBA.
   E. NiA. SSM

18. •A very long, straight wire carries a constant current. The magnetic field a distance d from the wire and far from its ends varies with distance d according to
   A. d^{-3}.
   B. d^{-2}.
   C. d^{-1}.
   D. d.
   E. d^2.

19. •A solenoid carries a current. If the radius of the solenoid were doubled, and all other quantities remained the same, the magnetic field inside the solenoid would
   A. remain the same.
   B. be twice as strong as initially.
   C. be half as strong as initially.
   D. be one-quarter as strong as initially.
   E. be four times as strong as initially.

20. ••Two parallel wires carry currents in opposite directions as shown in **Figure 19-28**. Which of the following statements is correct?

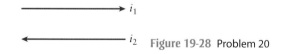

Figure 19-28 Problem 20

   A. The force on the i_2 wire is upward, and the force on the i_1 wire is upward.
   B. The force on the i_2 wire is downward, and the force on the i_1 wire is upward.
   C. The force on the i_2 wire is upward, and the force on the i_1 wire is downward.
   D. The force on the i_2 wire is downward, and the force on the i_1 wire is downward.
   E. Neither wire experiences a net force.

21. ••Two current-carrying wires are perpendicular to each other. One wire lies horizontally with the current directed toward the east. The other wire is vertical with the current directed upward.

What is the direction of the net magnetic force on the horizontal wire due to the vertical wire?

    A. east
    B. west
    C. south
    D. north
    E. zero force SSM

Estimation/Numerical Analysis

22. ••Give an estimate of the force per unit length acting on a pair of high-voltage transmission lines that carry current from a power plant to a distant substation for commercial use (assume that the wires carry current in the same direction).

23. •Estimate the magnetic field at the ground level beneath standard power lines in your neighborhood.

24. •Estimate the magnetic field needed to separate two uranium ions (one has a mass of 235 u, one has a mass of 238 u) into circular paths with radii separated by 1 cm. Assume the two ions enter the magnetic field with the same velocity and the same charge.

25. •Estimate the time required for an electron to complete a circular orbit of 10 cm radius in Earth's magnetic field when it moves at a speed that is of the order of 10^5 m/s. SSM

26. ••Estimate the maximum torque on a coil in a typical electric drill motor.

27. •The magnetic field due to a current-carrying cylinder of radius 1 cm is measured at various points ($r < 1$ cm and $r > 1$ cm). Graph the magnitude of the magnetic field B as a function of r and use this graph to find the functional relationship between the magnetic field and the radial distance. *Hint:* The magnetic field will be described by two different functions—one for inside the cylinder and one for outside the cylinder. After making an initial graph, you may want to graph the data for the inside and outside separately.

r(m)	B(T)	r(m)	B(T)
0.001	0.00050	0.015	0.00353
0.002	0.00100	0.020	0.00250
0.003	0.00152	0.025	0.00200
0.004	0.00200	0.030	0.00180
0.005	0.00252	0.035	0.00143
0.006	0.00300	0.040	0.00125
0.007	0.00350	0.045	0.00110
0.008	0.00401	0.050	0.00103
0.009	0.00453	0.100	0.000502
0.010	0.00500		

Problems

19-1 Magnetic forces are interactions between two magnets

19-2 Magnetism is an interaction between moving charges

19-3 A moving point charge can experience a magnetic force

28. •Convert the units for the following expressions for magnetic fields as directed:

    A. 5.00 T = _____ G
    B. 25,000 G = _____ T
    C. 7.43 mG = _____ μT
    D. 1.88 mT = _____ G

29. •Determine the directions of the magnetic forces that act on positive charges moving in the magnetic fields as shown in Figure 19-29. SSM

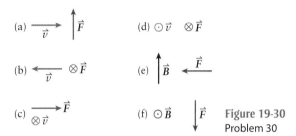

Figure 19-29
Problem 29

30. •Determine the direction of the missing vector, $\vec{v}$, $\vec{B}$, or $\vec{F}$, in the scenarios shown in Figure 19-30. All moving charges are positive.

(a) $\vec{v}$ → $\vec{F}$ ↑ (d) $\odot\vec{v}$ $\otimes\vec{F}$

(b) ← $\vec{v}$ $\otimes\vec{F}$ (e) $\vec{B}$ ↑ ← $\vec{F}$

(c) $\otimes\vec{v}$ → $\vec{F}$ (f) $\odot\vec{B}$ $\vec{F}$ ↓ Figure 19-30 Problem 30

31. •A +1 C charge moving at 1 m/s makes an angle of 45° with a uniform, 1-T magnetic field. What is the magnitude of the magnetic force that the charge experiences?

32. •An electron is moving with a speed of 18 m/s in a direction parallel to a uniform magnetic field of 2.0 T. What are the magnitude and direction of the magnetic force on the electron?

33. •A proton P travels with a speed of 18 m/s toward the top of the page through a uniform magnetic field of 2.0 T directed into the page as shown in Figure 19-31. What are the magnitude and direction of the magnetic force on the proton? SSM

Figure 19-31 Problem 33

34. •A proton is propelled at 2×10^6 m/s perpendicular to a uniform magnetic field. If it experiences a magnetic force of 5.8×10^{-13} N, what is the strength of the magnetic field?

35. •An electron moves with a velocity of 10^7 m/s in the x–y plane at an angle of 45° to both the $+x$ and $+y$ axes. There is a magnetic field of 3.0 T in the $+y$ direction. Calculate the magnetic force (magnitude and direction) on the electron. SSM

36. ••There is a uniform magnetic field of magnitude 2.2 T in the $+z$ direction. Find the force on a particle of charge -1.2 nC if its velocity is (a) 1.0 km/s in the y–z plane in a direction that makes an angle of 40° with the z axis and (b) 1.0 km/s in the x–y plane in a direction that makes an angle of 40° with the x axis.

19-4 A mass spectrometer uses magnetic forces to differentiate atoms of different masses

37. ••A beam of protons is directed in a straight line along the $+z$ direction through a region of space in which there are

crossed electric and magnetic fields. If the electric field is 500 V/m in the $-y$ direction and the protons move at a constant speed of 10^5 m/s, what must be the magnitude and direction of the magnetic field such that the beam of protons continues along its straight-line trajectory?

38. ••A beam of particles (each particle has a charge $q = -2e$ and a kinetic energy of 4.00×10^{-13} J) is deflected by the magnetic field of a bending magnet as shown in Figure 19-32. The radius of curvature of the beam is 20.0 cm, and the strength of the magnetic field is 1.50 T. (a) What is the mass of the particles making up the beam? (b) Sketch the path for the given ions in the given magnetic field as a reference path. Then sketch a path for a more massive doubly ionized negative ion and a less massive doubly ionized positive ion, both with the same speed as the given ions, for comparison.

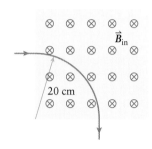

Figure 19-32 Problem 38

39. •A proton moves in a circle with a speed of 280 m/s through a uniform magnetic field of 2.0 T directed perpendicular to the circular path. What is the orbital radius of the proton?

19-5 Magnetic fields exert forces on current-carrying wires

40. •A 1.5-m length of straight wire experiences a maximum force of 2.0 N when in a uniform magnetic field that is 1.8 T. What current must be passing through it?

41. •A straight segment of wire 35.0 cm long carrying a current of 1.40 A is in a uniform magnetic field. The segment makes an angle of 53° with the direction of the magnetic field. If the force on the segment is 0.200 N, what is the magnitude of the magnetic field? SSM

42. •A straight wire of length 0.50 m is conducting a current of 2.0 A and makes an angle of 30° with a 3.0-T uniform magnetic field. What is the magnitude of the force exerted on the wire?

43. •A wire of length 0.50 m is conducting a current of 8.0 A in the $+x$ direction through a 4.0-T uniform magnetic field directed parallel to the wire. What are the magnitude and direction of the magnetic force on the wire?

44. •A wire of length 0.50 m is conducting a current of 8.0 A toward the top of the page and through a 4.0-T uniform magnetic field directed into the page as shown in Figure 19-33. What are the magnitude and direction of the magnetic force on the wire?

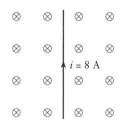

Figure 19-33 Problem 44

45. •A straight wire is positioned in a uniform magnetic field so that the maximum force on it is 4.0 N. If the wire is 80 cm long and carries a current that is 2 A, what is the magnitude of the magnetic field?

19-6 A magnetic field can exert a torque on a current loop

46. ••A square loop 10 cm on a side consists of 100 turns of wire that experiences a minimum torque of zero and a maximum

torque of 0.045 N · m. If the current in the loop is 2.82 A, calculate the magnetic field strength that the loop rests in.

47. ••What is the torque on a round loop of wire that carries a current of 100 A, has a radius of 10 cm, and whose plane makes an angle of 30° with a magnetic field of 0.244 T? Describe how the answer changes if the angle decreases to 10° and increases to 50°. SSM

19-7 Ampère's law describes the magnetic field created by current-carrying wires

48. ••A long, straight wire carries current in the $+z$ direction (out of the page). Determine the direction of the magnetic field due to the current at the points O, P, Q, and R (Figure 19-34).

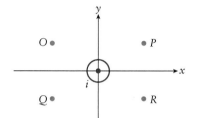

Figure 19-34 Problem 48

49. ••A long, straight wire carries current in the $+x$ direction. Determine the direction of the magnetic field due to the current at the points O, P, Q, and R (Figure 19-35).

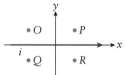

Figure 19-35 Problem 49

50. ••Using Figure 19-36, derive an expression for the magnetic field at the point C located at the center of the two circular, current-carrying arcs and the connecting radial lines. Assume the radii of the small and large arcs are r_1 and r_2, respectively.

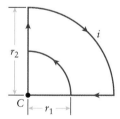

Figure 19-36 Problem 50

51. ••Derive an expression for the magnetic field at the point C at the center of the circular, current-carrying wire segments shown in Figure 19-37. SSM

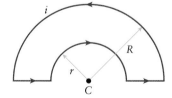

Figure 19-37 Problem 51

52. ••Calculate the magnitude of the magnetic field at a perpendicular distance of 2.2 m from a long, copper pipe that has a diameter of 2 cm and carries a current of 20 A.

53. ••Jerry wants to predict the magnetic field that a high-voltage line creates in his apartment. A current of 100 A passes through a wire that is 5 m from his window. Calculate the magnetic field.

How does the field compare to the magnitude of Earth's magnetic field of about 5×10^{-5} T in New York City? SSM

54. •A solenoid with 25 turns per centimeter carries a current of 25 mA. What is the magnetic field in the interior of the coils?

55. •You want to wind a solenoid that is 3.5 cm in diameter, is 16 cm long, and will have a magnetic field of 0.0250 T when a current of 3.0 A is in it. What total length of wire do you need?

56. •••A coaxial cable consists of a solid inner conductor of radius R_i, surrounded by a concentric outer conducting shell of radius R_o. Insulating material fills the space between the conductors. The inner conductor carries current to the right, and the outer conductor carries the same current to the left down the outer surface of the cable, as shown in Figure 19-38. Using Ampère's law, derive an expression for the magnetic field in three separate regions of space: inside the inner conductor, between the two conductors, and outside of the outer conductor.

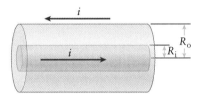

Figure 19-38 Problem 56

19-8 Two current-carrying wires exert magnetic forces on each other

57. •If wire 1 carries 2.00 A of current north, wire 2 carries 3.60 A of current south, and the two wires are separated by 1.40 m, calculate the force (magnitude and direction) acting on a 1.00-cm section of wire 1 due to wire 2. SSM

58. •What is the net force on the rectangular loop of wire that is 2.00 cm wide, is 6.00 cm long, and is located 2.00 cm from a long, straight wire that carries $i = 40.0$ A of current as shown in Figure 19-39? Assume a current of 20.0 A is in the loop.

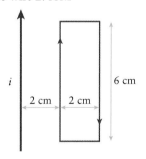

Figure 19-39 Problem 58

59. ••The fasteners on overhead power lines are 50.0 cm long. What force must they be able to withstand if two high-voltage lines are 2.00 m apart, each carrying 2500 A in the same direction?

General Problems

60. ••Horizontal electric power lines supported by vertical poles can carry large currents. Assume that Earth's magnetic field runs parallel to the surface of the ground from south to north with a magnitude of 0.50×10^{-4} T and that the supporting poles are 32 m apart. Find the magnitude and direction of the force that Earth's magnetic field exerts on a 32-m segment of wire carrying 95 A if the current runs (a) from north to south, (b) from east to west, or (c) toward the northwest making an angle of 30° north

of east. (d) Are any of the above forces large enough to have an appreciable effect on the power lines?

61. ••A levitating train is three cars long (180 m) and has a mass of 100 metric tons (1 metric ton = 1000 kg). The current in the superconducting wires is about 500 kA, and even though the traditional design calls for many small coils of wire, assume for this problem that there is a 180-m-long wire carrying the current. Find the size of the magnetic field needed to levitate the train.

62. ••An electron and a proton have the same kinetic energy upon entering a region of constant magnetic field and their velocity vectors are perpendicular to the magnetic field. Suppose the magnetic field is strong enough to allow the particles to circle in the field. What is the ratio of the radii of their circular paths r_p/r_e?

63. •••In Figure 19-40, which shows a mass spectrometer, a particle with charge $-e = -1.60 \times 10^{-19}$ C enters a region of magnetic field that has a strength of 0.00242 T, into the page. The velocity of the particle is confirmed with a velocity selector. The electric field is 90,000 V/m (down) and the magnetic field is 0.0053 T (into page) in the velocity selector. If the radius of curvature of the particle is 4.00 cm, use the information in this problem to calculate the mass of the particle. SSM

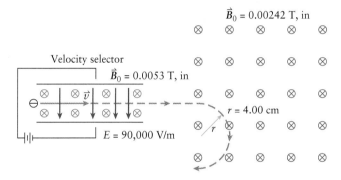

Figure 19-40 Problem 63

64. ••Biology The National High Magnetic Field Laboratory holds the world record for creating the largest magnetic field. Their long-pulse magnet produced a magnetic field of 60 T. To see if such a strong magnetic field could pose health risks for nearby workers, calculate the maximum acceleration the field could produce on Na⁺ ions (of mass 3.8×10^{-26} kg) in blood traveling through the aorta. The speed of blood is highly variable, but 50 cm/s is reasonable in the aorta. Does your result indicate that it would be dangerous to expose workers to such a large magnetic field?

65. ••During electrical storms, a bolt of lightning can transfer 10 C of charge in 2.0 μs (the amount and time can vary considerably). We can model such a bolt as a very long current-carrying wire. (a) What is the magnetic field 1.0 m from such a bolt? What is the field 1.0 km away? How do the fields compare with Earth's magnetic field? (b) Compare the fields in part (a) with the magnetic field produced by a typical household current of 10 A in a very long wire at the same distances from the wire as in (a). (c) How close would you have to get to the wire in part (b) for its magnetic field to be the same as the field produced by the lightning bolt at 1.0 km from the bolt?

66. •A straight wire carries a current of 8.00 A toward the top of the page. What are the magnitude and direction of the magnetic field at point P, which is 8.00 cm to the right of the wire as shown in **Figure 19-41**?

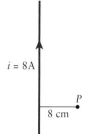

$i = 8A$

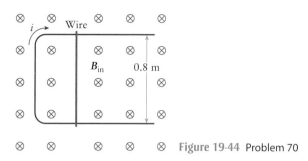

P

8 cm

Figure 19-41 Problem 66

67. ••A long, straight wire carries a current as shown in **Figure 19-42**. A charged particle moving parallel to the wire experiences a force of 0.80 N at point P. Assuming the same charge and same velocity, what would be the magnitude of the magnetic force on the charge at point S?

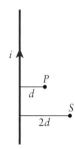

i

P

d

S

$2d$

Figure 19-42 Problem 67

68. ••Two long, straight wires parallel to the x axis are at $y = \pm 2.5$ cm (**Figure 19-43**). Each wire carries a current of 16 A in the $+x$ direction. Calculate the magnetic field on the y axis at (a) $y = 0$, (b) $y = 1.0$ cm, and (c) $y = 4.0$ cm.

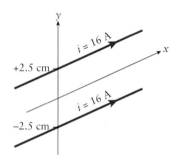

y

$i = 16$ A

x

+2.5 cm

$i = 16$ A

−2.5 cm

Figure 19-43 Problem 68

69. ••**Medical** Transcranial magnetic stimulation (TMS) is a noninvasive method to stimulate the brain using magnetic fields. It is used in treating strokes, Parkinson's disease, depression, and other physical conditions. In the procedure, a circular coil is placed on the side of the forehead to generate a magnetic field inside the brain. Although values can vary, a typical coil would be about 15 cm in diameter and contain 250 thin circular windings. The magnetic field in the cortex (3.0 cm from the coil measured along a line perpendicular to the coil at its center) is typically 0.50 T. (a) What current in the coil is needed to produce the desired magnetic field inside the brain? (b) What is the magnetic field at the center of the coil at the forehead? (c) If the current needed in part (a) seems too large, how could you easily achieve the same magnetic field with a smaller current?

70. ••A wire of mass 40 g slides without friction on two horizontal conducting rails spaced 0.8 m apart. A steady current of 100 A is in the circuit formed by the wire and the rails. A uniform magnetic field of 1.2 T, directed into the plane of the drawing, acts on it. (a) In which direction in **Figure 19-44** will the wire accelerate? (b) What is the magnetic force on the wire? (c) How long must the rails be if the wire, starting from rest, is to reach a speed of 200 m/s? (d) If the magnetic field were directed out of the page, how would your answers differ?

(e) What if the magnetic field were directed toward the top of the drawing?

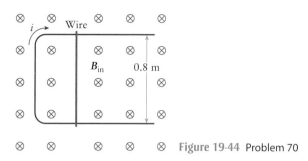

Wire

B_{in} 0.8 m

Figure 19-44 Problem 70

71. ••A small 20-turn current loop with a 4.00-cm diameter is suspended in a region with a magnetic field of 1000 G, with the plane of the loop parallel with the magnetic field direction. (a) What is the current in the loop if the torque exerted by the magnetic field on the loop is 4.00×10^{-5} N · m? (b) Describe the subsequent motion of the loop if it is allowed to rotate.

72. ••A long, straight wire carries a current of 1.2 A toward the south. A second, parallel wire carries a current of 3.8 A toward the north and is 2.8 cm from the first wire. What is the magnetic force per unit length each wire exerts on the other?

73. •••Two straight conducting rods, which are 1.0 m long, exactly parallel, and separated by 0.85 mm, are connected by an external voltage source and a 17-Ω resistance, as shown in **Figure 19-45**. The 0.5-Ω rod "floats" above the 2.5-Ω rod, in equilibrium. If the mass of each rod is 25 g, what must be the potential of the voltage source? SSM

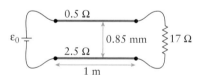

0.5 Ω

ε_0 0.85 mm   17 Ω

2.5 Ω

1 m

Figure 19-45 Problem 73

74. ••**Medical** When operated on a household 110-V line, typical hair dryers draw about 1650 W of power. We can model the current as a long, straight wire in the handle. During use, the current is about 3.0 cm from the user's head. (a) What is the current in the dryer? (b) What is the resistance of the dryer? (c) What magnetic field does the dryer produce at the user's head? Compare the field with Earth's magnetic field to decide if we should have health concerns about the magnetic field created when using a hair dryer.

75. ••Three very long, straight wires lie at the corners of a square of side d, as shown in **Figure 19-46**. The magnitudes of the currents in the three wires are the same, but the two diagonally opposite currents are directed into the page while the other one is directed outward. Derive an expression for the magnetic field (magnitude and direction) at the fourth corner of the square.

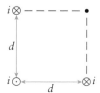

i

d

i d i

Figure 19-46 Problem 75

76. ••A 2.0-m lamp cord leads from the 110-V outlet to a lamp having a 75-W lightbulb. The cord consists of two insulated parallel wires 4.0 mm apart and held together by the insulation. One wire carries the current into the bulb, and the other carries it out. What is the magnitude of the magnetic field the cord produces (a) midway between the two wires and (b) 2.0 mm from one of the wires in the same plane in which the two wires lie? (c) Compare each of the fields in parts (a) and (b) with Earth's magnetic field (0.5×10^{-4} T). (d) What magnetic force (magnitude and direction) do the two wires exert on one another? Is the force large enough to stress the insulation holding the wires together?

77. ••Some people have raised concerns about the magnetic fields produced by current-carrying high-voltage lines in residential neighborhoods. Currents in such lines can be up to 100 A. Suppose you have such a line near your house. If the wires are supported horizontally 5.0 m above the ground on vertical poles and your living room is 12 m from the base of the poles, what magnetic field strength does the wire produce in your living room if it carries 100 A? Express your answer in teslas and as a multiple of Earth's magnetic field. Does the magnetic field from such wires seem strong enough to cause health concerns? SSM

78. ••**Medical** Magnetoencephalography (MEG) is a technique for measuring changes in the magnetic field of the brain caused by external stimuli such as touching the body or viewing images of food. Such a change in the field occurs due to electrical activity (current) in the brain. During the process, magnetic sensors are placed on the skin to measure the magnetic field at that location. Typical field strengths are a few femtoteslas (1 femtotesla = 1 fT = 10^{-15} T). An adult brain is about 140 mm wide, divided into two sections (called hemispheres although the brain is not truly spherical) each about 70 mm wide. We can model the current in one hemisphere as a circular loop, 65 mm in diameter, just inside the brain. The sensor is placed so that it is along the axis of the loop 2.0 cm from the center. A reasonable magnetic field is 5.0 fT at the sensor. According to this model, (a) what is the current in the brain and (b) what is the magnetic field at the center of the hemisphere of the brain?

79. ••Helmholtz coils are composed of two coils of wire that have their centers on the same axis, separated by a distance that is equal to the radius of the coils (**Figure 19-47**). The coils have N turns of wire that carry a current of i in the same direction. If one coil is centered at the origin, and the other at $x = R$, derive expressions for the net magnetic field due to the coils at the points (a) $x = R/2$ and (b) $x = 2R$.

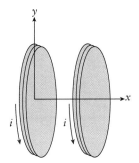

Figure 19-47 Problem 79

80. ••Geophysicists may use the gauss unit for magnetic field (10^4 G = 1 T). Earth's magnetic field at the equator can be taken as 0.7 G directed north. At the center of a flat circular coil that has 10 turns of wire and is 1.4 m in diameter, the coil's magnetic field exactly cancels Earth's field. (a) What must be the current in the coil? (b) How should the coil be oriented?

81. •••**Biology** Migratory birds use Earth's magnetic field to guide them. Some people are concerned that human-caused magnetic fields could interfere with bird navigation. Suppose that a pair of parallel high-voltage lines, each carrying 100 A, are 3.00 m apart and lie in the same horizontal plane. Find the magnitude and direction of the magnetic field the lines produce at a point 15.0 m above them equidistant from both lines in each of the following cases. (a) The lines run in the north–south direction and both currents run from north to south. (b) Both lines run in the north–south direction and the current in the eastern line runs northward while the current in the western line runs southward. (c) The lines run in the east–west direction, and both currents run from west to east. (d) Is it reasonable to think that the fields caused by the wires are likely to interfere with bird migration?

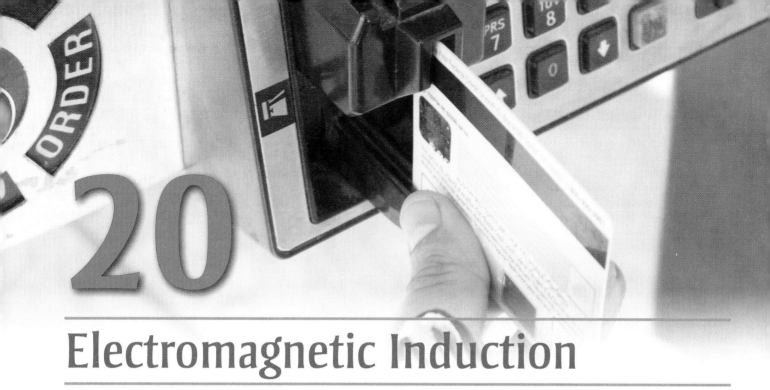

20
Electromagnetic Induction

What do you think?

The stripe on the back of a credit card is magnetized in a pattern that encodes your account information. A credit card reader contains a loop of wire, and when you swipe the card through the reader, the magnetized card's motion generates an electric current in the wire that sends a signal to the credit card company. To make this current flow, what kind of force must act on electrons in the wires of the card reader? (a) A magnetic force; (b) an electric force; (c) a combination of electric and magnetic forces.

In this chapter, your goals are to:

- (20-1) Explain the importance of electromagnetic induction.
- (20-2) Describe what is meant by a motional emf and an induced emf.
- (20-3) Explain what determines the magnitude and direction of an emf in a circuit with a changing magnetic flux.
- (20-4) Define the key properties of an ac generator.

To master this chapter, you should review:

- (6-2) The work done by a constant force.
- (12-3) Simple harmonic motion.
- (16-6) The electric flux through an area.
- (18-6) The electric power for a resistor.
- (19-3, The magnetic force on a moving charged particle and on a
  19-5) current-carrying wire.
- (19-7) The magnetic field produced by a current-carrying loop.

20-1 The world runs on electromagnetic induction

We've discussed electric circuits in which the current is caused by the emf provided by a battery. But many of the electric circuits around you are *alternating-current* circuits in which the current constantly changes direction. That includes the current in light fixtures, toasters, electric fans, and other devices plugged into wall sockets. (Alternating current also indirectly powers mobile devices like cell phones and laptop computers. These devices have batteries, but the batteries are recharged by plugging them into a wall socket.) What kind of emf produces an alternating current?

The answer to this question comes from a remarkable discovery made by physicists around 1830: *If the magnetic field in a region of space changes, the change gives rise to an electric field*. This electric field, called an *induced* field, is very different in character from the electric field produced by point charges that we described in Chapter 16: An induced electric field does not point away from positive charges and toward negative charges, but instead has field lines that form closed loops like magnetic field lines. That's just what's needed for that induced electric field to push charges around a loop of wire and generate an electric current. As we'll see later in this chapter, it's easy to make this induced electric field flip its direction back and forth, which makes a current that flips back and forth—in other words, an alternating current. The vast amount of electric current used by our technological civilization is produced in this way (Figure 20-1a).

We use the term **electromagnetic induction** for the process whereby a changing magnetic field induces an electric field. (The word *electromagnetic* shows that this

An electric generator produces current by electromagnetic induction: Coils of wire move relative to a magnetic field, which generates an emf in the coils. The motion can be powered by the wind, as in these wind turbines.

(a)

A changing magnetic field applied to the brain induces an electric field there, causing electric currents. Areas in red are where the currents are strongest.

(b)

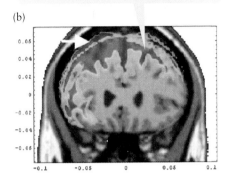

Figure 20-1 Electromagnetic induction Two examples of the phenomenon of electromagnetic induction, in which electric currents are induced by the presence of a changing magnetic flux.

process involves both electric and magnetic fields.) Electromagnetic induction has many applications beyond producing an alternating current to be delivered to wall sockets. It's how a credit card reader decodes the information on the card's magnetized strip (see the photo that opens this chapter). It's also at the heart of a relatively new medical technique called *transcranial magnetic stimulation* (TMS), which allows physicians to stimulate electrical activity in the brain without sticking electrodes to the scalp or inserting them through the skull. In TMS, a time-varying magnetic field is produced inside the brain by current-carrying coils around the head. This causes an induced electric field, which in turn causes currents to flow within the brain (Figure 20-1b). TMS has been used with some success to treat cases of depression that have not responded to more conventional therapy.

In this chapter we'll begin by describing the relationship between a changing magnetic flux and the electric field that it induces. We'll introduce two important laws that describe electromagnetic induction. The first of these, Faraday's law, will tell us how the emf that appears in a closed loop (such as an electric circuit) due to an induced electric field is related to the rate of change of the magnetic flux through the loop. The second, Lenz's law, will tell us the direction of this induced emf. We'll see how induced emf makes possible the important device called a *generator*, which converts mechanical energy into electric energy and creates an alternating emf. (Each of the wind turbines shown in Figure 20-1 uses its spinning blades to run a generator.) In Chapter 21 we'll see how all of these ideas explain the behavior of alternating-current circuits.

Take-Home Message for Section 20-1

✔ In electromagnetic induction, a time-varying magnetic flux in a certain region gives rise to an electric field in that same region.

✔ Electromagnetic induction is used to produce an alternating current.

20-2 A changing magnetic flux creates an electric field

Figure 20-2 shows an experiment that we can understand with the physics we already know. A loop of wire with an attached ammeter (a device for measuring the current in the loop) is moved toward the south pole of a stationary magnet. No current flows if the loop is held stationary. That's not surprising, since there's no source of emf connected to the loop. But a current *does* flow in the loop when it is moved toward the magnet's south pole (Figure 20-2a), and flows in the opposite direction when the loop is moved away from the magnet's south pole (Figure 20-2b). What's happening is that the mobile charges within the loop are moving along with the loop through the magnetic field of the bar magnet and so experience a magnetic force that pushes the charges around the loop (Figure 20-2c). Reversing the direction in which the loop and its charges move also reverses the direction of the magnetic force, so the charges are pushed in the opposite direction and the current direction reverses (Figure 20-2d). In either case we say that the magnetic force on the mobile charges is equivalent to an emf that makes the current flow. There is no magnetic force, and hence no emf, if the loop and its charges are at rest. (Recall from Section 19-2 that magnetic forces only act on *moving* charges.) Because the loop must be in motion in order for the emf to appear, we call it a **motional emf.**

motional emf

Figure 20-2 A loop of wire moving with respect to a magnet If a loop of wire moves toward or away from a magnet, a current flows in the loop. The current is caused by magnetic forces.

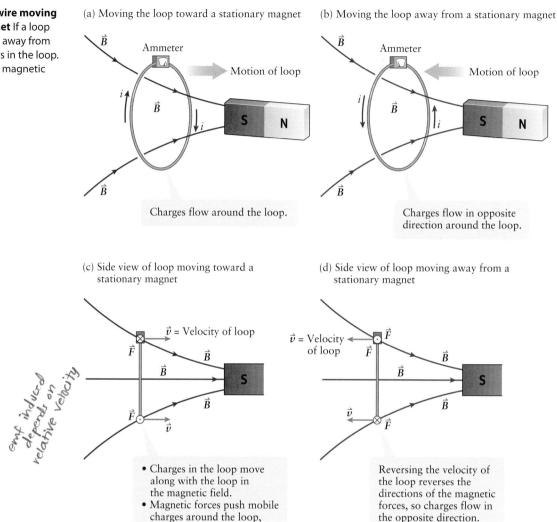

(a) Moving the loop toward a stationary magnet

Charges flow around the loop.

(b) Moving the loop away from a stationary magnet

Charges flow in opposite direction around the loop.

(c) Side view of loop moving toward a stationary magnet

- Charges in the loop move along with the loop in the magnetic field.
- Magnetic forces push mobile charges around the loop, causing a current.

emf induced depends on relative velocity

(d) Side view of loop moving away from a stationary magnet

Reversing the velocity of the loop reverses the directions of the magnetic forces, so charges flow in the opposite direction.

Figure 20-3 shows a very similar experiment that involves entirely new and different physics. Now we hold the loop stationary and move the south pole of the magnet either toward the loop (**Figure 20-3a**) or away from the loop (**Figure 20-3b**). In this case there can be no magnetic force on the mobile charges within the loop, because those charges are at rest in the stationary loop. Nonetheless, there is an emf in the loop and a current flows around the loop in response, but only when the magnet is moving relative to the loop. Since there is no magnetic force in this situation, it must be that the emf is due to an *electric* force on the mobile charges (**Figures 20-3c** and **20-3d**). What's happening is that when the magnet is moving, the magnetic field strength at the location of the loop is changing: It increases when the magnet's south pole moves toward the loop (Figure 20-3a) and decreases when the magnet's south pole moves away from the loop (Figure 20-3b). So this experiment shows that an electric field is *induced* by the changing magnetic field. For this reason we call the emf in the experiment of Figure 20-3 an **induced emf**. We use the term *electromagnetic induction* for any situation in which a changing magnetic field causes, or induces, an electric field.

Although the experiments in Figures 20-2 and 20-3 are different, they have the *same* result: Whether the loop moves toward the stationary magnet at 1 m/s as in Figure 20-2a or the magnet moves toward the stationary loop at 1 m/s as in Figure 20-3a, the same emf appears in the loop. In fact, the same emf appears if the magnet and loop are both moving, as long as the magnet and loop approach each other at a relative speed of 1 m/s. Since the result is the same in each of these cases, we should be able to describe all of these effects in terms of a single equation. But what equation is that?

induced emf

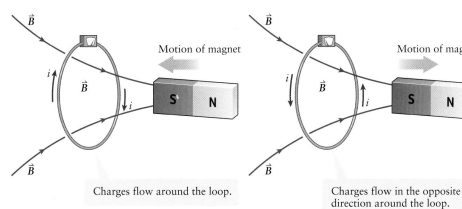

(a) Moving the magnet toward a stationary loop

(b) Moving the magnet away from a stationary loop

Motion of magnet

Motion of magnet

Charges flow around the loop.

Charges flow in the opposite direction around the loop.

Figure 20-3 A magnet moving with respect to a loop of wire If a magnet moves toward or away from a loop of wire, a current flows in the loop. Magnetic forces cannot explain why this happens, so electric forces must be present to produce the current.

(c) Side view of magnet moving toward a stationary loop

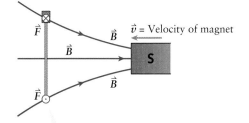

$\vec{v}$ = Velocity of magnet

(d) Side view of magnet moving away from a stationary loop

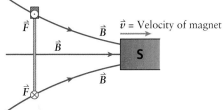

$\vec{v}$ = Velocity of magnet

- Charges in the loop experience electric forces.
- These forces push mobile charges around the loop, causing a current.

Reversing the velocity of the magnet reverses the directions of the electric forces, so charges flow in the opposite direction.

It turns out that a simple way to describe the emf in any of these situations is in terms of the change in *magnetic flux* through the loop in Figures 20-2 and 20-3. We define this in the same way that we defined *electric* flux in Section 16-6: It's the area A of the surface outlined by the loop, multiplied by $B \cos \theta$, the component of the magnetic field that's perpendicular to that surface (see part (a) of **Figure 20-4**). In equation form, the **magnetic flux** Φ_B ("phi-sub-B") through the loop is

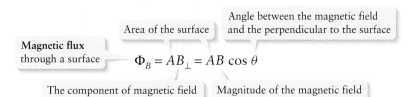

Angle between the magnetic field and the perpendicular to the surface

Area of the surface

Magnetic flux through a surface

$$\Phi_B = AB_\perp = AB \cos \theta$$

The component of magnetic field perpendicular to the surface

Magnitude of the magnetic field

Magnetic flux
(20-1)

(The subscript B reminds us that this is the flux of the magnetic field $\vec{B}$.) As parts (b) and (c) of Figure 20-4 show, the flux Φ_B can be positive or negative. Note that the choice of positive x direction is arbitrary; in Figure 20-4 we chose the positive x direction to be upward, so Φ_B is positive for the case shown in Figure 20-4b and negative for the case shown in Figure 20-4c. Had we chosen the positive x direction to be downward, we would have had $\Phi_B < 0$ in Figure 20-4b and $\Phi_B > 0$ in Figure 20-4c. It doesn't matter which one we choose, since the physics will turn out to be the same in either case. (It's much as in Chapter 2, where we had to make a choice of positive x direction for analyzing motion in a straight line. The actual motion didn't depend on which direction we chose to be positive and which to be negative.)

$\sqrt{x}$ *See the Math Tutorial for more information on trigonometry*

Figure 20-4 Magnetic flux (a) The magnetic flux through a loop. The flux is (b) most positive when $\vec{B}$ points along the perpendicular to the loop and (c) most negative when $\vec{B}$ points opposite to the perpendicular.

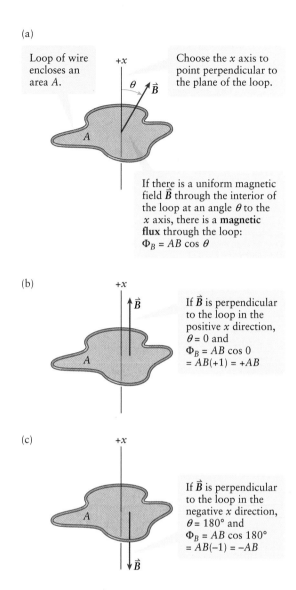

(a)

Loop of wire encloses an area A.

Choose the x axis to point perpendicular to the plane of the loop.

If there is a uniform magnetic field $\vec{B}$ through the interior of the loop at an angle θ to the x axis, there is a **magnetic flux** through the loop:
$\Phi_B = AB \cos \theta$

(b)

If $\vec{B}$ is perpendicular to the loop in the positive x direction, $\theta = 0$ and
$\Phi_B = AB \cos 0$
$= AB(+1) = +AB$

(c)

If $\vec{B}$ is perpendicular to the loop in the negative x direction, $\theta = 180°$ and
$\Phi_B = AB \cos 180°$
$= AB(-1) = -AB$

In Figures 20-2 and 20-3 the magnetic field is not uniform over the area enclosed by the loop, and so the perpendicular component $B_\perp = B \cos \theta$ has different values at different points on this area. In such a case $B_\perp$ in Equation 20-1 is the perpendicular component of the magnetic field *averaged* over the area A enclosed by the loop. Note that if the loop is actually a coil with N turns of wire, the net magnetic flux through the coil is N multiplied by the flux through one turn of the coil.

If the magnet and loop in Figures 20-2 and 20-3 are not moving with respect to each other, the magnetic flux through the loop remains the same. In this case there is no emf and no current in the loop. The flux changes, however, when either the loop moves relative to the magnet (Figure 20-2) or the magnet moves relative to the loop (Figure 20-3). In these cases there *is* an emf in the loop and the current. This suggests that *an emf appears in a loop when the magnetic flux through that loop changes*. This observation is known as **Faraday's law of induction**, named for the nineteenth-century English physicist Michael Faraday:

Faraday's law of induction
(20-2)

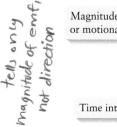

tells only magnitude of emf, not direction

Magnitude of the induced or motional emf in a loop

Change in the magnetic flux through the surface outlined by the loop

$$|\varepsilon| = \left| \frac{\Delta \Phi_B}{\Delta t} \right|$$

Time interval over which the change in magnetic flux takes place

This law states that the magnitude of the emf that appears in a loop is equal to the magnitude of the *rate of change* of the magnetic flux through the loop. If a large change in flux $\Delta\Phi_B$ happens in a short time interval Δt, the resulting emf has a large magnitude; if the change in flux is relatively small and happens over a long time interval, the resulting emf has a small magnitude.

Note that Equation 20-2 tells us only the *magnitude* of the emf, not its direction. In the following section we'll see how the direction is determined.

! Watch Out! It's not the magnetic flux that causes an emf, but the rate at which the flux changes.

The mere presence of magnetic flux through a loop does not cause an emf to appear in the loop. If a flux is present but does not change, such as what happens when a magnet and loop are held stationary with respect to each other, there is *no* resulting emf. An emf only appears when the flux *changes*, such as when the magnet and loop in Figures 20-2 and 20-3 move either toward or away from each other.

In the following example we'll check Faraday's law. We'll do this by considering a situation in which we can use our knowledge of magnetic forces to calculate the emf, then compare this to the emf calculated using Equation 20-2.

 Go to Interactive Exercise 20-1 for more practice dealing with motional emf

---

Example 20-1 Changing Magnetic Flux I: A Sliding Bar in a Magnetic Field

A copper bar of length L slides at a constant speed v along stationary, U-shaped copper rails (Figure 20-5). A uniform magnetic field of magnitude B is directed perpendicular to the plane of the bar and rails. The moving bar and stationary rails form a closed circuit, and an emf is produced in this circuit because the wire is moving in a magnetic field. Determine the emf in the circuit (a) by using the expression for the magnetic force on a charge in the moving wire and (b) by using Faraday's law of induction, Equation 20-2.

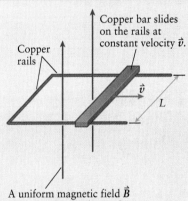

Copper bar slides on the rails at constant velocity $\vec{v}$.

A uniform magnetic field $\vec{B}$ points perpendicular to the plane of the rails.

Figure 20-5 **A sliding copper bar** What emf is generated in the bar as it slides in the presence of a magnetic field $\vec{B}$?

Set Up

For a battery, the magnitude of the emf equals the change in electric potential (potential energy per charge) for a charge that traverses the battery. That is, it's equal to the *work per charge* that the battery does on charges that travel from one terminal to the other. We'll use the same idea in part (a) to calculate the emf in terms of the work done by the magnetic force on a charged particle that travels the length of the moving bar. In part (b) we'll find the emf by instead using Equation 20-2. While the magnetic field doesn't change, the area of the loop outlined by the moving bar and the rails *does* change, and so the magnetic flux through this loop changes.

Magnetic force on a moving charged particle:

$$F = |q|vB \sin \theta \quad (19\text{-}1)$$

Work done by a constant force that points in the same direction as the straight-line displacement:

$$W = Fd \quad (6\text{-}1)$$

Magnetic flux:

$$\Phi_B = AB_\perp = AB \cos \theta \quad (20\text{-}1)$$

Faraday's law of induction:

$$|\varepsilon| = \left| \frac{\Delta\Phi_B}{\Delta t} \right| \quad (20\text{-}2)$$

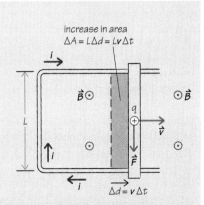

increase in area
$\Delta A = L\Delta d = Lv\Delta t$

Solve

(a) Find the magnetic force on a charged particle moving along with the copper bar.

For a positive charge q moving with the bar, the direction of the velocity $\vec{v}$ is perpendicular to the direction of the magnetic field $\vec{B}$. So $\theta = 90°$ in Equation 19-1, and the magnetic force $\vec{F}$ on such a charge has magnitude

$$F = qvB \sin 90° = qvB (1) = qvB$$

The direction of $\vec{F}$ is perpendicular to the directions of both $\vec{v}$ and $\vec{B}$, and so is directed along the length of the moving bar.

Use the magnetic force on a charged particle to find the emf produced in the bar.

The magnetic force $\vec{F}$ on a charge q causes it to move along the length L of the bar. Since $\vec{F}$ is in the same direction as the displacement of the charge, the work done on the charge as it travels this length is

$$W = FL = qvBL$$

The magnitude of the emf in the bar equals the work done per charge:

$$|\varepsilon| = \frac{W}{q} = \frac{qvBL}{q} = vBL$$

(b) Find the emf using Faraday's law of induction.

The magnetic field $\vec{B}$ points perpendicular to the plane of the loop outlined by the moving copper bar and the copper rails. If the area of this loop is A and we take the positive x direction to point out of the plane of the above figure (in the same direction as $\vec{B}$), then $\theta = 0$ in Equation 20-1. The magnetic flux through the loop is then

$$\Phi_B = AB \cos 0 = AB(1) = AB$$

The magnetic field is constant, but the area A changes with time because the bar moves. The speed v of the bar is just the distance Δd that the bar moves divided by the time Δt that it takes to move that distance, so

$$v = \frac{\Delta d}{\Delta t} \quad \text{and} \quad \Delta d = v\,\Delta t$$

During time Δt the area A of the loop outlined by the moving bar and rails increases by an amount $\Delta A = L\,\Delta d = Lv\,\Delta t$. Therefore the change in magnetic flux through the loop during this time is

$$\Delta\Phi_B = (\Delta A)B = (Lv\,\Delta t)B = vBL\,\Delta t$$

From Equation 20-2, the magnitude of the emf in the loop is

$$|\varepsilon| = \left|\frac{\Delta\Phi_B}{\Delta t}\right| = \left|\frac{vBL\,\Delta t}{\Delta t}\right| = vBL$$

Reflect

We find the same expression for the emf in both parts (a) and (b), as we must. We certainly haven't *proved* that Equation 20-2 is correct in all cases where an emf appears due to a wire moving in a magnetic field (a motional emf) or due to a changing magnetic field (an induced emf). But we do have added confidence that Equation 20-2 is valid, and a host of experiments backs up this conclusion.

Example 20-2 Changing Magnetic Flux II: A Varying Magnetic Field

A uniform magnetic field of magnitude $B = 1.50$ T is directed at an angle of 60.0° to the plane of a circular loop of copper wire. The loop is 3.50 cm in diameter. (a) What is the magnetic flux through the loop? What is the induced emf in the loop if the magnetic field decreases to zero (b) in 10.0 s or (c) in 0.100 s?

Set Up

The magnetic flux is given by Equation 20-1. Note that θ in this equation is the angle between the direction of magnetic field $\vec{B}$ and the *perpendicular* to the loop, so $\theta = 90.0° - 60.0° = 30.0°$. The magnetic flux through the loop changes when the field magnitude changes, so an emf will be induced in the loop. We'll use Equation 20-2 to calculate the magnitude of this induced emf.

Magnetic flux:

$$\Phi_B = AB_\perp = AB \cos \theta \qquad (20\text{-}1)$$

Area of a circle of radius r:

$$A = \pi r^2$$

Faraday's law of induction:

$$|\varepsilon| = \left| \frac{\Delta \Phi_B}{\Delta t} \right| \qquad (20\text{-}2)$$

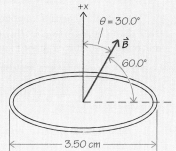

Solve

(a) Find the area of the loop, then use Equation 20-1 to calculate the magnetic flux through the loop.

The radius r of the loop is one-half of the diameter:

$$r = \frac{1}{2}(3.50 \text{ cm}) = 1.75 \text{ cm} = 1.75 \times 10^{-2} \text{ m}$$

The area of the loop is

$$A = \pi r^2 = \pi (1.75 \times 10^{-2} \text{ m})^2 = 9.62 \times 10^{-4} \text{ m}^2$$

From Equation 20-1, the magnetic flux through the loop is

$$\Phi_B = AB \cos \theta = (9.62 \times 10^{-4} \text{ m}^2)(1.50 \text{ T}) \cos 30.0°$$
$$= 1.25 \times 10^{-3} \text{ T} \cdot \text{m}^2$$

(b) The change in magnetic flux is the final value (zero) minus the initial value that we found in (a). Equation 20-2 tells us that to find the magnitude of the induced emf, we divide this change by the time $\Delta t = 10.0$ s over which the flux change takes place.

The change in magnetic flux is

$$\Delta \Phi_B = (\text{final flux}) - (\text{initial flux})$$
$$= 0 - 1.25 \times 10^{-3} \text{ T} \cdot \text{m}^2 = -1.25 \times 10^{-3} \text{ T} \cdot \text{m}^2$$

If the flux decreases to zero in $\Delta t = 10.0$ s, the magnitude of the induced emf is

$$|\varepsilon| = \left| \frac{\Delta \Phi_B}{\Delta t} \right| = \left| \frac{-1.25 \times 10^{-3} \text{ T} \cdot \text{m}^2}{10.0 \text{ s}} \right|$$
$$= 1.25 \times 10^{-4} \text{ T} \cdot \text{m}^2/\text{s} = \underline{1.25 \times 10^{-4} \text{ V}}$$

(c) Repeat part (b) with $\Delta t = 0.100$ s.

If the flux decreases to zero in just $\Delta t = 0.100$ s, the magnitude of the induced emf is

$$|\varepsilon| = \left| \frac{\Delta \Phi_B}{\Delta t} \right| = \left| \frac{-1.25 \times 10^{-3} \text{ T} \cdot \text{m}^2}{0.100 \text{ s}} \right|$$
$$= 1.25 \times 10^{-2} \text{ T} \cdot \text{m}^2/\text{s} = \underline{1.25 \times 10^{-2} \text{ V}}$$

Reflect

The induced emf is 100 times greater in part (c) than in part (b) because the same flux change takes place in 1/100 as much time. The faster the flux change, the greater the induced emf that results. Note that the emf is induced *only* during the time when the magnetic flux is changing. There is zero emf when the magnetic field is at its original value of 1.50 T, and there is zero emf when the magnetic field has stabilized at its final value of zero.

The emf in both cases is quite small because there is only a single turn of wire in the loop. If we replace the loop by a coil of the same diameter with 500 turns of wire, the induced emf is 500 times greater: $500 \times 1.25 \times 10^{-4}$ V $= 0.0625$ V in part (b), $500 \times 1.25 \times 10^{-2}$ V $= 6.25$ V in part (c). The key to generating a large induced emf is to have many turns of wire and a rapid change in magnetic flux.

In this example we chose the positive x direction to be upward, so that the angle θ between the magnetic field and the perpendicular to the loop was 30.0°. Can you show that we would have found the same results for emf had we chosen the positive x direction to be downward, so that $\theta = 150.0°$?

❓ Got the Concept? 20-1   A Wooden Loop

Suppose the loop in Example 20-2 were made out of wood rather than copper wire, but the magnetic field changes in the same manner as in part (b) of Example 20-2. Compared to the emf calculated in part (b) of Example 20-2, the emf induced in the wooden loop would be (a) zero; (b) much smaller, but not zero; (c) slightly less; (d) the same; (e) greater.

Take-Home Message for Section 20-2

✔   A motional emf appears in a conductor that moves in a magnetic field. The force that produces the emf is a magnetic one.

✔   An induced emf appears in any loop subjected to a changing magnetic field. The force that produces the emf is an electric one.

✔   Both motional emfs and induced emfs can be described by Faraday's law of induction: The magnitude of the emf in a loop is equal to the absolute value of the change in magnetic flux through the loop, divided by the time over which the change takes place.

20-3 Lenz's law describes the direction of the induced emf

Equation 20-2 tells us the *magnitude* of the emf that appears in a loop when there is a change in magnetic flux through that loop: $|\varepsilon| = |\Delta\Phi_B/\Delta t|$. It does not, however, tell us the *direction* of that emf—that is, the direction in which the emf tends to make current flow around that loop. As we'll see, there's a simple rule for determining this direction that works for both motional emfs (caused by a conductor moving in a magnetic field) and induced emfs (caused by a conductor being exposed to a changing magnetic field).

To learn about this rule, let's think again about the loop of wire in Figure 20-2. We saw in Section 20-2 that when the loop is moved toward or away from a bar magnet an emf is set up in the loop and a current flows. That's because the mobile charges in the wire experience magnetic forces as they move with the wire in the presence of the magnet's field $\vec{B}_{\mathrm{magnet}}$. That's not the only magnetic field present in this situation, however. As we learned in Section 19-7, a current-carrying loop produces a magnetic field of its own. The direction of the field $\vec{B}_{\mathrm{loop}}$ due to the loop depends on the direction of the current around the loop, and is given by a right-hand rule: Curl the fingers of your right hand around the loop in the direction of the current, and the extended thumb of your right hand will point in the direction of $\vec{B}_{\mathrm{loop}}$ in the interior of the loop (see part (a) of Figure 20-6). So whenever an emf appears in a loop—either a motional emf as in Figure 20-2 or an induced emf as in Figure 20-3—the current produced by that emf generates a magnetic field $\vec{B}_{\mathrm{loop}}$ whose direction depends on the direction of the current and emf. We call $\vec{B}_{\mathrm{loop}}$ an **induced magnetic field**.

The field $\vec{B}_{\mathrm{loop}}$ itself produces a magnetic flux through the loop, and it's the sense of this flux that will tell us the direction of the emf in the loop. Let's choose the positive x direction for the loop in Figure 20-6 to point to the right, perpendicular to the plane of the loop. If the loop is close to the south pole of a bar magnet as in Figure 20-6b, the field of the magnet causes a positive magnetic flux through the loop (the field $\vec{B}_{\mathrm{magnet}}$ points generally to the right, in the positive x direction). If the loop moves toward the magnet as in the left-hand side of Figure 20-6b, the field $\vec{B}_{\mathrm{magnet}}$ inside the loop increases and the positive flux increases. Experiment shows that the current induced in the loop gives rise to an induced magnetic field $\vec{B}_{\mathrm{loop}}$ within the loop which points in the *opposite* direction to $\vec{B}_{\mathrm{magnet}}$. So while the flux due to $\vec{B}_{\mathrm{magnet}}$ becomes more positive, $\vec{B}_{\mathrm{loop}}$ gives rise to a negative flux that opposes the change in the flux of $\vec{B}_{\mathrm{magnet}}$. If instead the loop moves away from the magnet as in the right-hand side of Figure 20-6b, the field $\vec{B}_{\mathrm{magnet}}$ inside the loop decreases and the positive flux decreases. In this case the direction of the induced current in the loop is reversed, as is the direction of the induced magnetic field $\vec{B}_{\mathrm{loop}}$: Now $\vec{B}_{\mathrm{loop}}$ inside the loop points in the *same* direction as $\vec{B}_{\mathrm{magnet}}$. The magnetic flux due to $\vec{B}_{\mathrm{loop}}$ is now positive, which opposes the negative change (decrease) in the flux of $\vec{B}_{\mathrm{magnet}}$.

In both cases shown in Figure 20-6b the induced magnetic field is in a direction opposite to the *change* in flux of the external magnetic field (in this case, the field due to the bar magnet). Many experiments show that this is always the case, no matter whether the induced magnetic field is due to a motional emf, an induced emf, or a combination

(a)

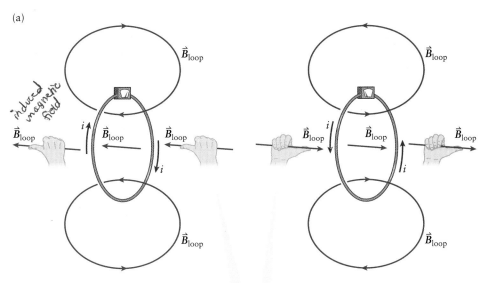

Figure 20-6 **Lenz's law** The current induced in a loop by a change in flux always acts to oppose the flux change.

induced magnetic field

causes a magnetic flux

- A current-carrying loop generates a magnetic field $\vec{B}_{loop}$.
- To find the direction of $\vec{B}_{loop}$, curl the fingers of your right hand around the loop in the direction of the current i. Your extended right thumb points in the direction of $\vec{B}_{loop}$ in the interior of the loop.
- $\vec{B}_{loop}$ itself causes a magnetic flux through the loop.

(b)

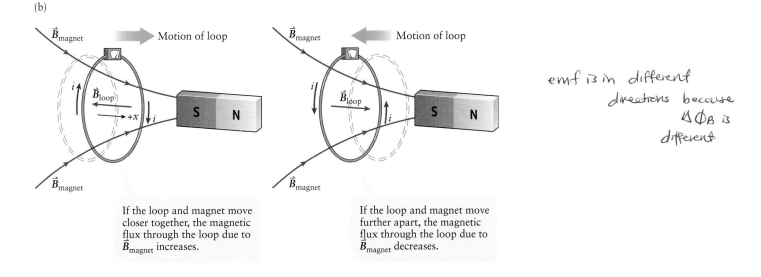

If the loop and magnet move closer together, the magnetic flux through the loop due to $\vec{B}_{magnet}$ increases.

If the loop and magnet move further apart, the magnetic flux through the loop due to $\vec{B}_{magnet}$ decreases.

emf is in different directions because $\Delta \Phi_B$ is different

of the two. The nineteenth-century Russian physicist Heinrich Lenz summarized these observations in a principle that we now call **Lenz's law**:

> *The direction of the magnetic field induced within a conducting loop opposes the change in magnetic flux that created it.*

It's common to combine Faraday's law and Lenz's law into a single equation:

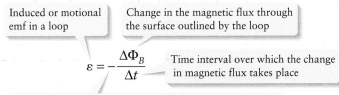

Induced or motional emf in a loop

Change in the magnetic flux through the surface outlined by the loop

$$\varepsilon = -\frac{\Delta \Phi_B}{\Delta t}$$

Time interval over which the change in magnetic flux takes place

The minus sign indicates that the current caused by the emf induces a magnetic field which opposes the change in flux.

Faraday's law and Lenz's law for induction
(20-3)

> **⚠ Watch Out!** Like Faraday's law, Lenz's law is about the *change* in flux.
>
> ● Notice that Lenz's law refers to the direction of the *change* in magnetic flux ("Is the flux increasing or decreasing?"), not to the direction of the field that causes the flux. The field $\vec{B}_{\mathrm{magnet}}$ points in the same direction in both of the situations shown in Figure 20-6b, but the flux change is different in the two situations and so the emf is in different directions as well.

We can check Lenz's law by revisiting the sliding copper bar from Example 20-1 (Section 20-2). The upward external magnetic field of magnitude B causes a positive magnetic flux through the loop formed by the sliding bar and the rails on which it slides (Figure 20-7). The area enclosed by this loop increases as the bar slides to the right, so the positive flux increases as well. By Lenz's law an induced current will flow in the loop in order to generate an induced magnetic field that opposes this change in flux. So this induced magnetic field must point downward, and to produce that induced field the current in the loop must be clockwise as seen from above the loop. That's just the direction of current flow that we depicted in the figure that accompanies Example 20-1. So this situation is consistent with Lenz's law.

The sliding copper bar in Figure 20-7 illustrates another aspect of Lenz's law. Once the induced current is flowing in the bar, the external magnetic field exerts a force on that current. Using the right-hand rule for this force (see Section 19-5), we see that this force points opposite to the direction in which the bar is moving. In other words, this force *opposes* the motion that gives rise to the change in flux through the loop made up of the bar and rails. That's always the case when a conductor moves through a magnetic field: A current is induced in the conductor, and the magnetic field exerts a force on the current that opposes the motion of the conductor. We can summarize this in an alternative statement of Lenz's law:

> *When the magnetic flux through a loop changes, current flows in a direction that opposes that change.*

In Example 20-1 we assumed that the bar slides on the rails at a constant speed. Our discussion shows that the bar won't do this on its own: Since a magnetic force opposes the motion of the bar, we need to apply an external force to keep the bar in motion. If we make the bar move faster, the magnetic flux through the loop changes more rapidly, the emf and resulting current in the loop are greater, and the magnetic force opposing the motion of the bar is greater. (The magnetic force on the bar turns out to be proportional to the speed of the bar, just like the drag force on a microscopic object moving through a fluid; see Section 5-5.) So we must apply a greater force to make the bar slide at a faster speed.

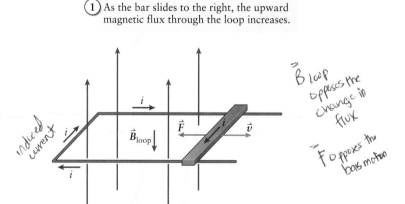

① As the bar slides to the right, the upward magnetic flux through the loop increases.

Figure 20-7 **A sliding copper bar revisited** Lenz's law helps explain the direction in which current flows in this situation.

② The induced current i produces an induced magnetic field $\vec{B}_{\mathrm{loop}}$ that opposes the change in flux.

③ The magnetic force on the current in the moving bar opposes the bar's motion.

This same effect explains the phenomenon of *magnetic braking*. If you try to make a magnet move past a conductor or a conductor move past a magnet, currents appear in the conductor. (These are called *eddy currents*, since their pattern resembles that of eddies in a body of water. The conductor does *not* need to be in the form of a loop for these currents to appear.) The magnetic force that the magnet exerts on the eddy currents opposes the motion of the conductor relative to the magnet, and so by Newton's third law there is a force that opposes the motion of the magnet relative to the conductor. One important application of this is to amusement park rides where it's necessary to slow down a fast-moving object such as a roller coaster car. When the car enters the part of the ride where it's supposed to slow down, a copper fin on the car passes through powerful permanent magnets mounted on the track. Eddy currents arise in the fin, and the interaction between the eddy currents and the field of the permanent magnets causes a force that smoothly brings the car to a slow speed. The car is then stopped by conventional mechanical braking.

Eddy currents are also used in an *electromagnetic flowmeter*, a device that can measure the rate of blood flow in an artery. Blood is an electrical conductor; eddy currents are induced in the blood as it flows past magnets in the flowmeter. The device records the small but measurable magnetic fields due to these currents and uses them to determine the rate of flow. The advantage of an electromagnetic flowmeter over other methods for measuring blood flow rate is that it is noninvasive: No component of the device need be surgically introduced into the body.

Another application of eddy currents is magnetic induction tomography, a relatively new experimental technique for medical imaging. In this technique, changing magnetic fields created by coils placed near a part of the body induce eddy currents. Observing the fields produced by these eddy currents is a way to monitor brain swelling. Eddy currents can also be used for the controlled, repeated delivery of medication. A capsule containing the drug is implanted in the body; the capsule is made from a gel that heats up slightly when there are eddy currents, opening pores through which the medication is released. The advantage of this approach is that no implanted electronics are required.

? Got the Concept? 20-2 Induced Current I

In Figure 20-8 a rectangular loop of wire moves to the right into a region of constant, uniform magnetic field. The field points into the plane of the figure, in a direction perpendicular to the plane of the loop. When the loop is entering the field region as in Figure 20-8a, what is the direction of the current around the loop? (a) clockwise; (b) counterclockwise; (c) the current is zero; (d) not enough information given to decide.

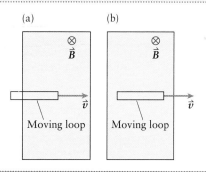

Figure 20-8 **A moving rectangular loop of wire** In each situation, what is the direction of the induced current in the moving loop of wire?

? Got the Concept? 20-3 Induced Current II

In Figure 20-8 a rectangular loop of wire moves to the right into a region of constant, uniform magnetic field. The field points into the plane of the figure, in a direction perpendicular to the plane of the loop. When the loop is moving and completely inside the field region as in Figure 20-8b, what is the direction of the current around the loop? (a) clockwise; (b) counterclockwise; (c) the current is zero; (d) not enough information given to decide.

Take-Home Message for Section 20-3

✔ An emf induced by a changing magnetic flux tends to cause a current to flow. This current generates a magnetic field of its own, called the induced magnetic field.

✔ The induced magnetic field is in a direction that opposes the change in flux that created the emf that gave rise to the induced field.

✔ Eddy currents arise whenever a conducting material, even a nonmagnetic one, moves relative to a magnetic field.

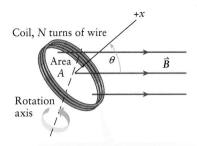

Coil, N turns of wire

+x

Area A

θ

$\vec{B}$

Rotation axis

As the coil rotates with angular speed ω, the angle θ between the magnetic field $\vec{B}$ direction and the perpendicular to the coil changes: $\theta = \omega t + \phi$.

Figure 20-9 **An ac generator** As the coil rotates, an oscillating emf is generated in the turns of wire that make up the coil.

√x *See the Math Tutorial for more information on trigonometry*

20-4 Faraday's law explains how alternating currents are generated

We learned in Section 18-2 about the importance of alternating current in technology. (If you're reading these words in a room lit by electric light, the light bulbs are powered by alternating current. If you're reading on a mobile device such as a tablet, the device's battery was charged by plugging it into a wall socket and using the alternating current delivered by that socket.) We now have the physics we need to understand how alternating current is produced.

Let's look at a coil of wire with N turns, each of which has area A. As **Figure 20-9** shows, this coil is free to rotate around an axis. We place the coil in a region of uniform magnetic field $\vec{B}$, then rotate the coil at a constant angular speed ω. What happens? As the coil rotates, the magnetic flux through each turn of the coil changes, and so an emf is generated. The total emf is N times greater than that due to a single turn of wire. As we will see, this is an alternating emf of just the sort required to generate an alternating current. That's why a rotating coil of the sort shown in Figure 20-9 is called an **ac generator**.

We begin by writing an equation for the magnetic flux through the rotating coil. The angle θ between the magnetic field $\vec{B}$ and the perpendicular to the coil changes as the coil rotates:

(20-4)
$$\theta = \omega t + \phi$$

In Equation 20-4 ϕ is the value of the angle at $t = 0$. From Equation 20-1, the magnetic flux through the N turns of the coil is

(20-5)
$$\Phi_B = NAB \cos \theta = NAB \cos(\omega t + \phi)$$

When $\theta = \omega t + \phi = 0$ so $\cos \theta = 1$, the perpendicular to the coil is in the same direction as $\vec{B}$ and the flux has its most positive value $\Phi_B = NAB$; when $\theta = \omega t + \phi = \pi/2$ so $\cos \theta = 0$, the coil is edge-on to the magnetic field and the flux is zero; when $\theta = \omega t + \phi = \pi$ and $\cos \theta = -1$, the perpendicular to the coil points opposite to $\vec{B}$ and the flux has its most negative value $\Phi_B = -NAB$; and so on. So the magnetic flux varies with time, and it follows that there will be an emf in the coil. Faraday's law and Lenz's law (Equation 20-3) tell us that the emf is equal to the negative of the rate of change of Φ_B. We actually know how to find the rate of change of a cosine function like that in Equation 20-5. In Section 12-3 we saw that the position of an object undergoing simple harmonic motion with amplitude A is given by

(12-6)
$$x = A \cos(\omega t + \phi)$$

The rate of change of position x is just the velocity v_x, which we found was equal to

(12-7)
$$v_x = -\omega A \sin(\omega t + \phi)$$

You can see that Equation 20-5 for magnetic flux is identical to Equation 12-6 for position, with amplitude A replaced by NAB (note that A in Equation 20-5 denotes area, not amplitude). Making the same replacement in Equation 12-7 tells us that the rate of change of magnetic flux through the rotating coil is

(20-6)
$$\frac{\Delta \Phi_B}{\Delta t} = -\omega NAB \sin(\omega t + \phi)$$

Substituting Equation 20-6 into Equation 20-3 then gives us the emf in the rotating coil:

Emf in an ac generator
(20-7)

Angular speed of the rotating coil   Time   Angle of the coil at $t = 0$

Emf produced by an ac generator

$$\varepsilon = \omega NAB \sin(\omega t + \phi)$$

Number of turns in the coil   Area of the coil   Magnitude of the magnetic field to which the coil is exposed

The emf alternates with angular frequency ω, the same as the angular speed of the rotating coil. The maximum value of the emf is $\varepsilon_{max} = \omega NAB$, which shows that we can increase the maximum emf by increasing the angular speed ω, the number of turns N, the coil area A, the magnetic field magnitude B, or a combination of these.

While the emf produced by an ac generator changes from positive to negative, the power delivered by the generator does not. As an example, suppose an ac generator is connected to a circuit device (such as a light bulb or a toaster) that we can represent as a resistor with resistance R. If we ignore the internal resistance of the coil, the emf is equal to the voltage drop across the resistor: $\varepsilon = iR$. The current in the resistor is therefore

$$i = \frac{\varepsilon}{R} = \frac{\omega NAB}{R}\sin(\omega t + \phi) \tag{20-8}$$

The current in the resistor alternates with the same angular frequency ω as the emf. From Section 18-6, the power into such a resistor is

$$P = i^2 R \tag{18-24}$$

If we substitute Equation 20-8 into Equation 18-24, we get

$$P = \left(\frac{\omega NAB}{R}\sin(\omega t + \phi)\right)^2 R = \frac{\omega^2 N^2 A^2 B^2}{R}\sin^2(\omega t + \phi) \tag{20-9}$$

Figure 20-10 shows graphs of the emf ε (Equation 20-7) and resistor power P (Equation 20-9) as functions of time. The power P is *never* negative, which means that energy always flows from the ac generator into the resistor, never the other way. The average value of the function $\sin^2(\omega t + \phi)$ is $\frac{1}{2}$, so the average power into the resistor is

$$P_{\text{average}} = \frac{\omega^2 N^2 A^2 B^2}{2R} \tag{20-10}$$

It may seem like the power given by Equation 20-10 comes "for free": You let the coil rotate, and an emf is generated that makes power flow into the resistor. Alas, this power comes at a price. As we discussed in Section 20-3, whenever a conductor (such as the coil of an ac generator) moves in the presence of a magnetic field, the conductor experiences a magnetic force that opposes its motion. So left to itself, the coil would quickly slow to a halt. To keep the coil in motion, you must apply a torque that just balances the effects of this magnetic force. At an electric generating station, this torque is applied to the blades of a turbine that is connected to the coil. The blades can be turned by the force of the wind (Figure 20-1a), by the force of flowing water at a hydroelectric plant, or by the force of fast-moving steam at a coal-fired or nuclear power plant (where heat from burning fossil fuels or radioactive decay is used to boil water and produce steam). Part of the mechanical power used to make the turbine spin goes

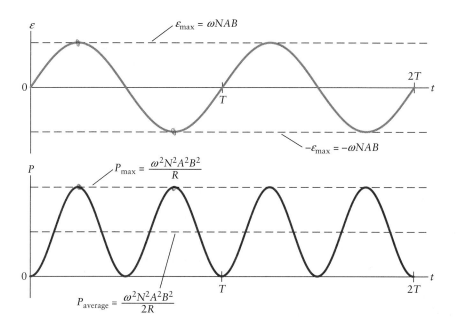

Figure 20-10 An ac generator: Emf and power These graphs show the emf ε generated in the coil shown in Figure 20-9, and the power P that this emf delivers to a resistor R connected to the coil. We assume $\phi = 0$ in Equations 20-7 and 20-9. Note that T is the time it takes for the coil to complete one rotation.

▶ Go to Interactive Exercise 20-2 for more practice dealing with coils

into the electric power provided by the generator; the rest is lost due to friction in the turbine and generator.

The ac generator is just part of what's needed for a system of power delivery based on alternating current. In Chapter 21 we'll explore in more detail the physics of alternating current in circuits.

Example 20-3 Lighting the Gym with a Bicycle

You attach the coil of an ac generator to an exercise bicycle, so that as you work the pedals the coil turns and an emf is generated. The generator is geared so that it makes 10 rotations for each rotation of the pedals. The coil has an area of 6.40×10^{-3} m^2, has 1200 turns of wire, and is in a magnetic field of magnitude 0.100 T. How many times a second must you turn the pedals in order to deliver an average power of 60.0 W to a light bulb with a resistance of 80.0 Ω?

Set Up

The situation is as shown in Figure 20-9. We'll use Equation 20-10 to solve for the angular speed ω of the generator coil. Since the coil makes 10 rotations for every rotation of the pedals, the angular speed of the pedals is equal to ω divided by 10.

Average power delivered by an ac generator to a resistor:

$$P_{average} = \frac{\omega^2 N^2 A^2 B^2}{2R} \tag{20-10}$$

Solve

Rewrite Equation 20-10 to solve for ω.

Find the angular speed of the generator coil from Equation 20-10:

$$\omega^2 = \frac{2RP_{average}}{N^2 A^2 B^2}$$

$$\omega = \frac{\sqrt{2RP_{average}}}{NAB} = \frac{\sqrt{2(80.0\ \Omega)(60.0\ \mathrm{W})}}{(1200)(6.40 \times 10^{-3}\ \mathrm{m}^2)(0.100\ \mathrm{T})}$$

$$= 128\ \mathrm{rad/s}$$

Convert this from radians per second to revolutions per second:

$$\omega = \left(128\ \frac{\mathrm{rad}}{\mathrm{s}}\right)\left(\frac{1\ \mathrm{rev}}{2\pi\ \mathrm{rad}}\right) = 20.3\ \mathrm{rev/s}$$

The generator turns 10 times faster than the pedals, so the pedals must turn at a rate of

$$(20.3\ \mathrm{rev/s})/10 = 2.03\ \mathrm{rev/s} = 122\ \mathrm{rev/min}$$

Reflect

A cycling cadence of 122 rev/min isn't difficult for an amateur cyclist, so you can certainly power a light bulb in this way. Exercise bicycles with generators of this kind are commercially available and are used to return power to the electrical grid in the same manner as residential solar panels.

❓ Got the Concept? 20-4 A Flickering Fluorescent Lamp

Certain types of fluorescent lamps flicker rapidly. That's because these lamps emit a pulse of light every time a burst of electric power is provided to the lamp. If such a lamp is powered by a source of emf that oscillates at 60 Hz, what is the frequency at which the lamp will flicker? (a) 30 Hz; (b) 60 Hz; (c) 120 Hz; (d) 240 Hz; (e) 3600 Hz.

Take-Home Message for Section 20-4

✔ An ac generator consists of a coil that rotates in a magnetic field.

✔ The emf produced by an ac generator oscillates at an angular frequency that equals the angular speed of the rotating coil.

Key Terms

ac generator
electromagnetic induction
Faraday's law of induction

induced emf
induced magnetic field
Lenz's law

magnetic flux
motional emf

Chapter Summary

Topic	Equation or Figure

Motional emf: An emf appears in a loop when that loop moves in the presence of a magnetic field. The emf is a result of magnetic forces on the mobile charges within the loop.

(c) Side view of loop moving toward a stationary magnet

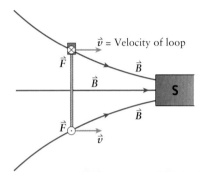

(Figure 20-2c)

- Charges in the loop move along with the loop in the magnetic field.
- Magnetic forces push mobile charges around the loop, causing a current.

Induced emf: An emf also appears in a loop when the magnetic field within the loop changes. Here the forces that create the emf are not magnetic, but electric; an electric field is produced by the changing magnetic field.

(c) Side view of magnet moving toward a stationary loop

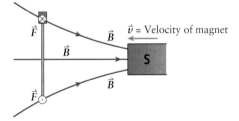

(Figure 20-3c)

- Charges in the loop experience electric forces.
- These forces push mobile charges around the loop, causing a current.

Magnitude of the induced or motional emf in a loop   Change in the magnetic flux through the surface outlined by the loop

$$|\varepsilon| = \left| \frac{\Delta \Phi_B}{\Delta t} \right|$$

(20-2)

Time interval over which the change in magnetic flux takes place

Lenz's law: The direction of an emf caused by a change in magnetic flux is such as to oppose the flux change. The induced current causes its own flux (which helps compensate for the change), and a conductor and magnet that move relative to each other experience magnetic forces that oppose this relative motion.

Induced or motional emf in a loop

Change in the magnetic flux through the surface outlined by the loop

$$\varepsilon = -\frac{\Delta\Phi_B}{\Delta t}$$

Time interval over which the change in magnetic flux takes place

(20-3)

The minus sign indicates that the current caused by the emf induces a magnetic field which opposes the change in flux.

Alternating current and an ac generator: To produce an alternating emf (the sort needed to cause an alternating current), rotate a coil in the presence of a constant magnetic field. The changing flux causes an emf that oscillates sinusoidally.

Angular speed of the rotating coil   Time   Angle of the coil at $t = 0$

Emf produced by an ac generator

$$\varepsilon = \omega NAB \sin(\omega t + \phi)$$

(20-7)

Number of turns in the coil

Area of the coil

Magnitude of the magnetic field to which the coil is exposed

Answer to What do you think? Question

(b) The electrons in the wire are initially at rest and so do not experience a magnetic force. (Recall that a magnetic field exerts a force only on a charged object that is in motion.) So it must be an *electric* force that sets the electrons into motion. This is a consequence of electromagnetic induction: As you swipe the credit card through the reader, the wire loop is exposed to a varying magnetic field from the stripe on the back of the card. This causes a changing magnetic flux through the loop, which induces an electric field and an emf. Electrons move in the loop in response to this emf.

Answers to Got the Concept? Questions

20-1 (d) The induced emf does *not* depend on what the loop is made of. (Note that Equation 20-2 makes no reference to the properties of the loop material.) So the emf will be exactly the same whether the loop is made of copper, wood, silver, rubber, or even air. The difference is that because copper is a good conductor while wood is a very poor conductor, a current will be generated in the copper loop but not in the wooden loop.

20-2 (b) As the loop in Figure 20-8a moves into the field region, there is an increasing magnetic flux through the loop due to the magnetic field directed into the plane of the figure. According to Lenz's law, current will flow in the loop to induce a magnetic field $\vec{B}_{\text{loop}}$ that will oppose this increase, so $\vec{B}_{\text{loop}}$ in the interior of the loop must point out of the plane in Figure 20-8a. The right-hand rule depicted in Figure 20-6a tells us that to induce such a field, current must flow counterclockwise around the loop. (You can confirm this by using the right-hand rule for the magnetic force on a moving charge. A positive charge in the right-hand leg of the moving loop feels an upward magnetic force, and this force drives current in a counterclockwise direction around the loop. There are also magnetic forces on charges in the top and bottom legs of the loop, but these forces have zero component along the length of the wire and so do not induce a current.)

20-3 (c) Although the loop in Figure 20-8b is moving, the magnetic flux through the loop remains constant because it is moving through a region of constant, uniform magnetic field. Since there is no flux change, Faraday's law tells us that no emf is induced and so no current will be generated. (You can confirm this by using the right-hand rule for the magnetic force on a moving charge. A positive charge in the right-hand leg of the moving loop feels an upward magnetic force, and this force by itself would drive current in a counterclockwise direction around the loop. But a positive charge in the left-hand leg of the moving loop also feels an upward magnetic force, which by itself would drive current in a clockwise direction around the loop. There are also magnetic forces on charges in the top and bottom legs of the loop, but these forces have zero component along the length of the wire and so do not induce a current. The net effect is that there is *zero* current in the loop.)

20-4 (c) Figure 20-10 shows that the power delivered by an ac generator goes through two up-and-down cycles during the time T required for the emf to go through a single cycle. So if the emf varies at 60 Hz, the power delivered to the fluorescent lamp varies at 2×60 Hz = 120 Hz.

Questions and Problems

In a few problems, you are given more data than you actually need; in a few other problems, you are required to supply data from your general knowledge, outside sources, or informed estimate.

Interpret as significant all digits in numerical values that have trailing zeros and no decimal points.

For all problems, use $g = 9.80 \text{ m/s}^2$ for the free-fall acceleration due to gravity. Neglect friction and air resistance unless instructed to do otherwise.

• Basic, single-concept problem
•• Intermediate-level problem, may require synthesis of concepts and multiple steps
••• Challenging problem
SSM *Solution is in Student Solutions Manual*

Conceptual Questions

1. •A common physics demonstration is to drop a small magnet down a long, vertical aluminum pipe. Describe the motion of the magnet and the physical explanation for the motion. SSM

2. ••Figure 20-11 depicts an electron in between the poles of an electromagnet. Explain how the electron is accelerated if the magnetic field is gradually being increased.

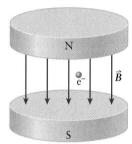

Figure 20-11 Problem 2

3. ••In a popular demonstration of electromagnetic induction, a metal plate is suspended in midair above a large electromagnetic coil, as shown in Figure 20-12. (a) How does this work? (b) If your professor does the demonstration, one thing you'll notice is that the plate gets quite hot. (In fact, you can end the demonstration by frying an egg on the plate!) Why does the plate become hot? (c) Would the trick work if the plate were made of an insulating material?

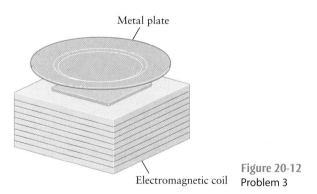

Metal plate

Figure 20-12
Electromagnetic coil Problem 3

4. •**Medical** In hospitals with magnetic resonance imaging facilities and at other locations where large magnetic fields are present, there are usually signs warning people with pacemakers and other electronic medical devices not to enter. Why?

5. •Two conducting loops with a common axis are placed near each other, as shown in Figure 20-13. Initially the currents in both loops are zero. If a current is suddenly set up in loop *a* in the direction shown, is there also a current in loop *b*? If so, in which direction? What is the direction of the force, if any, that loop *a* exerts on loop *b*? Explain your answer.

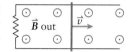

Figure 20-13 Problem 5

6. •A conducting rod slides without friction on conducting rails in a magnetic field as shown in Figure 20-14. The rod is given an initial velocity $\vec{v}$ to the right. Describe its subsequent motion and justify your answer.

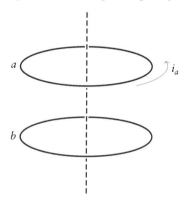

Figure 20-14 Problem 6

Multiple-Choice Questions

7. •Figure 20-15 shows a sequence of sketches depicting a rectangular loop passing from left to right through a region of constant magnetic field. The field points out of the page and perpendicular to the plane of the loop. In which one of the sequences is the magnetic flux through the loop decreasing? SSM
    A. from left to right approaching the magnetic field
    B. entering the magnetic field
    C. inside the magnetic field
    D. leaving the magnetic field
    E. from left to right moving away from the magnetic field

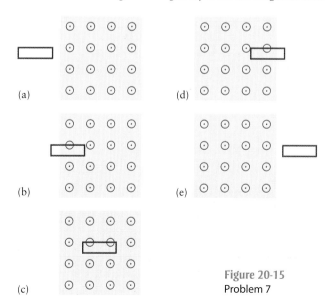

Figure 20-15
Problem 7

8. •On which variable does the magnetic flux depend?
   A. the magnetic field
   B. the area of a region through which the magnetic field passes
   C. the orientation of the field with respect to the region through which it passes
   D. all of the above
   E. none of the above

9. •Two metal rings with a common axis are placed near each other, as shown in **Figure 20-16**. If current i_a is suddenly set up and is increasing in ring a as shown, the current in ring b is
   A. zero.
   B. parallel to i_a.
   C. antiparallel to i_a.
   D. alternatively parallel and antiparallel to i_a.
   E. perpendicular to i_a. SSM

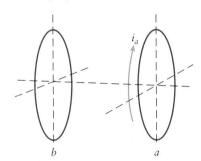

Figure 20-16 Problem 9

10. ••**Figure 20-17** shows two coils wound around an iron ring, which directs the magnetic field of each coil around the ring. Current appears in the second coil
    A. the moment the battery is connected by closing the switch.
    B. the entire time the battery is connected with the switch closed.
    C. the moment the battery is disconnected by opening the switch.
    D. the moment the battery is connected by closing the switch and the moment the battery is disconnected by opening the switch.
    E. the entire time the battery is disconnected with the switch open.

Figure 20-17 Problem 10

11. •The copper ring of radius R in **Figure 20-18** lies in a magnetic field pointed into the page. The field is uniformly decreasing in magnitude. The induced current in the ring is
    A. clockwise and constant.
    B. clockwise and changing.
    C. zero.
    D. counterclockwise and constant.
    E. counterclockwise and changing.

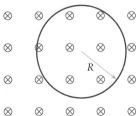

Figure 20-18 Problem 11

12. •A conducting loop moves at a constant speed parallel to a long, straight, current-carrying wire, as shown in **Figure 20-19**.
    A. The induced current in the loop will be clockwise.
    B. The induced current in the loop will be only parallel to the current i.
    C. The induced current in the loop will be counterclockwise.
    D. The induced current in the loop will be alternately clockwise and then counterclockwise.
    E. There will be no induced current in the loop.

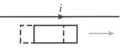

Figure 20-19 Problem 12

Estimation/Numerical Analysis

13. •Estimate the electric current in a generator in a large hydroelectric dam (such as Hoover Dam or the Oahe Dam).

14. •Estimate how many turbines (such as the ones at Hoover Dam) would be required to supply enough energy to power the United States.

15. •Give an estimate of the induced voltage created in trans-Atlantic communication cables due to fluctuations in Earth's magnetic field.

16. •Estimate the magnitude of the fluctuations in Earth's geomagnetic field during times of solar flares.

17. ••The induced voltage versus time for a coil that has 100 circular turns of wire with radii 25 cm is given in the table. Plot $V(t)$ and use this graph to predict the graph of the magnetic field as a function of time $B(t)$ that is passing through the loop (assume that the angle between the magnetic field and the area of the loop is 0°).

t(s)	V(V)	t(s)	V(V)
0	0	9	2
1	2	10	4
2	4	11	2
3	2	12	0
4	0	13	−2
5	−2	14	−4
6	−4	15	−2
7	−2	16	0
8	0		

Problems

20-1 The world runs on electromagnetic induction

20-2 A changing magnetic flux creates an electric field

18. •A single-turn circular loop of wire that has a radius of 5.0 cm lies in the plane perpendicular to a spatially uniform magnetic field. During a 0.12-s time interval, the magnitude of the field increases uniformly from 0.2 T to 0.4 T. Determine the magnitude of the emf induced in the loop during the time interval.

19. •A circular coil that has 100 turns and a radius of 10.0 cm lies in a magnetic field that has a magnitude of 0.0650 T directed perpendicular to the coil. (a) What is the magnetic flux through the coil? (b) The magnetic field through the coil is increased steadily to 0.100 T over a time interval of 0.500 s. What is the magnitude of the emf induced in the coil during the time interval?

20. •A 30-turn coil with a diameter of 6.00 cm is placed in a constant, uniform magnetic field of 1.00 T directed perpendicular to the plane of the coil. Beginning at time $t = 0$ s, the field is increased at a uniform rate until it reaches 1.30 T at $t = 10.0$ s. The field remains constant thereafter. What is the magnitude of the induced emf in the coil at (a) $t < 0$ s, (b) $t = 5.00$ s, and (c) $t > 10.0$ s? (d) Plot the magnetic field

and the induced emf as functions of time for the range $-5.00\ s < t < 15.0\ s$.

20-3 Lenz's law describes the direction of the induced emf

21. •Determine the direction of the induced current in the loop for each case shown in Figure 20-20.

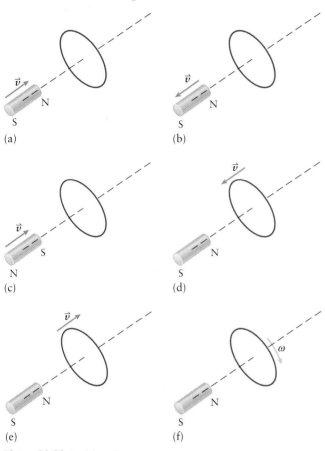

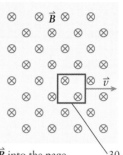

Figure 20-20 Problem 21

22. •A bar magnet is moved steadily through a wire loop as shown in Figure 20-21. Make a qualitative sketch of the induced emf in the loop as a function of time (be sure to include the times t_1, t_2, and t_3). Consider the direction of positive emf to be as indicated in the figure.

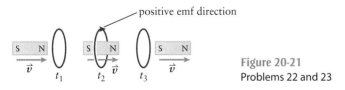

Figure 20-21
Problems 22 and 23

23. •A bar magnet is moved steadily through a wire loop as shown in Figure 20-21, except that the leading edge of the magnet is the south pole instead of the north pole. Make a qualitative sketch of the induced voltage in the loop as a function of time (be sure to include the times t_1, t_2, and t_3). Consider the direction of positive emf to be as indicated in the figure. SSM

24. ••A square, 30-turn coil 10.0 cm on a side with a resistance of 0.820 Ω is placed between the poles of a large electromagnet. The electromagnet produces a constant, uniform magnetic field of 0.600 T directed into the page. As suggested by Figure 20-22,

the field drops sharply to zero at the edges of the magnet. The coil moves to the right at a constant velocity of 2.00 cm/s. What is the current through the wire coil (a) before the coil reaches the edge of the field, (b) while the coil is leaving the field, and (c) after the coil leaves the field? (d) What is the total charge that flows past a given point in the coil as it leaves the field? (e) Plot the induced current in the loop as a function of the horizontal position of the right side of the current loop. Let the right-hand edge of the magnetic field region be $x = 0$. Your plot should be in the range of $-5.00\ cm < x < 20.0\ cm$.

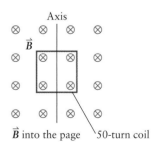

$\vec{B}$ into the page   30-turn coil   Figure 20-22 Problem 24

25. ••(a) Determine the magnitude and direction of the force on each side of the coil in Problem 24 for situations (a) through (c). (b) As the loop enters the field region from the left, what is the direction of the induced current and the resulting force on each segment of the coil?

20-4 Faraday's law explains how alternating currents are generated

26. ••A rectangular coil with sides 0.10 m by 0.25 m has 500 turns of wire. It is rotated about its long axis in a magnetic field of 0.58 T. At what frequency must the coil be rotated for it to generate a maximum potential of 110 V?

27. ••An electromagnetic generator consists of a coil that has 100 turns of wire, has an area of 400 cm², and rotates at 60 rev/s in a magnetic field of 0.25 T. What is the magnitude of the emf induced in the coil? SSM

28. •••A 50-turn square coil with a cross-sectional area of 5.00 cm² has a resistance of 20.0 Ω. The plane of the coil is perpendicular to a uniform magnetic field of 1.00 T. The coil is suddenly rotated about the axis shown in Figure 20-23 through an angle of 60° over a period of 0.200 s. (a) What charge flows past a point in the coil during that time? (b) If the loop is rotated a full 360° around the axis, what is the net charge that passes the point in the loop? Explain your answer.

Axis

$\vec{B}$

$\vec{B}$ into the page   50-turn coil

Figure 20-23 Problem 28

29. ••Perhaps it has occurred to you that we could tap Earth's magnetic field to generate energy. One way to do this would be to spin a metal loop about an axis perpendicular to Earth's magnetic field. Suppose that the metal loop is a square that is 45.0 cm on each side and that we want to generate an electric potential in the loop of amplitude 120 V at a place where Earth's magnetic field is 0.50×10^{-4} T. At what angular

speed (in rev/s) would we have to spin the coil? Does this appear to be a feasible method to extract energy from Earth's magnetic field?

General Problems

30. •••A long, rectangular loop of width w, mass m, and resistance R is being pushed into a magnetic field by a constant force $\vec{F}$ (Figure 20-24). Derive an expression for the speed of the loop while it is entering the magnetic field.

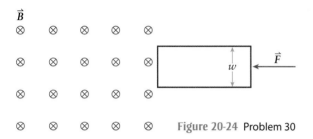

Figure 20-24 Problem 30

31. ••A magnetic field of 0.45 G is directed straight down, perpendicular to the plane of a circular coil of wire that is made up of 250 turns and has a radius of 20 cm. (a) If the coil is stretched, in a time of 15 ms, to a radius of 30 cm, calculate the emf induced in the coil during the process. (b) Assuming the resistance of the coil is a constant 25 Ω, what is the induced current in the coil during the process? (c) What is the direction of the induced current in the coil (clockwise or counterclockwise, as viewed from above)? SSM

32. ••**Astronomy** Activity on the Sun, such as solar flares and coronal mass ejections, hurls large numbers of charged particles into space. When the particles reach Earth, they can interfere with communications and the power grid by causing electromagnetic induction. As one example, a current of millions of amps (known as the *auroral electrojet*) that runs about 100 km above Earth's surface can be perturbed. The change in the current causes a change in the magnetic field it produces at Earth's surface, which induces an emf along Earth's surface and in the power grid (which is grounded). Induced electric fields as high as 6.0 V/km have been measured. We can model the circuit at Earth's surface as a rectangular loop made up of the power lines completed by a path through the ground beneath. We can treat the magnetic field created by the electrojet as being uniform (but not constant). Consider a 1.0-km-long stretch of power line that is 5.0 m above the surface of Earth. If the induced emf in the Earth–power line loop is 6.0 V, at what rate must the magnetic field through the loop be changing?

33. ••**Medical** During transcranial magnetic stimulation (TMS) treatment, a magnetic field typically of magnitude 0.50 T is produced in the brain using external coils. During the treatment, the current in the coils (and hence the magnetic field in the brain) rises from zero to its peak in about 75 μs. Assume that the magnetic field is uniform over a circular area of diameter 2.0 cm inside the brain. What is the magnitude of the average induced emf around this area in the brain during the treatment?

34. •A permanent bar magnet with the north pole pointing downward is dropped into a solenoid. (a) Determine the direction of the induced current that would be measured in the ammeter shown in Figure 20-25. (b) If the magnet is suddenly pulled upward through the solenoid, what is the direction of the induced current that would be measured in the ammeter?

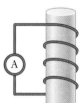

Figure 20-25 Problem 34

35. ••A pair of parallel conducting rails that are 12 cm apart lies at right angles to a uniform magnetic field of 0.8 T directed into the page, as shown in Figure 20-26. A 15-Ω resistor is connected across the rails. A conducting bar is moved to the right at 2 m/s across the rails. (a) What is the current in the resistor? (b) What direction is the current in the bar (up or down)? (c) What is the magnetic force on the bar? SSM

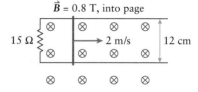

Figure 20-26 Problem 35

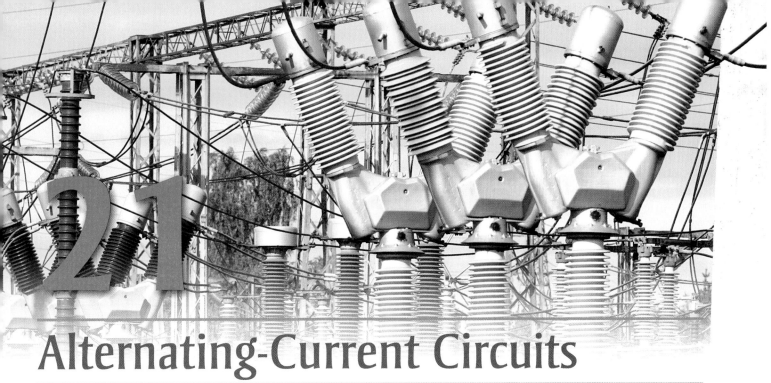

Alternating-Current Circuits

In this chapter, your goals are to:

- (21-1) Explain the importance of alternating current.
- (21-2) Describe what is meant by the root mean square value.
- (21-3) Calculate the voltage change produced by a transformer.
- (21-4) Describe why an inductor opposes changes in the current passing through it.
- (21-5) Explain the flow of energy in an *LC* circuit.
- (21-6) Describe what happens in a driven series *LRC* circuit when the driving frequency changes.
- (21-7) Discuss why current can flow in only one direction in a *pn* junction diode.

To master this chapter, you should review:

- (12-3, 12-4) Simple harmonic motion and the energy of a mass–spring system.
- (12-8) How a damped oscillator responds to different driving frequencies.
- (17-5, 17-6) The definition of capacitance and the electric energy stored by capacitors.
- (18-3, 18-6) The current through and voltage across a resistor, and the power into a resistor.
- (18-5) Using Kirchhoff's loop rule to analyze circuits.
- (19-7) The magnetic field created by a solenoid.
- (20-2, 20-3) How Faraday's law and Lenz's law describe the induced emf that opposes a change in magnetic flux through a loop.
- (20-4) How ac generators work and the power they generate.

What do you think?

These transformers use Faraday's law to raise and lower the voltage of alternating current. Can they also be used to raise or lower the voltage of *direct* current? (a) Yes; (b) yes, but only to raise the voltage; (c) yes, but only to lower the voltage; (d) no.

21-1 Most circuits use alternating current

In Chapter 20 we learned the principles of electromagnetic induction and how they can be used to *generate* an alternating current. In this chapter we'll learn how to *manipulate* and *use* alternating current. Most of the electrical power on our planet is transmitted and used in the form of an alternating current, so by studying this chapter you'll learn an important aspect of how the world around you works.

We'll see how *mutual inductance*—in which a changing emf in one coil makes it possible to induce an emf in another coil—is key for understanding how electric power can be transmitted efficiently over long distances. This same principle explains how the relatively high voltage of 120 to 240 V available from a wall socket can be used to charge a cell phone that operates at low voltage, typically 5 V or less (Figure 21-1a). We'll also introduce *self-inductance*, an effect in which a changing emf in a coil produces an emf reaction in the coil itself. Self-inductance is at the heart of an important device called an *inductor* that has many applications in circuits.

Figure 21-1 Alternating current
When you (a) use the transformer in your mobile phone charger or (b) tune your television to a different channel, you are using the physics of alternating current.

(a)

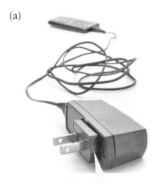

(b)

This transformer uses mutual inductance—in which a changing emf in one coil induces an emf in a second coil nearby—to reduce the voltage from a wall socket to a lower value suitable for charging a battery.

When you change the channel on a television, you're telling the TV tuning circuit to change its capacitance. This changes the natural frequency of the circuit so that it matches the carrier frequency of the channel you want to watch.

Many circuits that are connected to an ac source (such as the ac voltage provided by a wall socket) can be modeled as a combination of resistors, capacitors, and inductors. We'll see what happens when any one of these circuit elements by itself is connected to an ac source. We'll then go on to examine what happens in an *LRC series circuit* in which all three of these circuit elements are present. We'll find that such circuits hold the key to understanding what happens when you tune a radio or television to receive a particular station (Figure 21-1b). Finally, we'll take a brief look at *semiconductors*, a class of material that plays an important role in many circuits. Semiconductors are essential for the operation of two other important classes of circuit elements, called *diodes* and *transistors*.

Take-Home Message for Section 21-1

✔ The same principles of electromagnetic induction that explain how to produce an alternating current (ac) also tell us how to manipulate and use such currents.

✔ Many circuits that include an ac source can be modeled as a combination of resistors, capacitors, and inductors (a third type of circuit element).

21-2 We need to analyze ac circuits differently than dc circuits

Current arises in a circuit as a result of an applied voltage. The batteries that power your mobile phone or your flashlight are sources of (approximately) constant voltage. A 9-V battery, for example, introduces a roughly constant potential difference of 9 V between the two points at which it connects to a circuit. By convention, we refer to a circuit driven by a fixed voltage source, or one that does not change direction, as **a direct current**, or **dc** circuit. However, when you plug an electrical device into a wall socket, you are accessing an **alternating current**, or **ac** source.

The voltage from an ac source varies with time in a sinusoidal fashion, as we discussed in our description of ac generators in Section 20-4. In that section we wrote the emf delivered by an ac generator as

Emf in an ac generator
(20-7)

Angular speed of the rotating coil   Time   Angle of the coil at $t = 0$

Emf produced by an ac generator

$$\varepsilon = \omega N A B \sin(\omega t + \phi)$$

Number of turns in the coil   Area of the coil   Magnitude of the magnetic field to which the coil is exposed

The value V of this ac voltage oscillates between V_0 (the voltage amplitude) and $-V_0$.

The period T of this ac voltage equals the reciprocal of the frequency f:

$$T = \frac{1}{f} = \frac{2\pi}{\omega}$$

Figure 21-2 An ac voltage This figure graphs the ac voltage given by Equation 21-1. Note that at $t = 0$ the voltage is zero and increasing.

Let's choose the time $t = 0$ to be when the angle of the coil is $\phi = 0$, use the symbol V_0 for the combination of factors ωNAB, and use the symbol $V(t)$ for the time-varying emf. (We use V since emf is measured in volts.) If the generator is connected to two terminals, like the two terminals of a wall socket, then $V(t)$ represents the voltage between those terminals. We can then rewrite Equation 20-7 as a general equation that we'll use for ac sources of all kinds:

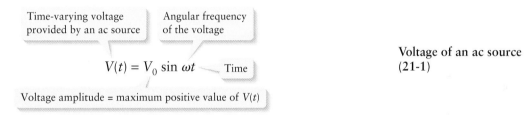

Time-varying voltage provided by an ac source

Angular frequency of the voltage

$$V(t) = V_0 \sin \omega t$$

Time

Voltage amplitude = maximum positive value of $V(t)$

Voltage of an ac source
(21-1)

Figure 21-2 graphs the voltage $V(t)$ as a function of time. The angular frequency ω of the voltage (in rad/s) is related to the frequency f of the voltage (in Hz) by the same relationship we used in Section 12-3 for simple harmonic motion:

$$\omega = 2\pi f \qquad (21\text{-}2)$$

The period T of the oscillation is equal to $1/f$. For example, in the United States and Canada ac voltage is applied at a frequency $f = 60$ Hz, so the period of oscillation is $T = 1/f = 1/(60\,\text{Hz}) = 0.017\,\text{s}$ and the angular frequency is $\omega = 2\pi f = (2\pi\,\text{rad})(60\,\text{Hz}) = 3.8 \times 10^2\,\text{rad/s}$.

If we attach a source of ac voltage described by Equation 21-1 to a resistor of resistance R, we can still use Equation 18-24 to calculate the power that flows into the resistor:

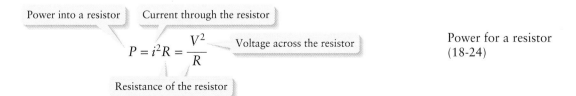

Power into a resistor

Current through the resistor

$$P = i^2 R = \frac{V^2}{R}$$

Voltage across the resistor

Resistance of the resistor

Power for a resistor
(18-24)

The only difference between a resistor in a dc circuit and one in an ac circuit is that because the voltage varies with time, the power into the resistor also varies with time. The *instantaneous* power into the resistor at time t is

$$(21\text{-}3) \qquad P(t) = \frac{V^2(t)}{R} = \frac{V_0^2 \sin^2(\omega t)}{R}$$

The value of $P(t)$ oscillates between zero (when $\sin \omega t = 0$) and V_0^2/R (when $\sin \omega t = 1$ or -1).

Often it's convenient to talk not about the instantaneous power $P(t)$ but the *average* power P_{average}. For example, if an appliance is designed to be powered by an ac voltage, its power rating in watts (such as a 1000-W microwave oven or a 1500-W hair dryer) is always stated in terms of the average power delivered to that appliance when in operation. The voltage amplitude V_0 and the resistance R are constants, so to find P_{average} we only have to figure out the average value of $\sin^2(\omega t)$. As we discussed in Section 20-4, this average value is $\frac{1}{2}$ (see Figure 20-10). So from Equation 21-3, the average power into the resistor is

$$(21\text{-}4) \qquad P_{\text{average}} = \frac{1}{2}\frac{V_0^2}{R}$$

The average power given by Equation 21-4 is one-half of the maximum value of the instantaneous power given by Equation 21-3.

Equation 21-4 is the basis of an alternative way to describe the voltage provided by an ac source. Comparing this equation to Equation 21-3, we can write

$$\left(\frac{V^2(t)}{R}\right)_{\text{average}} = \frac{1}{2}\frac{V_0^2}{R}$$

If we multiply both sides of this equation by the resistance R, we get

$$(V^2(t))_{\text{average}} = \frac{V_0^2}{2}$$

The left-hand side of this equation is called the *mean square value* of the ac voltage: We take the square of the voltage, then calculate the average (or *mean*) of that quantity. The square root of the mean square is called the **root mean square value**, or **rms value**, of the voltage:

The root mean square (rms) value of $V(t)$... ... is the square root of the average value of $V^2(t)$.

$$V_{\text{rms}} = \sqrt{(V^2(t))_{\text{average}}} = \frac{V_0}{\sqrt{2}}$$

If $V(t)$ is a sinusoidal function, its rms value equals the maximum value of $V(t)$ (the amplitude) divided by $\sqrt{2}$.

As an example, in the United States the standard value for the ac voltage supplied by a wall socket is 120 V; this is actually the rms voltage, so $V_{\text{rms}} = 120$ V. The voltage amplitude, or peak voltage, is larger by a factor of $\sqrt{2}$, so $V_0 = V_{\text{rms}}\sqrt{2} = (120\text{ V})\sqrt{2} = 170$ V.

In terms of the rms voltage, we can write the average power from Equation 21-4 as

$$(21\text{-}6) \qquad P_{\text{average}} = \frac{1}{2}\frac{V_0^2}{R} = \left(\frac{V_0^2}{2}\right)\frac{1}{R} = \left(\frac{V_0}{\sqrt{2}}\right)^2\frac{1}{R} = \frac{V_{\text{rms}}^2}{R}$$

This says that Equation 18-24, $P = V^2/R$, tells us the *average* power if we replace V by the rms value V_{rms} of the voltage.

Watch Out! **The average voltage is zero, but the average power is not.**

The average voltage of an ac source described by Equation 21-1 is zero. But the average power delivered by that source is *not* zero because power is proportional to the square of the voltage, which is always positive.

Example 21-1 Traveling Abroad with Your Hair Dryer

Seasoned travelers know that an electric device that works in one country may not work in another. A certain hair dryer made in the United States, where the rms wall voltage is 120 V, is rated at 1.5 kW. Find the power output of the hair dryer when connected to a wall outlet in Australia, where the voltage amplitude is 325 V. Treat the resistance of the hair dryer as the same regardless of the wall voltage. (In reality, the resistance depends somewhat on the temperature of the dryer's heating elements, which depends on the operating voltage.)

Set Up

We'll use Equation 21-6 to find the resistance of the hair dryer from the given values of $P_{average}$ and V_{rms} in the United States. We'll then use this same equation to find the average power in Australia. To do this, we'll need to know the rms voltage in Australia; we'll find this from the given value of the voltage amplitude V_0 by using Equation 21-5.

Average power into a resistor:

$$P_{average} = \frac{V_{rms}^2}{R} \qquad (21\text{-}6)$$

Root mean square (rms) voltage:

$$V_{rms} = \sqrt{(V^2(t))_{average}} = \frac{V_0}{\sqrt{2}} \qquad (21\text{-}5)$$

Solve

Use Equation 21-6 to determine the resistance of the hair dryer. We assume this has the same value no matter what the operating voltage.

Solve Equation 21-6 for resistance R:

$$R = \frac{V_{rms}^2}{P_{average}}$$

Substitute the values in the United States, $V_{rms} = 120$ V and $P_{average} = 1.5$ kW $= 1.5 \times 10^3$ W:

$$R = \frac{(120\,\text{V})^2}{1.5 \times 10^3\,\text{W}} = 9.6\,\Omega$$

(Recall that $1\,\text{W} = 1\,\text{V}^2/\Omega$.)

Find the value of V_{rms} in Australia; then use this to determine the average power into the hair dryer when used in Australia.

In Australia the voltage amplitude is $V_0 = 325$ V, so the rms voltage is

$$V_{rms} = \frac{V_0}{\sqrt{2}} = \frac{325\,\text{V}}{\sqrt{2}} = 230\,\text{V}$$

With this value of V_{rms}, the average power into the hair dryer is

$$P_{average,Australia} = \frac{V_{rms,Australia}^2}{R} = \frac{(230\,\text{V})^2}{9.6\,\Omega} = 5.5 \times 10^3\,\text{W} = 5.5\,\text{kW}$$

Reflect

Our result of 5.5 kW is *considerably* more power than the hair dryer is designed to take in. You may know how hot the air from a 1500-W hair dryer can be, or perhaps you've felt how hot a 100-W bulb can get after it's been on for a while. So you can probably imagine how hot 5500 W would make the heating element of the hair dryer. To radiate that much power, the required temperature of the heating element might well exceed the melting point of the material! We'll see in the following section how this hair dryer can be used safely even with the higher voltage provided in Australia.

? Got the Concept? 21-1 Average and Instantaneous Power

A resistor is connected to a source of ac voltage described by Equation 21-1. At which of the following times is the instantaneous power into the resistor equal to the average power into the resistor? (a) $t = 0$; (b) $t = T/8$; (c) $t = T/4$; (d) more than one of these; (e) none of these.

Take-Home Message for Section 21-2

✔ The arithmetic mean of a sinusoidally varying voltage is zero because it is negative as often as it is positive. The average power is not zero because it is proportional to the square of the voltage, which cannot be negative.

✔ It is common to characterize AC voltage by its root mean square (rms) value. The peak voltage is $\sqrt{2}$ times the rms voltage.

21-3 Transformers allow us to change the voltage of an ac power source

All ac circuits are intended to transfer energy from one location to another. For example, if you plug your mobile phone's charger into a wall socket, energy from an ac generator that may be hundreds of kilometers away flows through power lines (Figure 21-3) into your home and into your phone. The *voltage* associated with this energy has different values at different places in the circuit, however; the voltage in the power lines is typically hundreds of kilovolts, the voltage provided by a wall socket in the United States or Canada is 120 V, and the voltage supplied to your phone by the charger is 5 V. In this section we'll see the reason why different voltages are used in this way, and we'll learn about an important device called a *transformer* that makes it possible to raise or lower an alternating-current voltage.

To understand why high voltages are used in power lines, recall this relationship from Section 18-6:

Figure 21-3 Power lines When electric energy is transmitted over long distances in the form of an alternating current, it is most efficient to use very high voltage (typically 110 kV or higher). The power lines shown here are elevated so that people and vehicles cannot touch them, which would present a safety hazard.

Power for a circuit element (18-23)

Power produced by or transferred into a circuit element

$$P = iV$$

Current through the circuit element

Voltage (absolute value) across the circuit element

 Source of emf: Power flows out of the source and into the moving charges.
Resistor: Power flows out of the charges and into the resistor.

Equation 18-23 tells us that the same amount P of electric power can be delivered at any voltage V, provided that the product of current and voltage remains the same. This is true for ac circuits as well as dc circuits. For a power line, it's best to use high voltage and low current. The reason is that a long power line has a substantial resistance R, and so some of the energy being carried by the current goes into heating the power line rather than being transmitted to the end user. The rate at which energy is dissipated in a resistor with resistance R is given by Equation 18-24 (see Section 21-2), which we can write as $P = i^2R$. To minimize this energy loss, the current i should be as small as possible. It then follows from Equation 18-23 that the voltage V must be as large as possible.

! Watch Out! In power lines, don't confuse the voltage *between* adjacent cables with the voltage *across* the length of a single cable.

You might look at Equation 18-24 and think "It also says that $P = V^2/R$ for a resistor. So why isn't it preferable to have the voltage V be small to minimize energy loss?" The explanation is that in Equation 18-23, $P = iV$, V represents the voltage between the *two* cables that make up a power line. (You can see in Figure 21-3 that the cables come in pairs.) You need both cables to make a complete circuit, just as the power cord for a household appliance has two wires in it (one wire is connected to one prong of the plug at the end of the cord,

and the other wire to the other prong). This is the voltage that has to have a large value in order to make the current small. By contrast, in Equation 18-24 the quantity V is the potential difference between two ends of a *single* cable and is related to the current in the cable by the relationship $V = iR$ for resistors (Equation 18-9) that we learned in Section 18-3. This voltage is low if the current is small, so the power $P = i^2R = V^2/R$ that goes into heating the resistor is small.

In the home, however, the voltage must be kept small. That's because there is a risk that a person could be exposed to that voltage (for instance, if a person were to touch an appliance whose internal wiring had failed). If a voltage V is applied between two parts of your body, the resulting current through your body is proportional to V.

② The alternating current in the primary coil creates an alternating magnetic field of the same angular frequency ω. The field lines are "trapped" by the iron core and all pass through the secondary coil.

③ The alternating magnetic flux through the windings of the secondary coil induces an alternating emf, or output voltage, of the same angular frequency ω but of amplitude V_s.

Figure 21-4 **A transformer** This device uses two coils and a piece of magnetic material such as iron to raise or lower the voltage from an ac source.

① An input ac voltage with amplitude V_p and angular frequency ω causes an alternating current in the primary coil.

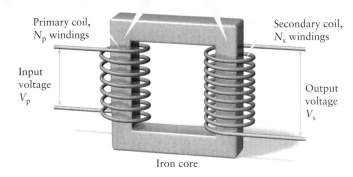

Primary coil, N_p windings

Secondary coil, N_s windings

Input voltage V_p

Output voltage V_s

Iron core

The amplitude V_s of the output voltage equals (N_s/N_p) times the amplitude V_p of the input voltage.
• If $N_s > N_p$, then $V_s > V_p$. This is a step-up transformer.
• If $N_s < N_p$, then $V_s < V_p$. This is a step-down transformer.

A current of just 0.1 A can cause ventricular fibrillation, and a current of 0.2 A or more can cause severe burns and stop the heart altogether. To reduce the risk of accident, the voltage supplied to the home has to be kept at a much lower value (120 to 240 V) than that used in power lines (more than 110 kV, or about a thousand times higher than what is safe for home use).

A device that can raise or lower an ac voltage to a desired value is called a **transformer**. The transformer shown in Figure 21-4 consists of two coils formed by winding separate wires around a piece of iron. The coil on the left, called the **primary coil**, is connected to the input voltage. The coil on the right, the **secondary coil**, is the source of the output voltage, even though there's no direct electric connection between the two coils. Let's see how this works.

When an ac voltage is applied to the primary coil in the transformer, the current in the windings gives rise to a magnetic field. This field oscillates with the same angular frequency ω as the ac voltage and current, and so creates a time-varying magnetic flux through the windings of the primary coil. Because iron is a magnetic material, the iron core confines the magnetic field lines so that the same flux that passes through each winding of the primary coil also passes through each winding of the secondary coil. As a result, there is a time-varying magnetic flux through the secondary coil. Faraday's law (see Sections 20-2 and 20-3) then tells us that a time-varying emf is induced in the secondary coil, which means this coil acts as a voltage source. The emf in the secondary coil oscillates with the same angular frequency as the current in the primary coil, so the output voltage produced by the secondary coil oscillates at the same rate as the input voltage that feeds into the primary coil. This effect, in which a change in the current in one coil induces an emf and current in a second coil, is called **mutual inductance**.

The *amplitude* of the output voltage in Figure 21-4 will be different from that of the input voltage—that is, the amplitude will be "transformed" by the transformer—if there are different numbers of windings in the primary and secondary coils (N_p and N_s, respectively, in Figure 21-4). To see how this can be, first notice that the time-varying magnetic flux Φ_B in the primary coil induces an emf there. From Equation 20-2, the magnitude of the emf induced in each of the windings of the primary coil is $|\Delta\Phi_B/\Delta t|$, the magnitude of the rate of change of the magnetic flux through that winding. There are N_p windings in the primary coil, so the total emf in that coil has magnitude $N_p|\Delta\Phi_B/\Delta t|$. By Kirchhoff's loop rule (Section 18-5), the sum of the voltage drops around the left-hand circuit that includes the primary coil must be zero, so the emf in the primary coil must have the same magnitude as the input voltage $V_p(t)$. (We're ignoring any voltage drops due to the resistance of the wires in this circuit.) So we can write

$$(21\text{-}7) \qquad |\text{input voltage}| = |V_p(t)| = N_p \left| \frac{\Delta \Phi_B}{\Delta t} \right|$$

Since the iron core ensures that there is the same magnetic flux Φ_B through each winding of the secondary coil as through each winding of the primary coil, the rate of change of the flux is the same and so the emf per winding is the same as well. There are N_s windings in the secondary coil, so the total emf in that coil has magnitude $N_s |\Delta \Phi_B / \Delta t|$. This emf acts as the voltage source for the circuit attached to the secondary coil (the right-hand circuit in Figure 21-4), so the magnitude of the output voltage $V_s(t)$ is

$$(21\text{-}8) \qquad |\text{output voltage}| = |V_s(t)| = N_s \left| \frac{\Delta \Phi_B}{\Delta t} \right|$$

If we divide Equation 21-8 by Equation 21-7, the factors of $|\Delta \Phi_B / \Delta t|$ cancel and we are left with

$$(21\text{-}9) \qquad \frac{|\text{output voltage}|}{|\text{input voltage}|} = \frac{|V_s(t)|}{|V_p(t)|} = \frac{N_s}{N_p}$$

Equation 21-9 is valid for the *instantaneous* values of the input and output voltages, so it must also be true for the *maximum* values. These are the voltage amplitudes, which we call V_p and V_s for the input (primary) voltage and output (secondary) voltage, respectively. Equation 21-9 is also true for the *rms* values of these voltages. So we are left with the result

Input and output voltages for a transformer (21-10)

Amplitudes of the input (p) and output (s) voltages

$$\frac{V_s}{V_p} = \frac{V_{s,\text{rms}}}{V_{p,\text{rms}}} = \frac{N_s}{N_p}$$

Number of windings in the primary coil (p) and secondary coil (s)

Rms values of the input (p) and output (s) voltages

In other words, in a transformer the ratio of the number of windings in the secondary coil (N_s) to the number of windings in the primary coil (N_p) determines the output voltage relative to the input voltage. In a **step-up transformer**, N_S is greater than N_P, so the output voltage is greater than the input voltage. In a **step-down transformer**, N_S is less than N_P, which results in an output voltage less than the input voltage. A step-up transformer is used at an electric power generation station to raise the voltage provided by the generator (which is connected to the primary coil) to a much larger value in the transmission lines (which are connected to the secondary coil). Close to where the electric power is to be used in home or industry, step-down transformers are used to reduce the voltage to a safe value. The charger that you use for recharging your mobile phone or other electronic device is also a step-down transformer that lowers the voltage provided by the wall socket to a much lower value suitable for the battery to be recharged.

The *current* in the secondary coil is also different from that in the primary coil. If we neglect any energy losses in the transformer (for instance, due to heating of the iron core by eddy currents), conservation of energy requires that the rate at which energy is delivered to the primary coil by the input voltage must equal the rate at which energy is transferred from the primary to the secondary coil. The rate of energy delivery is power and power equals current times voltage (Equation 18-23), so

input power in primary coil = output power into secondary coil

$$(21\text{-}11) \qquad i_p(t) V_p(t) = i_s(t) V_s(t)$$

In Equation 21-11 $i_p(t)$ and $V_p(t)$ are the current and voltage in the primary coil at time t, and $i_s(t)$ and $V_s(t)$ are the corresponding quantities in the secondary coil. If we rearrange Equation 21-11 and compare to Equation 21-9, we see that

$$\frac{|\text{output current}|}{|\text{input current}|} = \frac{|i_s(t)|}{|i_p(t)|} = \frac{|V_p(t)|}{|V_s(t)|} = \frac{N_p}{N_s} \qquad \text{(21-12)}$$

Compare Equation 21-12 with Equation 21-10. For a step-up transformer, for which N_s is greater than N_p, the output voltage is greater than the input voltage but the output current is smaller than the input current. The reverse is true for a step-down transformer.

Go to Interactive Exercise 21-1 for more practice dealing with coils and current

---

Example 21-2 A High-Voltage Transformer

Each of the 17 generators employed in the hydroelectric power plant at Hoover Dam in Colorado can generate up to 133 MW of average power. This power is delivered at 8.00 kV rms to a transformer that connects to a long-distance transmission line that operates at 5.00×10^2 kV rms. (a) What is the ratio of the number of windings in the secondary coil of the transformer to the number in the primary coil? (b) What is the rms current in the high-voltage transmission line?

Set Up

In part (a) we'll use Equation 21-10 to relate the ratio N_s/N_p (the number of windings in the secondary coil divided by the number in the primary coil) to the input and output rms voltages. In part (b) we'll assume that the transmission line delivers its power to a resistor, so we can relate the current, voltage, and power using Equations 18-9 and 18-23.

Input and output voltages for a transformer:

$$\frac{V_s}{V_p} = \frac{V_{s,rms}}{V_{p,rms}} = \frac{N_s}{N_p} \qquad \text{(21-10)}$$

Relationship among potential difference, current, and resistance:

$$V = iR \qquad \text{(18-9)}$$

Power for a circuit element:

$$P = iV \qquad \text{(18-23)}$$

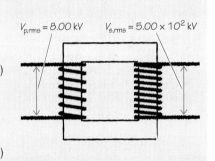

Solve

(a) We know both the primary and secondary rms voltages, so we can rearrange Equation 21-10 to solve for the ratio of the number of windings.

From Equation 21-10,

$$\frac{N_s}{N_p} = \frac{V_{s,rms}}{V_{p,rms}} = \frac{5.00 \times 10^2 \text{ kV}}{8.00 \text{ kV}} = 62.5$$

There are 62.5 times as many windings in the secondary coil as in the primary coil (or, equivalently, 125 windings in the secondary for every two windings of the primary).

(b) Since the resistance R is a constant, the current and voltage both have the same $\sin \omega t$ time dependence. We use this to relate the average power $P_{average} = 133$ MW to the rms values of the current and voltage in the transmission line.

If the voltage is $V(t) = V_0 \sin \omega t$, from Equation 18-9 the current is

$$i(t) = \frac{V(t)}{R} = \frac{V_0 \sin \omega t}{R} = \frac{V_0}{R} \sin \omega t = i_0 \sin \omega t$$

In this expression i_0 is the amplitude of the current. From Equation 18-23 the instantaneous power is

$$P(t) = i(t)V(t) = (V_0 \sin \omega t)(i_0 \sin \omega t) = i_0 V_0 \sin^2 \omega t$$

To find the average power $P_{average}$, replace $\sin^2 \omega t$ by its average value $\frac{1}{2}$:

$$P_{average} = \frac{i_0 V_0}{2} = \left(\frac{i_0}{\sqrt{2}}\right)\left(\frac{V_0}{\sqrt{2}}\right) = i_{rms} V_{rms}$$

The average power is the product of the rms current and the rms voltage. Since $i(t)$ has the same time dependence as $V(t)$, the rms value of current is equal to its maximum value divided by $\sqrt{2}$, just as for the voltage. From this expression, the rms current in the transmission line is

$$i_{s,\text{rms}} = \frac{P_{\text{average}}}{V_{s,\text{rms}}} = \frac{133\text{ MW}}{5.00 \times 10^2\text{ kV}} = \frac{133 \times 10^6\text{ W}}{5.00 \times 10^5\text{ V}} = 266\text{ A}$$

Reflect

In the process of delivering power to the end user, the voltage must ultimately be stepped down to 120 V. This process is not normally done with a single transformer. One reason is that if only one transformer were used for this stepping-down process, there would need to be more than 4000 windings in the primary coil for every one in the secondary.

We found a current of 266 A in the transmission line. Although that sounds large, it is only about a factor of 10 more than the current in household wiring. In addition, the power loss in the transmission wire is relatively low at this current. A few hundred kilometers of transmission wire might have a resistance of 25 Ω. The loss in such a line would be about 1.8 MW. Although in absolute terms this is a significant power loss, it represents only 1.4% of the total power transmitted.

Ratio of coil windings to step down 5.00×10^2 kV to 120 V:

$$\frac{N_p}{N_s} = \frac{V_p}{V_s} = \frac{5.00 \times 10^2\text{ kV}}{120\text{ V}} = 4170$$

The voltage between the two ends of the transmission line is $V_{\text{ends}} = iR$, where $R = 25$ Ω is the resistance of the line. Use the formula we derived in part (b) to find the power lost in the transmission line:

$$P_{\text{lost,average}} = i_{\text{rms}}V_{\text{ends,rms}} = i_{\text{rms}}(i_{\text{rms}}R)$$
$$= i_{\text{rms}}^2 R = (266\text{ A})^2(25\text{ }\Omega) = 1.8 \times 10^6\text{ W} = 1.8\text{ MW}$$

Express this as a percentage of the power that is transmitted:

$$\text{percentage} = \frac{\text{power lost}}{\text{power transmitted}} \times 100\%$$
$$= \frac{1.8\text{ MW}}{133\text{ MW}} \times 100\% = 1.4\%$$

❓ Got the Concept? 21-2 Transforming a Transformer

A certain step-up transformer has 100 windings in its primary coil and 250 windings in its secondary coil. You reverse the connections to the transformer so that the primary coil becomes the secondary and vice versa. If you now apply an ac input voltage of 120 V rms to the reversed transformer, what will be the rms output voltage? (a) 40 V; (b) 48 V; (c) 100 V; (d) 250 V; (e) 300 V.

Take-Home Message for Section 21-3

✔ A transformer uses mutual inductance to raise or lower an ac voltage.

✔ In a step-up transformer, there are more windings in the secondary coil than in the primary coil. The output voltage is greater than the input voltage, and the output current is less than the input current.

✔ In a step-down transformer, there are fewer windings in the secondary coil than in the primary coil. The output voltage is less than the input voltage, and the output current is greater than the input current.

21-4 An inductor is a circuit element that opposes changes in current

We saw in Section 21-3 that a transformer works by the mutual inductance of two coils: A change in current in one coil causes a change in the magnetic flux through the other coil, which induces an emf. The same physics applies equally well to individual windings within a *single* coil. You can think of each turn of the coil as a separate loop; as a changing current passes through any loop, the changing magnetic field that arises induces an emf in the other loops. This induced emf, according to Lenz's law, opposes the change in flux that created it. So if the current is increasing in the coil, the induced emf will oppose an increase in current. If the current is decreasing, the induced emf will be directed so that the current is augmented. This **self-inductance** has the net effect of opposing a change in current in a coil. Let's see how to determine the emf induced in this way.

Inductance of a Coil

Consider a coil of N windings in which there is a current i. This current causes the coil to produce a magnetic field, and the flux of this field through each winding of the coil is Φ_B. The total flux through all N windings is then $N\Phi_B$. We define the **inductance** of the coil (symbol L) as the total flux divided by the current that produces it:

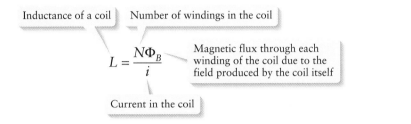

Definition of inductance
(21-13)

The units of inductance are henrys, abbreviated as H: $1\,\text{H} = 1\,\text{T} \cdot \text{m}^2/\text{A}$.

The magnetic field is directly proportional to the current i that produces it, and the same is true of the flux of that magnetic field. So the numerator in Equation 21-13 is proportional to the current i. Since the denominator in Equation 21-13 is equal to i, the current will cancel out when we calculate the inductance L. The inductance therefore does not depend on i, but only on geometrical factors such as the dimensions of the coil and physical constants. As an example, let's calculate the inductance of a long, straight coil or solenoid. We learned in Section 19-7 that the magnetic field of an ideal solenoid (one whose length is much greater than its diameter) is uniform, and has a magnitude given by Equation 19-17:

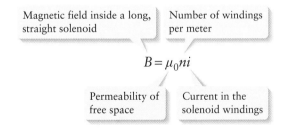

Magnetic field inside a long, straight solenoid
(19-17)

If the solenoid has length ℓ, the quantity n in Equation 19-17 equals the total number of windings N divided by ℓ. We can then write the magnetic flux through the cross-sectional area A of each winding of the solenoid as

$$\Phi_B = BA = (\mu_0 ni)A = \left(\frac{\mu_0 Ni}{\ell}\right)A$$

If we substitute this into Equation 21-13, we get the following expression for the inductance of the solenoid:

$$L = \frac{N\Phi_B}{i} = \frac{N}{i}\left(\frac{\mu_0 Ni}{\ell}\right)A = \frac{\mu_0 N^2 A}{\ell} \qquad (21\text{-}14)$$

(inductance of a solenoid)

For the solenoid, as for any coil, the inductance depends only on its dimensions and a physical constant (μ_0, the permeability of free space); inductance does not depend on current.

The quantity $\mu_0 = 4\pi \times 10^{-7}\,\text{T}\cdot\text{m/A}$ appears in all expressions for the inductance of a coil. Since $1\,\text{H} = 1\,\text{T}\cdot\text{m}^2/\text{A}$, it follows that $1\,\text{T}\cdot\text{m/A} = 1\,\text{H/m}$. So an alternative way to express the permeability of free space is $\mu_0 = 4\pi \times 10^{-7}\,\text{H/m}$. This is useful for problems that involve inductance, like the following example.

Example 21-3 How Large Is One Henry?

You decide to create a solenoid with inductance of 1.00 H by tightly wrapping wire around a plastic pipe. The wire has a diameter of 2.60 mm, and the pipe has an outside radius of 3.00 cm. How long a pipe will you need? Is 1.00 H a large or a small amount of inductance?

Set Up

Winding the wire tightly around the pipe will create a solenoid. The length of the solenoid may likely be large compared to its diameter, so we'll treat it as ideal. Then we can use Equation 21-14 to calculate the inductance of this solenoid. The pipe has a circular cross section, so to find A we'll use the expression for the area of a circle.

Inductance of a solenoid:

$$L = \frac{\mu_0 N^2 A}{\ell} \qquad (21\text{-}14)$$

Area of a circle of radius r:

$$A = \pi r^2$$

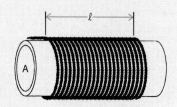

Solve

We don't know how many windings the solenoid will have. However, because the solenoid is tightly wrapped we can write the number of windings as the total length of the solenoid divided by the diameter of the wire. We substitute this expression for N into the inductance equation, and then solve for the length of the pipe.

The number of windings that will fit equals the length ℓ of the pipe divided by the diameter D_W of the wire:

$$N = \frac{\ell}{D_W}$$

Substitute this into Equation 21-14 for the inductance of the solenoid:

$$L = \frac{\mu_0 (\ell/D_W)^2 A}{\ell} = \frac{\mu_0 \ell A}{D_W^2}$$

The length of the pipe is therefore

$$\ell = \frac{L D_W^2}{\mu_0 A}$$

The cross-sectional area of the pipe of radius 3.00 cm is

$$A = \pi r^2 = \pi (3.00\,\text{cm})^2 = \pi (3.00 \times 10^{-2}\,\text{m})^2 = 2.83 \times 10^{-3}\,\text{m}^2$$

We are given $L = 1.00$ H, $D_W = 2.60 \times 10^{-3}$ m, and $\mu_0 = 4\pi \times 10^{-7}$ H/m, so

$$\ell = \frac{(1.00\,\text{H})(2.60 \times 10^{-3}\,\text{m})^2}{(4\pi \times 10^{-7}\,\text{H/m})(2.83 \times 10^{-3}\,\text{m}^2)}$$

$$= 1.90 \times 10^3\,\text{m} = 1.90\,\text{km}$$

Reflect

We would need a pipe nearly 2 km long to construct a 1.00-H inductor! Practical inductors have much smaller values of inductance. (Note that the length we determined is indeed much greater than the radius of the pipe, so we were justified in treating this as an ideal solenoid.)

The emf of an Inductor

An inductor in a circuit produces an emf if the current through the inductor changes. We can find an expression for this emf if we combine the definition of inductance,

Equation 21-13, with Faraday's and Lenz's laws (Section 20-3). From Equation 21-13, the total flux through an inductor is

$$N\Phi_B = Li$$

Faraday's and Lenz's laws tell us that the emf produced by an inductor is equal to the negative of the rate of change of the total flux through the inductor (Equation 20-3). So we can write the induced emf in the inductor as

$$\varepsilon = -\frac{\Delta(N\Phi_B)}{\Delta t} = -\frac{\Delta(Li)}{\Delta t}$$

(21-15)

Let's ignore the resistance of the wire that makes up the inductor, so this is an *ideal* inductor. Then ε in Equation 21-15 is also equal to the voltage V across the inductor (there is no additional voltage drop due to resistance, since there is zero resistance). Since the inductance L does not change with time, we can write Equation 21-15 as

Voltage across an inductor | Inductance of the inductor

$$V = -L\frac{\Delta i}{\Delta t}$$

Rate of change of the current i in the inductor

The negative sign means that the voltage opposes any change in the current.

Voltage across an inductor
(21-16)

Let's examine Equation 21-16 to see what it means. In **Figure 21-5a** an inductor (symbolized by a coil) carries a constant current i. The rate of change of this constant current is zero, so from Equation 21-16 there is zero voltage across the inductor: $V = 0$. (Remember, we are ignoring the resistance of the wires that make up the inductor.) In **Figure 21-5b** the current through the inductor increases, so in Equation 21-16 $\Delta i/\Delta t > 0$ and $V < 0$. The negative value of V means that there is a voltage *drop* from where the current enters the inductor to where it leaves the inductor. That's like the voltage drop for current that passes through a resistor, which means that the inductor is now resisting the current—or, more properly, resisting the increase in current. So in this case the voltage opposes the change in current. In **Figure 21-5c** the current through the inductor decreases, so in Equation 21-16 $\Delta i/\Delta t < 0$ and $V > 0$. The positive value of V means that there is a voltage *gain* from where the current enters the inductor to where it leaves. That's like the voltage gain for current directed through a battery (a source of emf) from the negative terminal to the positive terminal, so in this case the inductor is assisting the current and so is trying to prevent the current from decreasing.

(a) An inductor with constant current

(b) An inductor with increasing current

(c) An inductor with decreasing current

• The symbol for an inductor is a coil.
• An ideal inductor has zero resistance.

$i \longrightarrow$

If the current in the inductor is constant, there is zero voltage across the inductor: $V = 0$. This includes the case where there is no current.

The voltage *drops* from left to right along the inductor.

i at time $t \longrightarrow$
i at time $t + \Delta t \longrightarrow$

If current in the inductor is increasing so $\Delta i/\Delta t > 0$, there is negative voltage across the inductor: $V < 0$. This voltage opposes the increase in the current.

The voltage *rises* from left to right along the inductor.

i at time $t \longrightarrow$
i at time $t + \Delta t \longrightarrow$

If current in the inductor is decreasing so $\Delta i/\Delta t < 0$, there is positive voltage across the inductor: $V > 0$. This voltage opposes the decrease in the current.

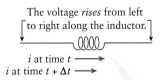

Figure 21-5 **Voltage across an inductor** The voltage across an ideal inductor (a coil with no resistance) depends on whether the current through the inductor is (a) constant, (b) increasing, or (c) decreasing.

Again, the voltage opposes the change in current. We can summarize these observations by saying that

> *The voltage across an inductor opposes any change in the current through the inductor. If the current is not changing, the voltage is zero.*

This observation is just what we would expect from Lenz's law, which says that the induced emf in any coil (including an inductor) always acts to oppose any changes. A useful rule is that V in Equation 21-16 represents the voltage *gain* across the inductor when traveling in the direction of the current. If V is negative, there is a voltage drop.

! Watch Out! Don't confuse inductors, capacitors, and resistors.

It's useful to contrast an inductor with two other devices used in circuits, a capacitor (Section 17-5) and a resistor (Section 18-3). The voltage across a capacitor is $V = q/C$, where q is the magnitude of the charge on either plate of the capacitor and C is the capacitance (a quantity that, like the inductance of an inductor, depends on the geometry of the capacitor). The greater the amount of charge on each capacitor plate, the greater the capacitor voltage. The voltage across a resistor that carries current i is $V = iR$, where R is the resistance of the resistor. This voltage is proportional to the rate at which charge moves through the resistor, that is, the current i. The greater the current, the greater the rate at which charge flows through the resistor and the greater the resistor voltage. By contrast, for an inductor Equation 21-16 tells us that the voltage is proportional to the rate of change of current—that is, to the rate of change of the rate at which charge flows through the inductor.

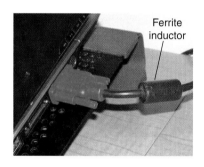

Ferrite inductor

Figure 21-6 Inductor on a cable The ferrite inductor on this monitor cable reduces the interference from external high-frequency electromagnetic signals.

Our observations show that an inductor in a circuit acts to oppose changes in the current within the circuit. The more rapid the change in current—that is, the greater the value of the quantity $\Delta i / \Delta t$ in Equation 21-16—the greater the voltage that will oppose the change. Some computer cables have a built-in inductor (Figure 21-6) that is designed to deal with high-frequency interference from signals radiated from other devices. These high-frequency signals would cause rapidly changing currents in the cables that could compromise the operation of the computer, so it's important to suppress them. If such a signal appears, the inductor will set up a voltage that opposes the signal. The inductor is made of a special material called a *ferrite* that also has a large resistance to high-frequency currents, so the energy in the stray signal is absorbed and dissipated as heat.

As we'll see in the following sections, inductors aren't just for suppressing currents. In the right circumstances, they can be used for *sustaining* an alternating current of just the right frequency. This will help us understand what's involved in tuning a television or radio receiver.

Example 21-4 Inductor Voltage

A 0.500-mH inductor is in a dc circuit that carries a current of 0.500 A. If the current increases to 0.900 A in 0.150 ms due to a fault in the circuit, how large is the voltage that appears across the inductor? From the perspective of a positive charge moving through the inductor, is there a voltage rise or drop across the inductor?

Set Up
We'll use Equation 21-16 to calculate the voltage induced across the inductor. To decide whether there's a voltage rise or drop, we'll use the idea that the voltage across the inductor acts to oppose the change in current.

Voltage across an inductor:

$$V = -L\frac{\Delta i}{\Delta t} \qquad (21\text{-}16)$$

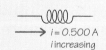

$i = 0.500\,A$
i increasing

Solve
Find the voltage that appears in the inductor.

The current increases from 0.500 A to 0.900 A in 0.150 ms = 0.150×10^{-3} s, so

$$\frac{\Delta i}{\Delta t} = \frac{0.900\,\text{A} - 0.500\,\text{A}}{0.150 \times 10^{-3}\,\text{s}} = 2.67 \times 10^3\,\text{A/s}$$

From Equation 21-16, the voltage across the inductor is

$$V = -L\frac{\Delta i}{\Delta t} = -(0.500 \times 10^{-3}\,\text{H})(2.67 \times 10^{3}\,\text{A/s})$$
$$= -1.33\,\text{H}\cdot\text{A/s} = -1.33\,\text{V}$$

The magnitude of the voltage is 1.33 V. The value of V is negative, which means that there is a voltage *drop* of 1.33 V across the inductor. Thus, the voltage opposes the flow of current and so opposes the increase in current (compare Figure 21-5b).

Reflect
The voltage induced in the inductor is comparable to that produced by a household AA or AAA battery. This voltage is only present during the time when the current is *changing*, however; when the current is constant, the inductor voltage is zero.

? Got the Concept? 21-3 Inductors vs. Capacitors vs. Resistors

For which of the following devices does the voltage across that device depend on the current that flows into that device? (a) An inductor; (b) a capacitor; (c) a resistor; (d) more than one of these; (e) none of these.

Take-Home Message for Section 21-4

✔ The inductance of a coil equals the net magnetic flux through the coil due to the current in the coil, divided by the current.

✔ An inductor is a circuit device that opposes any change in the current through that device. The voltage across an inductor is proportional to its inductance and to the rate of change of current through the inductor.

21-5 In a circuit with an inductor and capacitor, charge and current oscillate

Figure 21-7 shows a circuit made up of an inductor of inductance L and a capacitor of capacitance C connected by ideal, zero-resistance wires. This is called an **LC circuit**. With the switch open, positive charge $+Q_0$ is placed on the upper plate of the capacitor and negative charge $-Q_0$ is placed on the lower plate. This can be done by transferring a number of electrons from the upper plate to the lower plate. The switch is then closed, completing the circuit so that charge can flow. What happens? Remarkably, we'll see that the charge in this circuit behaves very much like the block attached to an ideal spring that we studied in Section 12-3: The charge *oscillates* back and forth in simple harmonic motion. Among other applications, this effect is used to generate the oscillating electric field that a microwave oven uses to cook food.

An LC Circuit and Simple Harmonic Motion

You might think that the excess electrons would move along the wires from the lower plate to the upper plate until the two plates are electrically neutral, at which point the motion of electrons would stop. Because there's an inductor in the circuit, however, what happens is dramatically different and far more interesting. **Figure 21-8** shows four steps in the behavior of the LC circuit. Once the switch is closed (Step 1 in Figure 21-8), a current is established around the circuit in the counterclockwise direction. (Remember that by convention, the direction of current is that in which positive charges would flow.) The current can't suddenly increase, however, because a voltage appears across the inductor to oppose the increase in current. Instead the current increases gradually and

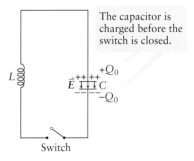

The capacitor is charged before the switch is closed.

Figure 21-7 An LC circuit An inductor (inductance L) and a capacitor (capacitance C) are connected as shown. Figure 21-8 shows what happens when we close the switch.

Figure 21-8 **Oscillations of an *LC*
circuit** When the switch in Figure
21-7 is closed, the capacitor charge
oscillates back and forth through the
circuit. These electrical oscillations are
directly analogous to the mechanical
oscillations of a block attached to an
ideal spring (Section 12-3).

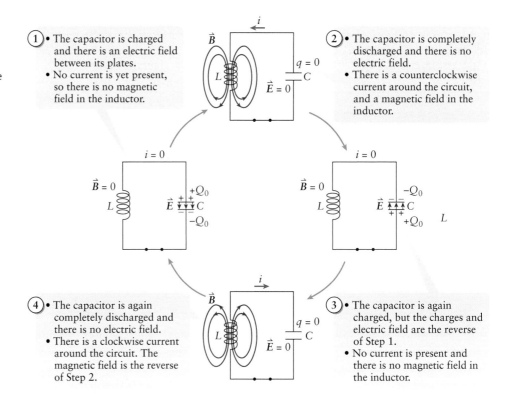

reaches a maximum value when the capacitor is fully discharged (Step 2 in Figure 21-8).
The current can't just come to a sudden stop, however; the inductor always opposes any
change in the current, so once the current starts to decrease a voltage will appear across
the inductor to slow that decrease. As a result the motion of charge continues. When the
current finally decreases to zero, the capacitor is again charged but with the opposite
polarity to its initial configuration (Step 3 in Figure 21-8). At this point a current is
established just as it was between Steps 1 and 2, but in the opposite direction: The current is clockwise, increasing gradually until the current is maximum and the capacitor
is again fully discharged (Step 4 in Figure 21-8). Again the inductor keeps the current
from coming to a sudden stop, and instead the current decreases gradually until the
capacitor is again fully charged and has returned to the configuration in Step 1. The
whole process then starts over!

You can see that what's happening here is an *oscillation* of charge and current. Let's
show that these oscillations are another example of *simple harmonic motion*, the special kind of oscillatory motion that we studied in Section 12-3. To see this, let's apply
Kirchhoff's loop rule to the *LC* circuit in Figure 21-8. This says that if we travel around
the circuit and measure the voltage across each element of the circuit, the sum of those
voltage changes must be zero.

We'll begin our trip around the circuit at the upper right-hand corner, and move
around the circuit clockwise so that we first pass downward through the capacitor. If we
let q be the charge on the upper plate of the capacitor at a given instant, the charge on
the lower plate at that same instant is $-q$. So if we start at the upper right-hand corner
of the circuit and travel downward through the capacitor, we'll encounter a voltage drop

(21-17)
$$V_C = -\frac{q}{C} \quad (LC \text{ circuit})$$

This is negative if q is positive, since then we travel from the positive charge to the
negative charge and so will encounter a decrease in electric potential (that is, a voltage
drop). If q is negative, so that there's negative charge on the upper plate and positive
charge on the lower plate, we'll encounter a voltage rise.

As we continue around the circuit, we'll encounter zero voltage across the
wires (which have zero resistance). We next pass upward through the inductor on

the left-hand side of the circuit. We'll take the current i to be positive if it also flows upward through the inductor (that is, clockwise around the circuit). If the current is positive, positive charge will flow onto the upper plate of the capacitor. The rate at which the charge q on the upper plate increases, measured in coulombs per second or amperes, is just equal to the current:

$$i = \frac{\Delta q}{\Delta t} \quad (LC \text{ circuit}) \tag{21-18}$$

From Equation 21-16, the voltage across the inductor is

$$V_L = -L\frac{\Delta i}{\Delta t} \quad (LC \text{ circuit}) \tag{21-19}$$

If the current through the inductor is positive (upward, or clockwise around the circuit) and increasing in magnitude (becoming more positive), then $\Delta i/\Delta t$ is positive and V_L is negative: We'll encounter a voltage drop as we move upward through the inductor. The same is true if the current through the inductor is negative (downward, or counterclockwise around the circuit) and decreasing in magnitude (becoming less negative). If instead the current through the inductor is either positive (clockwise) and decreasing or negative (counterclockwise) and becoming more negative, then $\Delta i/\Delta t$ is negative and V_L is positive: We'll encounter a voltage rise as we move upward through the inductor.

Kirchhoff's loop rule says that the sum of the voltages V_C from Equation 21-17 and V_L from Equation 21-19 must be zero:

$$V_C + V_L = -\frac{q}{C} - L\frac{\Delta i}{\Delta t} = 0$$

If we add $L(\Delta i/\Delta t)$ to both sides of this equation, we get

$$L\frac{\Delta i}{\Delta t} = -\frac{q}{C} \quad (LC \text{ circuit}) \tag{21-20}$$

On the face of it Equation 21-20 looks like a very difficult equation to solve. Both the capacitor charge q and the current i depend on time, and Equation 21-18 tells us that the current is equal to the rate of change of the capacitor charge. But in fact this is an equation that we've solved before, although with different symbols. Recall from Section 12-3 that if a block of mass m is attached to an ideal spring of spring constant k and displaced from equilibrium by x, the force that the spring exerts on the block is $F_x = -kx$. (Positive x means that the spring is stretched, and negative x means that the spring is compressed.) If the block is then released, it oscillates. The velocity of the block is the rate of change of the displacement x:

$$v_x = \frac{\Delta x}{\Delta t} \quad (\text{block and ideal spring}) \tag{21-21}$$

Newton's second law says that force $-kx$ on the block is equal to the mass of the block multiplied by its acceleration a_x. Acceleration is the rate of change of velocity, so we can write Newton's second law for the block as

$$m\frac{\Delta v_x}{\Delta t} = -kx \quad (\text{block and ideal spring}) \tag{21-22}$$

You can see that Equations 21-21 and 21-22 for a block and ideal spring are the *same* as Equations 21-18 and 21-20 for an LC circuit, with the names of the variables changed as listed in the first four rows of Table 21-1.

For a block attached to an ideal spring, the spring provides a restoring *force*: Whether the spring is compressed or stretched, this force tries to return the system to equilibrium. The block, however, has *inertia*: Once in motion, it tends to remain in motion, As a result, when the spring reaches the equilibrium position, it does not stop but instead overshoots. As a result, the block oscillates back and forth around its equilibrium position. In a similar way, the capacitor in an LC circuit provides a restoring *voltage* $-q/C$; whether q is positive or negative, this voltage drives charge around the circuit in a direction that tries to neutralize both plates. The inductor opposes changes in current; once a current is in motion, the inductor tends to keep it in motion. As a result, when the capacitor is in equilibrium with both of its plates neutral, the current "overshoots"

Table 21-1 Comparing a block on an ideal spring to an *LC* circuit

Quantity for a block on an ideal spring	Corresponding quantity for an *LC* circuit
spring displacement, x	capacitor charge, q
block velocity, $v_x = \Delta x / \Delta t$	current, $i = \Delta q / \Delta t$
spring constant of the spring, k	reciprocal of the capacitance, $1/C$
mass of the block, m	inductance, L
oscillation amplitude, A	maximum capacitor charge, Q_0
potential energy of the spring, $U_{spring} = \dfrac{1}{2}kx^2$	electric energy in the capacitor, $U_E = \dfrac{q^2}{2C}$
kinetic energy of the block, $K = \dfrac{1}{2}mv_x^2$	magnetic energy in the inductor, $U_B = \dfrac{1}{2}Li^2$

and the capacitor ends up charged again with reversed polarity. The result is an oscillation that's *exactly* like the simple harmonic motion of a block on an ideal spring. It's like watching the same play but with different actors in the cast: The inductor plays the role of the block, and the capacitor plays the role of the spring.

We can use Table 21-1 to tell us the angular frequency, period, and frequency for the oscillations of an *LC* circuit. From Equation 12-11, these quantities for a block on a spring are

Angular frequency, period, and frequency for a block attached to an ideal spring (12-11)

Angular frequency $\quad \omega = \sqrt{\dfrac{k}{m}}$

k = Spring constant of the spring

m = Mass of the object connected to the spring

Period $\quad T = \dfrac{2\pi}{\omega} = 2\pi\sqrt{\dfrac{m}{k}}$

Frequency $\quad f = \dfrac{1}{T} = \dfrac{1}{2\pi}\sqrt{\dfrac{k}{m}} = \dfrac{\omega}{2\pi}$

Table 21-1 tells us that k corresponds to $1/C$ and m corresponds to L. So the angular frequency, period, and frequency for the oscillations of an *LC* circuit are

Angular frequency, period, and frequency for an *LC* circuit (21-23)

Angular frequency $\quad \omega = \sqrt{\dfrac{1}{LC}}$

C = Capacitance

Period $\quad T = \dfrac{2\pi}{\omega} = 2\pi\sqrt{LC}$

L = Inductance

Frequency $\quad f = \dfrac{1}{T} = \dfrac{1}{2\pi}\sqrt{\dfrac{1}{LC}} = \dfrac{\omega}{2\pi}$

The smaller the product LC of inductance and capacitance, the greater the angular frequency and frequency of the oscillations and the shorter the oscillation period.

As an example, imagine an *LC* circuit made with a 2.00-μH inductor and a 10.0-nF capacitor. Then $L = 2.00 \times 10^{-6}$ H and $C = 10.0 \times 10^{-9}$ F, and from the last of Equations 21-23 the oscillation frequency is

$$f = \frac{1}{2\pi}\sqrt{\frac{1}{LC}} = \frac{1}{2\pi}\sqrt{\frac{1}{(2.00 \times 10^{-6}\,\text{H})(10.0 \times 10^{-9}\,\text{F})}}$$

$$= 1.13 \times 10^6 \text{ Hz} = 1.13 \text{ MHz}$$

The charges in this circuit oscillate back and forth 1.13 million times per second and complete one oscillation in a time $T = 1/f = 8.89 \times 10^{-7}$ s $= 0.889\,\mu$s. Such oscillations are much more rapid than those of a large mechanical oscillator such as a block attached to a spring.

The capacitor charge q is analogous to the displacement x for a block on a spring, so the maximum capacitor charge Q_0 is analogous to the amplitude A (the maximum displacement of the block on a spring). We can use this observation to write expressions for the capacitor charge q and current i as functions of time. In Section 12-3 we found that the displacement x and velocity v_x for a block on a spring are given by

$$x(t) = A\cos(\omega t + \phi) \quad \text{(block and ideal spring)} \qquad (12\text{-}6)$$
$$v_x(t) = -\omega A\sin(\omega t + \phi) \quad \text{(block and ideal spring)} \qquad (12\text{-}7)$$

Since A is analogous to Q_0, displacement x is analogous to q, and velocity v_x is analogous to i, we can write the following expressions for an LC circuit:

$$q(t) = Q_0\cos(\omega t + \phi) \quad \text{(LC circuit)} \qquad (21\text{-}24)$$
$$i(t) = -\omega Q_0\sin(\omega t + \phi) \quad \text{(LC circuit)} \qquad (21\text{-}25)$$

The angular frequency ω is given by the first of Equations 21-23: $\omega = \sqrt{1/LC}$. The cosine is maximum when the sine is zero and vice versa, so the capacitor charge is maximum when the current is zero and the capacitor charge is zero when the current is maximum. That's just what Figure 21-8 shows.

Since $\sin(\omega t + \phi)$ has values between $+1$ and -1, the maximum value of the current in Equation 21-25 is

$$i_{\max} = \omega Q_0 = \frac{Q_0}{\sqrt{LC}} \quad \text{(LC circuit)} \qquad (21\text{-}26)$$

As an example, consider the LC circuit we described above with $L = 2.00\,\mu$H $= 2.00 \times 10^{-6}$ H and $C = 10.0$ nF $= 10.0 \times 10^{-9}$ F. If the maximum capacitor charge is $Q_0 = 0.400\,\mu$C $= 4.00 \times 10^{-7}$ C, the maximum current in the LC circuit is

$$i_{\max} = \frac{Q_0}{\sqrt{LC}} = \frac{4.00 \times 10^{-7}\,\text{C}}{\sqrt{(2.00 \times 10^{-6}\,\text{H})(10.0 \times 10^{-9}\,\text{F})}} = 2.83\,\text{A}$$

Even though only a tiny amount of charge oscillates back and forth between the capacitor plates, the oscillation is so rapid that the maximum current is appreciable.

Energy in an *LC* Circuit

An electric circuit is first and foremost a means of transferring energy from one part of the circuit to another. The same is true for an LC circuit. To see what kinds of energy are involved, let's first look at the total mechanical energy E in a system made up of a block attached to an ideal spring. From Section 12-4, this is the sum of the kinetic energy K of the moving block and the elastic potential energy U_{spring} in the spring:

$$E = K + U_{\text{spring}} = \frac{1}{2}mv_x^2 + \frac{1}{2}kx^2$$

The kinetic and potential energies change, but the total energy E is constant and equal to $(1/2)kA^2$ (A is the amplitude, or maximum displacement).

Table 21-1 tells us how to find the analogous expression for the total energy of an LC circuit: Just replace m with L, k with $1/C$, x with q, and v_x with i. The result is

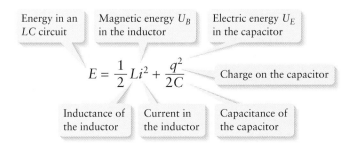

Energy in an *LC* circuit | Magnetic energy U_B in the inductor | Electric energy U_E in the capacitor

$$E = \frac{1}{2}Li^2 + \frac{q^2}{2C}$$

Charge on the capacitor

Inductance of the inductor | Current in the inductor | Capacitance of the capacitor

Magnetic and electric energies in an *LC* circuit

(21-27)

We recognize the second term on the right-hand side of Equation 21-27, $q^2/2C$, from Section 17-6: It's just the electric potential energy stored in the capacitor. Since a capacitor plays the same role in an LC circuit as the spring does in a block–spring combination, this electric potential energy is analogous to the elastic potential energy of a spring. We use the symbol U_E for this electric energy. The subscript E reminds us that an $\vec{E}$-field is involved; you can think of U_E as the energy required to take an uncharged capacitor and move charges $+q$ and $-q$ to the plates of the capacitor, thereby setting up an electric field $\vec{E}$ between the plates.

The first term on the right-hand side of Equation 21-27, $(1/2)Li^2$, is one that we haven't seen before. The presence of the inductance L in this term tells us that it represents energy stored in the inductor as a result of the presence of current. To see how this energy arises, recall from Section 21-4 that an inductor sets up an emf that opposes any change in the current through the inductor. If we want to make current flow through an inductor where none was flowing before, we have to do work against that emf. The quantity $(1/2)Li^2$ is exactly equal to the amount of work we have to do. By building up the current from zero to i, we also create a magnetic field $\vec{B}$, so we can think of $(1/2)Li^2$ as the energy required to set up a magnetic field in and around the inductor. That's why we call this **magnetic energy** and denote it by the symbol U_B (the subscript B reminds us that a $\vec{B}$-field is involved).

! Watch Out! **Magnetic energy is not the same as kinetic energy.**

Table 21-1 shows that the magnetic energy in an inductor, $U_B = (1/2)Li^2$, is analogous to the kinetic energy $K = (1/2)mv_x^2$ of the block in an oscillating block–spring system. This magnetic energy is only present if charges are in motion in the inductor, so i is nonzero. However, U_B is *not* the kinetic energy of the moving charges. That kinetic energy is very small because the moving electrons have very little mass. Instead, U_B is the energy stored in the magnetic field of the inductor. The inductor current i appears in the expression for U_B because it is this current that gives rise to the magnetic field.

In an LC circuit the energy oscillates between the electric and magnetic forms, just as the energy in a block–spring combination oscillates between the potential and kinetic forms (see Figure 12-10 in Section 12-4). In Figure 21-8 the energy is purely electric in Steps 1 and 3 (where the capacitor charge is maximum and the current is zero) and purely magnetic in Steps 2 and 4 (where the capacitor charge is zero and the current is maximum). If we substitute $q(t)$ from Equation 21-24 and $i(t)$ from Equation 21-25 into Equation 21-27, we get an expression for the energy of the LC circuit at any time:

$$E = U_B + U_E = \frac{1}{2}L\left[-\omega Q_0 \sin(\omega t + \phi)\right]^2 + \frac{\left[Q_0 \cos(\omega t + \phi)\right]^2}{2C}$$

(21-28)
$$= \frac{1}{2}L\omega^2 Q_0^2 \sin^2(\omega t + \phi) + \frac{Q_0^2}{2C}\cos^2(\omega t + \phi)$$

The first of Equations 21-23 tells us that $\omega = 1/\sqrt{LC}$, so $\omega^2 = 1/LC$ and $L\omega^2 = 1/C$. If we substitute this into the first term on the right-hand side of Equation 21-28 and recall that $\sin^2\theta + \cos^2\theta = 1$, we get

$$E = \frac{Q_0^2}{2C}\sin^2(\omega t + \phi) + \frac{Q_0^2}{2C}\cos^2(\omega t + \phi)$$

$$= \frac{Q_0^2}{2C}\left(\sin^2(\omega t + \phi) + \cos^2(\omega t + \phi)\right)$$

(21-29)
$$= \frac{Q_0^2}{2C} \quad (LC \text{ circuit})$$

The electric and magnetic energies in an oscillating LC circuit both vary with time, but the total energy $E = Q_0^2/2C$ remains constant.

Example 21-5 Analyzing an *LC* Circuit

An oscillating *LC* circuit is made of a 2.00–μH inductor and a 10.0–nF capacitor. The maximum charge on the capacitor is 0.400 μC. (a) Find the total energy in the circuit. (b) At an instant when one-quarter of the total energy is electric energy in the capacitor, find the absolute values of the capacitor charge and the current in the inductor.

Set Up

Equation 21-29 tells us the total energy of the *LC* circuit. This equals the sum of the magnetic and electric energies, which depend on the current *i* and capacitor charge *q* as Equation 21-27 tells us. We'll use this to find the absolute values of *i* and *q* at the instant in question.

Energy in an *LC* circuit:

$$E = \frac{Q_0^2}{2C} \qquad (21\text{-}29)$$

Magnetic and electric energies in an *LC* circuit:

$$E = U_B + U_E = \frac{1}{2}Li^2 + \frac{q^2}{2C} \quad (21\text{-}27)$$

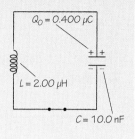

Solve

(a) Use Equation 21-29 to find the total energy.

The maximum capacitor charge is $Q_0 = 0.400\ \mu\text{C} = 0.400 \times 10^{-6}\ \text{C}$, and the capacitance is $C = 10.0\ \text{nF} = 10.0 \times 10^{-9}\ \text{F}$. The total energy of the oscillating circuit is

$$E = \frac{Q_0^2}{2C} = \frac{(0.400 \times 10^{-6}\ \text{C})^2}{2(10.0 \times 10^{-9}\ \text{F})} = 8.00 \times 10^{-6}\ \text{J} = 8.00\ \mu\text{J}$$

(b) The total energy remains constant. So if one-quarter of the energy is in the electric form, the other three-quarters must be in the magnetic form. Use this to determine the absolute values of the current and capacitor charge.

The magnetic energy is $\frac{3}{4}$ of the total energy:

$$U_B = \frac{1}{2}Li^2 = \frac{3}{4}E = \frac{3}{4}(8.00 \times 10^{-6}\ \text{J}) = 6.00 \times 10^{-6}\ \text{J}$$

The inductance is $2.00\ \mu\text{H} = 2.00 \times 10^{-6}\ \text{H}$. Solve for the absolute value of the current:

$$i^2 = \frac{2U_B}{L} = \frac{2(6.00 \times 10^{-6}\ \text{J})}{2.00 \times 10^{-6}\ \text{H}} = 6.00\ \text{A}^2$$

$$|i| = \sqrt{i^2} = \sqrt{6.00\ \text{A}^2} = 2.45\ \text{A}$$

The electric energy is $\frac{1}{4}$ of the total energy:

$$U_E = \frac{q^2}{2C} = \frac{1}{4}E = \frac{1}{4}(8.00 \times 10^{-6}\ \text{J}) = 2.00 \times 10^{-6}\ \text{J}$$

Solve for the absolute value of the capacitor charge:

$$q^2 = 2CU_E = 2(10.0 \times 10^{-9}\ \text{F})(2.00 \times 10^{-6}\ \text{J})$$
$$= 4.00 \times 10^{-14}\ \text{C}^2$$

$$|q| = \sqrt{q^2} = \sqrt{4.00 \times 10^{-14}\ \text{C}^2} = 2.00 \times 10^{-7}\ \text{C} = 0.200\ \mu\text{C}$$

Reflect

We can check our result for |q| by noting that U_E is proportional to q^2. If the electric energy U_E has its maximum value when $q = Q_0 = 0.400\ \mu\text{C}$, then U_E will have $\frac{1}{4}$ of this value when $q = Q_0/2 = 0.200\ \mu\text{C}$.

Note that all that we can determine in this problem are the absolute values of *i* and *q*. That's because the magnetic and electric energies are proportional to the squares of *i* and *q*, respectively. If the current were negative (counterclockwise in Figure 21-8) so that $i = -2.45$ A, the magnetic energy would be the same as if $i = +2.45$ A (clockwise current in Figure 21-8); if the capacitor charge were negative (so the upper capacitor plate in Figure 21-8 were negatively charged) so $q = -0.400\ \mu\text{C}$, the electric energy would be the same as if $q = +0.400\ \mu\text{C}$ (the upper capacitor plate were positively charged).

You use the physics of *LC* circuits whenever you use a microwave oven to heat food. At the heart of any microwave oven is a device called a *cavity resonator*, which is basically a hollow metal tube closed at both ends by metal caps. When equal amounts of positive and negative charges are placed on the caps, the charge flows back and forth between the caps along the inner surfaces of the tube. This gives rise to time-varying electric and magnetic fields inside the tube. The electric field is greatest when the magnetic field is zero and vice versa, just as in Figure 21-8. (The geometry of these fields is more complicated than in Figure 21-8, but the basic physics is the same.) The frequency at which the fields oscillate—typically 2.45 GHz, or 2.45×10^9 Hz, in a home microwave oven and 0.915 GHz in a large commercial oven—is determined by the size and shape of the cavity resonator. A small hole in the cavity resonator is connected to one end of a metal pipe, and the other end leads into the interior of the oven. The oscillating fields "flow" along the pipe into the oven, where the energy of the electric field is absorbed by water molecules in the food. This sets the molecules into vibration and raises the temperature of the food, cooking it.

There is resistance in a cavity resonator, just as there is in any real circuit. As a result, the oscillations will die away just as do the oscillations of a block on a spring when friction is present. To sustain the oscillations of a circuit, we need to *drive* the circuit with an alternating voltage, in much the same way you sustain the back-and-forth oscillations of a child on a playground swing by pushing on the child once per cycle. In the following section, we'll analyze what happens in a circuit that contains an inductor, a capacitor, a resistor, and a source of alternating voltage.

? Got the Concept? 21-4
Doubling the Charge in an *LC* Circuit

A certain *LC* circuit oscillates at frequency f when the maximum capacitor charge is Q_0. If you double the value of Q_0, the new frequency of oscillation is (a) $2f$; (b) $f\sqrt{2}$; (c) $f/\sqrt{2}$; (d) $f/2$; (e) none of these.

Take-Home Message for Section 21-5

✔ An *LC* circuit, made up of an inductor connected to a capacitor, is the electrical analog of a block attached to a spring. The inductor plays the role of the block, and the capacitor plays the role of the spring.

✔ In an *LC* circuit the capacitor charge and inductor current both oscillate with a frequency determined by the values of the inductance L and capacitance C.

✔ As the charge and current oscillate, the energy in the circuit oscillates between magnetic energy in the inductor and electric energy in the capacitor. If there is no resistance in the circuit, the energy remains constant.

21-6 When an ac voltage source is attached in series to an inductor, resistor, and capacitor, the circuit can display resonance

When you turn on the radio in a car, you expect to listen to only one station at a time. But the car antenna is being simultaneously bombarded by signals from *all* of the local radio stations. How does the radio "know" which signal to play through the car's speakers and which signals to ignore? The explanation is that the radio circuit is essentially a combination of an inductor, resistor, and capacitor in series, and this circuit is driven by an ac voltage coming from the radio signal. As we'll see, such a **series *LRC* circuit** can be designed to give a large current in response to a voltage at one particular frequency, while responding very little to voltages at other frequencies. When you tune the radio to a station with a particular frequency, you're adjusting the radio circuit so that the frequency at which it has the greatest response matches the frequency of the station you want to hear.

A driven series *LRC* circuit behaves in much the same way as a damped, driven mechanical oscillator like the one we studied in Section 12-8. (This would be a good time to review that section.) In order to understand the properties of a driven series *LRC* circuit, it's useful to first look at three simpler ac circuits: one with just an ac source and a resistor, one with just an ac source and a capacitor, and one with just an ac source and an inductor.

An ac Source and a Resistor

In **Figure 21-9a** an ac source is connected to a resistor of resistance R. The voltage provided by the source is

$$V(t) = V_0 \sin \omega t \qquad (21\text{-}30)$$

What current $i(t)$ is developed in the circuit when we close the switch, and how much power does the source deliver to the resistor? To answer these questions, let's apply Kirchhoff's loop rule, which states that the sum of the voltage drops around the circuit is zero. The voltage across the resistor is $i(t)$ multiplied by R, so the loop rule says that

$$V(t) - i(t)R = 0$$

 See the Math Tutorial for more information on trigonometry

(a) An ac source connected to a resistor

- Current i is in phase with source voltage V.
- Since i and V always have the same sign, the power $P = iV$ delivered by the source is always positive.

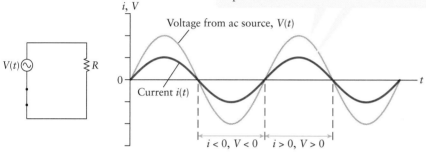

(b) An ac source connected to a capacitor

- Current i leads the source voltage V by 1/4 cycle.
- Half of the time i and V have the same sign, so the power $P = iV$ delivered by the source is positive; the other half of the time i and V have opposite signs and P is negative.

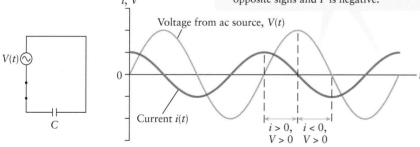

(c) An ac source connected to an inductor

- Current i lags the source voltage V by 1/4 cycle.
- Half of the time i and V have the same sign, so the power $P = iV$ delivered by the source is positive; the other half of the time i and V have opposite signs and P is negative.

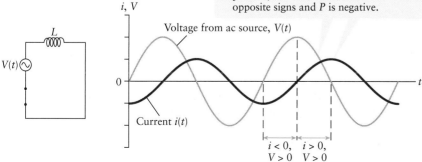

Figure 21-9 Three ac circuits The graphs show the source voltage and currents for three circuits: an ac source connected to (a) a resistor, (b) a capacitor, and (c) an inductor.

and so

(21-31)
$$i(t) = \frac{V(t)}{R} = \frac{V_0}{R} \sin \omega t \quad \text{(resistor and ac source)}$$

The graph of $V(t)$ and $i(t)$ versus time in Figure 21-9a shows that the current is in phase with the source voltage. From Equation 18-23, the power that the source delivers to the circuit (in this case, to the resistor) is equal to the product of the current through the source multiplied by the voltage across the source: $P(t) = i(t)V(t)$. The graph in Figure 21-9a shows that $i(t)$ and $V(t)$ always have the same sign (both positive or both negative), so the product $i(t)V(t)$ is always positive. Thus power is always being transferred from the ac source into the resistor.

An ac Source and a Capacitor

The circuit shown in **Figure 21-9b** is the same as in Figure 21-9a, except that we've replaced the resistor by a capacitor. The voltage provided by the source is still given by Equation 21-30, but the voltage across the capacitor is $q(t)/C$, where $q(t)$ is the capacitor charge. The loop rule now says that

$$V(t) - \frac{q(t)}{C} = 0$$

The capacitor charge is therefore

(21-32)
$$q(t) = CV(t) = CV_0 \sin \omega t \quad \text{(capacitor and ac source)}$$

The capacitor charge oscillates between $+CV_0$ and $-CV_0$.

To find the current in the circuit, two trigonometric identities that will be of use are $\sin \theta = \cos(\theta - \pi/2)$ and $\cos \theta = -\sin(\theta - \pi/2)$. Using the first of these, we can rewrite Equation 21-32 as

$$q(t) = CV_0 \cos\left(\omega t - \frac{\pi}{2}\right) \quad \text{(capacitor and ac source)}$$

The current $i(t)$ is the rate of change of the capacitor charge $q(t)$, just as it was for the LC circuit that we discussed in Section 21-5. In that section we saw that if $q(t) = Q_0 \cos(\omega t + \phi)$, then $i(t) = -\omega Q_0 \sin(\omega t + \phi)$ (see Equations 21-24 and 21-25). That's the same situation we have here, with Q_0 replaced by CV_0 and ϕ equal to $-\pi/2$. So the current in the circuit of Figure 21-9b is

$$i(t) = -\omega CV_0 \sin\left(\omega t - \frac{\pi}{2}\right)$$

Using the second trigonometric identity that we stated above, we can write this as

(21-33)
$$i(t) = \omega CV_0 \cos \omega t \quad \text{(capacitor and ac source)}$$

From Equation 21-32, the maximum charge on the capacitor is CV_0 no matter what the frequency. If the frequency is low and the period long, there's plenty of time for the charge to build up from zero to CV_0, so the required current will be low. Equation 21-33 shows the same thing: If the frequency is low, the angular frequency ω is likewise low and the current amplitude ωCV_0 is small. So a capacitor connected to an ac source acts to reduce the current amplitude at low frequency.

The graph in Figure 21-9b shows the source voltage and the current versus time. We say that the current *leads* the voltage by 1/4 cycle: The graph of current has its first peak at $t = 0$, but the graph of voltage reaches its first peak 1/4 cycle later. As a result, for half of the time the current and voltage have the same sign, while for the other half of the time the signs of $i(t)$ and $V(t)$ are opposite. Therefore for half of the time there is positive power $P(t) = i(t)V(t)$ out of the source and into the capacitor, while for the other half of the time the power is negative and energy is flowing out of the capacitor back into the source. What's happening is that the capacitor is alternately charging and discharging; when it's charging the electric energy U_E in the capacitor is increasing, and when it's discharging the electric energy is decreasing. Over one complete cycle, there's *zero* net flow of energy out of the source.

An ac Source and Inductor

Figure 21-9c shows a third circuit, one that includes only an ac source and an inductor. Again the source voltage is given by Equation 21-30. The voltage drop across the inductor is $L(\Delta i/\Delta t)$, so the loop rule says that

$$V(t) - L\frac{\Delta i(t)}{\Delta t} = 0$$

We can solve this for $\Delta i(t)/\Delta t$, which is the rate of change of the current:

$$\frac{\Delta i(t)}{\Delta t} = \frac{V(t)}{L} = \frac{V_0}{L}\sin \omega t \quad \text{(inductor and ac source)} \tag{21-34}$$

Given Equation 21-34 for the rate of change of current, what is the current itself? As we did for the capacitor circuit shown in Figure 21-9b, let's think for a minute about Equation 21-24, $q(t) = Q_0\cos(\omega t + \phi)$, and Equation 21-25, $i(t) = -\omega Q_0\sin(\omega t + \phi)$. In these equations $i(t)$ is the rate of change of $q(t)$. So these equations tell us that if the rate of change of a function is given by a constant multiplied by $\sin(\omega t + \phi)$, the function itself is given by the same constant divided by $-\omega$ and multiplied by $\cos(\omega t + \phi)$. Using this with Equation 21-34, in which the constant is V_0/L and $\phi = 0$, we conclude that the current for the circuit in Figure 21-9c is

$$i(t) = -\frac{V_0}{\omega L}\cos \omega t \quad \text{(inductor and ac source)} \tag{21-35}$$

Equation 21-35 says that the amplitude of the current is $V_0/(\omega L)$. Since ω is in the denominator, the current amplitude is small when the frequency, and hence the angular frequency ω, is high. That agrees with the idea that an inductor opposes changes in current. High frequency corresponds to rapid changes in current, which the inductor suppresses by making the current small. So an inductor connected to an ac source acts to reduce the current amplitude at high frequency. That's in sharp contrast to a capacitor, which acts to reduce the current amplitude at low frequency.

The graph in Figure 21-9c shows $V(t)$ and $i(t)$ versus time for the circuit with an inductor and an ac source. In this circuit the current *lags* behind the voltage by 1/4 cycle; if you look at any peak of the voltage, you'll see that a peak of the current occurs 1/4 cycle later. Just as for the capacitor circuit in Figure 21-9b, for half of the time the current and voltage have the same sign so power flows out of the source and adds energy to the inductor. In this case it's *magnetic* energy that's being added as the current through the inductor increases in magnitude. For the other half of the time, current and voltage have opposite signs, the source power is negative, and energy flows out of the inductor as the current decreases in magnitude. As for the capacitor, over one cycle there is zero net energy delivered to the inductor.

A Driven Series *LRC* Circuit

Let's now combine all of the devices shown in Figure 21-9 into a single circuit (Figure 21-10). Again the voltage of the ac source is given by Equation 21-30, $V(t) = V_0\sin \omega t$. Just as in the three simpler cases depicted in Figure 21-9, we expect that when the switch is closed there will be an oscillating current in the circuit with angular frequency ω. We can also expect that at low frequencies the presence of the capacitor will keep the current amplitude small, while at high frequencies the current amplitude will be small due to the presence of the inductor. At some frequency that's not too low and not too high, the current amplitude will be largest. That's the frequency where the voltage of the ac source produces the greatest response. One of our tasks is to find just what this frequency is.

If we apply Kirchhoff's loop rule to the circuit shown in Figure 21-10, we get the following equation:

$$V(t) - L\frac{\Delta i(t)}{\Delta t} - i(t)R - \frac{q(t)}{C} = 0 \quad \text{(driven series } LRC \text{ circuit)} \tag{21-36}$$

This is a complicated equation, since it involves the capacitor charge $q(t)$; the current $i(t)$, which is the rate of change of the capacitor charge; and the rate of change of the current, $\Delta i(t)/\Delta t$. It is possible to solve Equation 21-36 using some tricks of trigo-

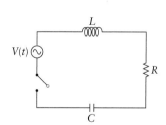

Figure 21-10 A driven series *LRC* circuit When the switch is closed, an oscillating current is set up in this circuit. The amplitude and phase of this current depend on the values of the inductance *L*, resistance *R*, and capacitance *C*, as well as the amplitude V_0 and angular frequency ω of the voltage provided by the ac source.

nometry, but it's a rather strenuous exercise. Instead, let's just present the result for the current $i(t)$ and see what it tells us:

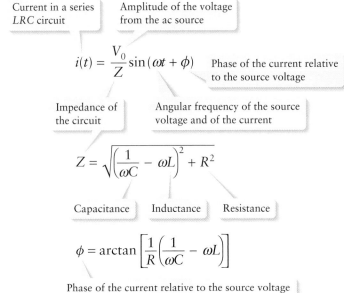

Current in a series LRC circuit

Amplitude of the voltage from the ac source

$$i(t) = \frac{V_0}{Z}\sin(\omega t + \phi)$$

Phase of the current relative to the source voltage

Impedance of the circuit

Angular frequency of the source voltage and of the current

Current in a driven LRC circuit
(21-37)

$$Z = \sqrt{\left(\frac{1}{\omega C} - \omega L\right)^2 + R^2}$$

Capacitance   Inductance   Resistance

$$\phi = \arctan\left[\frac{1}{R}\left(\frac{1}{\omega C} - \omega L\right)\right]$$

Phase of the current relative to the source voltage

The first of Equations 21-37 shows that the greater the value of the **impedance** Z, the smaller the amplitude V_0/Z of the current. Like resistance, impedance has units of ohms. So you can think of it as similar to an "effective resistance" of the circuit to an ac current. Unlike resistance, however, impedance depends on the angular frequency as given by the second of Equations 21-37. The quantity $[1/(\omega C) - \omega L]^2$ is large at both low and high angular frequencies; that's because the term $1/(\omega C)$ (due to the capacitor) gets very large for low values of ω, while the term ωL (due to the inductor) gets very large for high values of ω. So the impedance Z is large and the current amplitude V_0/Z is small for very low angular frequencies thanks to the capacitor, as well as for very high angular frequencies thanks to the inductor. That's just what we predicted above.

The impedance is smallest, and the current amplitude V_0/Z largest, when the quantity $[1/(\omega C) - \omega L]^2$ equals zero and so

$$\frac{1}{\omega C} - \omega L = 0$$

To solve for the angular frequency at which this happens, multiply this equation by ω/L:

$$\frac{\omega}{L}\left(\frac{1}{\omega C} - \omega L\right) = \frac{1}{LC} - \omega^2 = 0$$

So $\omega^2 = 1/(LC)$, or

(21-38)
$$\omega = \sqrt{\frac{1}{LC}}$$

We saw this same angular frequency in Equation 21-23: It's the angular frequency at which an LC circuit oscillates. We'll refer to $\sqrt{1/(LC)}$ as the **natural angular frequency** of the circuit and give it the symbol ω_0. It's the angular frequency at which the current would oscillate if the ac source and resistor weren't in the circuit at all!

In Section 12-8 we saw this same effect, called **resonance**, for an oscillating mechanical system. Equation 21-38 tells us that the current in a driven series LRC circuit is greatest, or in resonance, if the *driving* angular frequency ω equals the *natural* angular frequency $\omega_0 = \sqrt{1/(LC)}$. When $\omega = \omega_0 = \sqrt{1/(LC)}$, the impedance in Equation 21-37 is equal to R and the current amplitude is $V_0/Z = V_0/R$.

Why does resonance happen in a series LRC circuit?. The explanation is that at low frequencies the capacitor suppresses the current, and at high frequencies the inductor

(a) Amplitude of the current

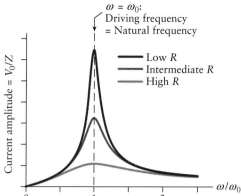

(b) Phase of the current

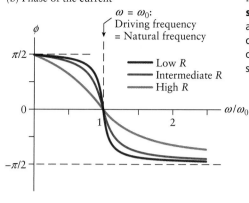

Figure 21-11 **Resonance in a driven series *LRC* circuit** (a) The amplitude and (b) the phase of the current in a driven series *LRC* circuit both depend on the angular frequency ω of the source voltage.

- The current amplitude is maximum at $\omega = \omega_0$ (resonance).
- The smaller the resistance, the greater the current amplitude at resonance.

- At $\omega = \omega_0$ (resonance), $\phi = 0$ and the current is in phase with the source voltage.
- If $\omega < \omega_0$, $\phi > 0$ and the current leads the source voltage. At very low frequencies, ϕ approaches $\pi/2$ (current leads voltage by 1/4 cycle).
- If $\omega > \omega_0$, $\phi < 0$ and the current lags behind the source voltage. At very high frequencies, ϕ approaches $-\pi/2$ (current lags voltage by 1/4 cycle).

suppresses the current. But at one special frequency—the natural frequency of the circuit—the effects of the capacitor and inductor cancel exactly so that the current in the circuit has its maximum amplitude. **Figure 21-11a** shows the current amplitude as a function of the driving frequency ω for three different values of the resistance R. For any value of R the current amplitude is maximum when $\omega/\omega_0 = 1$ and $\omega = \omega_0 = \sqrt{1/(LC)}$. As R decreases, the peak becomes higher (the maximum current amplitude increases).

The third of Equations 21-37 describes the phase ϕ of the current (see **Figure 21-11b**). The value of ϕ tells us how much the current leads the source voltage. At resonance, where $\omega/\omega_0 = 1$ and $\omega = \omega_0 = \sqrt{1/(LC)}$, we have $(1/(\omega C)) - \omega L = 0$ and $\phi = \arctan 0 = 0$. So at resonance the current is in phase with the source voltage, just as happens in a circuit with only an ac source and a resistor (Figure 21-9a).

For very low angular frequencies ω, the quantity $1/(\omega C)$ is large while ωL is small, so $(1/(\omega C)) - \omega L$ is large and positive. In this case ϕ is near $\pi/2$ radians, so at low frequencies the current leads the voltage by about 1/4 of a cycle (recall that there are 2π radians in a cycle). That's just like the behavior of a circuit with only an ac source and a capacitor (Figure 21-9b). By contrast, at very high angular frequencies the quantity $1/(\omega C)$ is small while ωL is large. Then $(1/(\omega C)) - \omega L$ is large and negative, and ϕ is near $-\pi/2$ radians or $-1/4$ cycle. The negative value of ϕ means that at high frequencies the current *lags* the source voltage by about 1/4 cycle. The circuit with just an ac source and an inductor in Figure 21-9c behaves in the same way.

As Figure 21-11b shows, changing the resistance doesn't affect the angular frequency at which $\phi = 0$. This always happens when $\omega = \omega_0 = \sqrt{1/(LC)}$. Changing the value of R does change the shape of the graph of ϕ versus ω, however; the smaller the resistance, the closer the driving angular frequency ω must be to the natural angular frequency ω_0 in order for ϕ to be close to zero.

These ideas show what's involved in tuning a radio. The signal from a broadcast station is at a particular frequency. The radio antenna detects that signal and converts it to an ac voltage with that same frequency. In order for the circuit of the radio receiver to respond to that frequency, the natural frequency of the receiver circuit must match the driving frequency of the signal from the radio station—that is, the radio receiver must be *tuned* to be in resonance with the station to which you want to listen. Other radio signals from other stations will be present in the circuit as well, but their driving frequencies won't match the natural frequency of the circuit. Hence these undesired signals will produce very little response in the radio (see Figure 21-11a), and you'll hear only the sound of the station to which you've tuned the radio. The following example illustrates this idea.

Go to Interactive Exercise 21-2 for more practice dealing with LRC circuits

Example 21-6 Tuning an FM Radio

The tuner knob on an FM radio moves the plates of an adjustable capacitor. This capacitor is in series with a 0.130-μH inductor and a net resistance of $755\ \Omega$. The peak current induced in this circuit by a radio wave becomes large when the natural frequency of the circuit matches the carrier frequency of the radio wave. (a) What is the frequency of the FM station that is tuned in when the capacitor in the radio is adjusted to 19.6 pF? (b) If the peak operating voltage in the tuning circuit is 9.00 V, what is the peak current?

Set Up

The tuning circuit resonates when driven by a radio wave with an angular frequency ω that equals the natural angular frequency ω_0 of the circuit. We'll use Equation 21-38 to determine the value of ω_0 for this circuit, and convert it to an ordinary frequency in Hz. We'll find the peak current, or current amplitude, using Equations 21-37.

Natural angular frequency of the tuning circuit:

$$\omega_0 = \sqrt{\frac{1}{LC}} \qquad (21\text{-}38)$$

Relationship between frequency and angular frequency:

$$f = \frac{\omega}{2\pi}$$

Amplitude of the oscillating current:

$$i_{0,\max} = \frac{V_0}{Z}$$

$$Z = \sqrt{\left(\frac{1}{\omega C} - \omega L\right)^2 + R^2} \qquad (21\text{-}37)$$

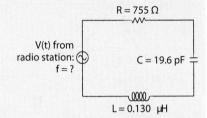

Solve

(a) Use the given values of capacitance and inductance to determine the natural frequency of the circuit.

We are given $L = 0.130\ \mu H = 0.130 \times 10^{-6}$ H and $C = 19.6$ pF $= 19.6 \times 10^{-12}$ F. From Equation 21-38 the natural angular frequency is $\omega_0 = \sqrt{1/(LC)}$, and the natural frequency is ω_0 divided by 2π:

$$f_0 = \frac{\omega_0}{2\pi} = \frac{1}{2\pi}\sqrt{\frac{1}{LC}} = \frac{1}{2\pi}\sqrt{\frac{1}{(0.130 \times 10^{-6}\ H)(19.6 \times 10^{-12}\ F)}}$$

$$= 99.7 \times 10^6\ Hz = 99.7\ MHz$$

(b) Find the amplitude of the oscillating current.

At resonance the term $(1/(\omega C)) - \omega L$ in the expression for impedance is equal to zero. Then Equation 21-37 tells us that $Z = R = 755\ \Omega$. If the voltage amplitude is 9.00 V, the current amplitude is

$$i_{0,\max} = \frac{V_0}{Z} = \frac{9.00\ V}{755\ \Omega} = 0.0119\ A = 11.9\ mA$$

Reflect

The carrier frequencies of FM radio stations lie in the megahertz (MHz) range. So the frequency we found, 99.7×10^6 Hz, is 99.7 MHz on your FM radio dial. The peak current, about 10 mA, is typical for a portable FM radio.

Note that you tune a television set in the same way. When you use the remote control to change the channel (Figure 21-1b), you're commanding a circuit in the television to change its capacitance. This adjusts the natural frequency of the circuit to match the carrier frequency of the channel you want to watch.

? Got the Concept? 21-5 Modifying a Series *LRC* Circuit

We've discussed the behavior of a series *LRC* circuit with $R = 1.00\ \Omega$, $C = 5.00\ \mu F$, $L = 20.0\ \mu H$, and $V_0 = 1.00$ V. If we increased the resistance to $2.00\ \Omega$ but kept all the other values the same, which of the following would be affected?

(a) The natural angular frequency of the circuit; (b) the current amplitude at resonance; (c) the current amplitude when the driving angular frequency is twice the natural angular frequency; (d) two of (a), (b), and (c); (e) all of (a), (b), and (c).

Take-Home Message for Section 21-6

✔ A driven series *LRC* circuit is made up of an inductor, capacitor, and resistor connected in series to a source of ac voltage.

✔ The response of a driven series *LRC* circuit depends on the frequency of the ac source. At low frequencies, the capacitor dominates the circuit and the current is small. The current is also small at high frequencies, where the inductor

dominates the circuit. When the frequency of the source equals the natural frequency of the circuit, the current is large and maximum power flows from the source into the resistor. This is resonance.

✔ As the charge and current oscillate, the energy in the circuit oscillates between magnetic energy in the inductor and electric energy in the capacitor. If there is no resistance in the circuit, the energy remains constant.

21-7 Diodes are important parts of many common circuits

We've just about come to the end of our discussion of electric circuits. We've seen two different types of voltage sources (batteries that provide a constant emf and ac sources that provide a time-varying emf) and three kinds of circuit elements (resistors, capacitors, and inductors). Before leaving the subject of circuits, however, let's take a look at another important circuit device called a *diode*. This device is possible because of *semiconductors*, a class of materials whose electrical properties are intermediate between those of conductors and insulators.

A common semiconductor is the element silicon (chemical symbol Si). Each silicon atom has four outer or *valence* electrons. In a conducting material such as copper or silver the valence electrons are free to move throughout the material, so the resistivity of the material is low. In silicon, however, the atoms are formed into a crystal structure in which all four of an atom's valence electrons are involved in chemical bonds with adjacent silicon atoms. As a result, the valence electrons are able to move only with great difficulty. That's why pure silicon has about 10^{10} times the resistivity of copper (though only about 10^{-13} the resistivity of an insulating material like rubber). That can change dramatically by **doping** the silicon—that is, by adding small amounts of a different kind of atom to solid silicon. Two common elements used for this purpose are phosphorous and boron.

Figure 21-12a shows a schematic illustration of a silicon crystal doped with a small amount of phosphorous (chemical symbol P), so that the phosphorous atoms replace a few of the silicon atoms. An atom of phosphorous has five valence electrons, four of which form bonds to the adjacent silicon atoms. The fifth electron can move relatively easily through the material, and so the doped silicon has much lower resistivity than does pure silicon. Adding just one part per million of phosphorous can decrease the resistivity by several powers of ten. Since the mobile charges in phosphorous-doped silicon are negatively charged electrons, we call such a material an ***n*-type semiconductor** (*n* for negative).

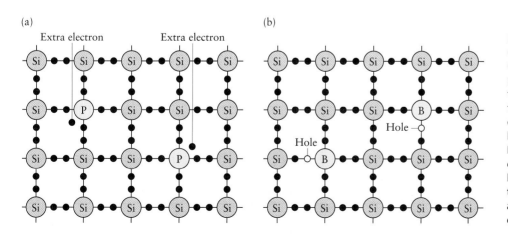

(a) | (b)

Figure 21-12 Doped semiconductors (a) Silicon (Si) doped with phosphorous (P) is an *n*-type semiconductor. Because phosphorous has five valence electrons, there is an extra, weakly bound electron that can contribute to electrical conduction. (b) Silicon doped with boron (B) is a *p*-type semiconductor. Because boron has only three valence electrons, there is a hole in one of its bonds. The hole can move through the semiconductor as though it were a positive charge, contributing to electrical conduction.

Figure 21-12b shows a different kind of doping in which some of the silicon atoms have been replaced with atoms of boron (chemical symbol B). Boron has only three valence electrons, so for each boron atom there is a "hole" in the electron population. These **holes** are actually free to travel through the boron-doped silicon. (A very rough analogy is to think of these holes as traveling like bubbles of carbon dioxide through a carbonated beverage.) Since a hole represents an absence of negative charge, it's equivalent to a *positive* charge. For this reason boron-doped silicon is called a ***p*-type semiconductor** (*p* for positive). Adding holes by doping the silicon with even a small amount of boron also causes a substantial decrease in resistivity.

We can make a novel kind of device if we take a crystal of silicon and dope one side of the crystal with boron and one side with phosphorous. The side with boron is the *p* side and the side with phosphorous is the *n* side (the two sides are *p*-type and *n*-type semiconductors, respectively). The region where the two sides meet is called a ***pn* junction**. Some of the mobile holes from the *p* side diffuse across the junction to the *n* side, and some of the mobile electrons from the *n* side diffuse to the *p* side. The semiconductor isn't a particularly good conductor, so neither the electrons nor the holes get very far from the junction. The net result is that there's a double layer of charge around the junction, a positive layer of holes on the *n* side of the junction and a negative layer of electrons on the *p* side of the junction (**Figure 21-13**).

To see what makes a *pn* junction useful, imagine placing it in a dc circuit with a source of emf ε and a resistor. If the polarity of the source is as shown in **Figure 21-14a**, so that the emf tries to make current flow from the *p* side to the *n* side, we find that a current does indeed flow through the junction and around the circuit. What happens is that the source creates an electric field within the *pn* junction that points from left to right, which helps to drive positively charged holes to the right across the junction and negatively charged electrons to the left across the junction. This enhanced diffusion of electrons and holes gives rise to a net current from left to right in the semiconductor. In this case the *pn* junction is said to be *forward biased*. If we reverse the polarity of the source as shown in **Figure 21-14b**, however, the source creates an electric field that points from right to left within the *pn* junction. This tends to push the holes leftward (back into the *p* side from which they came) and the electrons rightward (back into the *n* side from which they came). In this case the *pn* junction is said to be *reverse biased*: The emf suppresses the diffusion of electrons and holes, and there is little or no current. Essentially the junction conducts in only one direction! A single-junction semiconductor device like this is called a **diode**.

Diodes have many uses, one of which is converting alternating current into direct current. If we replace the source of emf in Figure 21-14 with an ac source, the diode will conduct electricity when the alternating voltage is positive (forward bias) but not when the alternating voltage is negative (reverse bias). In this case the current through the resistor in Figure 21-14 won't be constant, but it will be in one direction only.

Another use for *pn* junctions is in **photovoltaic solar cells**. Light shining on the *p* side of the junction can excite electrons and produce new holes in the process. If the electrons happen to migrate to the junction, they will be accelerated into the *n* side by the electric field between the double layer of charge at the junction. The result is an excess negative charge on the *n* side and an excess positive charge on the *p* side. This charge imbalance produces a potential difference between the two regions. If one terminal of a resistor is connected to the *p* side and the other terminal to the *n* side, a current will flow through the resistor due to the potential difference and energy will be transferred to the

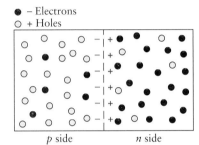

● – Electrons
○ + Holes

p side   *n* side

Figure 21-13 A *pn* junction The two sides of this semiconductor are doped with different atoms. Holes diffuse from the *p* side to the *n* side, and electrons diffuse from the *n* side to the *p* side. The result is a double layer of charge at the junction between the two sides.

Figure 21-14 A *pn*-junction diode This device is like a one-way valve for electric current. (a) Current will flow through the *pn* junction if the emf is applied as shown here (forward bias). (b) Little or no current will flow through the *pn* junction if the emf is reversed (reverse bias).

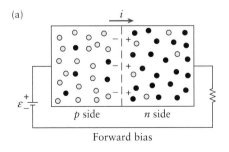

(a)

i

ε

p side   *n* side

Forward bias

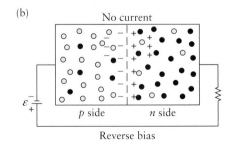

(b)

No current

ε

p side   *n* side

Reverse bias

resistor. The net effect is that the energy of sunlight is converted to electric energy. All solar cells designed for home and industrial use operate on this basic principle.

The basic process used in solar cells can also be run in reverse. In certain kinds of *pn* junctions, a forward bias like that shown in Figure 21-14a produces large concentrations of electrons on the *p* side and holes on the *n* side. When the electrons and holes diffuse across the barrier and recombine, energy is released in the form of light. Such a device is called a **light-emitting diode** or **LED**. LEDs are extremely efficient light sources: They produce much more light for a given power input than do incandescent light bulbs or fluorescent lamps. A television remote control (Figure 21-1b) sends commands to the television in the form of infrared light emitted by an LED.

Many other important semiconductor devices are used in circuits. In a *transistor*, a weak electrical current is used to modulate a stronger current. For example, the alternating current from an MP3 player carries all the musical information about the song you want to hear, but the current is too feeble to drive a speaker or even a pair of earbuds. By using transistors, however, this feeble current can be used to modulate a much stronger current from the MP3 player's battery. This modulated current *is* large enough to drive a speaker or earbuds, and that's what produces the music that you hear.

Take-Home Message for Section 21-7

✔ A semiconductor can be made a better conductor of electricity by doping it with a small amount of a different kind of atom. In an *n*-type semiconductor, there are extra electrons that are able to move through the material and carry a current. In a *p*-type semiconductor, the mobile charges are holes where there is an electron missing. Holes behave like positively charged objects.

✔ In a pn junction one side of the semiconductor is p-type and the other side is n-type. This has the property that it is much easier to establish a current through the junction in one direction compared to the opposite direction.

Key Terms

alternating current (ac)	magnetic energy	root mean square value
diode	mutual inductance	(rms value)
direct current (dc)	*n*-type semiconductor	secondary coil
doping	natural angular frequency	self-inductance
hole	*p*-type semiconductor	series *LRC* circuit
impedance	photovoltaic solar cell	step-down transformer
inductance	*pn* junction	step-up transformer
LC circuit	primary coil	transformer
light-emitting diode (LED)	resonance	

Chapter Summary

Topic	Equation or Figure
Root mean square values for alternating current: To find the root mean square or rms value of a quantity, first find the average (mean) value of the square of that quantity, then take the square root of that average. Alternating voltages are often described in terms of their rms values rather than their amplitudes.	The root mean square (rms) value of $V(t)$... ... is the square root of the average value of $V^2(t)$. $$V_{\mathrm{rms}} = \sqrt{(V^2(t))_{\mathrm{average}}} = \frac{V_0}{\sqrt{2}} \qquad (21\text{-}5)$$ If $V(t)$ is a sinusoidal function, its rms value equals the maximum value of $V(t)$ (the amplitude) divided by $\sqrt{2}$.

Transformers: A transformer uses electromagnetic induction to raise or lower the voltage of an alternating current. The ratio of the output voltage to the input voltage depends on the number of windings in the coils of the transformer. Raising the voltage lowers the current and vice versa. In an ideal transformer, no power is lost in this process.

Amplitudes of the input (p) and output (s) voltages

$$\frac{V_s}{V_p} = \frac{V_{s,rms}}{V_{p,rms}} = \frac{N_s}{N_p}$$

Number of windings in the primary coil (p) and secondary coil (s)

Rms values of the input (p) and output (s) voltages

(21-10)

Inductors: The simplest inductor is a coil of wire. If the current in the coil changes, there will be a change in the magnetic flux through the coil and an emf will be induced to oppose that change. So an inductor in a circuit always acts to oppose any change in the current in that circuit. The inductance of a coil depends on the coil's geometry.

Inductance of a coil   Number of windings in the coil

$$L = \frac{N\Phi_B}{i}$$

Magnetic flux through each winding of the coil due to the field produced by the coil itself

(21-13)

Current in the coil

Voltage across an inductor   Inductance of the inductor

$$V = -L\frac{\Delta i}{\Delta t}$$

Rate of change of the current i in the inductor

(21-16)

The negative sign means that the voltage opposes any change in the current.

LC circuits: A circuit made up of an inductor and capacitor is the electric analog of a block oscillating at the end of an ideal spring. The angular frequency of the oscillation is $\omega = \sqrt{1/(LC)}$. The energy in the LC circuit oscillates between electric energy in the capacitor and magnetic energy in the inductor.

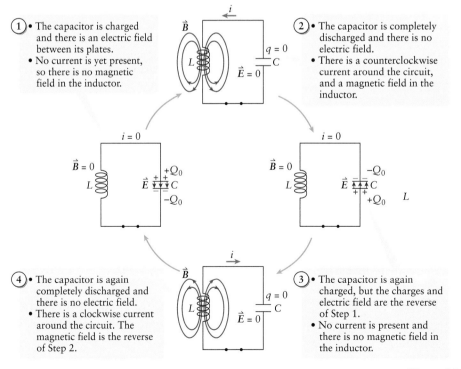

1. • The capacitor is charged and there is an electric field between its plates.
   • No current is yet present, so there is no magnetic field in the inductor.

2. • The capacitor is completely discharged and there is no electric field.
   • There is a counterclockwise current around the circuit, and a magnetic field in the inductor.

3. • The capacitor is again charged, but the charges and electric field are the reverse of Step 1.
   • No current is present and there is no magnetic field in the inductor.

4. • The capacitor is again completely discharged and there is no electric field.
   • There is a clockwise current around the circuit. The magnetic field is the reverse of Step 2.

(Figure 21-8)

Driven series *LRC* circuits: If an ac source is connected in series to an inductor, resistor, and capacitor to make a circuit, the current in the circuit oscillates at the angular frequency of the source. The voltage across the inductor leads the current by 1/4 cycle, the voltage across the resistor is in phase with the current, and the voltage across the capacitor lags the current by 1/4 cycle.

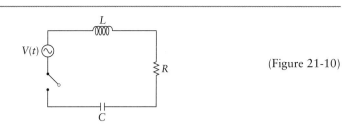

(Figure 21-10)

LRC circuit resonance: The current in a driven series *LRC* circuit has its maximum amplitude when the source (driving) angular frequency equals the natural angular frequency $\omega_0 = \sqrt{1/(LC)}$ of the circuit. When this happens, the circuit is in resonance. At resonance the current is in phase with the ac source voltage, and maximum power is delivered to the resistor. In addition, at resonance the voltages across the inductor and capacitor have equal amplitude but are 1/2 cycle out of phase with each other, so their voltages cancel.

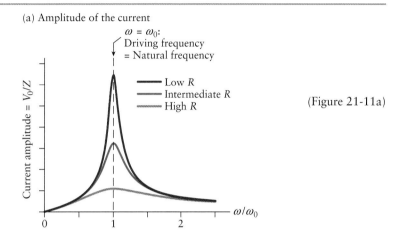

(a) Amplitude of the current

(Figure 21-11a)

Semiconductors and diodes: A semiconductor can be made a better conductor by adding impurities that contribute mobile electrons (an *n*-type semiconductor) or mobile holes (a *p*-type semiconductor). A semiconductor with one side that is *p*-type and one side that is *n*-type can act as a diode that conducts current easily from the *p* side to the *n* side, but conducts poorly in the opposite direction.

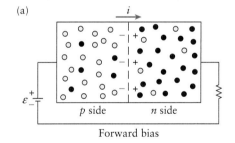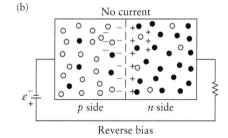

(Figure 21-14)

Answer to What do you think? Question

(d) As we learned in Chapter 20, Faraday's law only comes into play when currents are changing. If the current is constant, as in the case of direct current, Faraday's law has no effect. (See Section 21-3.)

Answers to Got the Concept? Questions

21-1 (b) The voltage provided by the ac source is $V(t) = V_0 \sin \omega t$, where the angular frequency ω is related to the period T by $\omega = 2\pi/T$. The instantaneous power into the resistor is given by Equation 21-3, $P(t) = V^2(t)/R$. At $t = 0$, $V(0) = V_0 \sin 0 = 0$, so $P(0) = 0$. At $t = T/8$, $\omega t = (2\pi/T)(T/8) = \pi/4$ so $V(T/8) = V_0 \sin(\pi/4) = V_0(1/\sqrt{2})$ and $P(T/8) = (V_0/\sqrt{2}^2)/R = V_0^2/(2R)$. At $t = T/4$, $\omega t = (2\pi/T)(T/4) = \pi/2$ so $V(T/4) = V_0 \sin(\pi/2) = V_0(1) = V_0$ and $P(T/4) = V_0^2/R$. Equation 21-4 tells us that the average power is $P_{average} = V_0^2/(2R)$, which is equal to the instantaneous power at $t = T/8$ and not at the other times. Note also that at $t = T/8$, the instantaneous voltage V is equal to the rms value $V_{rms} = V_0/\sqrt{2}$.

21-2 (b) With the connections reversed, the number of windings in the primary coil is $N_p = 250$ and the number of windings in the secondary coil is $N_s = 100$. With $V_{p,rms} = 120$ V, Equation 21-10 tells us that the output rms voltage is $V_{s,rms} = V_{p,rms}(N_s/N_p) = (120 \text{ V})(100/250) = 48$ V.

21-3 (c) The voltage across an inductor depends on the rate at which the current through the inductor is changing, not on the value of current itself. The voltage across a capacitor depends on the amount of charge on either plate, not on the current. (The current determines how rapidly the capacitor charge is changing and so how rapidly the voltage is changing, but does not affect the present value of the voltage.) The voltage $V = iR$ across a resistor is directly proportional to the current i, so this does depend on the value of current.

21-4 (e) As Equations 21-23 show, the oscillation frequency of an LC circuit depends on the values of inductance and capacitance alone. It does not depend on the amount of charge placed on the capacitor plates. This is the direct analog of the behavior of a block attached to an ideal spring, which we studied in Section 12-3: The oscillation frequency depends on the spring constant and the mass of the block, but does not depend on the amplitude of the oscillation.

21-5 (d) The natural angular frequency $\sqrt{1/(LC)}$ does not depend on the resistance. The other two quantities, however, do. You can see this from Figure 21-11a: Increasing the resistance decreases the current amplitude at *all* angular frequencies, not just at resonance.

Questions and Problems

In a few problems, you are given more data than you actually need; in a few other problems, you are required to supply data from your general knowledge, outside sources, or informed estimate. Interpret as significant all digits in numerical values that have trailing zeros and no decimal points. For all problems, use $g = 9.80 \text{ m/s}^2$ for the free-fall acceleration due to gravity. Neglect friction and air resistance unless instructed to do otherwise.
• Basic, single-concept problem
•• Intermediate-level problem, may require synthesis of concepts and multiple steps
••• Challenging problem
SSM *Solution is in Student Solutions Manual*

Conceptual Questions

1. •What does this statement mean: The voltage drop across an inductor leads the current by 1/4 cycle? SSM

2. •Why does a transformer whose primary coil has 10 times as many turns as its secondary coil normally deliver about 10 times more voltage than it receives?

3. •Does Equation 18-9 apply to alternating currents? If not, is there any manner in which it can be amended so that it will be valid?

4. •Explain why ac voltage is often described using the root mean square value rather than the average voltage.

5. •Search the Internet to determine the root mean square value of voltage in a common wall receptacle for five countries that are not mentioned in this chapter.

6. •Give a simple explanation as to why an electric appliance, which is supposed to be operated with a certain root mean square voltage, operated with a slightly increased root mean square voltage will lead to catastrophic results.

7. •Examine the label on the power converter for a laptop computer, digital camera, or cell phone. If you took the converter to a country where the rms voltage is 240 V, would you need a power transformer? Explain your answer.

8. •(a) What is a transformer? (b) How does it change the voltage input to some different voltage output? (c) Will a transformer work with a dc input?

9. •**Chemistry** Some of the most toxic substances that have been widely used in the United States (and many other countries in the world) are known as polychlorinated biphenyls or PCBs. (PCBs are a class of organic compounds that have two to 10 chlorine atoms attached to biphenyl, a molecule composed of two benzene rings.) Liquid PCBs were often used to fill the interior of transformers. Research PCBs and explain why they were used in transformers.

10. •Why is electric power for domestic use in the United States, Canada, and most of the Western Hemisphere transmitted at very high voltages and stepped down to 120 V by a transformer near the point of consumption?

11. •Discuss the storage of energy in an ideal LC circuit with no losses. SSM

12. •What change must you make in the current in an inductor to double the energy stored in it?

13. •Compare the expressions for the energy stored in an inductor and the energy stored in a capacitor and explain the similarities.

14. •Is the current through a resistor in an ac circuit always in phase with the potential applied to the circuit? Why or why not?

15. •A given length of wire is wound into a solenoid. How will its self-inductance be changed if it is rewound into another coil of (a) twice the length or (b) twice the diameter? SSM

16. •When the switch S is opened in the RL circuit shown in **Figure 21-15**, a spark jumps between the switch contacts. Why?

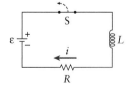

Figure 21-15 Problem 16

17. •Define the concept of impedance for an LRC circuit. What are the units of impedance?

Multiple-Choice Questions

18. •The most important advantage of ac over dc electrical signals is that
   A. electric power can't be delivered by a dc source.
   B. dc could only be used in the early days of electrical power distribution.
   C. dc results in more power loss in wire than ac.

D. it is relatively straightforward to change the voltage delivered by an ac source.

E. ac is safer.

19. •In a sinusoidal ac circuit with rms voltage V, the peak-to-peak voltage equals

A. $\sqrt{2}V$.

B. $2V$.

C. $V/\sqrt{2}$.

D. $2\sqrt{2}V$.

E. $V/2$. SSM

20. •The common electrical receptacle voltage in North America is often referred to as "120 volts ac." One hundred twenty volts is

A. the arithmetic mean of the voltage as it varies with time.

B. the root mean square (rms) average of the voltage as it varies with time.

C. the peak voltage from an ac wall receptacle.

D. the average voltage over many weeks of time.

E. one-half the peak voltage.

21. •If a power utility were able to replace an existing 500 kV transmission line with one operating at 1 MV, it would change the amount of heat produced in the transmission line to

A. 1/4 of the previous value.

B. 1/2 of the previous value.

C. 2 times the previous value.

D. 4 times the previous value.

E. 0.

22. •Two solenoids have the same cross-sectional area and length, but the first one has twice as many turns per unit length as the second one. The ratio of the inductance of the second to the first is

A. 1:1.

B. 1:$\sqrt{2}$.

C. 1:1/2.

D. 1:4.

E. 2:1.

23. •For an LC circuit, when the charge on the capacitor is 1/2 of the maximum charge, the energy stored in the capacitor is

A. the total energy.

B. 1/2 of the total energy.

C. 1/4 of the total energy.

D. 1/8 of the total energy.

E. twice the total energy. SSM

24. •If the current through an inductor were doubled, the energy stored in the inductor would be

A. halved.

B. the same.

C. doubled.

D. quadrupled.

E. 1/4 as much.

25. •The voltage leads the current

A. in circuits with an ac source and a resistor.

B. in circuits with an ac source and a capacitor.

C. in circuits with an ac source and an inductor.

D. in any ac circuit.

E. in circuits with an ac source and both a capacitor and an inductor.

Estimation/Numerical Analysis

26. •Estimate the power lost in a length of wire when the same power is transmitted by dc compared to ac.

27. •Estimate the ac current in an average household appliance. How much current does an average U.S. household draw? SSM

28. •Estimate the number of turns of the secondary compared to the number of turns of the primary for a transformer that is used to power your hair dryer that you purchased in the United States and used in Europe.

29. •The voltage applied to a series LRC circuit is shown in accompanying table. Predict and plot the current as a function of time if $L = 100$ mH, $C = 133$ μC, and $R = 50$ Ω.

t(s)	V(V)	t(s)	V(V)
1	12	14	12
2	12	15	12
3	12	16	−12
4	12	17	−12
5	12	18	−12
6	−12	19	−12
7	−12	20	−12
8	−12	21	12
9	−12	22	12
10	−12	23	12
11	12	24	12
12	12	25	12
13	12		

Proble.ms

21-1 Most circuits use alternating current

21-2 We need to analyze ac circuits differently than dc circuits

30. •A sinusoidally varying voltage is represented by $V(t) = (75.0$ V$)\sin(120\pi t)$. What are its frequency and peak voltage?

31. •Write an expression for the instantaneous voltage delivered by an ac generator supplying 120 V (rms) at 60 Hz?

32. •What is the rms current provided to a 60-W light bulb that is plugged into a 120-V rms wall receptacle?

33. •The maximum potential difference across the terminals of a 60.0-Hz sinusoidal ac source is +17.0 V at $t = 1/4$ cycle. If at $t = 0$ the potential difference is zero, calculate the potential difference at $t = 2.00$ ms. SSM

34. •An ac voltage is represented by $V(t) = (200$ V$) \sin(120\pi t)$. What is its rms voltage?

35. •The rms current passing through a 50.0-Ω resistor in a sinusoidal ac circuit is 12.0 A. What is the maximum voltage drop across the resistor?

36. •The peak-to-peak current passing through a 150-Ω resistor is 24.0 A. Find the maximum voltage across the resistor and the rms current through the resistor.

37. ••Derive a relationship between the rms current through and the maximum voltage across a resistor R. Apply your result to the following situations: (a) $R = 100$ Ω, $V_{max} = 50.0$ V, $i_{rms} = ?$, (b) $R = 200$ Ω, $i_{rms} = 2.50$ A, $V_{max} = ?$, and (c) $V_{max} = 28.0$ V, $i_{rms} = 127$ mA, $R = ?$

21-3 Transformers allow us to change the voltage of an ac power source

38. •The primary coil of a transformer makes 240 turns around its core, and the secondary coil of that transformer makes

80 turns. If the primary voltage is 120 V (rms), what is the secondary voltage?

39. •The 400-turn primary coil of a step-down transformer is connected to an ac line that is 120 V (rms). The secondary coil voltage is 6.50 V (rms). Calculate the number of turns in the secondary coil. SSM

40. •The primary coil of a transformer makes 240 turns, and the secondary coil of that transformer makes 80 turns. If an alternating current of 3.00 A (rms) passes through the primary coil of the transformer, what current passes through the secondary coil of that transformer? Assume the transformer is ideal.

41. •An ideal transformer produces an output voltage that is 500% larger than the input voltage. If the input current is 10 A (rms), what is the output current?

42. •A neon sign in a shop operates at around 12 kV (rms). A transformer is to step up the 120 V (rms) line voltage to that value. The secondary coil of the transformer has 20,000 turns. How many turns must be placed into the primary coil?

43. ••The 400-turn primary coil of a step-down transformer is connected to an ac line that is 120 V (rms). The secondary coil is to supply 15.0 A at 6.30 V (rms). Assuming no power loss in the transformer, calculate (a) the number of turns in the secondary coil and (b) the current in the primary coil. SSM

44. ••The ratio of the number of turns in the primary to the number of turns in the secondary of a step-down transformer is 25:1. If the input voltage across the primary coil is 750 V (rms) and the current in the output circuit is 25.0 A (rms), how much power is delivered to the secondary? Assume the transformer is 100% efficient.

45. ••**Chemistry** A high-voltage discharge tube is often used to study atomic spectra. The tubes require a large voltage across their terminals to operate. To get the large voltage, a step-up transformer is connected to a line voltage (120 V rms) and is designed to provide 5000 V (rms) to the discharge tube and to dissipate 75.0 W. (a) What is the ratio of the number of turns in the secondary to the number of turns in the primary? (b) What are the rms currents in the primary and secondary coils of the transformer? (c) What is the effective resistance that the 120-V source is subjected to?

21-4 An inductor is a circuit element that opposes changes in current

46. •How much voltage is produced by an inductor of value 25 μH if the time rate of change of the current is 58 mA/s?

47. ••Uncle Leo tunes an old-fashioned radio that has an antenna that is made from a 3.0-cm-long solenoid with a cross-sectional area of 0.50 cm^2, composed of 300 turns of fine copper wire. Calculate the inductance of the coil assuming it is air-filled. SSM

48. •What energy is stored in a 250-mH inductor with a current of 0.055 A?

49. ••A tightly wound solenoid of 1600 turns, cross-sectional area of 6.00 cm^2, and length of 20.0 cm carries a current of 2.80 A. (a) What is its inductance? (b) If the cross-sectional area is doubled, does anything happen to the value of the inductance? Explain your answer.

21-5 In a circuit with an inductor and capacitor, charge and current oscillate

50. •You have a 1.0-mH inductor. What size capacitor should you choose to make an oscillator with a natural frequency of 980 kHz?

51. •An LC circuit is formed with a 15-mH inductor and a 1000-μF capacitor. Calculate the frequency of oscillation f for the circuit.

52. •A 200-pF capacitor is charged to 120 V and then quickly connected to an inductor. Calculate the maximum energy stored in the magnetic field of the inductor as the circuit oscillates.

53. •LC circuits are used for filters in electronics. The selectivity of an LC circuit is defined as the ratio L/C. Calculate the selectivity of the bandwidth for an LC circuit composed of an inductor ($L = 0.250$ H) and a capacitor ($C = 875$ μF).

54. ••If the ratio of the energy stored in a capacitor compared to the total energy stored in an LC circuit is 0.5, calculate the ratio of the charge stored on the capacitor compared to the maximum charge stored on the capacitor in that circuit.

55. ••When the charge on the capacitor in an LC circuit is one-half of the maximum stored charge, calculate the ratio of the energy stored in the capacitor compared to the total energy in both the inductor and the capacitor.

21-6 When an ac voltage source is attached in series to an inductor, resistor, and capacitor, the circuit can display resonance

56. •An LRC circuit contains a 1.00-μF capacitor, a 5.00-mH inductor, and a 100-Ω resistor. What is its resonant frequency?

57. •An LRC circuit contains a 500-Ω resistor, a 5.00-H coil, and an unknown capacitor. The circuit resonates at 1000 Hz. What is the value of the capacitance? SSM

58. •The resonant frequency of an LRC circuit is 250 Hz. If the resistance is 200 Ω and the capacitance is 125 nF, what is the value of the inductance?

59. •A sinusoidal voltage of 120 V (rms) and a frequency of 60 Hz are applied to a 50.0-μF capacitor. Calculate the peak value of the current.

60. •A sinusoidal voltage that is 120 V (rms) and has a frequency of 60 Hz is applied to a 0.20 H inductor. Calculate the peak value of the current.

61. •A sinusoidal voltage of 50.0 V (peak) at a frequency of 400 Hz is applied to a capacitor of unknown capacitance. The current in the circuit is 400 mA (rms). (a) What is the capacitance? (b) If the frequency of the voltage is increased, what, if anything, will happen to the rms value of the current in the circuit? Why?

62. •A 100-Ω resistor is connected across a 120-V rms, 60-Hz ac power line. Calculate (a) i_{rms}, (b) i_{max}, and (c) the average power dissipated in the resistor.

63. ••A sinusoidal voltage of 40.0 V rms and a frequency of 100 Hz is applied to (a) a 100-Ω resistor, (b) a 0.200-H inductor, and (c) a 50.0-μF capacitor. Calculate the peak value of the current and the average power delivered in each case. SSM

64. •A potential of 40.0 V (rms) and a frequency of 100 Hz is applied to (a) a 0.200-H inductor and (b) a 50.0-μF capacitor. In each case, find the peak value of the current.

65. ••In an *LRC* series circuit, the inductance is 250 mH, the resistance is 20 Ω, and the capacitance is 350 μF. If the ac voltage applied across the circuit is given by $V(t) = (10 \text{ V}) \sin(12\pi t)$, calculate the voltage across each element at a time of $t = 0.04$ s. SSM

66. ••A 35-mH inductor with 0.20-Ω resistance is connected in series to a 200-μF capacitor and a 60-Hz, ac, 45-V source. Calculate the (a) rms current and (b) phase angle for the circuit.

67. ••Assume that the circuit in Figure 21-16 has *L* equal to 0.60 H, *R* equal to 250 Ω, and *C* equal to 3.5 μF. At a frequency of 60 Hz, what are the impedance and the phase angle between the current and voltage?

68. ••Assume that the circuit in Figure 21-16 has *L* equal to 0.60 H, *R* equal to 280 Ω, and *C* equal to 3.5 μF. The amplitude of the driving voltage is 150 V (rms). At a frequency of 60 Hz, what is the rms current in the circuit?

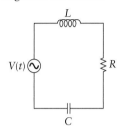

Figure 21-16
Problems 67 & 68

69. ••In an *LRC* circuit, the voltage amplitude and frequency of the source are 100 V and 500 Hz, respectively. The resistance has a value of 500 Ω, the inductance has a value of 0.20 H, and the capacitance has a value of 2.0 μF. (a) What is the impedance of the circuit? (b) What is the amplitude of the current from the source? (c) If the voltage of the source is given by $V(t) = (100 \text{ V}) \sin(1000\pi t)$, how does the current vary with time?

70. ••Compare the rms current that is created in a circuit with only a 75-nF capacitor if the frequency of the 120 V rms source is 60 Hz versus 100 Hz.

71. ••The tuner knob on an FM radio moves the plates of an adjustable capacitor that is in series with a 0.400-μH inductor and a net resistance of 1000 Ω. The peak current induced in the circuit by a radio wave of a specific carrier frequency becomes large when the natural frequency of the circuit matches the carrier frequency. What is the frequency of the FM station that is tuned in when the capacitor in the radio is adjusted to 5.80 pF? If the peak operating voltage in the tuning circuit is 9.00 V, what is the peak current?

72. ••A series *LRC* tuning circuit in a TV receiver resonates at 58 MHz. The circuit uses an 18-pF capacitor. What is the inductance of the circuit?

73. ••An *LRC* series circuit contains a 500-Ω resistor, a 5.00-H inductor, and a capacitor. What value of capacitance will cause the circuit to resonate at 1000 Hz? SSM

General Problems

74. ••A 150-Ω resistor connected across a 60-cycle ac power supply of voltage amplitude 75 V produces heat in the resistor at a certain rate. If you want to replace the ac source with a dc power supply and still produce the same rate of heating, what should be the voltage of the dc source?

75. •• You construct an *LRC* ac circuit using a 125-Ω resistor, a 12.5-mH inductor, and a parallel plate capacitor having a plate separation of 2.10 mm with a plastic material completely filling the region between the plates. The rectangular capacitor plates each measure 4.25 cm by 6.20 cm. (a) If you want the maximum rms current through the circuit at an ac frequency of 55.0 Hz, what should be the dielectric constant of the plastic in the capacitor? (b) Consult Table 17-1 to see if it appears feasible to achieve the desired results for the circuit, and explain why or why not. Would the use of an ultracapacitor (which can have a capacitance of up to several farads) allow you to achieve the desired results?

76. ••In Europe, the standard voltage is 240 V (rms), ac, and 60 cycles/s. Suppose you take your 5.00-W electric razor to Rome and plug it into the receptacle (with an adapter to fit the receptacle but not to change the voltage). (a) What power will it draw in Rome? (b) What rms current will run through the razor in the United States and in Rome? Is the razor in danger of being damaged by using it in Rome without a voltage adapter? (c) If you want to use the razor in Rome without damaging it, what type of transformer would you need? Be as quantitative as you can. (d) What is the resistance of your razor?

77. ••**Medical** A dc current of 60 mA can cause paralysis of the body's respiratory muscles and hence interfere with breathing, but only 15 mA (rms) of ac current will do the same thing. Suppose a person is working with electrical power lines on a warm humid day and therefore has a low body resistance of 1000 Ω. What dc and what ac (amplitude and rms) potentials would it take to cause respiratory paralysis?

78. ••A power cord has a resistance of 8.00×10^{-2} Ω and is used to deliver 1500 W of power. (a) If the power is delivered at 12.0 V (rms), how much power is dissipated in the power cord (assuming the current and voltage are in phase)? (b) If the power is delivered at 120 V rms, how much power is dissipated in the power cord (again assuming the current and voltage are in phase)? (c) Which voltage would you prefer to use to power your electrical device? Why?

79. •••An ac electrical generator is made by turning a flat coil in a uniform constant magnetic field of 0.225 T. The coil consists of 33 square windings and each winding is 15.0 cm on each side. It rotates at a steady rate of 745 rpm about an axis perpendicular to the magnetic field passing through the middle of the coil and parallel to two of its opposite sides. An 8.50-Ω light bulb is connected across the generator. (a) Find the voltage and current amplitudes for the light bulb. (b) At what average rate is heat generated in the light bulb? (c) How much energy is consumed by the light bulb every hour? SSM

80. ••A small ac heating coil consisting of 750 windings is 8.50 cm long. It has a diameter of 1.25 cm and a resistance of 2.15 Ω. The coil is connected in series with a 2240-μF ultracapacitor, and the combination is plugged into a household outlet (60.0 Hz and 120 V rms). (a) What is the impedance of the circuit? (b) What is the rms current through the resistor? (c) What is the maximum (or peak) current through the resistor?

81. ••Using the series *LRC* circuit in Figure 21-17, determine the maximum voltages V_R, V_C, and V_L across the resistor, capacitor, and inductor, respectively, (a) at a frequency of 100 Hz and (b) at resonance. (c) Explain why the sum of the maximum voltages in each case is larger than the maximum voltage of the applied

potential. (d) Why is the maximum voltage across the inductor equal to the maximum voltage across the capacitor at resonance?

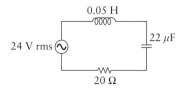

24 V rms
0.05 H
22 μF
20 Ω

Figure 21-17 Problem 81

82. ••An *LRC* circuit consists of a 15.0-μF capacitor, a resistor, and an inductor connected in series across an ac power source of variable frequency having a voltage amplitude of 25.0 V. You observe that when the power source frequency is adjusted to 44.5 Hz, the rms current through the circuit has its maximum value of 65.0 mA. What will be the rms current if you change the frequency of the power source to 60.0 Hz?

83. ••The circuit in Figure 21-18 is known as a low-pass filter because it allows low-frequency signals to pass and it attenuates

higher frequencies. Suppose its input is an ac signal that is composed of a broad range of frequencies. In this case, the voltage across the capacitor is the output that is detected. Show that the ratio of the output voltage to the input voltage is

$$\frac{V_o}{V_i} = \frac{1}{\sqrt{1 + (\omega RC)^2}}.$$ SSM

Figure 21-18 Problem 83

84. ••The circuit in Figure 21-19 is known as a high-pass filter because it allows high-frequency signals to pass and it attenuates lower frequencies.

V_i C R V_o

Figure 21-19 Problem 84

Suppose its input is an ac signal that is composed of a broad range of frequencies. The voltage across the resistor is the output that is detected. Show that the ratio of the output voltage to the input voltage is $\frac{V_o}{V_i} = \frac{1}{\sqrt{1 + 1/(\omega RC)^2}}.$

22

Electromagnetic Waves

Next sapling 22.1 - 23.4

In this chapter, your goals are to:

- (22-1) Define an electromagnetic wave.
- (22-2) Discuss how speed, frequency, and wavelength are related for electromagnetic waves, and describe the structure of an electromagnetic plane wave.
- (22-3) Explain what Maxwell's equations are and what they tell us about electromagnetic waves.
- (22-4) Calculate the energy density and intensity of an electromagnetic wave, and the energy of a photon.

To master this chapter, you should review:

- (13-2, 13-3, and 13-4) The properties of mechanical waves
- (16-6) Electric flux and Gauss's law for the electric field
- (17-5, 17-6) Capacitors
- (19-7) Ampère's law
- (20-2, 20-3) Magnetic flux and Faraday's law
- (21-2) Root-mean-square (rms) values
- (21-4, 21-5, and 21-6) Inductors

What do you think?

Our eyes are sensitive to the light from the setting sun, while the radio telescope in this photograph is sensitive to radio waves coming from distant objects in space. Both visible light and radio waves are kinds of electromagnetic waves. Compared to radio waves, visible light has (a) much higher frequency and much faster speed; (b) much lower frequency and much slower speed; (c) about the same frequency and about the same speed; (d) much higher frequency and about the same speed; (e) much lower frequency and about the same speed.

22-1 Light is just one example of an electromagnetic wave

What is light? The answer to this question was not discovered until the nineteenth century, when the Scottish physicist James Clerk Maxwell realized that light is an **electromagnetic wave**—a traveling disturbance that, unlike a sound wave, does not require a physical material through which to propagate. Instead, an electromagnetic wave involves oscillating electric and magnetic fields. These can exist even in the vacuum of space, which is why we can see the light from distant stars (**Figure 22-1a**).

In this chapter we'll study the properties of electromagnetic waves. We'll examine the broad variety of electromagnetic waves, which also includes x rays, microwaves, and radio waves. We'll see how wavelength, frequency, and speed are related for electromagnetic waves that propagate in a vacuum. We'll also look at the inner workings of a particularly simple kind of electromagnetic wave called a *sinusoidal plane wave*. We'll discover how such waves are possible by examining the fundamental equations that govern electric and magnetic fields. We'll then use these equations to help us understand

905

Figure 22-1 **Electromagnetic waves** Unlike sound waves or water waves, electromagnetic waves do not require the oscillation of any material substance. Instead, what oscillates are electric and magnetic fields. This explains (a) why these waves can propagate in a vacuum and (b) what determines their intensity.

(a)

We can see the light from this cluster of stars even though it is separated from us by 100,000 light years (about 10^{18} km) of nearly empty space. This is because electromagnetic waves, including visible light, can propagate in a vacuum.

(b)

The intensity of an electromagnetic wave—including this laser beam used in ophthalmic surgery—depends on the amplitudes of the electric and magnetic fields that make up the wave.

Take-Home Message for Section 22-1

✔ Visible light is an example of an electromagnetic wave. The properties of electromagnetic waves are explained by the equations of electricity and magnetism.

the amount of energy carried by an electromagnetic wave and the wave intensity (Figure 22-1b). We'll find that the energy of an electromagnetic wave comes in small packets called *photons*, whose properties help explain why certain kinds of electromagnetic waves can be harmful while others are not.

22-2 In an electromagnetic plane wave, electric and magnetic fields both oscillate

Experiment shows that in a vacuum, all electromagnetic waves—including radio waves, x rays, and others—propagate at the same speed. This is the **speed of light**, to which we give the symbol c:

$$c = 2.99792458 \times 10^8 \text{ m/s } (= 3.00 \times 10^8 \text{ m/s to 3 significant figures})$$

Different kinds of electromagnetic waves have different frequencies and wavelengths. In Section 13-3 we learned that for a mechanical wave, the frequency f and wavelength λ are related to the propagation speed of the wave v_p by $v_\mathrm{p} = f\lambda$ (Equation 13-2). The same relationship holds for electromagnetic waves in a vacuum with v_p related by c:

Handwritten note in margin:
Mechanical wave
$$v_p = f\lambda \qquad (22\text{-}1)$$

Propagation speed, frequency, and wavelength of an electromagnetic wave
(22-2)

Speed of light in a vacuum ⌐ Frequency of an electromagnetic wave

$$c = f\lambda$$

Wavelength of the wave in vacuum

Equation 22-2 tells us that the product of frequency f and wavelength λ has the same value, c, for *all* electromagnetic waves in a vacuum. The longer the wavelength, the lower the frequency; the shorter the wavelength, the higher the frequency.

Electromagnetic waves of any wavelength are possible. **Figure 22-2a** shows the names given to different wavelength ranges, which we refer to collectively as the **electromagnetic spectrum**. The human eye is sensitive to only a very narrow range of wavelengths known as **visible light** (**Figure 22-2b**). Visible light encompasses wavelengths from about $\lambda = 380$ nm to about $\lambda = 750$ nm (1 nm = 1 nanometer = 10^{-9} m). We perceive light of different wavelengths as having different colors; the shortest-wavelength light we

Handwritten note in margin:
Visible light
$\lambda = 380$ nm to
$\lambda = 750$ nm

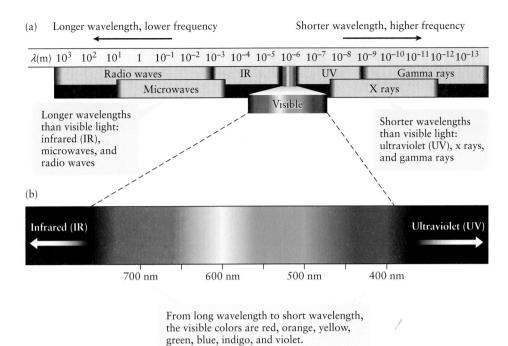

(a) Longer wavelength, lower frequency   Shorter wavelength, higher frequency

Longer wavelengths than visible light: infrared (IR), microwaves, and radio waves

Shorter wavelengths than visible light: ultraviolet (UV), x rays, and gamma rays

(b)

700 nm   600 nm   500 nm   400 nm

From long wavelength to short wavelength, the visible colors are red, orange, yellow, green, blue, indigo, and violet.

Figure 22-2 The electromagnetic spectrum (a) We classify electromagnetic waves according to their wavelength. (b) Visible light makes up a tiny portion of the entire electromagnetic spectrum.

can see is violet, and the longest-wavelength light we can see is red. At wavelengths shorter than visible light are ultraviolet light (UV), x rays, and gamma rays. At wavelengths longer than visible light are infrared light (IR), microwaves, and radio waves.

Other species can detect wavelengths longer or shorter than those visible to humans. Certain snakes (including pythons and rattlesnakes) have special pit organs on their heads that can sense infrared light. This enables these snakes to detect the radiation that both predators and prey emit due to their body temperature (see Section 14-7). Other species, such as damselfish, are able to sense ultraviolet light (**Figure 22-3**).

Note that in a medium other than a vacuum, electromagnetic waves propagate at speeds slower than c. For example, visible light travels at about 2.2×10^8 m/s $(0.73c)$ in water and at about $0.9998c$ in air. The propagation speed in a medium other than vacuum also depends on the wave frequency: For example, in ordinary glass blue light travels slightly slower than does red light. (We'll see in Chapter 23 that this explains why a prism is able to break white light into colors.) In a vacuum, however, electromagnetic waves of all frequencies propagate at c.

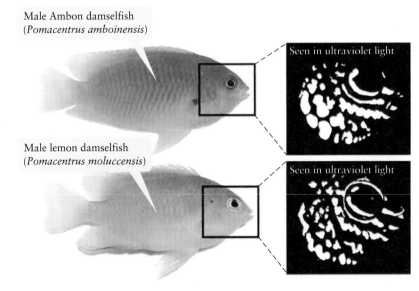

Figure 22-3 Using ultraviolet light for face recognition The Ambon damselfish (*Pomacentrus amboinensis*, a reef fish native to the western Pacific) has the ability to detect ultraviolet light. Where you see only dark and light bands on a fish's face, an Ambon sees an intricate pattern. This enables the territorial male Ambon damselfish (top) to identify and attack another Ambon in order to defend its territory, but ignore a male lemon damselfish (*Pomacentrus moluccensis*, bottom) with its slightly different facial pattern.

The structure of electromagnetic waves can be quite complex. For example, the waves that make up the beam of laser light shown in Figure 22-1b are strong near the center of the beam, then taper off in strength toward the edge of the beam. An idealized kind of wave that's much simpler to analyze but has all of the key properties of a more general electromagnetic wave is a **sinusoidal plane wave** (Figure 22-4a). As the name suggests, the disturbance in such a wave oscillates in a sinusoidal fashion. There are actually two disturbances in this wave, an electric field $\vec{E}$ with amplitude E_0 and a magnetic field $\vec{B}$ with amplitude B_0. Both of these fields are *transverse*: They are perpendicular to the direction of propagation, just like the disturbance for waves propagating along a stretched rope or string (Sections 13-3 and 13-4). The fields are also perpendicular to each other. In a snapshot of the wave as shown in Figure 22-4a, the $\vec{E}$ and $\vec{B}$ fields repeat over the same distance and so have the same wavelength λ. At a given point in space, as the wave passes by that point the $\vec{E}$ and $\vec{B}$ fields both oscillate with the same frequency f. Equation 22-2 then tells us that both fields propagate at the same speed $c = f\lambda$.

Figure 22-4 A sinusoidal electromagnetic plane wave
(a) The characteristics of a simple electromagnetic wave. This illustration shows two complete wavelengths of the wave. (b) How the electric field $\vec{E}$ and magnetic field $\vec{B}$ extend through space.

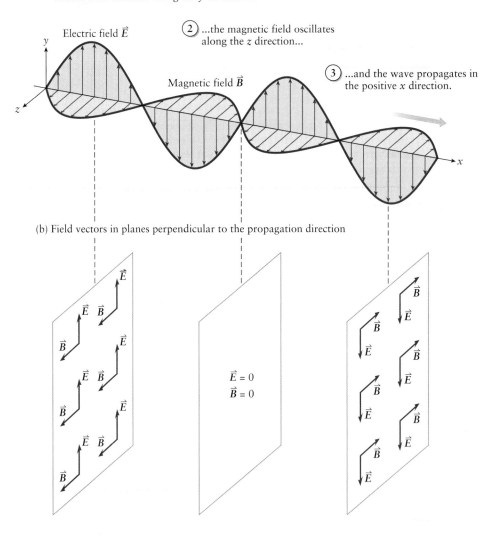

(a) An electromagnetic plane wave propagating in the positive x direction

(1) In this sinusoidal electromagnetic plane wave, the electric field oscillates along the y direction...

Electric field $\vec{E}$

(2) ...the magnetic field oscillates along the z direction...

Magnetic field $\vec{B}$

(3) ...and the wave propagates in the positive x direction.

(b) Field vectors in planes perpendicular to the propagation direction

$\vec{E} = 0$
$\vec{B} = 0$

• An electromagnetic wave is transverse: The electric and magnetic fields (the wave disturbances) are perpendicular to the propagation direction.
• In this sinusoidal plane wave, the electric and magnetic fields oscillate in phase.
• This is a plane wave because on any plane perpendicular to the direction in which the wave propagates, $\vec{E}$ has the same value at all points and $\vec{B}$ has the same value at all points.

(4) At any instant, the electric field vector has the same value at all points on a plane perpendicular to the direction in which the wave propagates (for this wave, any plane perpendicular to the x axis). The same is true for the magnetic field vector.

! Watch Out! **The fields of an electromagnetic plane wave extend beyond the axis of propagation.**

Figure 22-4a may give you the misleading impression that the $\vec{E}$ and $\vec{B}$ fields are present only along the x axis (the axis along which the wave propagates). Such a wave would be like an infinitely narrow laser beam. In fact, in a plane wave the fields have the same value at *all* points on a plane perpendicular to the direction of propagation (Figure 22-4b). That's the origin of the term *plane wave*. A true plane wave would extend infinitely far beyond the x axis in Figure 22-4a, so this is an idealization like a frictionless incline or a massless rope. But the fields shown in Figure 22-4 are a good approximation of the actual fields found near the center of the laser beam in Figure 22-1b.

In Section 13-3 we wrote a *wave function* that describes the disturbance associated with a wave on a rope. In the same fashion we can write wave functions for the electric and magnetic fields depicted in Figure 22-4a:

$$E_y(x,t) = E_0\cos(kx - \omega t + \phi)$$
$$B_z(x,t) = B_0\cos(kx - \omega t + \phi)$$
(22-3)

(sinusoidal electromagnetic plane wave)

Wave functions for electromagnetic waves

$k = 2\pi/\lambda$
$\omega = 2\pi f$

In Equations 22-3 we use the same symbols that we used for waves on a rope in Section 13-3: $k = 2\pi/\lambda$ is the angular wave number, $\omega = 2\pi f$ is the angular frequency, and ϕ is the phase angle (which tells us what point in the oscillation cycle corresponds to $x = 0, t = 0$). Note that $E_y(x, t)$ and $B_z(x, t)$ both depend on position x and time t in the same way, which tells us that the electric and magnetic fields oscillate in phase with the same wavelength and frequency. In addition, the electric field amplitude E_0 and the magnetic field amplitude B_0 in a plane wave are directly proportional to each other:

$$B_0 = \frac{E_0}{c}$$
(22-4)

(sinusoidal electromagnetic plane wave)

Example 22-1 A Radio Wave

A certain FM radio station broadcasts at a frequency of 98.7 MHz (1 MHz = 10^6 Hz). In the wave that reaches the radio in your car, the electric field amplitude is 6.00×10^{-2} V/m. Calculate the wavelength of the wave and the amplitude of the magnetic field.

Set Up

We are given the wave frequency f and the electric field amplitude E_0. We use Equation 22-2 to find the wavelength λ and Equation 22-4 to find the magnetic field amplitude B_0.

Propagation speed, frequency, and wavelength of an electromagnetic wave:

$$c = f\lambda$$
(22-2)

Relation between the electric and magnetic field amplitudes in an electromagnetic wave:

$$B_0 = \frac{E_0}{c}$$
(22-4)

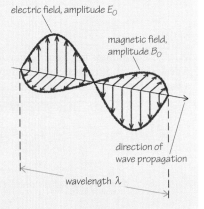

electric field, amplitude E_0

magnetic field, amplitude B_0

direction of wave propagation

wavelength λ

Solve

Use Equation 22-2 and the given frequency $f = 98.7$ MHz $= 98.7 \times 10^6$ Hz to solve for the wavelength.

From Equation 22-2,

$$\lambda = \frac{c}{f} = \frac{3.00 \times 10^8 \text{ m/s}}{98.7 \times 10^6 \text{ Hz}} = 3.04 \frac{\text{m}}{\text{s} \cdot \text{Hz}}$$

Recall that 1 Hz = 1 s^{-1}, so the units of s and Hz cancel:

$$\lambda = 3.04 \text{ m}$$

Use Equation 22-4 and the value $E_0 = 6.00 \times 10^{-2}$ V/m to solve for the magnetic field amplitude.

From Equation 22-4,

$$B_0 = \frac{E_0}{c} = \frac{6.00 \times 10^{-2}\,\text{V/m}}{3.00 \times 10^8\,\text{m/s}} = 2.00 \times 10^{-10}\left(\frac{\text{V}}{\text{m}}\right)\left(\frac{\text{s}}{\text{m}}\right)$$

We learned in Section 17-3 that 1 V/m = 1 N/C, and in Section 19-3 we learned that 1 T = 1 (N·s)/(C·m). So

$$B_0 = 2.00 \times 10^{-10}\left(\frac{\text{N}}{\text{C}}\right)\left(\frac{\text{s}}{\text{m}}\right) = 2.00 \times 10^{-10}\,\frac{\text{N·s}}{\text{C·m}}$$

$$= 2.00 \times 10^{-10}\,\text{T}$$

Reflect

Because the speed of light c has such a large value in m/s, the magnetic field amplitude B_0 in tesla (T) is much smaller than the electric field amplitude E_0 in volts per meter (V/m). As we'll see later in the chapter, however, the electric and magnetic fields prove to be equally important in an electromagnetic wave in vacuum.

In this example we've seen how to relate the units of magnetic field (T) to those of electric field (V/m): 1 T = 1 (V/m) · (s/m) = 1 V·s/m². We'll make use of this result in later examples.

Why must an electromagnetic wave propagating in vacuum be transverse? Would it be possible to have a longitudinal component of the wave, with an oscillating electric or magnetic field along the direction of propagation? For that matter, why is it necessary for an electromagnetic wave to include both electric and magnetic fields? Couldn't there be a wave that included an oscillating electric field but no magnetic field, or an oscillating magnetic field but no electric field? To answer these questions, we need to look at the fundamental equations that govern electric and magnetic fields.

? Got the Concept? 22-1 Electromagnetic Wave Speeds

Four electromagnetic waves in vacuum have different frequencies. Which of these propagates at the fastest speed? (a) $f = 3.95 \times 10^6$ Hz; (b) $f = 2.44 \times 10^7$ Hz; (c) $f = 1.26 \times 10^{11}$ Hz; (d) $f = 2.26 \times 10^8$ Hz; (e) all have the same speed.

Take-Home Message for Section 22-2

✔ Electromagnetic waves in vacuum propagate at the speed of light.

✔ Different kinds of electromagnetic waves have different wavelengths and frequencies. The human eye can detect only a very narrow range of wavelengths.

✔ In a sinusoidal electromagnetic plane wave, the electric and magnetic fields both oscillate in phase with the same wavelength and frequency. The field directions are perpendicular to each other and perpendicular to the direction of propagation.

22-3 Maxwell's equations explain why electromagnetic waves are possible

By the middle of the nineteenth century, scientists understood a great deal about electricity and magnetism. The work of Gauss, Ampère, and Faraday had established fundamental relationships that describe electric and magnetic phenomena; for example, it was understood that electric charges give rise to electric fields and that electric currents give rise to magnetic fields. It was the Scottish physicist James Clerk Maxwell who added to these fundamental relationships and forged our understanding of electricity and magnetism into a unified theory. In this section we'll look at four basic equations, known as **Maxwell's equations**, which describe *all* electromagnetic phenomena. We'll then see how these equations help us understand the nature of electromagnetic waves like the sinusoidal plane wave that we discussed in Section 22-2.

Gauss's Laws for Electricity and Magnetism

The first of Maxwell's equations is **Gauss's law for the electric field,** which we first encountered in Section 16-6. It states that the net electric flux through a closed surface (called a *Gaussian surface*) is proportional to the total amount of electric charge enclosed within that surface:

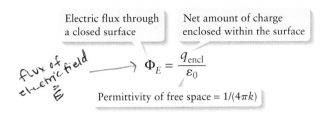

Electric flux through a closed surface

Net amount of charge enclosed within the surface

$$\Phi_E = \frac{q_{encl}}{\varepsilon_0}$$

flux of electric field $\vec{E}$

Permittivity of free space $= 1/(4\pi k)$

Gauss's law for the electric field (16-9)

In this equation we've added a subscript E to remind us that the quantity on the left-hand side is the flux of the electric field $\vec{E}$.

Figure 22-5a illustrates Gauss's law for the electric field. If a surface encloses a net positive charge, as for Gaussian surface 1 in Figure 22-5a, there is a net outward (positive) electric flux and electric field lines point out of the surface. If instead there is a net negative charge inside the surface, as for Gaussian surface 2 in Figure 22-5a, the

(a) Gauss's law for the electric field: $\Phi_E = q_{encl}/\varepsilon_0$

Gaussian surface 1 encloses positive charge. Field lines point out of this surface, and there is a net outward (positive) electric flux.

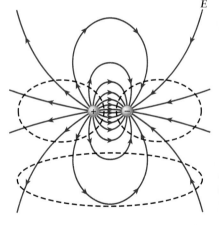

$\vec{E}$

Gaussian surface 2 encloses negative charge. Field lines point into this surface, and there is a net inward (negative) electric flux.

Gaussian surface 3 encloses zero charge. Each field line that points into this surface at one place points out at another place. There is zero net electric flux.

(b) Gauss's law for the magnetic field: $\Phi_B = 0$

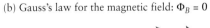

$\vec{B}$

Gaussian surface 4   Current loop   Gaussian surface 5

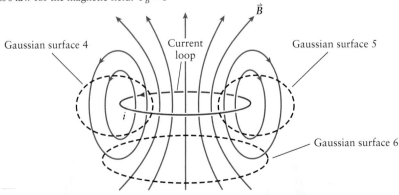

i

Gaussian surface 6

Gaussian surfaces 4 and 5 enclose part of the current loop. **Gaussian surface 6** enclose none of the current loop. For each of these surfaces, each field line that points into the surface at one place points out at another place. There is zero net magnetic flux for each surface.

Figure 22-5 Gauss's laws for the electric and magnetic fields (a) The electric flux through a closed surface depends on the charge enclosed by that surface. (b) The magnetic flux through any closed surface is zero.

- In a region of space where there are no electric charges, there is zero net electric flux through any Gaussian surface in that region.
- In any region of space, there is zero net magnetic flux through any Gaussian surface.

Margin note (handwritten): 1st equation – net electric flux and total electric charge net electric flux is inward (negative) and electric field lines point into the surface. If there is no charge at all inside the surface, as for Gaussian surface 3 in Figure 22-5a, there is zero net electric flux: Each field line that enters the surface at one point exits it at another point.

In addition to Gauss's law for the electric field, there is **Gauss's law for the magnetic field**. This states that for *any* closed Gaussian surface, there is *zero* net flux of the magnetic field $\vec{B}$ through that surface:

2nd law

There is zero magnetic flux through any closed surface, no matter what the size or shape of the surface or what it contains.

Magnetic flux through a closed surface

Gauss's law for the magnetic field
(22-5)

$$\Phi_B = 0$$

As an example, Figure 22-5b shows three Gaussian surfaces in the magnetic field of a current loop. For all three surfaces, each field line that enters the surface at one point exits it at another point. This is true whether there is current enclosed within the Gaussian surface (as for surfaces 4 and 5) or if there is no enclosed current (as for surface 6). Just as for the electric flux for Gaussian surface 3 in Figure 22-5a, this implies that there is zero net flux of magnetic field for all three surfaces. (There *would* be a nonzero net magnetic flux through a Gaussian surface if the surface enclosed an isolated north magnetic pole that had no associated south pole, or an isolated south pole that had no associated north pole. These would be the magnetic analogs of an isolated positive or negative electric charge. Physicists have searched for decades for such isolated poles, called *magnetic monopoles*. No confirmed observations have yet been made.)

What Gauss's Laws Tell Us about Electromagnetic Waves

Equations 16-9 and 22-5 are valid not just for static situations in which the electric and magnetic fields are constant, but also in situations where the $\vec{E}$ and $\vec{B}$ fields are changing. So Gauss's laws for the electric and magnetic fields must also hold true for electromagnetic waves. In Figure 22-6a we've drawn a Gaussian surface that encloses part of the sinusoidal plane wave shown in Figure 22-4. Because the wave is transverse, there is no electric or magnetic flux through the front or back face of the Gaussian surface. Although the fields change, at any instant there is as much outward flux of $\vec{E}$ or $\vec{B}$ on one edge of the Gaussian surface as there is inward flux on the opposite edge. So the net electric flux and the net magnetic flux through this surface is zero, and both of Gauss's laws are obeyed.

Gauss's laws also help us understand why electromagnetic waves in vacuum *must* be transverse. In Figure 22-6b we've sketched part of an electromagnetic wave with an oscillating *longitudinal* electric field (one that points along or opposite to the direction in which the wave propagates). For the Gaussian surface that we've chosen and at the instant of time shown, there is a net nonzero outward flux of electric field. But the wave is in a vacuum, so there can be no charge enclosed by the surface. Equation 16-9 then tells us that there *cannot* be a net electric flux through the surface in Figure 22-6b. We're forced to conclude that the longitudinal wave shown in Figure 22-6b is impossible because it contradicts Gauss's law for the electric field. Likewise, it would be impossible for an electromagnetic wave to have a longitudinal magnetic field, since this would violate Gauss's law for the magnetic field as given by Equation 22-5.

We see that Gauss's laws for the electric and magnetic fields explain why an electromagnetic wave must be transverse. Let's now examine the two remaining members of Maxwell's equations and see what they tell us about the nature of electromagnetic waves.

(a) Gauss's laws say that a transverse electromagnetic wave is permitted.

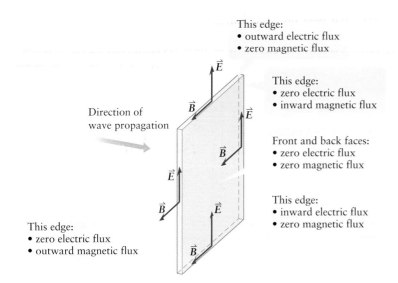

This edge:
• outward electric flux
• zero magnetic flux

This edge:
• zero electric flux
• inward magnetic flux

Direction of
wave propagation

Front and back faces:
• zero electric flux
• zero magnetic flux

This edge:
• inward electric flux
• zero magnetic flux

This edge:
• zero electric flux
• outward magnetic flux

• The net electric flux through this Gaussian
  surface is zero: $\Phi_E = 0$.
• The net magnetic flux through this Gaussian
  surface is zero: $\Phi_B = 0$.
• This agrees with Gauss's laws.

(b) Gauss's laws say that a longitudinal electromagnetic wave is not permitted.

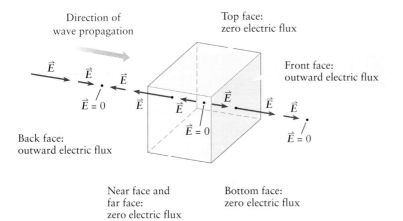

Direction of
wave propagation

Top face:
zero electric flux

Front face:
outward electric flux

$\vec{E} = 0$

$\vec{E} = 0$

$\vec{E} = 0$

Back face:
outward electric flux

Near face and
far face:
zero electric flux

Bottom face:
zero electric flux

*Because no charge can be
enclosed in a vacuum.*

• The net electric flux through this Gaussian
  surface is outward, so $\Phi_E \neq 0$.
• This disagrees with Gauss's law for the
  electric field.

Figure 22-6 **Gauss's laws applied to a sinusoidal electromagnetic plane wave** (a) A transverse
electromagnetic wave in vacuum is compatible with Gauss's laws, but (b) a longitudinal
electromagnetic wave in vacuum is not.

Faraday's Law: A Changing Magnetic Field Generates an Electric Field

The third of Maxwell's equations is **Faraday's law,** which we introduced in Sections 20-2
and 20-3. It states that an emf is induced in a loop if the magnetic flux through that
loop changes:

Induced or motional
emf in a loop

Change in the magnetic flux through
the surface outlined by the loop

$$\varepsilon = -\frac{\Delta\Phi_B}{\Delta t}$$

Time interval over which the change
in magnetic flux takes place

The negative sign indicates that the current caused by the emf
induces a magnetic field which opposes the change in flux.

Faraday's law and Lenz's law
for induction
(20-3)

Figure 22-7 A changing $\vec{B}$ produces a circulating $\vec{E}$ (a) Faraday's law says that an emf is induced in the loop when the magnetic field changes. (b) Fundamentally what happens when the magnetic field changes is that a circulating electric field is produced.

(a) A wire loop in a changing magnetic field

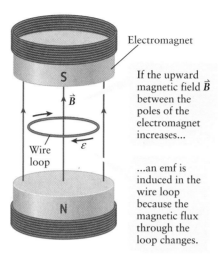

If the upward magnetic field $\vec{B}$ between the poles of the electromagnet increases...

...an emf is induced in the wire loop because the magnetic flux through the loop changes.

(b) A changing magnetic field produces a circulating electric field

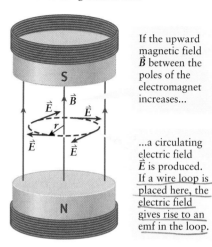

If the upward magnetic field $\vec{B}$ between the poles of the electromagnet increases...

...a circulating electric field $\vec{E}$ is produced. If a wire loop is placed here, the electric field gives rise to an emf in the loop.

As an example, Figure 22-7a shows a wire loop placed between the poles of an electromagnet. As the current in the electromagnet increases, the magnetic field increases and the increasing magnetic flux through the wire loop induces an emf in the loop. It's important to note that the emf is present whether the loop is made of a conductor like copper wire, a semiconductor like silicon, or an insulator like wood; the only difference is the amount of current that's established in response to the emf. The emf is even present if there is no material substance there at all!

Fundamentally, what happens is that the changing magnetic field generates a circulating electric field $\vec{E}$, as Figure 22-7b shows. The emf around a loop is just the *circulation* of the electric field around the loop. We define this in exactly the same way that we defined the circulation of magnetic field in Section 19-7. First imagine breaking the loop into a number of small segments of length $\Delta\ell$; then find the component $E_{\parallel}$ of the electric field that's tangent to each segment; then sum the products $E_{\parallel}\Delta\ell$:

(22-6)
$$\text{Circulation of the electric field} = \sum E_{\parallel}\Delta\ell$$

(A particularly simple case is the circulating electric field shown in Figure 22-7b. Here $\vec{E}$ points tangential to the circular loop of radius r, so the circulation is just the magnitude E multiplied by the total distance around the loop—that is, the circumference $2\pi r$ of the loop.)

We can now rewrite Faraday's law by replacing the induced emf ε in Equation 20-3 with the circulation of the electric field as given by Equation 22-6:

circulation $= E\, 2\pi r$

Faraday's law in terms of circulation
(22-7)

Circulation of electric field around a loop

Change in the magnetic flux through the surface outlined by the loop

$$\sum E_{\parallel}\Delta\ell = -\frac{\Delta\Phi_B}{\Delta t}$$

Time interval over which the change in magnetic flux takes place

Equation 22-7 tells us that a changing magnetic flux through a loop causes an electric field that circulates around that loop. This is true even if there is no material substance at the location of the loop. The minus sign in Equation 22-7 tells us the direction of the circulating electric field: If a conducting wire is present around the loop, the electric field will cause charges to flow, and the magnetic field due to those moving charges will oppose the change in magnetic field that caused the change in magnetic flux. (This is Lenz's law, which we introduced in Section 20-3.)

Example 22-2 An Electric Field Due to a Changing Magnetic Field

The uniform magnetic field shown in Figure 22-7b decreases in magnitude from 1.50 T to zero in a time t, inducing an electric field. What is the magnitude of this electric field around a loop 3.50 cm in diameter if (a) $t = 10.0$ s and (b) $t = 0.100$ s?

Set Up

Equation 20-1 tells us the magnetic flux through the loop of diameter 3.50 cm. The magnetic field $\vec{B}$ is perpendicular to the plane of the loop, so θ (the angle between the direction of $\vec{B}$ and the perpendicular to the loop) is $\theta = 0$.

The electric field in this case has the same magnitude E all the way around the circle and is tangent to the circle, so $E_{\parallel}$ in Equation 22-7 is equal to E.

Magnetic flux:

$$\Phi_B = AB_{\perp} = AB \cos\theta \quad (20\text{-}1)$$

Faraday's law in terms of circulation:

$$\sum E_{\parallel}\Delta\ell = -\frac{\Delta\Phi_B}{\Delta t} \quad (22\text{-}7)$$

Area of a circle of radius r:

$$A = \pi r^2$$

Circumference of a circle of radius r:

$$C = 2\pi r$$

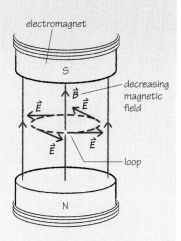

Solve

(a) First determine the initial and final values of magnetic flux and the change in magnetic flux for a circle of diameter 3.50 cm.

The radius of the loop is half the diameter:

$$r = \frac{1}{2}(3.50 \text{ cm}) = 1.75 \text{ cm} = 1.75 \times 10^{-2} \text{ m}$$

The area of the loop is

$$A = \pi r^2 = \pi (1.75 \times 10^{-2} \text{ m})^2 = 9.62 \times 10^{-4} \text{ m}^2$$

From Equation 20-1, the initial magnetic flux is

$$\Phi_B = AB \cos 0 = (9.62 \times 10^{-4} \text{ m}^2)(1.50 \text{ T})(1)$$
$$= 1.44 \times 10^{-3} \text{ T} \cdot \text{m}^2$$

The final magnetic field is zero, so the final magnetic flux is zero as well. The change in magnetic flux is

$$\Delta\Phi_B = (\text{final flux}) - (\text{initial flux}) = 0 - 1.44 \times 10^{-3} \text{ T} \cdot \text{m}^2$$
$$= -1.44 \times 10^{-3} \text{ T} \cdot \text{m}^2$$

Use Equation 22-7 to calculate the circulation of the electric field in the case where $\Delta t = 10.0$ s. Since the field magnitude E has the same value around the circle and $E_{\parallel} = E$, we can use this to calculate E.

The circulation of the electric field is equal to $-\Delta\Phi_B/\Delta t$:

$$\text{Circulation of the electric field} = \sum E_{\parallel}\Delta\ell = -\frac{\Delta\Phi_B}{\Delta t}$$

$$= -\frac{-1.44 \times 10^{-3} \text{ T} \cdot \text{m}^2}{10.0 \text{ s}} = 1.44 \times 10^{-4} \frac{\text{T} \cdot \text{m}^2}{\text{s}}$$

In Example 22-1 (Section 22-2) we saw that

$$1 \text{ T} = 1 \frac{\text{V} \cdot \text{s}}{\text{m}^2} \text{ so the magnitude of the circulation is}$$

$$1.44 \times 10^{-4} \frac{\text{T} \cdot \text{m}^2}{\text{s}} = 1.44 \times 10^{-4} \left(\frac{\text{V} \cdot \text{s}}{\text{m}^2}\right)\left(\frac{\text{m}^2}{\text{s}}\right) = 1.44 \times 10^{-4} \text{ V}$$

Since $E_{\parallel} = E$ and E has the same value at all points around the circle, we can write the circulation as

$$\sum E_{\parallel}\Delta\ell = \sum E\Delta\ell = E\sum\Delta\ell$$

The sum $\sum \Delta \ell$ is the total distance around the loop, equal to the loop circumference $2\pi r$. So

$$E\,(2\pi r) = 1.44 \times 10^{-4}\,\text{V}$$

$$E = \frac{1.44 \times 10^{-4}\,\text{V}}{2\pi r} = \frac{1.44 \times 10^{-4}\,\text{V}}{2\pi\,(1.75 \times 10^{-2}\,\text{m})} = 1.31 \times 10^{-3}\,\text{V/m}$$

(b) Repeat the calculation for the case where $\Delta t = 0.100$ s

Equation 22-7 tells us that the circulation of the electric field, and hence the electric field itself, is inversely proportional to the time Δt over which the magnetic flux changes. If the magnetic field drops to zero in $\Delta t = 0.100$ s rather than $t = 10.0$ s, the elapsed time is smaller by a factor of

$$\frac{0.100\,\text{s}}{10.0\,\text{s}} = 1.00 \times 10^{-2}$$

and so the induced field is larger by a factor of

$$\frac{1}{1.00 \times 10^{-2}} = 1.00 \times 10^{2}$$

Therefore the electric field in the case where $\Delta t = 0.100$ s is

$$E = (1.00 \times 10^{2})(1.31 \times 10^{-3}\,\text{V/m}) = 0.131\,\text{V/m}$$

Reflect

Our results show that the more rapid the change in magnetic field, the greater the magnitude of the electric field that is induced.

If a circular loop of conducting wire were placed along the circular path of diameter 3.50 cm, a current would be generated so as to produce a magnetic field that would oppose the change in magnetic flux. The upward magnetic field in the figure decreases, so the magnetic field produced in this way would have to be upward. The induced current that produces this magnetic field is in the same direction as the circulating electric field. So the electric field and current must both have the direction shown.

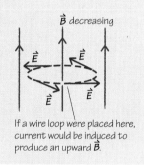

If a wire loop were placed here, current would be induced to produce an upward $\vec{B}$.

The Maxwell–Ampère Law: A Changing Electric Field Generates a Magnetic Field

Equation 22-7 tells us that a circulating electric field is produced by a magnetic field that changes over time. In Section 19-7 we learned that a circulating *magnetic* field is produced by electric charges in motion, that is, by a current. The mathematical expression of this statement is *Ampère's law*:

Ampère's law
(19-13)

Circulation of magnetic field around an Amperian loop

Permeability of free space

$$\sum B_{\parallel}\Delta \ell = \mu_0 i_{\text{through}}$$

Current through the Amperian loop

Figure 22-8a shows an application of Ampère's law that we introduced in Section 19-7: the magnetic field due to a long, straight, current-carrying wire. The current through each loop in the figure is equal to the current in the wire, so for each loop $i_{\text{through}} = i$. As a result, there is a magnetic field that circulates around each loop, and the circulation $\sum B_{\parallel}\Delta \ell$ of the magnetic field is equal to $\mu_0 i$.

Now suppose we break the wire in Figure 22-8a and insert two metal disks to form a parallel-plate capacitor (**Figure 22-8b**). If the same steady current exists as in Figure 22-8a, positive charge will build up on the lower plate and negative charge will build up on the upper plate. As a consequence, the electric field between the two plates will

(a) A current-carrying wire

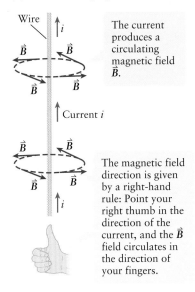

Wire

i

The current produces a circulating magnetic field $\vec{B}$.

$\vec{B}$ $\vec{B}$

$\vec{B}$ $\vec{B}$

Current i

$\vec{B}$ $\vec{B}$

$\vec{B}$ $\vec{B}$

i

The magnetic field direction is given by a right-hand rule: Point your right thumb in the direction of the current, and the $\vec{B}$ field circulates in the direction of your fingers.

(b) A current-carrying wire with a capacitor

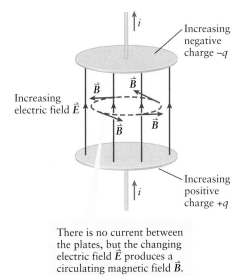

i

Increasing negative charge $-q$

Increasing electric field $\vec{E}$

$\vec{B}$ $\vec{B}$

$\vec{B}$ $\vec{B}$

i

Increasing positive charge $+q$

There is no current between the plates, but the changing electric field $\vec{E}$ produces a circulating magnetic field $\vec{B}$.

What produces a circulating $\vec{B}$ field?

Figure 22-8 A changing $\vec{E}$ produces a circulating $\vec{B}$ (a) Ampère's law says that a magnetic field circulates around a current-carrying wire. (b) A circulating magnetic field can also be produced by a changing electric field.

increase in magnitude. However, there is no current from the lower plate to the upper plate and so $i_{\text{through}} = 0$ for the loop between the plates in Figure 22-8b. So Equation 19-13 predicts that there should be no circulating magnetic field between the plates. Yet experiment shows that there *is* a circulating magnetic field between the plates! How can this be?

James Clerk Maxwell's great insight was to propose that if a changing magnetic field could produce a circulating electric field, then a changing *electric* field could produce a *magnetic* field. That's what's happening between the capacitor plates shown in Figure 22-8b: The electric field is increasing in magnitude as charge builds on the plates, and a circulating magnetic field results. To explain this effect, Maxwell expanded on Ampère's law as given by Equation 19-13 to include an additional term on the right-hand side. The result is called the **Maxwell-Ampère law**:

Analog of Faraday's Law in terms of circulation

Circulation of magnetic field around an Amperian loop

Permittivity of free space

Change in the electric flux through the surface outlined by the loop

$$\sum B_{\parallel} \Delta \ell = \mu_0 \left(i_{\text{through}} + \varepsilon_0 \frac{\Delta \Phi_E}{\Delta t} \right)$$

Permeability of free space

Current through the Amperian loop

Time interval over which the change in electric flux takes place

Maxwell-Ampère law
(22-8)

$\Phi = AE_{\perp} = AE\cos\theta$

$\varepsilon_0 = 8.85 \times 10^{-12}\,F/m$

The electric flux Φ_E through a loop, calculated using Equation 16-6, is defined in precisely the same manner as the magnetic flux Φ_B through a loop. The quantity $\varepsilon_0(\Delta\Phi_E/\Delta t)$ has units of amperes and is called the **displacement current**. Experiment confirms Equation 22-8: A circulating magnetic field can be produced by an electric current, by a changing electric field, or a combination.

We saw in Section 17-5 that we can write the permittivity of free space as $\varepsilon_0 = 8.85 \times 10^{-12}$ F/m. Since 1 F = 1 C/V and 1 A = 1 C/s, we can write this as

$$\varepsilon_0 = 8.85 \times 10^{-12}\,\frac{F}{m} = 8.85 \times 10^{-12}\,\frac{C}{V \cdot m} = 8.85 \times 10^{-12}\,\frac{A \cdot s}{V \cdot m} \qquad (22\text{-}9)$$

We'll make use of this equation in the following example.

Example 22-3 A Magnetic Field Due to a Changing Electric Field

A parallel-plate capacitor like that shown in Figure 22-8b has circular plates 5.00 cm in diameter. The electric field between the plates increases by 8.00×10^5 V/m in 1.00 s, inducing a magnetic field. What is the magnitude of that magnetic field around a loop 3.50 cm in diameter in the space between the capacitor plates?

Set Up

No charge actually moves through the loop in question, so $i_{\text{through}} = 0$ in Equation 22-8. However, the electric flux through the loop changes as the electric field changes, so the term $\Delta\Phi_E/\Delta t$ in Equation 22-8 is not zero and a circulating magnetic field will result. This example is very similar to Example 22-2, except that now we need to calculate the change in *electric* flux to determine the magnitude of the circulating *magnetic* field.

Equation 16-6 tells us the electric flux through the 3.50-cm diameter loop; the electric field $\vec{E}$ is perpendicular to the plane of the loop, so $\theta = 0$. The magnetic field has the same magnitude B all the way around the loop and is tangent to the loop, so $B_{\parallel} = B$ in Equation 22-8.

Electric flux:

$$\Phi_E = AE_{\perp} = AE\cos\theta \quad (16\text{-}6)$$

Maxwell-Ampère law:

$$\sum B_{\parallel}\Delta\ell = \mu_0\left(i_{\text{through}} + \varepsilon_0\frac{\Delta\Phi_E}{\Delta t}\right) \quad (22\text{-}8)$$

Permittivity of free space:

$$\varepsilon_0 = 8.85 \times 10^{-12}\,\frac{\text{A}\cdot\text{s}}{\text{V}\cdot\text{m}} \quad (22\text{-}9)$$

Permeability of free space:

$$\mu_0 = 4\pi \times 10^{-7}\,\text{T}\cdot\text{m/A}$$

Area of a circle of radius r:

$$A = \pi r^2$$

Circumference of a circle of radius r:

$$C = 2\pi r$$

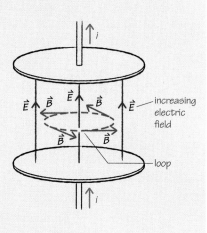

Solve

Find the displacement current $\varepsilon_0(\Delta\Phi_E/\Delta t)$ associated with the change in electric flux through the loop.

The change in electric flux in a time $\Delta t = 1.00$ s is equal to the area of the loop multiplied by the change in electric field in that time (recall that $\theta = 0$). The loop has radius $r = (1/2) \times (3.50\,\text{cm}) = 1.75\,\text{cm} = 1.75 \times 10^{-2}$ m, so

$$\begin{aligned}
\Delta\Phi_E &= A(\Delta E)\cos 0 = A(\Delta E)(1) \\
&= (\pi)(1.75 \times 10^{-2}\,\text{m})^2(8.00 \times 10^5\,\text{V/m})(1) \\
&= 7.70 \times 10^2\,\text{V}\cdot\text{m}
\end{aligned}$$

The displacement current through the loop is

$$\begin{aligned}
\varepsilon_0\frac{\Delta\Phi_E}{\Delta t} &= \left(8.85 \times 10^{-12}\,\frac{\text{A}\cdot\text{s}}{\text{V}\cdot\text{m}}\right)\left(\frac{7.70 \times 10^2\,\text{V}\cdot\text{m}}{1.00\,\text{s}}\right) \\
&= 6.81 \times 10^{-9}\,\text{A}
\end{aligned}$$

Use Equation 22-8 to determine the magnitude of the circulating magnetic field.

Since $i_{\text{through}} = 0$, the circulation of the magnetic field is equal to μ_0 times the displacement current:

$$\text{Circulation of the magnetic field} = \sum B_{\parallel}\Delta\ell = \mu_0\left(\varepsilon_0\frac{\Delta\Phi_E}{\Delta t}\right)$$

$$= \left(4\pi \times 10^{-7}\,\frac{\text{T}\cdot\text{m}}{\text{A}}\right)(6.81 \times 10^{-9}\,\text{A}) = 8.56 \times 10^{-15}\,\text{T}\cdot\text{m}$$

Since $B_{\parallel} = B$ and B has the same value at all points around the circle, we can write the circulation as

$$\sum B_{\parallel}\Delta\ell = \sum B\,\Delta\ell = B\sum\Delta\ell$$

The sum $\sum \Delta \ell$ is the total distance around the loop, equal to the loop circumference $2\pi r$. So

$$B(2\pi r) = 8.56 \times 10^{-15} \text{ T} \cdot \text{m}$$

$$B = \frac{8.56 \times 10^{-15} \text{ T} \cdot \text{m}}{2\pi(1.75 \times 10^{-2} \text{ m})} = 7.78 \times 10^{-14} \text{ T}$$

Reflect

The induced magnetic field is very weak (7.78×10^{-14} T) because the displacement current that produces it is very small (6.81×10^{-9} A). If the electric field between the plates were to change more rapidly, the displacement current would be greater and the induced magnetic field stronger.

The displacement current is in the same direction as the current that brings positive charge to the lower plate of the capacitor. The right-hand rule for using Ampère's law (Section 19-7) tells us that the induced magnetic field is in the direction shown.

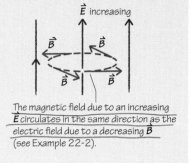

The magnetic field due to an increasing $\vec{E}$ circulates in the same direction as the electric field due to a decreasing $\vec{B}$ (see Example 22-2).

In a vacuum, where no electric currents are present, we can write Faraday's law and the Maxwell-Ampère law as

$$\sum E_\parallel \Delta \ell = -\frac{\Delta \Phi_B}{\Delta t}$$

$$\sum B_\parallel \Delta \ell = +\mu_0 \varepsilon_0 \frac{\Delta \Phi_E}{\Delta t} \qquad (22\text{-}10)$$

Notice that the first of Equations 22-10 (Faraday's law) has a minus sign, while the second of these equations (the Maxwell-Ampère law) does not. This says that the circulating electric field produced by a *decreasing* magnetic flux (so $\Delta \Phi_B / \Delta t < 0$) is in the same direction as the circulating magnetic field produced by an *increasing* electric flux (so $\Delta \Phi_E / \Delta t > 0$). That's why the electric field that we found in Example 22-2 due to a decreasing magnetic field circulates in the same direction as the magnetic field that we found in Example 22-3 due to an increasing electric field.

What Faraday's Law and the Maxwell–Ampère Law Tell Us about Electromagnetic Waves

Faraday's law tells us that a varying magnetic field gives rise to a circulating electric field, and the Maxwell-Ampère law tells us that a varying electric field gives rise to a circulating magnetic field. Taken together, these two laws tell us how electromagnetic waves are possible, and why they involve both electric and magnetic fields.

Imagine that you take an electric charge and oscillate it up and down. This will cause the electric field due to the charge to oscillate as well. By the Maxwell-Ampère law, it follows that a magnetic field will be produced which will circulate in one direction when the electric field is increasing and in the other direction when the electric field is decreasing. In other words, the magnetic field will itself oscillate. Faraday's law then tells us that this magnetic field will itself generate an electric field whose direction changes, circulating in one direction when the magnetic field is increasing and in the other direction when the magnetic field is decreasing. These two laws together tell us that a combination of an oscillating electric field and an oscillating magnetic field can sustain each other, and will continue to sustain each other even after the original source of the oscillating fields—the oscillating charge—has stopped moving.

Let's look at this in a little more detail for the electromagnetic plane wave depicted in Figure 22-4a. **Figure 22-9a** shows five snapshots of the wave as it propagates from left to right in the positive x direction. In each snapshot we're looking at the xy plane, and in each snapshot we've drawn a stationary rectangular loop one-half of a wavelength in

Figure 22-9 Flux and circulation in a sinusoidal electromagnetic plane wave (a) As an electromagnetic plane wave propagates past the dashed rectangular loop, the electric field circulates around the loop when the magnetic flux through the loop changes (Faraday's law). (b) For this loop, the magnetic field circulates around the loop when the electric flux through the loop changes (the Maxwell-Ampère law).

(a) Viewing the plane of the $\vec{E}$ field

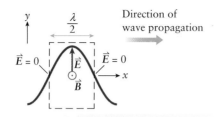

Magnetic flux through loop is maximum outward; zero circulation of $\vec{E}$.

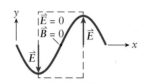

Magnetic flux through loop is changing from outward to inward; counterclockwise circulation of $\vec{E}$.

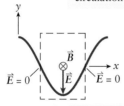

Magnetic flux through loop is maximum inward; zero circulation of $\vec{E}$.

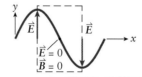

Magnetic flux through loop is changing from inward to outward; clockwise circulation of $\vec{E}$.

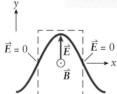

Again magnetic flux through loop is maximum outward; zero circulation of $\vec{E}$.

(b) Viewing the plane of the $\vec{B}$ field

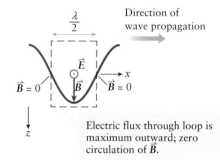

Electric flux through loop is maximum outward; zero circulation of $\vec{B}$.

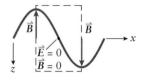

Electric flux through loop is changing from outward to inward; clockwise circulation of $\vec{B}$.

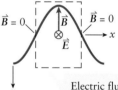

Electric flux through loop is maximum inward; zero circulation of $\vec{B}$.

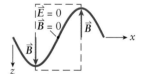

Electric flux through loop is changing from inward to outward; counterclockwise circulation of $\vec{B}$.

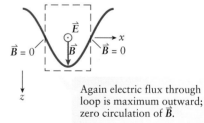

Again electric flux through loop is maximum outward; zero circulation of $\vec{B}$.

width. Because the magnetic field has only a z component, which is perpendicular to the plane of the loop, as the wave propagates there is a varying magnetic flux Φ_B through this loop. When the magnetic flux Φ_B is changing, there is a circulation of electric field around the loop; when the magnetic flux Φ_B has its maximum value either into or out of the xy plane, the flux is instantaneously not changing and there is zero circulation of electric field. That's just what Faraday's law tells us must be true.

Figure 22-9b shows five snapshots of the wave at the same instants as in Figure 22-9a, but in these snapshots we're looking at the xz plane. The rectangular loop in this part of the figure is the same size as those in part (a), but there's a varying *electric* flux Φ_E though this loop because the electric field has only a y component and so is perpendicular to the plane of this loop. When the magnetic flux Φ_E is changing, there is a circulation of magnetic field around the loop; when the magnetic flux Φ_E has its maximum value either into or out of the xz plane, the flux is instantaneously not changing and there is zero circulation of magnetic field. That's in perfect agreement with the Maxwell-Ampère law. Notice also that the circulation of the electric field in Figure 22-9a is counterclockwise when the magnetic flux Φ_B through the loop is changing from outward to inward, while the circulation of the magnetic field in Figure 22-9b is counterclockwise when the electric flux Φ_E is changing from inward to outward. That's a consequence of there being a minus sign in Faraday's law but no minus sign in the Maxwell-Ampère law, as we mentioned above in our discussion of Equations 22-10.

Faraday's law and the Maxwell-Ampère law show why the electric field and magnetic field are both necessary for the propagation of a wave: The varying electric field sustains the magnetic field, and the varying magnetic field sustains the electric field. Once an electromagnetic wave is started, it will continue to propagate through vacuum even across immense distances (Figure 22-1a). These laws also show why the electric and magnetic fields are naturally perpendicular to each other, and why the two fields oscillate together with the same frequency.

Using a more mathematical treatment of Faraday's law and the Maxwell-Ampère law, it's possible to calculate the speed c at which such a wave should propagate through a vacuum. It turns out that the speed is determined by the values of the permittivity of free space ε_0 and the permeability of free space μ_0, both of which appear in Equations 22-10:

$$c = \frac{1}{\sqrt{\mu_0 \varepsilon_0}}$$

If we substitute $\varepsilon_0 = 8.85 \times 10^{-12}\ (\text{A} \cdot \text{s})/(\text{V} \cdot \text{m})$ and $\mu_0 = 4\pi \times 10^{-7}\ \text{T} \cdot \text{m}/\text{A}$, we get

$$c = \frac{1}{\sqrt{\left(4\pi \times 10^{-7}\ \dfrac{\text{T} \cdot \text{m}}{\text{A}}\right)\left(8.85 \times 10^{-12}\ \dfrac{\text{A} \cdot \text{s}}{\text{V} \cdot \text{m}}\right)}} = 3.00 \times 10^8\ \sqrt{\frac{\text{V}}{\text{T} \cdot \text{s}}}$$

These are very odd units! Note, however, that $1\ \text{V} = 1\ \text{N} \cdot \text{m}/\text{C}$ and $1\ \text{T} = 1\ \text{N} \cdot \text{s}/(\text{C} \cdot \text{m})$, so

$$1\sqrt{\frac{\text{V}}{\text{T} \cdot \text{s}}} = 1\sqrt{\left(\frac{\text{N} \cdot \text{m}}{\text{C}}\right)\left(\frac{\text{C} \cdot \text{m}}{\text{N} \cdot \text{s}}\right)\left(\frac{1}{\text{s}}\right)} = 1\sqrt{\frac{\text{m}^2}{\text{s}^2}} = 1\ \frac{\text{m}}{\text{s}}$$

The above expression for the speed of an electromagnetic wave then becomes

Speed of light in a vacuum

$$c = \frac{1}{\sqrt{\mu_0 \varepsilon_0}} = 3.00 \times 10^8\ \text{m/s}$$

Permeability of free space   Permittivity of free space

Speed of light in a vacuum
(22-11)

This agrees with Equation 22-1 for the speed of light in a vacuum. The same mathematical analysis that leads to Equation 22-11 also relates the amplitudes E_0 and B_0 of the electric and magnetic fields in the wave: The result is that $B_0 = E_0/c$, the same result that we stated in Section 22-2 (Equation 22-4).

In the following section we'll use our insight into the nature of electromagnetic waves to help us analyze the energy carried by such waves.

! Watch Out! The speed of light in a medium other than a vacuum is less than *c*.

As we noted in Section 22-2, electromagnetic waves propagate at speed *c* only in a vacuum. We determined the value of *c* in Equation 22-11 using ε_0, the permittivity of free space, and μ_0, the permeability of free space; "free space" is equivalent to "in a vacuum." Different materials have different values of permittivity and permeability, which is why the speed of light is different in them.

? Got the Concept? 22-2 Maxwell's Equations

According to Maxwell's equations, which of the following situations is *possible*? (a) A closed surface that has a net outward magnetic flux through its surface. (b) A closed surface that has a net outward electric flux through its surface. (c) A loop that has zero electric flux through the interior of the loop and has a magnetic field that circulates around the loop. (d) More than one of (a), (b), and (c). (e) None of (a), (b), or (c).

Take-Home Message for Section 22-3

✔ Maxwell's equations are the four basic equations of electromagnetism.

✔ Gauss's law for the electric field says that the net electric flux through a closed surface is proportional to the charge enclosed by that surface. Gauss's law for the magnetic field says that the net magnetic flux through any closed surface must be zero.

✔ Gauss's laws explain why an electromagnetic wave in a vacuum must be a transverse wave.

✔ Faraday's law says that a circulating electric field is produced around a loop by a time-varying magnetic flux through the loop. The Maxwell-Ampère law states that a circulating magnetic field is produced around a loop by a current through the loop or a time-varying electric flux through the loop.

✔ Faraday's law and the Maxwell-Ampère law explain how an electromagnetic wave sustains itself, and why the electric and magnetic fields are mutually perpendicular.

22-4 Electromagnetic waves carry both electric and magnetic energy, and come in packets called photons

Electromagnetic waves carry energy. You can feel the energy delivered to your skin by sunlight (one kind of electromagnetic wave) on a sunny day. A microwave oven is useful for cooking because the water molecules found in food of all kinds absorb the energy in microwaves, which are at wavelengths between infrared and radio (Figure 22-2). And the energy in a laser beam is so tightly concentrated that it can be used as a surgical tool (Figure 22-1b).

The energy in an electromagnetic wave is actually contained within the electric and magnetic fields themselves. To see what this means, let's return to the physics of capacitors (Sections 17-5 and 17-6) and inductors (Sections 21-4 and 21-5). What we learn from these devices will give us insight into the energy content of electromagnetic waves.

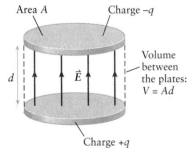

Figure 22-10 **Calculating electric energy density** The energy stored in this charged capacitor can be thought of as residing in the electric field $\vec{E}$ between the plates.

Energy in Electric and Magnetic Fields

Figure 22-10 shows a parallel-plate capacitor with plates of area *A* separated by a distance *d*. If the two plates are closely spaced, the electric field between the plates is approximately uniform and fills the volume between the plates (that is, there is very little field outside this volume). With a charge $+q$ on one plate and a charge $-q$ on the other plate, the magnitude of this field is

(17-10)

$$E = \frac{q}{\varepsilon_0 A}$$

between capacitors

It takes work to separate the charges $+q$ and $-q$, and the work that went into this goes into the electric potential energy U_E stored in the capacitor. The amount of stored energy is

$$U_E = \frac{q^2}{2C}$$
(17-17)

where C, the capacitance of the capacitor, is

$$C = \frac{\varepsilon_0 A}{d}$$
(17-13)

We can think of the electric energy in the capacitor as being stored in the electric field itself. To motivate this idea, let's substitute Equation 17-13 into Equation 17-17 and rearrange:

$$U_E = \frac{q^2}{2}\left(\frac{d}{\varepsilon_0 A}\right) = \frac{1}{2}\varepsilon_0\left(\frac{q^2}{\varepsilon_0^2 A^2}\right)Ad = \frac{1}{2}\varepsilon_0\left(\frac{q}{\varepsilon_0 A}\right)^2 Ad$$
(22-12)

The quantity $q/(\varepsilon_0 A)$ in parentheses on the far right-hand side of Equation 22-12 is just the electric field magnitude E from Equation 17-10. The quantity Ad is the volume V that the electric field occupies. So we can rewrite Equation 22-12 as

$$U_E = \left(\frac{1}{2}\varepsilon_0 E^2\right)V$$
(22-13)

Equation 22-13 tells us that the energy stored in the capacitor is equal to the volume V occupied by the electric field multiplied by a quantity $(1/2)\varepsilon_0 E^2$ that depends on the electric field magnitude E. This quantity has units of energy per volume (J/m^3) and is called the **electric energy density**:

Electric energy density   Electric field magnitude

$$u_E = \frac{1}{2}\varepsilon_0 E^2$$ units J/m^3

Permittivity of free space

Electric energy density
(22-14)

Equation 22-14 says that wherever there is an electric field, there is energy. This motivates the idea that the energy stored in a capacitor is stored in the field itself. We derived Equation 22-14 for the special case of a parallel-plate capacitor, but it turns out to be valid in any situation where an electric field is present.

We can come to a similar conclusion about the magnetic energy stored in an inductor like the one shown in Figure 22-11. Table 21-1 (Section 21-5) tells us that if an inductor has inductance L and carries a current i, the magnetic energy stored in the inductor is

magnetic energy stored in the inductor

$$U_B = \frac{1}{2}Li^2$$
(22-15)

For an inductor like that in Figure 22-11, which is a long solenoid with N turns of wire, length ℓ, and cross-sectional area A, the inductance is

inductance of inductor

$$L = \frac{\mu_0 N^2 A}{\ell}$$
(21-14)

Like the electric field inside the parallel-plate capacitor in Figure 22-10, the magnetic field inside the solenoid in Figure 22-11 is nearly uniform and confined to the volume inside the solenoid. From Equation 19-17, the magnitude of this magnetic field is $B = \mu_0 ni$, where n is the number of turns of wire per meter. This is just the total

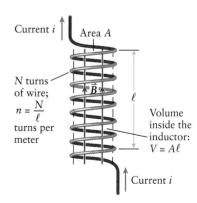

Current i Area A

N turns of wire;

$n = \dfrac{N}{\ell}$ turns per meter

$\vec{B}$ ℓ

Volume inside the inductor: $V = A\ell$

Current i

Figure 22-11 Calculating magnetic energy density The energy stored in this current-carrying inductor can be thought of as residing in the magnetic field $\vec{B}$ inside the inductor.

number of turns N divided by the length ℓ of the solenoid: $n = N/\ell$. So we can write the magnitude of the magnetic field inside the solenoid as

(22-16)
$$B = \frac{\mu_0 N i}{\ell}$$

Let's substitute Equation 21-14 for L into Equation 22-15 for the energy in the solenoid, then rearrange:

(22-17)
$$U_B = \frac{1}{2}\left(\frac{\mu_0 N^2 A}{\ell}\right)i^2 = \frac{1}{2\mu_0}\left(\frac{\mu_0^2 N^2 i^2}{\ell^2}\right)A\ell = \frac{1}{2\mu_0}\left(\frac{\mu_0 N i}{\ell}\right)^2 A\ell$$

The quantity $\mu_0 N i/\ell$ in parentheses on the far right of Equation 22-17 is the magnitude B of the magnetic field inside the solenoid. The quantity $A\ell$ is the volume V inside the solenoid, which is also the volume that the magnetic field occupies. So Equation 22-17 becomes

energy stored in inductor

(22-18)
$$U_B = \left(\frac{B^2}{2\mu_0}\right)V$$

Compare this to Equation 22-13 for the electric energy U_E. We see that according to Equation 22-18, the energy stored in the inductor equals the volume V occupied by the magnetic field multiplied by a quantity $B^2/(2\mu_0)$ that depends on the magnetic field magnitude B and has units J/m^3. We call $B^2/(2\mu_0)$ the **magnetic energy density**:

Magnetic energy density   Magnetic field magnitude

Magnetic energy density
(22-19)

J/m^3

$$u_B = \frac{B^2}{2\mu_0}$$

Permeability of free space

Just as Equation 22-14 tells us that there is electric energy wherever there is an electric field, Equation 22-19 tells us that there is magnetic energy wherever there is a magnetic field. This is true for an inductor, so we can think of the energy stored in a current-carrying inductor as being stored in the magnetic field within the inductor. But Equation 22-19 is valid wherever there is a magnetic field.

Energy in an Electromagnetic Plane Wave

Let's apply Equations 22-14 and 22-19 for the electric and magnetic energy densities to the sinusoidal electromagnetic plane wave that we introduced in Section 22-2. From Equation 22-3, in this plane wave the electric field has only a y component and the magnetic field has only a z component:

(22-3)
$$E_y(x,t) = E_0\cos(kx - \omega t + \phi)$$
$$B_z(x,t) = B_0\cos(kx - \omega t + \phi)$$

Substituting these into Equations 22-14 and 22-19, we find that the electric and magnetic energy densities in the plane wave are

(22-20)
$$u_E = \frac{1}{2}\varepsilon_0 E^2 = \frac{1}{2}\varepsilon_0 E_0^2\cos^2(kx - \omega t + \phi)$$
$$u_B = \frac{B^2}{2\mu_0} = \frac{B_0^2}{2\mu_0}\cos^2(kx - \omega t + \phi)$$

Note that u_E and u_B both depend on position x and time t in the same way. It may appear from Equation 22-20 that there are different amounts of energy in the electric and magnetic forms, since the coefficients $(1/2)\varepsilon_0 E_0^2$ and $B_0^2/2\mu_0$ are different. But we know from Equations 22-4 and 22-11 that $B_0 = E_0/c$ and $c = 1/\sqrt{\mu_0\varepsilon_0}$, so

(22-21)
$$\frac{B_0^2}{2\mu_0} = \frac{(E_0/c)^2}{2\mu_0} = \frac{(E_0\sqrt{\mu_0\varepsilon_0})^2}{2\mu_0} = \frac{1}{2}\left(\frac{\mu_0\varepsilon_0}{\mu_0}\right)E_0^2 = \frac{1}{2}\varepsilon_0 E_0^2$$

In other words, the coefficients of u_E and u_B in Equation 22-20 are equal. This means that at any position x and at any time t, an electromagnetic wave in vacuum has equal amounts of electric energy density and magnetic energy density:

$$u_E = u_B = \frac{1}{2}\varepsilon_0 E_0^2 \cos^2(kx - \omega t + \phi) = \frac{B_0^2}{2\mu_0}\cos^2(kx - \omega t + \phi) \qquad (22\text{-}22)$$

The *total* energy density u in the wave is the sum of u_E and u_B. Equation 22-22 tells us that $u_E = u_B$, so u is equal to $2u_E$ or $2u_B$:

$$u = u_E + u_B = \varepsilon_0 E_0^2 \cos^2(kx - \omega t + \phi) = \frac{B_0^2}{\mu_0}\cos^2(kx - \omega t + \phi) \qquad (22\text{-}23)$$

The value of u at any position varies with time, so it's often more useful to state its *average* value. The average value of the cosine function squared is $1/2$, so

Average energy density in an electromagnetic wave | Electric field magnitude | Electric field rms value | Magnetic field magnitude | Magnetic field rms value

$$u_{\text{average}} = \frac{1}{2}\varepsilon_0 E_0^2 = \varepsilon_0 E_{\text{rms}}^2 = \frac{B_0^2}{2\mu_0} = \frac{B_{\text{rms}}^2}{\mu_0}$$

Permittivity of free space | Permeability of free space

[handwritten: Sum of $u_E + u_B = u$ total energy density in the wave]

Average energy density in an electromagnetic wave
(22-24)

In Equation 22-24 we've used the root-mean-square (rms) values of the oscillating electric and magnetic fields, $E_{\text{rms}} = E_0/\sqrt{2}$ and $B_{\text{rms}} = B_0/\sqrt{2}$ (see Section 21-2).

An even more useful way to express the energy carried by an electromagnetic wave is in terms of the wave **intensity**, or average power per unit area. **Figure 22-12** shows a portion of a wave that has cross-sectional area A and length ℓ. The energy in this portion of the wave is u_{average} from Equation 22-24 multiplied by the volume $A\ell$ of this portion. This entire portion of the wave moves at speed c through the cross-sectional area A in a time t. The power equals the energy in the wave portion divided by the time $t = \ell/c$ that it takes this portion of the wave to travel at speed c through the cross-sectional area A. The intensity, to which we give the symbol S_{average}, equals the power divided by the area A. So

$$S_{\text{average}} = \frac{\text{energy}}{\text{time}} \times \frac{1}{\text{area}} = \frac{(u_{\text{average}} A\ell)}{\ell/c} \times \frac{1}{A} = u_{\text{average}} c$$

Using Equation 22-24, we can rewrite this as

Intensity of an electromagnetic wave | Speed of light | Magnetic field rms value

$$S_{\text{average}} = c\varepsilon_0 E_{\text{rms}}^2 = \frac{cB_{\text{rms}}^2}{\mu_0} = \frac{E_{\text{rms}}B_{\text{rms}}}{\mu_0}$$

Permittivity of free space | Electric field rms value | Permeability of free space

[handwritten: W/m²]

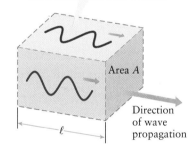

The electromagnetic wave energy in this volume moves with the wave at speed c.

Area A

Direction of wave propagation

ℓ

Figure 22-12 Calculating wave intensity The intensity of an electromagnetic wave equals the amount of wave energy that crosses an area A per unit time, divided by the area A.

Intensity of an electromagnetic wave
(22-25)

[handwritten: way to express energy carried by electromagnetic wave (Avg power / unit area)]

(To write the last expression in Equation 22-25, we used the result that since $B_0 = E_0/c$, it follows that $B_{\text{rms}} = E_{\text{rms}}/c$.)

The units of intensity are watts per square meter, or W/m^2. An everyday example is the intensity of sunlight that reaches Earth, about $1.36 \times 10^3 \text{ W/m}^2$. The following example shows that intensities of other common electromagnetic waves can be much smaller.

Example 22-4 Energy Density and Intensity in a Radio Wave

In Example 22-1 (Section 22-2) we considered an FM radio wave of frequency 98.7 MHz with an electric field amplitude 6.00×10^{-2} V/m. We found that the magnetic field has amplitude 2.00×10^{-10} T. Calculate the rms values of the electric and magnetic field, the average energy density in the wave, and the wave intensity.

Set Up

Each rms value is just equal to the amplitude divided by $\sqrt{2}$. Given the rms values, we'll use Equation 22-24 to calculate the energy density and Equation 22-25 to calculate the intensity.

Root-mean-square values:

$$E_{\mathrm{rms}} = \frac{E_0}{\sqrt{2}}$$

$$B_{\mathrm{rms}} = \frac{B_0}{\sqrt{2}}$$

Average energy density in an electromagnetic wave:

$$u_{\mathrm{average}} = \varepsilon_0 E_{\mathrm{rms}}^2 = \frac{B_{\mathrm{rms}}^2}{\mu_0} \quad (22\text{-}24)$$

Intensity of an electromagnetic wave:

$$S_{\mathrm{average}} = c\varepsilon_0 E_{\mathrm{rms}}^2 = \frac{cB_{\mathrm{rms}}^2}{\mu_0} = \frac{E_{\mathrm{rms}}B_{\mathrm{rms}}}{\mu_0} \quad (22\text{-}25)$$

Permittivity of free space:

$$\varepsilon_0 = 8.85 \times 10^{-12} \frac{\mathrm{C}}{\mathrm{V} \cdot \mathrm{m}} \quad (22\text{-}9)$$

Permeability of free space:

$$\mu_0 = 4\pi \times 10^{-7} \, \mathrm{T} \cdot \mathrm{m/A}$$

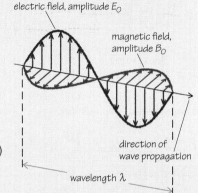

electric field, amplitude E_O

magnetic field, amplitude B_O

direction of wave propagation

wavelength λ

Solve

Calculate the rms values of the electric and magnetic fields.

We are given $E_0 = 6.00 \times 10^{-2}$ V/m and $B_0 = 2.00 \times 10^{-10}$ T. The corresponding rms values are

$$E_{\mathrm{rms}} = \frac{E_0}{\sqrt{2}} = \frac{6.00 \times 10^{-2}\,\mathrm{V/m}}{\sqrt{2}} = 4.24 \times 10^{-2}\,\mathrm{V/m}$$

$$B_{\mathrm{rms}} = \frac{B_0}{\sqrt{2}} = \frac{2.00 \times 10^{-10}\,\mathrm{T}}{\sqrt{2}} = 1.41 \times 10^{-10}\,\mathrm{T}$$

Use the value of E_{rms} to calculate the average energy density in the wave.

From Equation 22-24,

$$u_{\mathrm{average}} = \varepsilon_0 E_{\mathrm{rms}}^2 = \left(8.85 \times 10^{-12} \frac{\mathrm{C}}{\mathrm{V} \cdot \mathrm{m}}\right)\left(4.24 \times 10^{-2}\,\frac{\mathrm{V}}{\mathrm{m}}\right)^2$$

$$= 1.59 \times 10^{-14} \frac{\mathrm{C} \cdot \mathrm{V}}{\mathrm{m}^3}$$

A coulomb times a volt is a joule: $1\,\mathrm{C} \cdot \mathrm{V} = 1\,\mathrm{J}$. So

$$u_{\mathrm{average}} = 1.59 \times 10^{-14}\,\mathrm{J/m}^3$$

Find the intensity of the wave.

Comparing Equations 22-24 and 22-25 shows that the wave intensity is c times the average energy density:

$$S_{\mathrm{average}} = c\varepsilon_0 E_{\mathrm{rms}}^2 = c u_{\mathrm{average}}$$

$$= (3.00 \times 10^8\,\mathrm{m/s})(1.59 \times 10^{-14}\,\mathrm{J/m}^3)$$

$$= 4.78 \times 10^{-6}\,\frac{\mathrm{J}}{\mathrm{m}^2 \cdot \mathrm{s}}$$

A joule per second is a watt: $1\,\mathrm{J/s} = 1\,\mathrm{W}$. So

$$S_{\mathrm{average}} = 4.78 \times 10^{-6}\,\mathrm{W/m}^2$$

Reflect

The energy density and intensity are both very small quantities. It's a testament to the sensitivity of radio receivers that such a wave is quite easy to detect.

We can check our results by using the alternative expressions for $u_{average}$ and $S_{average}$ given in Equations 22-24 and 22-25. As an example, here's a check on the value of $S_{average}$.

From Equation 22-25,

$$S_{average} = \frac{E_{rms}B_{rms}}{\mu_0} = \frac{(4.24 \times 10^{-2}\,\text{V/m})(1.41 \times 10^{-10}\,\text{T})}{4\pi \times 10^{-7}\,\text{T}\cdot\text{m/A}}$$

$$= 4.78 \times 10^{-6}\,\frac{\text{V}\cdot\text{A}}{\text{m}^2}$$

One volt times one ampere is one watt ($1\,\text{V}\cdot\text{A} = 1\,\text{W}$), so

$$S_{average} = 4.78 \times 10^{-6}\,\text{W/m}^2$$

This agrees with our calculation above, as it must.

Photons

Equation 22-25 says that the intensity of an electromagnetic wave depends on the strength of the electric and magnetic fields that make up the wave, but not on the wave frequency. (The frequency f doesn't appear anywhere in this equation.) But everyday experience suggests that wave frequency *does* play a role in the energy carried by an electromagnetic wave. As an example, ultraviolet light can trigger a chemical reaction in the skin that causes a suntan or sunburn, but visible light cannot. (That's why sunscreen contains a substance that allows visible light to pass but blocks ultraviolet light.) X rays, with even higher frequency than ultraviolet light, can pierce soft tissue but not bone; as a result, an x-ray image allows a physician to diagnose a broken bone and a dentist to see cavities in your teeth. Gamma rays, with higher frequency than x rays, can damage DNA, cause cancer, and even kill cells. How can we explain these differences?

The explanation is that the energy of an electromagnetic wave propagates as small, individual packets of energy called **photons**. The energy of an individual photon is proportional to the wave frequency, and the proportionality constant h is called **Planck's constant**:

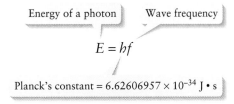

Energy of a photon   Wave frequency

$$E = hf$$

Planck's constant $= 6.62606957 \times 10^{-34}\,\text{J}\cdot\text{s}$

Energy of a photon
(22-26)

To three significant figures, $h = 6.63 \times 10^{-34}\,\text{J}\cdot\text{s}$.

Equation 22-26 explains why ultraviolet light causes a suntan or sunburn, but light in the visible spectrum does not. In order for tanning or burning to take place in your skin, individual molecules must absorb a certain minimum amount of energy from light to trigger a chemical change. A given molecule must absorb this energy in the form of a single photon, and a visible-light photon lacks sufficient energy to trigger this chemical change. An ultraviolet photon, by contrast, can trigger the change because it has a shorter wavelength, a higher frequency, and more energy per photon. An x-ray photon is more energetic still, which is why it is able to penetrate soft tissue.

Gamma-ray photons, x-ray photons, and short-wavelength, high-frequency ultraviolet photons have enough energy that they can dislodge an electron from an atom. Such **ionizing radiation** breaks apart molecules by pulling electrons from the chemical bonds that hold atoms together. Although ionizing radiation can directly break DNA molecules in living tissue, it's more likely to disrupt some other, more common molecule, such as water, to create highly reactive free radicals that then damage DNA. Depending on the severity of the damage, the cell may be able to recover. However, if the damage cannot be repaired, or if it is repaired incorrectly, the resulting mutations may be lethal. These effects generally go unnoticed until the next time the cell tries to divide. Because cancerous cells divide more frequently than most other cells in the body, they are more susceptible to radiation damage than most healthy cells.

(handwritten margin notes:) $f = c/\lambda$ $E = hf$

Planck's constant is very small, so a single photon carries only a miniscule amount of energy. As an example, a photon of red light with wavelength $\lambda = 750$ nm $= 7.50 \times 10^{-7}$ m has frequency $f = c/\lambda = (3.00 \times 10^8 \text{ m/s})/(7.50 \times 10^{-7} \text{ m}) = 4.00 \times 10^{14}$ Hz and energy $E = hf = (6.63 \times 10^{-34} \text{ J} \cdot \text{s})(4.00 \times 10^{14} \text{ Hz}) = 2.65 \times 10^{-19}$ J. (Recall that 1 Hz $= 1$ s^{-1}.) That's so small that you don't notice individual photons in the light from a lamp, just as you don't notice individual air molecules in a breeze against your face.

It's common to express photon energies in electron volts (eV). We introduced this unit in Section 17-6: 1 eV $= 1.60 \times 10^{-19}$ J. For a red photon of wavelength 750 nm, the energy is $E = (2.65 \times 10^{-19} \text{ J})/(1.60 \times 10^{-19} \text{ J/eV}) = 1.66$ eV. As the following example shows, radio photons have even less energy.

Example 22-5 Photons in a Radio Wave

For the radio wave of Examples 22-1 and 22-4, calculate (a) the energy per photon, (b) the number of photons per cubic meter, and (c) the number of photons per second that strike a receiver antenna of area 10.0 cm^2.

Set Up

The radio wave has frequency 98.7 MHz $= 98.7 \times 10^6$ Hz. We'll use this and Equation 22-26 to determine the energy of a single radio photon. From Example 22-4 we know that the energy density (energy per unit volume) of the wave is $u_{\text{average}} = 1.59 \times 10^{-14}$ J/m^3 and the intensity (energy per area per time) is $S_{\text{average}} = 4.78 \times 10^{-6}$ W/m^2. We'll use these and our calculated value of the photon energy to determine the number of photons per unit volume and the number of photons striking the antenna per time.

Energy of a photon:

$$E = hf \tag{22-26}$$

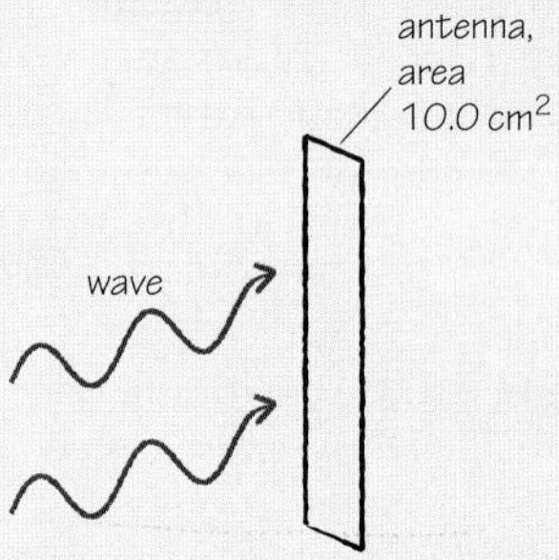

antenna, area 10.0 cm^2

wave

Solve

(a) Calculate the energy of an individual photon of frequency 98.7 MHz.

From Equation 22-26,

$$E = hf = (6.63 \times 10^{-34} \text{ J} \cdot \text{s})(98.7 \times 10^6 \text{ Hz})$$
$$= 6.54 \times 10^{-26} \text{ J} \cdot \text{s} \cdot \text{Hz}$$

Since 1 Hz $= 1$ s^{-1}, 1 J $\cdot$ s $\cdot$ Hz $= 1$ J and so

$$E = 6.54 \times 10^{-26} \text{ J or}$$

$$E = \frac{6.54 \times 10^{-26} \text{ J}}{1.60 \times 10^{-19} \text{ J/eV}} = 4.08 \times 10^{-7} \text{ eV}$$

This is much smaller than the energy of a visible-light photon (2.65×10^{-19} J $= 1.66$ eV for red light) because the frequency of the radio wave is much less than the frequency of visible light (4.00×10^{14} Hz for red light).

(b) Calculate the photon density (number of photons per unit volume).

The energy density in the wave is $u_{\text{average}} = 1.59 \times 10^{-14}$ J/m^3 and the energy per photon is $E = 6.54 \times 10^{-26}$ J/photon. The number of photons per unit volume is

$$\frac{\text{photons}}{\text{volume}} = \frac{\text{energy}}{\text{volume}} \times \frac{\text{photon}}{\text{energy}} = \frac{\left(\dfrac{\text{energy}}{\text{volume}}\right)}{\left(\dfrac{\text{energy}}{\text{photon}}\right)}$$

$$= \frac{1.59 \times 10^{-14} \text{ J/m}^3}{6.54 \times 10^{-26} \text{ J/photon}} = 2.43 \times 10^{11} \text{ photons/m}^3$$

Each cubic meter of this wave contains 2.43×10^{11} (243 billion) photons.

(c) Calculate the number of photons that strike an area of 10.0 cm² in 1.00 s.

The intensity (energy per area per time) of the wave is $S_{average} = 4.78 \times 10^{-6} \text{ W/m}^2$, so the rate at which energy arrives at the antenna of area $A = 10.0 \text{ cm}^2$ is

$$\frac{\text{energy}}{\text{area} \cdot \text{time}} \times \text{area} = S_{average} A$$

$$= (4.78 \times 10^{-6} \text{ W/m}^2)(10.0 \text{ cm}^2)\left(\frac{1 \text{ m}}{100 \text{ cm}}\right)^2$$

$$= 4.78 \times 10^{-9} \text{ W} = 4.78 \times 10^{-9} \text{ J/s}$$

The rate at which photons arrive at the antenna is

$$\frac{\text{photons}}{\text{time}} = \frac{\text{energy}}{\text{time}} \times \frac{\text{photon}}{\text{energy}} = \frac{\left(\dfrac{\text{energy}}{\text{time}}\right)}{\left(\dfrac{\text{energy}}{\text{photon}}\right)}$$

$$= \frac{4.78 \times 10^{-9} \text{ J/s}}{6.54 \times 10^{-26} \text{ J/photon}} = 7.30 \times 10^{16} \text{ photons/s}$$

In one second, 7.30×10^{16} (73 quadrillion) photons arrive at the antenna.

W = J/s

Reflect

Even this relatively low-intensity wave contains a tremendous number of photons per cubic meter and delivers an astronomical number of photons per second to a receiver.

Our results give us insight into the concerns that some people have expressed about mobile phones causing cancer. The frequencies used by mobile phones are about 7 to 27 times higher (about 700 to 2700 MHz), but even those frequencies correspond to very low photon energies (about 2.9×10^{-6} to 1.1×10^{-5} eV). As we described above, the sort of ionizing radiation that can cause cancer has very high frequency and very high photon energy of several electron volts. The photon energies associated with mobile phones are millions of times smaller, which should be far too low to have any carcinogenic effect.

! Watch Out! Photons are both particles and waves.

A common incorrect way to think about photons is to visualize them as small particles like miniature marbles, and to imagine that a large number of photons acting together behave like a wave. The reality is far different! Each individual photon has aspects of *both* wave and particle, and those particles are very different in character from ordinary objects such as marbles. We'll learn more about the curious properties of photons in Chapter 26.

? Got the Concept? 22-3 Photons   *same (energy/second) power*

Three lasers have equal power output. The first emits a pure violet light, the second emits a pure green light, and the third emits a pure red light. Which laser emits the greater number of photons per second? (a) The violet laser; (b) the green laser; (c) the red laser; (d) all emit the same number of photons per second; (e) answer depends on the value of the power output.

Take-Home Message for Section 22-4

✔ Energy is associated with both electric and magnetic fields. An electromagnetic wave in a vacuum has equal amounts of electric energy and magnetic energy per volume.

✔ The intensity of an electromagnetic wave is the average power per unit area. In a vacuum, the intensity equals the average energy per volume in the wave multiplied by the speed of light c.

✔ The energy of an electromagnetic wave comes in packets called photons. The energy of a single photon is proportional to the wave frequency.

Key Terms

displacement current
electric energy density
electromagnetic spectrum
electromagnetic wave
Faraday's law
Gauss's law for the electric field

Gauss's law for the magnetic field
intensity
ionizing radiation
magnetic energy density
Maxwell-Ampère law
Maxwell's equations

photon
Planck's constant
sinusoidal plane wave
speed of light
visible light

Chapter Summary

Topic	Equation or Figure	
Speed of electromagnetic waves: In a vacuum, all electromagnetic waves propagate at the same speed $c = 3.00 \times 10^8$ m/s. The shorter the wavelength, the higher the frequency of the wave. Our eyes are sensitive to only a narrow band of wavelengths known as the visible spectrum.	Speed of light in a vacuum Frequency of an electromagnetic wave $$c = f\lambda$$ Wavelength of the wave in vacuum	(22-2)

Electromagnetic plane waves: The simplest electromagnetic wave in vacuum is a sinusoidal plane wave. The electric and magnetic fields oscillate in phase, are perpendicular to each other, and are transverse (both are perpendicular to the direction of wave propagation). The amplitude B_0 of the magnetic field equals the amplitude E_0 of the electric field divided by the speed of light c.

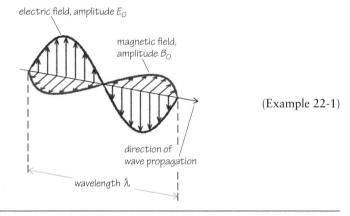

electric field, amplitude E_0
magnetic field, amplitude B_0
direction of wave propagation
wavelength λ

(Example 22-1)

Maxwell's equations: Four equations govern the behavior of electric and magnetic fields in all situations. The two Gauss's laws explain why electromagnetic waves in vacuum are transverse. Faraday's law and the Maxwell-Ampère law explain how the oscillations of the magnetic field produce the electric field, and how the oscillations of the electric field produce the magnetic field. As a result, the wave must have both electric and magnetic aspects, and the electric and magnetic fields are naturally perpendicular to each other. Maxwell's equations also successfully predict the value of the speed of light in terms of the permittivity of free space ε_0 and the permeability of free space μ_0.

Electric flux through a closed surface   Net amount of charge enclosed within the surface

$$\Phi_E = \frac{q_{\text{encl}}}{\varepsilon_0}$$

Permittivity of free space $= 1/(4\pi k)$

(16-9)

Magnetic flux through a closed surface   There is zero magnetic flux through any closed surface, no matter what the size or shape of the surface or what it contains.

$$\Phi_B = 0$$

(22-5)

Circulation of electric field around a loop   Change in the magnetic flux through the surface outlined by the loop

$$\sum E_{\parallel}\Delta\ell = -\frac{\Delta\Phi_B}{\Delta t}$$

Time interval over which the change in magnetic flux takes place

(22-7)

| Circulation of magnetic field around an Amperian loop | Permittivity of free space | Change in the electric flux through the surface outlined by the loop |

$$\sum B_\parallel \Delta \ell = \mu_0 \left(i_{\text{through}} + \varepsilon_0 \frac{\Delta \Phi_E}{\Delta t} \right) \qquad (22\text{-}8)$$

| Permeability of free space | Current through the Amperian loop | Time interval over which the change in electric flux takes place |

Electric and magnetic field energy in electromagnetic waves: The energy density (energy per volume) associated with an electric field is $u_E = (1/2)\varepsilon_0 E^2$, and the energy density associated with a magnetic field is $u_B = B^2/2\mu_0$. In an electromagnetic wave in vacuum, there are equal amounts of electric energy and magnetic energy. The average energy density in an electromagnetic wave can be expressed in terms of either the field amplitudes or their rms values. The intensity of the wave is the average power per unit area of the wave.

| Average energy density in an electromagnetic wave | Electric field magnitude | Electric field rms value | Magnetic field magnitude | Magnetic field rms value |

$$u_{\text{average}} = \frac{1}{2}\varepsilon_0 E_0^2 = \varepsilon_0 E_{\text{rms}}^2 = \frac{B_0^2}{2\mu_0} = \frac{B_{\text{rms}}^2}{\mu_0} \qquad (22\text{-}24)$$

| Permittivity of free space | Permeability of free space |

| Intensity of an electromagnetic wave | Speed of light | Magnetic field rms value |

$$S_{\text{average}} = c\varepsilon_0 E_{\text{rms}}^2 = \frac{cB_{\text{rms}}^2}{\mu_0} = \frac{E_{\text{rms}}B_{\text{rms}}}{\mu_0} \qquad (22\text{-}25)$$

| Permittivity of free space | Electric field rms value | Permeability of free space |

Photons: The energy of an electromagnetic wave comes in packets called photons that have properties of both wave and particle. The higher the wave frequency, the more energy there is per photon.

| Energy of a photon | Wave frequency |

$$E = hf \qquad (22\text{-}26)$$

Planck's constant = $6.62606957 \times 10^{-34}$ J $\cdot$ s

Answer to What do you think? Question

(d) As we describe in Section 22-2, in a vacuum all varieties of electromagnetic wave propagate at the same speed. Visible light has a much shorter wavelength (λ = about 400 to 700 nm) than radio waves (λ = a few centimeters to several meters) and so has a much higher frequency.

Answers to Got the Concept? Questions

22-1 (e) No matter what the frequency of an electromagnetic wave, its propagation speed in a vacuum is the same: the speed of light $c = 3.00 \times 10^8$ m/s.

22-2 (d) Situation (a) is impossible: Gauss's law for the magnetic field states that the net magnetic flux through any closed surface must be zero. Situation (b) is possible; according to Gauss's law for the electric field, there will be a net outward electric flux through a closed surface that encloses positive charge. Situation (c) is also possible: One example is the loop around the current-carrying wire in Figure 22-8a. There is zero electric field and hence zero electric flux through the interior of the loop, but the current gives rise to a magnetic field that circulates around the loop.

22-3 (c) All three lasers emit the same amount of electromagnetic wave energy per second. However, the energy per photon is different because each laser emits light of a different wavelength and frequency. Figure 22-2 shows that violet has the shortest wavelength, and so the highest frequency and highest photon energy; red has the longest wavelength, and so the lowest frequency and lowest photon energy. To emit the same amount of energy per second, the red laser must emit more of its low-energy photons per second than the other lasers.

Questions and Problems

In a few problems, you are given more data than you actually need; in a few other problems, you are required to supply data from your general knowledge, outside sources, or informed estimate.

Interpret as significant all digits in numerical values that have trailing zeros and no decimal points.

For all problems, use $g = 9.80 \text{ m/s}^2$ for the free-fall acceleration due to gravity. Neglect friction and air resistance unless instructed to do otherwise.

- • Basic, single-concept problem
- •• Intermediate-level problem, may require synthesis of concepts and multiple steps
- ••• Challenging problem

SSM *Solution is in Student Solutions Manual*

Conceptual Questions

1. •(a) Rank the following electromagnetic waves from the lowest to the highest wavelength: (a) microwaves, (b) red light, (c) ultraviolet light, (d) infrared light, and (e) gamma rays. (b) Which wavelength has the highest energy per photon? Which has the lowest?

2. •James Clerk Maxwell is credited with compiling the three laws of electricity and magnetism (Gauss's law, Ampère's law, and Faraday's law), adding his own law (also called Gauss's law for magnetism), modifying Ampère's law, and understanding the connections between electricity, magnetism, and optics. Were Maxwell's efforts more important than the individual discoveries of Gauss, Ampère, and Faraday? Explain your answer.

3. •Changing electric fields create changing magnetic fields. These oscillating fields propagate at the speed of light. Describe the orientation (directions) of the fields and the velocity of the electromagnetic wave. SSM

4. •Does a wire connected to a DC source, such as a battery, emit an electromagnetic wave?

5. •Describe how the frequency at which the changing electric and magnetic fields oscillate in electromagnetic waves is related to the speed of light.

6. •Describe how the frequency of an electromagnetic wave is related to the energy of the photons of that wave.

7. •The energy of an ultraviolet light photon is unrelated to the speed of the fundamental electromagnetic waves that make up such radiation. Explain how this is possible. SSM

8. •Match the equations that were conceived of by Carl Friedrich Gauss, André Ampère, Michael Faraday, and James Clerk Maxwell with the corresponding written statements:

i. $\sum B_{\parallel} \Delta \ell = \mu_0 \left(i_{\text{through}} + \varepsilon_0 \dfrac{\Delta \Phi_E}{\Delta t} \right)$

ii. $\sum E_{\parallel} \Delta \ell = -\dfrac{\Delta \Phi_B}{\Delta t}$

iii. $\Phi_B = 0$

iv. $\Phi_E = \dfrac{q_{\text{encl}}}{\varepsilon_0}$

A. The source of an electric field is an electric charge.

B. The source of a magnetic field is an electric current.

C. Changing magnetic fields induce changing electric fields.

D. Changing electric fields induce changing magnetic fields.

9. • Name three types of electromagnetic energy that you used today.

Multiple-Choice Questions

10. •In comparison to x rays, visible light has
   A. a speed that is faster.
   B. wavelengths that are longer.
   C. wavelengths that are equal.
   D. wavelengths that are shorter.
   E. frequencies that are equal.

11. •X rays and gamma rays
   A. have the same frequency.
   B. have the same wavelength.
   C. have the same speed.
   D. have the same "color."
   E. None of the above SSM

12. •In comparison to radio waves, visible light has
   A. a speed that is faster.
   B. wavelengths that are longer.
   C. wavelengths that are equal.
   D. wavelengths that are shorter.
   E. frequencies that are equal.

13. •Which of the following requires a physical medium through which to travel?
   A. radio waves
   B. light
   C. x rays
   D. sound
   E. gamma rays SSM

14. •In an *RC* circuit, the capacitor begins to discharge. In the region of space between the plates of the capacitor,
   A. there is an electric field but no magnetic field.
   B. there is a magnetic field but no electric field.
   C. there are both electric and magnetic fields.
   D. there are no electric and magnetic fields.
   E. there is an electric field whose strength is one-half that of the magnetic field.

15. •Maxwell's equations apply
   A. to both electric fields and magnetic fields that are constant over time.
   B. only to electric fields that are time-dependent.
   C. only to magnetic fields that are constant over time.
   D. to both electric fields and magnetic fields that are time-dependent.
   E. to both time-independent and time-dependent electric and magnetic fields. SSM

16. •The phase difference between the electric and magnetic fields in an electromagnetic wave is
   A. 90°.
   B. 180°.
   C. 0°.
   D. alternately 90° and 180°.
   E. alternately 0° and 90°.

Estimation/Numerical Analysis

17. •Estimate how long it would take for a bird to circumnavigate Earth traveling at 0.1c. (This would be one fast bird!)

18. •Estimate the average wavelength of radio waves that are received by (a) an AM radio and (b) an FM radio.

19. •Estimate the photon energy associated with the visible light (green) to which the human eye is most sensitive. SSM

20. •Estimate a simple conversion factor between joules and electron volts.

21. •Estimate the wavelength of electromagnetic radiation that would potentially be classified as ionizing radiation.

22. •Estimate the number of photons that are emitted in the 1000-h lifetime of a 100-W light bulb.

23. •An important news announcement is transmitted by radio waves to people who are 300 km away and sitting next to their radios and also by sound waves to people sitting 3 m from the newscaster in a newsroom. Who receives the news first? Explain your answer. SSM

24. •••Use a computer-based, graphical program to plot the graphs of the electric field (E) versus time (t) and the magnetic field (B) versus time (t) on a three-dimensional graph (see the table). Let t be on the x axis, E be on the y axis, and B be on the z axis. Explain the shape of the graph.

E (N/C)	B (×10⁻⁹ T)	t(s)
100	333	0
70.7	236	$\pi/4$
0	0	$\pi/2$
−70.7	−236	$3\pi/4$
−100	−333	π
−70.7	−236	$5\pi/4$
0	0	$3\pi/2$
70.7	236	$7\pi/4$
100	333	2π
70.7	236	$9\pi/4$
0	0	$5\pi/2$
−70.7	−236	$11\pi/4$
−100	−333	3π
−70.7	−236	$13\pi/4$
0	0	$7\pi/2$
70.7	236	$15\pi/4$
100	333	4π
70.7	236	$17\pi/4$
0	0	$9\pi/2$
−70.7	−236	$19\pi/4$
−100	−333	5π
−70.7	−236	$21\pi/4$
0	0 11	$\pi/2$
70.7	236	$23\pi/4$
100	333	6π

Problems

22-1 Light is just one example of an electromagnetic wave

22-2 In an electromagnetic plane wave, electric and magnetic fields both oscillate

25. ••Calculate the wavelengths of the electromagnetic waves with the following frequencies and classify the electromagnetic radiation of each (x ray, radio, etc.).
   A. $f = 4.14 \times 10^{15}$ Hz
   B. $f = 7.00 \times 10^{14}$ Hz
   C. $f = 8.00 \times 10^{16}$ Hz
   D. $f = 3.00 \times 10^{13}$ Hz
   E. $f = 9.00 \times 10^{12}$ Hz
   F. $f = 3.44 \times 10^{17}$ Hz
   G. $f = 8.23 \times 10^{15}$ Hz
   H. $f = 6.00 \times 10^{15}$ Hz

26. •Calculate the wavelengths of the electromagnetic waves with the following frequencies.
   A. $f = 7.50 \times 10^{15}$ Hz
   B. $f = 6.00 \times 10^{14}$ Hz
   C. $f = 5.00 \times 10^{14}$ Hz
   D. $f = 4.29 \times 10^{14}$ Hz
   E. $f = 7.50 \times 10^{16}$ Hz
   F. $f = 2.66 \times 10^{16}$ Hz
   G. $f = 8.23 \times 10^{17}$ Hz
   H. $f = 6.00 \times 10^{18}$ Hz

27. •Calculate the frequencies of the electromagnetic waves that have the following wavelengths:
   A. $\lambda = 700$ nm
   B. $\lambda = 600$ nm
   C. $\lambda = 500$ nm
   D. $\lambda = 400$ nm
   E. $\lambda = 100$ nm
   F. $\lambda = 0.0333$ nm
   G. $\lambda = 500$ μm
   H. $\lambda = 63.3$ pm SSM

28. •Calculate the frequencies of the electromagnetic waves that have the following wavelengths:
   A. $\lambda = 800$ nm
   B. $\lambda = 650$ nm
   C. $\lambda = 550$ nm
   D. $\lambda = 450$ nm
   E. $\lambda = 2.22$ nm
   F. $\lambda = 1.10 \times 10^{-8}$ m
   G. $\lambda = 50.0$ μm
   H. $\lambda = 33.4$ mm

29. •Calculate the range of wavelengths that are received by the typical FM radio in a car.

30. •The antenna for an AM radio station is a 75-m-high tower whose height is equivalent to one-quarter wavelength. At what frequency does the station transmit?

31. •How far does light travel in a vacuum in 10 ns? SSM

32. •How long does it take light to travel 300 km in a vacuum?

33. •How long does it take a radio signal from Earth to reach the Moon, which has an orbital radius of approximately 3.84×10^8 m?

34. •The basic formula relating the frequency and wavelength of an electromagnetic wave to the speed of light is given by

$$c = f\lambda$$

Starting with this expression, derive a similar relationship between the speed of light, the angular frequency ($\omega = 2\pi f$), and the wave number ($k = 2\pi/\lambda$).

35. •• (a) Calculate the wave number of an electromagnetic wave that has an angular frequency of 6.28×10^{15} rad/s. (b) Calculate the angular frequency, the frequency, and the wavelength of a photon that has a wave number of $k = 4\pi \times 10^6$ rad/m. SSM

36. ••The magnetic field of an electromagnetic wave is given by

$$B(x,t) = (0.7\,\mu\text{T})\sin\left[(8\pi \times 10^6\,\text{m}^{-1})x \right.$$
$$\left. - (2.40\pi \times 10^{15}\,\text{s}^{-1})t\right]$$

Calculate (a) the amplitude of the electric field, (b) the speed, (c) the frequency, (d) the period, and (e) the wavelength.

37. ••Suppose the electric field associated with the radio transmissions of a medical helicopter is given by

$$E(x,t) = (400\,\mu\text{N/C})\cos\left[(40\pi\,\text{nm}^{-1})x - (12\pi\,\text{s}^{-1})t\right]$$

Determine (a) the wave number, (b) the angular frequency, (c) the wavelength, and (d) the frequency of the electromagnetic wave associated with the electric field.

22-3 Maxwell's equations explain why electromagnetic waves are possible

38. ••The electric field in a region of space increases from 0 to 3000 N/C in 5.00 seconds. What is the magnitude of the induced magnetic field around a circular area with a diameter of 1.00 m oriented perpendicular to the electric field?

39. •••A parallel-plate capacitor has closely spaced plates. Charge is flowing onto the positive plate and off the negative plate at the rate $i = \Delta q/\Delta t = 2.8$ A. What is the displacement current through the capacitor between the plates? SSM

40. ••An 8-cm-diameter parallel-plate capacitor has a 1.0-mm gap. The electric field between the plates is increasing at the rate 1.0×10^6 V/(m·s). What is the magnetic field strength between the plates of the capacitor a distance of 5.0 cm from the axis of the capacitor?

41. ••Charge flows onto the positive plate of a 6.0-cm-diameter parallel-plate capacitor at the rate $i = \Delta q/\Delta t = 1.5$ A. What is the magnetic field between the plates at a distance of 3.0 cm from the axis of the plates?

22-4 Electromagnetic waves carry both electric and magnetic energy, and come in packets called photons

42. ••An electromagnetic plane wave has an intensity $S_{\text{average}} = 200$ W/m². (a) What are the rms values of the electric and magnetic field? (b) What are the amplitudes of the electric and magnetic fields?

43. ••The amplitude of an electromagnetic wave's electric field is 200 V/m. Calculate (a) the amplitude of the wave's magnetic field and (b) the intensity of the wave.

44. ••The rms value of an electromagnetic wave's magnetic field is 400 T. Calculate (a) the amplitude of the wave's electric field and (b) the intensity of the wave.

45. ••Calculate the wavelengths and frequencies of photons that have the following energies.
    A. $E_{\text{photon}} = 2.33 \times 10^{-19}$ J
    B. $E_{\text{photon}} = 4.50 \times 10^{-19}$ J
    C. $E_{\text{photon}} = 3.20 \times 10^{-19}$ J
    D. $E_{\text{photon}} = 8.55 \times 10^{-19}$ J
    E. $E_{\text{photon}} = 63.3$ eV
    F. $E_{\text{photon}} = 8.77$ eV
    G. $E_{\text{photon}} = 1.98$ eV
    H. $E_{\text{photon}} = 4.55$ eV

46. ••Calculate the wavelengths and frequencies of the photons that have the following energies.
    A. $E_{\text{photon}} = 3.45 \times 10^{-19}$ J
    B. $E_{\text{photon}} = 4.80 \times 10^{-19}$ J
    C. $E_{\text{photon}} = 1.28 \times 10^{-18}$ J
    D. $E_{\text{photon}} = 4.33 \times 10^{-20}$ J
    E. $E_{\text{photon}} = 931$ MeV
    F. $E_{\text{photon}} = 2.88$ keV
    G. $E_{\text{photon}} = 7.88$ eV
    H. $E_{\text{photon}} = 13.6$ eV

General Problems

47. ••**Biology** A recent study found that electrons having energies between 3.0 and 20 eV can cause breaks in a DNA molecule even though they do not ionize the molecule. If the energy were to come from light, (a) what range of wavelengths (in nanometers) could cause DNA breaks, and (b) in what part of the electromagnetic spectrum does the light lie?

48. ••**Medical** A dental x ray typically affects 200 g of tissue and delivers about 4.0 μJ of energy using x rays that have wavelengths of 0.025 nm. What is the energy (in electron volts) of such x ray photons, and how many photons are absorbed during the dental x ray? Assume the body absorbs all of the incident x rays.

49. ••A HeNe laser is made up of a cylindrical beam of light with a diameter of 0.750 cm. The energy is pulsed, lasting for 1.50 ns, and each burst contains an energy of 2.00 J. (a) What is the length of each pulse of laser light? (b) What is the average energy per unit volume for each pulse? SSM

50. •• (a) What is the energy of a photon of green light that has a wavelength of 525 nm? Give your answer in joules and electron volts. (b) What is the wave number of the photon?

23

Wave Properties of Light

next sapling 23.5 - 24.8

In this chapter, your goals are to:

- (23-1) Describe some key properties of light.
- (23-2) Explain Huygens' principle and what it tells us about the laws of reflection and refraction.
- (23-3) Recognize the special circumstances under which total internal reflection can take place.
- (23-4) Explain how a prism is able to break white light into its component colors.
- (23-5) Calculate how the intensity of light is affected by passing through a polarizing filter.
- (23-6) Use the idea of path length difference to calculate what happens in thin-film interference.
- (23-7) Explain why light spreads out when it passes through a narrow opening.
- (23-8) Calculate how the angular resolution of an optical device is limited by diffraction.

To master this chapter, you should review:

- (13-5) Constructive and destructive interference of waves
- (22-2) Plane waves and the electromagnetic nature of light
- (22-4) Photons

What do you think?

Honeybees are able to navigate with respect to the Sun, even though the Sun may not be directly visible to them. Which aspect of the wave nature of light do they exploit to do this? (a) The wavelength of light is shorter in air than in vacuum; (b) light is a transverse wave; (c) light spreads out when it passes through a narrow opening; (d) all of these; (e) none of these.

23-1 The wave nature of light explains much about how light behaves

In Chapter 22 we explored how light is an electromagnetic wave, with electric and magnetic fields that oscillate in phase and that carry equal amounts of energy in vacuum. In this chapter we'll explore several of the consequences of the wave nature of light. For most of this exploration we won't need the details about electric and magnetic fields; what's important is simply that in many cases light can be treated as a wave. As a result, many of the properties of light waves that we'll encounter apply equally well to sound and other types of waves.

We'll begin our discussion by introducing *Huygens' principle*, a simplified model that will help us understand how waves propagate through space. We'll use Huygens' principle to understand the laws of reflection (Figure 23-1a) and of *refraction*—the bending of light when it moves from one transparent material to another. We'll see that in certain circumstances light can be trapped inside a transparent material, just as if that material had mirrored surfaces. This effect, called *total internal reflection*, is essential for the medical technique of endoscopy. We'll see that the speed of light not only

935

Figure 23-1 Light waves in nature
(a) Reflection, (b) dispersion, and
(c) interference are among the many
phenomena that light waves exhibit in
the natural world.

Cats can see even in very
low light levels thanks to
reflection by a layer called
the *tapetum lucidum* at the
back of each eye. Incoming
light that isn't absorbed by
the retina reflects straight
back, and some of the
reflected light is detected
on the second pass.

The colors of the rainbow
are caused by dispersion:
The speed of light in water
depends on the frequency
of the light. As a result, each
color of sunlight follows a
different path as it enters a
raindrop and undergoes
refraction, reflects off the
back of the raindrop, and
undergoes refraction again
as it exits the raindrop.

The colors of this soap film
are caused by interference.
Some light reflects from the
front surface of the film,
and some enters the film
and reflects from the back
surface. If the wavelength is
just right, the two halves
interfere constructively and
produce a bright band.

(a)    (b)    (c)

Take-Home Message for Section 23-1

✔ Many of the key properties of light can be understood simply by using the idea that light can be treated as a wave.

depends on the material through which the light is traveling, but also on the wavelength of the light. This phenomenon, called *dispersion*, explains the vivid colors of a rainbow (Figure 23-1b).

One aspect of light that depends on its being a transverse wave is its *polarization*, which describes how the electric field vector of the wave is oriented. We'll see how light can become polarized by scattering or reflection, and we'll examine how polarizing filters work and why they're used in sunglasses.

We'll look at the phenomenon of *interference*, in which two light waves can add together constructively or destructively. Interference explains the colors seen in a thin film of soapy water (Figure 23-1c), as well as why cats and other animals have reflective eyes (see Figure 23-1a). We'll finish with a discussion of *diffraction*, an important effect in which light waves spread out when they pass through a small aperture.

23-2 Huygens' principle explains the reflection and refraction of light

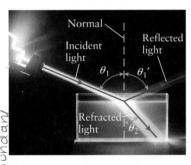

[handwritten margin note: All measured relative to the normal to the boundary]

Figure 23-2 Reflection and refraction When a beam of incident light strikes the surface of water at an angle θ_1 from the normal, some light is reflected at the same angle ($\theta_1' = \theta_1$) and some light goes into the water at a different angle θ_2.

We're used to the idea that light travels in straight lines. This is true if the material through which the light travels, called the medium, is uniform in its properties. But the direction in which light propagates changes when the light strikes a *boundary* between two different media, such as that between air and water (Figure 23-2). In general, some of the light reflects off the boundary, while the remainder travels into the second material at a different angle. The **law of reflection** states that the angle of the reflected light is the same as the angle of the incoming, or **incident**, light: $\theta_1' = \theta_1$. We use the term **refraction** to refer to the change in direction of the light that travels into the second medium so that the angle θ_2 for the refracted light is not equal to the angle θ_1 for the incident light.

! Watch Out! In reflection and refraction, angles are always measured from the normal to the boundary between the one medium and another.

Note that in Figure 23-2 the angles θ_1, θ_1', and θ_2 are all measured relative to the **normal** to the boundary, which is a line in the direction perpendicular to the boundary. A common mistake is to measure these angles relative to the boundary itself. If you make that mistake, you'll end up getting the wrong answer when you use the formulas that we'll derive in this section!

Why does the law of reflection hold true? And what determines how the direction of the light changes when it travels from one medium into another? We'll answer both of these questions using a model of waves introduced by the seventeenth-century Dutch scientist Christiaan Huygens. (Huygens' model precedes by two centuries Maxwell's complete description of light as an electromagnetic wave, but is consistent with it.) In particular, we'll see that what determines the difference between the incident and refracted angles are the *speeds* at which light travels in the two media. Refraction isn't just for visible light, but occurs for waves of all kinds: Radio waves, sound waves, and water waves may refract when crossing from one medium to another under the right circumstances.

Huygens' Principle and Reflection

Huygens considered waves that travel in two dimensions (like ripples on the surface of a pond) or in three dimensions (like light waves). He suggested that each point on a wave crest, or **front**, at time t can be treated as a source of tiny **wavelets** that themselves move at the speed of the wave (**Figure 23-3**). The wave front at a later time $t + \Delta t$ is then the superposition of all of these wavelets emitted at time t, and is tangent to the leading edges of the wavelets. This idea is called **Huygens' principle**.

As Figure 23-3 shows, Huygens' principle helps explain how circular waves in water retain their shape as they propagate outward from a splash in a pond, and how light waves spread out in spherical wave fronts from a light source. To analyze what's happening to the light beams in Figure 23-2, let's apply Huygens' principle to a *plane* wave like the ones we introduced in Section 22-2. A plane wave propagates in a single direction, and so it is a good description of the light beams shown in Figure 23-2. For each beam in Figure 23-2, the **ray** is an arrow that points in the direction of light propagation. Note that the ray is always perpendicular to the wave front.

Figure 23-4a shows how to apply Huygens' principle to understand the law of reflection. A plane wave with wave front *ABC* is directed at an angle toward the boundary between medium 1 (say, air) and medium 2 (say, water or glass). At time t point A on the wave front has just arrived at the boundary. A time Δt later, the wavelets from points A, B, and C have each spread outward by a distance $v_1\Delta t$, where v_1 is the speed at which waves propagate in medium 1. The wavelets from points B and C propagate forward through medium 1, while the wavelet from point A is reflected at the boundary and so propagates *backward* from the boundary into medium 1. (We haven't drawn the wavelets that propagate into medium 2. We'll return to those a little later to help us understand refraction. For now, we're concentrating on the light that's reflected back into medium 1.)

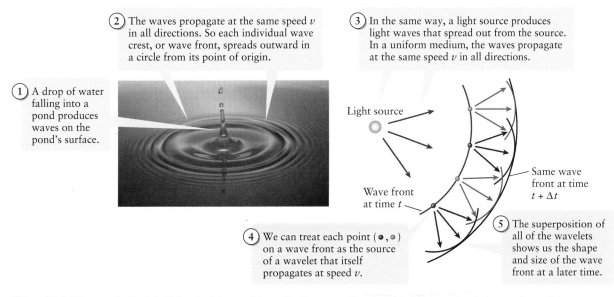

② The waves propagate at the same speed v in all directions. So each individual wave crest, or wave front, spreads outward in a circle from its point of origin.

③ In the same way, a light source produces light waves that spread out from the source. In a uniform medium, the waves propagate at the same speed v in all directions.

① A drop of water falling into a pond produces waves on the pond's surface.

Light source

Wave front at time t

Same wave front at time $t + \Delta t$

④ We can treat each point (●, ○) on a wave front as the source of a wavelet that itself propagates at speed v.

⑤ The superposition of all of the wavelets shows us the shape and size of the wave front at a later time.

Figure 23-3 Huygens' principle This principle provides a simple way to visualize wave propagation in terms of wavelets.

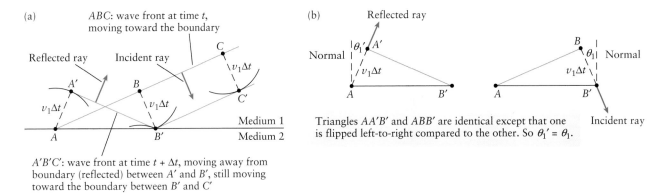

(a) ABC: wave front at time *t*, moving toward the boundary

A'B'C': wave front at time *t* + Δt, moving away from boundary (reflected) between A' and B', still moving toward the boundary between B' and C'

(b) Triangles AA'B' and ABB' are identical except that one is flipped left-to-right compared to the other. So $\theta_1' = \theta_1$.

Figure 23-4 **Huygens' principle and reflection** (a) A wave front reflected at a boundary between two media. (b) Finding the law of reflection.

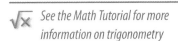

See the Math Tutorial for more information on trigonometry

If we draw a new wave front that's tangent to the leading edges of the wavelets that emanate from points *A*, *B*, and *C*, the result is A'B'C'. The wave front from B' to C' is still propagating toward the boundary; this represents the light still incident on the boundary. But the wave front from A' to B' is propagating away from the boundary and so represents the light reflected from the boundary.

The line BB' in Figure 23-4a is perpendicular to the incident wave front and so points in the direction of the incident ray. Likewise, the line AA' is perpendicular to the reflected wave front and so points in the direction of the reflected ray. We can determine how these directions are related to each other by noticing that the triangles ABB' and AA'B' are both right triangles, both have the same hypotenuse of length AB', and both have one side of length $v_1 \Delta t$ (Figure 23-4b). You can see that these two right triangles are identical, except that triangle AA'B' has been flipped left-to-right compared to triangle ABB'. So the angle θ_1' of the line AA' measured from the vertical (that is, from the normal to the boundary) must be the same as the angle θ_1 of the line BB' measured from the vertical. We conclude that

**The law of reflection
(23-1)**

When light reflects at the boundary between two media, the angle of the reflected ray from the normal...

$$\theta_1' = \theta_1$$

...is equal to the angle of the incident ray from the normal.

This is just the law of reflection that we mentioned above. We'll use Equation 23-1 extensively in Chapter 24 when we study the properties of mirrors.

Huygens' Principle and Refraction

We can also use Huygens' principle to determine the direction of the refracted ray in Figure 23-2. Figure 23-5a is similar to Figure 23-4a, except that for point *A* we've drawn only the wavelet that emanates from that point and propagates into medium 2. (We've already dealt with the reflected wave, so we don't need to draw that in Figure 23-5a.) We've assumed that the wave speed v_2 in medium 2 is slower than in medium 1, so in a time Δt the wavelet that propagates into medium 2 travels a shorter distance than those propagating in medium 1.

If we again draw a new wave front that's tangent to the leading edges of the wavelets from *A*, *B*, and *C*, the result is A″B'C'. As in Figure 23-4a, the wave front from B' to C' represents incident light in medium 1 that is still propagating toward the boundary at speed v_1. The wave front from A″ to B' represents *refracted* light that is propagating in medium 2 at speed v_2. The angle of the wave front, and hence the angle of the ray that's perpendicular to the wave front, has changed because the wave speed has changed.

Refraction

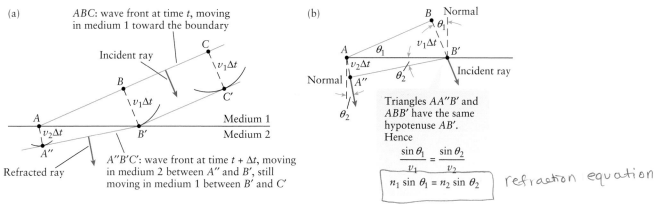

Figure 23-5 **Huygens' principle and refraction** (a) A wave front refracted at a boundary between two media. (b) Finding the law of refraction.

In Figure 23-5b we've redrawn the right triangles ABB' and $AA''B'$ from Figure 23-5a. Both triangles have the same hypotenuse of length AB', but the angles of the two triangles are different. Side BB' of triangle ABB' has length $v_1\Delta t$, points in the direction of the incident ray, and is at an angle θ_1 from the normal to the boundary. Because ABB' is a right triangle, you can see that the angle of side AB from the horizontal is also θ_1, the same as the angle of side BB' from the vertical (the normal to the boundary). If you look now at triangle $AA''B'$, you'll see that side AA'' has length $v_2\Delta t$, points in the direction of the refracted ray, and is at a different angle θ_2 from the normal to the boundary. And because $AA''B'$ is a right triangle, the angle of side $A''B'$ from the horizontal is θ_2, the same as the angle of side AA'' from the normal.

Recall that the sine of an angle in a right triangle equals the length of the side opposite to that angle divided by the length of the hypotenuse. For the angle θ_1 in triangle ABB' this statement about the sine function tells us that

$$\sin\theta_1 = \frac{\text{length of side } BB'}{\text{length of side } AB'} = \frac{v_1\Delta t}{\text{length of side } AB'} \quad \text{so} \quad \frac{\sin\theta_1}{v_1} = \frac{\Delta t}{\text{length of side } AB'} \qquad (23\text{-}2)$$

The same idea applied to the angle θ_2 in triangle $AA''B'$ tells us that

$$\sin\theta_2 = \frac{\text{length of side } AA''}{\text{length of side } AB'} = \frac{v_2\Delta t}{\text{length of side } AB'} \quad \text{so} \quad \frac{\sin\theta_2}{v_2} = \frac{\Delta t}{\text{length of side } AB'} \qquad (23\text{-}3)$$

If you compare Equations 23-2 and 23-3, you can see that

$$\frac{\sin\theta_1}{v_1} = \frac{\sin\theta_2}{v_2} \qquad (23\text{-}4)$$

Equation 23-4 tells us that the relationship between the angles θ_1 and θ_2 is determined by the speeds v_1 and v_2 of the wave in medium 1 and medium 2, respectively.

It's common to express the speed of light in a given medium in terms of a quantity called the **index of refraction**:

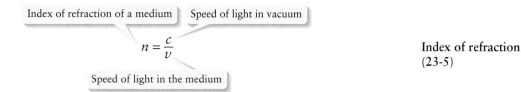

Index of refraction
(23-5)

The index of refraction of vacuum is 1, since light travels at the speed of light c so $v = c$ and $n = c/c = 1$. In any material medium, light travels slower than c, so $v < c$ and $n > 1$. The greater the value of the index of refraction n in a given medium, the slower the speed v at which light propagates in that medium. Table 23-1 lists the index of refraction of some common materials. Note that the index of refraction of air is equal to one to three significant digits, so we'll often take $n_{\text{air}} = 1$ in calculations.

Equation 23-5 tells us that $1/v = n/c$, so we can rewrite Equation 23-4 as

$$\left(\frac{n_1}{c}\right) \sin \theta_1 = \left(\frac{n_2}{c}\right) \sin \theta_2$$

If we cancel the factors of c on both sides of this equation we get

Angle of the incident ray from the normal

Angle of the refracted ray from the normal

Snell's law of refraction (23-6)

$$n_1 \sin \theta_1 = n_2 \sin \theta_2$$

Index of refraction for the medium with the incident light

Index of refraction for the medium with the refracted light

> Go to Interactive Exercise 23-1 for more practice dealing with refraction

Equation 23-6 is known as **Snell's law of refraction**. (The law is so named because it was deduced by the seventeenth-century Dutch scientist Willebrord Snellius. In fact, Equation 23-6 was first discovered by the Persian scientist Ibn Sahl in 984, more than 600 years before Snellius.)

Snell's law tells us that when a ray of light crosses from one medium to another, the product of the index of refraction and the sine of the angle the ray makes to the normal remains constant. When light passes into a material of higher index of refraction—for example, from air into glass—so that the speed of light is slower in the second medium and $n_2 > n_1$, the sine of the refracted angle and the angle itself both decrease. In this case $\theta_2 < \theta_1$ and the light bends closer to the normal (Figure 23-6a). When light instead passes into a material of lower index of refraction—for example, from glass into air—so that the speed of light is faster in the second medium and $n_2 < n_1$, the sine of the refracted angle and the angle itself both increase. In this situation $\theta_2 > \theta_1$ and the light bends away from the normal (Figure 23-6b).

The fraction of incident light that is reflected and the fraction that is refracted depend in part on the indices of refraction of the two media. (They also depend on the incident angle and on how the electric field vectors in the light wave are oriented relative to the boundary.) The index of refraction of a medium is often a function of the medium's density; one example is blood plasma, the density and index of refraction of which depend on the concentration of dissolved protein. Veterinarians can use this to estimate protein levels in livestock at the clinic or on the farm by measuring how much light refracts as it passes through a sample of an animal's plasma. Winemakers determine the amount of sugar in their grapes by using the same technique.

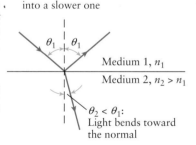

(a) Refraction from one medium into a slower one

θ_1 θ_1

Medium 1, n_1

Medium 2, $n_2 > n_1$

$\theta_2 < \theta_1$: Light bends toward the normal

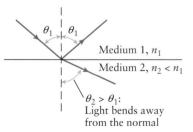

(b) Refraction from one medium into a faster one

θ_1 θ_1

Medium 1, n_1

Medium 2, $n_2 < n_1$

$\theta_2 > \theta_1$: Light bends away from the normal

Figure 23-6 Refraction toward and away from the normal Which way the refracted ray bends depends on whether the speed of light in the second medium is (a) slower or (b) faster than in the first medium.

Is the speed of light slower or faster in the second medium?

Table 23-1 Indices of Refraction

Material	Index of Refraction
vacuum	1.00000
air at 20 °c, 1 atm pressure	1.00029
ice	1.31
water at 20 °c	1.33
acetone	1.36
ethyl alcohol	1.36
eye, cornea	1.38
eye, lens	1.41
sugar water (high concentration)	1.49
Plexiglas	1.49
typical crown glass	1.52
sodium chloride	1.54
sapphire	1.77
diamond	2.42

Example 23-1 Seeing under Water

A surveyor looking at an aqueduct is just able to see the underwater edge where the far wall meets the bottom (Figure 23-7). If the aqueduct is 4.2 m wide and her line of sight to the near, top edge is 25° above the horizontal, find the depth of the aqueduct.

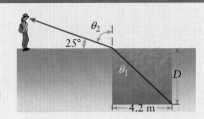

Figure 23-7 **A refracted view** If the surveyor just sees the bottom edge of the far wall of the aqueduct, how deep is the aqueduct?

Set Up

The light beam from the far edge of the bottom of the aqueduct refracts when it reaches the boundary between the water (medium 1) and the air (medium 2). Figure 23-7 shows that the angle of the refracted ray from the *normal* (shown as a vertical dashed line) is $\theta_2 = 90° - 25° = 65°$. We'll first use Snell's law to determine the angle θ_1 of the incident ray, and then use trigonometry and the given width of the aqueduct (4.2 m) to find the depth D.

Snell's law of refraction:

$$n_1 \sin \theta_1 = n_2 \sin \theta_2 \qquad (23\text{-}6)$$

Solve

Find the angle θ_1 of the incident ray using Snell's law.

Solve Equation 23-6 for the sine of the incident angle θ_1:

$$\sin \theta_1 = \frac{n_2}{n_1} \sin \theta_2$$

Figure 23-7 shows that $\theta_2 = 90° - 25° = 65°$, and from Table 23-1 we see that $n_1 = n_{\text{water}} = 1.33$ and $n_2 = n_{\text{air}} = 1.00$. So

$$\sin \theta_1 = \frac{1.00}{1.33} \sin 65° = 0.68$$

$$\theta_1 = \sin^{-1} 0.68 = 43°$$

Use trigonometry to determine the depth D of the aqueduct.

The incident ray that travels from the far edge of the aqueduct's bottom to the near, top edge is the hypotenuse of a right triangle with vertical dimension D and horizontal dimension 4.2 m. The side of length 4.2 m is the side opposite the angle θ_1, and the side of length D is the side adjacent to this angle. The tangent of θ_1 equals the opposite side divided by the adjacent side:

$$\tan \theta_1 = \frac{4.2 \text{ m}}{D}$$

Solve for the distance D:

$$D = \frac{4.2 \text{ m}}{\tan \theta_1} = \frac{4.2 \text{ m}}{\tan 43°} = 4.5 \text{ m}$$

The aqueduct is 4.5 m deep.

Reflect

The refracted angle $\theta_2 = 65°$ is larger than the incident angle $\theta_1 = 43°$, just as in Figure 23-6b. This makes sense since $n_2 < n_1$ (the index of refraction for air is smaller than the index for water).

Our brains are used to the idea that light travels in straight lines. If we extend the refracted ray backwards, we see that to the surveyor's eye the light from the far edge of the bottom of the aqueduct appears to be coming from a shallower depth than $D = 4.5$ m. You can easily see this effect in a swimming pool.

The red line represents a ray of light that bounces off the bottom of the pond and arrives at your eyes.

Your brain traces the light ray back along the blue dashed line...

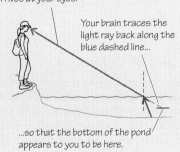

...so that the bottom of the pond appears to you to be here.

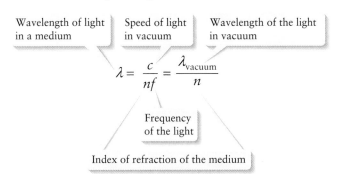

A chopstick in an empty glass

The same chopstick and glass, with water added

Figure 23-8 A refracted chopstick
Due to refraction, the submerged part of the chopstick on the right appears displaced from its actual position.

The situation shown in Figure 23-8 involves the same effect that makes the aqueduct in Example 23-1 appear shallower than it really is. When light from the submerged part of the chopstick passes from water to air at the boundary between the two, the light rays refract. As a result, it appears to our eyes that the submerged part is in a different position than its true location.

Frequency and Wavelength in Refraction

When a wave travels from one medium to another, the frequency of the wave remains the same. (In a given time interval, as many crests arrive at the boundary as leave the boundary. If this were not true, there would be a "traffic jam" of wave crests at the boundary.) However, since the wave speed is different in the two media, the wavelength must change. This follows from the relationship among the propagation speed v of the wave, the frequency f, and the wavelength λ. From Equation 13-2

$$v = f\lambda \quad \text{so} \quad \lambda = \frac{v}{f}$$

In a vacuum light waves travel at speed $v = c$, so the wavelength is

(23-7)
$$\lambda_{\text{vacuum}} = \frac{c}{f}$$

In a medium with index of refraction n, the wave speed from Equation 23-5 is $v = c/n$. Then we can write the wavelength of light in a medium as

Wavelength of light in a medium
(23-8)

Wavelength of light in a medium

Speed of light in vacuum

Wavelength of the light in vacuum

$$\lambda = \frac{c}{nf} = \frac{\lambda_{\text{vacuum}}}{n}$$

Frequency of the light

Index of refraction of the medium

Equation 23-8 says that the wavelength is shorter in a medium with a higher index of refraction, where the propagation speed is slower. For example, red light that has a wavelength $\lambda_{\text{vacuum}} = 750$ nm in vacuum has a wavelength in water ($n = 1.33$) equal to $\lambda = \lambda_{\text{vacuum}}/n = (750 \text{ nm})/(1.33) = 564$ nm. The frequency of this light is the same in both media: $f = c/\lambda_{\text{vacuum}} = (3.00 \times 10^8 \text{ m/s})/(750 \times 10^{-9} \text{ m}) = 4.00 \times 10^{14}$ Hz in vacuum and $f = v/\lambda = c/(n\lambda) = (3.00 \times 10^8 \text{ m/s})/((1.33)(564 \times 10^{-9} \text{ m})) = 4.00 \times 10^{14}$ Hz in water.

We learned in Section 22-4 that the energy of an electromagnetic wave comes in packets called *photons*. The energy of an individual photon is given by Equation 22-26, $E = hf$, where h is a constant called Planck's constant. This equation shows that the energy of a photon does not change when it passes from one medium to another, even though the speed of the photon changes: The frequency f remains constant, so the photon energy $E = hf$ is unchanged. This demonstrates one way that photons are very different from ordinary particles such as marbles, whose kinetic energy changes when their speed changes.

Speed changes but energy of photon stays the same

? Got the Concept? 23-1 Refraction

A beam of light in air travels into a transparent, flat-walled container made of Plexiglas. The incident angle is 5.00°. The light then travels from the Plexiglas into the water inside the container. In each refraction, does the ray bend closer to the normal or farther away from it? (a) Closer in both refractions; (b) farther away in both refractions; (c) closer going from air to Plexiglas, farther away going from Plexiglas to water; (d) farther away going from air to Plexiglas, closer going from Plexiglas to water; (e) in at least one of the refractions, the light does not bend at all.

Take-Home Message for Section 23-2

✔ Huygens' principle says that each point on a wave front acts as a source of wavelets. The new wave front is the superposition of the individual wavelets.

✔ When waves encounter a boundary between two media in which the wave speed is different, the waves can bounce back into the first medium (reflect) or pass into the second medium at a different angle (refract).

✔ The angle of the reflected ray is the same as the angle of the incident ray (law of reflection).

✔ Snell's law describes the direction of the refracted ray. If the wave speed is lower in the second medium than in the first, the refracted ray bends toward the normal. If the wave speed is faster in the second medium, the refracted ray bends away from the normal.

✔ The index of refraction is a measure of the speed of light in a medium relative to the speed of light in vacuum. The slower light travels in a medium, the larger its index of refraction.

23-3 In some cases light undergoes total internal reflection at the boundary between media

In the photograph shown in Figure 23-2, light in air encounters a boundary with water on the other side. Some of the incident light is reflected at the boundary, and some of it is refracted from the first medium (air) into the second medium (water). There are situations, however, in which *none* of the light is refracted into the second medium. Figure 23-9 shows a light beam in glass that encounters a boundary with air on the other side. As the photograph shows, 100% of the light is reflected back into the glass. This effect is called **total internal reflection**.

Figure 23-10 shows how this effect arises for the case of light in glass that reaches a boundary with air on the other side. Light travels more slowly in glass than air, so the index of refraction of glass is higher than the index of refraction of air. As light crosses the boundary from glass to air, it is bent away from the normal as in Figure 23-10a. As the incident angle of the light increases (Figure 23-10b), the refracted light gets farther from the normal and decreases in intensity. When the incident angle equals the **critical angle** θ_c, the refracted light lies exactly in the plane of the surface (Figure 23-10c). At this angle the intensity of the refracted light is zero. If the incident angle is greater than the critical angle, as in Figure 23-10d, the light is completely reflected back into the glass. This is total internal reflection.

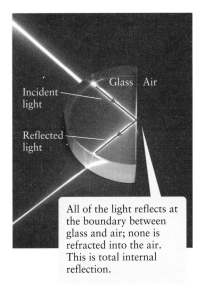

All of the light reflects at the boundary between glass and air; none is refracted into the air. This is total internal reflection.

Figure 23-9 Total internal reflection There is nothing unusual about the place where the light beam strikes the glass, yet none of the light escapes into the air on the other side.

(a) Light refracts as it passes from glass into air. The index of refraction of glass is greater than the index of refraction of air, so $\theta_2 > \theta_1$.

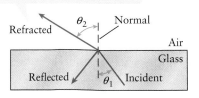

(b) As the angle θ_1 of the incident light increases, so does the angle θ_2 of the refracted light.

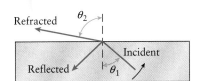

if $\theta_i > \theta_c$, then total internal reflection

(c) When the angle θ_1 of the incident light equals the critical angle θ_c, the refracted angle is $\theta_2 = 90°$. The intensity of the refracted light becomes zero.

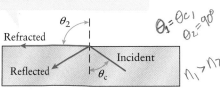

$\theta_1 = \theta_{c1}$ $\theta_2 = 90°$

$n_1 > n_2$

(d) When the angle θ_1 of the incident light is greater than the critical angle θ_c, the light is completely reflected back into the glass. This is total internal reflection.

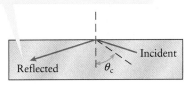

Figure 23-10 Approaching total internal reflection (a) and (b) Light approaches a boundary between glass and air at an incident angle θ_1 less than the critical angle θ_c. (c) When $\theta_1 = \theta_c$, the light ray is refracted along the surface. (d) Total internal reflection occurs for incident angles greater than the critical angle.

Total internal reflection is possible only when the first medium (in Figure 23-10, glass) has a higher index of refraction than the second medium (in Figure 23-10, air), so $n_1 > n_2$. Then the refracted angle θ_2 is greater than the incident angle θ_1, and the refracted angle can reach 90° as in Figure 23-10c. If the first medium has a lower index of refraction than the second medium, so $n_1 < n_2$, the refracted angle θ_2 is less than the incident angle θ_1, and the refracted angle can never reach 90°.

We can calculate the critical angle θ_c using Snell's law of refraction, Equation 23-6. When the incident angle θ_1 equals the critical angle θ_c, the refracted angle θ_2 equals 90°. If we substitute these into Equation 23-6 and recall that sin 90° = 1, we get

(23-9)
$$n_1 \sin \theta_c = n_2 \sin 90° = n_2 \quad \text{so} \quad \sin \theta_c = \frac{n_2}{n_1}$$

Note that the sine of an angle between 0 and 90° is between 0 and 1. If $n_1 > n_2$, the ratio n_2/n_1 is less than 1, and there will be some angle θ_c for which Equation 23-9 is satisfied. In this case total internal reflection is possible. But if $n_1 < n_2$, the ratio n_2/n_1 is greater than 1 and Equation 23-9 has no solution. This is another way of seeing that total internal reflection is possible only if $n_1 > n_2$.

If we solve Equation 23-9 for the critical angle θ_c, we get

Critical angle for total internal reflection

(23-10)

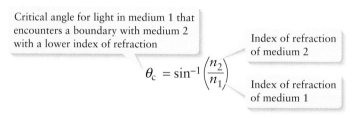

Critical angle for light in medium 1 that encounters a boundary with medium 2 with a lower index of refraction

Index of refraction of medium 2

$$\theta_c = \sin^{-1}\left(\frac{n_2}{n_1}\right)$$

Index of refraction of medium 1

 If the incident angle is greater than the critical angle, there is total internal reflection.

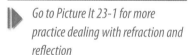

Go to Picture It 23-1 for more practice dealing with refraction and reflection

Notice that the critical angle depends on the indices of refraction of the media on *both* sides of a boundary.

Example 23-2 Critical Angles

(a) A laser is aimed from under the water toward the surface, as in Figure 23-11. Find the critical angle of the light incident in the water beyond which total internal reflection occurs.
(b) Find the critical angle if the liquid in the tank were replaced by water containing a high concentration of dissolved sugar.

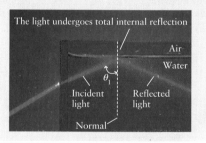

The light undergoes total internal reflection

Air
Water
θ_1
Incident light
Reflected light
Normal

Figure 23-11 Total internal reflection in water Light from a laser aimed from under the water toward the surface is totally internally reflected. What is the minimum incident angle for which total internal reflection will occur?

Set Up
Before hitting the surface (the boundary between water and air), the light is propagating in water, so this is medium 1 with index of refraction n_1. In part (a), medium 1 is ordinary water with $n_1 = 1.33$; in part (b), medium 1 is sugar water with $n_1 = 1.49$. In both parts, medium 2 on the other side of the surface is air with index of refraction $n_2 = 1.00$. We'll use Equation 23-10 to solve for the critical angle in each case.

Critical angle for total internal reflection:

$$\theta_c = \sin^{-1}\left(\frac{n_2}{n_1}\right) \tag{23-10}$$

Solve

✓ (a) Find the critical angle if medium 1 is water.

With $n_1 = 1.33$ and $n_2 = 1.00$, the critical angle is

$$\theta_c = \sin^{-1}\left(\frac{1.00}{1.33}\right) = \sin^{-1} 0.752 = 48.8°$$

If the angle of incidence θ_1 is 48.8° or greater, the light will undergo total internal reflection and no light will go into the air. (Note that θ_1 in Figure 23-11 is approximately 60°, which is indeed greater than 48.8°.)

✓ (b) Find the critical angle if medium 2 is sugar water.

With $n_1 = 1.49$ and $n_2 = 1.00$, the critical angle is

$$\theta_c = \sin^{-1}\left(\frac{1.00}{1.49}\right) = \sin^{-1} 0.671 = 42.2°$$

Reflect

Adding sugar to water results in a lower speed of light, which increases the index of refraction. The minimum angle of incidence for which total internal reflection occurs is therefore smaller for sugar water than for pure water. The smaller the critical angle, the larger the range of angles at which light experiences total internal reflection.

Total internal reflection explains what happens when you look down the highway on a hot day and see what appear to be puddles of water on the road. You might even see the reflection of cars and other objects in these "puddles," as in Figure 23-12. The explanation is that air sits in layers above the road, each layer a bit warmer, less dense, and with a lower index of refraction than the one above it. Light from the blue sky that propagates down toward the road is refracted as it encounters the boundary between one layer of air and the next. Because the index of refraction of the layer of air closer to the road is lower, the light is bent farther from the normal. The normal direction in these refractions is vertical, so the light is refracted closer to horizontal, as Figure 23-13 shows. Light that strikes the boundary between layers at a large angle with respect to the normal (a grazing angle as measured from the air layer boundary) can experience total internal reflection and reflect back up into the higher layer. Your eyes trace the light rays back along straight lines, so that the light appears to be coming from a point close to the road. What looks like a puddle is actually a refracted image of the blue sky.

Endoscopy is a medical procedure used to see inside the body. It relies on a light fiber, an optical device used to carry light and sometimes data encoded in pulses of light, from one place to another. A beam of light sent down a light fiber experiences total internal reflection at the surface of the fiber, which results in multiple reflections that keep the beam inside the fiber. The bent bar of Plexiglas in Figure 23-14 carries light in

Figure 23-12 **A mirage** What causes these shimmering patches on the road that look like puddles?

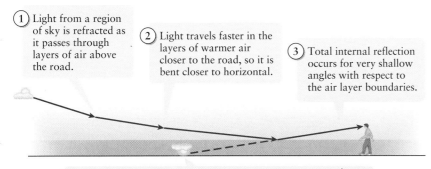

① Light from a region of sky is refracted as it passes through layers of air above the road.

② Light travels faster in the layers of warmer air closer to the road, so it is bent closer to horizontal.

③ Total internal reflection occurs for very shallow angles with respect to the air layer boundaries.

④ An image of the region of sky is formed close to the ground that gives the appearance of shimmery, blue puddles on the road.

Figure 23-13 **Explaining a mirage** The "puddles" in Figure 23-12 are an illusion caused by total internal reflection.

Figure 23-14 Light trapped by total internal reflection Light propagates through the interior of a curved bar of Plexiglas by a series of total internal reflections.

a similar way. Because light experiences total internal reflection when it encounters the surface, no light is lost as the beam is reflected back. Notice in Figure 23-14 that no light leaks out of the bar at the points where the beam hits the surface. Most light fiber is made by surrounding a central core with either one or two layers of a "cladding" made from a material of lower index of refraction than the core. The index of refraction of the core and the cladding is chosen to adjust the critical angle at the core-cladding boundary in order to optimize the light-carrying properties of the light fiber. Many endoscopes actually use bundles of several fibers. Light is sent down some fibers to illuminate the subject. Other fibers carry the image back to a camera.

As the following example shows, light fibers can also be used to deliver light energy to the interior of a patient's body in order to relieve the pain of kidney stones.

Example 23-3 Laser Lithotripsy

In a common technique used by urologists to remove kidney stones from the urinary tract, light from a powerful laser is directed down a light fiber inserted into a patient's body through a catheter. The energy of the light is absorbed by the stone, pulverizing it so that the small fragments can safely be passed by the patient. A commonly used light fiber consists of a glass-like core of diameter 2.00 mm and an index of refraction of 1.55. It is surrounded by a cladding of index of refraction 1.46. What is the minimum radius of curvature at which the fiber can be bent before light, initially traveling parallel to the axis of the fiber, will begin to leak out of the fiber (Figure 23-15)?

The ray of light colored blue strikes the surface at an angle greater than the critical angle, and experiences total internal reflection.

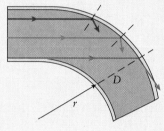

The ray of light colored green strikes the surface at an angle greater than, but closer to, the critical angle and experiences total internal reflection.

The ray of light colored red strikes the surface at an angle less than the critical angle, so it does not experience total internal reflection and leaves the light fiber.

Figure 23-15 A bent light fiber What is the minimum radius of curvature r of this light fiber to keep light from leaking out?

Set Up

Light leaks out of the fiber when it strikes the surface of the fiber at an angle less than the critical angle. As Figure 23-15 shows, leakage is more likely to happen the deeper into the bend that light strikes the surface of the fiber. So when the fiber is bent to the radius at which light just begins to leak out, the light that strikes the surface farthest into the bend hits at an incident angle just equal to the critical angle. To understand the geometry of this ray, we consider the triangle formed by it and the radius of the bend, as the figure shows. We labeled the radius r_{min} to indicate that light would leak out were the light fiber bent at a smaller radius. We'll use Equation 23-10 to calculate the critical angle.

Critical angle for total internal reflection:

$$\theta_c = \sin^{-1}\left(\frac{n_2}{n_1}\right) \qquad (23\text{-}10)$$

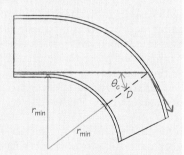

Solve

First use Equation 23-10 to find the critical angle θ_c for light in the core of the fiber reflecting off the boundary with the cladding.

The core in which the light travels has index of refraction $n_1 = 1.55$. The light reflects from the boundary with the cladding, which has index of refraction $n_2 = 1.46$. The critical angle is

$$\theta_c = \sin^{-1}\left(\frac{1.46}{1.55}\right) = \sin^{-1} 0.942 = 70.4°$$

If the incident angle at which light strikes the cladding is more than 70.4°, the light undergoes total internal reflection. If the incident angle is less than 70.4°, light will leak out of the fiber.

Use trigonometry and the calculated value of θ_c to determine the minimum radius of curvature r_{min}.

The figure above shows a right triangle with one side of length r_{min} and a hypotenuse of length $r_{min} + D$. The angle opposite the side of length r_{min} is the critical angle θ_c. The sine of this angle equals the length r_{min} divided by the length of the hypotenuse:

$$\sin \theta_c = \frac{r_{min}}{r_{min} + D}$$

Solve this equation for r_{min}:

$$(r_{min} + D) \sin \theta_c = r_{min}$$
$$r_{min} \sin \theta_c + D \sin \theta_c = r_{min}$$
$$r_{min}(1 - \sin \theta_c) = D \sin \theta_c$$
$$r_{min} = D\left(\frac{\sin \theta_c}{1 - \sin \theta_c}\right)$$

Substitute $D = 2.00$ mm and the calculated value $\theta_c = 70.4°$:

$$r_{min} = (2.00 \text{ mm})\left(\frac{\sin 70.4°}{1 - \sin 70.4°}\right) = (2.00 \text{ mm})\left(\frac{0.942}{1 - 0.942}\right)$$
$$= 32.4 \text{ mm} = 3.24 \text{ cm}$$

Reflect

Our calculation shows that such a light fiber can be bent with a radius of curvature as small as 3.25 cm before light leaks out. This radius is more than adequate to accommodate the twists and turns a fiber would make under normal circumstances in the process of being inserted into a patient.

? Got the Concept? 23-2 Total Internal Reflection

It's possible for a beam of light in Plexiglas to undergo total internal reflection at a boundary if the material on the other side of the boundary is (a) ethyl alcohol; (b) sapphire; (c) diamond; (d) more than one of these; (e) none of these.

Take-Home Message for Section 23-3

✔ When light traveling in a medium with index of refraction n_1 reaches a boundary with a second medium of index of refraction n_2, total internal reflection can happen if $n_1 > n_2$.

✔ Total internal reflection takes place only if the incident angle measured from the normal at the boundary is greater than the critical angle θ_c given by Equation 23-10. In this case, none of the light is refracted into the second medium.

23-4 The dispersion of light explains the colors from a prism or a rainbow

White light is a mixture of all the colors of the visible spectrum. You can see these colors by allowing sunlight to pass through a glass prism, as in Figure 23-16: The different colors emerge in different directions. This happens because the speed of light in a medium other than vacuum, such as the glass in a prism, is different for different frequencies of light. (In vacuum the speed is equal to c for all frequencies.) This is a result of how a light wave interacts with the atoms of a transparent material through which the wave passes. This variation of speed with frequency is called dispersion.

Since the speed v of light waves in a medium depends on the frequency, the index of refraction $n = c/v$ (Equation 23-5) depends on frequency as well. In most transparent materials the speed decreases with increasing frequency, from red to yellow to violet (see Figure 22-2): Yellow light travels slightly more slowly than red light, and blue light travels slightly more slowly than yellow light. This means that the index of refraction n increases with increasing frequency. Note that Equation 23-7 tells us that the wavelength in vacuum is inversely proportional to frequency: $\lambda_{vacuum} = c/f$. So we

Figure 23-16 **Dispersion** Light of different frequencies, and hence different colors, propagates at different speeds through the glass of which this prism is made. As a result, different colors refract along slightly different paths.

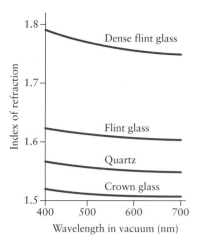

Figure 23-17 **Dispersion in different materials** The index of refraction in glass varies with the vacuum wavelength of light and with the type of glass.

can also say that the value of n decreases with increasing vacuum wavelength. (We specify *vacuum* wavelength since the wavelength in the medium depends on the value of n; see Equation 23-8.) Figure 23-17 shows how the index of refraction varies with vacuum wavelength for four different transparent materials. Note that the indices of refraction given in Table 23-1 are for yellow light, near the middle of the visible spectrum.

Snell's law of refraction, Equation 23-6, tells us that the angle at which light refracts as it crosses the boundary between two transparent media depends on their indices of refraction. So it follows that different colors of light, with different vacuum wavelengths, refract at different angles. Figure 23-18 shows this for light passing from vacuum into glass. The higher the index of refraction of the second medium, the more the refracted light is bent toward the normal. In common glass, the index of refraction is about 1.51 for red light, about 1.52 for yellow light, and about 1.53 for blue light. Because the index of refraction of short-wavelength blue light is higher than for long-wavelength red light, the blue light bends more toward the normal than does the red light. This causes the different colors of light to be spread out or dispersed. (This is the origin of the term *dispersion*.)

The same effect explains the appearance of a rainbow (Figure 23-1b). When raindrops in midair are illuminated by the Sun, sunlight enters each raindrop, is partly reflected off the back of the drop, and then exits out the front of the drop. The index of refraction of water is different for different wavelengths, so each color of light emerges in a slightly different direction to form a rainbow.

The variation of the index of refraction n over the visible spectrum characterizes the dispersion of light in that material. In glass, for example, the index of refraction varies by 0.02 from red ($n = 1.51$) to blue ($n = 1.53$), while for diamond the index of refraction varies by 0.04 from red ($n = 2.41$) to blue ($n = 2.45$). As a result, the colors of white light are spread out over a wider angle by a cut diamond than by a piece of glass cut to the same shape. This high value of dispersion contributes to the "sparkly" character of a cut diamond.

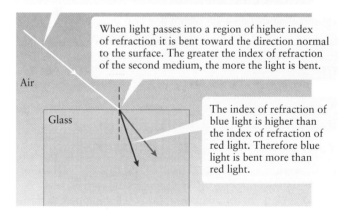

White light is composed of light across all wavelengths (colors) in the visible spectrum. The colors spread apart when light crosses the boundary between two light transmitting media.

When light passes into a region of higher index of refraction it is bent toward the direction normal to the surface. The greater the index of refraction of the second medium, the more the light is bent.

Air

Glass

The index of refraction of blue light is higher than the index of refraction of red light. Therefore blue light is bent more than red light.

Figure 23-18 **Analyzing dispersion** The difference in refracted angle between the red and blue light is greatly exaggerated for clarity.

Example 23-4 Dispersion in Dense Flint Glass

A narrow beam of white light enters a rectangular block of dense flint glass at an angle of 60.0° from the normal. The block is 0.500 m on a side. Calculate how far apart the red and blue parts of the visible spectrum will be when the light leaves the glass. The index of refraction of dense flint glass is 1.75 for red light and 1.79 for blue light.

Set Up

Both colors of light are incident on the block at the same angle: $\theta_1 = 60.0°$. However, the red and blue refract at different angles because the index of refraction is different for the two colors. We'll use Snell's law, Equation 23-6, to calculate the angles θ_{red} and θ_{blue}, then use trigonometry to find the distances d_{red} and d_{blue} shown in the figure. The difference between these distances tells us how far apart the points are where the two colors emerge from the glass.

Snell's law of refraction:

$$n_1 \sin \theta_1 = n_2 \sin \theta_2 \qquad (23\text{-}6)$$

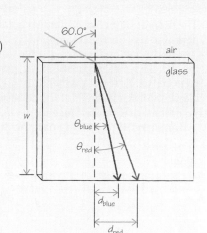

Solve

Find the refracted angles θ_{red} and θ_{blue} for the two colors.

For both colors $n_1 = n_{air} = 1.00$ and $\theta_1 = 60.0°$. For red light $n_2 = 1.75$, so Equation 23-6 becomes

$$1.00 \sin 60.0° = 1.75 \sin \theta_{red}$$

$$\sin \theta_{red} = \frac{1.00 \sin 60.0°}{1.75} = 0.4948$$

$$\theta_{red} = \sin^{-1} 0.4948 = 29.66°$$

(We've kept an extra significant digit in our result; we'll round off at the end of the calculation.) Do the same calculation for blue light, for which $n_2 = 1.79$:

$$1.00 \sin 60.0° = 1.79 \sin \theta_{blue}$$

$$\sin \theta_{blue} = \frac{1.00 \sin 60.0°}{1.79} = 0.4838$$

$$\theta_{blue} = \sin^{-1} 0.484 = 28.93°$$

Find the distances d_{red} and d_{blue} for the two colors of light, and from these find the separation between the two colors as they exit the glass.

For each color of light, the ray that extends from where the light enters the block of glass to where it exits the block forms the hypotenuse of a right triangle. The other two sides are the vertical dimension of the block, $w = 0.500$ m, and the distance d_{red} or d_{blue} that the light is displaced horizontally. In each case the side of length w is adjacent to the angle θ_{red} or θ_{blue}, and the side of length d_{red} or d_{blue} is opposite to that angle. In a right triangle, the length of the opposite side divided by the length of the adjacent side equals the tangent of the angle, so

$$\tan \theta_{red} = \frac{d_{red}}{w}$$

$$\tan \theta_{blue} = \frac{d_{blue}}{w}$$

Solve for the distances d_{red} and d_{blue} using the angles that we calculated above:

$$d_{red} = w \tan \theta_{red} = (0.500 \text{ m}) \tan 29.66° = 0.2847 \text{ m}$$
$$d_{blue} = w \tan \theta_{blue} = (0.500 \text{ m}) \tan 28.93° = 0.2764 \text{ m}$$

The distance between where the red light exits the glass and where the blue light exits the glass is

$$d_{red} - d_{blue} = 0.2847 \text{ m} - 0.2764 \text{ m} = 0.0083 \text{ m} = 8.3 \text{ mm}$$

Reflect

The separation between the two colors is fairly substantial, so this block of glass does a good job of spreading white light into its constituent colors.

In this example we used a block made of flint glass, sometimes called "lead glass" because lead is often added during the glass-making process. We invite you to repeat this calculation for crown glass, the kind of glass commonly used to make windows, for which $n_{red} = 1.51$ and $n_{blue} = 1.53$. Since crown glass has a smaller index of refraction than flint glass, and because the difference between the values of the two indices is smaller for crown glass than for flint glass, you'll find that the distance $d_{red} - d_{blue}$ is smaller for crown glass. Which type of glass would be a better choice for a prism intended to spread apart the different colors of light, as in Figure 23-16?

? Got the Concept? 23-3 **Dispersion**

A flash of white light (containing all of the colors of the visible spectrum) in air shines straight down on the surface of a pond of water. Which color of light from the flash reaches the bottom of the pond first? (a) The blue light; (b) the yellow light; (c) the red light; (d) all reach the bottom at the same time.

red → blue; index of refraction increases

Take-Home Message for Section 23-4

✔ The speed of light in a material medium depends on the frequency of the light. This is called dispersion. In most materials, the index of refraction for visible light increases from the red end of the spectrum (low frequency, long wavelength) to the blue end of the spectrum (high frequency, short wavelength).

✔ When light crosses a boundary into a medium of different index of refraction, different colors refract by different angles. This causes the colors to spread apart by an amount that depends on how strongly the index of refraction varies with wavelength.

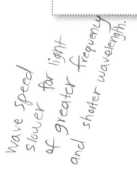

Wave speed slower for light of greater frequency and shorter wavelength.

23-5 In a polarized light wave, the electric field vector points in a specific direction

When a honeybee finds nectar, it communicates the location to other bees in the hive. In the 1940s Austrian ethologist Karl von Frisch established that if bees can see even a small patch of blue sky, they can use the position of the Sun to describe the path back to the food from the hive. (For his work on animal behavior, von Frisch received the Nobel Prize in Physiology or Medicine in 1973.) How is it that bees can know the position of the Sun even if they can't see it directly?

The explanation is that light is a transverse electromagnetic wave, with an oscillating electric field $\vec{E}$ and magnetic field $\vec{B}$ that are perpendicular to each other and to the direction of wave propagation (see Figure 22-4). The orientation of the $\vec{E}$ field is called the **polarization** of the light wave. (We don't need to separately state the orientation of $\vec{B}$, since we know that it's perpendicular to both the direction of propagation and the orientation of $\vec{E}$.) Natural light such as that emitted by the Sun or an ordinary light bulb is **unpolarized**: The orientation of the electric field changes randomly from one moment to the next. For example, if the wave is propagating in the positive x direction, at one moment $\vec{E}$ may be oriented along the y axis, a short time later it may be oriented along the z axis, a short time after that it may be oriented at 23.7° to the y axis, and so on. The reason for this is that a source of natural light emits light in the form of a stream of photons (see Section 22-4), and the orientation of $\vec{E}$ varies randomly from one photon to another.

When sunlight scatters from molecules or small particles in the atmosphere, however, the scattered light that we see has its $\vec{E}$ field oriented predominantly in one direction (Figure 23-19). (Scattering is stronger for short-wavelength light than for long-wavelength light, which is why the color of the sky is dominated by short-wavelength blue light.) Light in which the orientation of the $\vec{E}$ field changes randomly,

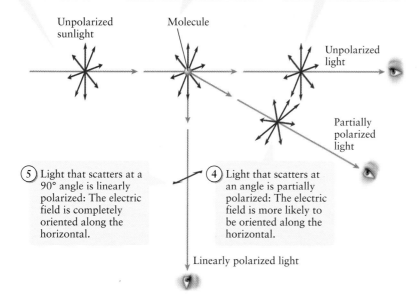

(1) The blue arrows indicate the orientation of the electric field. In unpolarized light, this orientation changes randomly.

(2) When light strikes a molecule in the atmosphere, the polarization and direction of propagation can both change.

(3) Light that scatters in the forward direction (that is, is not deflected) remains unpolarized.

Unpolarized sunlight

Molecule

Unpolarized light

Partially polarized light

(5) Light that scatters at a 90° angle is linearly polarized: The electric field is completely oriented along the horizontal.

(4) Light that scatters at an angle is partially polarized: The electric field is more likely to be oriented along the horizontal.

Linearly polarized light

Figure 23-19 **Polarization by scattering** The extent to which sunlight is polarized by scattering depends on the scattering angle.

but is more likely to be in one orientation than in other orientations, is called **partially polarized**. Light for which the $\vec{E}$ field is oriented *completely* along one direction is called **linearly polarized**. For example, the $\vec{E}$ field for the light wave shown in Figure 22-4 has only a y component, so we say this light is linearly polarized along the y axis. Bees have the ability to detect the polarization of light coming from the sky, and by using this they can infer the position of the Sun. (Human eyes, by contrast, are only weakly sensitive to polarization.)

Polarizing Light with a Polarizing Filter

A simple way to make polarized light from unpolarized light is by using a **polarizing filter.** This is a transparent sheet which contains long-chain molecules that are all oriented in the same direction. These molecules absorb light whose polarization direction is along the axis of the molecules, but have no effect on light that is polarized perpendicular to that direction. An equivalent way to think of a polarizing filter is as a series of slits that allow waves to pass that are polarized along the slits, but block waves that are polarized perpendicular to the slits (**Figure 23-20**). If we send unpolarized light into

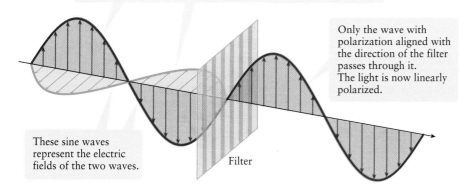

Two waves of different polarizations encounter a polarizing filter.

Only the wave with polarization aligned with the direction of the filter passes through it. The light is now linearly polarized.

These sine waves represent the electric fields of the two waves.

Filter

Figure 23-20 **A linear polarizing filter I** A filter of this kind allows only light with an electric field component aligned in a certain direction to pass through it.

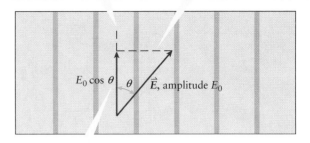

(1) The polarizing filter allows light polarized along this direction to pass. It blocks light polarized in the perpendicular direction.

(2) As a result, this component of the electric field of a light wave is blocked by the filter...

(3) ...and only this component is allowed to pass through the filter.

Figure 23-21 A linear polarizing filter II What happens to light that enters a polarizing filter with its electric field $\vec{E}$ at an angle to the polarization direction?

- The transmitted light is linearly polarized along the polarization direction of the filter.
- The electric field amplitude of the transmitted light equals the incident amplitude E_0 multiplied by cos θ.
- The intensity of the transmitted light equals the incident intensity I_0 multiplied by $\cos^2 \theta$.

the filter, we can think of the electric field $\vec{E}$ at any instant as having a component along the direction of the filter and a component perpendicular to that direction. Only the component of $\vec{E}$ along the direction of the filter will pass through, so the light that emerges from the filter will be linearly polarized.

Since a polarizing filter absorbs some of the light that falls on it, the light exiting the filter is in general less intense than the incident light that falls on it. Suppose the incident light has an oscillating electric field with magnitude E_0 that is oriented at an angle θ to the polarization direction of the filter (Figure 23-21). The component of this field that will be allowed to pass through is the component aligned with the filter direction:

$$(23\text{-}11) \qquad\qquad E_{\text{aligned}} = E_0 \cos \theta$$

The intensity of the light is proportional to the square of the electric field amplitude. The amplitude is E_0 for the incident light and (from Equation 23-11) $E_0 \cos \theta$ for the transmitted light, so the intensity I of the transmitted light equals $\cos^2\theta$ times the intensity I_0 of the incident light:

$$(23\text{-}12) \qquad\qquad I = E_0^2 \cos^2 \theta = I_0 \cos^2 \theta$$

If the incident light is unpolarized, the value of the angle θ will vary randomly between 0 and 360°. The average value of $\cos^2\theta$ over this range is $1/2$, so if unpolarized light incident on a polarizing filter has intensity I_0, the intensity of the light that emerges from the filter will be $I = I_0/2$.

Polarizing Light by Reflection

Another way to convert unpolarized light into light that is at least partially polarized is by *reflection*. When light strikes the boundary between two media, a fraction of the light is reflected and the remainder is refracted. The fraction that is reflected depends not only on the incident angle of the light, but also on the orientation of the electric field of the incident light—that is, on the polarization. In general, light polarized parallel to the boundary surface is reflected more strongly than light polarized in the perpendicular direction. If the incident light is unpolarized, that means the reflected light will be at least partially polarized in the orientation parallel to the boundary surface.

As an example, sunlight that reflects from the windshield and the hood of the car in Figure 23-22a is partially polarized parallel to the reflecting surface, so the electric field of the reflected light has a strong horizontal component and a weak vertical component. In Figure 23-22b we've placed a polarizing filter in front of the camera

(a) Photographed without a filter

(b) Photographed with a polarizing filter oriented vertically, to block light with horizontal polarization

Figure 23-22 Reducing reflections with a polarizing filter Reflected light is linearly polarized. The polarizing lenses often used for sunglasses can dramatically reduce glare from reflected light.

with its polarizing direction in the vertical direction, so light with horizontal polarization is blocked. You can see that the reflected light is greatly suppressed. Polarizing sunglasses use this same principle to minimize reflections from the road or the surface of a lake or ocean: Sunlight reflected from these horizontal surfaces is predominantly polarized in the horizontal direction, so the filters that make up the sunglass lenses are oriented to block light with a horizontal polarization. If you hold a pair of polarizing sunglasses in front of your eyes and slowly rotate them by 90° until the left lens is above the right lens, you'll see that the reflections reappear because the filter direction now aligns with the polarization direction of sunlight reflected from the ground.

For incident unpolarized light at one particular value of the angle of incidence, the reflected light is *completely* polarized in the direction parallel to the boundary between the two media. The angle of incidence θ_1 that results in this special condition is called **Brewster's angle** θ_B. When $\theta_1 = \theta_B$, it turns out that the sum of the incident angle θ_1 and the refracted angle θ_2 equals 90° (Figure 23-23):

$$\theta_1 + \theta_2 = 90° \quad \text{when } \theta_1 = \theta_B \tag{23-13}$$

We can use this to solve for the value of Bewster's angle θ_B in terms of the indices of refraction n_1 and n_2 of the two media. Snell's law of refraction, Equation 23-6, gives us this relationship between θ_1 and θ_2:

$$n_1 \sin \theta_1 = n_2 \sin \theta_2 \tag{23-6}$$

If θ_1 equals Brewster's angle θ_B, Equation 23-13 tells us that

$$\theta_2 = 90° - \theta_1 = 90° - \theta_B$$

If we substitute this into Equation 23-6, we get

$$n_1 \sin \theta_B = n_2 \sin (90° - \theta_B)$$

We know from trigonometry that $\sin (90° - \theta_B) = \cos \theta_B$. So

$$n_1 \sin \theta_B = n_2 \cos \theta_B$$

Divide both sides of this equation by n_1, then divide both sides by $\cos \theta_B$. The result is

$$\frac{\sin \theta_B}{\cos \theta_B} = \frac{n_2}{n_1}$$

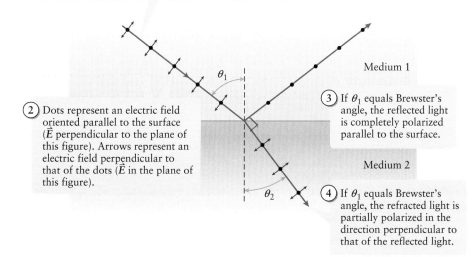

① Unpolarized light propagating in medium 1 strikes the boundary with medium 2 at an incident angle θ_1.

θ_1

Medium 1

② Dots represent an electric field oriented parallel to the surface ($\vec{E}$ perpendicular to the plane of this figure). Arrows represent an electric field perpendicular to that of the dots ($\vec{E}$ in the plane of this figure).

③ If θ_1 equals Brewster's angle, the reflected light is completely polarized parallel to the surface.

Medium 2

θ_2

④ If θ_1 equals Brewster's angle, the refracted light is partially polarized in the direction perpendicular to that of the reflected light.

Figure 23-23 **Polarization by reflection** Light reflected from a surface is partially polarized and becomes completely polarized when the incident angle equals Brewster's angle.

The sine of an angle divided by its cosine equals the tangent of the angle. So $\tan\theta_B = n_2/n_1$, or

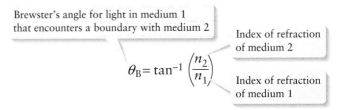

Brewster's angle for light in medium 1 that encounters a boundary with medium 2

Index of refraction of medium 2

$$\theta_B = \tan^{-1}\left(\frac{n_2}{n_1}\right)$$

Index of refraction of medium 1

Brewster's angle for polarization by reflection (23-14)

If the incident angle equals Brewster's angle, the reflected light is completely polarized in the direction parallel to the surface of the boundary.

Go to Interactive Exercise 23-2 for more practice dealing with polarization

If the angle of incidence for unpolarized light is something other than θ_B, the reflected ray is partly polarized. If unpolarized light strikes the boundary perpendicular to it, so $\theta_1 = 0$, there is no change in polarization of the reflected light.

Example 23-5 Brewster's Angle for Air to Water

At what incident angle must light strike the surface of a pond so that the reflected light is completely polarized?

Set Up

This is the situation shown in Figure 23-23, with air as medium 1 and water as medium 2. The reflected light will be completely polarized if the incident angle θ_1 is equal to Brewster's angle θ_B given by Equation 23-14. Table 23-1 tells us the value of the indices of refraction: $n_1 = n_{air} = 1.00$ and $n_2 = n_{water} = 1.33$.

Brewster's angle for polarization by reflection:

$$\theta_B = \tan^{-1}\left(\frac{n_2}{n_1}\right) \qquad (23\text{-}14)$$

Solve

Use Equation 23-14 to find Brewster's angle for this situation.

With $n_1 = 1.00$ and $n_2 = 1.33$, Brewster's angle is

$$\theta_B = \tan^{-1}\left(\frac{1.33}{1.00}\right) = 53.1°$$

The reflected light will be completely polarized if the incident angle θ_1 is equal to $\theta_B = 53.1°$.

Reflect

The boundary between the two media (air and water) is the horizontal surface of the pond, so the normal to this surface is vertical. Our result tells us that if light from the sky strikes the surface at an angle of 53.1° from the vertical, the reflected light will be completely polarized. Light striking close to this angle will be strongly polarized but not 100% polarized.

You should repeat this calculation for light coming from below the water (say, from a diver's flashlight) and striking the surface of the pond from below. Can you show that in this case, the light that reflects back into the water will be completely polarized if the incident light is at an angle of 36.9° to the vertical?

? Got the Concept? 23-4   Polarizing Filters I

Two polarizing filters, *A* and *B*, are placed one behind the other with their transmission directions perpendicular to each other. A beam of unpolarized light is directed at filter *A*. What fraction of the original light intensity will remain after the light passes through both filters *A* and *B*? (a) 1/2; (b) 1/4; (c) 1/8; (d) zero; (e) none of these.

? Got the Concept? 23-5 Polarizing Filters II

Two polarizing filters, A and B, are placed one behind the other with their transmission directions perpendicular to each other. A third polarizing filter, C, is placed in between them. The transmission direction of C is halfway between those of filters A and B (that is, at 45° to that of filter A and at 45° to that of filter B). A beam of unpolarized light is directed at filter A. What fraction of the original light intensity will remain after the light passes through filters A, B, and C? (a) 1/2; (b) 1/4; (c) 1/8; (d) zero; (e) none of these.

Take-Home Message for Section 23-5

✔ The orientation of the oscillating electric field in a light wave tells you the polarization of that wave.

✔ In natural light, the direction of the electric field changes randomly and is equally likely to be in any direction perpendicular to the propagation direction of the wave. Such light is called unpolarized.

✔ Light that has its electric field oriented in one specific direction is called linearly polarized. Unpolarized light can be polarized by scattering, by passing it through a polarizing filter, or by reflection.

23-6 Light waves reflected from the layers of a thin film can interfere with each other, producing dazzling effects

The wings of the butterfly *Morpho menelaus* (Figure 23-24) show brilliant, almost glowing colors. Yet the material of which the wings are made is colorless! The explanation for this seeming contradiction is that the colors are produced by the interference of light, a process similar to the interference of sound waves we investigated in Chapter 13.

In Section 13-5 we saw that sound waves interfere constructively or destructively depending on how the peaks and troughs of the two waves align. The same is true for light waves. Let's consider the process that occurs when light of a single wavelength strikes a layer of a transparent material. (Light that has a single definite wavelength is called **monochromatic**.) An example of a transparent layer of this sort is the thin *tapetum lucidum* ("shining carpet") that lines the back of some animals' eyes behind the retina (see Figure 23-1a). The material behind the *tapetum lucidum* is opaque (light cannot pass through it), so we can think of the *tapetum lucidum* as a thin layer against an opaque backing. This thin layer, often called a **thin film**, is of thickness D.

As Figure 23-25a shows, some of the light waves that strike the outer surface of the thin film are reflected. The remaining light enters the thin film (Figure 23-25b), then strikes the boundary between the film and the opaque backing, where it is reflected back up. (Some light is also absorbed by the backing.) As Figure 23-25c shows, the light that is reflected from the front of the thin film and the light that is reflected from the opaque backing both end up traveling upward in the same direction.

What was originally a single light wave striking a thin layer of a transparent material has been split into two waves that have traveled different paths. The light that enters the layer and reflects off the inner surface travels farther than the light reflected off the outer surface. For this reason the two light waves, which had to be in phase initially because they were part of the same wave before striking the upper surface, may not be in phase when they recombine. The interference that occurs depends on the number of wave cycles that fit into the extra distance traveled by the light that enters the thin layer. If an integer number of cycles (1, 2, 3, ... cycles) fit into that extra distance, then the two outgoing waves are in phase and constructively interfere. The surface of the film appears bright. If an odd number of half cycles (1/2, 3/2, 5/2, ... cycles) fit into the extra distance, however, the two outgoing waves are 180° out of phase and destructively interfere. (We specify an *odd* number of half cycles because an *even* number of half cycles is the same as an integer number of full cycles: for example,

Figure 23-24 **Butterfly interference** The iridescent colors on the wings of this *Morpho menelaus* butterfly result from the interference of light waves that reflect from the wing surfaces.

(a)
Some of the light that strikes the surface of a thin transparent layer is reflected.

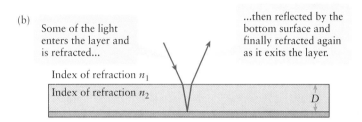

(b)
Some of the light enters the layer and is refracted...

...then reflected by the bottom surface and finally refracted again as it exits the layer.

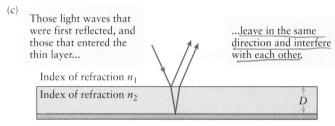

(c)
Those light waves that were first reflected, and those that entered the thin layer...

...leave in the same direction and interfere with each other.

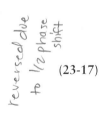

Figure 23-25 Interference from a thin film For clarity in this figure we've drawn light rays that reach the surface of a thin film at an angle. In our calculations we'll assume that the light hits the surface face-on.

4/2 = 2. This case gives constructive, not destructive, interference.) When destructive interference occurs, the two outgoing waves cancel each other out and the surface of the film appears dark.

Let's consider the case in which the incident light is normal to the surface of the thin film (that is, the light strikes the film face-on). Then the path length difference Δ_{pl} for the two waves—one that reflects off the front of the film and the other that reflects off the opaque backing—is twice the thickness of the film:

$$(23\text{-}15) \qquad \Delta_{pl} = 2D$$

Constructive interference occurs when an integer number of wavelengths of the light exactly fit into this path length difference of 2D. Destructive interference occurs when an odd number of half-wavelengths fit into the distance 2D.

However, the wavelength of light inside the film is not the same as the wavelength outside the film. Recall from Section 23-2 that the wavelength of light in a medium with index of refraction n is

$$(23\text{-}8) \qquad \lambda = \frac{\lambda_{vacuum}}{n}$$

As in Figure 23-25, let's say that the material outside the film has index of refraction n_1, and the material of which the film is made has index of refraction n_2. Then Equation 23-8 tells us that the wavelengths in the two materials are

$$\lambda_1 = \frac{\lambda_{vacuum}}{n_1} \text{ outside the film}$$

$$\lambda_2 = \frac{\lambda_{vacuum}}{n_2} \text{ inside the film}$$

Comparing these two, we see that

$$(23\text{-}16) \qquad \lambda_2 = \frac{n_1 \lambda_1}{n_2}$$

We now know everything we need to determine how the light reflected from the outer surface of the thin film interferes with the light that enters the film and is reflected from the inner surface. Constructive interference occurs when the path difference equals an integer number of wavelengths λ_2 (the wavelength of light inside the film):

reversed due to 1/2 phase shift

$$(23\text{-}17) \qquad 2D = m\lambda_2 = m\frac{n_1\lambda_1}{n_2}, \quad m = 1, 2, 3, \ldots$$

(constructive interference, thin film with an opaque backing)

The case m = 1 in Equation 23-17 tells us the minimum thickness of a thin film that results in constructive interference when light of wavelength λ_1 strikes the surface face-on:

$$(23\text{-}18) \qquad D_{min} = \frac{n_1\lambda_1}{2n_2}$$

(minimum thickness for constructive interference, thin film with an opaque backing)

Destructive interference occurs when the path difference equals an odd number of one-half the wavelength λ_2 inside the film:

reversed due to 1/2 phase shift

$$(23\text{-}19) \qquad 2D = (2m-1)\frac{\lambda_2}{2} = (2m-1)\frac{n_1\lambda_1}{2n_2}, \quad m = 1, 2, 3, \ldots$$

(destructive interference, thin film with an opaque backing)

The case $m = 1$ in Equation 23-19 tells us the minimum thickness of a thin film that results in destructive interference when light of wavelength λ_1 strikes the surface face-on:

$$D_{min} = \frac{n_1 \lambda_1}{4 n_2} \tag{23-20}$$

(minimum thickness for destructive interference,
thin film with an opaque backing)

The *tapetum lucidum* of an animal's eye appears shiny because it preferentially reflects certain wavelengths—to be specific, those that satisfy Equation 23-17, where n_2 is the index of refraction of the material of which the *tapetum lucidum* is made. This gives rise to the phenomenon of *eyeshine* that you can see in Figure 23-1a. The *tapetum lucidum* for the cat in that photo preferentially reflects green light because an integer number of wavelengths of that color fit into the path difference (twice the thickness of the *tapetum lucidum*). Other colors with other wavelengths do not satisfy that relationship perfectly, so they are not reflected as strongly. The eyes of different animals shine with different colors depending on the thickness of the *tapetum lucidum*.

The same effect explains the colors of the *Morpho* butterfly shown in Figure 23-24. The wings of *Morpho* are covered with microscopic scales that act like a thin film. The thickness of these scales is such that there is constructive interference for reflected light at wavelengths in the blue-green part of the spectrum. As a result, *Morpho* appears to glow at those wavelengths.

Go to Interactive Exercise 23-3 for more practice dealing with films

Example 23-6 A Soapy Film

Monochromatic light that has wavelength 560 nm in air strikes a layer of soapy water, which has an index of refraction of 1.40 and rests on a bathroom tile. (a) If the layer of soapy water is 700 nm thick, does constructive interference, destructive interference, or neither occur when the light strikes the surface close to the normal? (b) What is the minimum thickness of soapy water that would result in no (or minimum) reflection from the surface?

Set Up

The situation is the same as in Figure 23-25. Interference occurs between (i) light that is reflected from the top surface of the soapy layer and (ii) light that enters the soapy layer and is eventually reflected back out. In part (a) we'll compare the given values of wavelength and film thickness to Equations 23-17 and 23-19 to decide whether the interference is constructive, destructive, or something in between. In part (b) we'll use Equation 23-20 to find the minimum thickness for destructive interference. In both parts, Equation 23-16 will help us relate the wavelength of the light in air to its wavelength in the soapy water.

Constructive interference:

$$2D = m\lambda_2 = m\frac{n_1 \lambda_1}{n_2}, \quad m = 1, 2, 3, \ldots \tag{23-17}$$

Destructive interference:

$$2D = (2m - 1)\frac{\lambda_2}{2} = (2m - 1)\frac{n_1 \lambda_1}{2 n_2}, \quad m = 1, 2, 3, \ldots \tag{23-19}$$

Minimum thickness for destructive interference:

$$D_{min} = \frac{n_1 \lambda_1}{4 n_2} \tag{23-20}$$

Wavelength in two different media:

$$\lambda_2 = \frac{n_1 \lambda_1}{n_2} \tag{23-16}$$

Solve

(a) In this situation medium 1 is air ($n_1 = 1.00$) and medium 2, of which the film is made, is soapy water ($n_2 = 1.40$). The wavelength in the soapy water is therefore shorter than in air.

Wavelength in air: $\lambda_1 = 560$ nm

Wavelength in soapy water:

$$\lambda_2 = \frac{n_1 \lambda_1}{n_2} = \frac{(1.00)(560 \text{ nm})}{1.40} = 400 \text{ nm}$$

Find how many wavelengths of the wavelength λ_2 in the soapy water fit into the path length difference $\Delta_{pl} = 2D$, where $D = 700$ nm is the film thickness.

The path length difference between the waves that reflect off the top and bottom surfaces of the film is

$$\Delta_{pl} = 2D = 2(700 \text{ nm}) = 1400 \text{ nm}$$

The number of wavelengths that fit into this path length difference equals Δ_{pl} divided by λ_2:

$$\frac{\Delta_{pl}}{\lambda_2} = \frac{1400 \text{ nm}}{400 \text{ nm}} = 3.5 = \frac{7}{2}$$

The path length difference is an odd number of half wavelengths. We conclude that there is destructive interference between the light that reflects from the top surface of the soapy water and the light that reflects from the bottom surface (where the soapy water touches the tile).

(b) The minimum thickness required for destructive interference is such that the path length difference $2D$ is one half-wavelength, so $2D = \lambda_2/2$ and $D = \lambda_2/4$.

From Equation 23-20, the minimum thickness for destructive interference is

$$D_{min} = \frac{n_1\lambda_1}{4n_2} = \frac{(1.00)(560 \text{ nm})}{4(1.40)} = 100 \text{ nm}$$

Alternatively, since $\lambda_2 = n_1\lambda_1/n_2$ from Equation 23-16,

$$D_{min} = \frac{\lambda_2}{4} = \frac{400 \text{ nm}}{4} = 100 \text{ nm}$$

Reflect

When light of wavelength 560 nm in air strikes the soapy layer close to the normal, destructive interference occurs both when the layer is 100 nm thick (so twice the thickness is 200 nm, or 1/2 the wavelength in the soapy water) and when the layer is 700 nm thick (so twice the thickness is 1400 nm, or 7/2 the wavelength in the soapy water). For these thicknesses, reflections from the surface would be minimized, and the surface would look dark.

Destructive interference always occurs when twice the thickness of the layer is an odd multiple of one-half of the wavelength. Can you see that destructive interference would also occur if the film of soapy water were either 300 nm thick or 500 nm thick?

Figure 23-1c shows another example of thin-film interference. This film was made by dipping an open ring into soapy water, then holding the ring vertically. Some of the light that strikes the soap film is reflected from the front surface, while some passes into the film before being reflected at the back surface. When these two light waves recombine, the wavelength of light (the color) that results in constructive interference appears bright. The thickness of the film increases from top to bottom, so the path difference for the two waves is different at different places on the film. That's why the brightest color you see (corresponding to the wavelength for which there is constructive interference) is different at different positions. The thickness of the film is relatively constant *across* the film, however, so the colors appear in bands.

Note that the very top of the soap film in Figure 23-1c appears dark. That may come as a surprise because the top of the film is very thin (far thinner than a wavelength of visible light) so the path difference between light that reflects from the front and back surfaces should be negligible. As a result, there should be constructive interference at the top of the film, and the top should appear bright rather than dark. Why is our prediction incorrect?

To see the explanation, we need to go back to our discussion in Chapter 13 of how waves are reflected from a boundary. In Section 13-6 we saw that a wave pulse on a rope is inverted as it reflects from a fixed boundary. This inversion is equivalent to a phase shift of one-half of a wavelength; the position along the wave that arrived at the boundary as a peak has been reflected as a trough. In general, any wave is inverted when reflected, either partially or completely, from a boundary going from a material of higher wave speed to one of lower wave speed. (For the rope, the wave speed is zero on the other side of the boundary, which is definitely lower than the speed along the rope.) So when light traveling in air is partially reflected from the surface of the soap film, the reflected light wave is inverted. This does not happen, however, to the light that enters the soap film and is reflected from the back surface. At that boundary, light is moving from a medium of lower speed (the soap film) to one of higher speed (air), so no inversion occurs. Thus, even if there were no path length difference between the light wave

important

(low n to high index)

inverts

inverts

$2D = m\lambda_2$
Constructive if 0 or 2 inversions

$2D = (2m-1)\frac{\lambda_2}{2}$
Destructive for 0 or 2 inversions
Constructive for 1 inversion

An inversion is $1/2 \lambda$ difference

in phase.

$\lambda_2 = \dfrac{n_1 \lambda_1 \text{ vac}}{n_2}$

that reflects from the front surface than the one that reflects from the back surface, these two waves would be 180° out of phase because one is inverted upon reflection but the other is not. This would result in destructive interference. That's just what we see at the top of the soap film in Figure 23-1c, where the film is very thin compared to the wavelength of the light. This explains the dark band across the top of the film.

Thanks to the additional half-wavelength phase shift for a film like that in Figure 23-1c, Equation 23-17 no longer tells us the condition for constructive interference and Equation 23-19 no longer tells us the conditions for destructive interference. Instead, the roles of these equations are reversed: Equation 23-17 is now the condition for *destructive* interference, and Equation 23-19 is now the condition for *constructive* interference.

Example 23-7 Reducing the Reflection

The glass in an LCD display is sometimes given a thin coating to reduce glare. Reflection of light incident on the display, particularly light that strikes close to the normal to the plane of the display, is minimized by means of thin-film interference. Zirconium acrylate, a plastic doped with zirconium that has an index of refraction of 1.54, is often used in antireflective coatings. What is the minimum thickness of zirconium acrylate that will accomplish the desired reduction in reflection for light that has wavelength 560 nm in air? Note that the index of refraction of zirconium acrylate is higher than that of glass.

Set Up

We want the light reflected from the surface of the coating to interfere destructively with the light reflected from the zirconium acrylate-to-glass boundary. This requires that when these two light waves recombine they are shifted by an odd number of half-wavelengths. One half-wavelength shift occurs because the light in air is inverted when it reflects off the surface of the zirconium acrylate, but the light in zirconium acrylate is not inverted when it reflects off the glass. (That's because light travels more slowly in zirconium acrylate than in air, but faster in glass than in zirconium acrylate.) Any additional shift arises from the path difference of $2D$ between the two waves. For the total shift to be equivalent to an odd number of half-wavelengths, the shift due to the path difference must be an *integer* number of wavelengths. The minimum thickness corresponds to $2D$ equal to one wavelength λ_2, where λ_2 is the wavelength in the zirconium acrylate as given by Equation 23-16.

Wavelength in two different media:

$$\lambda_2 = \frac{n_1\lambda_1}{n_2} \qquad (23\text{-}16)$$

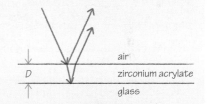

Solve

First calculate the wavelength of the light in zirconium acrylate.

We are given the wavelength in air ($n_1 = 1.00$): $\lambda_1 = 560$ nm. The wavelength in zirconium acrylate ($n_2 = 1.54$) is given by Equation 23-16:

$$\lambda_2 = \frac{n_1\lambda_1}{n_2} = \frac{(1.00)(560\text{ nm})}{1.54} = 364\text{ nm}$$

The minimum film thickness D_{min} for destructive interference is such that $2D_{min}$ equals λ_2. Use this to solve for D_{min}.

The condition for the minimum thickness that leads to destructive interference is

$$2D_{min} = \lambda_2 = 364\text{ nm}$$

$$D_{min} = \frac{364\text{ nm}}{2} = 182\text{ nm}$$

Reflect

For most people, light sensitivity peaks in the range of 555 nm to 565 nm, which we perceive as yellow. Antireflective coatings are usually optimized for yellow light (which includes the 560-nm wavelength we've used here) for that reason.

In this section we've emphasized interference due to light reflecting from a *thin* film. Why is it important to specify this? The explanation is that in this section we've assumed that a steady, continuous train of light waves is incident on the film. However, the light from ordinary sources such as light bulbs and the Sun is emitted in a sequence of short bursts, each of which is a segment of wave no more than a few micrometers to about a millimeter in length. This is called the *coherence length* of the light. The phase of the wave changes randomly from one burst to the next. If the thickness of the film is small compared to the coherence length, then the two waves that interfere—the light that reflects from the back of the film and the light that reflects from the front of the film—are part of the same burst, and so the phase relationships we have developed in this section (which assumed a steady train of waves) are valid. But if the thickness of the film is large compared to the coherence length, the two waves are likely to be from different wave bursts and will differ in phase by a random and rapidly changing amount. As a result, the interference between the waves will be neither always constructive nor always destructive, and any interference effects will be wiped out. That's why you won't see interference effects like those we've described in this section from a thick film like an ordinary pane of glass, which is several millimeters deep. (However, you *can* see interference effects for a glass pane if the light source is a laser. A laser produces light in a very different way from an ordinary light bulb, and the coherence length can be several meters.)

? Got the Concept? 23-6 Inversion on Reflection

Figure 23-26 shows light shining on a thin layer of oil that has an index of refraction of 1.4. The oil layer is atop a piece of glass that has an index of refraction of 1.5. On the underside of the glass is air. At which boundary or boundaries does the reflected light undergo an inversion? (a) The air–oil boundary; (b) the oil–glass boundary; (c) the glass–air boundary; (d) more than one of these; (e) none of these.

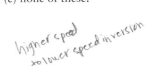

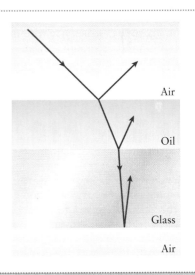

$n = 1.4$

$n = 1.5$

Air

Oil

Glass

Air

Figure 23-26 Layers of reflection At which boundary or boundaries does the reflected light undergo an inversion?

Take-Home Message for Section 23-6

✔ Light waves can interfere constructively or destructively depending on how the peaks and troughs of two waves align when they combine.

✔ When a single beam of light strikes a thin layer of a transparent material, some of the light is immediately reflected while some enters the layer and is reflected off its inner, lower surface. Because these two light rays travel a different distance, they may not be in phase when they recombine, resulting in interference. Whether there is destructive interference, constructive interference, or neither depends on the number of wave cycles that fit into the extra distance traveled by the light that enters the thin layer.

23-7 Diffraction is the spreading of light when it passes through a narrow opening

We saw in Section 23-2 that Huygens' principle—the notion that each point on a wave front itself acts as a source of waves—helped us explain the laws of reflection and refraction. Huygens' principle will also allow us to understand **diffraction,** in which waves tend to spread out when they pass through a narrow opening or near the sharp edge of an object.

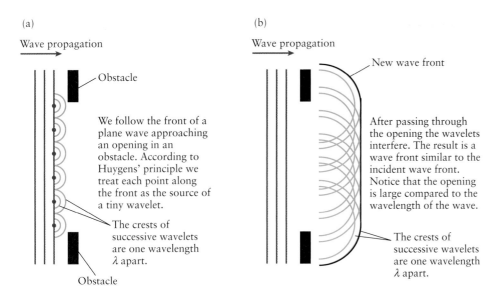

Figure 23-27 **Diffraction I: A wide opening** Huygens' principle predicts that if the width of the opening is large compared to the wavelength, most of the wave continues straight ahead through the opening. A slight amount diffracts to the sides.

Consider a wave front of a plane wave that passes through an opening in an obstacle, as in **Figure 23-27a**. The opening is wide compared to the wavelength of the wave. Many of the wavelets pass through the opening, as in **Figure 23-27b**. The resulting wave on the other side of the obstacle is mostly a new plane wave, though the ends of the wave front are curved. This means that most of the wave energy continues straight through the opening, with only a small fraction "leaking" to the sides.

Something rather different happens when the opening is comparable in size to or smaller than the wavelength. That's the situation in **Figure 23-28a**, in which the opening is much narrower than in Figure 23-27a. (It's not necessary for the opening to be exactly the same size as the wavelength, or smaller than the wavelength, or to bear any particular relationship to the wavelength other than the two being *about* the same size.) Because the opening is narrow, only a very few of the wavelets that comprise each wave front pass through it. And because so few wavelets get through, they cannot reproduce the straight wave front that was incident on the opening. After passing through the narrow opening, the wave spreads out; it has undergone diffraction (**Figure 23-28b**). Diffraction is also present in Figure 23-27b, as shown by the waves that "leak" to the side of the opening, but to a much smaller extent.

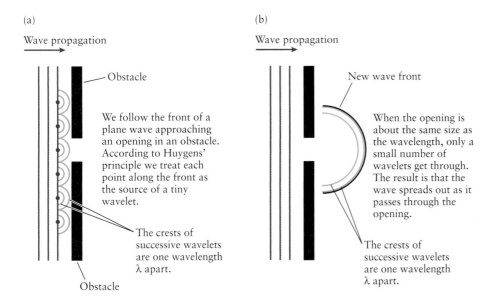

Figure 23-28 **Diffraction II: A narrow opening** If the width of the opening is comparable to the wavelength, the wave spreads out substantially after exiting the opening. (Compare Figure 23-27.)

Figure 23-29 **Water wave diffraction**
Compare these photographs to
Figures 23-27 and 23-28.

(a) Water waves pass through a wide opening

(b) Water waves pass through a narrow opening

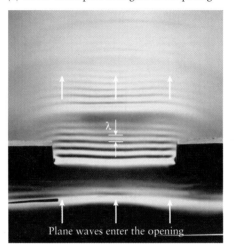

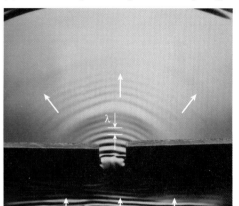

Plane waves enter the opening

Plane waves enter the opening

- The width of the opening is large compared to the wavelength λ.
- Most of the wave energy continues straight ahead, so there is very little diffraction.

- The width of the opening is comparable in size to the wavelength λ.
- The wave spreads out after passing through the opening; diffraction is important.

Figure 23-29 shows these effects for water waves passing through an opening in an obstacle. If the width of the opening is large compared to the wavelength, as in Figure 23-29a, there are almost no effects of diffraction. But if the opening is comparable in size to the wavelength as in Figure 23-29b, the wave spreads out. This explains why you hear someone talking on the other side of an open door, even if you're not directly in front of it. The wavelengths of sound used in human speech are in the range of a few meters, comparable in size to the width of a typical door (about one meter). As a result sound waves emerging from a door spread out like the water waves in Figure 23-29b, making it easy to *eavesdrop* on conversations. However, you can't *see* through an open door if you stand to one side because the wavelengths of visible light (around 550 nm, or 5.5×10^{-7} m) are very small compared to the width of the door. So light waves passing through the door behave like the water waves in Figure 23-29a.

There's much more to diffraction than how much waves spread out after passing through an opening. To consider diffraction quantitatively, imagine that we send a beam of light of a single wavelength λ through a long, narrow opening or *slit*. The slit has width w. After passing through the slit, the light falls on a screen a distance L away (Figure 23-30). You might expect that the pattern on the screen would be a single blob of light. But as the photo in Figure 23-30 shows, what actually appears is a series of bright and dark patches called **fringes**. This **diffraction pattern** arises because the light that reaches a given spot on the screen is a superposition of

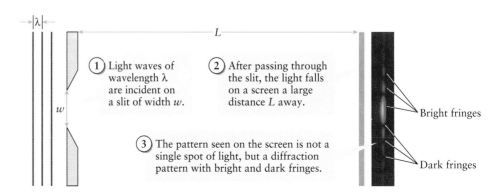

Figure 23-30 **Diffraction through a slit** A diffraction pattern arises when monochromatic light passes through a narrow slit.

① Light waves of wavelength λ are incident on a slit of width w.

② After passing through the slit, the light falls on a screen a large distance L away.

③ The pattern seen on the screen is not a single spot of light, but a diffraction pattern with bright and dark fringes.

Bright fringes

Dark fringes

Huygens wavelets from each part of the slit. At some locations, these wavelets interfere destructively; these are the locations of the dark fringes. At locations between the dark fringes, the wavelets interfere more or less constructively, so at these locations we see bright fringes.

Let's see how to determine the positions of the dark fringes. The first dark fringe is found where light waves from the upper half of the slit, of width $w/2$, and the lower half, also of width $w/2$, interfere destructively. This means that as these waves travel from the slit to the screen, the waves from one half of the slit must travel one half-wavelength farther than the waves from the other half, so the path length difference Δ_{pl} equals $\lambda/2$. We can use this to easily determine the position of the first dark fringe if we assume that the distance L to the screen is much greater than the slit width w. With this assumption, **Figure 23-31** shows that for a given position on the screen, Δ_{pl} is related to the angle θ between the normal to the slit and a line to that position:

$$\Delta_{\mathrm{pl}} = \left(\frac{w}{2}\right) \sin \theta \qquad (23\text{-}21)$$

$\sqrt{x}$ *See the Math Tutorial for more information on trigonometry*

To find the angle for the first dark fringe, substitute $\Delta_{\mathrm{pl}} = \lambda/2$ (the condition for destructive interference) into Equation 23-21:

$$\frac{\lambda}{2} = \left(\frac{w}{2}\right) \sin \theta$$

so

$$\sin \theta = \frac{\lambda}{w} \qquad (23\text{-}22)$$

(first dark fringe)

There are two of these first dark fringes, one on either side of the center of the pattern (see Figure 23-31).

To see how the second dark fringe arises, imagine breaking the slit into quarters, each of width $w/4$. Destructive interference occurs when light from the first quarter arrives at the screen out of phase with light from the second quarter, and light from the third quarter arrives at the screen out of phase with light from the fourth quarter.

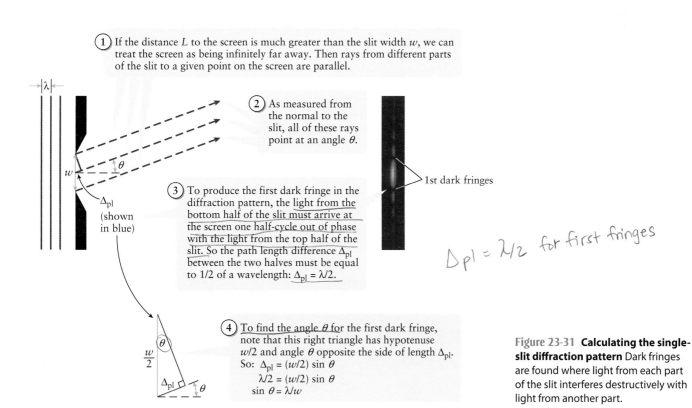

① If the distance L to the screen is much greater than the slit width w, we can treat the screen as being infinitely far away. Then rays from different parts of the slit to a given point on the screen are parallel.

② As measured from the normal to the slit, all of these rays point at an angle θ.

1st dark fringes

③ To produce the first dark fringe in the diffraction pattern, the light from the bottom half of the slit must arrive at the screen one half-cycle out of phase with the light from the top half of the slit. So the path length difference Δ_{pl} between the two halves must be equal to 1/2 of a wavelength: $\Delta_{\mathrm{pl}} = \lambda/2$.

④ To find the angle θ for the first dark fringe, note that this right triangle has hypotenuse $w/2$ and angle θ opposite the side of length Δ_{pl}.
So: $\Delta_{\mathrm{pl}} = (w/2) \sin \theta$
$\lambda/2 = (w/2) \sin \theta$
$\sin \theta = \lambda/w$

Figure 23-31 Calculating the single-slit diffraction pattern Dark fringes are found where light from each part of the slit interferes destructively with light from another part.

The path length difference Δ_{pl} between light from adjacent quarters is given by Equation 23-21 with $w/2$ replaced by $w/4$. The condition that $\Delta_{pl} = \lambda/2$ then tells us that

$$\frac{\lambda}{2} = \left(\frac{w}{4}\right)\sin\theta$$

If we solve this for $\sin\theta$, we get

(23-23)
$$\sin\theta = \frac{2\lambda}{w}$$

(second dark fringe)

This gives a larger value for $\sin\theta$, and hence for θ, than for the first dark fringe given by Equation 23-22. For the third dark fringe the factor of 2 in Equation 23-23 is replaced by a 3, for the fourth dark fringe it is replaced by a 4, and so on. In general, the angle of the mth dark fringe is given by

Angle between the normal to the slit and the location of the mth dark fringe

Number of the dark fringe: $m = 1, 2, 3, ...$

**Dark fringes in diffraction through a slit
(23-24)**

Θ from Fig 23-31

$$\sin\theta = \frac{m\lambda}{w}$$

Wavelength of the light

Width of the slit

Equations 23-22 and 23-23 are special cases of Equation 23-24, with $m = 1$ and $m = 2$ respectively.

The middle of the central *bright* fringe (at the center of the diffraction pattern in Figure 23-30 or Figure 23-31) is where all of the waves from the slit arrive in phase, since they all travel essentially the same distance to this point. This point, which corresponds to $\theta = 0$, is where the intensity is maximum. The next bright fringe, located between the first and second dark fringes, is fainter than the central bright fringe. Roughly speaking, at this bright fringe the light from the upper third of the slit interferes destructively with the light from the middle third of the slit, canceling it out. What remains is the light from the lower third of the slit. This wave has one-third the amplitude and hence $(1/3)^2 = 1/9$ the intensity of the wave that reaches the middle of the central bright fringe. (Recall from Section 22-4 that the intensity of an electromagnetic wave is proportional to the square of the amplitude of the electric field.) Successive bright fringes are even fainter. Note that the point of greatest intensity in the central bright fringe is at its center; for the other bright fringes, the point of greatest intensity is close to (but slightly displaced from) a point halfway between the adjacent dark fringes.

Diffraction of light by a slit is compelling evidence that light is a wave. If light were a stream of Newtonian particles, like miniature billiard balls, sending the light through a narrow slit would simply produce a narrower stream. The presence of bright and dark fringes like those shown in Figures 23-30 and 23-31 can only be explained by the wave nature of light.

Go to Interactive Exercise 23-4 for more practice dealing with diffraction

---

Example 23-8 Diffraction through a Slit

A green laser pointer emits light at a wavelength of 532 nm. You aim the beam from this laser at a slit 1.50 μm wide. Find the angles of the first, second, and third dark fringes in the diffraction pattern.

Set Up

We are given the wavelength λ and the slit width w, and we want to find the value of the angle θ for the $m = 1, 2$, and 3 dark fringes. We'll use Equation 23-24 for this purpose.

Dark fringes in diffraction through a slit:

$$\sin\theta = \frac{m\lambda}{w} \qquad (23\text{-}24)$$

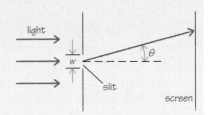

Solve

Find the angle of the first dark fringe ($m = 1$).	We have $\lambda = 532$ nm $= 5.32 \times 10^{-7}$ m and $w = 1.50$ μm $= 1.50 \times 10^{-6}$ m. From Equation 23-24 with $m = 1$,

$$\sin \theta_1 = \frac{\lambda}{w} = \frac{5.32 \times 10^{-7} \text{ m}}{1.50 \times 10^{-6} \text{ m}} = 0.355$$

$$\theta_1 = \sin^{-1} 0.355 = 20.8°$$

Find the angle of the second dark fringe ($m = 2$).	From Equation 23-24 with $m = 2$,

$$\sin \theta_2 = \frac{2\lambda}{w} = \frac{2(5.32 \times 10^{-7} \text{ m})}{1.50 \times 10^{-6} \text{ m}} = 0.709$$

$$\theta_2 = \sin^{-1} 0.709 = 45.2°$$

Find the angle of the third dark fringe ($m = 3$).	From Equation 23-24 with $m = 3$,

$$\sin \theta_3 = \frac{3\lambda}{w} = \frac{3(5.32 \times 10^{-7} \text{ m})}{1.50 \times 10^{-6} \text{ m}} = 1.06$$

This equation has *no* solution! The sine of an angle cannot be greater than 1, so there is no value of θ_3 that satisfies this equation. We are forced to conclude that the diffraction pattern in this situation has a first dark fringe at $\theta_1 = 20.8°$ and a second dark fringe at $\theta_2 = 45.2°$, but there is no third dark fringe before the end of the pattern at $\theta = 90°$.

Reflect

The number of dark fringes present in the diffraction pattern of a slit depends on the relative sizes of the wavelength λ and the slit width w. We explore this further below.

Figure 23-32 shows the intensity in the diffraction pattern of a slit as a function of the angle θ. As the width of the slit is increased from $w = \lambda$ (Figure 23-32a) to $w = 4\lambda$ (Figure 23-32b) to $w = 8\lambda$ (Figure 23-32c), the pattern becomes narrower. Note that if $w = \lambda$ as in Figure 23-32a, the first dark fringe corresponds to $\sin \theta = \lambda/w = 1$ so $\theta = \sin^{-1} 1 = 90°$; there are no dark fringes at smaller angles. The light intensity is

(a) Slit width w = wavelength λ

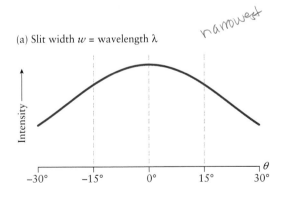

(b) Slit width w = 4 × wavelength λ

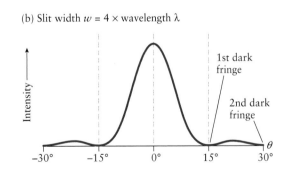

(c) Slit width w = 8 × wavelength λ

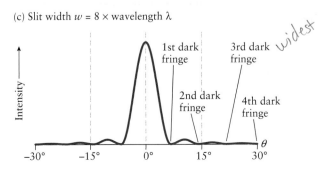

Figure 23-32 **Intensity in single-slit diffraction** The intensity in the diffraction pattern from a narrow slit depends on the relative size of the slit width w and the wavelength λ.

spread out very broadly over all angles, much like what happens to the water waves in Figure 23-29b. If the slit width is large compared to the wavelength, as in Figure 23-32c, most of the intensity goes into the central bright maximum centered at $\theta = 0$. In this case, the vast majority of the light emerging from the slit goes straight ahead or very nearly so. That's just like what happens to the water waves in Figure 23-29a, for which there is very little diffraction.

In this section we've concentrated on the diffraction that takes place when waves, including light waves, pass through a narrow opening. But diffraction can also happen when waves encounter an obstacle. One example is a sound wave coming from one side of your head. As we mentioned above, sound waves used in speech have wavelengths of a meter or more, which is large compared to the diameter of a typical human head. As a result, these sound waves are able to diffract around your head, so you can hear the sound with both ears. Because the sound wave must travel a greater distance to one ear than to the other, there will be a phase difference between the waves that the two ears detect. Your brain detects and processes this information about phase, and uses this to help determine the direction from which the sound is coming.

? Got the Concept? 23-7 Comparing Three Slits

The photographs in Figure 23-33 show the diffraction pattern that is created when red laser light passes through a narrow slit. The wavelength of the light is the same in all three photographs, but the width of the slit is different. Order the photographs from the widest slit to the narrowest one.

Figure 23-33 **Three diffraction patterns** Which pattern was produced by the widest slit? By the narrowest slit?

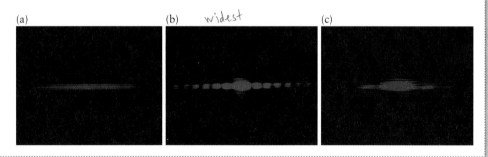

(a)   (b) *widest*   (c)

Take-Home Message for Section 23-7

✔ Waves passing through a narrow opening tend to spread out, or diffract. The diffraction is more important the smaller the size of the opening.

✔ The diffraction pattern caused by waves passing through a narrow slit has bright and dark fringes. The positions of these depend on the relative size of the wavelength and the slit width.

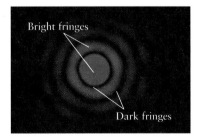

Figure 23-34 **Diffraction by a circular aperture I** Compare the diffraction pattern from a circular aperture with that for a narrow slit (Figure 23-30).

23-8 The diffraction of light through a circular aperture is important in optics

An important real-life application of diffraction is the case of light passing through a *circular* aperture. That's what happens whenever light enters the lens of a microscope, the circular mirror of an astronomical telescope, or the pupil of a human eye. Figure 23-34 shows the diffraction pattern produced by red laser light passing through a circular aperture. Like the diffraction pattern of a slit (Section 23-7), there are bright and dark fringes. But because the aperture is circular, the diffraction pattern has circular symmetry.

Figure 23-35 shows how we define the angle θ of a point in the diffraction pattern of a circular aperture of diameter D through which passes light of wavelength λ. For a narrow slit, we found that the diffraction pattern depends on the relative sizes of the wavelength λ and the slit width w; in the same way, the diffraction pattern of a circular

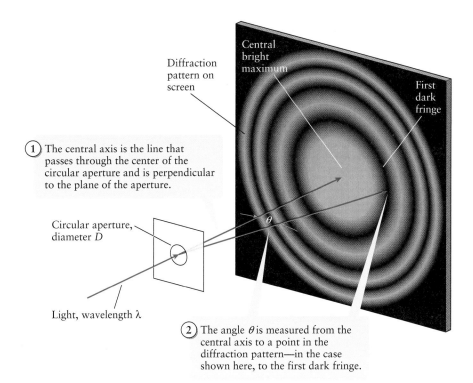

Figure 23-35 Diffraction by a circular aperture II The size of the first dark fringe—and hence the size of the central bright maximum that it surrounds—depends on the ratio of wavelength λ to aperture diameter D.

Diffraction pattern on screen

Central bright maximum

First dark fringe

(1) The central axis is the line that passes through the center of the circular aperture and is perpendicular to the plane of the aperture.

Circular aperture, diameter D

Light, wavelength λ

(2) The angle θ is measured from the central axis to a point in the diffraction pattern—in the case shown here, to the first dark fringe.

aperture depends on the relative sizes of λ and D. In particular, the location of the center of the first dark fringe is given by

$$\sin \theta = 1.22 \frac{\lambda}{D} \qquad\qquad \text{(23-25)}$$

(circular aperture, first dark fringe)

This is similar to Equation 23-22 for the first dark fringe formed by light passing through a narrow slit of width w, with w replaced by the diameter D. The factor of 1.22 results from the different geometry of a circular opening versus a rectangular one.

Figure 23-36 shows the diffraction pattern made by two pointlike objects as the objects are moved closer and closer together. In Figure 23-36a the two objects are so far apart we see only one object and its associated diffraction pattern in the field of view. In Figure 23-36b a second object has been brought close to the first; the diffraction patterns of the two objects overlap but are still distinct. In Figure 23-36c, however, the two objects are very close together. Their diffraction patterns overlap so much that it is barely possible to tell the two objects apart. We have run into the limit on our ability to **resolve**, or optically distinguish, the two objects.

(a)   (b)   (c)

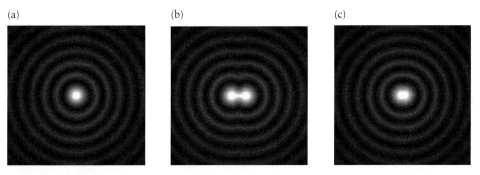

Figure 23-36 Angular resolution (a) Light from a single point source gives rise to a diffraction pattern when it passes through a circular aperture. (b) These diffraction patterns from two distant point sources partially overlap but are distinct. (c) At the limit of our ability to resolve two distant point sources, the diffraction patterns formed as their light passes through a circular aperture overlap and are barely distinguishable.

The nineteenth-century English physicist John William Strutt, 3rd Baron Rayleigh, proposed the following criterion for resolvability: Two distant, pointlike objects observed through a circular aperture can be resolved when the central maximum of one coincides with the center of the first dark fringe of the other. The angle θ_R that separates two point objects that are just barely resolved through a circular aperture, known as the **angular resolution** of the aperture, is then just the angle given by Equation 23-25:

Rayleigh's criterion for resolvability (23-26)

Angle between two pointlike objects that can barely be resolved through an optical device

Wavelength of the light

$$\sin \theta_R = 1.22 \frac{\lambda}{D}$$

Diameter of the circular aperture of the device

A physician's eye chart is just a device for measuring the value of θ_R for each of your eyes. If you have normal vision, your unaided eye can distinguish objects (such as the lines that make up the letter "E" on an eye chart) that are separated by an angle of as small as 1/60 of a degree. So $\theta_R = (1/60)°$ for a person with normal vision. The smaller the value of θ_R, the better the resolution and the smaller the details that can be resolved.

Example 23-9 The Hubble Space Telescope

The Hubble Space Telescope (HST) has a circular aperture 2.4 m in diameter. What is the theoretical angular resolution of the HST for light of wavelength 550 nm?

Set Up
Equation 23-26 gives the angular resolution, the smallest angular separation of two objects that can be resolved as a result of diffraction effects.

Rayleigh's criterion for resolvability:

$$\sin \theta_R = 1.22 \frac{\lambda}{D} \qquad (23\text{-}26)$$

Solve
We are given $\lambda = 550$ nm and $D = 2.4$ m. We calculate θ_R from these using Equation 23-26.

From Equation 23-26,

$$\sin \theta_R = 1.22 \frac{(550 \times 10^{-9}\,\text{m})}{(2.4\,\text{m})} = 2.8 \times 10^{-7}$$

$$\theta_R = \sin^{-1}(2.8 \times 10^{-7}) = 1.6 \times 10^{-5} \text{ degree}$$

Reflect
The photograph shown here is a Hubble Space Telescope image of the minor planet Pluto and its largest moon, Charon. Pluto and Charon are separated by about 25×10^{-5} degrees in this image. They are easily resolved by HST, for which $\theta_R = 1.6 \times 10^{-5}$ degrees. By contrast, a telescope of the same diameter on Earth would have a much poorer resolution of about 0.03 degrees due to the blurring effects of the atmosphere. Such a telescope would not be able to resolve Pluto and Charon. This illustrates one important rationale for orbiting telescopes, which operate high above the atmosphere.

Pluto

Charon

❓ Got the Concept? 23-8 Angular Resolution

Rank the following telescopes from best to worst angular resolution. Assume that diffraction is the only limiting factor. (a) A radio telescope with an effective diameter of 27 km that observes waves at wavelength 21 cm. (b) A telescope in orbit that has a mirror of 0.85 m diameter and that observes infrared light of wavelength 4.5 μm. (c) An inexpensive telescope for amateur astronomers with a lens of 5.0 cm diameter that observes visible light of wavelength 550 nm.

Take-Home Message for Section 23-8

✔ The angular resolution of an optical device is limited by diffraction, which blurs the images of even pointlike objects.

✔ Rayleigh's criterion states that two pointlike objects can just be resolved if the center of the bright maximum for one object coincides with the first dark fringe for the second object.

Key Terms

angular resolution
Brewster's angle
critical angle
diffraction
diffraction pattern
dispersion
fringe
front
Huygens' principle

incident
index of refraction
law of reflection
linearly polarized
monochromatic
normal
partially polarized
polarization
polarizing filter

ray
refraction
resolve
Snell's law of refraction
thin film
total internal reflection
unpolarized
wavelet

Chapter Summary

Topic	Equation or Figure
Speed of light in a medium: In a medium (transparent material) other than vacuum, the speed of light is less than c. This is described in terms of the index of refraction n of the material. The value of n is different for different materials. The wavelength also changes when light enters a medium.	Index of refraction of a medium Speed of light in vacuum $$n = \frac{c}{v}$$ (23-5) Speed of light in the medium
	Wavelength of light in a medium Speed of light in vacuum Wavelength of the light in vacuum $$\lambda = \frac{c}{nf} = \frac{\lambda_{vacuum}}{n}$$ (23-8) Frequency of the light Index of refraction of the medium

Huygens' principle, reflection, and refraction: Each point on a wave front (or wave crest) acts as a source of spherical waves called wavelets. This principle helps us explain the laws of reflection and refraction, which describe what happens when light encounters the boundary between two media. The angle of the reflected light is always equal to the angle θ_1 of the incident light, and the angle θ_2 of the refracted light is given by Snell's law.

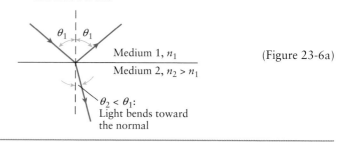

(a) Refraction from one medium into a slower one

Medium 1, n_1
Medium 2, $n_2 > n_1$

$\theta_2 < \theta_1$:
Light bends toward the normal

(Figure 23-6a)

Angle of the incident ray from the normal

Angle of the refracted ray from the normal

$$n_1 \sin \theta_1 = n_2 \sin \theta_2$$

(23-6)

Index of refraction for the medium with the incident light

Index of refraction for the medium with the refracted light

Total internal refraction: If light in one medium reaches a boundary with a second medium of lower index of refraction ($n_1 > n_2$), total internal reflection is possible. It will only take place if the incident angle is greater than the critical angle for the two media.

Critical angle for light in medium 1 that encounters a boundary with medium 2 with a lower index of refraction

Index of refraction of medium 2

$$\theta_c = \sin^{-1}\left(\frac{n_2}{n_1}\right)$$

Index of refraction of medium 1

(23-10)

 If the incident angle is greater than the critical angle, there is total internal reflection.

Dispersion: For a given material, the index of refraction depends on the frequency of the light (or, equivalently, the wavelength in vacuum). This is the reason why a prism or water droplets can break white light into its constituent colors.

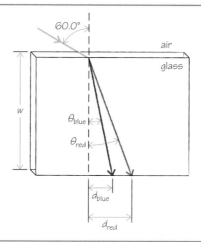

$60.0°$

air
glass

w

θ_{blue}
θ_{red}

d_{blue}
d_{red}

(Example 23-4)

Polarization: The polarization of a light wave is a description of the orientation of its electric field vector. In unpolarized light, this orientation changes randomly. In linearly polarized light, the orientation is in a fixed direction. Unpolarized light can become polarized by scattering from the atmosphere, by passing through a polarizing filter, or by reflection at Brewster's angle.

Brewster's angle for light in medium 1 that encounters a boundary with medium 2

Index of refraction of medium 2

$$\theta_B = \tan^{-1}\left(\frac{n_2}{n_1}\right)$$

Index of refraction of medium 1

(23-14)

 If the incident angle equals Brewster's angle, the reflected light is completely polarized in the direction parallel to the surface of the boundary.

Thin-film interference: If a thin, transparent film is illuminated with monochromatic light, the light that reflects from the back surface of the film interferes with light that reflects from the front surface. If the two waves emerge in phase, the interference is constructive; if they are out of phase, it is destructive. The details of the interference depend on whether there is an inversion of the wave on each reflection.

(Example 23-7)

Diffraction: When monochromatic light illuminates a narrow slit, a pattern of bright and dark fringes results. The dark fringes appear where light from any one part of the slit interferes destructively with light from some other part of the slit, so that zero net electromagnetic wave reaches the position of the dark fringe. The narrower the width of the slit, the broader the diffraction pattern.

Angle between the normal to the slit and the location of the mth dark fringe

Number of the dark fringe: $m = 1, 2, 3, \ldots$

$$\sin \theta = \frac{m\lambda}{w}$$

Wavelength of the light (23-24)

Width of the slit

Diffraction by a circular aperture: Light also diffracts when it passes through a circular aperture. This sets a fundamental limit on the angular resolution of an optical device: If two objects are closer together than an angle θ_R, it will be impossible to tell with the device whether they are two objects or a single object.

Angle between two pointlike objects that can barely be resolved through an optical device

Wavelength of the light

$$\sin \theta_R = 1.22 \frac{\lambda}{D}$$

(23-26)

Diameter of the circular aperture of the device

Answer to What do you think? Question

(b) Honeybees use the polarization of sunlight scattered by the atmosphere to determine the position of the Sun in the sky. Light is a transverse wave, with the field vectors perpendicular to the propagation direction; the polarization of the wave tells us the particular orientation of the electric field vector $\vec{E}$ in the wave. If light were a longitudinal wave like sound waves, $\vec{E}$ would always be along the propagation direction and the idea of polarization would not apply. See Section 23-5.

Answers to Got the Concept? Questions

23-1 (c) For the first refraction, medium 1 is air and medium 2 is Plexiglas. Table 23-1 shows that the index of refraction of Plexiglas (1.49) is greater than that of air (1.00), so $n_2 > n_1$. Snell's law of refraction (Equation 23-6) tells us that in this situation $\theta_2 < \theta_1$, so the angle θ_2 that the light makes to the normal in Plexiglas is less than the angle θ_1 the light makes in air. So the light bends toward the normal as it crosses the boundary from air into Plexiglas. For the second refraction, medium 1 is Plexiglas and medium 2 is water. Since water has a lower index of refraction (1.33) than Plexiglas (1.49), in this case $n_2 < n_1$ and Snell's law tells us that in this situation $\theta_2 > \theta_1$. The angle θ_2 that the light makes to the normal in water is greater than the angle θ_1 the light makes in Plexiglas, so the light bends away from the normal as it crosses the boundary from Plexiglas into water. Can

you show that the angle is 3.35° in Plexiglas and 3.76° in water?

23-2 (a) Total internal reflection at a boundary between media is possible only if the second medium has a lower index of refraction than the first medium. Here the first medium is Plexiglas, so from Table 23-1 $n_1 = 1.49$. Total internal reflection is possible for ethyl alcohol, for which $n_2 = 1.36$ is less than n_1. It is not possible for sapphire ($n_2 = 1.77$) or diamond ($n_2 = 2.42$), since both of these have an index of refraction greater than $n_1 = 1.49$.

23-3 (c) For water, like most materials, the index of refraction n is greater, and so the wave speed $v = c/n$ is slower for light of greater frequency and shorter wavelength. This means that the blue light is slowest and the red light is fastest, so the red light reaches the bottom of the pond slightly before the other colors.

23-4 (d) Once the light has passed through filter A, it will have one-half the intensity of the incident unpolarized light and will be completely polarized in the transmission direction of that filter. Filter B is set so that its transmission direction is perpendicular to that of filter A. So the angle between the polarization direction of the light that reaches filter B and the transmission direction of filter B is $\theta = 90°$. So from Equation 23-12 the intensity of the light that emerges from filter B will be the intensity of light reaching it from filter A multiplied by $\cos^2 90° = 0$. In other words, *no* light emerges from filter B.

23-5 (c) Once the light has passed through filter A, it will have one-half the intensity of the incident unpolarized light and will be completely polarized in the transmission direction of that filter. The transmission direction of filter C is at $45°$ to that of filter A, so from Equation 23-12 the light that emerges from filter C is reduced in intensity by an additional factor of $\cos^2 45° = (1/\sqrt{2})^2 = 1/2$. This light is polarized in the transmission direction of filter C. This light now enters filter B, which has a transmission direction at $45°$ to the direction of filter C. So the light that emerges from filter B is polarized in the transmission direction of filter B and has $\cos^2 45° = (1/\sqrt{2})^2 = 1/2$ the intensity of the light that reached it from filter C. So compared to the unpolarized light that was incident on filter A, the polarized light that emerges from filter B has $(1/2) \times (1/2) \times (1/2) = 1/8$ the intensity.

23-6 (d) When light is reflected from a boundary going from a material of higher wave speed to one of lower wave speed, the wave is inverted. No inversion occurs when light is reflected from a boundary where the transition is from a medium of lower speed to one of higher speed. The higher a material's index of refraction, the lower the speed of light is in that material.

Another way to state the rule is that a light wave is inverted when it is reflected from a boundary going from a material that has a lower index of refraction to one that has a higher index of refraction. So here, the speed of light is highest in air, lower in the oil, and lowest in glass. Light striking the air-to-oil boundary is therefore inverted upon reflection, as is light striking the oil-to-glass boundary. In both cases, the boundary separates a material of higher light speed from one of lower light speed. No inversion occurs at the glass-to-air boundary, however, because the speed of light in glass is lower than the speed in air.

23-7 (b), (c), (a) The narrower the slit, the more the diffracted light is spread out (see Figure 23-32). A good way to compare the three photographs is by looking at the width of the central bright maximum. The central bright maximum is widest in (a), so this is the narrowest slit. Of the other two photos, (c) has the wider central maximum, so the slit width is narrower in photo (c) than in photo (b). Hence the order from the widest to the narrowest slit is (b), (c), (a).

23-8 (b), (a), (c) The smaller the value of θ_R as given by Equation 23-26, the better the diffraction-limited angular resolution. For (a) $\lambda = 21$ cm $= 0.21$ m and $D = 27$ km $= 2.7 \times 10^4$ m, so $\sin \theta_R = 1.22\lambda/D = 9.5 \times 10^{-6}$ and $\theta_R = 5.4 \times 10^{-4}$ degrees. (Note that 27 km is not the diameter of a single telescope. To improve the angular resolution, multiple radio telescopes many kilometers apart are linked together so that they act like a single, gigantic telescope.) For (b) $\lambda = 4.5$ μm $= 4.5 \times 10^{-6}$ m and $D = 0.85$ m, so $\sin \theta_R = 6.5 \times 10^{-6}$ and $\theta_R = 3.7 \times 10^{-4}$ degrees. For (c) $\lambda = 550$ nm $= 550 \times 10^{-9}$ m and $D = 5.0$ cm $= 0.050$ m, so $\sin \theta_R = 1.3 \times 10^{-5}$ and $\theta_R = 7.7 \times 10^{-4}$ degrees. (The actual resolution would be worse due to atmospheric turbulence.) So (b) is the best, (a) is second, and (c) is worst.

Questions and Problems

In a few problems, you are given more data than you actually need; in a few other problems, you are required to supply data from your general knowledge, outside sources, or informed estimate. Interpret as significant all digits in numerical values that have trailing zeros and no decimal points. For all problems, use $g = 9.80$ m/s² for the free-fall acceleration due to gravity. Neglect friction and air resistance unless instructed to do otherwise.

• Basic, single-concept problem

•• Intermediate-level problem, may require synthesis of concepts and multiple steps

••• Challenging problem

SSM *Solution is in Student Solutions Manual*

Conceptual Questions

1. •What is Huygens' principle, and why is it necessary to understand Snell's law of refraction?

2. •Why do you expect the last color of the sunset to be on the red end of the visible spectrum?

3. •Explain why the Moon appears to change colors during a total lunar eclipse (when Earth's shadow completely blocks the light coming from the Sun). SSM

4. •Does the depth of a pool determine the critical angle that a light ray will have as it travels from the bottom of the pool and heads toward the air above the water? Explain your answer.

5. •Does the refraction of light make a swimming pool seem deeper or shallower? Explain your answer.

6. •Give two common uses of total internal reflection.

7. •In your own words, explain why the phenomenon of total internal reflection only occurs when light moves from a medium with a larger index of refraction toward a medium with a smaller index of refraction.

8. •Describe the physical interactions that take place when unpolarized light is passed through a polarizing filter. Be sure to describe the electric field of the light before and after the filter as well as the incident and transmitted intensities of the light source.

9. •Recently, researchers have created materials with a negative index of refraction. Explain what happens to the angle of refraction if light enters such a material from air. SSM

10. •Describe how polarized sunglasses work. Why do such sunglasses have *vertically* polarized lenses (as opposed to *horizontally* polarized lenses)?

11. •Give two or three examples of thin-film interference.

12. •A thin layer of gasoline floating on water appears brightly colored in sunlight. From where do the colors come?

13. •Does the phenomenon of diffraction apply to wave sources

other than light? Give an example if it does.

14. •Why was diffraction discovered in the lab after the theory that explains it was developed?

15. •Sunlight striking a diamond throws rainbows of color in every direction. From where do the colors come? SSM

16. •Linearly polarized light is incident at Brewster's angle on the surface of an optical medium. What can be said about the refracted and reflected beams if the incident beam is polarized (a) parallel to the plane of the surface and (b) perpendicular to the plane of the surface?

Multiple-Choice Questions

17. •Which kind of wave can refract when crossing from one medium to another with the wave having a different speed in the two media?
   A. electromagnetic waves
   B. sound waves
   C. water waves
   D. electromagnetic, sound, and water waves
   E. only electromagnetic and sound waves

18. •When light enters a piece of glass from air with an angle of θ with respect to the normal to the boundary surface,
   A. it bends with an angle larger than θ with respect to the normal to the boundary surface.
   B. it bends with an angle smaller than θ with respect to the normal to the boundary surface.
   C. it does not bend.
   D. it bends with an angle equal to two times θ with respect to the normal to the boundary surface.
   E. it bends with an angle equal to one-half θ with respect to the normal to the boundary surface.

19. •Which phenomenon would cause monochromatic light to enter the prism and follow along the path as shown in Figure 23-37?

Figure 23-37 Problem 19

   A. reflection
   B. refraction
   C. interference
   D. diffraction
   E. polarization

20. •Which color of light, red or blue, travels faster in crown glass?
   A. red
   B. blue
   C. their speeds are the same
   D. it depends on the material surrounding the glass
   E. blue if the glass is thin

21. •Two linear polarizing filters are placed one behind the other, so that their transmission directions are parallel to one another. A beam of unpolarized light of intensity I_0 is directed at the two filters. What fraction of the light will pass through both filters?
   A. 0
   B. $(1/2)I_0$
   C. I_0
   D. $(1/4)I_0$
   E. $2I_0$ SSM

22. •Two linear polarizing filters are placed one behind the other, so that their transmission directions form an angle of 45°. A beam of unpolarized light of intensity I_0 is directed at the two filters. What fraction of the light will pass through both?
   A. 0
   B. $(1/2)I_0$
   C. I_0
   D. $(1/4)I_0$
   E. $2I_0$

23. •A monochromatic light passes through a narrow slit and forms a diffraction pattern on a screen behind the slit. As the wavelength of the light decreases, the diffraction pattern
   A. shrinks with all the fringes getting narrower.
   B. spreads out with all the fringes getting wider.
   C. remains unchanged.
   D. spreads out with all the fringes getting alternately wider and then narrower.
   E. becomes dimmer.

24. •A monochromatic light passes through a narrow slit and forms a diffraction pattern on a screen behind the slit. As the slit width increases, the diffraction pattern
   A. shrinks with all the fringes getting narrower.
   B. spreads out with all the fringes getting wider.
   C. remains unchanged.
   D. spreads out with all the fringes getting alternately wider and then narrower.
   E. becomes dimmer.

25. •Figure 23-38 shows two single-slit diffraction patterns. The distance between the slit and the viewing screen is the same in both cases. Which of the following is true about the width of the slits, w_a and w_b?

(a)

(b)

Figure 23-38 Problem 25

   A. $w_a > w_b$
   B. $w_a < w_b$
   C. $w_a = w_b$
   D. $w_a = (1/2)w_b$
   E. $w_a = (1/4)w_b$ SSM

Estimation/Numerical Analysis

26. •Estimate the speed of light in (a) air, (b) water, and (c) glass.

27. •Estimate the range of values for the index of refraction that you will use in this class.

28. •Estimate the time required for light to travel from (a) the Moon to Earth, (b) the Sun to Earth, and (c) Earth to Alpha Centauri (the nearest star to our solar system).

29. •Estimate the distance that light travels in (a) one second, (b) one minute, and (c) one year. SSM

30. •Estimate the range of wavelengths of visible light in water.

31. •A beam of light travels from medium 1 to medium 2 in the x–y plane. Medium 1 is found in quadrants 2 and 3; medium 2 is in quadrants 1 and 4. The beam touches each of the points in the x–y plane given in the table below. Calculate the ratio of the index of refraction of medium 2 to medium 1.

x (cm)	y (cm)	x (cm)	y (cm)
−4	−2.00	+1	+0.296
−3	−1.52	+2	+0.595
−2	−1.02	+3	+0.901
−1	−0.514	+4	+1.20
0	0		

Problems

23-1 The wave nature of light explains much about how light behaves

23-2 Huygens' principle explains the reflection and refraction of light

32. •The speed of light in a newly developed plastic is 1.97×10^8 m/s. Calculate the index of refraction.

33. ••The index of refraction for a vacuum is 1.00000. The index of refraction for air is 1.00029. Determine the ratio of time required for light to travel through 1000 m of air to the time required for light to travel through 1000 m of vacuum.

34. •Calculate the speed of light for each of the following materials:
   A. ice
   B. acetone
   C. Plexiglas
   D. sodium chloride
   E. sapphire
   F. diamond
   G. water
   H. crown glass

35. ••The speed of light in methylene iodide is 1.72×10^8 m/s. The index of refraction of water is 1.33. Through what distance of methylene iodide must light travel such that the time to travel through the methylene iodide is the same as the time required for light to travel through 1000 km of water? SSM

36. ••Light travels from air toward water. If the angle that is formed by the light beam in air with respect to the normal line between the two media is 27°, calculate the angle of refraction of the light in the water.

37. •Determine the unknown angle in each of the situations in Figure 23-39.

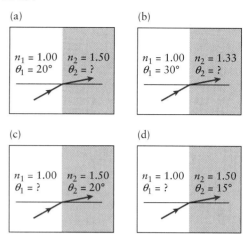

Figure 23-39 Problem 37

38. •Determine the unknown index of refraction in each of the situations in Figure 23-40.

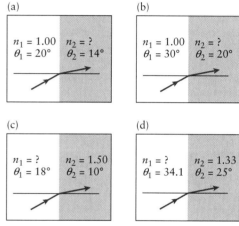

Figure 23-40 Problem 38

23-3 In some cases light undergoes total internal reflection at the boundary between media

39. •Calculate the critical angle for the following:
   A. Light travels from plastic ($n = 1.50$) to air ($n = 1.00$).
   B. Light travels from water ($n = 1.33$) to air ($n = 1.00$).
   C. Light travels from glass ($n = 1.56$) to water ($n = 1.33$).
   D. Light travels from air ($n = 1.00$) to glass ($n = 1.55$). SSM

40. •For each of the critical angles given, calculate the index of refraction for the optical materials that light travels out of toward air ($n = 1.00$):
   A. $\theta_c = 48.5°$
   B. $\theta_c = 47.0°$
   C. $\theta_c = 42.6°$
   D. $\theta_c = 35.0°$
   E. $\theta_c = 55.7°$
   F. $\theta_c = 38.5°$
   G. $\theta_c = 22.2°$
   H. $\theta_c = 75.0°$

41. •What is the critical angle for light traveling from sapphire to air?

42. ••What is the largest angle θ_1 that will ensure that light is totally internally reflected in the fiber-optic pipe made of acrylic ($n = 1.50$) shown in Figure 23-41?

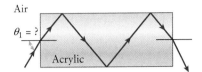

Figure 23-41 Problem 42

43. ••At what angle with respect to the vertical must a scuba diver look in order to see her friend standing on the very distant shore? Take the index of refraction of the water to be $n = 1.33$. SSM

44. ••A point source of light is 2.50 m below the surface of a pool. What is the diameter of the circle of light that a person above the water will see? Assume the water has an index of refraction of $n = 1.33$.

45. ••A block of glass that has an index of refraction of 1.55 is completely immersed in water ($n = 1.33$). What is the critical angle for light traveling from the glass to the water?

23-4 The dispersion of light explains the colors from a prism or a rainbow

46. ••For a certain optical medium, the speed of light varies from a low value of 1.90×10^8 m/s for violet light to a high value of 2.00×10^8 m/s for red light. (a) Calculate the range of the index of refraction of the material for visible light. (b) A white light is incident on the medium from air, making an angle of 30.0° with the normal. Compare the angles of refraction for violet light and red light. (c) Repeat the previous part when the incident angle is 60.0°.

47. •••A beam of light shines on an equilateral glass prism at an angle of 45° to one face (Figure 23-42a). (a) What is the angle at which the light emerges from the opposite face given that $n_{glass} = 1.57$? (b) Now consider what happens when dispersion is involved (Figure 23-42b). Assume the incident ray of light spans the spectrum of visible light between 400 nm and 700 nm (violet to red, respectively). The index of refraction for violet light in the glass prism is 1.572, and it is 1.568 for red light in the glass prism. Find the distance along the right face of the prism between the points where the red light and violet light emerge back into air. Assume the prism is 10.0 cm on a side and the incident ray hits the midpoint of the left face.

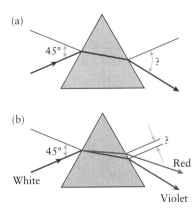

Figure 23-42 Problem 47

48. ••A light beam strikes a piece of glass with an incident angle of 45.0°. The beam contains two colors: 450 nm and an unknown wavelength. The index of refraction for the 450 nm light is 1.482. Determine the index of refraction for the unknown wavelength if the angle between the two refracted rays is 0.275°. Assume the glass is surrounded by air.

49. •••Blue light (500 nm) and yellow light (600 nm) are incident on a 12-cm-thick slab of glass as shown in Figure 23-43. In the glass, the index of refraction for the blue light is 1.545, and for the yellow light it is 1.523. What distance along the glass slab (side AB) separates the points at which the two rays emerge back into air? SSM

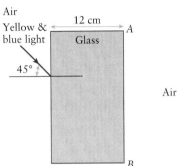

Figure 23-43 Problem 49

23-5 In a polarized light wave, the electric field vector points in a specific direction

50. •Unpolarized light is passed through an optical filter that is oriented in the vertical direction. If the incident intensity of the light is 78 W/m², what are the polarization and intensity of the light that emerges from the filter?

51. •What angle(s) does vertically polarized light make relative to a polarizing filter that diminishes the intensity of the light by 25%? SSM

52. ••Light that passes through a series of three polarizing filters emerges from the third filter horizontally polarized with an intensity of 250 W/m². If the polarization angle between the filters increases by 25° from one filter to the next, find the intensity of the incident beam of light, assuming it is initially unpolarized.

53. ••Vertically polarized light that has an intensity of 400 W/m² is incident on two polarizing filters. The first filter is oriented 30.0° from the vertical while the second filter is oriented 75.0° from the vertical. Predict the intensity and polarization of the light that emerges from the second filter.

54. •What is Brewster's angle when light in water is reflected off a glass surface? Assume $n_{water} = 1.33$ and $n_{glass} = 1.55$.

55. •The critical angle between two optical media is 60.0°. What is Brewster's angle at the same interface between the two media?

56. •(a) What would Brewster's angle be for reflections off the surface of water when the light source is beneath the surface? (b) Compare that answer to the angle for total internal reflection when light starts in water and reflects off air.

57. ••At what angle θ above the horizontal is the Sun when a person observing its rays reflected off water finds them linearly polarized along the horizontal (Figure 23-44)? SSM

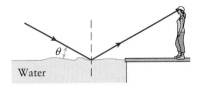

Figure 23-44 Problem 57

23-6 Light waves reflected from the layers of a thin film can interfere with each other, producing dazzling effects

58. ••A ray of light is reflected from a thin film back into air. If the film is actually a coating on a slab of glass ($n_{film} < n_{glass}$), describe the phase changes that the reflected ray undergoes (a) as it reflects off the front surface of the film and (b) as it reflects off the back surface of the film (the film–glass interface).

59. •What is the wavelength of red light (700 nm in air) when it is inside a glass slab with $n = 1.55$?

60. ••When white light illuminates a thin film with normal incidence, it strongly reflects both indigo light (450 nm in air) and yellow light (600 nm in air) (Figure 23-45). Calculate the minimum thickness of the film if it has an index of refraction of 1.28 and it sits atop a slab of glass that has $n = 1.50$.

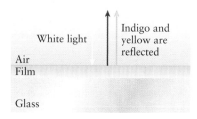

Figure 23-45 Problem 60

61. ••When white light illuminates a thin film normal to the surface, it strongly reflects both blue light (500 nm in air) and red light (700 nm in air) (Figure 23-46). Calculate the minimum thickness of the film if it has an index of refraction of 1.35 and it "floats" on water with $n = 1.33$. SSM

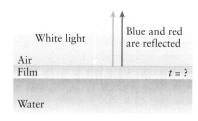

Figure 23-46 Problem 61

62. ••A soap bubble is suspended in air. If the thickness of the soap is 625 nm and both blue light (500 nm in air) and red light (700 nm in air) are *not* observed to reflect from the soap film, what is the index of refraction of the thin film?

63. ••A thin film of cooking oil ($n = 1.38$) is spread on a puddle of water ($n = 1.33$). What are the minimum and the next three thicknesses of the oil that will strongly reflect blue light having a wavelength in air of 518 nm?

64. ••What is the minimum thickness of a nonreflective coating of magnesium chloride ($n = 1.39$) so that no light centered around 550 nm in air will reflect back off a glass lens ($n = 1.56$)?

65. ••Water ($n = 1.33$) in a shallow pan is covered with a thin film of oil that is 450 nm thick and has an index of refraction of 1.45. What visible wavelengths will *not* be present in the reflected light when the pan is illuminated with white light and viewed from straight above? SSM

23-7 Diffraction is the spreading of light when it passes through a narrow opening

66. •Light that has a wavelength of 550 nm is incident on a single slit that is 10.0 μm wide. Determine the angular location of the first three dark fringes that are formed on a screen behind the slit.

67. •Light that has a wavelength of 475 nm is incident on a single slit that is 800 nm wide. Calculate the angular location of the first three dark fringes that are formed on a screen behind the slit.

68. •What is the highest order dark fringe that is found in the diffraction pattern for light that has a wavelength of 633 nm and is incident on a single slit that is 1500 nm wide?

69. •The highest order dark fringe found in a diffraction pattern is 6. Determine the wavelength of light that is used with the single slit that has a width of 3500 nm.

70. •When blue light ($\lambda = 500$ nm) is incident on a single slit, the central bright spot has a width of 8.75 cm. If the screen is 3.55 m distant from the slit, calculate the slit width.

71. ••A He–Ne laser illuminates a narrow, single slit that is 1850 nm wide. The first dark fringe is found at an angle of 20.0° from the central peak. Determine the wavelength of the light from the laser. SSM

72. ••Yellow light that has a wavelength of 625 nm produces a central maximum peak that is 24.0 cm wide on a screen that is 1.58 m from a single slit. Calculate the width of the slit.

23-8 The diffraction of light through a circular aperture is important in optics

73. •Light from a helium–neon laser with a wavelength of 633 nm passes through a 0.180-mm-diameter hole and forms a diffraction pattern on a screen 2.0 m behind the hole. Calculate the diameter of the central maximum.

74. •**Biology** The average pupil is 5.0 mm in diameter, and the average normal-sighted human eye is most sensitive at a wavelength of 555 nm. What is the eye's angular resolution in radians?

75. •**Astronomy** The telescope at Mount Palomar has an objective mirror that has a diameter of 508 cm. What is the angular limit of resolution for 560-nm light in degrees and radians?

76. •**Astronomy** The Hubble Space Telescope has a diameter of 2.4 m. What is the angular limit of resolution due to diffraction when a wavelength of 540 nm is viewed?

77. •The distance from the center of a circular diffraction pattern to the first dark ring is 15,000 wavelengths on a screen that is 0.85 m away. What is the size of the aperture? SSM

78. ••**Biology** Assume your eye has an aperture diameter of 3.00 mm at night when bright headlights are pointed at it. At what distance can you see two headlights separated by 1.50 m as distinct? Assume a wavelength of 550 nm, near the middle of the visible spectrum.

General Problems

79. ••One way of describing the speed of light in an optical material is to specify the ratio of the time that is required for light to travel through a vacuum to the time required for light to travel through the same length of the optical material. For example, if light travels through a material in 150% of the time for light to travel through a vacuum, the speed of light in the material would be $2/3 = 1/1.5$ that in a vacuum. Complete the table by giving the speed of light and the index of refraction

for each of the following optical materials, listed with the corresponding percentage.

Optical material with percentage of time required for light to pass through compared to an equal length of vacuum	Speed of light	Index of refraction
100%		
125%		
150%		
200%		
500%		
1000%		

80. ••(a) Determine the index of refraction for medium 2 if the distance between points B and C in Figure 23-47 is 0.75 cm. Assume the index of refraction in medium 1 is 1.00. (b) Suppose $n_2 = 1.55$. Calculate the distance between points B and C.

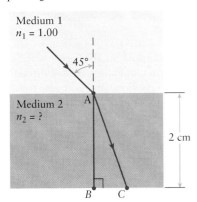

Figure 23-47 Problem 80

81. ••Prove that in the case where there are more than two optically different media sandwiched together, with air on the left and air on the right, the angle at which light returns to air is independent of the indices of refraction of the interior media (Figure 23-48). In other words, the refraction angle, θ_n, is *only* dependent on n_1, θ_1, and n_i (not $n_2, n_3, n_4, \ldots$).

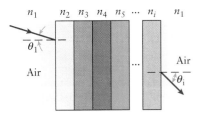

Figure 23-48 Problem 81

82. ••One of the world's largest aquaria is the Monterey Bay Aquarium. The viewing wall is made of acrylic and is 0.33 m thick, and the tank holds 1.2 million gallons of water (Figure 23-49). If a ray of light is directed into the plastic from air at an angle of 40°, calculate the angle that the ray will make (a) when it enters the plastic and (b) when it enters the seawater. The indices of refraction for air, acrylic, and seawater are listed on the figure.

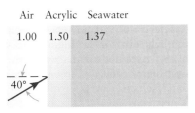

Figure 23-49 Problem 82

83. ••A flat glass surface ($n = 1.54$) has a layer of water ($n = 1.33$) of uniform thickness directly above the glass. At what minimum angle of incidence must light in the glass strike the glass–water interface for the light to be totally internally reflected at the water–air interface? SSM

84. •••Light rays fall normally on the vertical surface of a glass prism ($n = 1.55$) as shown in Figure 23-50. (a) What is the largest value of θ such that the ray is totally internally reflected at the slanted face? (b) Repeat the calculation if the prism is immersed in water with $n = 1.33$.

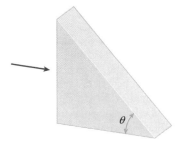

Figure 23-50 Problem 84

85. ••The object in Figure 23-51 is a depth $d = 0.85$ m below the surface of clear water. (a) How far from the end of the dock, distance D in the figure, must the object be if it cannot be seen from any point on the end of the dock? The index of refraction of water is 1.33. (b) If you could change the index of refraction of the water, how would you change it so that the object could be seen at any distance from the dock?

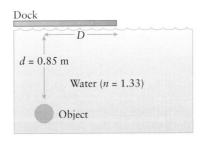

Figure 23-51 Problem 85

86. ••The polarizing angle for light that passes from water ($n = 1.33$) into a certain plastic is 61.4°. What is the critical angle for total internal reflection of the light passing from the plastic into air?

87. •••A baseball is hit into a round pool of water that is 4.00 m deep and 17.0 m across (Figure 23-52). It lands right in the center of the pool. A large round raft shaped like a lily pad is floating in the pool, concentrically on top of the location of the ball. What minimum diameter must the raft have in order to completely obscure the ball from sight? Assume that water has an index of refraction of 1.33.

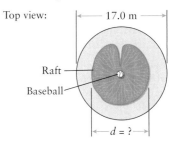

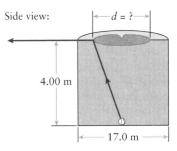

Figure 23-52 Problem 87

88. ••A glass lens that has an index of refraction equal to 1.57 is coated with a thin layer of transparent material that has an index of refraction equal to 2.10. If white light strikes the lens at near-normal incidence, light of wavelengths 495 nm in air and 660 nm in air are absent from the reflected light. What is the thinnest possible layer of material for which this can be accomplished?

89. ••Unpolarized light of intensity 100 W/m² is incident on two ideal polarizing sheets that are placed with their transmission axes perpendicular to each other. An additional polarizing sheet is then placed between the two, with its transmission axis oriented at 30° to that of the first. (a) What is the intensity of the light passing through the stack of polarizing sheets? (b) What orientation of the middle sheet enables the three-sheet combination to transmit the greatest amount of light? SSM

90. ••Unpolarized light that has an intensity of 850 W/m² is incident on a series of polarizing filters as shown in Figure 23-53. If the intensity of the light after the final filter is 75 W/m², what is the orientation of the second filter relative to the x axis? *Hint:* cos(90 − θ) = sin θ and 2 sin θ cos θ = sin 2θ.

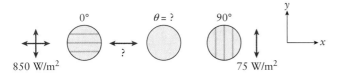

Figure 23-53 Problem 90

91. ••A thin film of soap solution (n = 1.33) has air on either side and is illuminated normally with white light. Interference minima are visible in the reflected light only at wavelengths of 400, 480, and 600 nm in air. What is the minimum thickness of the film?

92. ••A brass sheet has a thin slit scratched in it. At room temperature (22.0 °C) a laser beam is shined on the slit and you observe that the first dark diffraction spot occurs at ± 25.0° on either side of the central maximum. The brass sheet is then immersed in liquid nitrogen at 77.0 K until it reaches the same temperature as the nitrogen. It is removed and the same laser is shined on the slit. At what angle will the first dark spot now occur? Consult Table 14-1 for the thermal properties of brass.

93. ••A wedge-shaped air film is made by placing a small slip of paper between the edges of two thin plates of glass 12.5 cm long. Light of wavelength 600 nm in air is incident normally on the glass plates. If interference fringes with a spacing of 0.200 mm are observed along the plate, how thick is the paper? This form of interferometry is a very practical way of measuring small thicknesses. SSM

94. ••A thin layer of SiO, having an index of refraction of 1.45, is used as a coating on certain solar cells. The refractive index of the cell itself is 3.5. (a) What is the minimum thickness of the coating needed to cancel visible light of wavelength 400 nm in the light reflected from the top of the coating in air? Are any other visible wavelengths also canceled? (b) Suppose that technological limitations require you to make the coating 3.0 times as thick as in part (a). Which, if any, visible wavelengths in the reflected light will be canceled in air and which, if any, will be reinforced?

95. ••You want to coat a pane of glass that has an index of refraction of 1.54 with a 155-nm-thick layer of material that has an index of refraction greater than that of the glass. The purpose of the coating is to cancel light reflected off the top of the film having a wavelength (in air) of 550 nm. The coating material is very expensive, so you want to use the thinnest possible layer. (a) What should be the index of refraction of the coating? (b) If, due to technological difficulties, you cannot achieve a uniform coating at the desired thickness, what are the next three thicknesses of the coating you could use?

96. ••Two point sources of light each of which has a wavelength of 500 nm are photographed from a distance of 100 m using a camera with a 50.0-mm focal length lens. The camera aperture is 1.05 cm in diameter. What is the minimum separation of the two sources if they are to be resolved in the photograph, assuming the resolution is diffraction limited?

97. ••In 2009, researchers reported on evidence that a giant tsunami had hit the eastern coast of the Mediterranean Sea (present-day Lebanon and Israel) around 1600 BCE, causing huge damage to the civilizations located there. It is believed that the tsunami was caused by the eruption of the Thera volcano near the island of Crete. The waves would have passed through the 100-mile-wide opening between Crete and Rhodes, which would cause them to diffract and spread out. Satellite observations of tsunamis show that the waves measure about 250 mi from a crest to the adjacent trough, and the time between successive crests is typically 60 min. (a) How fast do tsunami waves travel? (b) How long after the eruption of Thera would the tsunami reach the eastern shore of the Mediterranean Sea, 600 mi from Thera? (c) For these waves, could we apply the formula w sin θ = mλ to find the angles at which the waves cancel after passing through the 100-mile "slit" between Crete and Rhodes? Explain why or why not. SSM

98. ••**Biology** The pupil (the opening through which light enters the lens) of a house cat's eye is round under low light but ciliary muscles narrow it to a thin vertical slit in very bright light. Assume that bright light of wavelength 550 nm in air is entering the eye perpendicular to the lens and that the pupil has narrowed to a slit that is 0.500 mm wide. What are the three smallest angles on either side of the central maximum at which no light will reach the cat's retina (a) if we imagine the eye is filled with air, and (b) if we take into consideration that in reality the eye is filled with a fluid having index of refraction of approximately 1.4? (c) Would the cat be aware of the pattern of alternating dark and light fringes? Why or why not?

99. ••If you peek through a 0.75-mm-diameter hole at an eye chart, you will notice a decrease in visual acuity. Calculate the angular limit of resolution if the wavelength is taken as 575 nm. Compare the result to a 4-mm pupil of the eye that has an angular resolution of 1.75×10^{-4} rad.

100. ••**Biology** The pupil of the eye is the circular opening through which light enters. Its diameter can vary from about 2.0 mm to about 8.0 mm to control the intensity of the light reaching the retina. (a) Calculate the angular resolution, θ_R, of the eye for light that has a wavelength of 550 nm in both bright light and dim light. In which light can you see more sharply, dim or bright? (b) You probably have noticed that when you squint, objects that were a bit blurry suddenly become somewhat clearer. In light of your results in part (a), explain why squinting helps you see an object more clearly.

101. ••**Biology** Under bright light, the pupil of the eye (the circular opening through which light enters) is typically 2.0 mm in diameter. The diameter of the eye is about 25 mm. Suppose you are viewing something with light of wavelength 500 nm. Ignore the effect of the lens and the vitreous humor in the eye. (a) At what angles (in radians and degrees) will the first three diffraction dark rings occur on either side of the central bright spot on the retina at the back of the eye? (b) Approximately how far (in millimeters) from the central bright spot would the dark rings in part (a) occur? (c) Explain why we do not actually observe such diffraction effects in our vision.

102. •••**Astronomy** Sometime around 2022, astronomers at the European Southern Observatory hope to begin using the E-ELT (European Extremely Large Telescope), which is planned to have a primary mirror 42 m in diameter. Let us assume that the light it focuses has a wavelength of 550 nm. (a) What is the most distant Jupiter-sized planet the telescope could resolve, assuming it operates at the diffraction limit? Express your answer in meters and light-years. (b) The nearest known exoplanets (planets beyond the solar system) are around 20 light-years away. What would have to be the minimum diameter of an optical telescope to resolve a Jupiter-sized planet at that distance using light of wavelength 550 nm? (1 light-year $= 9.461 \times 10^{15}$ m)

103. ••**Astronomy** Under the best atmospheric conditions at the premium site for land-based observing (Mauna Kea, Hawaii, elevation ~4.27 km), an optical telescope can resolve celestial objects that are separated by one-fourth of a second of arc (arcsec). The viewing never gets any better than this because of atmospheric turbulence, which makes the images jitter.

(a) What minimum diameter aperture is necessary to provide arcsec resolution due to diffraction? (b) Is there ever any point in building a telescope much bigger than this? Explain your answer. SSM

104. ••**Astronomy** The Herschel infrared telescope, launched in 2009, made observations from 2010 to 2013. Its primary mirror is 3.5 m in diameter, and the telescope focuses infrared light in the range of 55 μm to 672 μm. (a) What wavelength in its observing range will give the maximum angular resolution? What is that maximum resolution (in radians and seconds, $1° = 60'$ and $1' = 60'$)? (b) To achieve the same resolution as in part (a) using visible light of wavelength 550 nm, what should be the mirror diameter of an optical telescope? (c) What is the smallest infrared source that the Herschel infrared telescope can resolve at a distance of 150 light-years? (A light year is the distance that light travels in one year—about 9.461×10^{15} m.)

105. ••**Astronomy** The world's largest refracting telescope is at Yerkes Observatory in Williams Bay, Wisconsin. Its objective is 1.02 m in diameter. Suppose you could mount the telescope on a spy satellite 200 km above the ground. (a) Assuming that the resolution is diffraction limited, what minimum separation of two objects on the ground could it resolve? Take 550 nm as a representative wavelength for visible light. (b) Because of atmospheric turbulence, objects on the surface of Earth can be distinguished only if their angular separation is at least 1.00 arcsec. How far apart would two objects on Earth's surface be if they subtended an angle of 1.00 arcsec as measured from the satellite? Compare this with your answer to part (a).

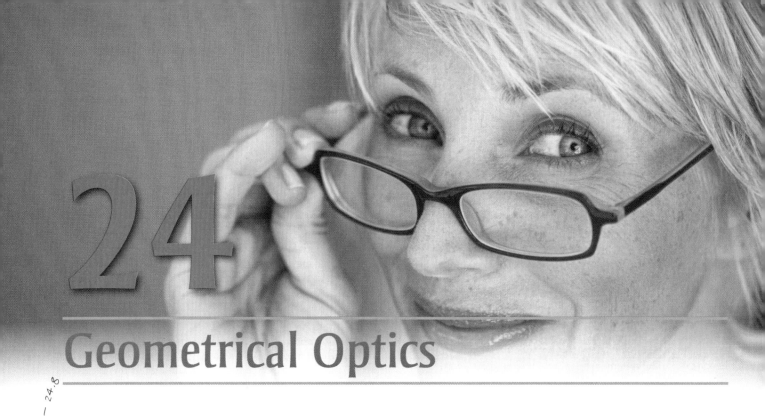

24

Geometrical Optics

What Do You Think?

Many older adults can see distant objects clearly but must wear corrective eyeglasses to see nearby objects (for example, to read a book). Are the lenses of these eyeglasses (a) thicker at the middle, (b) thicker at the edges, or (c) of uniform thickness?

In this chapter, your goals are to:

- (24-1) Explain the importance of optical devices.
- (24-2) Describe how a plane mirror forms an image.
- (24-3) Use ray diagrams to explain how the image formed by a concave mirror depends on the position of the object.
- (24-4) Calculate the position and height of an image made by a concave mirror.
- (24-5) Explain the differences between the images made by a convex mirror and a concave mirror.
- (24-6) Calculate the position and height of an image made by a concave mirror.
- (24-7) Describe how the curved surfaces of a lens make light rays converge or diverge.
- (24-8) Calculate the focal length of a lens based on its composition and shape.

To master this chapter, you should review:

- (3-8) Calculating the length of a circular arc.
- (23-2) The law of reflection and Snell's law of refraction.

24-1 Mirrors or lenses can be used to form images

You use optical devices every time you see something. The simplest optical device is a piece of reflective material (a *mirror*) or transparent material (a *lens*) shaped so that it changes the direction of light in a regular way. Mirrors and lenses can be used to change the apparent size of objects (Figure 24-1a); the lens of your eye is an essential part of vision and needs to be replaced if it becomes clouded with age (Figure 24-1b).

To understand the physics of such optical devices, all we need are the laws of reflection and of refraction that we learned in Chapter 23 along with a little bit of geometry. We won't need to refer at all to the wave properties of light. (Indeed, much of the basic physics of mirrors and lenses was deduced before it was understood that light is a wave.) For this reason we use the term **geometrical optics** to refer to the science of mirrors and lenses.

We'll begin by considering a simple plane mirror. We'll see how the law of reflection explains the kind of image that it forms. We'll then use similar ideas to understand

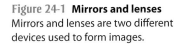

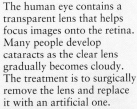

(a)

This dentist uses a curved mirror to make a magnified image of a patient's tooth. He can also get a magnified view by using the lenses attached to his eyeglasses.

(b)

The human eye contains a transparent lens that helps focus images onto the retina. Many people develop cataracts as the clear lens gradually becomes cloudy. The treatment is to surgically remove the lens and replace it with an artificial one.

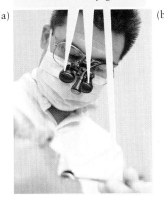
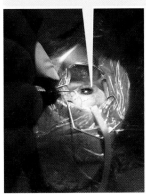

Figure 24-1 Mirrors and lenses Mirrors and lenses are two different devices used to form images.

Take-Home Message for Section 24-1

✔ A mirror forms images by the reflection of light from the mirror's surface.

✔ A transparent lens forms images by the refraction of light as it enters and exits the lens.

---

the kind of images formed by a mirror with a surface that's curved either inward or outward. In contrast to mirrors, lenses use the refraction of light to form an image: Light rays can change direction when they enter a lens and again when they exit the lens. Happily, we'll find that many of the same ideas that we'll develop by considering curved mirrors apply equally well to lenses.

24-2 A plane mirror produces an image that is reversed back to front

Our visual system can detect only objects that emit or reflect light. (We can't see in the dark!) Although it's easy to find examples of luminous things, such as the Sun, a light bulb, or the screen of a mobile device, most of what we see only reflects light. As we learned in Section 23-2, a ray of light that strikes a surface always reflects in such a way that the angle of incidence equals the angle of reflection. (This is the law of reflection.) However, reflected light is usually **diffuse** because it reflects from an object's surface in many random directions. This is because most surfaces are uneven (**Figure 24-2a**).

If an object has a flat surface, however, light rays that strike that surface are all reflected in the same general direction (**Figure 24-2b**). Such a flat, reflecting surface is called a **plane mirror,** and reflections from such a surface are called specular (from the Latin word for mirror, *speculum*). Your reflection in a bathroom mirror is an example of **specular reflection.** Let's look more closely at how a plane mirror creates an image of an object.

Figure 24-3 shows how your eye and brain interpret light coming from an **object** (a term that refers to anything that acts as a source of light rays for an optical device). Some of the light rays coming from the object go to your eyes. Your brain traces those rays backward to a common origin and interprets that point as the location of the object. **Figure 24-4** shows a similar situation, except that we have added a plane mirror. Some of the light from the object strikes the mirror and is reflected toward your eye. These rays appear to be coming from a point behind the mirror, so your brain interprets the light as coming from that point. We say that the mirror has formed an **image** of the object, and this image lies behind the mirror.

As you can see in Figure 24-4, no light rays actually pass through the location of the image. For this reason, the image formed by a plane mirror is said to be a **virtual image.** We will shortly encounter some optical devices that cause light rays to bend toward each other, so that the image forms where light rays do actually meet. An image formed by light rays coming together is called a **real image.**

(a)

Light that reflects from an object's surface in many random directions is called diffuse. Uneven surfaces produce diffuse light even when the incident rays come from the same direction.

(b)

Light that reflects from a smooth surface is called specular. Light rays coming from the same general direction all reflect in the same general direction.

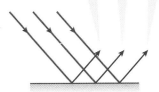

Figure 24-2 Diffuse and specular reflection Light reflecting from (a) an uneven surface and (b) a smooth surface.

Figure 24-3 Locating an object
When light strikes our eyes, we trace the rays of light back along straight lines to an apparent common source.

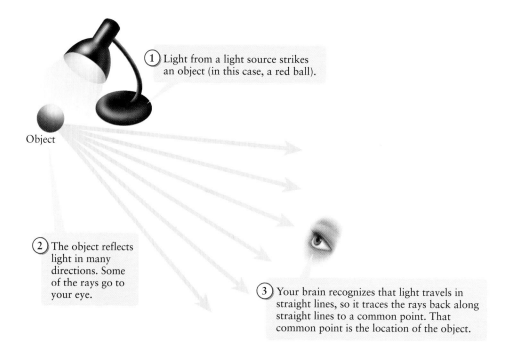

① Light from a light source strikes an object (in this case, a red ball).

Object

② The object reflects light in many directions. Some of the rays go to your eye.

③ Your brain recognizes that light travels in straight lines, so it traces the rays back along straight lines to a common point. That common point is the location of the object.

We can determine the position of the image made by a plane mirror by using a **ray diagram** like that in Figure 24-5. In such a diagram, we draw a few light rays coming from the object and show how they reflect from the mirror. We've drawn the object as a red arrow of height h_O (the **object height**) located a distance d_O (the **object distance**) from the mirror. To determine the position of the image, we must draw at least two rays that emanate from the tip of the object arrow. (We've drawn three in Figure 24-5.) When each ray strikes the mirror, it obeys the law of reflection: The angle of the reflected ray equals the angle of the incident ray. Note that the horizontal ray, which strikes the mirror face-on, is reflected back in a horizontal direction toward the tip of the object arrow.

The reflected rays diverge from each other and never actually meet. But if we trace these rays back to where they would meet, as shown by the dashed lines in Figure 24-5,

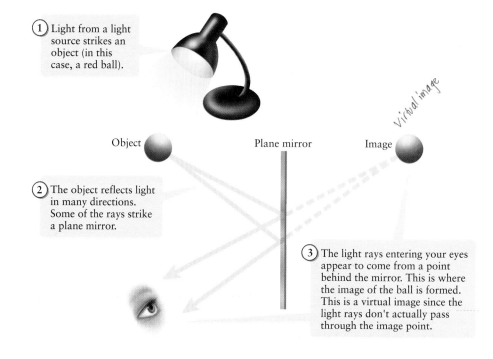

① Light from a light source strikes an object (in this case, a red ball).

Object   Plane mirror   Image

Virtual image

② The object reflects light in many directions. Some of the rays strike a plane mirror.

③ The light rays entering your eyes appear to come from a point behind the mirror. This is where the image of the ball is formed. This is a virtual image since the light rays don't actually pass through the image point.

Figure 24-4 Locating an image
When light from a mirror strikes our eyes, we use the same technique as in Figure 24-3 to determine the position of the image made by the mirror.

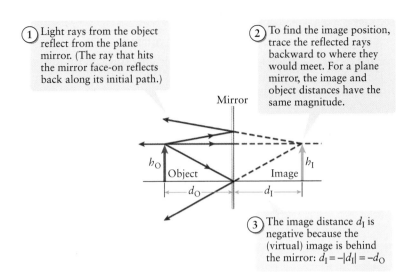

① Light rays from the object reflect from the plane mirror. (The ray that hits the mirror face-on reflects back along its initial path.)

② To find the image position, trace the reflected rays backward to where they would meet. For a plane mirror, the image and object distances have the same magnitude.

③ The image distance d_I is negative because the (virtual) image is behind the mirror: $d_I = -|d_I| = -d_O$

Figure 24-5 **A ray diagram for a plane mirror** This diagram helps us determine the position, orientation, and size of the image.

we find the position of the tip of the image arrow. The geometry of the light rays requires that the image be as far behind the front of the mirror as the object is in front of the mirror. In other words, the **image distance** d_I has the same magnitude as the object distance d_O. We'll use the convention that a point on the reflective side of the mirror (to the left of the mirror in Figure 24-5) is at a positive distance, while a point on the back side of the mirror (to the right of the mirror in Figure 24-5) is at a negative distance. So in Figure 24-5 the object distance d_O is positive but the image distance d_I is negative. We can write the relationship between the image and object distances for a plane mirror as

The image distance is negative.   The object distance is positive.

$$d_I = -|d_I| = -d_O$$

The negative value of d_I indicates that the image is on the opposite side of the mirror from the object.

Image distance for a plane mirror
(24-1)

For example, if you stand a distance $d_O = 1.0$ m in front of a plane mirror, your image is at $d_I = -1.0$ m—that is, 1.0 m behind the mirror.

Figure 24-5 also shows that the **image height** h_I is the same as the object height h_O. We define the **lateral magnification** m as the ratio of the image height to the object height. (We will often refer to m simply as the "magnification.")

Lateral magnification   Image height

$$m = \frac{h_I}{h_O}$$

Object height

Lateral magnification
(24-2)

For a plane mirror, $h_I = h_O$, so the magnification is $m = 1$. For an optical device that results in magnification greater than 1, the image is larger than the object. Such a device is commonly called a *magnifier*; it forms a magnified image of the object. When m is less than one, the image formed by the optical device is smaller than the object. As we'll see in later sections, a curved mirror can form an image that is either larger or smaller than an object placed in front of it.

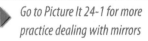
Go to Picture It 24-1 for more practice dealing with mirrors

! **Watch Out!** The image formed by a plane mirror appears *behind* the mirror.

The image does not form "on" a plane mirror, but rather behind it. This is evident from the ray diagrams in Figure 24-5. But if this diagram doesn't convince you, you can prove it to yourself by taping a bit of paper to a plane mirror, then looking at your reflection in the mirror from about 1 m away.

You'll find it difficult, likely impossible, to focus your eyes on both your image in the mirror and the piece of paper at the same time. That's because your image in the mirror is twice as far from your eyes as the paper on the surface of the mirror. The image is behind the mirror, not on it.

! **Watch Out!** The image formed by a plane mirror is reversed back to front, *not* left to right.

It's a common misconception that your image in a plane mirror is reversed left to right. A better description is that your image is reversed back to front. As an example, in Figure 24-6 a rectangular box *ABCDEFGH* sits in front of a plane mirror, making an image *A'B'C'D'E'F'G'H'*. Note that the face *A'B'C'D'* of the image has the same orientation as the face *ABCD* of the object and is the same distance from the mirror; the same is true of the face *E'F'G'H'* of the image and the corresponding face *EFGH* of the object. You can see that the net result is that the image is identical to the object but flipped from back to front.

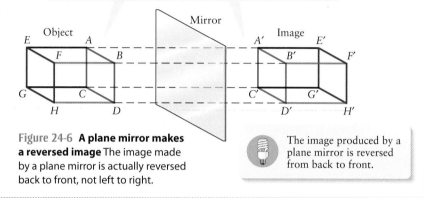

Each point on the image behind the mirror is directly opposite the corresponding point on the object in front of the mirror: *A'* opposite *A*, *B'* opposite *B*, and so on.

Figure 24-6 A plane mirror makes a reversed image The image made by a plane mirror is actually reversed back to front, not left to right.

The image produced by a plane mirror is reversed from back to front.

? **Got the Concept? 24-1** See You, See Me I

You look into a mirror hanging near the corner of a hallway and see the eyes of someone standing around the corner. Is she able to see you? (a) Yes; (b) no; (c) not enough information given to decide.

? **Got the Concept? 24-2** See You, See Me II

You look into a mirror hanging near the corner of a hallway and see the right hand of someone standing around the corner but not his eyes. Is he able to see you? (a) Yes; (b) no; (c) not enough information given to decide.

Take-Home Message for Section 24-2

✔ When light rays coming from the same general direction hit a plane mirror, they tend to be reflected in the same general direction.

✔ If an object is placed in front of a plane mirror, the image is as far behind the mirror as the object is in front.

The image is the same size as the object, but reversed from back to front.

✔ A plane mirror makes a virtual image; the light rays coming from the image do not actually pass through the image position.

24-3 A concave mirror can produce an image of a different size than the object

Figure 24-7 shows a jalapeño pepper placed in front of two curved mirrors. The mirror in Figure 24-7a is **convex** (its reflective surface is curved outward) and produces an image that is smaller than the object. The mirror in Figure 24-7b is **concave** (its reflective surface

(a) Convex mirror   (b) Concave mirror

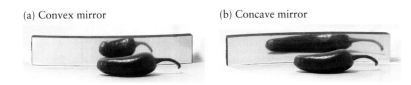

Figure 24-7 Images from curved mirrors A curved mirror can produce an image that is a different size than the object.

is curved inward, or "caved in") and produces an image that is larger than the object. This is in contrast to the plane mirror that we studied in Section 24-2, which always produces an image of the same size as the object. What accounts for the difference?

Figure 24-8 shows ray diagrams for parallel light rays striking two *spherical* mirrors—that is, mirrors that are shaped like a section cut from a complete sphere. Figure 24-8a shows that parallel light rays converge after they strike a concave mirror, while Figure 24-8b shows that parallel light rays diverge after they strike a convex mirror. By comparison, parallel light rays that strike a plane mirror remain parallel after striking it (see Figure 24-2b). It's important to consider parallel light rays as we examine various types of mirrors and lenses because light rays that come from distant objects are either parallel or nearly so. The Sun's rays, for example, are essentially parallel when they strike Earth. You can see this from the crisp shadows cast by an object placed in the Sun, such as the hanging frame in Figure 24-9a. In contrast, the shadow cast by a light bulb in Figure 24-9b is fuzzy because the light rays from the light bulb are not parallel.

The concave mirror is of particular interest because it has many practical uses. For example, a bathroom mirror used for applying makeup or for shaving is concave so that it gives an enlarged image of your face. Telescopes used by professional and amateur astronomers have a large concave mirror that brings the light from distant objects to a focus, forming an image. (The dish of the radio telescope in the photograph that opens Chapter 22 is a concave mirror for radio waves. The waves are brought to a focus at a radio receiver, mounted on struts above the telescope dish.) Automobile headlights use the same principle in reverse: A light bulb is placed in front of a concave mirror, and the mirror reflects the light forward to illuminate the road ahead. In this section and the next, we'll look at the concave mirror in detail; we'll return to the convex mirror in Sections 24-5 and 24-6.

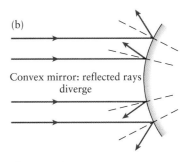

(a)

Concave mirror: reflected rays converge

(b)

Convex mirror: reflected rays diverge

Figure 24-8 Concave and convex mirrors (a) Parallel light rays converge after reflecting from a concave mirror, but (b) diverge after reflecting from a convex mirror.

(a)

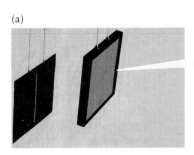

Rays of light from the distant Sun strike the hanging frame. Notice the sharp edges on all of the shadows, including those of the strings holding the frame.

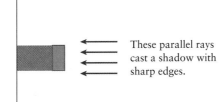

These parallel rays cast a shadow with sharp edges.

(b)

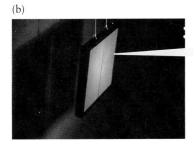

Light from a lamp strikes the frame. Because the light bulb is close to the frame, the central region of the shadow is darker than the outer regions.

These non-parallel rays cast a shadow with fuzzy edges.

Figure 24-9 Parallel and non-parallel light rays These photographs show the difference between light rays that are (a) parallel and (b) non-parallel.

• A concave, spherical mirror causes parallel light rays to nearly converge along the principal axis of the mirror.

• If we consider only rays close to the principal axis, or if the mirror is only a small arc of the complete sphere, the light rays all converge to essentially a single point.

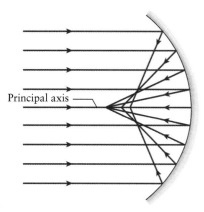

Figure 24-10 Reflection from a spherical mirror How parallel light rays behave when they strike a concave, spherical mirror.

Focal length of a spherical mirror
(24-3)

Figure 24-10 shows what happens when a series of parallel light rays strike a concave mirror formed from a fairly large section of a sphere. The incoming rays are parallel to each other and also parallel to the **principal axis** of the mirror, the axis that runs through the center of the sphere and also the center of the mirror. The law of reflection tells us that the angles from the normal to the mirror of the incident and reflected rays are equal at the point where each ray strikes. Notice that while the reflected rays generally converge along the principal axis of the mirror, they do not converge to the exact same point. This characteristic of spherical mirrors is referred to as *spherical aberration*. Only those incoming rays that are relatively close to the axis reflect through a point close enough together for us to say that they are focused on a point. Therefore, we limit the reflective surface of the mirrors we consider to a relatively small section of a sphere. Here "relatively small" means that the size of the reflective surface is small compared to the radius of the sphere. There is no specific cut-off value; rather, the smaller the mirror compared to the radius, the more tightly focused the reflected rays will be. For the rest of this chapter, we will deal with spherical mirrors small enough that we will treat them as if all rays parallel to the principal axis are focused to a single point.

Figure 24-11 defines many of the variables we use to describe the physics of a concave, spherical mirror. The **focal point** of the mirror is the point *F* along the principal axis at which incident rays parallel to the principal axis converge and come to a common focus when they reflect off the mirror. Although the mirror is only a small section of a full sphere, the point C—the **center of curvature** of the mirror—corresponds to the center of that sphere, and the distance from C to any point on the mirror is *r*, the radius of the sphere. The distance *r* is also called the **radius of curvature** of the mirror. The distance from the focal point to the center of the mirror is *f*, the **focal length**.

A glance at Figure 24-11 suggests that the focal length is about half as great as the radius of curvature. In fact, for a concave mirror small enough that all rays parallel to the principal axis are focused at the focal point, the focal length is *exactly* half of the radius:

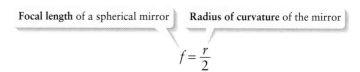

Focal length of a spherical mirror | Radius of curvature of the mirror

$$f = \frac{r}{2}$$

Equation 24-3 says that the tighter the curve of a spherical mirror and hence the smaller the radius of curvature *r*, the shorter the focal length *f*. (We'll prove the relationship $f = r/2$ in Section 24-4.)

To see how a concave mirror forms an image, let's put our standard arrow at a point far from the mirror, as in Figure 24-12a. "Far" in this case means that the object distance d_O is greater than the radius of the mirror *r*; in other words, the base of the arrow is farther from the mirror, along the principal axis, than the center of curvature C. Now let's trace two rays of light from the tip of the arrow, as they strike and then are reflected from the mirror. The image of the arrow's tip forms where those two rays intersect. Although any two light rays that originate at the tip of the arrow and that strike the mirror will work, we chose the two used in Figure 24-12 that are particularly convenient. In Figure 24-12b, we trace a ray that starts parallel to the principal axis because all light rays parallel to the principal axis are reflected through the focal point. In Figure 24-12c we add the trace of a ray that strikes the center of the mirror. Because the normal to the surface at this point lies along the principal axis, it is easy to apply the law of reflection; the incident and reflected rays are symmetric around the axis of the mirror.

Where does the image of the base of the arrow form? The base of the arrow is on the principal axis, which is by definition the normal to the mirror's surface at the center of the mirror. So a light ray coming from the base of the arrow is reflected straight back from the center of the mirror, which means the image of the base of the arrow forms along the principal axis. Figure 24-12d shows the final image of the arrow. This is an **inverted image**: It's flipped upside down compared to the object. The image is also smaller than the object. Finally, the image is real; that is, it forms in front of the mirror where light rays reflected from different parts of the mirror's surface meet.

r is the radius of curvature of the full sphere. | *f* is the focal length.

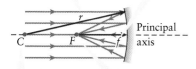

C is the center of curvature of the full sphere. | *F* is the focal point of the mirror.

Figure 24-11 Mirror nomenclature This drawing defines many of the variables we use to describe a concave, spherical mirror.

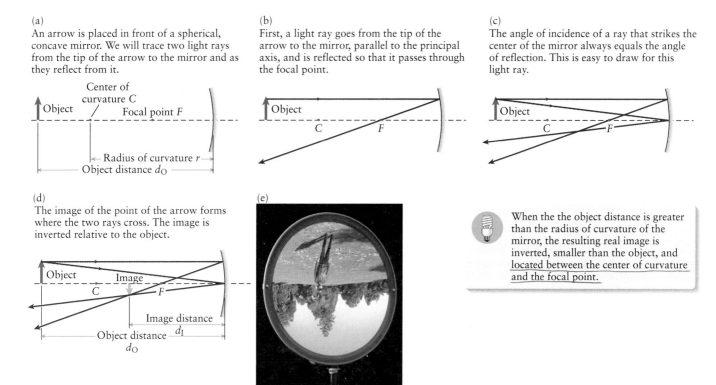

(a)
An arrow is placed in front of a spherical, concave mirror. We will trace two light rays from the tip of the arrow to the mirror and as they reflect from it.

(b)
First, a light ray goes from the tip of the arrow to the mirror, parallel to the principal axis, and is reflected so that it passes through the focal point.

(c)
The angle of incidence of a ray that strikes the center of the mirror always equals the angle of reflection. This is easy to draw for this light ray.

(d)
The image of the point of the arrow forms where the two rays cross. The image is inverted relative to the object.

(e)

When the the object distance is greater than the radius of curvature of the mirror, the resulting real image is inverted, smaller than the object, and located between the center of curvature and the focal point.

Figure 24-12 **Ray diagram for a concave mirror I** (a) A distant object in front of a concave mirror. (b), (c), (d) Locating the image of this object. (e) The image of a person standing far from a concave mirror.

Figure 24-12d shows both the object distance d_O and the image distance d_I; both distances are positive, since the object and image are both on the reflective side of the mirror. It also shows that if d_O is large, the image is closer to the mirror than the object is, so $d_I < d_O$. Note that the image distance is greater than the focal length (the distance from the mirror to the focal point F). This is how a concave mirror is used in a telescope: The object is very far away, while the image is formed very close to the mirror and is much smaller than the object. (For example, the Moon is 3.84×10^5 km distant and 1738 km in radius, but the Moon's image made by an amateur astronomer's telescope is formed only a meter or so from the mirror and is less than a centimeter in radius.) Figure 24-12e shows the inverted image of one of the authors of this book standing far away from a concave mirror.

Let's see what happens if we move the object closer and closer to the mirror. Figure 24-13a shows the arrow object placed at the center of curvature C, so the object distance equals the radius of curvature r. From Equation 24-3 the focal length $f = r/2$, so for the case shown in Figure 24-13 $d_O = r = 2f$. We trace the same two light rays as in Figure 24-12; now where the two rays meet—that is, where the image forms—is farther from the mirror. This image is the same size as the object and, as in the previous case, is both real (the light rays pass through the image) and inverted. Notice that when the object is placed at the center of curvature, the image forms at the center of curvature, so $d_I = d_O$. In Figure 24-13b one of the authors of this book is standing at the center of curvature of a concave mirror; note that his inverted image is larger than that shown in Figure 24-12e.

In Figure 24-14a we've moved the arrow closer still to the mirror, so that it sits between the center of curvature C and the focal point F. In this case the

Figure 24-13 **Ray diagram for a concave mirror II** (a) We move the object from Figure 24-12 to the center of curvature of the concave mirror. (b) The image of a person standing at the center of curvature of a concave mirror.

(a)

(b)

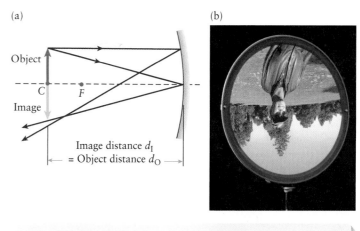

When the object is located at the radius of curvature of the mirror, the resulting real image is inverted, the same size as the object, and located at the radius of curvature.

(a)

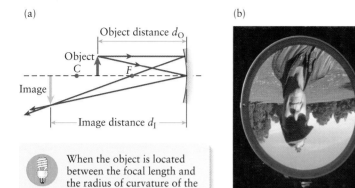

Object distance d_O

Object

C F

Image

Image distance d_I

> When the object is located between the focal length and the radius of curvature of the mirror, the resulting real image is inverted, larger than the object, and located outside the radius of curvature.

(b)

Figure 24-14 **Ray diagram for a concave mirror III** (a) We move the object from Figures 24-12 and 24-13 to a point between the center of curvature and the focal point of the concave mirror. (b) The image of a person standing at such a point.

(a)

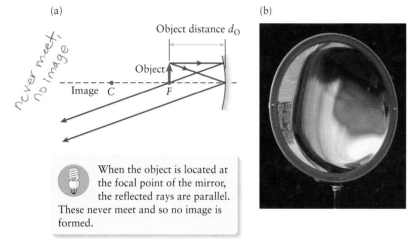

never meet,
no image

Object distance d_O

Object

Image C F

> When the object is located at the focal point of the mirror, the reflected rays are parallel. These never meet and so no image is formed.

(b)

Figure 24-15 **Ray diagram for a concave mirror IV** (a) We move the object from Figures 24-12, 24-13, and 24-14 to the focal point of the concave mirror. (b) The image of a person standing near the focal point.

object distance is between $2f$ and f: $2f > d_O > f$. Again we've traced the same two light rays that we considered in the previous two cases. The image is still both real and inverted but has moved farther still from the mirror, with the net result that it is now farther from the mirror than the object (so $d_I > d_O$) and larger than the object. **Figure 24-14b** shows such an image of one of the authors of this book; compare Figures 24-12e and 24-13b.

Comparing Figures 24-12, 24-13, and 24-14 shows that the image is getting larger and moving farther from the mirror as we move the object closer to the mirror. **Figure 24-15a** presents the limiting case: We place the object at the focal point so that the object distance equals the focal length, or $d_O = f$. After reflection, light rays that emanate from the tip of the arrow are parallel. (This is how a concave mirror is used in an automobile headlight: The lamp itself is placed close to the focal point of the curved mirror behind it, and the reflected light forms a beam of nearly parallel light rays.) Because the reflected rays are parallel, they never meet; therefore, no image forms when the object is placed at the focal point. When the object is slightly farther from the mirror than the focal point, the image formed by the mirror is extremely large and extremely far from the mirror. The author in **Figure 24-15b** is standing close to the focal point of the concave mirror; there is no sharp image.

What happens when the object is placed even closer to the mirror, inside the focal point? Then the object distance is less than the focal length ($d_O < f$). As **Figure 24-16a** shows, in this case the reflected rays never actually meet but rather appear to meet at a point behind the mirror. This means that the image is virtual, like the image formed by the plane mirror of Section 24-2, and the image distance d_I is negative: $d_I = -|d_I| < 0$. The image is larger than the object and is an **upright image** (its orientation is the same as that of the object). That's the kind of enlarged image you see when you look in a curved bathroom mirror for shaving or applying makeup. **Figure 24-16b** shows

(a)

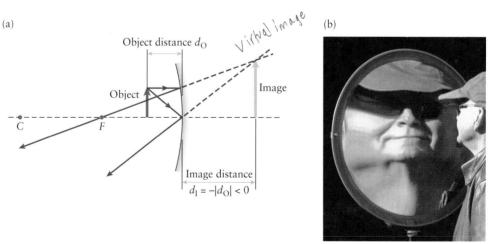

Object distance d_O

Virtual image

Object

C F

Image

Image distance $d_I = -|d_O| < 0$

> When the object is closer to the mirror than the focal point, the resulting virtual image is upright, larger than the object, and located behind the mirror.

(b)

Figure 24-16 **Ray diagram for a concave mirror V** (a) We move the object from Figures 24-12, 24-13, 24-14, and 24-15 to a point inside the focal point of the concave mirror. (b) The image of a person standing at such a point.

such an enlarged image of one of the authors of this book (compare Figures 24-12e, 24-13b, 24-14b, and 24-15b). The photograph in Figure 24-7b shows another such image.

So far we've investigated the formation of images by concave mirrors entirely by drawing light rays. In the following section we'll take a more quantitative look at how the position of an object placed in front of a concave mirror determines the position and size of the resulting image.

? Got the Concept? 24-3 A Soup Spoon

A shiny spoon is not so different in shape than a spherical mirror: It's concave on one side and convex on the other. Your reflection from the concave side of the spoon, when held at arm's length, is upside down and appears to float in front of the spoon. When you hold the spoon about 6 cm from your eye you see only a blur reflected in the spoon, but when you hold it about 4 cm from your eye your reflection is right side up and appears to be behind the spoon. What is the approximate focal length of the spoon? (a) More than 6 cm; (b) 6 cm; (c) between 4 and 6 cm; (d) 4 cm; (e) less than 4 cm.

? Got the Concept? 24-4 Covering a Concave Mirror

You place a light bulb oriented vertically (with its base on the bottom) just outside the focal point of a concave mirror. The resulting real image of the light bulb falls on a wall far from the mirror. This image is inverted and larger than the light bulb. If you were to paint the bottom half of the mirror black, so that light rays that strike this half of the mirror would not be reflected, the image would show (a) only the top of the light bulb; (b) only the bottom of the light bulb; (c) only the left side of the light bulb; (d) only the right side of the light bulb; (e) the entire light bulb. *but image is dimmer, because half as much light is reflected.*

Take-Home Message for Section 24-3

✔ A concave spherical mirror is in the shape of the inner surface of a section of a sphere.

✔ When an object is outside the focal point of a concave mirror, the image is outside the focal point, real and inverted.

✔ As the object is moved closer to the focal point, the image becomes larger and forms farther from the mirror.

✔ When the object is at the focal point, the reflected rays are parallel. Effectively, the image is at infinity and infinitely large.

✔ If the object is placed inside the focal point, the image is virtual and upright.

24-4 Simple equations give the position and magnification of the image made by a concave mirror

Let's return to the concave spherical mirror that we considered in the previous section. We'll see that given the radius of the mirror and the object distance, we can determine exactly how far from the mirror the image forms and how much magnification the mirror provides. All we need are a little geometry and the law of reflection.

The Mirror Equation for a Concave Mirror

To find the mathematical relationship among the object distance d_O, image distance d_I, and the radius of curvature r, it's convenient to imagine that the object is a point located on the principal axis of the mirror as in **Figure 24-17**. Then the image will also lie along the principal axis. (The explanation is the same one we used in Section 24-3 to show why the image of the base of the arrow in Figure 24-12 must lie on the principal axis.) Figure 24-17 also shows an arbitrarily chosen ray of light coming from the object and reflecting off the mirror. The image I forms where this reflected ray intercepts the principal axis.

Figure 24-17 **Analyzing a concave spherical mirror** A ray diagram for a point object on the principal axis of a concave spherical mirror.

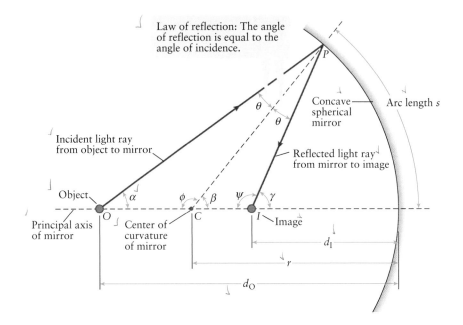

Watch Out! The path from object to mirror to image isn't just for one special light ray.

It's tempting to look at Figure 24-17 and think that there's something special about the light ray we've drawn. But *any* ray that emanates from the object point O in Figure 24-17 will reflect off the mirror and pass through the *same* image point I, provided the ray is at a shallow enough angle α to the principal axis of the mirror. So every part of the mirror contributes to forming the image. (Note that in Figure 24-17 we've drawn the angle α as fairly large. We've done this just to make the figure easier to see. If we drew α as a realistically small angle, the incident and reflected rays would be so close to the principal axis that it would be hard to see the angles in the figure.)

As Figure 24-17 shows, the light ray we've drawn goes from the object point O to a point P on the mirror, then to the image point I. We can find a relationship among d_O, d_I, and r by first finding a relationship among the angles in Figure 24-17 labeled α (the angle between the principal axis and the light ray from O to P), β (the angle between the principal axis and a line from the center of curvature C to P), and γ (the angle between the principal axis and the light ray from P to I). Since we're considering α to be a small angle, then necessarily β and γ are small angles as well. For that reason we can treat each of the three regions formed by one of these angles and the arc length s of the mirror as a sector (a pie-like slice) of a circle. The arc length is common to all three sectors. The length of the arc of a circle of radius r subtended by an angle θ equals $r\theta$ (see Section 3-8), provided the angle θ is in radians. So in Figure 24-17

See the Math Tutorial for more information on geometry

$$(24\text{-}4) \qquad s = d_O\alpha; \quad s = r\beta; \quad s = d_I\gamma$$

Note also that the sum of the angles in any triangle equals 180° or π radians. For the triangle OPC we have

$$(24\text{-}5) \qquad \alpha + \theta + \phi = \pi$$

However, the angles ϕ and β in Figure 24-17 must add to 180°, or π radians (together they make up half a circle). So $\phi + \beta = \pi$ and $\phi = \pi - \beta$, and Equation 24-5 becomes

$$(24\text{-}6) \qquad \alpha + \theta + \pi - \beta = \pi \quad \text{or} \quad \theta = \beta - \alpha$$

Similarly, for the triangle OPI the sum of the angles is π:

$$(24\text{-}7) \qquad \alpha + 2\theta + \psi = \pi$$

Figure 24-17 shows that $\psi + \gamma = \pi$, so $\psi = \pi - \gamma$. Triangle CPI shows that $\beta + \theta + \psi = \pi$. Comparing with Equation 24-7, this means it must be true that $\alpha + 2\theta = \beta + \theta$. Equation 24-7 can then be written as

$$(24\text{-}8) \qquad \beta + \theta + \pi - \gamma = \pi \quad \text{or} \quad \theta = \gamma - \beta$$

Equations 24-6 and 24-8 are two different expressions for the angle θ in Figure 24-17. If we set these equal to each other, we get

$$\beta - \alpha = \gamma - \beta \quad \text{or} \quad \alpha + \gamma = 2\beta \tag{24-9}$$

From Equations 24-4, we have $\alpha = s/d_O$, $\beta = s/r$, and $\gamma = s/d_I$. Substituting these into Equation 24-9 gives

$$\frac{s}{d_O} + \frac{s}{d_I} = 2\frac{s}{r}$$

To simplify we divide through by the arc length s, giving us the relationship we've been seeking between the object distance, image distance, and radius of curvature of the mirror:

$$\frac{1}{d_O} + \frac{1}{d_I} = \frac{2}{r} \tag{24-10}$$

To help interpret Equation 24-10, recall from Figure 24-11 that if the incident light rays are parallel, they come to a focus at the focal point a distance f from the mirror. So in this case $d_I = f$. The incident rays will be parallel if the object is infinitely far away, so $d_O \rightarrow \infty$ and $1/d_O \rightarrow 0$. Then Equation 24-10 becomes

$$0 + \frac{1}{f} = \frac{2}{r} \quad \text{or} \quad f = \frac{r}{2}$$

This justifies Equation 24-3: The focal length of a concave spherical mirror is one-half of the radius of curvature.

If we replace $2/r$ with $1/f$ in Equation 24-10, we get the final form of the **mirror equation**:

$$\underset{\text{Object distance}}{\frac{1}{d_O}} + \underset{\text{Image distance}}{\frac{1}{d_I}} = \frac{1}{f} \quad \text{Focal length}$$

Mirror equation and lens equation
(24-11)

We also call Equation 24-11 the *lens equation* because, as we shall see in Section 24-8, it's also applicable to the image formed by a lens.

The focal length f of a concave mirror is taken to be positive because the center of curvature C of the mirror is in front of the mirror, that is, on its reflective side (to the left in Figure 24-17). Likewise, since the object point O is in front of the mirror, the object distance d_O is positive. The image distance d_I, however, can be positive if the image is real (in front of the mirror) or negative if the image is virtual (behind the mirror). To see when the image distance is positive and when it is negative, let's rewrite Equation 24-11 to solve for d_I:

$$\frac{1}{d_I} = \frac{1}{f} - \frac{1}{d_O} = \frac{d_O}{d_O f} - \frac{f}{d_O f} = \frac{d_O - f}{d_O f} \quad \text{or} \quad d_I = \frac{d_O f}{d_O - f} \tag{24-12}$$

In Equation 24-12 d_O and f are both positive for a concave mirror, so the numerator $d_O f$ is positive. However, the denominator $d_O - f$ can be positive or negative depending on whether d_O is larger or smaller than f.

If the object is outside the focal point so that $d_O > f$, then the denominator $d_O - f$ in Equation 24-12 is positive and the image distance is positive. Therefore, in this case the image made by the mirror is real. As the object moves closer to the focal point, the difference between d_O and f gets smaller and so d_I given by Equation 24-12 gets larger. Hence the image moves father away from the mirror. If the object is inside the focal point, however, then $d_O < f$ and the denominator $d_O - f$ is negative. Then d_I is negative as well, and the image is virtual. That's exactly the behavior that we deduced in Section 24-3 by analyzing ray diagrams.

As a further check on Equations 24-11 and 24-12, note that if the object is exactly two focal lengths from the mirror so $d_O = 2f$, the image distance is

$$d_I = \frac{d_O f}{d_O - f} = \frac{(2f)f}{2f - f} = \frac{2f^2}{f} = 2f$$

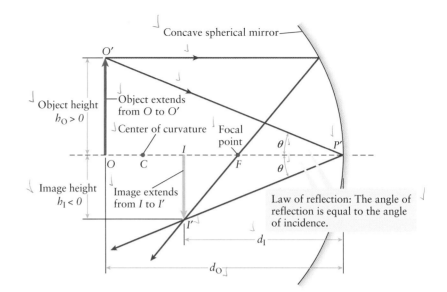

• The right triangles $OO'P'$ and $II'P'$ both include the same angle θ.
• So $\tan \theta = h_O/d_O = |h_I|/d_I$ and $|h_I|/h_O = d_I/d_O$.
• Since h_I is negative, $h_I = -|h_I|$ and $m = h_I/h_O = -|h_I|/h_O = -d_I/d_O$.

So when the object is a distance $2f$ from the mirror—which, because $r = 2f$, is at the center of curvature—the image is at the same position. That's the same conclusion we came to by using the ray diagram in Figure 24-13.

Magnification for a Concave Mirror

We can also use simple geometry to find an expression for the lateral magnification of the image produced by a concave spherical mirror. In **Figure 24-18** we've replaced the point object at O with an upright arrow of height h_O that extends from O (on the principal axis) to O'. Just as in Figure 24-12, we draw two rays coming from the tip of the object arrow at O' to determine the position I' of the tip of the image arrow.

Since the object in Figure 24-18 is outside the focal point F, the image is real and inverted and so the height of the image is negative: $h_I = -|h_I|$. By the law of reflection, the ray from O' to the point P' at the center of the mirror makes the same angle θ with the mirror's principal axis (shown as a dashed line in Figure 24-18) as does the reflected ray from P' to I'. If you look at the right triangle $OO'P'$, you'll see that the tangent of θ (the opposite side divided by the adjacent side) is $\tan \theta = h_O/d_O$; if you do the same for the right triangle $II'P'$, you'll see that $\tan \theta = |h_I|/d_I$. Setting these two expressions equal to each other, we see that

$$\frac{h_O}{d_O} = \frac{|h_I|}{d_I} \quad \text{or} \quad \frac{|h_I|}{h_O} = \frac{d_I}{d_O}$$

Since $h_I = -|h_I|$, we can rewrite this as

(24-13)
$$-\frac{h_I}{h_O} = \frac{d_I}{d_O} \quad \text{or} \quad \frac{h_I}{h_O} = -\frac{d_I}{d_O}$$

From Equation 24-2, the lateral magnification is $m = h_I/h_O$. So Equation 24-13 tells us that

Lateral magnification for a mirror or lens
(24-14)

$m+$ image upright
$m-$ image inverted

Lateral magnification	Image height	Image distance

$$m = \frac{h_I}{h_O} = -\frac{d_I}{d_O}$$

Object height	Object distance

 Go to Interactive Exercise 24-1 for more practice dealing with concave mirrors

A negative value of the magnification m means that the image is inverted, as in Figure 24-18. If m is positive, the image is upright. (As we'll see in Section 24-8, the same equation also applies to lenses.)

We've derived Equation 24-14 for the case in which the image is in front of the mirror (the same side as the object) and hence real, but it's also true when the image is on the back side of the mirror and hence virtual. When the object is far from the mirror and the image is close to the focal point, d_I is positive and small compared to d_O, so m is small and negative; the mirror produces a reduced, inverted image (see Figure 24-12). As the object distance decreases, the image distance increases and the ratio $m = -d_I/d_O$ increases in absolute value. As we saw above, when $d_O = 2f$ the image distance d_I is also equal to $2f$; then $m = -1$ and the inverted image is as large as the object (see Figure 24-13). If we move the object even closer to the focal point, but still outside it, the image distance is greater than the object distance and the absolute value of the magnification is greater than 1. Hence the inverted image is larger than the object (see Figure 24-14). If we move the object inside the focal point so that $d_O < f$, the image distance is negative (the image is behind the mirror) and the image is virtual. In this case Equation 24-14 tells us that m is positive, so the virtual image is upright (see Figure 24-16). We see that Equation 24-14 gives us the same results as we deduced from the ray diagrams in Section 24-3.

Table 24-1 summarizes when the radius of curvature r, focal length f, image distance d_I, image height h_I, and magnification m are positive and when they are negative.

Table 24-1 Sign Conventions for Mirrors

mirror radius of curvature r	• positive for a concave mirror • negative for a convex mirror
focal length f	• positive for a concave mirror • negative for a convex mirror
image distance d_I	• positive if on the reflective side of the mirror (the same side as the object); the image is then a real image • negative if on the non-reflective side of the mirror (the opposite side from the object); the image is then a virtual image
image height h_I	• positive if the image is upright (the same orientation as the object) • negative if the image is inverted (flipped upside down compared to the object)
lateral magnification m	• positive if the image is upright (the same orientation as the object) • negative if the image is inverted (flipped upside down compared to the object)

Example 24-1 Images Made by a Concave Mirror

An object is 1.50 cm tall and is placed in front of a spherical concave mirror that has a radius of curvature equal to 20.0 cm. Find the image distance and height if the object is (a) 14.0 cm from the mirror; (b) 6.00 cm from the mirror.

Set Up

Our mathematical tools are the expression for the focal length of the mirror, the mirror equation, and the equation for lateral magnification. We're given the radius of curvature $r = 20.0$ cm and the object height $h_O = 1.50$ cm; our goal is to find the values of the image distance d_I and image height h_I for the cases $d_O = 14.0$ cm and $d_O = 6.00$ cm.

Focal length of a spherical mirror:

$$f = \frac{r}{2} \qquad (24\text{-}3)$$

Mirror equation:

$$\frac{1}{d_O} + \frac{1}{d_I} = \frac{1}{f} \qquad (24\text{-}11)$$

Lateral magnification of a mirror

$$m = \frac{h_I}{h_O} = -\frac{d_I}{d_O} \qquad (24\text{-}14)$$

Solve

First calculate the focal length of the mirror.

From Equation 24-3,

$$f = \frac{r}{2} = \frac{20.0 \text{ cm}}{2} = 10.0 \text{ cm}$$

(a) The object distance $d_O = 14.0$ cm is greater than the focal length $f = 10.0$ cm but less than $2f = 20.0$ cm. That is, the object is inside the center of curvature but outside the focal point. We've drawn a ray diagram that's similar to Figure 24-14. This tells us to expect that the image will be real, inverted, farther from the mirror than the object is, and larger than the object.

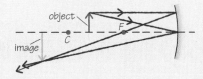

Calculate the image distance using the mirror equation.

From Equation 24-11,

$$\frac{1}{d_I} = \frac{1}{f} - \frac{1}{d_O} = \frac{1}{10.0 \text{ cm}} - \frac{1}{14.0 \text{ cm}}$$

$$= 0.1000 \text{ cm}^{-1} - 0.0714 \text{ cm}^{-1} = 0.0286 \text{ cm}^{-1} \text{ so}$$

$$d_I = \frac{1}{0.0286 \text{ cm}^{-1}} = 35.0 \text{ cm}$$

The image distance is positive, so the image is 35.0 cm in front of the mirror. An image that forms in front of the mirror is a real image.

Calculate the image height using the magnification equation.

From Equation 24-14,

$$m = \frac{h_I}{h_O} = -\frac{d_I}{d_O} = -\frac{35.0 \text{ cm}}{14.0 \text{ cm}} = -2.50$$

The image is 2.50 times larger than the object and, as the minus sign shows, inverted. Solve for the image height h_I:

$$h_I = mh_O = (-2.50)(1.50 \text{ cm}) = -3.75 \text{ cm}$$

The inverted image is 3.75 cm high.

(b) Now the object distance $d_O = 6.00$ cm is less than the focal length $f = 10.0$ cm, so the object is inside the focal point. We've drawn a ray diagram that's similar to Figure 24-16. This tells us to expect that the image will be virtual, upright, farther from the mirror, and larger than the object.

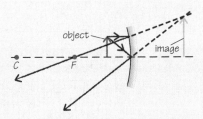

Calculate the image distance using the mirror equation.

From Equation 24-11,

$$\frac{1}{d_I} = \frac{1}{f} - \frac{1}{d_O} = \frac{1}{10.0 \text{ cm}} - \frac{1}{6.00 \text{ cm}}$$

$$= 0.1000 \text{ cm}^{-1} - 0.1667 \text{ cm}^{-1} = -0.0667 \text{ cm}^{-1} \text{ so}$$

$$d_I = \frac{1}{-0.0667 \text{ cm}^{-1}} = -15.0 \text{ cm}$$

The image distance is negative, so the image is 15.0 cm behind the mirror. An image that forms behind the mirror is a virtual image.

Calculate the image height using the magnification equation.

From Equation 24-14,

$$m = \frac{h_I}{h_O} = -\frac{d_I}{d_O} = -\frac{(-15.0 \text{ cm})}{6.00 \text{ cm}} = +2.50$$

Again the image is 2.50 times larger than the object, but is now upright as shown by the plus sign. Solve for the image height h_I:

$$h_I = mh_O = (+2.50)(1.50 \text{ cm}) = +3.75 \text{ cm}$$

The upright image is 3.75 cm high.

Reflect

In both parts, the position and size of the image are consistent with our ray diagrams. It's always a good idea to draw such diagrams as a check on your calculations using the mirror equation and magnification equation.

? Got the Concept? 24-5 A Concave Mirror

If an object is placed 12.0 cm from a concave mirror, the resulting real image is 2.00 times as large as the object. What is the focal length of the mirror? (a) 6.00 cm; (b) 8.00 cm; (c) 16.0 cm; (d) 20.0 cm; (e) 24.0 cm.

Take-Home Message for Section 24-4

✔ The mirror equation relates the focal length of a concave mirror and the positions of the object and image.

✔ A positive value of the image distance d_I indicates that the image is in front of the mirror and is real.

A negative value of d_I indicates that the image is behind the mirror and virtual (the light rays never actually go there).

✔ The magnification of the image is positive if the image is upright and negative if the image is inverted.

24-5 A convex mirror always produces an image that is smaller than the object

Let's now turn our attention to images produced by a convex mirror. As Figure 24-7a shows, if an object is held next to a convex mirror, the resulting image is smaller than the object. The same is true if an object is far away from a convex mirror. Figure 24-19 shows the reflection of a Ferris wheel in the convex surface of a person's eye. Although the Ferris wheel is tens of meters in diameter, its image is only about a centimeter in diameter—smaller than the iris of the eye. In this section we'll use ray diagrams to understand the nature of the images formed by a convex mirror.

Recall from Figure 24-8 the key difference between concave and convex mirrors: Parallel light rays that reflect from a concave mirror converge toward a point in front of the mirror, while parallel light rays that reflect from a convex mirror diverge from a point on the back side of the mirror. (If the mirrors are spherical, the rays don't truly converge on or diverge from a single point. But this is a good description of what happens if the rays are all close to the principal axis of the mirror. We'll make this assumption—the same that we made in Sections 24-3 and 24-4 for concave mirrors—throughout this section.)

Because the focal point of a convex mirror is behind the mirror, we say that the focal length f is negative. For a convex mirror, we call the focal point *virtual* because parallel light rays that reflect from the mirror seem to emanate from that point but don't actually pass through it.

Figure 24-20 shows an object placed a distance d_O in front of a convex mirror. The focal point F lies a distance $|f|$ behind the mirror. (Since the focal length f is negative, the distance is the absolute value of f.) The center of curvature C is also behind the mirror, so the radius of curvature r is also negative. It turns out that the focal length f is equal to one-half of the radius of curvature r, just as for a concave mirror:

$$f = \frac{r}{2}$$ (24-3)

(We'll justify this statement in Section 24-6.) For a concave mirror, f and r are both positive; for a convex mirror, f and r are both negative.

To find the location of the image in Figure 24-20, we draw two rays that emanate from the tip of the object arrow, just as we did for the ray diagrams in Section 24-3 for a concave mirror. The image of the arrow tip forms where the two reflected rays appear to meet. The image of the base of the arrow must lie along the principal axis. (A light ray coming from the base of the arrow and traveling along the principal axis strikes the mirror normal to the surface, and so is reflected straight back. The image of the base of the arrow must therefore form somewhere on the principal axis.) As Figure 24-20 shows, the image is upright. The image is virtual because it forms behind the mirror, just like the image formed by a plane mirror (Section 24-2): The reflected rays don't actually

Figure 24-19 **Reflections in a golden eye** A Ferris wheel can be seen reflected from the surface of this person's eye. The eye is relatively spherical, so its outer surface acts like a convex mirror.

Figure 24-20 Ray diagram for a convex mirror How the image is formed for an object in front of a convex mirror.

- A convex mirror has a negative focal length: $f = -|f|$.
- For any object distance d_O, a convex mirror produces a virtual image behind the mirror.
- The image distance d_I is negative: $d_I = -|d_I|$.
- The virtual image is upright and smaller than the object.

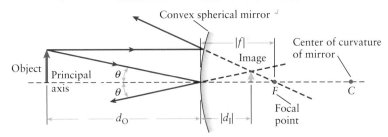

① A light ray that starts off parallel to the principal axis is reflected so that when we trace the ray back behind the mirror, it appears to pass through the virtual focal point F.

Convex spherical mirror

Object Principal axis θ θ

$|f|$ Image Center of curvature of mirror

F C Focal point

d_O $|d_I|$

② A light ray that strikes the center of the mirror at an angle θ from the principal axis reflects at the same angle.

③ Trace the two reflected rays to where they appear to cross behind the mirror. This is where the virtual image forms.

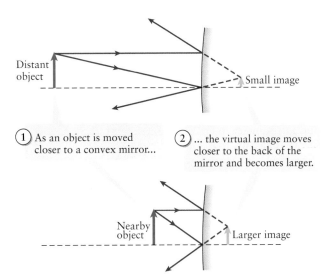

Distant object

Small image

① As an object is moved closer to a convex mirror...

② ... the virtual image moves closer to the back of the mirror and becomes larger.

Nearby object

Larger image

Figure 24-21 Moving closer to a convex mirror As the object far from a convex mirror is moved closer to the mirror, the image increases in size.

cross there. Hence the image distance d_I is negative. We saw in Section 24-3 that a concave mirror produces a virtual image only if the object is inside the focal point; <u>a convex mirror produces a virtual image for *any* position of the object</u>.

The image in Figure 24-20 is smaller than the object, no matter what the object distance is. To show this, note that the height of the image arrow is determined in part by the angle of the ray that emerges from the tip of the object arrow moving parallel to the principal axis. The dashed line from where this ray hits the mirror to the focal point F slopes downward toward F, so the image must be smaller than the object. Figure 24-20 also shows that the image formed by the convex mirror will always be upright, always be smaller than the object, and always be closer to the mirror than the virtual focal point.

Figure 24-21 shows how the image size and position change as the object is moved closer to the mirror. As the object distance d_O decreases, the image becomes larger and closer to the mirror, but remains virtual and upright. If the object were moved all the way to the surface of the mirror, the image would be exactly the same size as the object.

Because a convex mirror makes objects appear smaller, they provide a wide-angle view. Rear-view mirrors on automobiles and trucks often include a convex mirror to allow the driver to see as much of the area behind the vehicle as possible. (Any mirror with the label "Objects in mirror are closer than they appear" is a convex mirror.)

In the following section, we'll look at these ideas more quantitatively and see how to calculate the image size and position for a convex mirror.

? Got the Concept? 24-6 Solar Cooking

The Sun is so far from Earth that rays of sunlight are effectively parallel. Sunlight reflected from a mirror can be focused onto a pot, raising its temperature enough to pasteurize water in the pot or cook food. What kind of spherical mirror could be used for this purpose? (a) A convex mirror; (b) a concave mirror; (c) either a convex or concave mirror.

Take-Home Message for Section 24-5

✔ Parallel light rays that strike a convex mirror diverge and appear to emanate from a virtual focal point behind the mirror.

✔ If an object is placed anywhere in front of a convex mirror, the resulting image is virtual, upright, closer to the mirror than the object is, and smaller than the object.

24-6 The same equations used for concave mirrors also work for convex mirrors

Just as we did for the concave mirror in Section 24-4, we can use geometry and the law of reflection to find an equation that relates the object and image distances and the radius of curvature for a convex mirror. We'll also find an equation that relates the sizes of the image and object for a convex mirror. Remarkably, we'll see that these equations are exactly the same as those for a concave mirror, provided we're careful with the signs of the radius of curvature, focal length, and image distance for a convex mirror.

The Mirror Equation for a Convex Mirror

In Figure 24-22 we've placed a point object on the principal axis of a convex mirror with the (negative) radius of curvature r. This is analogous to Figure 24-17, in which we placed a point object on the principal axis of a concave mirror. The object is at position O, a distance d_O from the mirror. We've drawn an arbitrarily chosen light ray that travels away from the object at an angle α from the principal axis and reflects off the mirror at P; if we extend the reflected ray backward, the extension (shown as a dashed blue line) crosses the principal axis at I. A second light ray (not shown) that travels away from the object along the principal axis will be reflected straight back along that axis. If we extend this reflected ray backward, it will meet the extension of the first reflected ray at I. Hence I is the position of the image made by the mirror. As we expect, this image is behind the mirror, so the image distance is negative. The center of curvature C of the mirror also lies behind the mirror, so the radius of curvature is also negative. Mathematically, $d_I = -|d_I| < 0$ and $r = -|r| < 0$.

Note that the red dashed line CP in Figure 24-22 is normal to the mirror at P, so by the law of reflection the incident ray and the reflected ray are both at the same angle θ relative to this normal. It follows that θ is also the angle between the line CP and the backward extension PI of the reflected ray.

To relate the object distance d_O, image distance d_I, and radius of curvature r, we'll first find a relationship among α, the angle between the principal axis and the light ray incident on the mirror; β, the angle between the principal axis and the reflected ray; and γ, the angle between the principal axis and the red dashed line CP that defines the normal at the point P where the light ray is reflected. Notice that each of these three angles is part of a right triangle for which one side is the thick black line PQ in Figure 24-22: triangle OPQ for angle α, triangle IPQ for angle β, and triangle CPQ for angle γ. In each case the line PQ, of length h, is the side opposite the angle. Strictly speaking, we're only considering light rays that are very close to the principal axis, or equivalently

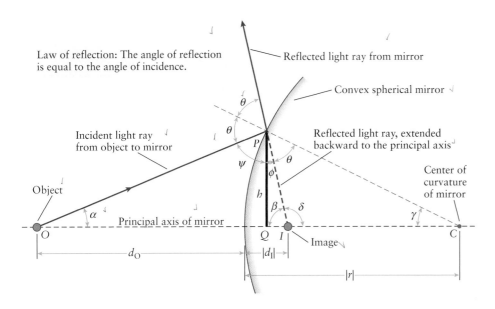

Figure 24-22 **Analyzing a convex spherical mirror** A ray diagram for a point object on the principal axis of a convex spherical mirror.

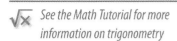

See the Math Tutorial for more information on trigonometry

a mirror of very large radius so that we can regard the mirror's curved surface as nearly flat. As a result, we can treat d_O as the base of triangle OPQ, $|d_I|$ as the base of triangle IPQ, and $|r|$ as the base of triangle CPQ. For each triangle the tangent of the angle equals the length h of the side opposite the angle, divided by the length of the triangle's base. So

(24-15)
$$\tan \alpha = \frac{h}{d_O}, \quad \tan \beta = \frac{h}{|d_I|}, \quad \tan \gamma = \frac{h}{|r|}$$

If the incident ray OP is at a small angle to the principal axis, then α, β, and γ are all small angles. (We've drawn the angle α fairly large in Figure 24-22 to make the geometry easier to visualize. In reality, it must be quite small to conform to the approximation that the rays are nearly parallel to the principal axis.) The tangent of a small angle is approximately equal to the angle in radians, so Equation 24-15 becomes

(24-16)
$$\alpha = \frac{h}{d_O}, \quad \beta = \frac{h}{|d_I|}, \quad \gamma = \frac{h}{|r|}$$

To relate the angles α, β, and γ, we'll do as in Section 24-4 and use the result that the sum of the angles of any triangle must be 180°, or π radians. For the triangle IPC that connects the image point I, reflection point P, and center of curvature C, the three angles are γ, θ, and δ. So

(24-17)
$$\gamma + \theta + \delta = \pi$$

Figure 24-22 shows that the angles β and δ must add to π radians (together they make up half a circle around the point I). So $\beta + \delta = \pi$, $\delta = \pi - \beta$, and Equation 24-17 becomes

(24-18)
$$\gamma + \theta + \pi - \beta = \pi \quad \text{or} \quad \theta = \beta - \gamma$$

For the triangle OPC that connects the object point O, reflection point P, and center of curvature C, the angles are α, $(\psi + \theta)$, and γ, so we have

(24-19)
$$\alpha + (\psi + \theta + \phi) + \gamma = \pi$$

We can simplify Equation 24-19 by noting from Figure 24-22 that the angles θ, ψ, ϕ and θ form a half-circle around the point P, so their sum is π radians: $\theta + \psi + \phi + \theta = \pi$, or $\psi + \theta + \phi = \pi - \theta$. If we substitute this expression for $(\psi + \theta + \phi)$ into Equation 24-19, we get

(24-20)
$$\alpha + \pi - \theta + \gamma = \pi \quad \text{or} \quad \theta = \alpha + \gamma$$

Equations 24-18 and 24-20 are both expressions for the angle θ. If we set these equal to each other, we get

(24-21)
$$\beta - \gamma = \alpha + \gamma \quad \text{or} \quad \alpha - \beta = -2\gamma$$

Equation 24-21 is the relationship among the angles α, β, and γ we've been looking for. We can now get a relationship among the object distance d_O, image distance d_I, and radius of curvature r by substituting the expressions for α, β, and γ from Equation 24-16 into Equation 24-21:

$$\frac{h}{d_O} - \frac{h}{|d_I|} = -\frac{2h}{|r|}$$

If we divide through by the height h of the black line in Figure 24-22 and recall that the image distance and radius of curvature are both negative, so that $d_I = -|d_I|$ and $r = -|r|$, this becomes

(24-22)
$$\frac{1}{d_O} + \frac{1}{d_I} = \frac{2}{r}$$

Equation 24-22 is *exactly* the same as Equation 24-10, which we derived for a *concave* mirror. It's reassuring that we get the same expression for both kinds of mirrors.

Note that if the object is infinitely far away, so that $d_O \to \infty$ and $1/d_O \to 0$, the rays from the object to the mirror will all be parallel to the axis and the virtual image will be formed at the focal point F, a distance $|f|$ behind the mirror. Then $d_I = -|f| = f$ (recall that the focal length f is negative). For this situation, Equation 24-22 becomes

$$\frac{1}{f} = \frac{2}{r} \quad \text{or} \quad f = \frac{r}{2}$$

This result is exactly the same as that for a concave spherical mirror: The focal length f is equal to one-half of the radius of curvature r (Equation 24-3). The only difference is that for a convex mirror, f and r are both negative.

If we substitute $1/f = 2/r$ into Equation 24-22, we get the mirror equation for a convex mirror:

$$\frac{1}{d_O} + \frac{1}{d_I} = \frac{1}{f} \tag{24-11}$$

This is Equation 24-11, the same mirror equation that we derived for a *concave* mirror in Section 24-4. This equation works equally well whether the focal length is positive (for a concave mirror) or negative (for a convex mirror).

We can use Equation 24-11 to explore the properties of the image made by a convex mirror. We saw in Section 24-4 that this equation can be rewritten as

$$d_I = \frac{d_O f}{d_O - f} \tag{24-12}$$

Using $f = -|f|$ for the negative focal length of a convex mirror, Equation 24-12 becomes

$$d_I = \frac{d_O(-|f|)}{d_O - (-|f|)} = -\left(\frac{d_O}{d_O + |f|}\right)|f| \tag{24-23}$$

The right-hand side of Equation 24-23 is always negative, so the image distance d_I for a convex mirror will always be negative (the image will always be behind the mirror). Note that the fraction in parentheses is always less than or equal to 1, so the image distance is always between 0 and $-f$. That is, the image always forms somewhere between the mirror and the focal point.

Magnification for a Convex Mirror

Figure 24-23 again shows the image made by a convex spherical mirror, but now an upright arrow of height h_O is the object. This arrow is a distance d_O in front of the mirror. Just as for a concave mirror (see Figure 24-18 in Section 24-4), we draw two rays coming from the tip of the object arrow at O'. One ray is parallel to the principal axis of the mirror; after reflection, this travels away from the mirror as though it had been emitted from the focal point F. The other ray bounces off the center of the mirror at P', and the reflected ray makes the same angle θ with the principal axis as the incident ray from O' to P'. If we extend this reflected ray backward, the extension is at the same angle θ to the principal axis. This extension meets the extension of the other reflected ray at I', the position of the tip of the image arrow. The base of the image arrow is at I, directly underneath I' on the principal axis. The image is behind the mirror, so the image distance d_I is negative. This image is also upright, so the image height h_I is positive.

To relate the image and object heights, note that the right triangles $OO'P'$ and $II'P'$ both include the same angle θ. For $OO'P'$ the tangent of θ (the opposite side divided by the adjacent side) equals h_O/d_O; for $II'P'$ the tangent of θ equals $h_I/|d_I|$. These two expressions for $\tan \theta$ must be equal, so

$$\frac{h_O}{d_O} = \frac{h_I}{|d_I|} \quad \text{or} \quad \frac{h_I}{h_O} = \frac{|d_I|}{d_O} = -\frac{d_I}{d_O} \quad (\text{since } d_I = -|d_I|)$$

The lateral magnification m equals h_I/h_O, so it follows that for a convex mirror

$$m = \frac{h_I}{h_O} = -\frac{d_I}{d_O} \tag{24-14}$$

Figure 24-23 **Magnification for a convex spherical mirror** We replace the point object in Figure 24-22 with an object arrow.

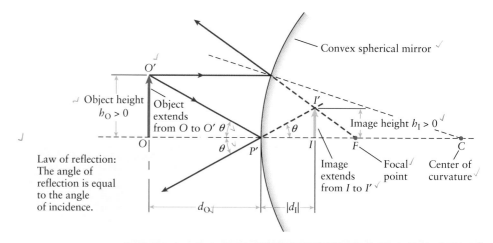

Figure 24-23 **Magnification for a convex spherical mirror** We replace the point object in Figure 24-22 with an object arrow.

- The right triangles $OO'P'$ and $II'P'$ both include the same angle θ.
- So $\tan \theta = h_O/d_O = h_I/|d_I|$ and $h_I/h_O = |d_I|/d_O$.
- Since d_I is negative, $d_I = -|d_I|$ and $m = h_I/h_O = |d_I|/d_O = -d_I/d_O$.

 Go to Interactive Exercise 24-2 for more practice dealing with concave and convex mirrors

This is the same expression for magnification as for a concave mirror. Since the image distance d_I is negative for a convex mirror, Equation 24-14 tells us that the magnification is positive and so the image is upright. In addition, for a convex mirror the image is closer to the mirror than is the object, so $|d_I| < d_O$ and the value of m is less than 1. That is, the image made by a convex mirror is always smaller than the object, just as we saw in Section 24-5.

! **Watch Out!** **Positive magnification indicates that an image is upright.**

Remember that the sign of the magnification of a mirror doesn't indicate whether the image is larger or smaller than the object, only whether the image is upright or inverted. An upright image, for example, is always associated with a positive magnification, regardless of whether the image is larger than or smaller than the object.

Example 24-2 An Image Made by a Convex Mirror

An object is 1.50 cm high and is placed in front of a spherical convex mirror with a radius of curvature of magnitude 48.0 cm. Find the image distance and height if the object is (a) 68.0 cm from the mirror; (b) 3.00 cm from the mirror.

Set Up
This is similar to Example 24-1 in Section 24-4. The key difference is that the mirror is now convex, so the radius of curvature is negative: $r = -48.0$ cm. We're given the object height $h_O = 1.50$ cm, and want to find the values of the image distance d_I and image height h_I for the cases $d_O = 68.0$ cm and $d_O = 3.00$ cm. Our tools are Equations 24-3, 24-11, and 24-14, which apply to convex mirrors as well as concave ones.

Focal length of a spherical mirror:

$$f = \frac{r}{2} \qquad (24\text{-}3)$$

Mirror equation:

$$\frac{1}{d_O} + \frac{1}{d_I} = \frac{1}{f} \qquad (24\text{-}11)$$

Lateral magnification of a mirror

$$m = \frac{h_I}{h_O} = -\frac{d_I}{d_O} \qquad (24\text{-}14)$$

Solve

Use Equation 24-3 to calculate the focal length of the mirror.

From Equation 24-3,

$$f = \frac{r}{2} = \frac{-48.0\text{ cm}}{2} = -24.0\text{ cm}$$

The negative value of focal length means that the focal point is behind the mirror.

In both cases, we expect that the image will be virtual (behind the mirror), smaller than the object, and closer to the mirror than the object is. The specific position and size of the image will be different in the two cases, however.

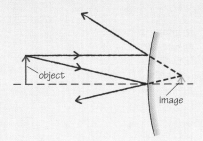

(a) With an object distance $d_O = 68.0$ cm, we use Equation 24-11 to find the image distance and Equation 24-14 to find the image size.

From the mirror equation (Equation 24-11),

$$\frac{1}{d_I} = \frac{1}{f} - \frac{1}{d_O} = \frac{1}{(-24.0\text{ cm})} - \frac{1}{68.0\text{ cm}}$$

$$= -0.0417\text{ cm}^{-1} - 0.0147\text{ cm}^{-1} = -0.0564\text{ cm}^{-1}\text{ so}$$

$$d_I = \frac{1}{(-0.0564\text{ cm}^{-1})} = -17.7\text{ cm}$$

The image distance is negative, so the image is 17.7 cm behind the mirror. An image that forms behind the mirror is a virtual image. The magnification of the image is, from Equation 24-14,

$$m = \frac{h_I}{h_O} = -\frac{d_I}{d_O} = -\frac{(-17.7\text{ cm})}{68.0\text{ cm}} = +0.261$$

Since $m > 0$, the image is upright. The image is 0.261 as tall as the object, so its height is

$$h_I = mh_O = (+0.261)(1.50\text{ cm}) = 0.391\text{ cm}$$

(b) Repeat the calculations of part (a) with $d_O = 3.00$ cm.

Calculate the image distance:

$$\frac{1}{d_I} = \frac{1}{f} - \frac{1}{d_O} = \frac{1}{(-24.0\text{ cm})} - \frac{1}{3.00\text{ cm}}$$

$$= -0.0417\text{ cm}^{-1} - 0.333\text{ cm}^{-1} = -0.375\text{ cm}^{-1}\text{ so}$$

$$d_I = \frac{1}{(-0.375\text{ cm}^{-1})} = -2.67\text{ cm}$$

Again the image distance is negative. Note that with a smaller object distance than in (a), the image distance is also smaller.

The magnification of the image is

$$m = \frac{h_I}{h_O} = -\frac{d_I}{d_O} = -\frac{(-2.67\text{ cm})}{3.00\text{ cm}} = +0.889$$

Again $m > 0$ and the image is upright. The image is 0.889 as tall as the object; its height is

$$h_I = mh_O = (+0.889)(1.50\text{ cm}) = +1.33\text{ cm}$$

Reflect

As the object is moved closer to the convex mirror, the image moves closer to the mirror and increases in size.

? Got the Concept? 24-7 A Convex Mirror

If the image made by a convex mirror of focal length f is 1/2 the height of the object, the distance from the object to the mirror must be equal to (a) $4|f|$; (b) $2|f|$; (c) $|f|$; (d) $|f|/2$; (e) $|f|/4$.

Take-Home Message for Section 24-6

✔ The mirror equation that relates the image distance, object distance, and focal length of a convex mirror is the same as for a concave mirror. The same is true for the equation that relates the heights of the image and the object to the image and object distances, and the equation that then relates the focal length and radius of curvature.

✔ The fundamental difference between concave and convex mirrors is that a convex mirror has a negative radius of curvature and so a negative focal length.

24-7 Convex lenses form images like concave mirrors and vice versa

A curved mirror forms images by reflection. A **lens**—a piece of glass or other transparent material with a curved surface on its front side, back side, or both sides—forms images by the *refraction* of light as the light enters and leaves the lens. Just as for a curved mirror, the images made by a lens can be larger or smaller than the object (Figure 24-24). In this section we'll explore how lenses form images.

The key idea that we need to understand lenses is the law of refraction. As we learned in Section 23-2, a light ray changes direction when it moves from one transparent material or medium to a second medium in which light travels at a different speed. The ray bends toward the normal if the speed of light is slower in the second medium (Figure 24-25a); this happens, for example, when light moves from air into glass. The ray bends away from the normal if the speed of light is faster in the second medium, as when light moves from glass into air (Figure 24-25b).

Figure 24-24 Lenses can magnify or shrink Different types of lenses can make (a) large or (b) small images.

This rodent appears larger when viewed through a magnifying glass, a lens that is convex on both sides.

The physicist appears smaller than actual size when viewed through this lens, which is concave on both sides.

(a)
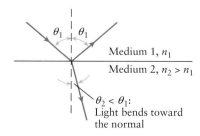

(b)
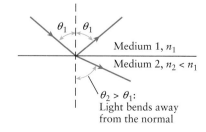

Figure 24-25 The law of refraction A light ray crossing the boundary between two transparent media can refract either (a) toward or (b) away from the normal, depending on how the speed of light compares in the two media.

(a) Refraction from one medium into a slower one

Medium 1, n_1
Medium 2, $n_2 > n_1$

$\theta_2 < \theta_1$:
Light bends toward the normal

(b) Refraction from one medium into a faster one

Medium 1, n_1
Medium 2, $n_2 < n_1$

$\theta_2 > \theta_1$:
Light bends away from the normal

(a) Parallel light rays entering a glass sphere

Each dashed line represents the normal to the surface at the point where a light ray enters the sphere.

The incident angle θ_1 and the refracted angle θ_2 obey the law of refraction.

θ_2 is less than θ_1 for light crossing from air to glass, so the rays are bent toward the principal axis of the sphere. Hence the light rays converge as they enter the sphere.

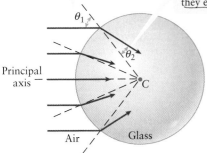

(b) Parallel light rays exiting a glass sphere

θ_2 is greater than θ_1 for light crossing from glass to air, so the rays are bent toward the principal axis of the sphere. Hence the light rays converge as they exit the sphere.

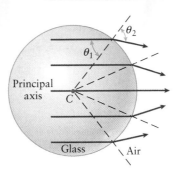

Figure 24-26 **Refraction by a glass sphere** If a glass sphere is surrounded by air, parallel light rays converge whether they (a) enter the sphere or (b) exit the sphere.

Figure 24-26 shows how parallel light rays refract when they enter or exit a glass sphere (in which the speed of light is relatively slow) surrounded by air (in which the speed of light is almost as fast as in vacuum). We say that both the front and back surfaces of the sphere are convex (they bulge outward). Part (a) of this figure shows that light rays converge when they enter through the convex surface of the sphere; part (b) shows that parallel rays also converge when they exit through the sphere's convex surface. Figure 24-27 shows what happens when parallel light rays enter or exit a piece of glass with a concave surface (one that bulges inward). In part (a) the parallel rays diverge as they cross the concave surface into the glass, and in part (b) the rays diverge as they exit through the concave surface of the glass.

In Figures 24-26 and 24-27 we've considered only what happens when light enters or leaves a glass object with a curved surface. In a lens the light enters from outside, passes through the lens, then exits. So each ray of light refracts *twice* as it travels into and out of the lens. Figure 24-26 shows that the refractions will make the rays converge if each surface is convex, and Figure 24-27 shows that the refractions will make the rays diverge if each surface is concave. So a lens with two convex surfaces—called a *convex lens*—will be a **converging lens** that takes incoming parallel light rays and makes them

(a) Parallel light rays entering a piece of glass with a spherical cutout

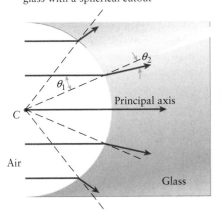

θ_2 is less than θ_1 for light crossing from air to glass, so the rays are bent away from the principal axis of the glass. Hence the light rays diverge as they enter the glass.

(b) Parallel light rays exiting a piece of glass with a spherical cutout

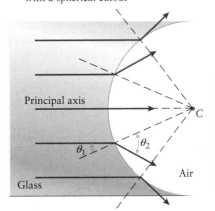

θ_2 is greater than θ_1 for light crossing from glass to air, so the rays are bent away from the principal axis of the glass. Hence the light rays diverge as they exit the glass.

Figure 24-27 **Refraction by glass with a spherical cutout** If a piece of glass with a spherical cutout is surrounded by air, parallel light rays diverge whether they (a) enter the glass through the cutout or (b) exit the glass through the cutout.

Figure 24-28 **Converging and diverging lenses** Two examples of converging lenses and two examples of diverging lenses.

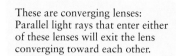

These are converging lenses: Parallel light rays that enter either of these lenses will exit the lens converging toward each other.

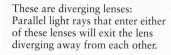

These are diverging lenses: Parallel light rays that enter either of these lenses will exit the lens diverging away from each other.

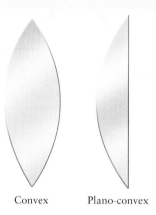

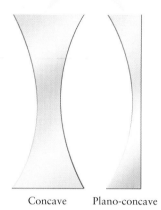

Convex   Plano-convex   Concave   Plano-concave

converge toward the principal axis. The same is true for a lens with one convex surface and one flat surface, called a *plano-convex* lens. The refraction at the flat surface by itself causes neither convergence nor divergence by itself. Similarly, a lens with two concave surfaces—called a *concave* lens—will be a **diverging lens** that takes incoming parallel light rays and makes them diverge away from the principal axis. A lens with one concave surface and one flat surface, called a *plano-concave* lens, will also be a diverging lens (**Figure 24-28**).

> **! Watch Out! The curvature of a piece of glass has the opposite effect to the same curvature in a mirror.**
>
> We saw in Section 24-3 that a concave mirror causes light rays to converge on reflection. By contrast, a concave lens causes light rays to diverge as they pass through. Likewise, while we saw in Section 24-5 that a convex mirror causes light rays to diverge on reflection, a convex lens causes light rays to converge as they pass through. Mirrors and lenses are different!

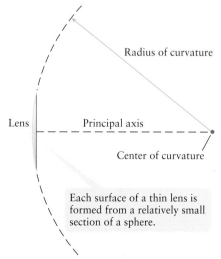

Radius of curvature

Lens   Principal axis

Center of curvature

Each surface of a thin lens is formed from a relatively small section of a sphere.

Figure 24-29 **A thin lens** Analyzing image formation by a lens is much easier if we assume that the lens is thin and has spherical surfaces.

A lens with spherical surfaces suffers from the same spherical aberration that we noted in Section 24-3 for spherical mirrors: Parallel light rays do not all focus to the same point for a converging lens, or appear to originate from a single point for a diverging lens. However, we can neglect this effect if the rays of light are all close to the principal axis. We can assure this by using only a small section of a large spherical surface to form each surface of the lens (**Figure 24-29**). In addition we'll assume that there is very little thickness of material between the front and back surfaces of the lens. The result is called a **thin lens**. We'll consider only thin lenses for the rest of this chapter, so that we can neglect spherical aberration and treat each lens as if parallel rays are focused to a single point. Eyeglasses and contact lenses are everyday examples of thin lenses.

Ray Diagrams for Converging Lenses

We saw earlier that ray diagrams are powerful tools for visualizing how a curved mirror produces an image. Let's see how to draw a ray diagram to help us locate the position of the image made by a converging lens.

Figure 24-30 shows a thin convex lens, created by placing two thin spherical sections such as the one shown in Figure 24-29 back-to-back. Notice that we have marked two points as a focal point *F*. For a thin lens the focal length, the distance from the center of the lens to the focal point, is the same on both sides of the lens, even if the two surfaces have a different radius of curvature. That's why we've drawn

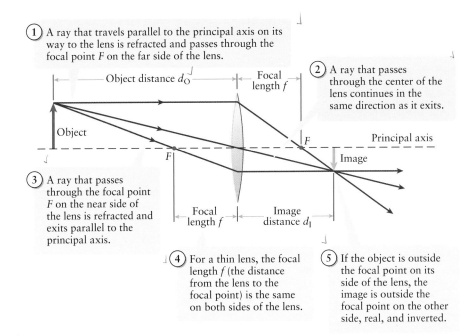

(1) A ray that travels parallel to the principal axis on its way to the lens is refracted and passes through the focal point F on the far side of the lens.

Object distance d_O

Focal length f

(2) A ray that passes through the center of the lens continues in the same direction as it exits.

Object

Principal axis

F

F

Image

(3) A ray that passes through the focal point F on the near side of the lens is refracted and exits parallel to the principal axis.

Focal length f

Image distance d_I

(4) For a thin lens, the focal length f (the distance from the lens to the focal point) is the same on both sides of the lens.

(5) If the object is outside the focal point on its side of the lens, the image is outside the focal point on the other side, real, and inverted.

▼ **Figure 24-30 Ray diagram for a converging lens I** How the image is formed for an object placed outside the focal point of a converging lens.

outside focal point

the two focal points F in Figure 24-30 the same distance from the geometrical center of the lens.

In Figure 24-30 we've drawn a red arrow as our object on the left-hand side of the lens, and placed this arrow outside the focal point. We've also drawn three representative light rays emanating from the tip of the object arrow. We've drawn the rays as if they refract only once, along the centerline of the lens (the line perpendicular to the principal axis that runs through the center of the lens). Notice how these three rays behave:

▶ *Go to Picture It 24-2 for more practice dealing with converging lenses*

(1) A ray that arrives at the lens traveling parallel to the principal axis is refracted so that it passes through the focal point F on the far (right-hand) side of the lens.

(2) A ray that strikes the lens directly at its center continues in a straight line as it exits. This is only an approximation because it assumes that the ray undergoes no refraction on entering or leaving the lens. But this is a good approximation for a thin lens, which is nearly flat at its very center.

(3) A ray that passes through the near (left-hand) focal point F on its way to the lens exits the lens traveling parallel to the principal axis. To see why this is the case, imagine that we could record a video of light traveling along this path, then run the video backward. Then we would see a light ray coming to the far right in Figure 24-30 and traveling parallel to the principal axis. After this ray passes through the lens from right to left, it would naturally pass through the focal point F on the left-hand side. If we now run that same video forward, we see light from the object following the path shown in Figure 24-30.

The image of the tip of the arrow in Figure 24-30 forms where the rays coming from the tip cross. The image of the base of the arrow must form along the principal axis, because a ray that comes from the base and travels along the axis strikes the center of the lens and therefore continues in the same direction. You can see that the image of the arrow is real (the light rays actually cross at the image point) and is inverted.

Note that for the case shown in Figure 24-30, the object is relatively far from the lens (the object distance d_O is greater than twice the focal length f) and the resulting image is real, inverted, *and* smaller than the object. That's exactly the same result we found for a concave mirror in Section 24-3 (see Figure 24-12). This reinforces the idea that a convex lens behaves similarly to a concave mirror. **Figure 24-31** shows an example of a small, inverted image formed when light from a distant object passes through a

Figure 24-31 A real, inverted image made by a converging lens The lens in the person's hand makes a real, inverted image of a park bench.

convex lens. You can see the same phenomenon in Figure 11-39 (Section 11-12), in which a sphere of water acts as a convex lens.

Watch Out! If a lens forms a real image, it must be on the side of the lens opposite to the object.

We've defined a real image as one that forms where reflected or refracted light rays converge. A concave mirror can cause light rays to converge only on the reflective side of the mirror, which is where we would put an object, so this is the only side on which a real image can form. A lens, however, can only cause light rays to converge on the side of the lens *opposite* to the object, as in Figure 24-30. As we'll see below, only a converging lens can produce a real image, and only if the object is outside the focal point as in Figure 24-30.

Recall that for mirrors our convention was that the image distance is positive if the image is real and so is on the same side of the mirror as the object. For a lens we'll also say that the image distance is positive if the image is real. This means that for a lens, a positive image distance d_I implies that the image is on the *opposite* side of the lens from the object. For example, the image distance is positive in Figure 24-30.

In Figure 24-32 we've moved the standard object to a point between the lens and the focal point, so the object distance is less than the focal length: $d_O < f$. To locate the image, we've drawn just two light rays from the tip of the arrow. One ray moves parallel to the principal axis of the lens and therefore passes through the focal point after being refracted by the lens. The other ray we have drawn strikes the center of the lens and so continues straight. (In Figure 24-30 we drew a third ray through the focal point on the same side of the lens as the object. We don't draw that ray here, since it leads away from the lens rather than toward it.) Notice that these two rays do not meet. But if we extend the rays backward as shown by the dashed lines, they appear to meet on the same side of the lens as the object. Where they meet is the location of the image. The rays do not actually cross there, so this is a *virtual* image. Because the image is on the same side of the lens as the object, the image distance is negative: $d_I < 0$.

As in Figure 24-30, the base of the image arrow must lie on the principal axis of the lens. It follows that the image is upright. Note that the image is also larger than the object. This is the kind of image shown in Figure 24-24a; when we view a rodent by holding a magnifier (a convex lens) close to the rodent (so that the rodent is inside the focal point of the lens), we see an enlarged, upright image of the rodent. This is very similar to the image made by a concave mirror when an object is placed inside the focal point of the mirror (see Figure 24-16).

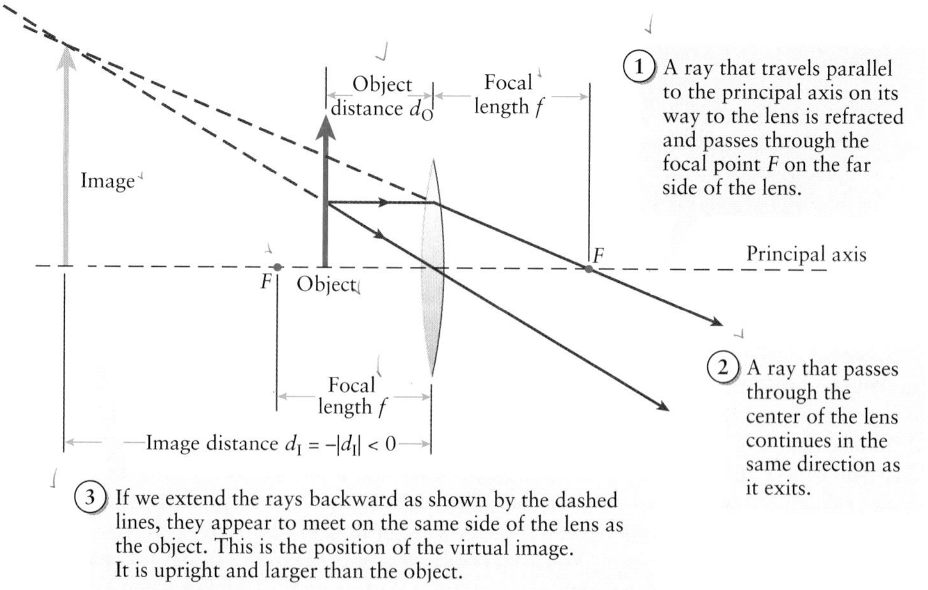

(handwritten margin note: Real image: converging lens & object outside focal point)

(handwritten margin note: inside focal point)

① A ray that travels parallel to the principal axis on its way to the lens is refracted and passes through the focal point F on the far side of the lens.

② A ray that passes through the center of the lens continues in the same direction as it exits.

③ If we extend the rays backward as shown by the dashed lines, they appear to meet on the same side of the lens as the object. This is the position of the virtual image. It is upright and larger than the object.

Figure 24-32 Ray diagram for a converging lens II How the image is formed for an object placed inside the focal point of a converging lens.

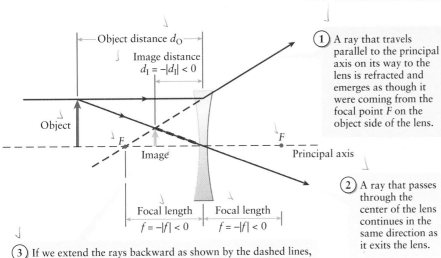

Figure 24-33 **Ray diagram for a diverging lens** How the image is formed for an object placed in front of a diverging lens.

① A ray that travels parallel to the principal axis on its way to the lens is refracted and emerges as though it were coming from the focal point *F* on the object side of the lens.

② A ray that passes through the center of the lens continues in the same direction as it exits the lens.

③ If we extend the rays backward as shown by the dashed lines, they appear to meet on the same side of the lens as the object. This is the position of the virtual image. It is upright and smaller than the object.

Ray Diagrams for Diverging Lenses

We can also use ray diagrams to learn about the image made by a diverging lens. In Figure 24-33 we've placed an object arrow in front of a thin lens with two concave surfaces. Since this lens causes parallel light rays to diverge, we say that the focal length is negative: $f = -|f| < 0$. (The focal length of a convex mirror, which also causes parallel light rays to diverge, is also negative.)

We've drawn two light rays to determine where the image forms, and as we did for the thin, convex lens, we've drawn the rays as if they refract only once along the center-line of the lens. The refracted rays don't actually meet, but if we extend these rays backward, we find the location where the extensions meet. The image of the tip of the arrow forms at this point; the image is virtual (since the rays don't actually meet there), upright, and smaller than the object. This is the same behavior we saw in Section 24-5 for convex mirrors. As for a convex mirror, the image is virtual, upright, and smaller no matter what the object distance.

Figure 24-24b shows an image made by a concave lens. Just like the image in Figure 24-33, this image is upright and smaller than the object (the physicist standing behind the lens).

❓ Got the Concept? 24-8   A Lens in Sugar Water

You make a lens with two convex surfaces out of ice ($n = 1.31$). You then submerge the lens in a large tank of concentrated sugar water ($n = 1.49$). If you place an object in the tank and in front of the lens, an image is formed at a position inside the tank. What kind of image is this? (a) A real image; (b) a virtual image that is larger than the object; (c) a virtual image that is smaller than the object; (d) either (a) or (b), depending on the distance from the object to the lens; (e) any of (a), (b), or (c), depending on the distance from the object to the lens.

Take-Home Message for Section 24-7

✔ When parallel light rays enter one side of a convex lens, they exit the lens converging toward the focal point on the other side of the lens.

✔ When parallel light rays enter a concave lens, they exit the lens diverging away from the focal point on the same side of the lens that the rays entered.

✔ A convex lens produces a real image if the object is outside the focal point, and produces a virtual image if the object is inside the focal point.

✔ A concave lens produces a virtual image no matter what the position of the object.

24-8 The focal length of a lens is determined by its index of refraction and the curvature of its surfaces

As we did for concave mirrors and convex mirrors, we'd like to find equations for the focal length of a lens; for the relationship among the focal length of a lens, the object distance, and the image distance; and for the magnification of an image produced by a lens. As we'll see, the latter two equations turn out to be identical to those for curved mirrors. The expression for the focal length, however, is a little more complicated and requires a bit of interpretation.

Focal Length of a Thin Lens

The focal length f of a thin lens depends on the index of refraction n of the material of which it is made. The greater the value of n, the more sharply a light ray is bent as it passes either from the surrounding air (for which the index of refraction is essentially 1) into the lens or from the lens into the air. The value of f also depends on how the front and back surfaces of the lens are curved. The mathematical expression of these relationships for a lens in air is called the **lensmaker's equation:**

Lensmaker's equation for the focal length of a thin lens (24-24)

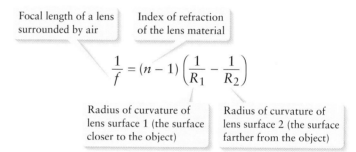

Focal length of a lens surrounded by air   Index of refraction of the lens material

$$\frac{1}{f} = (n - 1)\left(\frac{1}{R_1} - \frac{1}{R_2}\right)$$

Radius of curvature of lens surface 1 (the surface closer to the object)   Radius of curvature of lens surface 2 (the surface farther from the object)

Equation 24-24 can be derived by applying Snell's law of refraction (Equation 23-6) to a ray of light refracted by both surfaces of a lens. The derivation is beyond our scope, however.

The values of the radii of curvature R_1 and R_2 depend on how sharply and in what direction the two surfaces of the lens are curved. As Figure 24-34 shows, we take a radius to be positive if the center of curvature is on the other side of the lens from the object, but negative if the center of curvature is on the same side as the object. For example, in Figure 24-34 the radius R_1 of surface 1 is positive, but the radius R_2 of surface 2 is negative. The following examples illustrate how to use Equation 24-24.

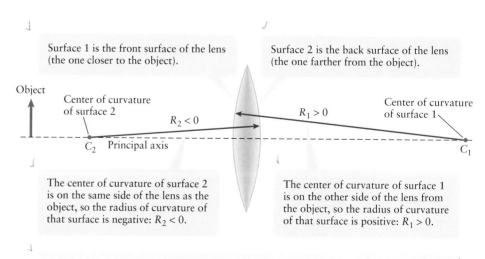

Surface 1 is the front surface of the lens (the one closer to the object).

Surface 2 is the back surface of the lens (the one farther from the object).

Object

Center of curvature of surface 2

$R_2 < 0$

$R_1 > 0$

Center of curvature of surface 1

C_2 Principal axis

C_1

The center of curvature of surface 2 is on the same side of the lens as the object, so the radius of curvature of that surface is negative: $R_2 < 0$.

The center of curvature of surface 1 is on the other side of the lens from the object, so the radius of curvature of that surface is positive: $R_1 > 0$.

Figure 24-34 Interpreting the lensmaker's equation For each surface of a thin lens, the sign of the radius of curvature depends on where the center of curvature is located.

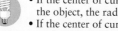

- If the center of curvature of a lens surface is on the other side of the lens from the object, the radius of curvature R of that surface is positive.
- If the center of curvature of a lens surface is on the same side of the lens as the object, the radius of curvature R of that surface is negative.
- A flat surface has an infinite radius of curvature R, so $1/R = 0$ for that surface.

Example 24-3 Calculating Focal Length for a Convex Lens

In Figure 24-34 the front surface (surface 1) of the lens has a radius of curvature of magnitude 15.0 cm, and the back surface (surface 2) has a radius of curvature of magnitude 25.0 cm. The lens is made of crown glass with an index of refraction 1.520. (a) Calculate the focal length of the lens. (b) Next flip the orientation of the lens so that the front surface is now the one with a radius of curvature of magnitude 25.0 cm, and the back surface is now the one with a radius of curvature of magnitude 15.0 cm. Calculate the focal length of the lens in this case.

Set Up

For each situation we'll draw the lens and determine on which side of the lens the centers of curvature lie; that will tell us whether R_1 and R_2 are positive or negative. We'll then use Equation 24-24 to calculate the focal length.

Lensmaker's equation for the focal length of a thin lens:

$$\frac{1}{f} = (n - 1)\left(\frac{1}{R_1} - \frac{1}{R_2}\right) \tag{24-24}$$

Solve

(a) Both surfaces are convex. Hence the center of curvature of each surface is on the other side of the lens from that surface. The center of curvature C_1 of the front surface (on the left in the figure) is on the far side of the lens, so R_1 is positive: $R_1 = +15.0$ cm. The center of curvature C_2 of the back surface (on the right in the figure) is on the near side of the lens, so R_2 is negative: $R_2 = -25.0$ cm.

Calculate the focal length.

From Equation 24-24,

$$\frac{1}{f} = (1.520 - 1)\left(\frac{1}{(+15.0 \text{ cm})} - \frac{1}{(-25.0 \text{ cm})}\right)$$

$$= 0.520[(0.0667 \text{ cm}^{-1}) - (-0.0400 \text{ cm}^{-1})]$$

$$= 0.0555 \text{ cm}^{-1}$$

$$f = \frac{1}{0.0555 \text{ cm}^{-1}} = 18.0 \text{ cm}$$

Although R_2 is negative, we subtract rather than add $1/R_2$ in Equation 24-24, so both the $1/R_1$ term and the $1/R_2$ term—that is, both surface 1 and surface 2—contribute to giving $1/f$ a positive value. As a result, the focal length is positive, as we expect for a convex lens.

(b) Because we have flipped the lens around, we have interchanged surfaces 1 and 2. The object is still to the left of the lens, however. As in part (a), R_1 is positive and R_2 is negative, but now $R_1 = +25.0$ cm and $R_2 = -15.0$ cm.

Calculate the focal length.

From Equation 24-24,

$$\frac{1}{f} = (1.520 - 1)\left(\frac{1}{(+25.0 \text{ cm})} - \frac{1}{(-15.0 \text{ cm})}\right)$$

$$= 0.520[(0.0400 \text{ cm}^{-1}) - (-0.0667 \text{ cm}^{-1})]$$

$$= 0.0555 \text{ cm}^{-1}$$

$$f = \frac{1}{0.0555 \text{ cm}^{-1}} = 18.0 \text{ cm}$$

Again both the $1/R_1$ term and the $1/R_2$ term contribute to giving $1/f$ a positive value. The focal length has the same positive value as in part (a).

Reflect

The focal length of this thin lens stays the same after we flip it back to front. That's consistent with the statement we made in the previous section that the focal points on either side of the lens are both the same distance (that is, the same focal length) from the lens.

Example 24-4 Calculating Focal Length for a Concave Lens

A certain lens made of crown glass ($n = 1.520$) has concave front and back surfaces. The front surface has a radius of curvature of magnitude 15.0 cm, and the back surface has a radius of curvature of magnitude 25.0 cm. Calculate the focal length of the lens.

Set Up

As in the previous example, we'll first draw the lens and use our drawing to decide whether R_1 and R_2 are positive or negative. Equation 24-24 will then allow us to calculate the focal length.

Lensmaker's equation for the focal length of a thin lens:

$$\frac{1}{f} = (n - 1)\left(\frac{1}{R_1} - \frac{1}{R_2}\right) \tag{24-24}$$

Solve

Both surfaces are concave. Hence the center of curvature of each surface is on the same side of the lens as that surface. The center of curvature C_1 of the front surface (on the left in the figure) is on the near side of the lens, so R_1 is negative: $R_1 = -15.0$ cm. The center of curvature C_2 of the back surface (on the right in the figure) is on the far side of the lens, so R_2 is positive: $R_2 = +25.0$ cm.

Calculate the focal length.

From Equation 24-24,

$$\frac{1}{f} = (1.520 - 1)\left(\frac{1}{(-15.0\text{ cm})} - \frac{1}{(+25.0\text{ cm})}\right)$$
$$= 0.520[(-0.0667\text{ cm}^{-1}) - (0.0400\text{ cm}^{-1})]$$
$$= -0.0555\text{ cm}^{-1}$$

$$f = \frac{1}{(-0.0555\text{ cm}^{-1})} = -18.0\text{ cm}$$

The $1/R_1$ term is negative, and we subtract from it the positive $1/R_2$ term. So both terms contribute to giving $1/f$ a negative value so that the focal length is negative.

Reflect

The focal length is negative, which is just what we expect from a concave lens. We encourage you to repeat the calculation with the lens reversed back to front, as in part (b) of Example 24-3; you should get the same result for the focal length. Do you?

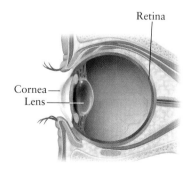

Retina

Cornea—
Lens—

Figure 24-35 The human eye Both the cornea and the lens cause light to refract at their curved surfaces. Together, these parts of the eye cause light to be focused on the retina.

Two structures in the human eye focus images on the retina, allowing us to see clearly: the cornea, which makes up the outer surface of the eye, and the lens, which is suspended inside the eye behind the pupil (Figure 24-35). The lens is made of a flexible material. As tiny muscles around the lens contract and relax, the lens becomes more or less round. As the shape of the lens changes, the radius of curvature of its surfaces changes, and this in turn changes the focal length of the lens. This is important because no matter how far away an object is, the image distance is always the same—the distance from the lens to the retina, where images are formed and detected. So the focal length has to change in order to have the image remain on the retina.

By the time we reach middle age, the lenses in our eyes eventually lose some of their elasticity. As a result, the front and back surfaces of the lenses cannot curve as sharply, and light from objects close to the eyes cannot be focused on the retina. This makes it difficult to see objects at close range. This condition, called *presbyopia*, is treated by wearing reading glasses. These glasses are made from convex lenses that help the eye focus the light from a nearby object onto the retina. The lenses of the eyes can also become clouded by cataracts, in which case they are surgically replaced by artificial lenses (see Figure 24-1b).

Image Position and Magnification for a Thin Lens

To see how the image distance is related to the object distance and the focal length for a thin, convex lens, let's look again at a ray diagram like Figure 24-30. We've drawn such a diagram in Figure 24-36. Let's see how to use trigonometry to find relationships between the distances of interest.

The shaded regions in Figure 24-36a are similar triangles. That's because the ray from the tip of the object at O' to the tip of the image at I' passes through the center of the lens at C without deflection, so the angle θ is the same in both triangle $O'C'C$ and triangle ICI'. The ratio of the heights of the two triangles is therefore equal to the ratio of their bases; that is

$$\frac{|h_I|}{h_O} = \frac{d_I}{d_O} \tag{24-25}$$

(The height h_I of the image in Figure 24-36a is negative because the image is inverted. That's why we've used $|h_I|$ in Equation 24-25 for the distance from I to I'.) The two shaded regions in Figure 24-36b—the right triangles $FC'C$ and $FI'I$—are also similar triangles. That's because the straight ray from C' to I' passes through the focal point, so the angle ϕ is the same on either side of F. So for these two triangles as well, the ratio of their heights is equal to the ratio of their bases:

$$\frac{|h_I|}{h_O} = \frac{d_I - f}{f} \tag{24-26}$$

If we combine Equation 24-25 and Equation 24-26, we find a relationship between the image distance d_I, the object distance d_O, and the focal length f:

$$\frac{d_I}{d_O} = \frac{d_I - f}{f} \quad \text{or} \quad \frac{d_I}{d_O} = \frac{d_I}{f} - 1 \quad \text{or} \quad \frac{d_I}{d_O} + 1 = \frac{d_I}{f}$$

Divide both sides by d_I:

$$\frac{1}{d_O} + \frac{1}{d_I} = \frac{1}{f} \quad \text{(lens equation)} \tag{24-11}$$

This equation should look familiar; it's the same equation that we deduced for spherical mirrors in Section 24-4. It applies just as well to thin, spherical convex lenses. Now

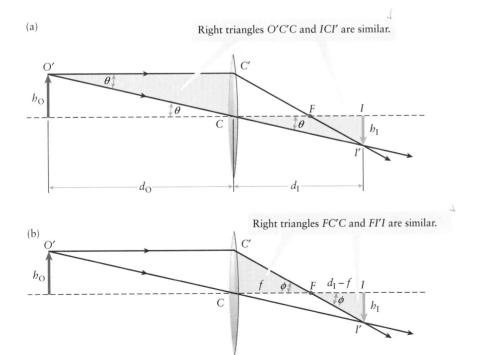

Figure 24-36 Analyzing a converging lens Ray diagrams for an object arrow on the principal axis of a converging lens. The similar triangles in (a) and (b) help us determine the position and magnification of the image.

we see why we were justified in calling Equation 24-11 the mirror and lens equation—it works for both.

The lens equation is expressed in terms of the reciprocals of the object distance d_O, image distance d_I, and focal length f. For that reason it's common, especially among optometrists, to characterize lenses (and also mirrors) in terms of the reciprocal of the focal length:

(24-27)
$$P = \frac{1}{f}$$

The quantity P is called the **power** of the lens or mirror. The units of P are m^{-1}; $1\ m^{-1}$ is known as 1 **diopter**. The larger the power of a thin, convex lens, the closer to the lens parallel light rays are brought to a focus. A 2-diopter lens brings parallel rays to a focus $1/2$ m (0.5 m) from the lens, while a 5-diopter lens brings parallel rays to a focus $1/5$ m (0.2 m) from the lens.

Equation 24-25 above also tells us the magnification of the image. Since the image height h_I is negative in Figure 24-36a (the image is inverted), h_I is equal to $-|h_I|$ and $|h_I| = -h_I$. If we substitute this into Equation 24-25, we get

$$\frac{(-h_I)}{h_O} = \frac{d_I}{d_O} \quad \text{or} \quad \frac{h_I}{h_O} = -\frac{d_I}{d_O}$$

Lateral magnification is equal to the ratio of image height to object height: $m = h_I/h_O$ (Equation 24-2). So we can write the magnification of the image produced by a thin lens as

(24-14)
$$m = \frac{h_I}{h_O} = -\frac{d_I}{d_O} \quad \text{(lateral magnification)}$$

That's the same Equation 24-14 that we derived in Section 24-4 for a curved mirror.

We've derived Equations 24-11 and 24-14 for a thin convex lens with a positive focal length, but they turn out to be equally valid for a thin concave lens with a negative focal length. As for mirrors, it's important to keep track of the signs of quantities in these equations. Table 24-2 summarizes when the quantities in the lens-maker's equation (Equation 24-24), the lens equation (Equation 24-11), and the magnification equation (Equation 24-14) are positive and when they are negative.

Because the equations for mirrors and thin lenses are effectively identical, many of the same conclusions that we came to for images made by a curved mirror also apply to images made by a thin lens:

- If the lens has a positive focal length (a converging lens) and an object is placed outside the focal point, the image is real and inverted. Depending on the object distance, the image can be smaller, larger, or the same size as the object.

- If the lens has a positive focal length (a converging lens) and an object is placed inside the focal point of a converging lens, the image is virtual, upright, and larger than the object.

- If the lens has a negative focal length (a diverging lens), the image is virtual, upright, and smaller than the object. This is true for any object distance.

One key difference between mirrors and lenses is in the position of the image. For a mirror a real image is on the same side of the mirror as the object, and a virtual image is on the other side of the mirror. For a lens a real image is on the opposite side of the lens from the object, and a virtual image is on the same side of the lens as the object.

Go to Interactive Exercise 24-3 for more practice dealing with lenses

Table 24-2 Sign Conventions for Lenses

lens surface radius of curvature R_1 or R_2	• positive if the center of curvature is on the side of the lens opposite from the object • negative if the center of curvature is on the same side of the lens as the object
focal length f	• positive for a converging lens • negative for a diverging lens
image distance d_I	• positive if on the side of the lens opposite from the object; the image is then a real image • negative if on the same side of the lens as the object; the image is then a virtual image
image height h_I	• positive if the image is upright (the same orientation as the object) • negative if the image is inverted (flipped upside down compared to the object)
lateral magnification m	• positive if the image is upright (the same orientation as the object) • negative if the image is inverted (flipped upside down compared to the object)

Example 24-5   An Image Made by a Convex Lens

A thin convex lens has a focal length of 15.0 cm. How far from the lens does the image form when an object is placed 9.00 cm from the center of the lens? Is the image virtual or real? Is the image inverted or upright? By what factor is the image magnified relative to the object?

Set Up

We'll begin by drawing a ray diagram to help us visualize the kind of image that will be produced. We'll use Equation 24-11 to calculate the position of this image, and Equation 24-14 to calculate the image height compared to the object height.

Lens equation:

$$\frac{1}{d_O} + \frac{1}{d_I} = \frac{1}{f} \qquad (24\text{-}11)$$

Magnification:

$$m = \frac{h_I}{h_O} = -\frac{d_I}{d_O} \qquad (24\text{-}14)$$

Solve

We begin by drawing a ray diagram. We've drawn one ray from the object that enters the lens parallel to the principal axis, exits the lens, and passes through the focal point F on the other side of the lens. We've also drawn a ray that passes through the center of the lens. The rays never meet on the other side of the lens, but their extensions do meet on the same side of the lens as the object. So the image will be virtual. As the diagram shows, the image is also upright and larger than the object.

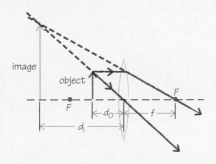

Find the image distance using the lens equation.

We are given $d_O = 9.00$ cm and $f = 15.0$ cm. From Equation 24-11,

$$\frac{1}{d_I} = \frac{1}{f} - \frac{1}{d_O} = \frac{1}{15.0 \text{ cm}} - \frac{1}{9.00 \text{ cm}}$$

$$= 0.06667 \text{ cm}^{-1} - 0.111 \text{ cm}^{-1}$$

$$= -0.0444 \text{ cm}^{-1}$$

$$d_I = \frac{1}{(-0.0444 \text{ cm}^{-1})} = -22.5 \text{ cm}$$

The negative value of d_I means that the image is on the same side of the lens of the object, just as the ray diagram shows. It must therefore be a virtual image.

Find the image height using the magnification equation.

From Equation 24-14,

$$m = \frac{h_I}{h_O} = -\frac{d_I}{d_O} = -\frac{(-22.5 \text{ cm})}{9.00 \text{ cm}} = +2.50$$

The plus sign means the image is upright, in agreement with the ray diagram. The image is 2.50 times as large as the object.

Reflect

Although we didn't try to draw the ray diagram to exact scale, by eye the ratio of the focal length to the object distance looks to be about 2 to 1, which is certainly consistent with the actual values $f = 15.0$ cm and $d_O = 9.00$ cm. So we expect that the image distance in the diagram is about one and half times the focal length (22.5 cm compared to 15.0 cm)—and it is. Can you verify that the height of the image in the diagram is about 2.5 times the height of the object?

❓ Got the Concept? 24-9   Focusing a Camera

Light rays from a distant object are essentially parallel. That's why to focus a digital camera on a distant object, the lens is adjusted so that the distance from the lens to the electronic image sensor (which records the image) is equal to the focal length f. To focus on a nearby object, the lens must be positioned so that the distance from the lens to the image sensor is (a) less than f; (b) more than f; (c) f.

? Got the Concept? 24-10 A Magnifying Glass

By placing an object closer to your eyes, it looks larger and you can see it in more detail. The object has an angular size θ_0 (Figure 24-37a). If you're typical, however, you can focus on an object only if it's about 25 cm or farther from your eyes, which may not allow you to see small details clearly. A magnifying glass, a convex lens with a relatively short focal length, is the solution. When a magnifying glass is placed in between the object and your eye and the object is closer to the lens than the focal point, the image forms with angular size θ_M, which is larger than θ_0. In addition, the image forms farther from your eye than the 25 cm limit so you can focus clearly on it. Draw a ray diagram that shows the principle of a magnifying glass, starting from the sketch in Figure 24-37b.

(a)   When an object is placed close to the eye it spans an angular size θ_0.

(b)   A magnifying glass is placed between the object and the eye, in order to see more detail without moving the object closer to the eye. Where does the image form?

Figure 24-37 **A magnifying glass** (a) The angular size of an object. (b) Use this setup to draw a ray diagram that demonstrates the principle of a magnifying glass.

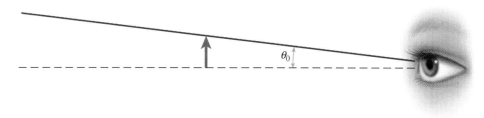

? Got the Concept? 24-11 Microscopes

You've used microscopes many times, but do you know how they work? By applying what you've learned about lenses, you can figure it out for yourself! A simple compound microscope has two lenses, the objective and the eyepiece. An object is placed just beyond the focal point of the objective in order to form a magnified image. Using the sketch in Figure 24-38, determine the location and size of the image by tracing rays from the tip of the arrow to the eye. Is the image real or virtual?

Figure 24-38 **A compound microscope** Use this setup to draw a ray diagram that demonstrates the principle of a compound microscope.

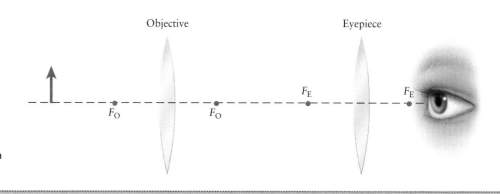

Take-Home Message for Section 24-8

✔ The focal length of a thin lens is determined by the index of refraction of the lens material and the radii of curvature of the front and back surfaces of the lens. These radii can be positive or negative.

✔ The same equation that relates object distance, image distance, and focal length for a spherical mirror also applies to thin lenses. The magnification equation is also the same for spherical mirrors and thin lenses.

✔ Image distance d_I is positive when the image is real and negative when the image is virtual. The focal length f is positive for a thin, convex lens and negative for a thin, concave lens.

Key Terms

center of curvature
concave
converging lens
convex
diffuse light
diopter
diverging lens
focal length
focal point
geometrical optics
image

image distance
image height
inverted image
lateral magnification
lens equation
lensmaker's equation
mirror equation
object
object distance
object height
plane mirror

power (of a lens or mirror)
principal axis
radius of curvature
ray diagram
real image
specular reflection
thin lens
upright image
virtual image

Chapter Summary

Topic	Equation or Figure	
Plane mirrors: A plane mirror makes an upright, virtual image of any object. The image is the same size as the object, so the lateral magnification is $m = 1$. The image is reversed back to front, not side to side.	The image distance is negative. The object distance is positive. $$d_I = -|d_I| = -d_O$$ The negative value of d_I indicates that the image is on the opposite side of the mirror from the object.	(24-1)
	Lateral magnification Image height $$m = \frac{h_I}{h_O}$$ Object height	(24-2)
Spherical mirrors: If we consider only parallel light rays that are close to the principal axis of a spherical concave mirror, the reflected rays all converge at the focal point. The distance from the center of the mirror to the focal point is the focal length. If the mirror is convex rather than concave, parallel light rays diverge rather than converge after reflection. The radius of curvature and the focal length are both negative for a convex mirror.	**Focal length** of a spherical mirror **Radius of curvature** of the mirror $$f = \frac{r}{2}$$	(24-3)

Image formation by a concave mirror:
We can locate the image made by a concave mirror by drawing a ray diagram. If the object is outside the focal point, the image is real and inverted; its size depends on how far the object is from the mirror. If the object is inside the focal point, the image is virtual, upright, and larger than the object.

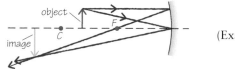

(Example 24-1, figure #1)

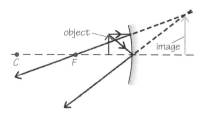

(Example 24-1, figure #2)

Image formation by a convex mirror:
A ray diagram also helps us locate the image made by a convex mirror. No matter where the object is placed, the image is virtual, upright, and smaller than the object.

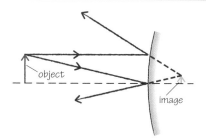

(Example 24-2, figure #1)

The mirror equation and lens equation:
The mirror equation and lens equation relates the object distance, image distance, and focal length for either a concave or convex mirror or a converging or diverging lens. The lateral magnification depends on the image and object distances, and can be positive (if the image is upright) or negative (if the image is inverted).

$$\frac{1}{d_O} + \frac{1}{d_I} = \frac{1}{f} \quad \text{Focal length}$$

(24-11)

Object distance | Image distance

Lateral magnification | Image height | Image distance

$$m = \frac{h_I}{h_O} = -\frac{d_I}{d_O}$$

(24-14)

Object height | Object distance

Lenses: Due to refraction, light rays converge as they enter or exit a converging glass lens and diverge as they enter or exit a diverging glass lens. A converging lens behaves similarly to a concave mirror, and a diverging lens behaves similarly to a convex mirror. The focal length of a lens is given by the lensmaker's equation, which involves the index of refraction of the lens material and the radii of curvature of the lens surfaces. This equation is valid if the lens is thin.

These are converging lenses: Parallel light rays that enter either of these lenses will exit the lens converging toward each other.

These are diverging lenses: Parallel light rays that enter either of these lenses will exit the lens diverging away from each other.

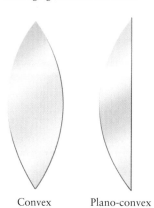
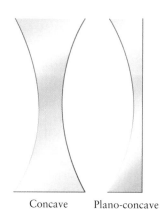

(Figure 24-28)

Convex   Plano-convex   Concave   Plano-concave

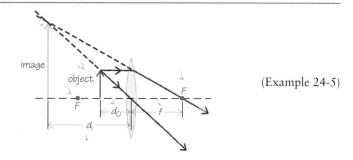

$$\frac{1}{f} = (n - 1)\left(\frac{1}{R_1} - \frac{1}{R_2}\right) \qquad \text{(24-24)}$$

Focal length of a lens surrounded by air

Index of refraction of the lens material

Radius of curvature of lens surface 1 (the surface closer to the object)

Radius of curvature of lens surface 2 (the surface farther from the object)

Image formation by lenses: A converging lens produces a real, inverted image if the object is outside the focal point, and a virtual, upright, enlarged image if the object is inside the focal point. A diverging lens always produces a virtual, upright, reduced image. The same equation that relates object distance d_O, image distance d_I, and focal length f for a spherical mirror also applies to thin lenses, as does the equation for lateral magnification.

(Example 24-5)

Answer to What do you think? Question

(a) These eyeglasses are used by people with presbyopia. Their eyes are able to take essentially parallel light rays from distant objects and make them converge onto the retina to form an image, but aren't able to do the same with the diverging light rays coming from a nearby object. The corrective eyeglasses take these diverging light rays and make them nearly parallel, so a converging lens is needed. A converging lens is thicker at the middle than it is at the edges. We discuss the physics of the eye and of lenses in Sections 24-7 and 24-8.

Answers to Got the Concept? Questions

24-1 (a) If you see someone's eyes in a mirror, she can see you as well. This must be true because if light rays can reflect off the mirror from her eyes to yours, then light can follow that path in the opposite direction and reflect from your eyes to hers, too.

24-2 (c) From the information given it's not possible to know whether the person around the corner can see you. We've sketched this situation in **Figure 24-39**. Light rays that reflect off the mirror from his right hand to your eyes, shown in red, enable you to see his hand. You can't see his eyes because the path that light would need to follow for this to happen, shown by the light blue, dashed line, does not intercept the mirror.

The light ray drawn in red leaves the right hand of the person around the corner, reflects off the mirror, and arrives at your eyes. You can see his right hand.

Mirror

The other person from above

If your right arm is extended, light can follow a path from your hand to the mirror and then to the other person's eyes.

The dashed light blue line indicates the path a light ray would have to travel for you to see his eyes, and for him to see yours. The mirror does not extend far enough for light to reflect in this way.

You from above

Figure 24-39

Whether the other person can see some part of you, however, depends on how far out you are holding your right hand. If it is far enough away from your body, light rays can follow the path drawn in dark blue, from your hand to his eyes.

24-3 (b) When an object is placed farther from a concave mirror than the focal point, the image is real and inverted. Note that a real image forms in front of a mirror; that is, it appears to be in front of the mirror. So when you hold the spoon at arm's length, the object (you!) must therefore be farther from the reflective surface than the focal point. When you hold the spoon 4 cm from your eye, however, you (your eye) must be closer to the spoon than its focal point. An object placed closer to a reflective surface than the focal point forms a virtual, inverted image. Because a virtual image is formed at the point where light rays from any point on an object appear to meet (but don't actually) the image forms behind the mirror and is upright. The focal point of the spoon must be about 6 cm from the spoon. At the focal point no image forms because light rays from any point on your face are reflected parallel to each other.

24-4 (e) A concave mirror makes an image because light rays from a point on the object that strike anywhere on the mirror are all reflected to the same point on the image. If you black out part of the mirror, there are fewer positions on the mirror that contribute to the image, but they still form an image at the same place. This is true for any point on the object, so the image will still show the entire light bulb. The only difference is that the image will be dimmer because only half as much light energy is reflected by the half-painted mirror.

24-5 (b) A real image made by a concave mirror is also an inverted image, so the magnification is $m = -2.00$. Equation 24-14 then tells us the image distance: $m = -d_I/d_O$, so $d_I = -md_O = -(-2.00)(12.0 \text{ cm}) = +24.0 \text{ cm}$. (The positive value of the image distance tells us that the image is in front of the mirror, as must be true for a real image.) The mirror equation, Equation 24-11, then tells us the focal length: $1/f = 1/d_O + 1/d_I = 1/(12.0 \text{ cm}) + 1/(24.0 \text{ cm}) = 0.0833 \text{ cm}^{-1} + 0.0417 \text{ cm}^{-1} = 0.125 \text{ cm}^{-1}$, so $f = 1/(0.125 \text{ cm}^{-1}) = 8.00 \text{ cm}$. Note that the object distance of 12.0 cm is greater than the focal length, as must be the case if the image is to be real.

24-6 (b) A concave mirror can focus parallel light rays; a convex mirror cannot, and so it cannot concentrate solar energy onto a pot for cooking. Another way to see this is to think of the mirror as making an image of the Sun that lies at the position of the pot so as to concentrate the Sun's light there. This is possible with a concave mirror, which makes a real image of a distant object. A convex mirror, by contrast, can only make a virtual image whose rays never actually cross. Hence a convex mirror is no good for concentrating solar energy.

24-7 (c) We can solve this problem using the mirror equation, Equation 24-11, and the expression for magnification, Equation 24-14. The image is one-half the height of the object, so the lateral magnification is $m = h_I/h_O = 1/2$. From Equation 24-14 the magnification m is also equal to $-d_I/d_O$, so $-d_I/d_O = 1/2$ and $d_I = -d_O/2$. That is, the image is half as far from the mirror as the object is and is behind the mirror (because d_I is negative). Substitute $d_I = -d_O/2$ into the mirror equation and solve for d_O:

$$\frac{1}{d_O} + \frac{1}{d_I} = \frac{1}{f} \quad \text{so} \quad \frac{1}{d_O} + \left(-\frac{2}{d_O}\right) = -\frac{1}{d_O} = \frac{1}{f}$$

$$\text{and} \quad d_O = -f = |f|$$

(The focal length of a convex mirror is negative, so $f = -|f|$.) The distance from the object to the mirror equals the distance from the mirror to the focal point.

24-8 (c) The lenses we described in Section 24-7 are made of a material in which light travels more slowly than in the surrounding air—that is, a material with a higher index of refraction than the surroundings. In this situation, however, the lens is made of a material with a *lower* index of refraction

than the surroundings. Hence light rays refract in the opposite sense as they enter and exit the lens, which means that this convex lens is actually a diverging lens. Like the diverging lens in Figure 24-33, the image made by this lens is virtual and smaller than the object.

24-9 (b) As the object approaches the lens, the object distance decreases and so the image distance increases. Hence the sensor must be placed farther from the lens, or equivalently the lens must be farther away from the sensor. On some digital cameras you can actually see the lens move outward from the camera body as the lens focuses on a nearby object.

24-10 Figure 24-40 shows the ray diagram.

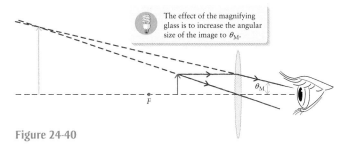

The effect of the magnifying glass is to increase the angular size of the image to θ_M.

Figure 24-40

24-11 Figure 24-41 shows the ray diagram.

Notice that the ray that passes through the focal point of the objective exits parallel to the principal axis of the objective, so that this ray enters the eyepiece parallel to the principal axis of the eyepiece. For that reason, this ray passes through the focal point of the eyepiece when it exits.

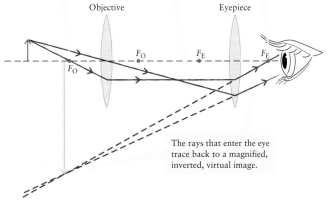

The rays that enter the eye trace back to a magnified, inverted, virtual image.

Figure 24-41

Questions and Problems

In a few problems, you are given more data than you actually need; in a few other problems, you are required to supply data from your general knowledge, outside sources, or informed estimate.

Interpret as significant all digits in numerical values that have trailing zeros and no decimal points.

For all problems, use $g = 9.80 \text{ m/s}^2$ for the free-fall acceleration due to gravity. Neglect friction and air resistance unless instructed to do otherwise.

• Basic, single-concept problem

•• Intermediate-level problem, may require synthesis of concepts and multiple steps

••• Challenging problem

SSM *Solution is in Student Solutions Manual*

Conceptual Questions

1. •A plane mirror seems to invert your image left and right but not up and down. Why is this? SSM

2. •Explain why reflected light is usually diffuse and scattered.

3. •What is the difference between a real image and a virtual image?

4. •When you view a car's side mirror, you see a smaller image than you would if the mirror were flat. Is the mirror concave or convex? Explain your answer.

5. •Explain the meaning, in terms of physics, of the phrase etched on the right side mirror of most cars: "Objects in mirror are closer than they appear."

6. •What is the radius of curvature of a plane mirror? Explain your answer.

7. •For a certain lens in air, both radii of curvature are positive. Is it a converging lens or a diverging lens, or do you need additional information to tell? Explain your answer. SSM

8. •If we move a glass lens from air into water, what will happen to the focal length of the lens? Explain your answer.

9. •A laptop computer is connected to a video projector which projects an image on a screen. If the lens of the projector is half covered, what happens to the image? Explain your answer.

10. •**Biology** The image focused on your retina is actually inverted (sketch a simple ray trace diagram showing this observation). What does this fact say about our definitions of "right side up" and "upside down"?

11. •**Biology** Experimental subjects who wear inverting lenses (glasses that invert all images) for several days adapt to their new perception of the world so well that they can even ride a bicycle. Several days after the glasses are removed, their perceptions return to normal. Discuss this phenomenon and comment.

12. •**Medical** Explain why converging lenses are used to correct farsightedness (hyperopia) while diverging lenses are used for nearsightedness (myopia).

13. •**Biology** Discuss why nearsightedness is not found in all people, but virtually everyone eventually becomes farsighted as they age.

14. ••Explain why looking through a small opening often provides visual acuity even to an extremely nearsighted person.

Multiple-Choice Questions

15. •Which is true when an object is moved farther from a plane mirror?
   A. The height of the image decreases and the image distance increases.
   B. The height of the image stays the same and the image distance increases.
   C. The height of the image increases and the image distance increases.
   D. The height of the image stays the same and the image distance decreases.
   E. The height of the image decreases and the image distance decreases. SSM

16. •A real image can form in front of
   A. a plane mirror.
   B. a concave mirror.
   C. a convex mirror.
   D. any type of mirror.
   E. no mirrors.

17. •When an object is placed a little farther from a concave mirror than the focal length, the image is
   A. magnified and real.
   B. magnified and virtual.
   C. smaller and real.
   D. smaller and virtual.
   E. smaller and reversed.

18. •If you want to start a fire using sunlight, which kind of mirror would be most efficient?
   A. a plane mirror
   B. a concave mirror

   C. a convex mirror
   D. any type of plane, concave, or convex mirror
   E. It is not possible to start a fire using sunlight and a mirror; you must use a concave lens.

19. •An object is placed at the center of curvature of a concave mirror. Which of the following is true about the image?
   A. real and upright
   B. real and inverted
   C. virtual and upright
   D. virtual and inverted
   E. real, inverted, and reversed SSM

20. •When an object is placed farther from a convex mirror than the focal length, the image is
   A. larger and real.
   B. larger and virtual.
   C. smaller and real.
   D. smaller and virtual.
   E. smaller and reversed.

21. •**Medical** When a dentist needs a mirror to see an enlarged, upright image of a patient's tooth, what kind of mirror should he use?
   A. a plane mirror
   B. a concave mirror
   C. a convex mirror
   D. either a plane mirror or a concave mirror
   E. either a plane mirror or a convex mirror

22. •A magnifier allows one to look at a very near object by forming an image of it farther away. The object appears larger. To create a magnifier, one would use a
   A. short focal length (< 1 m) converging lens.
   B. short focal length (< 1 m) diverging lens.
   C. long focal length (>1 m) converging lens.
   D. long focal length (>1 m) diverging lens.
   E. either a converging or a diverging lens.

23. •A compound microscope is a two-lens system used to look at very small objects. Which of the following statements is correct?
   A. The objective lens and the eyepiece both have the same focal length, and both serve as simple magnifiers.
   B. The objective lens is a short focal length, converging lens and the eyepiece functions as a simple magnifier.
   C. The objective lens is a long focal length, converging lens and the eyepiece functions as a simple magnifier.
   D. The objective lens is a short focal length, diverging lens and the eyepiece functions as a simple magnifier.
   E. The objective lens is a long focal length, diverging lens and the eyepiece functions as a simple magnifier. SSM

Estimation/Numerical Analysis

24. •Estimate the focal length (in cm) of the corrective lenses in an average pair of glasses.

25. •Estimate the focal length of a convex blind spot mirror that is often added to a vehicle's outside mirror.

26. •Estimate the radius of curvature of a spherical mirror that is typically found in the corridors of busy hospitals, for example, to help prevent collisions when going around a corner.

27. •Give three examples of spherical concave mirrors in your daily life and estimate their approximate focal lengths. SSM

28. •Give three examples of spherical convex mirrors in your daily life and estimate their approximate focal length.

29. •Determine the focal length for an unknown lens with the following object and image distances:

Object Distance (cm)	Image Distance (cm)	Object Distance (cm)	Image Distance (cm)
30	98	60	37
35	67	65	35
40	53	70	34
45	47	75	33
50	42	80	32
55	38		

Problems

24-1 Mirrors or lenses can be used to form images

24-2 A plane mirror produces an image that is reversed back to front

30. •The angle of incidence on a flat mirror is 0°; what is the angle of reflection?

31. •Two flat mirrors are perpendicular to each other. An incoming beam of light makes an angle of $\theta = 30°$ with the first mirror as shown in Figure 24-42. What angle will the outgoing beam make with respect to the normal of the second mirror? SSM

Figure 24-42 Problem 31

32. •A 1.8-m-tall man stands 2.0 m in front of a vertical plane mirror. How tall will the image of the man be?

33. •What must be the minimum height of a plane mirror in order for a 1.80-m-tall person to see a full image of himself?

34. ••A plane mirror is 10 m away from and parallel to a second plane mirror (Figure 24-43). Find the location of the first five images formed by each mirror when an object is positioned exactly in the middle between the two mirrors.

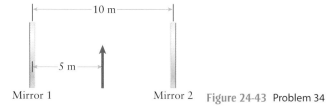

Mirror 1   Mirror 2   Figure 24-43 Problem 34

35. ••A plane mirror is 10 m away from and parallel to a second plane mirror (Figure 24-44). Find the location of the first five images formed by each mirror when an object is positioned 3 m from one of the mirrors. SSM

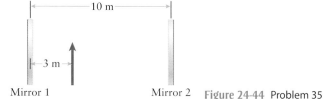

Mirror 1   Mirror 2   Figure 24-44 Problem 35

36. •At what point(s) A–E will the image of the face be visible in the plane mirror of length L (Figure 24-45)? The distance x

equals $L/2$. The points A–E are collinear and separated by a distance of $3L/4$, and point C lies on a line bisecting the mirror. Assume the face lies directly on point C.

Figure 24-45 Problem 36

37. •Using a ruler, a protractor, and the law of reflection, show the location of the image of your face when you stand a short distance in front of a plane mirror as shown in Figure 24-46.

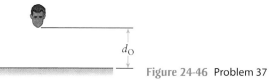

Figure 24-46 Problem 37

38. ••One person is looking in a plane mirror at the image of a second person (Figure 24-47). Using a ruler, a protractor, and the law of reflection, show the location of the image as seen by the first person.

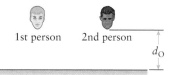

1st person   2nd person

Figure 24-47 Problem 38

39. ••Suppose an autofocusing camera is positioned 2 m in front of a plane mirror. Describe the self-portrait that is photographed when you snap a picture of yourself. An autofocus camera sends infrared waves from a transmitter and receives the reflected waves that bounce off of the objects in front of the camera to determine the distance at which to focus the lens.

24-3 A concave mirror can produce an image of a different size than the object

40. •Describe the difference between the images seen in a spherical, concave mirror when the object is "up close" (closer to the image than the focal length) compared to "far away" (outside the focal length).

41. •Are there any situations where a real image is formed in a spherical, concave mirror? Describe the location of the object relative to the focal point for such situations. SSM

42. •What is a good way to remember the general shape of a spherical concave mirror?

24-4 Simple equations give the position and magnification of the image made by a concave mirror

43. •An object is placed 8.0 cm in front of a concave mirror that has a 10-cm radius of curvature. Calculate the image distance and the magnification of the image. Determine if the image is real or virtual and whether it is inverted or upright by using (a) a ray trace diagram and (b) the mirror equation. SSM

44. •An object 1.0 cm tall is placed 3.0 cm in front of a spherical concave mirror with a radius of curvature equal to 10.0 cm. Calculate the image distance and height by using (a) a ray trace diagram and (b) the mirror equation.

45. •An object 1.0 cm tall is placed 6 cm in front of a spherical concave mirror with a radius of curvature equal to 10.0 cm. Calculate the image distance and height by using (a) a ray trace diagram and (b) the mirror equation.

46. ••The radius of curvature of a spherical concave mirror is 20 cm. Describe the image formed when a 10-cm-tall object is positioned (a) 5 cm from the mirror, (b) 20 cm from the mirror, (c) 50 cm from the mirror, and (d) 100 cm from the mirror. For each case give the image distance, the image height, the type of image (real or virtual), and the orientation of the image (upright or inverted).

47. ••The radius of curvature of a spherical concave mirror is 15 cm. Describe the image formed when a 20-cm-tall object is positioned (a) 10 cm from the mirror, (b) 20 cm from the mirror, and (c) 100 cm from the mirror. For each case give the image distance, the image height, the type of image (real or virtual), and the orientation of the image (upright or inverted). SSM

48. ••An object is positioned 24 cm from a spherical concave mirror of unknown focal length. The image that is formed is 30 cm from the mirror. (a) Calculate the focal length. (b) Is this answer unique? (c) Is the image real or virtual? (d) If the object is 10 cm tall, determine the height of the image(s).

49. ••Construct the ray trace diagrams to locate the images in the following cases. (a) A 10-cm-tall object is located 5 cm in front of a spherical concave mirror with a radius of curvature of 20 cm. (b) A 10-cm-tall object is located 10 cm in front of a spherical concave mirror with a radius of curvature of 20 cm. (c) A 10-cm-tall object is located 20 cm in front of a spherical concave mirror with a radius of curvature of 20 cm. In each case, start by estimating the height of the image from a sketch.

50. ••Derive a relationship between the radius of curvature of a spherical, concave mirror and the object distance that gives an upright image that is four times as tall as the object.

24-5 A convex mirror always produces an image that is smaller than the object

51. •Describe the difference between the images seen in a spherical, convex mirror when the object is "up close" (a shorter distance from the mirror than the focal distance of the mirror) compared to "far away" (a longer distance from the mirror than the focal distance of the mirror).

52. •Are there any situations where a real image is formed in a spherical, convex mirror? Why or why not?

53. •How can you remember the difference between the shapes of a spherical *concave* mirror and a spherical *convex* mirror?

54. ••Construct ray trace diagrams to locate the images and estimate the image height in each of the following cases. (a) A 10-cm-tall object located 5 cm in front of a spherical convex mirror with a radius of curvature of 20 cm. (b) A 10-cm-tall object located 10 cm in front of a spherical convex mirror with a radius of curvature of 20 cm. (c) A 10-cm-tall object located 20 cm in front of a spherical convex mirror with a radius of curvature of 20 cm.

24-6 The same equations used for concave mirrors also work for convex mirrors

55. ••The radius of curvature of a spherical convex mirror is 20 cm. Describe the image formed when a 10-cm-tall object is positioned (a) 20 cm from the mirror, (b) 50 cm from the mirror, and (c) 100 cm from the mirror. For each case, provide the image distance, the image height, the type of image (real or virtual), and the orientation of the image (upright or inverted). SSM

56. ••The radius of curvature of a spherical convex mirror is 15 cm. Describe the image formed when a 20-cm-tall object is positioned (a) 5 cm from the mirror, (b) 20 cm from the mirror, and (c) 100 cm from the mirror. For each case, provide the image distance, the image height, the type of image (real or virtual), and the orientation of the image (upright or inverted).

57. •A car's convex rearview mirror has a radius of curvature equal to 15 m. What are the magnification, type, and location of the image that is formed by an object that is 10 m from the mirror?

58. •An 18-cm-long pencil is placed beside a convex spherical mirror, and its image is 10.5 cm in length. If the radius of curvature of the mirror is 88.4 cm, find the image distance, the object distance, and the magnification of the pencil.

59. •A 1-cm-long horse fly hovers 1 cm from a shiny sphere with a radius of 25 cm. Describe the location of the image of the fly, its type (real or virtual), and its length.

60. •A spherical convex mirror is placed at the end of a driveway on a corner with a limited view of oncoming traffic. The mirror has a radius of curvature of 1.85 m. Where will the image of a car that is 12.6 m from the mirror appear?

61. ••Using the mirror equation, prove that all images in spherical convex mirrors are virtual. SSM

62. •A girl sees her image in a shiny glass sphere tree ornament that has a diameter of 10 cm. The image is upright and is located 1.5 cm behind the surface of the ornament. How far from the ornament is the child located?

63. •A shiny sphere, 30 cm in diameter, is placed in a garden for aesthetic purposes. Determine the type, location, and height of the image of a 6.00-cm-tall squirrel located 40 cm in front of the sphere.

24-7 Convex lenses form images like concave mirrors and vice versa

64. •Under what circumstances will the images formed by converging or diverging lenses be designated as "real"? Indicate the type or types of lenses and the required position of the object.

65. •A real image that is created due to reflection in a spherical mirror appears in front of the mirrored surface. Is this also the case for a real image that is created due to refraction in a lens? SSM

66. •Where does the bending of light physically take place in a typical biconcave or biconvex lens? Is this how we draw ray trace diagrams? Why or why not?

67. •Which type of lens is used in eyeglasses to correct (a) nearsightedness (myopia)? (b) farsightedness (hyperopia)?

68. •How many rays are required to trace out the image that is formed by a lens in a ray trace diagram?

24-8 The focal length of a lens is determined by its index of refraction and the curvature of its surfaces

69. •A 10-cm-tall object is located in front of a converging lens with a power of 5 diopters. Describe the image created (type, location, height) and draw the ray trace diagrams if the object is located (a) 5 cm from the lens, (b) 10 cm from the lens, (c) 20 cm from the lens, and (d) 50 cm from the lens. SSM

70. ••A 10-cm-tall object is located in front of a diverging lens with a power of −5 diopters. Describe the type, location, and height of the image created, and draw the ray trace diagrams if the object is located (a) 5 cm from the lens, (b) 10 cm from the lens, (c) 20 cm from the lens, and (d) 50 cm from the lens.

71. ••A 2.00-cm-tall object is 18.0 cm in front of a converging lens with a focal length of 30.0 cm. (a) Use the lens equation and (b) a ray trace diagram to describe the type, location, and height of the image that is formed.

72. ••A lens is formed from a plastic material that has an index of refraction of 1.55. If the radius of curvature of one surface is 1.25 m and the radius of curvature of the other surface is 1.75 m, use the lens maker's equation to calculate the focal length and the power of the lens.

73. ••A glass lens ($n = 1.60$) has a focal length of −31.8 cm and a plano-concave shape. (a) Calculate the radius of curvature of the concave surface. (b) If a lens is constructed from the same glass to form a plano-convex shape with the same radius of curvature, what will the focal length be? SSM

74. ••A 2.00-cm-tall object is 30.0 cm in front of a converging lens that has a focal length of 18.0 cm. (a) Use the lens equation and (b) a ray trace diagram to describe the type, location, and height of the image that is formed.

75. •••**Biology** Calculate the overall magnification of a compound microscope that uses an objective lens with a focal length of 0.50 cm, an eyepiece with a focal length of 2.50 cm, and a distance of 18 cm between the two.

General Problems

Note: In these problems "infinity" means a very large distance compared to the focal length of a lens.

76. ••The opposite walls of a barber shop are covered by plane mirrors, so that multiple images arise from multiple reflections, and you see many reflected images of yourself, receding to infinity. The width of the shop is 6.50 m, and you are standing 2.00 m from the north wall. (a) How far apart are the first two images of you behind the north wall? (b) What is the separation of the first two images of you behind the south wall? Explain your answer.

77. ••An object is 40.0 cm from a concave spherical mirror whose radius of curvature is 32.0 cm. Locate and describe the type and magnification of the image formed by the mirror (a) by calculating the image distance and lateral magnification and (b) by drawing a ray trace diagram. On the ray trace diagram draw an eye in a position from which it can view the image.

78. ••**Biology** A typical human eye is nearly spherical and usually about 2.5 cm in diameter. Suppose a person first looks at a coin that is 2.3 cm across, located 30.0 cm from her eye, and then looks up at her friend who is 1.8 m tall and 3.25 m away. (a) Find the approximate size of each image (coin and friend) on her retina. (*Hint:* Just look at rays from the top and bottom of the object that pass through the center of the lens.) (b) Are the images in part (a) upright or inverted, and are they real or virtual?

79. ••**Biology** A typical human lens has an index of refraction of 1.43. The lens has a double convex shape, but its curvature can be varied by the ciliary muscles acting around its rim. At minimum power, the radius of the front of the lens is 10.0 cm,

while that of the back is 6.00 mm. At maximum power the radii are 6.00 mm and 5.50 mm, respectively. (The numbers can vary somewhat.) If the lens were in air, (a) what would be the ranges of its focal length and its power (in diopters)? (b) At maximum power, where would the lens form an image of an object 25 cm in front of the front surface of the lens? (c) Would the image fall on the retina of a human eye? The retina is located approximately 2.5 cm from the lens.

80. ••**Biology** A typical person's eye is 2.5 cm in diameter and has a near point (the closest an object can be and still be seen in focus) of 25 cm, and a far point (the farthest an object can be and still be in focus) of infinity. (a) What is the range of the effective focal lengths of the focusing mechanism (lens plus cornea) of the typical eye? (b) Is the equivalent focusing mechanism of the eye a diverging or a converging lens? Justify your answer without using any mathematics, and then see if your answer is consistent with your result in part (a).

81. •••A geneticist looks through a microscope to determine the phenotype of a fruit fly. The microscope is set to an overall magnification of 400x with an objective lens that has a focal length of 0.60 cm. The distance between the eyepiece and objective lenses is 16 cm. Find the focal length of the eyepiece lens assuming a near point of 25 cm (the closest an object can be and still be seen in focus). SSM

82. ••A thin lens made of glass that has a refractive index equal to 1.60 has surfaces with radii of curvature that have magnitudes equal to 12.0 and 18.0 mm. What are the possible values for its focal length? Sketch a cross-sectional view of the lens for each possible combination, making sure to label the radii of curvature of each surface of the lens and the associated focal length of the entire lens.

83. ••You are designing lenses that consist of small double convex pieces of plastic having surfaces with radii of curvature of magnitudes 3.50 cm on one side and 4.25 cm on the other side. You want the lenses to have a focal length of 1.65 cm in air. What should be the index of refraction of the plastic to achieve the desired focal length?

84. •••A lens of focal length +15.0 cm is 10.0 cm to the left of a second lens of focal length −15.0 cm. (a) Where is the final image of an object that is 30.0 cm to the left of the positive lens? (b) Is the image real or virtual? (c) How do the image's size and orientation compare to those of the original object? Explain your answer. (d) How must you be oriented to see the image?

85. •••A thin, diverging lens having a focal length of magnitude 45.0 cm has the same principal axis as a concave mirror with a radius of 60.0 cm. The center of the mirror is 20.0 cm from the lens, with the lens in front of the mirror. An object is placed 15.0 cm in front of the lens. (a) Where is the final image due to the lens–mirror combination? (b) Is the final image real or virtual? Upright or inverted? (c) Suppose now that the concave mirror is replaced by a convex mirror of the same radius. Repeat parts (a) and (b) for the new lens–mirror combination. SSM

86. •••When you place a bright light source 36.0 cm to the left of a lens, you obtain an upright image 14.0 cm from the lens and also a faint inverted image 13.8 cm to the left of the lens that is due to reflection from the front surface of the lens. When the lens is turned around, a faint inverted image is 25.7 cm to the left of the lens. What is the index of refraction of the material?

87. •••A thin, converging lens having a focal length of magnitude 25.00 cm is placed 1.000 m from a plane mirror that

is oriented perpendicular to the principal axis of the lens. A flower, 8.400 cm tall, is 1.450 m from the mirror on the principal axis of the lens. (a) Where is the final image of the flower produced by the lens–mirror combination? Is it real or virtual? Upright or inverted? How tall is the image? (b) If the converging lens is replaced by a diverging lens having a focal length of the same magnitude as the original lens, what will be the answers to part (a)?

88. •••**Biology** A compound microscope has a tube length of 20.0 cm and an objective lens of focal length 8.0 cm. (a) If it is to have a magnifying power of 200x, what should be the focal length of the eyepiece? (b) If the final image is viewed at infinity, how far from the objective should the object be placed?

89. ••**Biology** You may have noticed that the eyes of cats appear to glow green in low light. This effect is due to the reflection of light by the *tapetum lucidum*, a highly reflective membrane just behind the retina of the eye. Light that has passed through the retina without hitting photoreceptors is reflected back to the retina, thus enabling the animal to see much better than humans in low light. The eye of a typical cat is about 1.25 cm in diameter. Assume that the light enters the eye traveling parallel to the principal axis of the lens. (a) If some of the light reflected off the *tapetum lucidum* escapes being absorbed by the retina, where will it be focused? (b) The refractive index of the liquid in the eye is about 1.4. How does this affect the location of the image in part (a)? SSM

90. ••**Medical** A nearsighted eye is corrected by placing a diverging lens in front of the eye. The lens will create a virtual image of a distant object at the far point (the farthest an object can be and still be in focus) of the myopic viewer where it will be clearly seen. In the traditional treatment of myopia, an object at infinity is focused to the far point of the eye. If an individual has a far point of 70 cm, prescribe the correct power of the lens that is needed.

91. ••**Medical** A farsighted eye is corrected by placing a converging lens in front of the eye. The lens will create a virtual image that is located at the near point (the closest an object can be and still be in focus) of the viewer when the object is held at a comfortable distance (usually taken to be 25 cm). If a person has a near point of 75 cm, what power reading glasses should be prescribed to treat this hyperopia?

92. •••**Medical** (a) Prove that when two thin lenses are pressed next to one another, the combined focal length (f combined) of the two lenses, acting together, is given by

$$\frac{1}{f_{\text{combined}}} = \frac{1}{f_1} + \frac{1}{f_2}$$

(b) Describe how this relates to a prescription for a contact lens which is placed directly on the eye? (Assume there is no significant separation between the contact lens and the lens of the eye.) (c) Why would an eyeglass prescription that is identical to a contact prescription give a very subtle difference in the image seen?

93. ••**Medical** An optometrist tests a person and finds that without glasses, he needs to have his eyes 15.0 cm from a book to read comfortably and can focus clearly only on distant objects up to 2.75 m away, but no farther. A typical normal eye should be able to focus on objects that are between 25.0 cm (the near point) and infinity (the far point) from the eye. (a) What type of correcting lenses does the person need: single focal length or bifocals? Why? (b) What should the optometrist specify as the focal length(s) of the correcting contact lens or lenses? (c) What is the power (in diopters) of the correcting lens or lenses? SSM

94. •••**Medical** One of the inevitable consequences of aging is a decrease in the flexibility of the lens. This leads to the farsighted condition called *presbyopia* (elder eye). Almost every aging human will experience it to some extent. However, for the myopic person, at some point, it is possible that far vision will be limited by a subpar far point *and* near vision will be hampered by an expanding near point. One solution is to wear bifocal lenses that are diverging in the upper half to correct the nearsightedness and converging in the lower half to correct the farsightedness.

Suppose one such individual asks for your help. The patient complains that he can't see far enough to safely drive (his far point is 112 cm) and he can't read the font of his smart phone without holding it beyond arm's length (his near point is 83 cm)! Prescribe the bifocals that will correct the visual issues for your patient.

95. ••A common zoom lens for a digital camera covers a focal length range of 18 mm to 200 mm. For the purposes of this problem, treat the lens as a thin lens. If the lens is zoomed out to 200 mm and is focused on a petroglyph that is 15.0 m away and 38 cm wide, (a) how far is the lens from the photosensor array of the camera and (b) how wide is the image of the petroglyph on the sensors? (c) If the closest that the lens can get to the sensors at its 18 mm focal length is 5.2 cm, what is the closest object it can focus on at that focal length? SSM

96. ••A macro lens is designed to take very close-range photographs of small objects such as insects and flowers. At its closest focusing distance, a certain macro lens has a focal length of 35.0 mm and forms an image on the photosensors of the camera that is 1.09 times the size of the object. (a) How close must the object be to the lens to achieve this maximum image size? (b) What is the magnification if the object is twice as far from the lens as in part (a)? For this problem, treat the lens as a thin lens.

97. •**Astronomy** A refracting astronomical telescope, or refractor, consists of an eyepiece lens at one end of a cylindrical tube and an objective lens at the other end. The objective lens gathers light from a distant object (such as a planet) and focuses it at the focal point of the eyepiece lens. The eyepiece basically acts as a simple magnifier to create a virtual image of the objective's image. The overall magnification, M, is found to be $M = -f_o/f_e$, where f_o is the focal length of the objective and f_e is the focal length of the eyepiece. (a) Calculate the magnification of the 36-in refractor at Lick Observatory on Mount Hamilton near San Jose, California. The focal length of the objective lens is 17.37 m, and the focal length of the eyepiece is 22 mm. (b) What is the significance of the negative sign in the magnification equation?

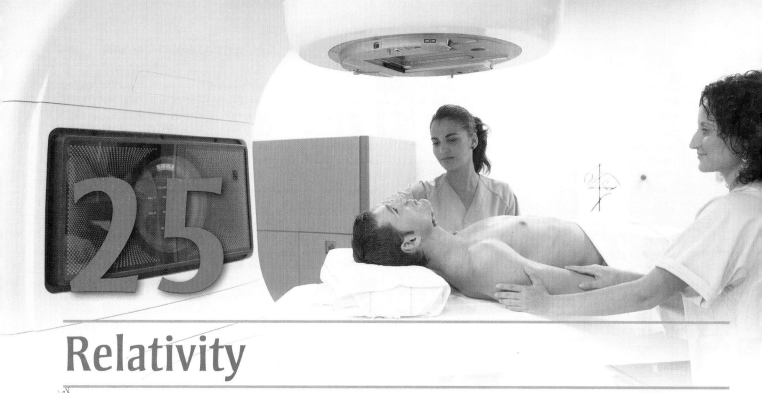

Relativity

All chapter

What Do You Think?

In cancer radiotherapy, a beam of electrons is accelerated to nearly the speed of light. The kinetic energy of the electrons is used to create intense X ray beams that can be accurately targeted on the location of the cancerous tissue. Which requires more energy: (a) accelerating an electron from rest to 90% of the speed of light, or (b) further accelerating that electron from 90 to 99% of the speed of light?

In this chapter, your goals are to:

- (25-1) Identify some circumstances under which the physics you know breaks down.
- (25-2) Describe how different observers view the same motion in Newtonian physics.
- (25-3) Explain how the Michelson-Morley experiment helped rule out the ether model of the propagation of light.
- (25-4) Describe how the time interval between two events can have different values in different frames of reference.
- (25-5) Calculate how the dimensions of an object change when it is in motion.
- (25-6) Explain why the speed of light in a vacuum is an ultimate speed limit.
- (25-7) Calculate the rest energy of an object with mass.
- (25-8) Explain what the principle of equivalence tells us about the nature of gravity.

To master this chapter, you should review:

- (2-7) The motion of objects in free fall.
- (3-6) Projectile motion.
- (4-2, 4-3) Newton's first and second laws.
- (6-3) Kinetic energy and the work-energy theorem.
- (7-4) Elastic collisions.
- (7-6) The relationship between external forces and momentum change.
- (22-3) The electromagnetic nature of light.

25-1 The concepts of relativity may seem exotic, but they're part of everyday life

Most of our everyday experiences involve objects that are either not moving with respect to us or are moving at speeds that are slow compared to the speed of light in a vacuum. In the late nineteenth century, scientists began to realize that the laws of physics that we've developed so far in this book don't properly describe light or objects moving at speeds near the speed of light. Albert Einstein first understood the way to extend physics into this regime of extremely high speed.

In this chapter we'll focus on Einstein's *special theory of relativity*. We'll see how discoveries concerning the nature of light helped motivate the central ideas of this theory. We'll also see how a simple postulate—that the speed of light does not depend on the motion of either the emitter or the observer of the light—leads to a radical transformation of our understanding of space and time. We'll discover that the speed of light is an ultimate speed limit, which means that it is impossible to accelerate an object beyond

(a)
In order for a GPS receiver to accurately determine its position, it must use relativity to adjust for time flowing at a different rate aboard a GPS satellite than on Earth.

(b)
Sunlight is a result of reactions deep inside the Sun that convert mass into energy—a process predicted by the special theory of relativity.

Figure 25-1 **Relativity in your world** (a) A GPS-equipped mobile phone and (b) the light from the Sun illustrate applications of the theory of relativity.

that speed. We'll also see that mass is simply another form of energy. We'll conclude with a look at Einstein's *general theory of relativity*, which provides new insights into the nature of gravity.

Although the effects of the special theory of relativity are most pronounced for objects moving at very high speeds, they can be seen in the world around you. Your mobile phone probably has the ability to determine its location using the Global Positioning System, or GPS (Figure 25-1a). A GPS receiver detects signals from a collection of satellites in Earth's orbit and calculates its position by timing those signals. However, the satellites move at about 28,000 km/h (18,000 mi/h) relative to Earth, and special relativity tells us that a moving clock or timekeeper runs at a different rate than a stationary one (see Section 25-4). Your smartphone has to be able to correct for this in order to give you accurate positioning information. Ordinary sunlight is also a consequence of relativity: The Sun shines by converting a fraction of its mass into electromagnetic energy, a direct application of the idea that objects have energy simply as a consequence of having mass (Figure 25-1b).

25-2 Newton's mechanics includes some ideas of relativity

You're on a train in the station, looking out the window at the train right next to yours. One of the trains is moving, but you can't tell which (Figure 25-2). Is your train moving and the other one stationary, or vice versa? Perhaps both are moving. How can you tell?

To address this question, let's return to Newton's first law, which we introduced in Section 4-3:

If the net external force on an object is zero...   ...the object does not accelerate...

$$\text{If } \sum \vec{F}_{\text{ext}} = 0, \text{ then } \vec{a} = 0 \text{ and } \vec{v} = \text{constant}$$

...and the velocity of the object remains constant. If the object is at rest, it remains at rest; if it is in motion, it continues in motion in a straight line at a constant speed.

Take-Home Message for Section 25-1

✔ The physics we have learned so far must be modified for objects moving at speeds comparable to the speed of light.

✔ These modifications lead us to new ideas about space, time, and energy.

Figure 25-2 **Relative motion** This is the view of one train as seen from the window of another. Which train is moving?

Newton's first law of motion (4-6)

Figure 25-3 Two observers
Observers on the platform in a train
station use coordinates x, y, and z and
measure time t. An observer aboard
the train uses coordinates x', y', and z'
and measures t'.

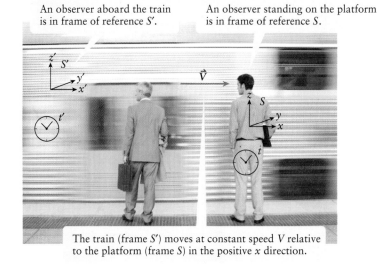

An observer aboard the train
is in frame of reference S'.

An observer standing on the platform
is in frame of reference S.

The train (frame S') moves at constant speed V relative
to the platform (frame S) in the positive x direction.

(a)
As measured in S

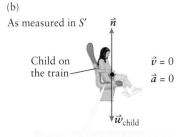

Net force on the child = 0.
Acceleration of the child = 0.

(b)
As measured in S'

Net force on the child = 0.
Acceleration of the child = 0.

Figure 25-4 Newton's first law in two frames of reference The free-body diagram for a child on the train as measured in the two frames of reference shown in Figure 25-3.

The first law states that an object that experiences no net force could *either* be at rest or in motion at a constant velocity. Newton's laws therefore treat both cases identically. So although we might make a distinction between an object at rest and one in (uniform) motion with respect to us, the laws of physics do not. In the case of the two trains, this tells us something profound: If you have no reference to the ground or the tracks on which the trains move, there is *no* way to design an experiment or make a measurement that would tell you whether the other train is moving at a constant velocity with respect to yours or your train is moving a constant velocity with respect to the other one.

A useful way to think about this idea is to introduce the concept of a **frame of reference** (also called a *reference frame* or simply *frame*). This is a coordinate system with respect to which we can make observations or measurements. Figure 25-3 shows two different frames of reference. A person standing on the train platform is in frame S and measures the positions of objects using the coordinates x, y, and z. This person has a watch, and the time of a certain event as measured on this watch is t. A person riding on the train is in frame S' and measures the positions of objects using the coordinates x', y', and z'. The time of an event as measured by this person is t'. If neither frame of reference is accelerating, the message of Newton's first law is that *both* frames are equally good for making measurements.

As an example, let's consider a child sitting in one of the seats on the train (Figure 25-4). The train is on a straight track and is moving relative to the platform at a constant speed V. As measured by a person on the platform (frame of reference S), the child is moving at a constant velocity with zero acceleration, so the net force on the child must be zero: The upward normal force exerted by the seat exactly balances the downward gravitational force that Earth exerts on the child. As measured by another passenger seated on board the train (frame of reference S'), the child is at rest (the child isn't moving relative to that passenger). So as measured in S', the net force on the child must again be zero. The two observers—one in frame of reference S and one in frame of reference S'—disagree about *how* the child moves. But they agree that the child's motion is in accordance with Newton's first law, Equation 4-6.

It's not just Newton's first law that applies in both frames of reference depicted in Figure 25-3. Newton's second law also applies in both frames:

If a **net external force** acts on an object...

...the object accelerates. The acceleration is in the same direction as the net force.

Newton's second law of motion
(4-2)

$$\sum \vec{F}_{ext} = m\vec{a}$$

The magnitude of acceleration that the net external force causes depends on the mass m of the object (the quantity of material in the object). The greater the mass, the smaller the acceleration.

As an illustration, suppose the child riding on the train is tossing a ball up and down (**Figure 25-5**). A passenger seated in the train (frame of reference S') sees the motion of the ball as purely vertical (We draw the ball's motion with a slight horizontal displacement to distinguish the up and down motion, but this motion is really only along a vertical line). By contrast, a person on the platform (frame of reference S) sees the ball following a parabolic path: The ball has the same horizontal component of velocity V as the train, and it maintains that horizontal velocity during its flight. As for the child in Figure 25-4, the observers in the two frames of reference disagree about how the ball moves: The motion is along a line as seen in S' but is in a plane as seen in S. Both observers, however, agree that the ball obeys Newton's second law. That's because as observed from both frames of reference, when the ball is in flight only the gravitational force acts on it, and so the acceleration is downward and has magnitude g. In frame S' the straight up-and-down motion is free fall, as we described in Section 2-7; in frame S the ball is in projectile motion, as we described in Section 3-6. Each description is correct for the frame of reference in which the motion is observed.

We refer to a frame of reference attached to an object that does not accelerate as an **inertial frame**. Newton's first law (also called the law of inertia) defines an inertial frame of reference. If, according to an observer, an object at rest tends to remain at rest, and an object in motion maintains the same velocity (speed and direction) unless acted on by a net force, then the object is in an inertial reference frame. If one frame of reference S is inertial, a second frame of reference S' is also inertial if it moves at a constant velocity relative to S. That's the case for the two frames of reference depicted in Figure 25-3.

By contrast, a frame of reference attached to an accelerated object is a **noninertial frame**. To an observer in a noninertial frame, Newton's first law does *not* hold true. An example is a frame of reference attached to a car that is accelerating forward. A ball sitting on the floor of this car has zero net force on it (the upward normal force exerted by the floor balances the downward gravitational force), yet the ball accelerates toward the back of the car. A rotating frame of reference, such as a carnival merry-go-round, is also noninertial because an object that follows a circular path is accelerating. Just like a ball in a car that accelerates forward, a ball placed on the merry-go-round floor has zero net force acting on it, yet this ball will tend to roll to the outside of the merry-go-round. In the frame of reference of a person riding on the merry-go-round, Newton's first law does not hold true.

Strictly speaking, an observer at rest on Earth's surface, such as the person in frame S standing on the platform in Figure 25-3, is in a noninertial frame. That's because Earth rotates on its axis like a merry-go-round and also moves along a roughly circular orbit around the Sun. However, the accelerations involved with those motions are so small (each is a small fraction of g) that for many purposes we can ignore them. As a result, we can safely regard the frame of reference S in Figure 25-3 as an effectively inertial one, and likewise for the frame S' attached to the train.

We've seen that Newton's first and second laws—the fundamental laws of motion—work equally well in both inertial frame S and inertial frame S'. This suggests that the same should be true for *any* inertial frame of reference. This statement is called the **principle of Newtonian relativity:**

> The laws of motion are the same in all inertial frames of reference.

The word *relativity* means that measurements made relative to one inertial frame of reference are just as valid as those made relative to another inertial frame. Since the laws of motion don't distinguish between two inertial frames S and S', it's meaningless to ask which frame is "really" moving and which frame is "really" at rest. (You may think that frame S is the one that's really at rest because it's stationary with respect to the platform. But remember that the platform is on Earth and that our entire planet is in motion through the solar system.) So another way to express the principle of Newtonian relativity is:

> There is no way to detect absolute motion. Only motion relative to a selected frame of reference can be detected.

Let's apply the principle of Newtonian relativity to the ball shown in Figure 25-5. Suppose you are standing on the platform as the train goes by, so you are in reference frame S. You see the ball following a parabolic path, and you make observations of the x, y, and z coordinates of the ball as functions of time t. The child riding in the train is in reference frame S' and sees the ball moving straight up and down relative to her. As

(a)
As measured in S'

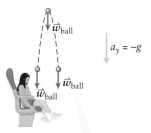

$a_y = -g$

Net force on ball = gravitational force
Ball experiences free fall.

(b)
As measured in S

Net force on ball = gravitational force
Ball experiences projectile motion.

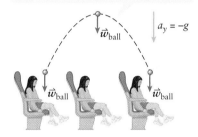

$a_y = -g$

Figure 25-5 A tossed ball in two frames of reference The child on the train from Figure 25-4 tosses a ball straight up and down relative to her.

the ball moves, she measures the x', y', and z' coordinates of the ball as functions of time t'. The two sets of coordinates have the same orientation, as Figure 25-6 shows. To calibrate the two clocks—one in frame S, the other in frame S'—you and the child both set your clocks to read zero at the instant that you pass each other, when the origins of your frames of reference coincide. So at this instant $t = t' = 0$.

Suppose you and the child both measure the same **event**—that is, the ball being at a certain point in its motion, such as being at the high point in its path. In your S frame the coordinates of this event in space and time are x, y, z, and t, and the coordinates of this same event as measured in the S' frame are x', y', z', and t'. (Note that we are thinking of time as a fourth coordinate of an event.) As Figure 25-6 shows, a simple set of equations relates the coordinates of the same event in the two frames:

Galilean coordinate transformation (25-1)

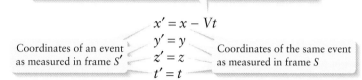

Inertial frame of reference S' moves at speed V in the positive x direction relative to inertial frame of reference S.

$$x' = x - Vt$$
$$y' = y$$
$$z' = z$$
$$t' = t$$

Coordinates of an event as measured in frame S' Coordinates of the same event as measured in frame S

This set of equations is known as the **Galilean transformation**. Note that the relative motion of the two frames of reference along the positive x axis does not impact the measured values of the y and z components, which are the same in both reference frames.

We can use the Galilean transformation to compare the velocity of the ball as measured in frame S' to its velocity as measured in frame S. To do this, we'll look at *two* events in the motion of the ball separated by a short time interval from t_1 to t_2 as measured in frame S. During this time interval, the coordinates of the ball as measured in frame S change from x_1, y_1, and z_1 to x_2, y_2, and z_2. The components of the ball's velocity $\vec{v}$ in frame S are

(25-2)

$$v_x = \frac{\Delta x}{\Delta t} = \frac{x_2 - x_1}{t_2 - t_1}$$
$$v_y = \frac{\Delta y}{\Delta t} = \frac{y_2 - y_1}{t_2 - t_1}$$
$$v_z = \frac{\Delta z}{\Delta t} = \frac{z_2 - z_1}{t_2 - t_1}$$

At time t, the coordinate of the ball in S' is $x' = x - Vt$.

At $t = 0$, the origins of frames S and S' coincide.

$t' = 0$

S, S' $y' = y$

$x' = x$

$z' = z$

$t = 0$

S y

x

z

t

S' y'

x'

z'

t'

Figure 25-6 Comparing coordinates in two frames of reference The coordinates of a ball as measured in the two frames of reference shown in Figure 25-3.

For the same two events as measured in frame S' the time interval is from t'_1 to t'_2, and the coordinates change from x'_1, y'_1, and z'_1 to x'_2, y'_2, and z'_2. So in frame S' the components of the object's velocity $\vec{v}'$ are

$$v'_x = \frac{\Delta x'}{\Delta t'} = \frac{x'_2 - x'_1}{t'_2 - t'_1}$$

$$v'_y = \frac{\Delta y'}{\Delta t'} = \frac{y'_2 - y'_1}{t'_2 - t'_1} \qquad (25\text{-}3)$$

$$v'_z = \frac{\Delta z'}{\Delta t'} = \frac{z'_2 - z'_1}{t'_2 - t'_1}$$

Now substitute the expressions for x', y', z', and t' from Equations 25-1 into Equations 25-3. We get

$$v'_x = \frac{(x_2 - Vt_2) - (x_1 - Vt_1)}{t_2 - t_1} = \frac{(x_2 - x_1) - V(t_2 - t_1)}{t_2 - t_1} = \frac{(x_2 - x_1)}{t_2 - t_1} - V$$

$$v'_y = \frac{y_2 - y_1}{t_2 - t_1} \qquad (25\text{-}4)$$

$$v'_z = \frac{z_2 - z_1}{t_2 - t_1}$$

If we now compare Equations 25-4 for the velocity components in frame S' to Equations 25-2 for the velocity components in frame S, we see that

Inertial frame of reference S' moves at speed V in the positive x direction relative to inertial frame of reference S.

Velocity components of an object as measured in frame S'

$$v_x' = v_x - V$$
$$v_y' = v_y$$
$$v_z' = v_z$$

Velocity components of the same object as measured in frame S

Galilean velocity transformation
(25-5)

Equations 25-5, which relate the velocity of an object in frame S' to the velocity of the same object in frame S, are called the **Galilean velocity transformation**.

As an example, think again of the ball shown in Figure 25-5. If the ball moves straight up and down as measured in frame S', then in that frame the ball is in free fall with only a y component of velocity. The other two components are zero: $v'_x = v'_z = 0$. As measured in frame S, the velocity of the ball has components

$$v_x = v'_x + V = V$$
$$v_y = v'_y$$
$$v_z = v'_z = 0$$

As measured in frame S, the ball moves up and down along the y direction with the same velocity as measured in frame S': $v_y = v'_y$. At the same time, as measured in frame S the ball maintains a constant velocity V in the x direction. That's just the behavior we expect for a projectile: Its motion is a combination of up-and-down free fall and constant-velocity horizontal motion (see Section 3-6).

> **! Watch Out!** There is nothing special about either the S frame or the S' frame.
>
> In Figures 25-5 and 25-6 we've chosen to think of frame S as at rest and frame S' as moving. This selection is purely our choice, because the principle of Newtonian relativity says that *all* inertial frames are equivalent. We could just as well say that frame S' is stationary and frame S is moving with speed V in the negative x direction.

You may wonder why we've spent so much time and effort explaining motion as seen from two different inertial frames of reference. As we'll discover in the next few sections, the reason is that something remarkable happens when the relative speed V

of the two frames is comparable to c, the speed of light in a vacuum. In that case we'll find that the Galilean transformations given by Equations 25-1 and 25-5 do not hold true. For example, as measured in the two different frames, the time interval between events can be different and objects can have different dimensions. These remarkable observations will radically transform our notions of the nature of time and space themselves.

Example 25-1 Two Cars

You observe two race cars approaching you. A red car is in one lane moving at 24 m/s relative to you. In a second lane, a blue car is moving at 36 m/s relative to you. (a) What is the velocity of the red car as measured by the driver of the blue car, relative to her frame of reference? (b) What is the velocity of the blue car as measured by the driver of the red car, relative to his frame of reference?

Set Up

We'll use Equations 25-5 to transform the velocity of a car as measured in one frame of reference to the velocity of the same car as measured in a different frame of reference. Note that these equations assume that frame S' is moving relative to frame S at speed V in the positive x direction. We'll have to keep this in mind when deciding which frame of reference corresponds to S and which to S'.

Galilean velocity transformation:

$$v'_x = v_x - V$$
$$v'_y = v_y$$
$$v'_z = v_z \qquad (25\text{-}5)$$

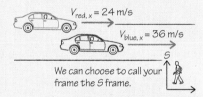

The given speeds of both cars are measured relative to you, that is, in your frame.

We can choose to call your frame the S frame.

Solve

(a) Let's take the positive x direction to be in the direction both cars are moving relative to you. Then we'll take S to be your frame of reference and S' to be the frame of reference of the driver of the blue car. The relative speed of these two frames is $V = 36$ m/s (the speed of the blue car relative to you). All of the motions in this example are along the x axis, so we don't need the y or z members of Equations 25-5.

Use the x equation from Equations 25-5 to relate the velocity of the red car as measured by you ($v_{red,x} = +24$ m/s) to its velocity as measured by the driver of the blue car ($v'_{red,x}$):

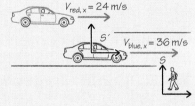

$$v'_{red,x} = v_{red,x} - V$$
$$= +24 \text{ m/s} - 36 \text{ m/s}$$
$$= -12 \text{ m/s}$$

As measured by the driver of the blue car, the red car is moving in the negative x direction—that is, backwards—at 12 m/s.

(b) Again we take S to be your frame of reference, but now S' is the frame of reference of the driver of the red car. The relative speed of these two frames is $V = 24$ m/s (the speed of the red car relative to you).

Use the x equation from Equations 25-5 to relate the velocity of the blue car as measured by you ($v_{blue,x} = +36$ m/s) to its velocity as measured by the driver of the red car ($v'_{blue,x}$):

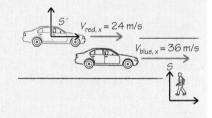

$$v'_{blue,x} = v_{blue,x} - V$$
$$= +36 \text{ m/s} - 24 \text{ m/s}$$
$$= +12 \text{ m/s}$$

As measured by the driver of the red car, the blue car is moving in the positive x direction—that is, forwards—at 12 m/s.

Reflect

The driver of the blue car sees the red car falling farther and farther behind her at a rate of 12 m/s, while the driver of the red car sees the blue car moving farther and farther in front of him at a rate of 12 m/s. Note that the *magnitude* of the two answers is the same: Both drivers agree that the other driver is moving at a relative speed of 12 m/s.

? **Got the Concept? 25-1** Groundspeed versus Airspeed

A typical jet airliner has a cruise airspeed—that is, its speed relative to the air through which it is flying—of 900 km/h. If the wind at the airliner's cruise altitude is blowing at 100 km/h from west to east, what is the speed of the airliner relative to the ground if the airplane is flying from west to east? From east to west? (a) 800 km/h west to east, 1000 km/h east to west; (b) 1000 km/h west to east, 800 km/h east to west; (c) 800 km/h in both directions; (d) 900 km/h in both directions; (e) 1000 km/h in both directions.

Take-Home Message for Section 25-2

✔ Newton's laws of motion treat all nonaccelerating objects identically, whether the objects are in motion or at rest.

✔ The laws of motion are the same in all inertial frames of reference. There is no way to detect absolute motion;

only motion relative to a selected frame of reference can be detected.

✔ The Galilean transformations allow you to convert the coordinates and velocity of an object observed in one inertial frame of reference to the coordinates and velocity of the same object observed in a different inertial frame.

25-3 The Michelson-Morley experiment shows that light does not obey Newtonian relativity

We saw in the preceding section that Newton's laws of motion are the same in all inertial frames. Is the same true for the other laws of physics? Physicists asked this very question during the second half of the nineteenth century, specifically about the laws of electromagnetism. As we will see, finding the answer to this question had profound consequences that led to fundamental changes in our understanding of nature.

We learned in Section 22-3 that the laws of electromagnetism explain how electromagnetic waves, including visible light, are possible. These laws also predict that the speed of electromagnetic waves in a vacuum is

Speed of light in a vacuum

$$c = \frac{1}{\sqrt{\mu_0 \varepsilon_0}} = 3.00 \times 10^8 \text{ m/s}$$

Permeability of free space   Permittivity of free space

Speed of light in a vacuum
(22-11)

Our discussion in Section 25-2 tells us that we can measure relative motion but not absolute motion. So when we say that the speed of electromagnetic waves in a vacuum is c, we are forced to ask this question: Relative to what inertial frame of reference is the speed of light equal to c?

In the nineteenth century, the most common answer to this question was to imagine a substance that fills all space. This substance, which was called the *luminiferous ether*, was thought to be the medium for electromagnetic waves, just as air is the medium for sound waves in our atmosphere. This substance must be of extraordinarily low density, so that its presence is almost undetectable. (The word *luminiferous* comes from the Latin for "light-bearing." *Ether* in this phrase has nothing to do with the organic compounds of the same name.) In this model, c is the speed of electromagnetic waves relative to the frame in which the luminiferous ether is at rest.

How can we test whether this model is correct? To see the answer, note that the speed of sound waves in dry air is 343 m/s, but you will measure a different speed if a wind is blowing (that is, the air is moving relative to you). If a sound wave is traveling from west to east and a wind is blowing past you at 10 m/s from west to east, the sound will move relative to you at 343 m/s + 10 m/s = 353 m/s. If the sound wave is traveling from west to east and a 10-m/s wind is blowing from east to west, the speed of the sound wave relative to you will be 343 m/s − 10 m/s = 333 m/s. The same should be true for light waves if the luminiferous ether is moving past you, so that there is an "ether wind." If a light wave is traveling from west to east and the ether is moving from west to east relative to you at 10 m/s, you would measure the speed of the wave to be $c + 10$ m/s; if the ether is instead moving from east to west relative to you at 10 m/s, you would measure the speed of the wave to be $c - 10$ m/s. If we can detect these small changes in the speed of light, that would be evidence that the luminiferous ether really exists.

Nineteenth-century scientists looked to Earth's motion around the Sun as a source of "ether wind." Our planet moves around its orbit at an average speed of 29.8 km/s $= 2.98 \times 10^4$ m/s, or about $10^{-4}\,c$. If the luminiferous ether is at rest relative to the solar system as a whole, we should experience an "ether wind" that blows past our moving planet at $10^{-4}\,c$. Depending on the direction of that "ether wind" relative to the direction of light propagation, we would expect the speed of light to vary between $(1 + 10^{-4})c$ and $(1 - 10^{-4})c$. The challenge is to design an experiment that can detect such small changes in the speed of light.

In 1887 the American scientists Albert Michelson and Edward Morley carried out the first definitive experiment of this kind. This seminal experiment is known as the **Michelson-Morley experiment**. Their apparatus, called an *interferometer*, split a beam of light into two, sent the two beams along perpendicular paths, and then allowed them to recombine at a viewing screen (**Figure 25-7**). What is seen on the viewing screen is an interference pattern between the waves in the two beams. The nature of this pattern depends on the difference in length between the two paths and also on whether the speeds at which light travels along each path are the same or different. If there is an "ether wind" that is more nearly aligned with one leg of the interferometer than the other, the speed of light should indeed be different along the two legs. Their apparatus was sensitive enough that Michelson and Morley should have been able to measure the effects of an "ether wind" due to Earth's motion around the Sun.

Michelson and Morley's results were striking: They found *no* effect due to an "ether wind." They and other scientists refined and repeated the experiment many times, but the results were always the same: There is no evidence for the existence of the luminiferous ether. The conclusion from this and other experiments is that electromagnetic waves do not require the presence of a material medium, but can propagate in a complete vacuum.

Under Newtonian relativity, to explain a constant value of the speed of light in a vacuum required that the equations of electromagnetism hold true only in a specific reference frame, one that is at rest relative to the medium for electromagnetic waves. But if there is no luminiferous ether, there is no such medium and hence no such special frame of reference. So physicists had no choice but to conclude that Newtonian relativity does *not* apply to electromagnetic waves. It was left to Albert Einstein to modify the ideas of Newtonian relativity and find a new way to look at the relationships between measurements made in different inertial frames of reference. We'll explore Einstein's simple yet radical ideas in the following section.

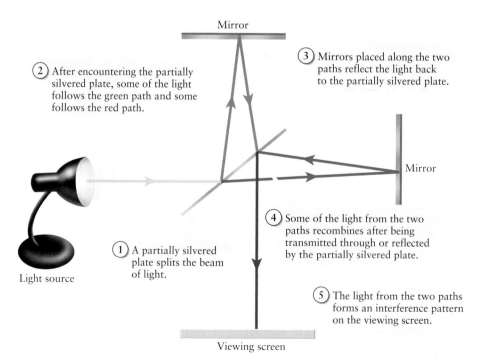

2 After encountering the partially silvered plate, some of the light follows the green path and some follows the red path.

Mirror

3 Mirrors placed along the two paths reflect the light back to the partially silvered plate.

Mirror

4 Some of the light from the two paths recombines after being transmitted through or reflected by the partially silvered plate.

1 A partially silvered plate splits the beam of light.

Light source

5 The light from the two paths forms an interference pattern on the viewing screen.

Viewing screen

Figure 25-7 The Michelson-Morley experiment simplified If the luminiferous ether exists and is moving relative to this apparatus, its presence will be apparent in the interference pattern on the viewing screen.

? Got the Concept? 25-2 Speed of Sound in Water

A bottlenose dolphin (*Tursiops truncates*) can swim at speeds up to about 10 m/s. These dolphins produce sounds used for social communication and for echolocation (using sound to detect objects around them while swimming in dark or murky waters). If a swimming dolphin travels at top speed and produces a sound wave that propagates forward, how fast does that wave travel relative to the dolphin? The speed of sound in water is 1500 m/s. (a) 10 m/s; (b) 1490 m/s; (c) 1500 m/s; (d) 1510/m/s; (d) 10 m/s; (e) answer depends on the frequency of the sound wave.

Take-Home Message for Section 25-3

✔ Experiment shows that electromagnetic waves do not require a material medium; they can easily propagate in a perfect vacuum.

✔ As a result, we conclude that electromagnetic waves do not obey Newtonian relativity.

25-4 Einstein's relativity predicts that the time between events depends on the observer

German-born theoretical physicist Albert Einstein published his **special theory of relativity**, or *special relativity* for short, in 1905. Einstein based his theory on two postulates:

- *First postulate:* All laws of physics are the same in all inertial frames.
- *Second postulate:* The speed of light in a vacuum is the same in all inertial frames, independent of both the speed of the source of the light and the speed of the observer.

The adjective "special" means that the theory applies to the special case of inertial frames and constant-velocity motion. Einstein's *general* theory of relativity, which we'll discuss in Section 25-8, extends these ideas to accelerating, noninertial frames.) The first postulate extends the principle of relativity beyond Newton's laws of motion to include thermodynamics and electricity and magnetism, including electromagnetic waves. Einstein's second postulate derives from our conclusion in Section 25-3 that there is no special frame of reference in which the speed of light in a vacuum is equal to $c = 3.00 \times 10^8$ m/s. The implications of this second postulate totally transformed early twentieth-century scientists' understanding of the natural world.

Perhaps the most astounding consequence of the second postulate of special relativity is that the time interval between two events is *not* an absolute. Rather, this time interval depends on the motion of the frame of reference from which time is measured. To see how this comes about, we'll do a thought experiment. Einstein was well known for his ability to solve problems by carrying out these *gedanken*, or thought experiments, some of which were technologically impossible to do as real physical experiments.

Imagine a *light clock*, a special clock that uses light to measure intervals of time (**Figure 25-8**). A laser fires an extremely brief burst, or pulse, of light straight downward. This is reflected by a mirror and arrives at a light-sensitive detector next to the laser. For simplicity, we can treat the laser and detector as being at the same position. One tick of the clock is the time interval between the pulse leaving the laser (event 1) and the pulse arriving at the detector (event 2). In the frame of reference of the clock, these two events occur at the same point in space. We use the term **proper time** for the time interval between two events that occur at the same place, and we denote it by the symbol Δt_{proper}. Another way to think of proper time is that it is the time interval as measured in a frame of reference in which the clock is at rest. You can think of the rest frame of the clock as being "attached" to it.

Figure 25-8 A light clock The duration of each "tick" of this clock is the time between the start of the pulse and the time when the reflected light arrives at the detector.

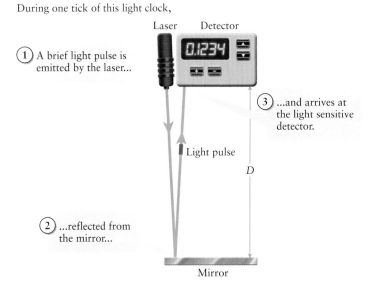

During one tick of this light clock,

Laser   Detector

0.1234

① A brief light pulse is emitted by the laser...

③ ...and arrives at the light sensitive detector.

Light pulse

D

② ...reflected from the mirror...

Mirror

! **Watch Out!** Every object—including a moving object—has a rest frame.

An object is always at rest with respect to itself, which means that an object is not moving in a reference frame that is attached to it. Notice, however, that the statement is a relative one—the object is not moving *relative* to its own rest frame.

We can still define any number of other frames with respect to which the object *is* in motion. A person standing on the sidewalk is at rest in her own rest frame, but is moving as measured from the frame of reference of a car driving past.

During one tick of our light clock, a light pulse travels the distance D to the mirror and then the same distance back to the detector, for a total distance of $2D$. We imagine that the clock is placed in a vacuum, so that the speed of the light pulse is c. The time interval for the tick is then

$$\Delta t_{\text{proper}} = \frac{2D}{c} \qquad (25\text{-}6)$$

We now let the light clock move at speed V relative to us. In keeping with the way we name frames in discussing Newtonian relativity, we attach a frame S' to the clock and consider ourselves in the S frame. The S' frame therefore moves at speed V relative to the S frame. Figure 25-9 shows the process of a single tick of the clock as observed from our S frame. During the time that it takes for the light pulse to get from the laser to the mirror, the entire clock has moved. And during the time between the reflection from the mirror and the arrival of the light pulse at the detector, the clock has moved again. The total distance L that the clock has moved during this time equals the product of the speed V and the time interval of one clock tick. Be careful, however: We *cannot* assume that the time interval for one tick as measured in S is Δt_{proper} because Δt_{proper} is measured in frame S' at rest with respect to the clock. Instead, we use the symbol Δt for the time interval of one tick as observed from the S frame. So

$$L = V\Delta t \qquad (25\text{-}7)$$

From our vantage point in the S frame, the path followed by the light pulse during time Δt traces out two sides of a triangle (Figure 25-9).

$\sqrt{}\!\!\times$ *See the Math Tutorial for more information on trigonometry*

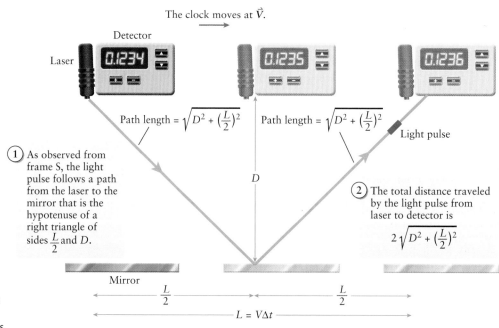

The clock moves at $\vec{V}$.

Detector

Laser

Path length $= \sqrt{D^2 + \left(\frac{L}{2}\right)^2}$ Path length $= \sqrt{D^2 + \left(\frac{L}{2}\right)^2}$

Light pulse

(1) As observed from frame S, the light pulse follows a path from the laser to the mirror that is the hypotenuse of a right triangle of sides $\frac{L}{2}$ and D.

D

(2) The total distance traveled by the light pulse from laser to detector is

$$2\sqrt{D^2 + \left(\frac{L}{2}\right)^2}$$

Mirror

$\frac{L}{2}$ $\frac{L}{2}$

$L = V\Delta t$

Figure 25-9 A moving light clock and time dilation When the light clock depicted in Figure 25-8 moves, the distance the light pulse travels is longer than twice the distance from the laser to the mirror.

(3) As observed in frame S, $L = V\Delta t$ is the total distance that the clock travels during the travel time Δt of the light pulse.

In our rest frame S the time interval is therefore equal to the sum of the lengths of these two sides—the distance the light pulse travels—divided by the speed at which the light pulse travels in frame S. The second postulate of special relativity assures us that the speed of the light pulse is c in *all* inertial frames, and so is the same in frame S as in frame S'. Thus

$$\Delta t = \frac{2\sqrt{D^2 + \left(\frac{L}{2}\right)^2}}{c} \tag{25-8}$$

We now have two expressions for the time interval between the light pulse leaving the laser and the same pulse arriving at the detector. As measured in the clock rest frame S', this time interval is the proper time Δt_{proper} as given by Equation 25-6; as measured in our rest frame S, the time interval is Δt as given by Equation 25-8. To compare these two time intervals, first note that the distance L does not appear in the expression for Δt_{proper}. To eliminate L from the expression for Δt, substitute Equation 25-7 into Equation 25-8:

$$\Delta t = \frac{2\sqrt{D^2 + \left(\frac{V\Delta t}{2}\right)^2}}{c}$$

Square both sides of this equation and rearrange to bring the two Δt terms together:

$$c^2\Delta t^2 = 4\left(D^2 + \left(\frac{V\Delta t}{2}\right)^2\right) = 4D^2 + V^2\Delta t^2$$

$$c^2\Delta t^2 - V^2\Delta t^2 = 4D^2 \quad \text{or} \quad (c^2 - V^2)\Delta t^2 = 4D^2$$

Solve for Δt:

$$\Delta t^2 = \frac{4D^2}{c^2 - V^2} = \frac{4D^2}{c^2\left(1 - \frac{V^2}{c^2}\right)} \quad \text{or} \quad \Delta t = \frac{2D}{c\sqrt{1 - \frac{V^2}{c^2}}} \tag{25-9}$$

If we compare Δt as given by Equation 25-9 to Δt_{proper} as given by Equation 25-6, we get

Time interval between two events as measured in frame S', in which those events occur at the same place.

Time interval between the same two events as measured in frame S

$$\Delta t = \frac{\Delta t_{proper}}{\sqrt{1 - \frac{V^2}{c^2}}}$$

Speed of S' relative to S

Speed of light in a vacuum

a moving clock runs slowly

Time dilation
(25-10)

An observer in S moving relative to the clock measures a time interval Δt for one tick of the light clock. An observer in S' who is at rest with respect to the clock measures the time interval for one tick to be Δt_{proper}, regardless of whether the observer and clock are moving with respect to some *other* observer. Equation 25-10 shows that Δt and Δt_{proper} are *not* equal: An observer in the S frame sees time running at a different rate than an observer in the S' frame. This is astounding!

For any nonzero value of V, the speed of one inertial frame relative to another, the denominator in Equation 25-10 is less than 1. It follows that Δt is greater than Δt_{proper}. That is, the time interval as measured in S is longer than as measured in S'. We say that the time interval as measured in S has been expanded or *dilated* (the same term used to refer to an increase in size of the pupil of the eye). That's why this effect of special relativity is known as **time dilation**. If $\Delta t_{proper} = 1$ s, the time interval Δt for one tick as measured in S will be greater than 1 s. Equivalently, if Δt as measured in S equals 1 s, Δt_{proper} as measured in S' will be less than 1 s. So an observer in S says that the light clock, which is moving past her at speed V, is ticking off time slowly: After 1 s has elapsed according to the observer in S, the moving clock has ticked off less than 1 s. So a shorthand way to express Equation 25-10 is that *a moving clock runs slowly*.

Our derivation of Equation 25-10 used the second postulate of relativity, the idea that the speed of light in a vacuum is the same in frame S as in frame S'. The difference between Δt and Δt_{proper} was then a result of the light pulse having to travel different distances in the two frames. As a result, it may seem that Equation 25-10 is valid only for time intervals measured using a light clock. But the first postulate of relativity says that *all* laws of physics are the same in every inertial frame of reference. This implies that time dilation is also valid for time intervals measured using a mechanical clock, such as a grandfather clock that keeps time with an oscillating pendulum or a wristwatch that uses an oscillating piece of quartz (Figure 12-1b). Imagine a pendulum clock, ticking once every time the pendulum makes one full oscillation. If the clock is on your desk, you see it swinging back and forth once every second. If the clock is moving, then the swing of the pendulum is slower, so the clock ticks more slowly. As we will see, however, the effect is very small unless V is very large.

! Watch Out! 🦅 Time dilation occurs whether or not a clock is present.

You don't need to have a clock per se for the effects of time dilation to occur. Time runs slowly in one frame moving with respect to another frame, even if there is no clock in the moving frame and even if there is no observer in the other frame. Imagine, say, that you can place a *Caenorhabditis elegans* worm (Figure 25-10) on a spacecraft that will fly past Earth at high speed. *C. elegans* is an organism popular among geneticists because it grows to adulthood in a series of easily identifiable developmental stages that all together take less than three days. Each stage lasts eight to twelve hours. Were you to observe a *C. elegans* worm as it moved past you at high speed, each of those stages might take days or even years, measured on a clock at rest with respect to you.

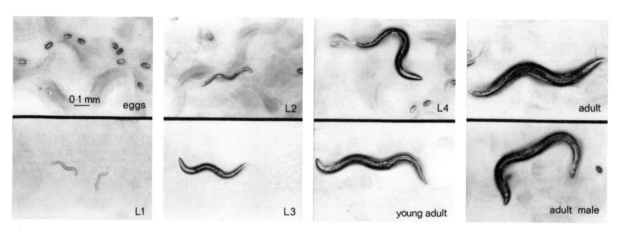

Figure 25-10 **A worm "clock"** The worm *Caenorhabditis elegans* is an organism popular among geneticists because it grows to adulthood in a series of easily identifiable developmental stages each of which lasts a well-defined period of time.

! Watch Out! The effect of time dilation only arises when events in one frame are viewed from a second frame in motion relative to the first.

We imagined, above, a pendulum clock that ticks once every time the pendulum makes one full oscillation. If the clock sits on your desk, you see it swinging back and forth once every second. If you're sitting at your desk and you see the clock move across your desk, you see the pendulum swing more slowly. But if you and your clock are moving together relative to some other specific object or frame, you will see no time dilation effect on your own clock, regardless of how fast you and the clock are moving relative to that other object or frame. You will still see the pendulum swinging back and forth once every second. Although you and the clock are moving relative to the other object or frame, you and the clock are not moving relative to each other.

We do not observe time dilation in everyday life because most objects travel very slowly compared to the speed of light. As an example, in 2006 the *New Horizons* spacecraft was launched toward Pluto at 16,260 m/s = 58,536 km/h = 36,373 mi/h relative to Earth. (This is the greatest launch speed ever given to a spacecraft.) Even this tremendous speed is only 5.42×10^{-5} of the speed of light c, and the factor $1/\sqrt{1 - V^2/c^2}$ in Equation 25-10 is larger than 1 by only 1.47×10^{-9}. Due to time dilation, a clock on board *New Horizons* does indeed run slowly as measured from Earth, but by only one second every 21.5 years! As the following two examples show, however, time dilation can be substantial if the speeds involved are **relativistic speeds**— that is, an appreciable fraction of the speed of light, so that the consequences of Einstein's special theory of relativity are apparent.

Example 25-2 A Moving Clock

A clock moves past you. What must be the speed of the clock relative to you so that you see it as running at one-half (0.500) the rate of the clock on your cell phone in your hand?

Set Up

For a moving clock to be observed as running at 0.500 the rate of a clock at rest with respect to you means making Δt (the time interval measured by you) equal to twice Δt_{proper} (the time interval measured in the rest frame of the moving clock). That implies the factor $1/\sqrt{1 - V^2/c^2}$ in Equation 25-10 must be equal to $1/0.500 = 2.00$. We'll use this to solve for the speed V of the moving clock relative to you.

Time dilation:

$$\Delta t = \frac{\Delta t_{proper}}{\sqrt{1 - \dfrac{V^2}{c^2}}} \qquad (25\text{-}10)$$

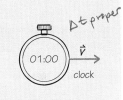

Solve

Determine the value of V that makes $\Delta t = 2.00 \Delta t_{proper}$.

From Equation 25-10, we want

$$\Delta t = \frac{\Delta t_{proper}}{\sqrt{1 - \dfrac{V^2}{c^2}}} = 2.00 \Delta t_{proper} \quad \text{so} \quad \frac{1}{\sqrt{1 - \dfrac{V^2}{c^2}}} = 2.00$$

Solve for the value of the speed V:

$$\sqrt{1 - \frac{V^2}{c^2}} = \frac{1}{2.00} = 0.500$$

$$1 - \frac{V^2}{c^2} = (0.500)^2 = 0.250$$

$$\frac{V^2}{c^2} = 1 - 0.250 = 0.750$$

$$\frac{V}{c} = \sqrt{0.750} = 0.866$$

$$V = 0.866c$$

$$= 0.866(3.00 \times 10^8 \text{ m/s}) = 2.60 \times 10^8 \text{ m/s}$$

Reflect

For you to observe the moving clock running a factor of 2.00 slower than the clock at rest with respect to you, the relative speed between you and the clock would need to be 86.6% of the speed of light in a vacuum. This is more than 10^4 times faster than the top speed of any craft built by humans.

Example 25-3 Muon Decay

Our planet is continually bombarded by fast-moving subatomic particles from space (mostly protons). When these particles collide with the atoms and molecules that make up Earth's upper atmosphere, they can produce a new subatomic particle called the *muon*. These muons travel at high speed, around $0.994c$. Muons naturally decay, however; if you create a number of muons in the lab, on average half of them will decay after $1.56\ \mu s$ ($1\ \mu s = 10^{-6}$ s). This time is known as the *half-life* of the muon. Of the remaining muons, another half will decay after another $1.56\ \mu s$, and so on. If 1.00×10^6 muons are created at an altitude of 15.0 km, (a) how many would you expect to strike Earth if time-dilation effects were ignored? (b) How many would you expect to strike Earth when time dilation is taken into account?

Set Up

The production and decay of the muon occur at the same place in the rest frame of the muon, so the half-life of $1.56\ \mu s$ is a proper time interval in the muon frame. We'll call this frame S'. This frame moves at speed $V = 0.994c$ relative to our frame on Earth, and so we measure a different half-life Δt as given by Equation 25-10. We'll use the idea that the muon population decreases by $1/2$ in each half-life to determine how many reach our planet's surface.

Time dilation:

$$\Delta t = \frac{\Delta t_{\text{proper}}}{\sqrt{1 - \dfrac{V^2}{c^2}}} \qquad (25\text{-}10)$$

muon produced

15.0 km

$V = 0.994c$

ground

Solve

(a) If there were no time dilation, the half-life of the muon in the Earth frame S would be $\Delta t = \Delta t_{\text{proper}} = 1.56\ \mu s$, the same as in the muon frame S'. Use this to calculate the number of muons that successfully reach Earth's surface.

Time for a muon to travel a distance $d = 15.0$ km at $V = 0.994\ c$:

$$T = \frac{d}{V} = \frac{15.0\ \text{km}}{0.994(3.00 \times 10^8\ \text{m/s})} \left(\frac{10^3\ \text{m}}{1\ \text{km}}\right)$$

$$= 5.03 \times 10^{-5}\ \text{s}\left(\frac{1\ \mu s}{10^{-6}\ \text{s}}\right) = 50.3\ \mu s$$

Express this as a multiple of the half-life:

$$\frac{T}{\Delta t_{\text{proper}}} = \frac{50.3\ \mu s}{1.56\ \mu s} = 32.2\ \text{half-lives}$$

After one half-life, the number of muons has decreased to $1/2$ of its initial value; after two-half-lives, to $(1/2) \times (1/2) = 1/2^2$ of its initial value; after three half-lives, to $(1/2) \times (1/2) \times (1/2) = 1/2^3$ of its initial value; and so on. So after 32.2 half-lives, the number of muons remaining would be the original number of 1.00×10^6 multiplied by $1/2^{32.2}$:

$$\text{Muons remaining} = (1.00 \times 10^6)\left(\frac{1}{2^{32.2}}\right)$$

$$= (1.00 \times 10^6)(1.97 \times 10^{-10})$$

$$= 1.97 \times 10^{-4}$$

Much less than 1 muon, on average, survives the trip to Earth's surface.

(b) With time dilation, we first calculate the half-life in the Earth frame using Equation 25-10. We then use the same method as in part (a) to calculate the number of muons that reach Earth's surface.

Accounting for time dilation, the half-life as measured in the Earth frame S is the half-life measured in the muon frame S' divided by $\sqrt{1 - V^2/c^2}$:

$$\Delta t = \frac{\Delta t_{\text{proper}}}{\sqrt{1 - \dfrac{V^2}{c^2}}} = \frac{1.56\ \mu s}{\sqrt{1 - \dfrac{(0.994c)^2}{c^2}}} = \frac{1.56\ \mu s}{\sqrt{1 - (0.994)^2}}$$

$$= \frac{1.56\ \mu s}{\sqrt{0.0120}} = \frac{1.56\ \mu s}{0.109}$$

$$= 14.3\ \mu s \quad \text{rs.} \quad 1.56\ \mu s$$

Due to time dilation, the half-life as measured in the Earth frame is about 9 times longer than the half-life as measured in the muon frame.

From part (a), the time for a muon to travel to the surface is 50.3 μs. Expressed as a multiple of the time-dilated half-life, this is

$$\frac{T}{\Delta t} = \frac{50.3\ \mu s}{14.3\ \mu s} = 3.53 \text{ half-lives}$$

The number of muons remaining at the surface is the original number of 1.00×10^6 multiplied by $1/2^{3.53}$:

$$\begin{aligned}
\text{Muons remaining} &= (1.00 \times 10^6)\left(\frac{1}{2^{3.53}}\right) \\
&= (1.00 \times 10^6)(0.0868) \\
&= 8.68 \times 10^4
\end{aligned}$$

About one in eleven of the muons produced in the upper atmosphere reaches Earth's surface, far more than would be the case were there no time dilation.

Reflect

One of the earliest confirmations of Einstein's special theory of relativity was a 1941 experiment that compared the number of muons observed in an hour at the top of Mount Washington in New Hampshire (about 2000 m above sea level) to the number observed at the base of the mountain (about 1100 m above sea level). Were there no time dilation, the number of muons observed at the base of the mountain would have been about 10% of the number at the top. The measured rate at the base was over 80%, precisely in accordance with Einstein's prediction. This experiment provided dramatic evidence that the bizarre phenomenon of time dilation is very real.

As we found in this example, the effect of time dilation causes time to run more slowly for the muon than we measure in our own frame, so fewer half-lives elapse for a given distance traveled. Because fewer half-lives have elapsed, fewer muons have decayed.

? Got the Concept? 25-3   Proper Time

A clock is placed aboard Starship *Alpha*, which Albert flies past Earth at half the speed of light relative to Earth. Barbara flies Starship *Beta* alongside *Alpha* at the same velocity. George pilots Starship *Gamma* past Earth at half the speed of light relative to Earth, but in the opposite direction. Elena observes from Earth. For which of these observers is the time interval between ticks of the clock equal to the proper time? (a) Albert; (b) Barbara; (c) George; (d) Elena; (e) more than one of these.

? Got the Concept? 25-4   Comparing Clocks

A clock is placed aboard Starship *Alpha*, which Albert flies past Earth at half the speed of light relative to Earth. Elena observes from Earth, where she has an identical clock. Which pair of words correctly fills in the blanks in this statement: "Elena measures Albert's clock as running _____, and Albert measures Elena's clock as running _____." (a) slow, fast; (b) slow, slow; (c) fast, fast; (d) fast, slow.

Take-Home Message for Section 25-4

✔ Einstein based his special theory of relativity on two postulates: first, that all laws of physics are the same in all inertial frames, and second, that the speed of light in a vacuum is the same in all frames and independent of both the speed of the source of the light and the speed of the observer.

✔ If two events happen at the same place in one frame, the time interval between these events as measured in that frame is called the proper time. As measured from a second frame moving relative to the first one, the time interval between those two events is longer than the proper time. This is called time dilation.

25-5 Einstein's relativity also predicts that the length of an object depends on the observer

We learned in the preceding section that the time interval between two events is not an absolute, but depends on the motion of the observer. As we will see in this section, the *length* of an object is also not an absolute: Different observers will measure the same object as having different dimensions. So the postulates of special relativity tell us that the nature of space and time is very different from what had been thought previous to the work of Einstein.

We'll conclude this section by looking at the *Lorentz transformation*, a set of equations that allows us to convert the space and time coordinates of an event in one inertial frame of reference to the coordinates in a second inertial frame. This transformation is an extension of the Galilean transformation that we explored in Section 25-2. We'll see that the Galilean transformation is actually just a special case of the Lorentz transformation valid for two frames that are moving with respect to each other at a speed far less than *c*.

Length Contraction

Figure 25-11 shows a thought experiment in which we look at the same motion in two different frames of reference, much as we did in Figures 25-8 and 25-9. In Figure 25-11a a rod is moving to the right at speed V relative to frame S, which you can think of as our frame of reference. To determine the length of the rod, we use our clock to measure the time interval Δt_S required for the rod to travel its own length. The right-hand end of the rod is at $x = 0$ at time $t = 0$, and the left-hand end of the rod is at $x = 0$ at a later time $t = \Delta t_S$. The length L of the rod in our frame is therefore the speed V of the rod multiplied by the time Δt_S needed to travel its own length:

(25-11)
$$L = V\Delta t_S$$
(length of the rod in frame S)

Figure 25-11b shows the same process as observed in the frame of reference S' in which the rod is at rest. In this frame the rod has length L_{rest}, and the frame S moves to the left at speed V. The point $x = 0$ on frame S travels from one end of the rod to the other in a time $\Delta t_{S'}$. The length of the rod equals the speed V of frame S multiplied by the time $\Delta t_{S'}$ that frame S needs to travel this length:

(25-12)
$$L_{rest} = V\Delta t_{S'}$$
(length of the rod in its own rest frame S')

(a) Motion of a rod as observed in frame S

The right-hand end of the rod coincides with the point $x = 0$ at $t = 0$.

As measured in frame S, the rod has length L. It moves at speed V in the positive x direction.

The left-hand end of the rod coincides with the point $x = 0$ at $t = \Delta t_S$.

As measured in S, the rod moves a distance L in time Δt_S. So $L = V\Delta t_S$.

(b) The same events as observed in the rod rest frame S'

The right-hand end of the rod coincides with the point $x = 0$ at $t' = 0$.

As measured in frame S', the rod is at rest and has length L_{rest}. Frame S moves at speed V in the negative x direction.

The left-hand end of the rod coincides with the point $x = 0$ at $t' = \Delta t_{S'}$.

As measured in the rod rest frame S', the frame S moves a distance L_{rest} in time $\Delta t_{S'}$. So $L_{rest} = V\Delta t_{S'}$.

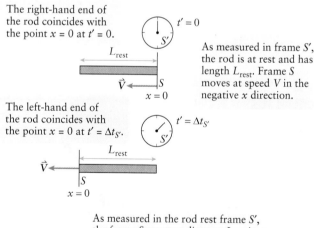

- The two events are (1) the right-hand end of the rod coinciding with $x = 0$ and (2) the left-hand end of the rod coinciding with $x = 0$.
- These two events happen at the same place in frame S, so Δt_S is the proper time interval between these events. The time interval $\Delta t_{S'}$ must therefore be greater than Δt_S (time dilation).
- The distance L_{rest} must therefore be greater than the distance L. So the length L of the moving rod is less than the length L_{rest} of the rod in a frame where it is at rest. This is length contraction.

Figure 25-11 **A moving rod and length contraction** A rod as observed in (a) a frame in which the rod is moving along its length and (b) a frame in which the rod is at rest.

To see how the lengths as measured in the two frames compare, note that the length measurements in S and S' both involve the same pair of events: the right-hand end of the rod coinciding with the point $x = 0$ in frame S, and the left-hand end of the rod coinciding with that same point. These two events happen at the same location in frame S, but at different locations in frame S'. So Δt_S is the proper time interval between these events. The time interval $\Delta t_{S'}$ measured in frame S' is therefore longer ("dilated") compared to the time interval Δt_S measured in frame S, and the two time intervals are related by Equation 25-10:

$$\Delta t_{S'} = \frac{\Delta t_S}{\sqrt{1 - \dfrac{V^2}{c^2}}} \qquad (25\text{-}13)$$

Multiply both sides of Equation 25-13 by V, then replace $V\Delta t_S$ by L in accordance with Equation 25-11 and replace $V\Delta t_{S'}$ by L_{rest} according to Equation 25-12:

$$V\Delta t_{S'} = \frac{V\Delta t_S}{\sqrt{1 - \dfrac{V^2}{c^2}}} \quad \text{or} \quad L_{\text{rest}} = \frac{L}{\sqrt{1 - \dfrac{V^2}{c^2}}}$$

We can rewrite this as

Length of an object in a frame of reference S' in which it is at rest.

Speed of S' relative to S

$$L = L_{\text{rest}}\sqrt{1 - \frac{V^2}{c^2}}$$

Length of the same object in a frame of reference S in which the object is moving along its length.

Speed of light in a vacuum

Length contraction
(25-14)

If the rod is moving, V is greater than zero and the factor $\sqrt{1 - V^2/c^2}$ is less than 1. So the length L of the moving rod is less than the length L_{rest} of the rod at rest. In other words, *a moving object is shortened along the direction in which it is moving.* This is called **length contraction**.

> **Watch Out!** **Length contraction occurs only along the direction of motion.**
>
> There is *no* change in length of a moving object in any direction other than the direction of motion. For example, if the rod in Figure 25-11 is moving in the x direction relative to frame S, an observer in S will measure the rod as having a shorter length in the x direction than will an observer in S' at rest with respect to the rod. But both observers will agree about the height and width of the rod (its dimensions in the y and z directions).

Length contraction gives us an alternative way to understand the results of Example 25-3 in the preceding section, in which we used time dilation to explain how short-lived muons produced at an altitude of 15.0 km are able to survive their trip to Earth's surface. Imagine a vertical rod that extends 15.0 km upward from the ground to where the muons are produced. This rod is stationary relative to Earth, so its length as measured by you on the ground is its rest length: $L_{\text{rest}} = 15.0\text{ km} = 1.50 \times 10^4$ m. But as seen from the frame of a descending muon, this rod is moving upward along its length at $V = 0.994c$. As measured by a muon, the length of this rod is contracted in accordance with Equation 25-14:

$$L = L_{\text{rest}}\sqrt{1 - \frac{V^2}{c^2}} = (15.0\text{ km})\sqrt{1 - \frac{(0.994c)^2}{c^2}}$$

$$= (15.0\text{ km})\sqrt{1 - (0.994)^2}$$

$$= (15.0\text{ km})\sqrt{0.0120} = (15.0\text{ km})(0.109)$$

$$= 1.64\text{ km} = 1.64 \times 10^3\text{ m} \quad \text{vs. } 15\text{ km}$$

Moving at $V = 0.994c$, this contracted rod travels past the muon in a time

$$\frac{1.64 \times 10^3 \text{ m}}{0.994c} = \frac{1.64 \times 10^3 \text{ m}}{(0.994)(3.00 \times 10^8 \text{ m/s})} = 5.50 \times 10^{-6} \text{ s} = 5.50 \ \mu\text{s}$$

The half-life of the muon in its own rest frame is $1.56 \ \mu\text{s}$, so from the muon's perspective it takes just $(5.50 \ \mu\text{s})/(1.56 \ \mu\text{s}) = 3.53$ half-lives for the contracted rod to move past it—that is, for the muon to move from where it is produced to Earth's surface. That's exactly the result that we found in Example 25-3 using the ideas of time dilation. The ideas of length contraction and time dilation are mutually consistent!

Example 25-4   A Flying Meter Stick I

A meter stick (length 1.00 m) hurtles through space at a speed of $0.800c$ relative to you, with its length aligned with the direction of motion. What do you measure as the length of the meter stick?

Set Up

The meter stick's length along its direction of motion is 1.00 m as measured in its own rest frame S', so $L_{\text{rest}} = 1.00$ m. Your frame is frame S, and the two frames are moving relative to each other at $V = 0.800c$. We'll use Equation 25-14 to find the length L as measured in your frame.

Length contraction:

$$L = L_{\text{rest}}\sqrt{1 - \frac{V^2}{c^2}} \qquad (25\text{-}14)$$

$L_{\text{rest}} = 1.00$ m

$V = 0.800c$
= velocity of the meter stick relative to frame S

Solve

Substitute $L_{\text{rest}} = 1.00$ m and $V = 0.800c$ into Equation 25-14 and calculate L.

From Equation 25-14,

$$L = L_{\text{rest}}\sqrt{1 - \frac{V^2}{c^2}} = (1.00 \text{ m})\sqrt{1 - \frac{(0.800c)^2}{c^2}}$$

$$= (1.00 \text{ m})\sqrt{1 - (0.800)^2} = (1.00 \text{ m})\sqrt{1 - 0.640}$$

$$= (1.00 \text{ m})\sqrt{0.360} = (1.00 \text{ m})(0.600)$$

$$= 0.600 \text{ m} = 60.0 \text{ cm} \quad \text{Vs. } 100 \text{ cm}$$

Reflect

As measured in your frame of reference, the meter stick is only 60.0 cm in length. This is not an optical illusion; the meter stick really is only 60.0% as long in your frame of reference as in the rest frame of the meter stick.

Example 25-5   A Flying Meter Stick II

The meter stick from the preceding example again hurtles through space at a speed of $0.800c$ relative to you, but now it is tilted at an angle of $30.0°$ as measured in its rest frame with respect to the direction of motion. Now what do you measure as the length of the meter stick?

Set Up

As measured in the meter stick's rest frame S', the meter stick is inclined at an angle $\theta' = 30.0°$ to its direction of motion relative to frame S. The x dimension of the stick undergoes length contraction given by Equation 25-14, but the y dimension does not. We'll calculate the x and y dimensions of the stick in your frame S, then use the Pythagorean theorem to find the length of the stick in frame S.

Length contraction:

$$L = L_{\text{rest}}\sqrt{1 - \frac{V^2}{c^2}} \qquad (25\text{-}14)$$

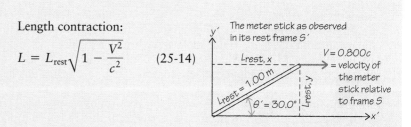

The meter stick as observed in its rest frame S'

$V = 0.800c$
= velocity of the meter stick relative to frame S

$L_{\text{rest},x}$

$L_{\text{rest}} = 1.00$ m

$\theta' = 30.0°$

$L_{\text{rest},y}$

Solve

First calculate the x and y dimensions of the meter stick in its rest frame.

In the rest frame S' of the stick, its dimensions are

$$L_{\text{rest},x} = L_{\text{rest}}\cos\theta' = (1.00 \text{ m})\cos 30.0° = 0.866 \text{ m}$$

$$L_{\text{rest},y} = L_{\text{rest}}\sin\theta' = (1.00 \text{ m})\sin 30.0° = 0.500 \text{ m}$$

Calculate the x and y dimensions of the meter stick in your frame, in which the stick is moving at $V = 0.800c$ along the x direction.

In your frame S, the x dimension of the stick is contracted:

$$L_x = L_{\text{rest},x}\sqrt{1 - \frac{V^2}{c^2}} = (0.866 \text{ m})\sqrt{1 - \frac{(0.800c)^2}{c^2}}$$

$$= (0.866 \text{ m})\sqrt{1 - 0.640} = (0.866 \text{ m})\sqrt{0.360}$$

$$= 0.520 \text{ m}$$

Length contraction occurs only along the direction of motion, so the y dimension of the stick is *not* contracted:

$$L_y = L_{\text{rest},y} = 0.500 \text{ m}$$

Use the Pythagorean theorem to calculate the length of the meter stick in your frame.

From the Pythagorean theorem, the length of the meter stick in your frame S is

$$L = \sqrt{L_x^2 + L_y^2}$$

$$= \sqrt{(0.520 \text{ m})^2 + (0.500 \text{ m})^2}$$

$$= \sqrt{(0.520 \text{ m})^2 + (0.500 \text{ m})^2}$$

$$= \sqrt{0.520 \text{ m}^2} = 0.721 \text{ m}$$

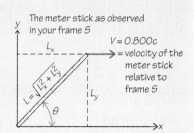

The meter stick as observed in your frame S

$V = 0.800c$ = velocity of the meter stick relative to frame S

Reflect

Because the meter stick is not aligned with the direction of motion, the amount of contraction is not as great as in Example 25-4, in which the meter stick was completely aligned with the direction of motion. Notice also that the angle θ of the meter stick in your frame is different from the angle $\theta' = 30.0°$ in the stick's rest frame. Can you show that $\theta = 43.9°$?

The Lorentz Transformation

The Galilean transformation that we presented in Section 25-2 is not consistent with the postulates of special relativity. A set of transformation equations that *is* consistent with relativity is the **Lorentz transformation**. As in Figure 25-3, we take frame S' to be moving at speed V in the positive x direction relative to frame S. The origins of the two frames coincide at $t = 0$ in frame S and $t' = 0$ in frame S'. If an event takes place at coordinates x, y, z, t in frame S, the coordinates of that same event in frame S' are

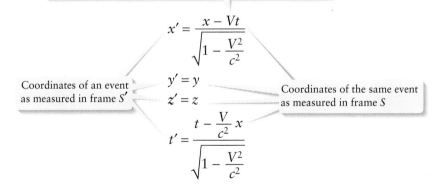

Inertial frame of reference S' moves at speed V in the positive x direction relative to inertial frame of reference S.

$$x' = \frac{x - Vt}{\sqrt{1 - \dfrac{V^2}{c^2}}}$$

$$y' = y$$
$$z' = z$$

$$t' = \frac{t - \dfrac{V}{c^2}x}{\sqrt{1 - \dfrac{V^2}{c^2}}}$$

Coordinates of an event as measured in frame S'

Coordinates of the same event as measured in frame S

Lorentz transformation
(25-15)

Note that at speeds that are far less than the speed of light, the ratio V/c is very small and can be treated as essentially zero. Then $\sqrt{1 - V^2/c^2}$ is essentially equal to 1, and Equations 25-15 become

$$x' = x - Vt, \quad y' = y, \quad z' = z, \quad t' = t$$

These are just the Galilean transformation equations that we presented in Section 25-2. So at speeds far slower than the speed of light, the equations of Einstein's special theory of relativity reduce to the equations of Newtonian relativity.

Equations 25-15 are useful for finding the coordinates of an event in frame S' if we know the event's coordinates in frame S. If we instead want to determine the coordinates of an event in frame S from the coordinates in frame S', it's most convenient to use the *inverse* Lorentz transformation:

> Inertial frame of reference S' moves at speed V in the positive x direction relative to inertial frame of reference S.

Inverse Lorentz transformation (25-16)

$$x = \frac{x' + Vt'}{\sqrt{1 - \dfrac{V^2}{c^2}}}$$

$$y = y'$$

$$z = z'$$

$$t = \frac{t' + \dfrac{V}{c^2}x'}{\sqrt{1 - \dfrac{V^2}{c^2}}}$$

> Coordinates of an event as measured in frame S

> Coordinates of the same event as measured in frame S'

It's possible to derive the equations for time dilation (Equation 25-10) and length contraction (Equation 25-14) from the Lorentz transformation equations. Instead, let's look at another remarkable result of the special theory of relativity that we can deduce from the Lorentz transformation equations.

Example 25-6 Simultaneity Is Relative

In your reference frame you have a meter stick that is oriented along the x axis, with one end at $x = 0$ and the other end at $x = 1.00$ m. There is a light bulb at each end of the stick, and you make the two bulbs flash simultaneously (as measured by you) at $t = 0$. A spacecraft flies past you at $V = 0.800c$ in the positive x direction. (a) According to an observer in the spacecraft, are the two light flashes simultaneous? If not, which flash happens first as measured by her? (b) On board the spacecraft is an identical meter stick with a light bulb at each end. The observer on board the spacecraft places the two ends of the stick at $x' = 0$ and $x' = 1.00$ m, and she makes the two bulbs flash simultaneously (as measured by her) at $t' = 0$. According to you, are the two light flashes simultaneous? If not, which flash happens first as measured by you?

Set Up

In part (a) we're given the coordinates in frame S of two events, the flash of the left-hand bulb at $x = 0$ and $t = 0$ and the flash of the right-hand bulb at $x = 1.00$ m and $t = 0$. We'll use the t' equation from the Lorentz transformation, Equations 25-15, to determine the times of these two events as measured in the spaceship frame S'. Similarly, in part (b) we're given the coordinates in frame S' of two other events, the flash of the left-hand bulb at $x' = 0$ and $t' = 0$ and the flash of the right-hand bulb at $x' = 1.00$ m and $t' = 0$. We'll find the times of these events as measured in your S frame using the t equation from the inverse Lorentz tranformation, Equations 25-16.

Time equation from the Lorentz transformation:

$$t' = \frac{t - \dfrac{V}{c^2}x}{\sqrt{1 - \dfrac{V^2}{c^2}}} \qquad (25\text{-}15)$$

Time equation from the inverse Lorentz transformation:

$$t = \frac{t' + \dfrac{V}{c^2}x'}{\sqrt{1 - \dfrac{V^2}{c^2}}} \qquad (25\text{-}16)$$

(a) Meter stick at rest in frame S: What does an observer in frame S' measure?

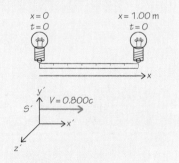

(b) Meter stick at rest in frame S': What does an observer in frame S measure?

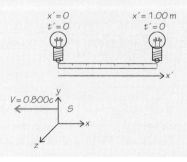

Solve

(a) Use the Lorentz transformation to calculate the times of the simultaneous flashes in S as measured in S'.

The left-hand bulb on the meter stick at rest in frame S flashes at $x = 0$, $t = 0$. In frame S' this bulb flashes at

$$t'_{\text{left}} = \frac{(0) - \dfrac{V}{c^2}(0)}{\sqrt{1 - \dfrac{V^2}{c^2}}} = 0$$

The right-hand bulb on the meter stick at rest in frame S flashes at $x = 1.00$ m, $t = 0$. In frame S' this bulb flashes at

$$t'_{\text{right}} = \frac{(0) - \dfrac{V}{c^2}(1.00\text{ m})}{\sqrt{1 - \dfrac{V^2}{c^2}}} = \frac{-\dfrac{(0.800c)}{c^2}(1.00\text{ m})}{\sqrt{1 - \dfrac{(0.800c)^2}{c^2}}}$$

$$= \frac{-0.800\text{ m}}{c\sqrt{1 - (0.800)^2}} = \frac{-0.800\text{ m}}{(3.00 \times 10^8\text{ m/s})(0.600)}$$

$$= -4.44 \times 10^{-9}\text{ s}$$

As observed from the spaceship frame S', the two events are *not* simultaneous: The right-hand bulb flashes 4.44×10^{-9} s before the left-hand bulb.

(b) Use the inverse Lorentz transformation to calculate the times of the simultaneous flashes in S' as measured in S.

The left-hand bulb on the meter stick at rest in frame S' flashes at $x' = 0$, $t' = 0$. In frame S this bulb flashes at

$$t_{\text{left}} = \frac{(0) + \dfrac{V}{c^2}(0)}{\sqrt{1 - \dfrac{V^2}{c^2}}} = 0$$

The right-hand bulb on the meter stick at rest in frame S' flashes at $x' = 1.00$ m, $t' = 0$. In frame S this bulb flashes at

$$t_{\text{right}} = \frac{(0) + \dfrac{V}{c^2}(1.00\text{ m})}{\sqrt{1 - \dfrac{V^2}{c^2}}} = \frac{+\dfrac{(0.800c)}{c^2}(1.00\text{ m})}{\sqrt{1 - \dfrac{(0.800c)^2}{c^2}}}$$

$$= \frac{+0.800\text{ m}}{c\sqrt{1 - (0.800)^2}} = \frac{+0.800\text{ m}}{(3.00 \times 10^8\text{ m/s})(0.600)}$$

$$= +4.44 \times 10^{-9}\text{ s}$$

As observed from your frame S, the two events are *not* simultaneous: The right-hand bulb flashes 4.44×10^{-9} s after the left-hand bulb.

Reflect

This example illustrates yet another counterintuitive consequence of Einstein's special theory of relativity: Two events that are simultaneous to one observer need not be simultaneous to another observer. So even the simple statement "Two things happen at the same time" has to be qualified by stating in which frame of reference it holds true. Time in relativity is not an absolute!

? Got the Concept? 25-5 No Contraction?

A rod with a rest length of 1.00 m whizzes past you at $0.995c$. Is it possible that you could measure its length to be equal to its rest length? (a) Yes; (b) no.

Take-Home Message for Section 25-5

✔   When an object moves relative to an observer, its length in the direction of motion is contracted compared to the length measured in the object's rest frame. This is called length contraction.

✔   The Lorentz transformation allows you to calculate the coordinates of an event measured in one inertial frame based on the coordinates of that event measured in another inertial frame. Unlike the Galilean transformation, the Lorentz transformation is consistent with the postulates of relativity.

25-6 The relative velocity of two objects is constrained by the speed of light, the ultimate speed limit

Suppose you are an outfielder running to catch a batted baseball (Figure 25-12a). The baseball is traveling at 30.0 m/s relative to the ground, and you are running toward the ball at 10.0 m/s relative to the ground. The Galilean velocity transformation that we learned in Section 25-2 says that, relative to you, the baseball travels at 30.0 m/s plus 10.0 m/s, or 40.0 m/s; the velocities simply add. But suppose that instead of being an outfielder running at 10.0 m/s, you are an astronaut flying in your spaceship at 1.00×10^8 m/s as in Figure 25-12b. Instead of moving toward a baseball, you are moving toward a light beam aimed at you by a stationary astronaut. The same idea that we applied to the baseball predicts that relative to you, the light beam travels at $c = 3.00 \times 10^8$ m/s (the speed of the light beam relative to the other astronaut) plus

(a)

As seen by the outfielder, the ball is approaching her at (30.0 m/s) + (10.0 m/s) = 40.0 m/s.

10.0 m/s   Ball   30.0 m/s

(b)

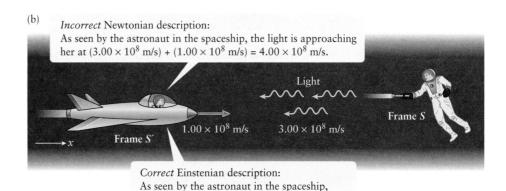

Incorrect Newtonian description:
As seen by the astronaut in the spaceship, the light is approaching her at $(3.00 \times 10^8$ m/s$) + (1.00 \times 10^8$ m/s$) = 4.00 \times 10^8$ m/s.

Light

Frame S

1.00×10^8 m/s 3.00×10^8 m/s

Frame S'

x

Correct Einsteinian description:
As seen by the astronaut in the spaceship, the light is approaching her at 3.00×10^8 m/s.

Figure 25-12 **Velocity addition—Newtonian and Einsteinian**
(a) An outfielder running toward a batted baseball. (b) An astronaut in her spaceship flying toward a light beam.

1.00×10^8 m/s (the speed of your spaceship relative to the other astronaut), or 4.00×10^8 m/s. But that *cannot* be correct: The second postulate of the special theory of relativity says that the speed of light in vacuum is the same to all inertial observers. So the light beam must also travel at speed $c = 3.00 \times 10^8$ m/s relative to you. Clearly, the Galilean transformation for velocities is inadequate. In this section we'll explore the *Lorentz velocity transformation*, which allows us to combine velocities in a way that is consistent with Einsteinian relativity.

It's possible to derive the transformation of velocities from the Lorentz transformation of coordinates that we introduced in Section 25-5 (see Equations 25-15 and 25-16). We'll skip over the derivation and just present the result for the special case in which all motions are along the same line, as in Figure 25-12. Suppose there are two inertial frames of reference, S and S', such that frame S' is moving at speed V in the positive x direction relative to frame S. An object is moving relative to frame S along the x direction, with an x component of velocity v_x. In frame S' the same object has an x component of velocity given by

Inertial frame of reference S' moves at speed V in the positive x direction relative to inertial frame of reference S.

x component of velocity of an object moving along the x axis as measured in frame S'

$$v_x' = \frac{v_x - V}{1 - \dfrac{V}{c^2}v_x}$$

x component of velocity of the same object as measured in frame S

Lorentz velocity transformation
(25-17)

This is called the **Lorentz velocity transformation**. Notice that if the speed V of one frame relative to the other is small compared to c, the ratio V/c is much less than 1 and the denominator is essentially equal to 1. Then Equation 25-17 becomes $v_x' = v_x - V$, which is just the Galilean velocity transformation (the first of Equations 25-5). The same thing happens if the velocity v_x of the object relative to frame S is small compared to c. So if either of the speeds involved is a small fraction of the speed of light, the Lorentz velocity transformation reduces to the Galilean velocity transformation. This justifies our use of the Galilean velocity transformation for slow-moving objects.

Equation 25-17 allows us to find the object's velocity relative to frame S' if we know its velocity relative to frame S. If instead we know the object's velocity relative to frame S' and want to calculate its velocity relative to frame S, we use the **inverse Lorentz velocity transformation**:

Inertial frame of reference S' moves at speed V in the positive x direction relative to inertial frame of reference S.

x component of velocity of an object moving along the x axis as measured in frame S

$$v_x = \frac{v_x' + V}{1 + \dfrac{V}{c^2}v_x'}$$

x component of velocity of the same object as measured in frame S'

Inverse Lorentz velocity transformation
(25-18)

Let's see what the Lorentz velocity transformation tells us about the situation shown in Figure 25-12b. The astronaut with the flashlight is in frame of reference S, and you and your spaceship are in frame of reference S'. You are moving to the right (in the positive x direction), which is just how frame S' must move relative to frame S in order to use Equation 25-17 or 25-18. The relative speed of the two frames is V. We know that the light travels relative to the astronaut with the flashlight at the speed of light in the negative x direction, so its velocity relative to S is $v_x = -c$. We can then use Equation 25-17 to calculate the velocity v_x' of the light relative to you in frame S':

$$v_x' = \frac{v_x - V}{1 - \dfrac{V}{c^2}v_x} = \frac{-c - V}{1 - \dfrac{V}{c^2}(-c)} = \frac{-(c + V)}{1 + \dfrac{V}{c}}$$

We can simplify this by factoring c out of the numerator:

$$v'_x = \frac{-c\left(1 + \dfrac{V}{c}\right)}{1 + \dfrac{V}{c}} = -c$$

This says that as measured by you in frame S', the light travels at the speed of light c in the negative x direction just as in frame S. Note that while $V = 1.00 \times 10^8$ m/s in Figure 25-12b, we didn't have to plug in a numerical value for V to get this result: No matter what the relative speed of the two frames, each observer will see light propagating in a vacuum at the same speed c.

A direct consequence of this calculation with the Lorentz velocity transformation is that *no object can move faster than c, the speed of light in a vacuum, in any inertial frame of reference*. If an object (light) is traveling at speed c in one inertial frame, it is traveling at c in all inertial frames; if an object is traveling slower than c in one inertial frame, it is traveling slower than c in all inertial frames. Thus the speed of light in a vacuum represents an ultimate speed limit.

Example 25-7 Baseball for Superheroes

Suppose the baseball game in Figure 25-12 is being played by superheroes. The outfielder has the power of super speed, and can run at $0.300c$ relative to the ground. The batter has the power of super strength, and can bat the ball with such force that the ball ends up traveling horizontally at $0.900c$ relative to the ground. (The bat and ball are made of super materials that can withstand the tremendous forces required.) How fast is the ball moving relative to the outfielder?

Set Up

This is nearly the same situation that we discussed above with the two astronauts and the beam of light, except that the astronauts have been replaced by (super) baseball players and the beam of light has become a baseball. We use Equation 25-17 to calculate the velocity of the ball relative to the outfielder in frame S'.

Lorentz velocity transformation:

$$v'_x = \frac{v_x - V}{1 - \dfrac{V}{c^2}v_x} \qquad (25\text{-}17)$$

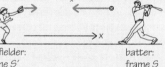

outfielder: frame S' batter: frame S

$V = 0.300c$ $v_x = -0.900c$

Solve

Use the velocity of the ball relative to the batter (v_x) and the velocity of the outfielder relative to the batter (V) to find the velocity of the ball relative to the outfielder (v'_x).

The outfielder (frame S') is moving in the positive x direction relative to the batter (frame S) at $V = 0.300c$. The velocity of the ball relative to the batter (frame S) is $v_x = -0.900c$ (negative because the ball is moving in the negative x direction). From Equation 25-17, the velocity of the ball relative to the outfielder is

$$v'_x = \frac{v_x - V}{1 - \dfrac{V}{c^2}v_x} = \frac{-0.900c - 0.300c}{1 - \dfrac{0.300c}{c^2}(-0.900c)}$$

$$= \frac{-1.200c}{1 + (0.300)(0.900)} = \frac{-1.200c}{1.27}$$

$$= -0.945c$$

Relative to the outfielder, the ball is moving at $0.945c$ in the negative x direction (to the left).

Reflect

In the Galilean velocity transformation, the velocity of the ball relative to the outfielder would have been $v'_x = v_x - V = -0.900c - 0.300c = -1.200c$, which is faster than the speed of light in a vacuum. Thanks to the $1 - (V/c^2)v_x$ term in the denominator of Equation 25-17, the actual speed of the ball relative to the outfielder is faster than $0.900c$ but still slower than c. The batter may be super-powered, but the ball cannot exceed the speed of light as measured by any observer.

? Got the Concept? 25-6 Returning a Super Baseball

Suppose the super outfielder in Example 25-7 catches the ball and throws it back toward the super batter, who is still standing at home plate. If the outfielder is still running at 0.300c and she throws the ball at 0.700c relative to her, what is the speed of the ball relative to the batter? (a) 0.331c; (b) 0.506c; (c) 0.826c; (d) 1.00c; (e) 1.26c.

Take-Home Message for Section 25-6

✔ At speeds that are an appreciable fraction of the speed of light, we must use the Lorentz velocity transformation to calculate relative velocities. This transformation respects the rule that the speed of light in a vacuum is the same in all inertial frames of reference.

✔ No object can move faster than c, the speed of light in a vacuum, in any inertial frame of reference.

25-7 The equations for kinetic energy and momentum must be modified at very high speeds

We have seen that the speed of light in a vacuum c is an ultimate speed limit: No object can travel faster than c in any inertial frame of reference. As we'll see in this section, this tells us that the expressions we learned earlier in this book for kinetic energy K (Chapter 6) and momentum $\vec{p}$ (Chapter 7) *cannot* be entirely correct, but fail at speeds that are a reasonable fraction of c. We'll see the reasons why this must be so, and encounter a new kind of energy called *rest energy* that is intrinsic to any object with mass.

In Section 6-3 we introduced kinetic energy through the work-energy theorem:

Work done on an object by the net force on that object

$$W_{net} = K_f - K_i$$

Kinetic energy of the object after the work is done on it

Kinetic energy of the object before the work is done on it

The work-energy theorem
(6-9)

This says that if an object starts at rest so that its initial kinetic energy K_i is zero, the more work the net force on an object does, the greater the final kinetic energy K_f of the object. In principle, there is no limit to how much kinetic energy an object can acquire. There is a problem, however: Using Newtonian physics, we found that the expression for the kinetic energy K of an object of mass m moving at speed v is

$$K = \frac{1}{2}mv^2$$

Since the speed of light in a vacuum c is the maximum speed an object can acquire, this expression says that the maximum kinetic energy that an object can acquire is $(1/2)mc^2$. This contradicts the idea that there should be no limit on an object's kinetic energy. Clearly, we need an improved equation for kinetic energy.

There is a similar problem with the Newtonian expression for momentum. We saw in Section 7-6 that the change in momentum of an object is determined by the external forces on the object and the duration of the time interval over which the forces act:

The sum of all external forces acting on an object

Duration of a time interval over which the external forces act

$$\left(\sum \vec{F}_{external\ on\ object} \right) \Delta t = \vec{p}_f - \vec{p}_i = \Delta\vec{p}$$

External force and momentum change for an object
(7-23)

Change in the momentum $\vec{p}$ of the object during that time interval

Suppose an object starts at rest so that its initial momentum $\vec{p}_i$ is zero and a constant net force acts on it. Then Equation 7-23 says that the longer the time interval Δt that the force acts, the greater the final momentum $\vec{p}_f$ of the object. In principle there is no limit on how long the force can act, so there should be no upper limit on an object's momentum. However, this can't be reconciled with the Newtonian expression for the momentum of an object of mass m with velocity $\vec{v}$:

$$\vec{p} = m\vec{v}$$

According to this expression, the maximum magnitude of momentum that an object of mass m can have is $p = mc$, which would be attained only when an object is moving at the speed of light. This directly contradicts the notion that there should be no upper limit on momentum. Just as for kinetic energy, we need a new expression for momentum that's consistent with the special theory of relativity.

It's possible to derive the correct expressions for K and $\vec{p}$ by looking closely at what happens in an elastic collision, in which both mechanical energy and momentum are conserved (see Section 7-5). The derivation is beyond our scope, so we'll just look at the results. For a particle moving with velocity $\vec{v}$, both the expression for the kinetic energy and the expression for the momentum involve a quantity called **relativistic gamma**:

**Relativistic gamma
(25-19)**

Relativistic gamma for a particle moving at speed v

$$\gamma = \frac{1}{\sqrt{1 - \dfrac{v^2}{c^2}}}$$

Speed of the particle

Speed of light in a vacuum

This dimensionless quantity is equal to 1 when $v = 0$ and becomes infinitely large as v approaches c (**Figure 25-13**). In terms of relativistic gamma, we can write the correct expressions for kinetic energy and momentum as

**Einsteinian expressions for
kinetic energy and momentum
(25-20)**

Kinetic energy of a particle of mass m and velocity $\vec{v}$

$$K = (\gamma - 1)mc^2$$

$$\vec{p} = \gamma m\vec{v}$$

Relativistic gamma for speed v

Momentum of a particle of mass m and velocity $\vec{v}$

 At speeds that are a small fraction of the speed of light c, these are approximately equal to the Newtonian expressions

$$K = \frac{1}{2}mv^2 \text{ and } \vec{p} = m\vec{v}$$

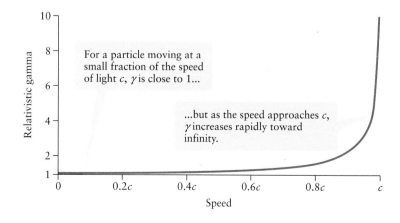

For a particle moving at a small fraction of the speed of light c, γ is close to 1...

...but as the speed approaches c, γ increases rapidly toward infinity.

Figure 25-13 Relativistic gamma
The quantity $\gamma = 1/\sqrt{1 - (v^2/c^2)}$ increases dramatically as speed v approaches the speed of light.

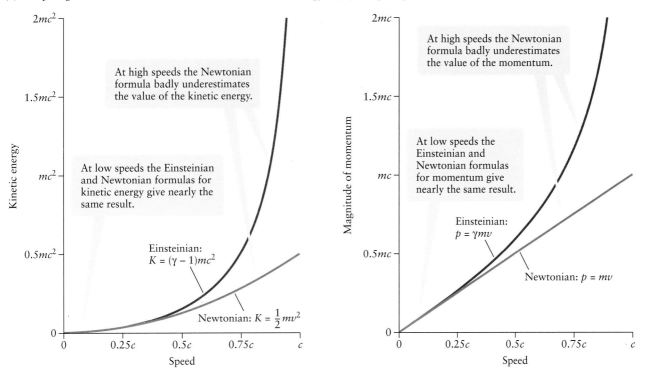

(a) Comparing Newtonian and Einsteinian formulas for kinetic energy

At high speeds the Newtonian formula badly underestimates the value of the kinetic energy.

At low speeds the Einsteinian and Newtonian formulas for kinetic energy give nearly the same result.

Einsteinian: $K = (\gamma - 1)mc^2$

Newtonian: $K = \frac{1}{2}mv^2$

(b) Comparing Newtonian and Einsteinian formulas for momentum

At high speeds the Newtonian formula badly underestimates the value of the momentum.

At low speeds the Einsteinian and Newtonian formulas for momentum give nearly the same result.

Einsteinian: $p = \gamma mv$

Newtonian: $p = mv$

Figure 25-14 **Kinetic energy and momentum in relativity** These graphs show how the Newtonian and Einsteinian expressions for (a) the kinetic energy of a particle and (b) the momentum of a particle depend on the speed of the particle.

Figure 25-14 compares the kinetic energy K and the magnitude of momentum p from Equations 25-20 to the Newtonian expressions for these quantities. The graphs show that if v is a small fraction of the speed of light, the Einsteinian and Newtonian expressions give essentially identical results. This justifies our use of these expressions in earlier chapters, in which we considered only relatively slow-moving objects. But as the speed v approaches c, the Einsteinian expressions from Equations 25-20 approach infinity. So even though the speed of light is an upper limit to the speed of an object, there is no upper limit on the kinetic energy or momentum of an object.

Example 25-8 The Energy and Momentum Costs of High Speed

Electrons can be accelerated to speeds very close to the speed of light. The mass of an electron is 9.11×10^{-31} kg.
(a) How much kinetic energy must be given to an electron to accelerate it from rest to $0.900c$? How much momentum?
(b) How much additional kinetic energy must be given to the electron to accelerate it from $0.900c$ to $0.990c$? How much additional momentum?

Set Up
We use Equations 25-20 to find the kinetic energy K and magnitude of momentum p for each speed. Equation 25-19 tells us the value of relativistic gamma for each speed.

Relativistic gamma:

$$\gamma = \frac{1}{\sqrt{1 - \dfrac{v^2}{c^2}}} \qquad (25\text{-}19)$$

Einsteinian kinetic energy and momentum:

$$K = (\gamma - 1)\, mc^2$$
$$\vec{p} = \gamma m \vec{v} \qquad (25\text{-}20)$$

Solve

(a) The kinetic energy that must be given to the electron is the difference between its kinetic energy at $v = 0.900c$ and its kinetic energy at $v = 0$, and similarly for the momentum.

At $v = 0$,

$$\gamma = \frac{1}{\sqrt{1 - \dfrac{v^2}{c^2}}} = \frac{1}{\sqrt{1 - 0}} = \frac{1}{1} = 1$$

$$K = (\gamma - 1)\,mc^2 = (1 - 1)\,mc^2 = 0$$
$$p = \gamma mv = (1)\,m\,(0) = 0$$

Just as in Newtonian physics, a particle at rest has zero kinetic energy and zero momentum.

To calculate K and p at nonzero speeds, it's useful to first find the values of mc and mc^2 for an electron:

$$mc = (9.11 \times 10^{-31}\ \text{kg})(3.00 \times 10^8\ \text{m/s})$$
$$= 2.73 \times 10^{-22}\ \text{kg} \cdot \text{m/s}$$

$$mc^2 = (9.11 \times 10^{-31}\ \text{kg})(3.00 \times 10^8\ \text{m/s})^2$$
$$= 8.20 \times 10^{-14}\ \text{kg} \cdot \text{m}^2/\text{s}^2$$
$$= 8.20 \times 10^{-14}\ \text{J}$$

At $v = 0.900c$,

$$\gamma = \frac{1}{\sqrt{1 - \dfrac{v^2}{c^2}}} = \frac{1}{\sqrt{1 - \dfrac{(0.900c)^2}{c^2}}} = \frac{1}{\sqrt{1 - (0.900)^2}}$$

$$= \frac{1}{\sqrt{0.190}} = 2.29$$

$$K = (\gamma - 1)mc^2 = (2.29 - 1)(8.20 \times 10^{-14}\ \text{J})$$
$$= 1.06 \times 10^{-13}\ \text{J}$$

$$p = \gamma mv = (2.29)m(0.900c) = (2.29)(0.900)mc$$
$$= (2.29)(0.900)(2.73 \times 10^{-22}\ \text{kg} \cdot \text{m/s})$$
$$= 5.64 \times 10^{-22}\ \text{kg} \cdot \text{m/s}$$

The electron begins with $K = 0$ and $p = 0$, so it must be given 1.06×10^{-13} J of kinetic energy and 5.64×10^{-22} kg $\cdot$ m/s of momentum to accelerate it from rest to $0.900c$.

(b) Repeat the calculation in part (b) for the additional kinetic energy and momentum that must be given to the electron to accelerate it from $0.900c$ to $0.990c$.

At $v = 0.990c$,

$$\gamma = \frac{1}{\sqrt{1 - \dfrac{v^2}{c^2}}} = \frac{1}{\sqrt{1 - \dfrac{(0.990c)^2}{c^2}}} = \frac{1}{\sqrt{1 - (0.990)^2}}$$

$$= \frac{1}{\sqrt{0.0199}} = 7.09$$

$$K = (\gamma - 1)\,mc^2 = (7.09 - 1)(8.20 \times 10^{-14}\ \text{J})$$
$$= 4.99 \times 10^{-13}\ \text{J}$$

$$p = \gamma mv = (7.09)\,m\,(0.990c) = (7.09)(0.990)mc$$
$$= (7.09)(0.990)(2.73 \times 10^{-22}\ \text{kg} \cdot \text{m/s})$$
$$= 1.92 \times 10^{-21}\ \text{kg} \cdot \text{m/s}$$

The difference between these values and the values of K and p at $v = 0.900c$ from part (a) tells us the additional kinetic energy and momentum that must be given to the electron:

$$\Delta K = 4.99 \times 10^{-13}\ \text{J} - 1.06 \times 10^{-13}\ \text{J}$$
$$= 3.93 \times 10^{-13}\ \text{J}$$

$$\Delta p = 1.92 \times 10^{-21}\ \text{kg} \cdot \text{m/s} - 5.64 \times 10^{-22}\ \text{kg} \cdot \text{m/s}$$
$$= 1.35 \times 10^{-21}\ \text{kg} \cdot \text{m/s}$$

Reflect

Compared to accelerating an electron from rest to $0.900c$, accelerating that same electron from $0.900c$ to $0.990c$ requires 3.70 times as much additional kinetic energy and 2.40 times as much additional momentum. As the speed gets closer and closer to c, it requires ever-greater amounts of kinetic energy and momentum to cause an ever-smaller speed increase. You can see that an object can never be accelerated from rest to the speed of light. This would require adding *infinite* amounts of kinetic energy and momentum to the object.

We mentioned above that the Einsteinian expressions for kinetic energy and momentum, Equations 25-20, come from an analysis of elastic collisions. In particular, we must demand that if energy and momentum are conserved in one inertial frame of reference, they must be conserved in *all* inertial frames. That's required if energy and momentum are to be consistent with the first of the postulates of special relativity. It turns out that in order for this to be the case, we must also include a term mc^2 in the total energy of a particle of mass m. This quantity is called the **rest energy** of a particle, since it is present even when the particle is not in motion.

Rest energy of a particle   Mass of the particle

$$E_0 = mc^2$$

Speed of light in a vacuum

Rest energy
(25-21)

Equation 25-21 is one of the most famous in science, and one of the most misunderstood. Rest energy is not potential energy; potential energy is associated with a force an object experiences. (For example, gravitational potential energy is associated with the gravitational force.) It is not kinetic energy; kinetic energy is associated with motion. Rest energy is energy that is intrinsic to an object because of its mass.

The physical interpretation of Equation 25-21 is that *mass and energy are equivalent*, or that this mass is simply one possible manifestation of energy. Stated another way, the mass of an object is a measure of its rest energy content. Figure 25-15 is visual evidence of the equivalence of mass and energy. This image shows the result of a head-on collision between two protons, each of which was moving at just under the speed of light. Dozens of new particles appear after the collision. These are not fragments of the colliding protons, but rather new particles that were created in the collision. This is possible because some of the kinetic energy of the colliding protons was converted into the rest energy of these new particles; another portion of this kinetic energy went into the kinetic energies of the new particles, which fly away from the collision site at high speeds.

Collision between protons happens here.

Each colored curve shows the track of a particle produced in the collision.

Figure 25-15 Converting kinetic energy to new particles This image from the Large Hadron Collider at CERN in Geneva, Switzerland shows dozens of new particles produced by a collision between energetic protons.

You can observe the equivalence of mass and energy in action whenever you go outside on a sunny day or look up at the stars on a clear night. The Sun and stars shine thanks to a process occurring in their interiors in which hydrogen nuclei are fused together to form nuclei of helium. The mass of the products of such a reaction is slightly less than the mass of the hydrogen nuclei present before the reaction. The "lost" mass is converted to energy in the form of electromagnetic radiation, and it is that radiation that we see in the form of sunlight and starlight.

Example 25-9   Converting Mass to Energy in the Sun

The Sun emits 3.84×10^{26} J of energy every second. (a) At what rate (in kg/s) is the Sun's mass decreasing? (b) How much mass has the Sun lost since it was formed 4.56×10^9 years ago? Assume that it has emitted energy at the same rate over its entire history. Compare to the present-day mass of the Sun, 1.99×10^{30} kg.

Set Up

The emitted energy comes from the rest energy of mass that is "lost" in nuclear reactions in the Sun's interior. We use Equation 25-21 to find the amount of mass equivalent to the energy emitted in one second. The total mass lost over the Sun's lifetime equals this amount of mass multiplied by the number of seconds that have elapsed since the Sun formed.

Rest energy:

$$E_0 = mc^2 \qquad\qquad (25\text{-}21)$$

Solve

(a) Calculate the amount of mass lost by the Sun per second.

Mass equivalent of 3.84×10^{26} J:

$$m = \frac{E_0}{c^2} = \frac{3.84 \times 10^{26}\,\text{J}}{(3.00 \times 10^8\,\text{m/s})^2} = 4.27 \times 10^9\,\text{J}\cdot\text{s}^2/\text{m}^2$$

Since $1\,\text{J} = 1\,\text{kg}\cdot\text{m}^2/\text{s}^2$, we can write this as

$$m = 4.27 \times 10^9\,\text{kg}$$

The mass of the Sun decreases at a rate of 4.27×10^9 kg/s.

(b) Calculate the total amount of mass lost by the Sun in its history.

The age of the Sun in seconds is

$$(4.56 \times 10^9\,\text{y})\left(\frac{365.25\,\text{d}}{1\,\text{y}}\right)\left(\frac{24\,\text{h}}{1\,\text{d}}\right)\left(\frac{60\,\text{min}}{1\,\text{h}}\right)\left(\frac{60\,\text{s}}{1\,\text{min}}\right) = 1.44 \times 10^{17}\,\text{s}$$

In this number of seconds, the total mass lost by the Sun is

$$\left(4.27 \times 10^9\,\frac{\text{kg}}{\text{s}}\right)(1.44 \times 10^{17}\,\text{s}) = 6.14 \times 10^{26}\,\text{kg}$$

As a fraction of the Sun's present-day mass, the amount of mass lost is $(6.14 \times 10^{26}\,\text{kg})/(1.99 \times 10^{30}\,\text{kg}) = 3.09 \times 10^{-4} = 0.0309$ percent of the total mass.

Reflect

In more than 4 billion years of producing energy at a prodigious rate, the Sun has lost only a tiny fraction of its total mass. This is a testament to how much energy can be released by converting even a small amount of mass.

? Got the Concept? 25-7   Can We Build a Starship?

The total amount of electric energy produced per year in the United States from all sources is about 1.5×10^{19} J. You propose to use all of this energy to accelerate a spacecraft to $0.990c$ in order to travel to other stars. About how massive could your proposed starship be? (a) About 3×10^6 kg (the mass of an ocean liner); (b) about 3×10^5 kg (the mass of a large airliner); (c) about 3×10^3 kg (the mass of a sport utility vehicle); (d) about 30 kg (the mass of a kayak or canoe); (e) about 3 kg (the mass of a skateboard).

Take-Home Message for Section 25-7

✔ The mathematical expressions for kinetic energy and momentum have to be modified to be consistent with special relativity. Both the kinetic energy and momentum

of an object increase without limit as the object's speed approaches *c*.

✔ Any object with mass has a kind of energy called rest energy.

25-8 Einstein's general theory of relativity describes the fundamental nature of gravity

Sitting in a chair in the patent office in Bern, Switzerland, in 1907, a young Albert Einstein had what he would call "the happiest thought of my life." He imagined a man falling freely from the roof of a house and realized that "at least in his immediate surroundings—there exists no gravitational field." If the man released an object, for example, it would accelerate at the same rate as the man accelerated, and because the man would "not feel his own weight," it would appear to him that neither he nor the object was experiencing a gravitational force.

Einstein's thought experiment led him to postulate a new principle called the **principle of equivalence**. This principle states that

A gravitational field is equivalent to an accelerated frame of reference in the absence of gravity.

The principle of equivalence dictates that it is not possible to distinguish experimentally between a system in an accelerating frame and a system under the influence of gravity. In other words, if you were to drop a ball in a windowless elevator car that makes no noise and doesn't shake, you could not tell from the motion of the ball whether the elevator car were sitting stationary on the surface of a planet and experiencing its gravity or accelerating in empty space, far from sources of gravity. For example, if you were standing in this elevator car and experienced an acceleration of 9.80 m/s^2, you wouldn't be able to tell whether the car was stationary on Earth or accelerating at 9.80 m/s^2 in the absence of Earth's gravity.

Let's explore physics in this imaginary elevator car further. In **Figure 25-16a** a ball is thrown horizontally while the elevator car is stationary near Earth's surface. Due to the force of gravity, the ball accelerates downward, following a familiar parabolic arc.

(a) This elevator car is stationary near the surface of Earth. A ball thrown horizontally follows a parabolic path as it falls.

(b) This elevator car is stationary, far from Earth or any other massive object. A ball thrown horizontally travels in a straight line.

Figure 25-16 **Elevator cars on Earth and in space** A ball is thrown horizontally in (a) an elevator car on Earth's surface and (b) in an elevator car far from any massive object.

The figure shows the positions of the ball at five instants, spanning four equal time intervals. What if the elevator car were far from Earth and from any other massive object that could exert a noticeable gravitational force on the ball? In that case the ball would travel along a straight line, as shown in Figure 25-16b. No gravity means no acceleration . . . or does it? Consider the situation shown in Figure 25-17. The elevator car is again far from any object that could exert a noticeable gravitational force on the ball, but now the car is accelerating. The direction of the acceleration is "up" (toward the top of the page). Because there is no discernible gravitational force, the ball travels in a straight line as in Figure 25-16b. However, because the elevator car is accelerating, an observer in the car sees the ball trace out a parabolic arc with respect to the walls and floor of the car. Remember that there are no windows in the elevator car, and it makes no noise and does not vibrate as it moves. An observer in the car cannot, therefore, detect its motion. The observer feels the effect of the principle of equivalence: It appears that the ball is falling under the influence of gravity.

Einstein's theory of gravitation is therefore a theory of accelerating frames of reference. Because this is more general than the case of inertial frames, the type that was at the heart of the special theory of relativity, this expanded theory is called the **general theory of relativity.**

To see the real power of Einstein's postulate about the equivalence of acceleration and gravity, substitute a beam of light for the ball in our elevator car thought experiment. We expect that a beam of light travels in a straight line, so in analogy to throwing the ball, shining the light horizontally should result in a horizontal beam of light. Certainly that is the case when the elevator is stationary and far from any massive objects. If the elevator were accelerating, however, as in the case of the ball depicted in Figure 25-17, an observer in the elevator car would *not* see the light move straight toward the far wall. You might need a precise measuring device to detect it, but the light would hit the opposing wall below the height at which it started. Did the light bend, or is it that the elevator car has moved during the time that the light travels from one side of the car to the other? The principle of equivalence tells us that the observer has no way of knowing whether the car is accelerating or whether it is at rest near an object that exerts a gravitational force on it. We are therefore justified in declaring that the car is stationary and that the light is deflected! In addition, we must conclude that the beam's parabolic path results from the influence of a gravitational force. Both the beam of light and the ball accelerate due to a gravitational field.

The deflection of light in the presence of massive objects is small, an effect you might imagine would have been impossible to measure in Einstein's time. Einstein realized, however, that because the Sun's mass is large, if the light from a distant star were to pass close to the Sun, the change in its direction would be measurable. Light passing close to the Sun would only be visible during a total eclipse, so in order to test Einstein's theory the English astronomer Sir Arthur Eddington led an expedition to an island off the west coast of Africa in 1919 to photograph a total eclipse of the Sun. Eddington compared the position relative to other stars of a star cluster that was optically close to the Sun during the eclipse to its position at other times. His measurements confirmed Einstein's theory that massive objects deflect light.

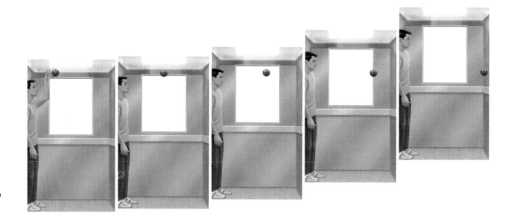

Figure 25-17 An accelerating elevator car in space When a ball is thrown horizontally in an elevator car, which is both far from any massive object and also accelerating, it follows a straight path. An observer in the elevator, however, would see the ball follow a parabolic path with respect to the floor and walls of the elevator car.

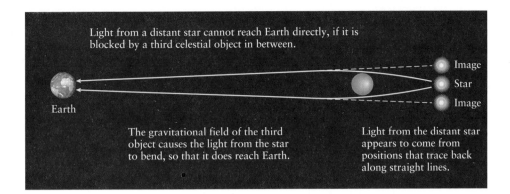

Light from a distant star cannot reach Earth directly, if it is blocked by a third celestial object in between.

Earth

The gravitational field of the third object causes the light from the star to bend, so that it does reach Earth.

Light from the distant star appears to come from positions that trace back along straight lines.

Image
Star
Image

Figure 25-18 Gravity deflects light
A massive object positioned between Earth and a distant star bends light from the star so that it can be seen on Earth. This is called gravitational lensing.

A more stunning phenomenon associated with gravitational bending of light occurs when multiple images of a distant star form as light is bent by a closer, massive celestial object, such as the black hole at the center of a galaxy. **Figure 25-18** depicts such *gravitational lensing*, in which a massive object is positioned directly between Earth and a distant star. Light from the star cannot reach Earth directly, but the lensing effect results in light initially not propagating toward Earth to be bent back toward us. In this way light from the star can approach Earth from many directions—for example, the two directions shown in the figure. **Figure 25-19a** presents an example of gravitational lensing showing four distinct images. When light is bent around the intervening object and reaches Earth from a full circle around it, the star's light is spread out into a ring around the lensing object, as in **Figure 25-19b**.

Earlier in this chapter we saw that Einstein's special theory of relativity addresses time dilation; a clock moving at constant speed relative to an observer is seen to be running slowly. Imagine a scenario in which *two* observers moving at constant speed relative to each other can each see a clock carried by the other. Each observes a moving clock, so each measures the other's clock to be running slowly *compared to* his own clock. Is this possible? If the two observers stopped their relative motion and brought the clocks together, certainly both clocks could not be slow relative to the other!

Einstein's general theory of relativity is required to resolve this puzzle. When the two observers are in relative motion at constant speed, their frames are symmetric; that is, either observer can consider the other to be in motion. To bring the two clocks together to compare their times, however, one of the observers must decelerate. The two frames are therefore no longer symmetric. One is inertial and the other, the frame that decelerates, is not. In addition, general relativity must be applied to understand the measurement of time from one frame to the other. Time runs more slowly in the frame that experiences the change in speed.

This issue of acceleration and deceleration is central to the so-called twins paradox, one of the most well-known puzzles in relativity. We'll state it this way: Bertha takes a

(a)   (b)

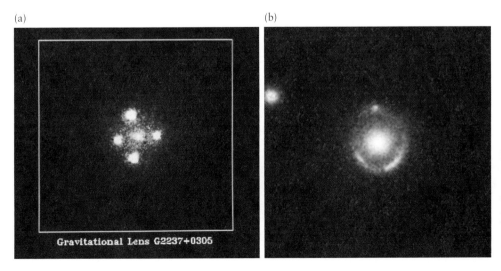

Gravitational Lens G2237+0305

Figure 25-19 Gravitational lenses
(a) These four separate images of a single star are formed by a massive gravitational lens. (b) Light leaving a distant star travels in every direction. When that light is bent toward Earth by a massive gravitational lens, the image formed can be a circular ring.

trip to the star Proxima Centauri on a fast spaceship, while her twin Eartha stays at home. Eartha's clock ticks off ten years in the time Bertha is gone. Special relativity suggests that were Eartha to observe Bertha's clock, she would see it running slowly compared to her own. After ten hours have elapsed on Eartha's clock, for example, she observes that only eight hours have elapsed on Bertha's. Eartha expects that when Bertha returns, she will have aged eight years, not ten—Bertha will be younger than Eartha at the end of the trip. But according to the puzzle, Bertha considers herself at rest and sees Eartha moving away and then returning. In the telling of the paradox, the claim is made that the two views are equally valid, so Bertha sees Eartha's clock running slowly and expects that Eartha will be younger than she is at the end of the trip. But the two views are not symmetric! Bertha must accelerate to leave Earth, decelerate when she reaches Proxima Centauri, and then accelerate and decelerate again in order to return home. For this reason her frame can be distinguished from Eartha's. It is simply not correct to declare that each twin can consider the other to be moving in an inertial frame; special relativity does not apply. Although the mathematical details are beyond the scope of this book, general relativity predicts that the clock that accelerates and decelerates (Bertha's clock) will run slowly, so Bertha will be younger than Eartha upon her return.

Take-Home Message for Section 25-8

✔ The principle of equivalence, a central postulate of Einstein's general theory of relativity, declares that a gravitational field is equivalent to an accelerated frame of reference in the absence of gravity.

✔ It is not possible to distinguish between a system in an accelerating frame and a system under the influence of gravity.

✔ One important prediction of general relativity is that light bends as it passes through a gravitational field.

Key Terms

event
frame of reference (reference frame)
Galilean transformation
Galilean velocity transformation
general theory of relativity
inertial frame
inverse Lorentz velocity transformation

length contraction
Lorentz transformation
Lorentz velocity transformation
Michelson-Morley experiment
noninertial frame
principle of equivalence
principle of Newtonian relativity

proper time
reference frame (frame of reference)
relativistic gamma
relativistic speed
rest energy
special theory of relativity
time dilation

Chapter Summary

Topic	Equation or Figure	
Newtonian relativity: The principle of Newtonian relativity states that the laws of motion are the same in all inertial frames of reference. The Galilean transformation relates the coordinates of an event in one frame to the coordinates of the same event in another frame, and the Galilean velocity transformation relates the velocity of an object relative to one frame to its velocity relative to another frame. These transformations are valid only for speeds that are small compared to the speed of light c.	Inertial frame of reference S' moves at speed V in the positive x direction relative to inertial frame of reference S. Coordinates of an event as measured in frame S' $$x' = x - Vt$$ $$y' = y$$ $$z' = z$$ $$t' = t$$ Coordinates of the same event as measured in frame S	(25-1)
	Inertial frame of reference S' moves at speed V in the positive x direction relative to inertial frame of reference S. Velocity components of an object as measured in frame S' $$v_x' = v_x - V$$ $$v_y' = v_y$$ $$v_z' = v_z$$ Velocity components of the same object as measured in frame S	(25-5)

The Michelson-Morley experiment:
In the nineteenth century it was hypothesized that space was filled with a material medium, called the luminiferous ether, that was required in order for light to propagate through space. The Michelson-Morley experiment provided evidence that the ether does not exist, and motivated Einstein's special theory of relativity.

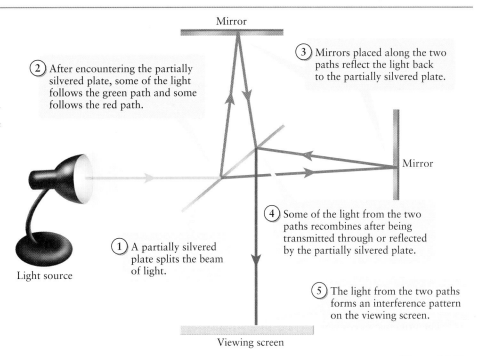

(2) After encountering the partially silvered plate, some of the light follows the green path and some follows the red path.

(3) Mirrors placed along the two paths reflect the light back to the partially silvered plate.

Mirror

Light source

(1) A partially silvered plate splits the beam of light.

(4) Some of the light from the two paths recombines after being transmitted through or reflected by the partially silvered plate.

(5) The light from the two paths forms an interference pattern on the viewing screen.

Viewing screen

(Figure 25-7)

The special theory of relativity:
Einstein's special theory of relativity is based on the postulates that all laws of physics are the same in all inertial frames, and the speed of light in a vacuum is the same in all inertial frames. Two consequences are that moving clocks run slow and moving objects are shortened along their direction of motion.

Time interval between two events as measured in frame S', in which those events occur at the same place.

Time interval between the same two events as measured in frame S

$$\Delta t = \frac{\Delta t_{proper}}{\sqrt{1 - \dfrac{V^2}{c^2}}}$$

Speed of S' relative to S

Speed of light in a vacuum

(25-10)

Length of an object in a frame of reference S' in which it is at rest.

$$L = L_{rest}\sqrt{1 - \frac{V^2}{c^2}}$$

Speed of S' relative to S

Speed of light in a vacuum

Length of the same object in a frame of reference S in which the object is moving along its length.

(25-14)

The Lorentz transformation:
The Lorentz transformation is a generalization of the Galilean transformation that is valid for all speeds. The Lorentz velocity transformation is a generalization of the Galilean velocity transformation: It shows that the speed of light c is an ultimate speed limit and that an object at rest can never be accelerated to c.

Inertial frame of reference S' moves at speed V in the positive x direction relative to inertial frame of reference S.

Coordinates of an event as measured in frame S'

$$x' = \frac{x - Vt}{\sqrt{1 - \dfrac{V^2}{c^2}}}$$

$$y' = y$$
$$z' = z$$

$$t' = \frac{t - \dfrac{V}{c^2}x}{\sqrt{1 - \dfrac{V^2}{c^2}}}$$

Coordinates of the same event as measured in frame S

(25-15)

Inertial frame of reference S' moves at speed V in the positive x direction relative to inertial frame of reference S.

x component of velocity of an object moving along the x axis as measured in frame S'

$$v_x' = \frac{v_x - V}{1 - \frac{V}{c^2} v_x}$$

x component of velocity of the same object as measured in frame S

(25-17)

Kinetic energy, momentum, and rest energy: The formulas for kinetic energy and momentum must be modified to account for c being the ultimate speed limit. The relativity postulates also show that an object with mass m has a rest energy $E_0 = mc^2$ even when it is not in motion.

Kinetic energy of a particle of mass m and velocity $\vec{v}$

$$K = (\gamma - 1)mc^2$$

$$\vec{p} = \gamma m\vec{v}$$

Relativistic gamma for speed v

Momentum of a particle of mass m and velocity $\vec{v}$

(25-20)

At speeds that are a small fraction of the speed of light c, these are approximately equal to the Newtonian expressions

$$K = \frac{1}{2} mv^2 \text{ and } \vec{p} = m\vec{v}$$

The general theory of relativity: The principle of equivalence states that a gravitational field is equivalent to an accelerated frame of reference in the absence of gravity. This principle shows that light is affected by gravity just as objects with mass are.

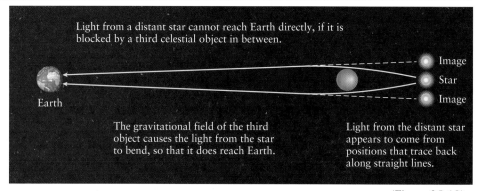

Light from a distant star cannot reach Earth directly, if it is blocked by a third celestial object in between.

Image
Star
Image

Earth

The gravitational field of the third object causes the light from the star to bend, so that it does reach Earth.

Light from the distant star appears to come from positions that trace back along straight lines.

(Figure 25-18)

Answer to What do you think? Question

(b)   As an object approaches the speed of light, it becomes increasingly difficult to further increase its speed. See Example 25-8 in Section 25-7.

Answers to Got the Concept? Questions

25-1   (b) Let S be the frame of reference of an observer on the ground, and let S' be the frame of reference of an observer moving with the air (for example, an observer floating in a balloon). Then S' is moving due east at $V = 100 \text{ km/h}$, so we take the positive x direction to be to the east. First suppose the airliner is flying west to east, so its velocity relative to the air is $v_x' = +900 \text{ km/h}$. From the first of Equations 25-5, the velocity of the airplane relative to the ground in this situation is $v_x = v_x' + V = +900 \text{ km/h} + 100 \text{ km/h} = +1000 \text{km/h}$; relative to the ground, the airliner is moving east at 1000 km/h. If instead the airliner is flying east to west, its velocity relative to the air is $v_x' = -900 \text{ km/h}$ and its velocity relative to the ground is $v_x = v_x' + V = -900 \text{ km/h} + 100 \text{ km/h} = -800 \text{ km/h}$; the airliner is moving west at 800 km/h relative

to the ground. In North America high-altitude winds typically blow from west to east as in this example, so a west-to-east airliner trip is typically faster than an east-to-west trip between the same two airports.

25-2   (b) This is similar to Got the Concept? Question 25-1. Let S be the frame of reference of the water, and let S' be the frame of reference of the dolphin. The dolphin swims in the positive x direction at speed $V = 10 \text{ m/s}$. As measured in frame S, the sound wave moves in the positive x direction at velocity $v_x = 1500 \text{ m/s}$. From the first of Equations 25-5, the velocity of the sound wave relative to S', the frame of reference of the dolphin, is $v_x' = v_x - V = 1500 \text{ m/s} - 10 \text{ m/s} = 1490 \text{ m/s}$. Because the dolphin is "chasing" the sound wave, the speed of the wave relative to the dolphin is less than its speed relative to the water. As we will see in Section 25-6, light waves do not behave like this: If you shoot a laser beam forward from a fast-moving vehicle, the speed of the light wave relative to the vehicle is $c = 3.00 \times 10^8$ m/s no matter how fast it is moving.

25-3 (e) The proper time is measured in any frame that is at rest with respect to the clock. Albert aboard Starship *Alpha* is at rest with respect to the clock, so he measures the clock's proper time. Although Barbara is not on the same craft as the clock, she is also not moving relative to it, so she also measures the clock's proper time. George's speed is the same as the clock's, but not in the same direction. So George is moving with respect to the clock and therefore does not observe proper time intervals. The same is true for Elena on Earth, who is moving relative to the clock. We conclude that the time interval between clock ticks is the proper time for both Albert and Barbara, but it is not the proper time for either George or Elena.

25-4 (b) The statement of time dilation is that moving clocks run slow. Albert's clock is moving relative to Elena, and Elena's clock is moving at the same speed relative to Albert. So *both* observers will see the other person's clock running slow. There is no contradiction, since the two observers are looking at different pairs of events: Elena is looking at the time interval between two clicks of Albert's clock, while Albert is looking at the time interval between two clicks of Elena's clock.

25-5 (a) The length you measure will be the same as the rest length if the rod is oriented so that its length is perpendicular to the direction of motion. There is no length contraction along

a perpendicular direction. (The *width* of the rod will be contracted along the direction of motion, but that's not what the question is about.)

25-6 (c) The outfielder's frame is S', so the velocity of the ball relative to her is $v'_x = +0.700c$ (positive since the ball is now moving in the positive x direction). She is still moving at $V = 0.300c$ relative to the batter. The velocity of the ball relative to the batter is v_x, given by the inverse Lorentz velocity transformation (Equation 25-18):

$$v_x = \frac{v'_x + V}{1 + \dfrac{V}{c^2}v'_x} = \frac{0.700c + 0.300c}{1 + \dfrac{(0.300c)}{c^2}(0.700c)}$$

$$= \frac{1.00c}{1 + 0.210} = 0.826c$$

25-7 (d) The kinetic energy of the spacecraft will be $K = (\gamma - 1)mc^2$. From Example 25-8, $\gamma = 7.09$ if $v = 0.990c$, so $K = (7.09 - 1)mc^2 = 6.09mc^2$. If all of the 1.5×10^{19} J goes into the kinetic energy K of the starship, the mass that can be accelerated to $0.990c$ will be

$$m = \frac{K}{6.09c^2} = \frac{1.5 \times 10^{19}\,\text{J}}{(6.09)(3.00 \times 10^8\,\text{m/s})^2}$$

$$= 27\,\text{J} \cdot \text{s}^2/\text{m}^2 = 27\,\text{kg}$$

That's about the mass of a kayak or canoe, not including any occupants. This calculation shows that it's simply not feasible with current technology to build a spacecraft that can carry humans and travel at speeds approaching the speed of light.

Questions and Problems

In a few problems, you are given more data than you actually need; in a few other problems, you are required to supply data from your general knowledge, outside sources, or informed estimate. Interpret as significant all digits in numerical values that have trailing zeros and no decimal points.
For all problems, use $g = 9.80$ m/s^2 for the free-fall acceleration due to gravity. Neglect friction and air resistance unless instructed to do otherwise.
• Basic, single-concept problem
•• Intermediate-level problem, may require synthesis of concepts and multiple steps
••• Challenging problem
SSM *Solution is in Student Solutions Manual*

Conceptual Questions

1. •Why do you think there was so much resistance from the established physics community when Einstein proposed his new theory of relativistic motion in 1905?

2. •How would you change the following Galilean transformation equations in the case where a frame was moving both in the x and the y direction?

$$x' = x - Vt$$
$$y' = y$$
$$z' = z$$
$$t' = t$$

3. •What is a frame of reference? What is an inertial frame of reference? SSM

4. •What kind of reference frame is Earth's surface? Explain your answer.

5. •What does the phrase "all motion is relative, there is no absolute motion" mean?

6. •Describe the Michelson and Morley experiment. Why do you think it was repeated so many times? (It may be the most often performed experiment in the history of physics.)

7. •Explain how the measurement of time enters into the determination of the length of an object. SSM

8. •What are the two postulates of the special theory of relativity?

9. •Is it possible for one observer to find that event A happens after event B and another observer to find that event A happens before event B? Explain your answer.

10. •Is it possible to accelerate an object to the speed of light in a real situation? Explain your answer.

11. •Gene Roddenberry created the popular TV and film series *Star Trek*. Once at a public lecture he was asked if he thought we would ever be able to travel faster than light. He responded, "No, and that's a good thing because it means that we can always do it in science fiction." How have science fiction writers gotten around the cosmic speed limit?

12. •What is the fundamental postulate of the general theory of relativity?

13. •Describe one of the predictions of general relativity.

Multiple-Choice Questions

14. •Which of the following statements are true?

A. The laws of motion are the same in all inertial frames of reference.

B. The laws of motion are the same in all reference frames.

C. There is no way to detect absolute motion.

D. All of the above statements are true.

E. Only two of the above statements are true.

15. •If we used radio waves to communicate with an alien spaceship approaching Earth at 10% of the speed of light, we would receive their signals at a speed of

A. $0.10c$.

B. $0.90c$.

C. $0.99c$.

D. $1.00c$.

E. $1.10c$. SSM

16. •Time dilation means that

A. the slowing of time in a moving frame of reference is only an illusion resulting from motion.

B. time really does pass more slowly in a frame of reference moving relative to a frame of reference at relative rest.

C. time really does pass more slowly in a frame of reference at rest relative to a frame of reference that is moving.

D. time is unchanging regardless of the frame of reference.

E. no two clocks at rest can ever read the same time.

17. •A meter stick hurtles through space at a speed of $0.95c$ with its length perpendicular to the direction of motion. You measure its length to be equal to

A. 0 m.

B. 0.05 m.

C. 0.95 m.

D. 1.00 m.

E. 1.05 m.

18. •A particle has an Einsteinian momentum of p. If its speed doubles, the Einsteinian momentum will be

A. greater than $2p$.

B. equal to $2p$.

C. less than $2p$

D. equal to $2p/c$.

E. equal to p.

19. •Consider two atomic clocks, one at the GPS ground control station near Colorado Springs (elevation 1830 m) and the other one in orbit in a GPS satellite (altitude 20,200 km). According to the general theory of relativity, which atomic clock runs slow?

A. The clock in Colorado runs slow.

B. The clock in orbit runs slow.

C. The clocks keep identical time.

D. The orbiting clock is 95% slower than the clock in Colorado.

E. The orbiting clock alternately runs slow and then fast depending on where the Sun is. SSM

Estimation/Numerical Analysis

20. •Estimate the difference in a 10,000-s time interval as measured by a proper observer and a relative observer traveling on a commercial jetliner.

21. •Estimate the minimum speed required to observe a 10% change between the proper time and the relative time.

22. •Estimate the percent error between the Newtonian momentum and the Einsteinian momentum at (a) typical macroscopic speeds and (b) speeds that are a substantial fraction of c.

23. •Are there any numerical values of v/c that lend themselves to relatively simple calculations of γ (without the use of a calculator)?

24. •If all effects due to general relativity are neglected, and the effects of time dilation are cumulative, how many trips into outer space would an astronaut have to make to experience a 10% change in the length of his life (as measured by an observer at rest on Earth)? Is the time longer or shorter by 10% according to the observer on Earth?

25. •Estimate the percentage of the world's population that has a basic understanding of Einstein's special theory of relativity (assume your understanding of the concepts in this chapter constitutes a basic understanding). SSM

26. ••Estimate the increase in rest energy of a copper sample [100 kg, $c = 387$ J/(kg·°C)] if its temperature is increased 100°C.

27. •This problem is intended to help you to understand the behavior of relativistic variables such as time, length, and momentum. To begin, construct a table (like the one shown below) of the following values for γ versus v/c, in which

$$\gamma = \frac{1}{\sqrt{1 - \left(\dfrac{v}{c}\right)^2}} \qquad c = 3.00 \times 10^8 \text{ m/s}$$

v/c	γ	v/c	γ
0	0.7	0.4	0.999
0.1	0.8	0.5	0.9999
0.2	0.9	0.6	
0.3	0.99		

Now, plot a graph of the variable γ versus v/c. As long as γ is close to unity, the effects of Einstein's special theory of relativity go unnoticed. (a) From your graph (or your chart) determine the maximum value of v/c for which γ is less than 1% larger than unity (1.0). (b) What numerical value does the slope take at $v/c = 0.1$, $v/c = 0.5$, $v/c = 0.9$, and $v/c = 0.999$?

Problems

25-1 The concepts of relativity may seem exotic, but they're part of everyday life

25-2 Newton's mechanics includes some ideas of relativity

28. ••A bicyclist rides at 8.00 m/s toward the north. A car is moving at 25.0 m/s, also toward the north, and is initially behind the rider. A truck is moving at 15.0 m/s toward the south, approaching the bicycle and the car. (a) Make a sketch of the three vehicles and label their respective velocity vectors, relative to the ground. (b) Calculate the relative velocities of each vehicle compared to the other two. Solve the problem both with vector diagrams and with equations.

29. ••A frame of reference, S, is fixed on the surface of Earth with the x axis pointing toward the east, the y axis pointing toward the north, and the z axis pointing up. A second frame of reference, S', is moving at a constant 4.00 m/s toward the east. (a) Describe mathematically the relationships between x' and x,

y' and y, and z' and z. (b) A picture is taken in the S frame of reference at $t = 4.00$ s at the point (2 m, 1 m, 0 m). Calculate the corresponding values of x', y', and z' for the same event in the S' frame. SSM

30. •A boat sails from the pier at Fisherman's Wharf in San Francisco at 4.00 m/s, directly toward Alcatraz. A kite rider heads directly away from Alcatraz toward the boat at a relative speed of 6.00 m/s, according to the skipper of the sailboat. Calculate the velocity of the kite rider relative to the pier.

31. ••A radio-controlled model car travels at 15.0 m/s, to the right, relative to the parking lot that it is driving on. A girl on her scooter chases after the model car at 4.00 m/s. The parking lot represents reference frame S, the model car is in frame S', and the girl is described by frame S''. Assume that the car and the girl are both at the origin of the parking lot at $t = 0$ s. Write down the Galilean transformation between (a) S and S', (b) S and S'', and (c) S' and S''.

32. ••Assume that the origins of S and S' coincide at $t = t' = 0$ s. An observer in inertial frame S measures the space and time coordinates of an event to be $x = 750$ m, $y = 250$ m, $z = 250$ m, and $t = 2.0$ μs. What are the space coordinates of the event in inertial frame S' which is moving in the $+x$ direction at a speed of $0.01c$ relative to S?

33. ••At time $t' = 4.00 \times 10^{-3}$ s, as measured in S', a particle is at the point $x' = 10$ m, $y' = 4$ m, and $z' = 6$ m. Compute the corresponding values of x, y, and z, as measured in S, for (a) a relative velocity between S' and S of $+500$ m/s and (b) a relative velocity between S' and S of -500 m/s. Assume that the origins of S and S' are coincident at $t = 0$ and that the motion lies along the x and x' axes.

34. ••Suppose that at $t = 6.00 \times 10^{-4}$ s, the space coordinates of a particle are $x = 100$ m, $y = 10$ m, and $z = 30$ m according to coordinate system S. Compute the corresponding values as measured in the frame S' if the relative velocity between S' and S is 150,000 m/s along the x and x' axes. The reference frames start together, with their origins coincident at $t = 0$.

35. •••A float plane lands on a river. Assume the velocity of the airplane relative to the air is 30 m/s, due east; the velocity of the wind is 20 m/s, due north; and the current in the river is 5 m/s, due south. Calculate the velocity of the plane just before it lands relative to the water. SSM

36. •A spaceship moves by Earth at 2.4×10^8 m/s. A satellite moves by Earth in the opposite direction at 1.6×10^8 m/s. Use the Galilean transformation to calculate the speed of the satellite relative to the spaceship and comment on your answer.

25-3 The Michelson-Morley experiment shows that light does not obey Newtonian relativity

37. •The Michelson-Morley experiment was performed hundreds of times in a futile attempt to find the luminiferous ether. Why was the experiment performed on an enormous slab of marble that was floated in a pool of mercury?

38. ••Consider an airplane traveling at 25 m/s airspeed between two points that are 2000 km apart. What is the round-trip time for the plane (a) if there is no wind? (b) if there is a wind blowing at 10 m/s, perpendicular to the line joining the two points? (c) if there is a wind blowing at 10 m/s along the line joining the two points?

39. ••Suppose that the length L in the Michelson-Morley experiment is 20,000 m. What would be the difference in light travel times on the two legs of the Michelson-Morley experiment *if the ether existed* and if Earth moved relative to it at the following speeds: (a) at its orbital speed around the Sun, (b) at $0.01c$, (c) at $0.1c$, (d) at $0.5c$, and (e) at $0.9c$?

25-4 Einstein's relativity predicts that the time between events depends on the observer

40. ••An observer in reference frame S observes that a lightning bolt strikes the origin and 10^{-4} s later a second lightning bolt strikes the same location. What is the time separation between the two lightning bolts determined by a second observer in reference frame S' moving at a speed of $0.8c$ along the collinear x–x' axis?

41. •A radioactive particle travels at $0.80c$ relative to the laboratory observers who are performing research. Calculate the half-life of the particle as measured in the laboratory frame compared to the half-life according to the proper frame of the particle.

42. •Muons at rest in the laboratory have an average lifetime of about 2.2 μs. What is the average lifetime, measured in the laboratory, of muons traveling at $0.99c$ with respect to the laboratory?

43. ••When muons traveled through the laboratory at a speed of $0.98c$, scientists obtained an average lifetime value of 11 μs before a muon decayed. If the muons were at rest in the laboratory, what would be the average lifetime of muons? SSM

44. ••The time dilation effect between two frames of reference is measured in the lab to have a 0.01% difference between the relative time and the proper time $[\Delta t = (t - t_{\text{proper}})/t_{\text{proper}} \times 100\% = 0.01\%]$. Calculate the relative speed of the two reference frames.

45. ••Astronomy The Andromeda galaxy is a spiral galaxy that is a distance of 2.54 million light-years from Earth. Is there any possible speed that a spaceship can achieve to deliver a human being to this galaxy? Consider the lifetime of a human to be 80 years.

46. ••Astronomy The nearest star to our own Sun is Proxima Centauri, 4.24 light-years away. If a spaceship travels at $0.75c$, how much time is required for a one-way trip (a) according to an Earth observer, and (b) according to the captain of the ship?

47. ••A subatomic particle is traveling with a "γ factor," $\gamma = 1/\sqrt{1 - (v/c)^2}$, of 20 as observed by a radiation monitor in a nuclear power plant. The particle is measured to "decay" 30 ns after it is observed by the plant. What is the proper lifetime of such a particle?

48. ••A spaceship travels at $0.95c$ toward Alpha Centauri. According to Earthlings, the distance is 4.37 light-years. (a) From the perspective of the space travelers, how long does it take to reach this star if the ship starts at Earth? (b) How long do Earthlings measure for the trip?

25-5 Einstein's relativity also predicts that the length of an object depends on the observer

49. ••A car travels in the positive x direction in the reference frame S. The reference frame S' moves at a speed of $0.80c$, along the x axis. The proper length of the car is 3.20 m. Calculate the length of the car according to observers in the S' frame. SSM

50. •The length of a spaceship is measured to have a relative length that is two-thirds of its proper length. What is the speed of the spaceship relative to the observer?

51. • A stick moves past an observer at a speed of $0.44c$. According to the observer the stick is oriented parallel to the direction of motion and is 0.88 m long. Determine the proper length of the stick.

52. ••A standard tournament domino is 1.5 in wide and 2.5 in long. Describe how you might orient a domino so that it will measure 1.5 in by 1.5 in as it moves by. What relative speed is required?

53. ••How fast must a pion be moving to travel 100 m (according to the laboratory frame) before it decays? The average lifetime, at rest, of a pion is 2.60×10^{-8} s. Give your answer in units of meters per second (m/s) and as a fraction of the speed of light (in other words, $v = ?$ and $v/c = ?$). SSM

25-6 The relative velocity of two objects is constrained by the speed of light, the ultimate speed limit

54. ••Spaceship A moves at $0.8c$ toward the right, while spaceship B moves in the opposite direction at $0.7c$ (both speeds are measured relative to Earth). (a) Calculate the velocity of Earth relative to spaceship A. (b) Calculate the velocity of Earth relative to spaceship B. (c) Calculate the velocity of spaceship A relative to spaceship B.

55. ••A (very fast) car traveling with a velocity of $+0.35c$ passes an observer sitting on the side of the road. A truck traveling with a velocity of $+0.25c$ passes the same observer. Determine the relative velocity between the car and the truck. Give your answer in terms of "the velocity of the truck relative to the car is . . ." and "the velocity of the car relative to the truck is. . . ."

56. ••A spaceship flies by Earth at $0.92c$. It fires a rocket at $0.75c$ in the forward direction, relative to the spaceship. What is the velocity of the rocket relative to Earth?

57. ••Suppose the spaceship in problem 56 continues to fly by Earth at $0.92c$. This time, however, it fires a rocket at $0.75c$ in the backward direction relative to the spaceship. What is the velocity of the rocket relative to Earth? SSM

58. ••Prove that the relative velocity of a laser fired from a spaceship that is moving at $0.92c$ past Earth will be c from the perspective of Earth.

59. ••A proton travels in the accelerator at Fermilab at 99.999954% of the speed of light. An anti-proton travels in the opposite direction at the same speed. What is the relative velocity of the two particles?

25-7 The equations for kinetic energy and momentum must be modified at very high speeds

60. ••A 2.00-kg object moves at 400,000 m/s. (a) Calculate the Newtonian momentum of the object. (b) Calculate the Einsteinian momentum of the object. (c) Which of the answers is correct? What is the percent difference?

61. ••An electron travels at $0.444c$. Calculate (a) its Einsteinian momentum, (b) its Einsteinian kinetic energy, (c) its rest energy, and (d) the total energy of the electron. SSM

62. •A proton ($m = 1.673 \times 10^{-27}$ kg) is traveling at $0.5c$. Calculate its Einsteinian momentum and its Einsteinian kinetic energy.

63. •A particle is traveling with respect to an observer such that its Einsteinian energy is twice its rest energy. How fast is it moving with respect to the observer?

64. ••A particle has a rest energy of 5.33×10^{-13} J and a total energy of 9.61×10^{-13} J. Calculate the momentum of the particle.

65. ••A proton has a rest energy of 1.50×10^{-10} J and a momentum of 1.07×10^{-19} kg·m/s. Calculate its speed. SSM

25-8 Einstein's general theory of relativity describes the fundamental nature of gravity

66. ••What would an observer measure for the free-fall acceleration in an elevator near the surface of Earth if the elevator accelerates downward at 8 m/s²?

67. ••What would an observer measure for the free-fall acceleration in an elevator near the surface of Earth if the elevator accelerates downward at 18 m/s²? SSM

General Problems

68. ••A super rocket car traverses a straight track 2.40×10^5 m long in 10^{-3} s as measured by an observer next to the track. (a) How much time elapses on a clock in the rocket car during the run? (b) What is the distance traveled in traversing the track as determined by the driver of the rocket car?

69. ••Observers in reference frame S see an explosion located at $x_1 = 580$ m. A second explosion occurs 4.5 μs later at $x_2 = 1500$ m. In reference frame S', which is moving along the $+x$ axis at speed v, the explosions occur at the same point in space. What is the separation in time between the two explosions as measured in S'? SSM

70. ••A spaceship departs from Earth for the star Alpha Centauri, which is 4.37 light-years away. The spaceship travels at $0.77c$. (a) What is the time required to get there as measured by a passenger on the spaceship? (b) How long does it take for the spaceship to arrive at Alpha Centauri as measured on Earth?

71. ••A radioactive nucleus traveling at a speed of $0.8c$ in a laboratory decays and emits an electron in the same direction as the nucleus is moving. The electron travels at a speed of $0.6c$ relative to the nucleus. (a) How fast is the electron moving according to an observer in the laboratory? (b) Rocket A travels away from Earth at $0.6c$, and rocket B travels away from Earth in exactly the opposite direction at $0.8c$. What is the speed of rocket B as measured by the pilot of rocket A? (c) Why did you get the same answer that you did for part (a)?

72. ••A beam of pions has a speed of $0.88c$. Their mean lifetime, as measured in the reference frame of the laboratory, is 2.6×10^{-8} s. What is the distance traveled by the laboratory, as measured by the pion, during its lifetime?

73. •••Muons have a proper lifetime of 2.20×10^{-6} s. Suppose a muon is formed at an altitude of 3000 m and travels at a speed of $0.950c$ straight toward Earth. (a) Does the muon reach Earth's surface before it decays? Complete the problem from both perspectives: the muon's point of view and Earth's reference frame. (b) Calculate the minimum speed of the muon so that it *just barely* reaches Earth's surface as it decays.

74. ••Two students, Nora and Allison, are both the same age when Allison hops aboard a flying saucer and blasts off to achieve a cruising speed of $0.800c$ for 20 years (according

to Allison). Neglecting the acceleration of the ship during blastoff, landing, and turnarounds, find the difference in age between Nora and Allison when they are reunited. Who is younger?

75. ••Suppose a jet plane flies at 300 m/s relative to an observer on the ground. Using only special relativity, determine the distance of the flight, as measured by an observer on the ground, before the clocks aboard the plane are 10 s behind clocks on the ground. Assume the two clocks were originally synchronized to start. SSM

76. ••At the end of the linear accelerator at the Stanford Linear Accelerator Center (SLAC), electrons have a speed of $0.99999999995c$. (a) Calculate the value of γ for an electron at SLAC. (b) What time interval would an observer at rest relative to the accelerator measure for a time interval of 1.66 μs measured from the electron's perspective?

77. ••The proper half-life of a certain subatomic particle is about 1×10^{-8} s. How fast is a beam of these particles moving if one-half of them decay in 6×10^{-8} s as measured in the laboratory?

78. ••Based on experiments in your lab, you know that a certain radioisotope has a half-life of 2.25 μs. As a high-speed spaceship passes your lab, you measure that the same isotope at rest inside the spaceship takes 3.15 s for one-half of it to decay. (a) What is the half-life of the isotope as measured by an astronaut working inside the spaceship? (b) How fast is the spaceship traveling relative to Earth?

79. ••Recent home energy bills indicate that a household used 411 kWh of electrical energy and 201 therms for gas heating and cooking in a period of one month. Given that 1.0 therm is equal to 29.3 kWh, how many milligrams of mass would need to be converted directly to energy each month to meet the energy needs for the home? SSM

80. ••We know 1 kg of trinitrotoluene (TNT) yields an energy of 4.2 MJ. The energy released comes from the chemical bonds in the material. How much mass would be required to create an explosion equivalent to 1.8×10^9 kg TNT? Assume all the energy comes simply from the mass of the material.

81. ••Astronomy High-speed cosmic rays strike atoms in Earth's upper atmosphere and create secondary showers. Suppose a particle in one of the showers is created 25.0 km above the surface traveling downward at 90.0% the speed of light. Consider the following two events: "a particle is created in the upper atmosphere" and "a particle strikes the ground." We can view the events from two reference frames, one fixed on Earth and one traveling with the created particle. (a) In which of the reference frames are the proper time and the rest length between the two events measured? Explain your reasoning. (b) In the particle's reference frame, how long after its creation does it take it to reach the ground? (c) In Earth's reference frame, how long after creation does it take the particle to reach the ground? (d) Show that the times in parts (b) and (c) are consistent with time dilation.

82. •••A rocket 642 m long is traveling parallel to Earth's surface at $0.5c$ from left to right. At time $t = 0$, a light flashes for an instant at the center of the rocket. Detectors at opposite ends of the rocket record the arrival of the light signal. Call event A the light striking the left detector and event B the light striking the right detector. Observers at rest in the rocket and on Earth record the events. (a) What is the speed of the light signal as measured by the observer (i) at rest in the rocket and (ii) at rest on Earth? (b) At what time after the flash do events A and B occur as measured by (i) the observer in the rocket and (ii) the observer at rest on Earth? Which event occurs first in each case? (c) Show that the results in part (b) are consistent with time dilation.

83. ••Biology Twin astronauts, Harry and Larry, have identical pulse rates of 70 beats/min on Earth. Harry remains on Earth, but Larry is assigned to a space voyage during which he travels at $0.75c$ relative to Earth. What will be Larry's pulse rate as measured by (a) Harry on Earth and (b) the doctor in Larry's rocket?

84. ••A rocket is traveling at speed v relative to Earth. Inside a lab in the rocket, two laser beams are turned on, one pointing in the forward direction and the other pointing in the backward direction relative to the rocket's velocity. (a) What is the speed of each laser beam relative to the laboratory in the rocket? (b) Use the Lorentz velocity transformation to find the velocity of each laser beam as measured by an observer at rest on Earth. (c) As observed from Earth, how fast are the two laser beams separating from *each other*?

85. •••A 1000-kg rocket is flying at $0.90c$ relative to your lab. Calculate the kinetic energy of the rocket using the Einsteinian formula and the ordinary Newtonian formula. What is the percent error if we use the Newtonian formula? Does the Newtonian formula overestimate or underestimate the kinetic energy?

86. •••A spaceship is traveling at $0.50c$ relative to Earth. Inside the ship, a cylindrical piston that is 50.0 cm long and 4.50 cm in diameter contains 1.25 mol of ideal gas at 25.0°C under 2.20 atm of pressure. The cylinder is oriented with its axis parallel to the direction in which the spaceship is flying. What is the particle density (in molecules per cubic meter) of the gas in the cylinder as measured by (a) an astronaut in the rocket ship's lab and (b) a scientist in an Earth lab?

87. •••Astronomy In December 2009, the discovery was announced of a planet that may have a large amount of water, and hence would be a good candidate for possible life. The planet, GJ 1214b, orbits a small star that is 42 light-years from Earth. In the future, we might decide to send some astronauts to explore the planet. When they arrive there, we want them to be young enough to perform tests. Suppose that the captain is 25 years old at launch time. (a) What is the minimum speed the spaceship will need for the captain to be no more than 60 years old at arrival? (b) As soon as the spaceship arrives at the planet, the captain has orders to send a radio signal to Earth to notify Mission Control that the trip was successful. How many years after launch from Earth will it be when the signal arrives at Earth? You can ignore acceleration times and any motion of Earth and GJ 1214b. SSM

26

Quantum Physics and Atomic Structure

What do you think?

Some of the stars in this photo are blue, while others are red. (The blue stars are surrounded by dust, which gives the appearance of a wispy cloud.) Based on the color alone, you can conclude that the blue stars are (a) hotter than the red stars; (b) cooler than the red stars; (c) made of different materials than the red stars; (d) both (a) and (c); (e) both (b) and (c).

In this chapter, your goals are to:

- (26-1) Recognize the limitations of classical physics for explaining the properties of light and matter.
- (26-2) Describe how the photoelectric effect and blackbody radiation provide evidence for the photon picture of light.
- (26-3) Explain why the wavelength of a photon increases if it scatters from an electron.
- (26-4) Calculate the wavelength of a particle such as an electron.
- (26-5) Describe why atoms absorb light at only certain wavelengths, and why they emit and abosrb light at the same wavelengths.
- (26-6) Explain how the Bohr model of the atom explains the spectrum of hydrogen.

To master this chapter, you should review:

- (3-8)   Uniform circular motion.
- (4-2)   Newton's second law.
- (8-8)   Angular momentum of a moving particle.
- (14-3)   The energy of molecules in an ideal gas.
- (14-7)   Heat transfer by radiation.
- (17-3)   Electric potential.
- (22-2, 22-3,   The electromagnetic nature of light.
  22-4)
- (25-7)   Relativistic kinetic energy and momentum.

26-1 Experiments that probe the nature of light and matter reveal the limits of classical physics

We have uncovered hints that light and other electromagnetic waves have a particle-like nature. Is there strong evidence that light actually comes in the form of photons? Can a photon strike a particle such as an electron and "bounce" or scatter the way a cue ball does when it strikes a billiard ball? If waves have a particle-like nature, do particles ever exhibit properties we associate with waves?

In this chapter we'll see that the answer to each of these questions is "yes." The photoelectric effect (Figure 26-1a), in which light striking a surface causes the surface to eject electrons, can only be understood if light comes in the form of photons. The

Night vision goggles "amplify light" using the photoelectric effect. Even faint light causes a surface to emit electrons, and the current of these electrons can be amplified in a circuit to generate a brighter image that the wearer of the goggles can see.

The characteristic color of neon lights is due to the structure of the neon atom. When excited by an electric current, neon atoms in the light tube jump to a higher energy level. When they return to the starting energy level, they emit photons of a specific frequency, wavelength, and color.

(a)

(b)

Figure 26-1 **Photons, electrons, and atoms** Two examples of the interaction between light and matter. These can only be understood if we use the ideas that light has particle aspects and matter has wave aspects.

Take-Home Message for Section 26-1

✔ Light waves have some of the characteristics of particles, and particles such as electrons have some of the characteristics of waves.

✔ The energies of atoms are quantized (restricted to certain specific values).

same is true for blackbody radiation, the light emitted by an object as a consequence of its temperature. We'll see direct evidence that photons really do behave like tiny "cue balls" when they scatter from electrons. And we'll learn that particles like electrons also have a wavelike aspect. This wave nature of matter will help us understand the structure of atoms and the manner in which atoms absorb and emit light (Figure 26-1b). We'll find that the energies of atoms are *quantized*—that is, they can have only certain very definite values. The lesson of this chapter is that the microscopic world of atoms and light is very different in character from the macroscopic world of ordinary-sized objects that we see around us.

26-2 The photoelectric effect and blackbody radiation show that light is absorbed and emitted in the form of photons

By the end of the nineteenth century, it was recognized that light was an electromagnetic wave. Maxwell's equations (Section 22-3) describe the properties of these waves in great detail, and so physicists were confident that they had a deep understanding of the nature of light. But experimental studies of two very different phenomena showed that the true nature of light is actually more complex. In one of these phenomena, the *photoelectric effect*, a material absorbs light and the absorbed energy is used to eject electrons; in the other, called *blackbody radiation*, a material emits electromagnetic radiation when it is heated. The discoveries made by studying these two phenomena radically altered how physicists answered the question, "What is light?"

The Photoelectric Effect

When light strikes certain materials, electrons can be ejected from the surface of those materials (Figure 26-2a). This **photoelectric effect**, discovered in 1886, plays an important role in biological research through a technique called *photoemission electron microscopy* or PEEM. In this technique a biological sample is illuminated with ultraviolet light or X rays. Different materials in the surface layer of the sample will emit fewer or greater numbers of electrons in response to this light.

(a)

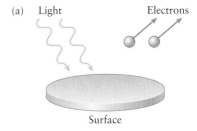

Light   Electrons

Surface

(b)   Fibrinogen   Substrate

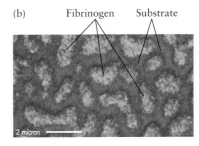

Figure 26-2 **The photoelectric effect** (a) In the photoelectric effect, electrons escape from a surface when the surface is illuminated with light. (b) An example of photoemission electron microscopy. Protein interactions with polymers are critical to understanding the compatibility of synthetic materials with blood (such as medical implants). Here the spatial distribution of fibrinogen, a blood protein, is mapped relative to a PS/PMMA polymer substrate, using photoemission electron microscopy (PEEM). Data measured at the Advanced Light Source, Lawrence Berkeley National Laboratory. This image could be made because fibrinogen and the substrate emit different numbers of electrons when illuminated with ultraviolet light.

By recording the differences in the number of emitted electrons, called **photoelectrons,** it's possible to construct an image of the sample surface that shows where the different materials are located. Figure 26-2b shows an example of an image made in this way.

What makes the photoelectric effect so remarkable is that it does not behave in accordance with the idea that light is an electromagnetic wave. It takes energy to liberate an electron from the surface of a material, and in the photoelectric effect this energy is provided by the electric and magnetic energy in the light absorbed by the surface. We learned in Section 22-4 that according to Maxwell's equations, the intensity of such a wave depends on the rms values of the electric and magnetic fields in that wave, but not on the frequency of the wave:

Intensity of an
electromagnetic wave
(22-25)

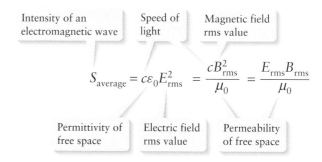

$$S_{\text{average}} = c\varepsilon_0 E_{\text{rms}}^2 \;=\; \frac{cB_{\text{rms}}^2}{\mu_0} \;=\; \frac{E_{\text{rms}}B_{\text{rms}}}{\mu_0}$$

Equation 22-25 leads us to predict that light of *any* frequency should be able to liberate an electron from the surface of a material, provided the light wave is sufficiently intense. Experiment shows that this is not the case. For example, if the biological sample shown in Figure 26-2 is illuminated with red light, no electrons are ejected no matter how intense the light. But if instead we illuminate the sample with X rays, which have a higher frequency than red light, electrons *are* ejected. (Figure 26-2b was made by using X rays.) This is impossible to understand on the basis of Maxwell's equations.

In 1905, the same year that he published his special theory of relativity, Albert Einstein proposed a simple but radical explanation for the strange behavior of the photoelectric effect. He suggested that light of frequency *f* comes in small packets, each with an energy *E* that is directly proportional to the frequency. Today these packets are called **photons.** We first encountered this idea in Section 22-4:

Energy of a photon
(22-26)

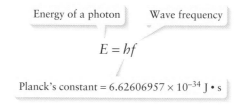

$$E = hf$$

Planck's constant $= 6.62606957 \times 10^{-34}$ J $\cdot$ s

We'll see shortly how the value of Planck's constant *h* is determined from the photoelectric effect.

Let's see how Einstein's idea explains the properties of the photoelectric effect. The minimum amount of energy required to remove a single electron from a material is called the **work function** Φ_0 of the material (Φ is the uppercase Greek letter phi). The value of Φ_0 varies from one material to another; it is small if electrons are easy to remove, and large if electrons are hard to remove. In a given material, some electrons will be more difficult to remove, but Φ_0 represents the energy required to remove the most easily dislodged electron. Einstein proposed that an electron can absorb only a single photon at a time. So for even the most easily dislodged electron to be ejected from the material, it must absorb a photon with an energy equal to or greater than Φ_0. If the energy of the absorbed photon is greater than Φ_0, the energy that remains after the electron is ejected goes into the kinetic energy of the electron as it flies away from the material. Electrons that require more energy to be ejected will emerge from the material with less kinetic energy, but those with *maximum* kinetic energy will be those that were the easiest to dislodge. So in Einstein's picture, the most energetic electrons ejected from the material will emerge with kinetic energy K_{max} given by

Maximum kinetic energy of an electron ejected from a material by the photoelectric effect

Work function of the material

$$K_{max} = hf - \Phi_0$$

Planck's constant

Frequency of the light used to illuminate the material

Maximum kinetic energy of an electron in the photoelectric effect (26-1)

Equation 26-1 tells us that that a graph of K_{max} as a function of the light frequency f should be a straight line of slope h (see Figure 26-3). Since kinetic energy can never be negative, Equation 26-1 also tells us that electrons will be emitted only if $hf - \Phi_0 > 0$, or

$$f > f_0 = \frac{\Phi_0}{h}$$

(26-2)

In words, Equation 26-2 says that electrons will be emitted from the surface only if the light frequency is greater than a threshold frequency f_0 equal to the work function Φ_0 divided by Planck's constant h. This agrees with the observation that no electrons are ejected if a surface is illuminated with light of too low a frequency, no matter how intense the light.

The graph shown in Figure 26-3 turns out to be an excellent match to experimental measurements of the maximum kinetic energy of ejected electrons for different frequencies f. The slope of the graph tells us the value of Planck's constant h. Equation 26-1 also shows that if we extend the graph of K_{max} as a function of f to (unphysical) values of K_{max} less than zero, the graph intercepts the vertical axis at $-\Phi_0$. If we repeat the experimental measurements for a second material with a different work function Φ_0, the straight line intercepts the vertical axis at a different point but has the same slope h as for the first material. This reinforces the idea that Planck's constant is a universal constant associated with the energy of the photons, not the specific properties of the material that absorbs the photons.

The remarkable fit of Einstein's theory to experiment is powerful evidence that light is indeed absorbed in the form of photons with energy $E = hf$ as given by Equation 22-26. In recognition of his achievement, Einstein was awarded the Nobel Prize in Physics in 1921 for his explanation of the photoelectric effect.

 Go to Interactive Exercise 26-1 for more practice dealing with the photoelectric effect

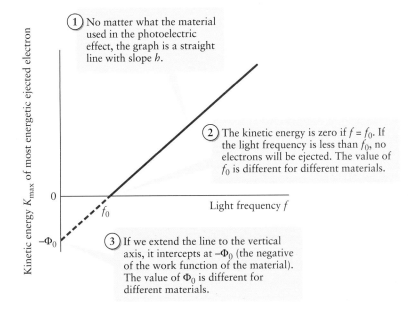

① No matter what the material used in the photoelectric effect, the graph is a straight line with slope h.

② The kinetic energy is zero if $f = f_0$. If the light frequency is less than f_0, no electrons will be ejected. The value of f_0 is different for different materials.

③ If we extend the line to the vertical axis, it intercepts at $-\Phi_0$ (the negative of the work function of the material). The value of Φ_0 is different for different materials.

Figure 26-3 **Electron kinetic energy versus frequency in the photoelectric effect** The kinetic energy of the most energetic electron ejected in the photoelectric effect depends on the frequency of the light. This cannot be explained using the wave model of light, but can be explained by the photon concept.

> **! Watch Out!** Some electrons require more energy than Φ_0 to be ejected.
>
> The work function Φ_0 is the smallest amount of energy required to eject an electron from a given material under the most favorable conditions. Other electrons in the material require more energy to eject and will emerge from the material with less kinetic energy than the value K_{max} given by Equation 26-1.

Example 26-1 The Photoelectric Effect with Cesium

The work function for a sample of cesium is 3.43×10^{-19} J. (a) What is the minimum frequency of light that will result in electrons being ejected from this sample by the photoelectric effect? (b) What is the maximum wavelength of light that will result in electrons being ejected from this sample by the photoelectric effect?

Set Up

Equation 26-1 tells us that the minimum energy required to eject an electron corresponds to having $K_{max} = 0$, so the electrons just barely make it out of the cesium. We'll use this to find the threshold frequency f_0 that just barely allows an electron to be ejected. We'll find the corresponding wavelength using Equation 22-2.

Maximum kinetic energy of an electron in the photoelectric effect:

$$K_{max} = hf - \Phi_0 \qquad (26\text{-}1)$$

Propagation speed, frequency, and wavelength of an electromagnetic wave:

$$c = f\lambda \qquad (22\text{-}2)$$

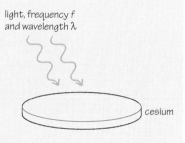

light, frequency f and wavelength λ

cesium

Solve

(a) Use Equation 26-1 to calculate the frequency f that corresponds to $K_{max} = 0$.

From Equation 26-1 with $K_{max} = 0$,

$$0 = hf - \Phi_0$$
$$hf = \Phi_0$$

This says that the photon energy hf is just enough to remove the most easily dislodged electron from the material (which requires energy Φ_0), with nothing left over to give the electron any kinetic energy. So $hf = \Phi_0$ is the minimum photon energy that will eject an electron, and the frequency f of this photon is the minimum (threshold) frequency that will do the job:

$$f_0 = f = \frac{\Phi_0}{h}$$
$$= \frac{3.43 \times 10^{-19} \text{ J}}{6.63 \times 10^{-34} \text{ J} \cdot \text{s}} = 5.17 \times 10^{14} \text{ s}^{-1} = 5.17 \times 10^{14} \text{ Hz}$$

(b) Find the wavelength that corresponds to the frequency that we calculated in part (a).

We can rewrite Equation 22-2 as

$$\lambda = \frac{c}{f}$$

In words, this says that wavelength is inversely proportional to frequency. So the *minimum* frequency of light that will eject an electron corresponds to the *maximum* wavelength that will eject an electron:

$$\lambda_{max} = \frac{c}{f_0} = \frac{3.00 \times 10^8 \text{ m/s}}{5.17 \times 10^{14} \text{ Hz}}$$
$$= 5.80 \times 10^{-7} \text{ m} = 580 \text{ nm}$$

(Recall that 1 nm $= 10^{-9}$ m.)

Reflect

Figure 22-2 shows that a wavelength of 580 nm is in the yellow-green part of the visible spectrum. If we illuminate cesium with light of higher frequency and shorter wavelength than this (for example, blue or violet light), the photons will have more energy than the minimum and electrons will be ejected from the cesium. If instead we illuminate cesium with light of lower frequency and longer wavelength (for example, orange or red light), the photons will have less energy than the required minimum and no electrons will be ejected.

Blackbody Radiation

The photoelectric effect shows that light is *absorbed* in the form of photons. If we are to fully believe the photon concept, however, it must also be true that light is *emitted* in the form of photons. We learned in Section 14-7 that ordinary objects emit electromagnetic radiation as a result of their temperature. If we can find evidence that this emission is in the form of photons, we will have further evidence that the photon description of light is the correct one. Let's take a closer look at the detailed properties of radiation of this kind.

Experiment shows that the rate at which an object emits radiation is proportional to its surface area A and to the fourth power of its Kelvin temperature T:

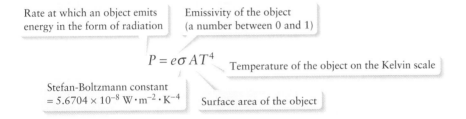

Rate at which an object emits energy in the form of radiation

Emissivity of the object (a number between 0 and 1)

$$P = e\sigma A T^4$$

Temperature of the object on the Kelvin scale

Stefan-Boltzmann constant
$= 5.6704 \times 10^{-8}\ \mathrm{W \cdot m^{-2} \cdot K^{-4}}$

Surface area of the object

Rate of energy flow in radiation (14-22)

The higher the temperature of an object of a given size, the greater the radiated power P and so the more brightly it glows.

Experiment also shows that the *color* of the radiation emitted by an object depends on its temperature T (Figure 26-4). A heated object emits light at all wavelengths, but emits most strongly at a particular frequency called the *frequency of maximum emission*. As the temperature increases, the frequency of maximum emission increases.

Equation 14-22 shows that the radiated power also depends on a quantity e called the *emissivity*, which depends on the properties of the object's surface. This has its greatest value ($e = 1$) for an idealized type of dense object called a **blackbody.** An ideal blackbody does not reflect any light at all, but absorbs all radiation falling on it. If a blackbody is in thermal equilibrium with its surroundings, it must emit energy at the

(1) This metal bar heated with a flame emits light at all frequencies, but glows most strongly at red frequencies.

(2) As the temperature of the bar increases, it glows most strongly at orange frequencies...

(3) ...and at even higher temperatures, it glows most strongly at yellow frequencies.

- A hot, dense object emits electromagnetic radiation. The idealized case is called **blackbody radiation.**
- The frequency of maximum emission is directly proportional to the Kelvin temperature T of the object: The higher the temperature T, the greater the frequency of maximum emission.

Figure 26-4 Radiation from heated objects The color of the light from a heated object depends on its temperature.

same rate that it absorbs it in order that its temperature *T* remain constant. So in addition to being a perfect absorber of energy, an ideal blackbody in thermal equilibrium with its surroundings is also a perfect emitter of energy because it emits as much energy as it absorbs.

Ordinary objects, such as tables, textbooks, and people, are not ideal blackbodies; they reflect light, which is why they are visible. (Even a piece of wood darkened with soot or painted a dull black reflects *some* light.) But it is possible to make a nearly ideal blackbody simply by building a box and drilling a small hole in one side (**Figure 26-5a**). Light that enters the hole will reflect around inside the box, with part of the light energy being absorbed by the walls on each reflection. Eventually all of the light energy will be absorbed, so the interior of the box acts like a perfect absorber and is effectively an ideal blackbody. You can see this effect if you look into another person's eye (**Figure 26-5b**). The pupil at the center of the iris appears black, even though the tissues that line the interior of the eye are pinkish in color. That's because after multiple reflections, those tissues almost completely absorb light that enters the eye through the pupil.

If the box in Figure 26-5a is in thermal equilibrium at temperature *T*, the rate at which the walls absorb energy in the form of radiation must be equal to the rate at which the walls emit energy. The cavity in the interior of the box will be filled with this radiation, which will itself be in thermal equilibrium with the walls of the box. Since the walls act as an ideal blackbody, the light that fills the cavity is effectively **blackbody radiation**—the kind of light that would be emitted by a perfect blackbody of emissivity $e = 1$ at temperature *T*. We can study this light by examining the small fraction of light that emerges from the hole in Figure 26-5a.

Figure 26-6 shows the experimentally observed *spectrum* of blackbody radiation—that is, the relative amount of light energy present at different frequencies—for two different temperatures. Note that the high-temperature curve lies above the low-temperature curve at all frequencies. This tells us that the higher the temperature of a blackbody, the greater the amount of radiation at all frequencies. Note also that as the blackbody temperature increases, the peak of the curve shifts to a higher frequency. This agrees with our observation about how the frequency of maximum emission varies with an object's temperature (Figure 26-4). Figure 26-6 explains why we can't see the radiation from objects at room temperature, about $T = 300$ K. At this relatively low temperature, the frequency of maximum emission is in the infrared, which our eyes cannot see. There is some emission at visible frequencies (in Figure 26-6, to the right of the peak of the curve), but the amount of emission is so low at $T = 300$ K that our eyes can't detect it.

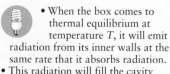

- When the box comes to thermal equilibrium at temperature *T*, it will emit radiation from its inner walls at the same rate that it absorbs radiation.
- This radiation will fill the cavity inside the box.
- Some of the radiation will leak out of the hole. Because the box acts as a perfect blackbody, the radiation that leaks out will be blackbody radiation.

(a)

① Light enters a hole in the box.

② The light reflects off the inner walls of the box many times. On each reflection, part of the light energy is absorbed by the walls.

③ Eventually all of the light energy is absorbed. Hence the interior of the box (the cavity) acts like a perfect absorber— that is, an ideal blackbody.

(b)

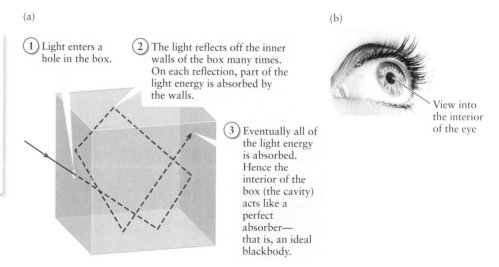

View into the interior of the eye

Figure 26-5 A blackbody cavity (a) A box with a small hole in one side is a good approximation to an ideal blackbody. (b) The interior of the eye has very similar properties to the box in (a).

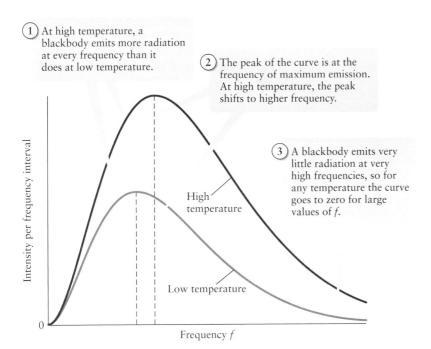

① At high temperature, a blackbody emits more radiation at every frequency than it does at low temperature.

② The peak of the curve is at the frequency of maximum emission. At high temperature, the peak shifts to higher frequency.

③ A blackbody emits very little radiation at very high frequencies, so for any temperature the curve goes to zero for large values of f.

High temperature

Low temperature

Intensity per frequency interval

Frequency f

Figure 26-6 Blackbody spectra
The spectrum of light emitted by an ideal blackbody depends on the temperature of the blackbody.

In the late nineteenth century, physicists tried to understand the shape of the black-body spectrum shown in Figure 26-6 using their knowledge of thermodynamics and electromagnetic waves. Their efforts ended in failure. To understand how they failed, we begin by noting that the electromagnetic waves inside the cavity in Figure 26-5a should be in the form of *standing* waves. That's because the waves will bounce back and forth between the walls of the cavity. We learned in Section 13-6 that when waves bounce back and forth between the ends of a string that's tied down at both ends, steady wave patterns arise for waves of certain wavelengths and frequencies. The same is true for electromagnetic waves in a cavity. A key difference is that while the patterns of nodes and antinodes in standing waves on a string are one-dimensional (along the length of the string), the corresponding patterns for standing waves in a cavity are three-dimensional and so are more complicated. Another difference is that a standing electromagnetic wave involves two varying quantities, the electric field and the magnetic field. For our purposes, however, all we need to know is that these standing waves exist.

The shape of the blackbody spectrum in Figure 26-6 should be indicative of how much energy is present in the standing waves in each frequency range. The general shape of these curves shows that there should be relatively little energy at very low frequencies, more energy at frequencies near the frequency of maximum emission, and again relatively little energy at high frequencies. However, nineteenth-century physics suggested that *every* possible standing wave in the cavity should contain on average the same amount of energy. This conclusion came from the equipartition theorem, which we first encountered in Section 14-3. This theorem states that a molecule in a gas at a Kelvin temperature T has, on average, an amount of energy $(1/2)kT$ for each degree of freedom of the molecule, where $k = 1.381 \times 10^{-23}$ J/K is the Boltzmann constant. Arguments from thermodynamics suggest that for the same reason, a standing wave inside a cavity in thermal equilibrium at temperature T should also possess an average amount of energy equal to $(1/2)kT$ per degree of freedom. There are two degrees of freedom per standing wave, one for the electric field and one for the magnetic field, so the total average energy per standing wave should be kT. It turns out that the number of standing waves in a given frequency interval increases with increasing frequency. So according to nineteenth-century physics, the total energy per frequency interval (equal to the energy kT per standing wave multiplied by the number of standing waves per frequency interval) should increase with increasing frequency and *never* decrease (**Figure 26-7**). This is in profound disagreement with the experimentally observed shape of the spectrum.

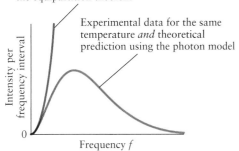

Theoretical prediction using the equipartition theorem

Experimental data for the same temperature *and* theoretical prediction using the photon model

Intensity per frequency interval

0

Frequency f

- A photon model accurately describes the spectrum of blackbody radiation.
- The model that does not use the photon concept fails to describe the spectrum.

Figure 26-7 **The photon model explains blackbody radiation** A model for blackbody radiation that does not use the photon concept predicts (incorrectly) that the intensity should increase without limit as the frequency increases.

If we now introduce the photon concept, however, we *can* match the experimentally observed spectrum. We still use the idea that the average energy available for a standing wave of frequency f is kT, but now this energy goes into photons of that frequency which each have energy $E = hf$. This energy fundamentally comes from the walls of the box in Figure 26-5a, since these walls are what emit the photons. At low frequencies the photon energy hf is small compared to the available energy kT, so there will be very many photons present for a low-frequency standing wave. But at high frequencies hf is much larger than kT, which means that the energy required to produce a photon is larger than the average energy available to produce one. Hence the average number of photons present for that standing wave will be very small. (It need not be zero, since kT is only the *average* energy available. From time to time, the available energy will be greater than kT, and some photons can be produced.) So even though energy is *available* for a high-frequency standing wave, that energy can't be used to create photons, and so the amount of energy *present* is quite small. That's just the effect we need to make the theoretical curve in Figure 26-7 decline at large frequencies.

Using a somewhat different version of this photon argument, in 1900 the German physicist Max Planck was able to make a theoretical prediction for the blackbody spectrum that was in excellent agreement with the experimental spectrum shown in Figure 26-7. Planck's theoretical formula was the first to involve the new quantity h that now bears his name, and the value of h given in Equation 22-26 is the one that gives the best match between this formula and the experimental data. Planck was awarded the 1918 Nobel Prize in Physics for his achievement.

The explanation of blackbody spectra in terms of photons is the evidence we were seeking that light is emitted in the form of photons. In the following section we'll see even more compelling evidence for the photon picture of light.

? Got the Concept? 26-1 Blackbody Radiation

Two objects of the same size are both perfect blackbodies. One is at a temperature of 3000 K, so its frequency of maximum emission is in the infrared part of the electromagnetic part of the spectrum; the other is at a temperature of 12,000 K, so its frequency of maximum emission is in the ultraviolet part of the spectrum. Compared to the object at 3000 K, the object at 12,000 K (a) emits more infrared light; (b) emits more visible light; (c) emits more ultraviolet light; (d) two of (a), (b), and (c); (e) all of (a), (b), and (c).

Take-Home Message for Section 26-2

✔ In the photoelectric effect, an electron in a material can absorb light energy that strikes the surface and as a result be ejected from the surface.

✔ For any material, the light must have a certain minimum frequency in order for electrons to be ejected. This is evidence that light comes in the form of photons, with an energy proportional to their frequency.

✔ A perfect blackbody is an ideal absorber of light and also an ideal emitter of light. The spectrum of light emitted by a blackbody depends on its temperature.

✔ The spectrum of light emitted by a blackbody can only be understood if we use the idea that light is emitted in the form of photons.

26-3 As a result of its photon character, light changes wavelength when it is scattered

Blackbody radiation and the photoelectric effect suggest that photons can be treated like tiny bundles. We have therefore made the claim that as quantized units of energy, photons have a particle-like nature. Are there other implications of this claim? For example, an electron absorbs the energy of a photon in the photoelectric effect. But if a photon is like a particle, is it possible that, as in the collisions we studied in Chapter 7, a photon could strike an electron and bounce off? If so, based on our experience with collisions, we ought to expect that linear momentum would be conserved in such an interaction and that the momentum of the photon should change as a result. This effect, called *Compton scattering*, is further evidence that light does indeed come in the form of photons.

Let's first see how to express the momentum of a photon. We saw in Section 25-7 that for a particle of mass m, we can write the kinetic energy and momentum as

Kinetic energy of a particle of mass m and velocity $\vec{v}$

$$K = (\gamma - 1)mc^2$$

Relativistic gamma for speed v

$$\vec{p} = \gamma m\vec{v}$$

Momentum of a particle of mass m and velocity $\vec{v}$

Einsteinian expressions for kinetic energy and momentum (25-20)

 At speeds that are a small fraction of the speed of light c, these are approximately equal to the Newtonian expressions

$$K = \frac{1}{2}mv^2 \text{ and } \vec{p} = m\vec{v}$$

In these expressions, the quantity γ (relativistic gamma) is

Relativistic gamma for a particle moving at speed v

$$\gamma = \frac{1}{\sqrt{1 - \dfrac{v^2}{c^2}}}$$

Speed of the particle

Speed of light in a vacuum

Relativistic gamma (25-19)

We also saw that an object of mass m has a rest energy E_0 that is present even when it is not moving:

Rest energy of a particle   Mass of the particle

$$E_0 = mc^2$$

Speed of light in a vacuum

Rest energy (25-21)

If we combine the first of Equations 25-20 with Equations 25-19 and 25-21, we find that the total energy of a particle (kinetic energy plus rest energy) is

$$E = K + E_0 = (\gamma - 1)mc^2 + mc^2 = \gamma mc^2 = \frac{mc^2}{\sqrt{1 - \dfrac{v^2}{c^2}}} \qquad (26\text{-}3)$$

From the second of Equations 25-20, the magnitude of the momentum of a particle is

$$p = \gamma mv = \frac{mv}{\sqrt{1 - \dfrac{v^2}{c^2}}} \qquad (26\text{-}4)$$

Comparing Equations 26-3 and 26-4, we see that

(26-5)
$$p = \frac{Ev}{c^2} \quad \text{(momentum of a particle of total energy } E\text{)}$$

A photon has zero mass, which is why it can travel at the speed of light. (Any object with nonzero mass would require an infinite amount of kinetic energy to reach $v = c$, as we described in Section 25-7.) As such, we can't apply Equation 26-3 or 26-4 directly to photons. But we can use the combination in Equation 26-5, in which mass does not appear explicitly. Setting $v = c$ in Equation 26-5, we get the following relationship for the momentum of a photon:

(26-6)
$$p = \frac{Ec}{c^2} = \frac{E}{c} \quad \text{(momentum of a photon)}$$

From Equation 22-26 we can write $E = hf$, and we know that for light waves $c = f\lambda$ (Equation 22-2). If we substitute these into Equation 26-6, we get an alternative expression for the momentum of a photon:

$$p = \frac{hf}{f\lambda}$$

or, simplifying,

**Momentum and energy of a photon
(26-7)**

Magnitude of the momentum of a photon · Energy of the photon

$$p = \frac{E}{c} = \frac{h}{\lambda}$$

Planck's constant

Speed of light in a vacuum · Wavelength of the photon

A photon's momentum is directly proportional to the energy that it carries and inversely proportional to its wavelength. Thus a violet photon of wavelength 400 nm has twice the momentum, as well as twice the energy, of an infrared photon of wavelength 800 nm.

In the early 1920s, American physicist Arthur Compton showed conclusively that photons have momentum and that the momentum is inversely proportional to the wavelength as stated in Equation 26-7. In his experiments, an X ray photon collided with an electron in a carbon atom, a process now called **Compton scattering**. Compton detected both the electron, which is knocked out of the atom, and the scattered photon. Compton could only account for the directions and energies of the electron and the scattered photon by associating a momentum vector with the incoming and outgoing photon and then requiring that momentum be conserved. In other words, he showed that the photon description of light applies not just to the absorption and emission of light, but also to what happens to light when it is scattered. For revealing this fundamental aspect of light, Compton was awarded the Nobel Prize in Physics in 1927.

Compton's results are most concisely expressed in terms of the change in wavelength of the photon as a result of the collision with the electron. **Figure 26-8** shows the collision of a photon (symbol γ_i) and a stationary electron (symbol e^-). The electron is scattered at angle ϕ relative to the initial direction of the photon, and the photon scatters at angle θ relative to its initial direction. Although we have labeled the photon γ_f after the collision, the scattered photon is the same photon that collided with the electron. The subscripts "i" for "initial" and "f" for "final" instead imply that the energy, wavelength, and other quantities associated with the photon have changed as a result of the collision. For example, because energy is conserved during

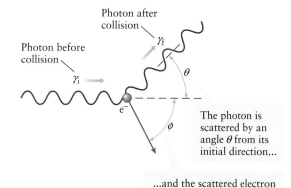

Photon after collision — γ_f

Photon before collision — γ_i

e^-

θ

ϕ

The photon is scattered by an angle θ from its initial direction...

...and the scattered electron moves off at an angle ϕ from the initial direction of the photon.

Figure 26-8 Compton scattering A photon undergoes a wavelength shift when it scatters from an electron that is initially at rest.

the collision and because some of the photon's energy is almost always transferred to the electron, the outgoing photon carries less energy than it had initially. From Equation 26-7, the energy of a photon is

$$E = pc = \frac{hc}{\lambda}$$

Because the photon has less energy after the collision than it had initially, E_f is less than E_i, and the final wavelength λ_f is greater than the initial wavelength λ_i. In other words, as the energy of the photon decreases, its wavelength increases. Compton found that the increase in wavelength $\Delta\lambda$ from λ_i to λ_f is a function of the angle θ at which the photon scatters:

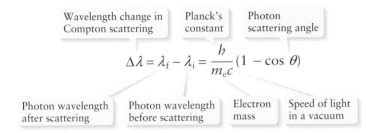

Wavelength change in Compton scattering   Planck's constant   Photon scattering angle

$$\Delta\lambda = \lambda_f - \lambda_i = \frac{h}{m_e c}(1 - \cos\theta)$$

Photon wavelength after scattering   Photon wavelength before scattering   Electron mass   Speed of light in a vacuum

Compton scattering equation (26-8)

The proportionality constant in Equation 26-8 is known as the Compton wavelength λ_C:

$$\lambda_C = \frac{h}{m_e c}$$

(26-9)

 See the Math Tutorial for more information on trigonometry

The scattering angle θ of the photon ranges from 0° (straight forward) to 180° (straight back). When the photon continues in the same direction after the collision, so that θ equals 0° and $\cos\theta$ equals 1, $\Delta\lambda$ equals zero. In other words, if $\theta = 0°$ there is no change in the photon's wavelength and no change in the photon's energy.

 Go to Picture It 26-1 for more practice dealing with Compton scattering

The maximum change in a photon's wavelength and energy when it undergoes Compton scattering occurs when it scatters straight back, in the direction opposite to the one in which it approached the electron. In this case $\theta = 180°$, so $\cos\theta = -1$ and the term in parentheses in Equation 26-8 equals 2. The maximum possible change in the photon's wavelength is therefore $2\lambda_C$, or twice the Compton wavelength. Using the known values of h, m_e, and c, we find

$$\lambda_C = \frac{h}{m_e c} = \frac{6.63 \times 10^{-34}\,\text{J}\cdot\text{s}}{(9.11 \times 10^{-31}\,\text{kg})(3.00 \times 10^8\,\text{m/s})}$$

$$= 2.43 \times 10^{-12}\,\text{m} = 2.43 \times 10^{-3}\,\text{nm} = 0.00243\,\text{nm}$$

(Recall that 1 nm = 10^{-9} m.) The small value of λ_C means that Compton scattering has a negligible effect on the wavelength of visible-light photons. The wavelength of visible light ranges between about 380 nm and 750 nm; the Compton wavelength is considerably smaller. For example, if a photon of wavelength 400 nm is scattered by an electron, its wavelength changes by a value on the order of 0.00243 nm. Such a tiny change is very difficult to measure, so we can ignore the wavelength change of a visible-light photon due to Compton scattering. But for an X ray photon with a wavelength on the order of 10^{-11} m = 0.01 nm, a wavelength shift of 0.00243 nm corresponds to a large percentage of the initial wavelength. That's why Compton first noticed this effect in the scattering of X rays.

If light did not have a photon aspect, we would expect *no* wavelength change on scattering. A light wave of frequency f and wavelength λ encountering an electron would make the electron oscillate at the same frequency f, and the electron would emit radiation with the same frequency f and hence the same wavelength λ as the initial light wave. The change in wavelength that Compton observed is unambiguous evidence that light does indeed have a particle character.

Example 26-2 Compton Scattering

A photon carries 2.00×10^{-14} J of energy. It undergoes Compton scattering in a block of carbon. What is the largest fractional change in energy the photon can undergo as a result?

Set Up

Given the initial photon energy $E_i = 2.00 \times 10^{-14}$ J, we can calculate its wavelength λ_i using Equation 26-7. We use Equation 26-8 to calculate the change in wavelength due to scattering; this will be maximum if $\theta = 180°$, so $\cos\theta = -1$. Once we know the final wavelength λ_f, we can use Equation 26-7 again to find the final photon energy. Comparing this to the initial photon energy tells us the fractional change in energy.

Momentum and energy of a photon:

$$p = \frac{E}{c} = \frac{h}{\lambda} \qquad (26\text{-}7)$$

Compton scattering equation:

$$\Delta\lambda = \lambda_f - \lambda_i = \frac{h}{m_e c}(1 - \cos\theta) \qquad (26\text{-}8)$$

Before:

photon   electron at rest

After:

scattered photon   recoiling electron

Solve

First calculate the wavelength of the initial photon using Equation 26-7.

From Equation 26-7, the wavelength of the initial photon is

$$\lambda_i = \frac{hc}{E_i} = \frac{(6.63 \times 10^{-34}\,\text{J}\cdot\text{s})(3.00 \times 10^{8}\,\text{m/s})}{2.00 \times 10^{-14}\,\text{J}}$$

$$= (9.95 \times 10^{-12}\,\text{m})\left(\frac{1\,\text{nm}}{10^{-9}\,\text{m}}\right) = 9.95 \times 10^{-3}\,\text{nm}$$

Calculate the wavelength shift using Equation 26-8.

The maximum wavelength shift is with $\theta = 180°$ and $\cos\theta = -1$:

$$\Delta\lambda_{max} = \lambda_f - \lambda_i = \frac{h}{m_e c}(1 - \cos 180°)$$

$$= (2.43 \times 10^{-3}\,\text{nm})[1 - (-1)] = 4.86 \times 10^{-3}\,\text{nm}$$

The wavelength of the final photon equals the wavelength of the initial photon plus the shift $\Delta\lambda$. Use this to find the energy of the final photon.

The final wavelength is

$$\lambda_f = \lambda_i + \Delta\lambda = 9.95 \times 10^{-3}\,\text{nm} + 4.86 \times 10^{-3}\,\text{nm}$$

$$= 1.481 \times 10^{-2}\,\text{nm} = 1.481 \times 10^{-11}\,\text{m}$$

The energy of the final photon is

$$E_f = \frac{hc}{\lambda_f} = \frac{(6.63 \times 10^{-34}\,\text{J}\cdot\text{s})(3.00 \times 10^{8}\,\text{m/s})}{1.481 \times 10^{-11}\,\text{m}}$$

$$= 1.34 \times 10^{-14}\,\text{J}$$

This is less than the energy of the initial photon. The lost energy has gone into the kinetic energy of the scattered electron.

Express the energy change as a fraction of the initial photon energy.

The fractional energy change is the energy change $E_f - E_i$ divided by the initial energy E_i:

$$\text{fractional energy change} = \frac{E_f - E_i}{E_i}$$

$$= \frac{1.34 \times 10^{-14}\,\text{J} - 2.00 \times 10^{-14}\,\text{J}}{2.00 \times 10^{-14}\,\text{J}}$$

$$= -0.328 = -32.8\%$$

Reflect

An initial photon of this high energy and short wavelength can lose as much as 32.8% (nearly one-third) of its initial energy when it undergoes Compton scattering.

Example 26-2 suggests why X rays are useful in cancer radiation therapy. If an X ray photon strikes an electron in a water molecule within a cancerous cell, the photon can scatter and transfer a substantial amount of energy to the electron. This energy is great enough that the electron escapes from the molecule, leaving the water molecule in an ionized state. These ionized water molecules damage the DNA of the cancerous cell and cause cell death.

? Got the Concept? 26-2 Compton Scattering

Suppose a photon has a wavelength equal to the Compton wavelength λ_C. If this photon collides with an electron, and the photon is scattered through an angle of 90°, what will be the wavelength of the photon after the collision? (a) zero; (b) $\lambda_C/2$; (c) λ_C; (d) $2\lambda_C$; (e) $3\lambda_C$.

Take-Home Message for Section 26-3

✔ A photon has zero mass but does have momentum. The magnitude of the momentum is proportional to the photon energy and inversely proportional to the wavelength.

✔ In Compton scattering, a photon scatters from an electron. The photon loses energy and momentum, and these are transferred to the electron. The change in wavelength of the photon depends on the angle through which it is scattered.

26-4 Matter, like light, has aspects of both waves and particles

We have seen that light, previously thought to be a wave, is actually a stream of particles. But these particles (photons) have a wave aspect: Associated with them is a frequency f, which determines the photon energy $E = hf$, and a wavelength λ, which determines the photon momentum $p = h/\lambda$. So light has a dual nature, with attributes of both wave and particle. Is it possible that the same could be true for ordinary matter, which we know is made of particles such as electrons, protons, and neutrons? Could these particles also have a wave aspect?

In 1924, French graduate student Louis de Broglie (pronounced "de broy'") proposed precisely that idea. In particular, he suggested that the relationship $p = h/\lambda$ between momentum p and wavelength λ that applies to photons should also apply to particles such as electrons (**Figure 26-9**). The wavelength of a particle is called its **de Broglie wavelength**:

A particle has a de Broglie wavelength...   ...equal to Planck's constant...

$$\lambda = \frac{h}{p}$$

...divided by the momentum of the particle. The greater the momentum, the shorter the de Broglie wavelength.

de Broglie wavelength
(26-10)

How large should we expect the wavelength of an electron to be? Let's examine the case of an electron of charge $q = -e$ that gains its momentum by moving through a potential difference of V, so the electron starts at position a where the potential is zero and moves to a position b where the potential has a positive value V. (You may want to review the discussion of electric potential in Section 17-3.) The electron potential energy then changes from $U_a = qV_a = (-e)(0)$ to $U_b = qV_b = (-e)V = -eV$. The change in electric potential energy is

$$\Delta U = U_b - U_a = (-eV) - 0 = -eV < 0$$

Figure 26-9 **Wave-particle duality**
Matter has both particle aspects
(speed and momentum) and wave
aspects (wavelength).

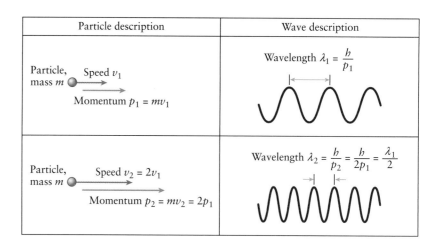

The de Broglie
wavelength is inversely
proportional to the
momentum. If the momentum is
doubled, the wavelength
decreases by 1/2.

The electric potential energy decreases by an amount eV. Mechanical energy is conserved if the only force acting on the electron is the (conservative) electric force, so the decrease in electric potential energy equals the gain in kinetic energy of the electron:

(26-11)
$$\Delta K = -\Delta U = -(-eV) = +eV > 0$$

If the electron starts at rest, its initial kinetic energy is zero and its final kinetic energy is $K = (1/2)mv^2$, so $\Delta K = (1/2)mv^2 - 0 = (1/2)mv^2$. (We're assuming that the electron is moving at a speed much slower than the speed of light c, so we don't have to use the Einsteinian expression for kinetic energy.) Then, from Equation 26-11, the final kinetic energy of the electron is

$$K = \frac{1}{2}mv^2 = eV$$

Solve for the final speed of the electron:

(26-12)
$$v^2 = \frac{2K}{m} = \frac{2eV}{m} \quad \text{so} \quad v = \sqrt{v^2} = \sqrt{\frac{2eV}{m}}$$

The final momentum of the electron is $p = mv$, and its final wavelength is $\lambda = h/p = h/mv$ from Equation 26-10. If we substitute v from Equation 26-12 into the formula for the wavelength of the electron, we get

(26-13)
$$\lambda = \frac{h}{mv} = \frac{h}{m}\sqrt{\frac{m}{2eV}} = \frac{h}{\sqrt{2meV}}$$

Suppose that the electron is accelerated through a potential difference $V = 50.0$ V. Substituting this into Equation 26-13 along with the values of Planck's constant h, the electron mass m, and the magnitude e of the electron charge, we get

$$\lambda = \frac{6.63 \times 10^{-34} \, \text{J} \cdot \text{s}}{\sqrt{2(9.11 \times 10^{-31} \, \text{kg})(1.60 \times 10^{-19} \, \text{C})(50.0 \, \text{V})}}$$
$$= 1.74 \times 10^{-10} \, \text{m} = 0.174 \, \text{nm}$$

To see how to measure such a short electron wavelength, note that a photon with this wavelength is in the X ray region of the electromagnetic spectrum (see Figure 22-2). It was known in the 1920s that X rays show interference effects when they reflect from adjacent atoms in a crystal: At certain angles, waves that reflect from one atom will interfere constructively (so the reflected intensity is high) with those that reflect from neighboring atoms, while at other angles they interfere destructively (so the reflected intensity is near zero). This can happen because the spacing between adjacent atoms in a crystal is around 0.1 nm, comparable to the wavelength of the X rays. So if electrons have a wave aspect, we expect that a beam of electrons that have been accelerated from rest through 50.0 V should display the same kind of interference effects as a beam of X rays.

In 1927 the American physicists Clinton Davisson and Lester Germer performed precisely this kind of experiment using a beam of electrons directed at a target of crystalline nickel. They found that the intensity of reflected electrons was greater for certain angles, just as for X rays. What's more, the angles at which this maximum intensity occurred were precisely those expected if the wavelength of electrons was given by the de Broglie relation, Equation 26-10. This groundbreaking result was quickly confirmed in experiments carried out by the British physicist G. P. Thomson. These results resoundingly confirmed de Broglie's remarkable hypothesis and showed that matter does indeed have a wave aspect. (The 1927 Nobel Prize in Physics went to de Broglie; the 1937 prize was shared by Davisson and Thomson.) The dual character of *both* light and matter, which have both wave and particle characteristics, is called **wave-particle duality.**

Wave-particle duality is a surprising and counterintuitive fact about the nature of our universe. It is also of tremendous practical use. One important application that has revolutionized biology is the *electron microscope.* A major limitation of ordinary microscopes is that the smallest detail that can be resolved is about the size of a wavelength of visible light (about 380 to 750 nm). This makes microscopes useless for seeing details of the structure of viruses, for example, which range in size from 5 to 300 nm. But as our above example of the 50.0-V electron shows, the wavelength of electron waves can be a fraction of a nanometer. So images made with an *electron microscope* can reveal details that are forever hidden from an ordinary visible-light microscope. Figure 26-10 is an electron microscope image of an influenza virus, in which details smaller than a nanometer across can be seen.

Because the de Broglie wavelength is inversely proportional to momentum, the wavelength is very small for objects with large momentum. That's why we don't see wave effects for objects in our environment that are large enough to see with the unaided eye: The wavelengths of such objects are infinitesimal. As an example, a dust mote (such as you might see floating in the air when a shaft of sunlight comes in the window) has mass of about 8×10^{-10} kg. If it drifts at a speed of 1 mm/s = 10^{-3} m/s, its momentum is

$$p = mv = (8 \times 10^{-10}\,\text{kg})(10^{-3}\,\text{m/s}) = 8 \times 10^{-13}\,\text{kg} \cdot \text{m/s}$$

and its de Broglie wavelength is

$$\lambda = \frac{h}{p} = \frac{6.63 \times 10^{-34}\,\text{J} \cdot \text{s}}{8 \times 10^{-13}\,\text{kg} \cdot \text{m/s}} = 8 \times 10^{-22}\,\text{m}$$

That's about 10^{-6} of the diameter of a proton! It's impossible to see wave effects from a wave with such a tiny wavelength: To see diffraction of such a wave, we would have to create a slit whose width is much smaller than the width of a single proton. Wave effects are even smaller for larger objects (greater mass *m*) moving faster (greater speed *v*). So the wave aspect of matter is generally only noticeable on the atomic or subatomic scale. Particles such as electrons exhibit noticeable wave properties; objects such as dust motes, baseballs, and humans do not.

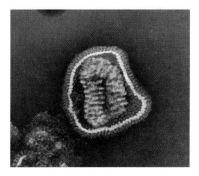

Figure 26-10 An electron micrograph When a beam of low-energy electrons is shot through a thin slice of a specimen in a transmission electron microscope (TEM), the pattern formed by the diffracted electrons forms an image. A TEM captured this (false-color) image of an influenza virus particle, which is only about 100 nm in diameter. Because the wavelengths of low-energy electrons are so much shorter than those of light, a TEM is capable of significantly better resolution than light microscopes (better than 0.005 nm for a TEM, compared to about 0.2 μm with the most powerful optical microscopes).

Example 26-3 Finding the Wavelength of a Room-Temperature Neutron

A nuclear reactor emits *thermal neutrons.* These are neutrons that behave as though they were particles of an ideal gas at Kelvin temperature *T*. We learned in Section 14-3 that the average kinetic energy of a particle in an ideal gas at temperature *T* is $(3/2)kT$, where $k = 1.381 \times 10^{-23}$ J/K is the Boltzmann constant. Calculate the de Broglie wavelength of an average neutron at 293 K (room temperature). The mass of a neutron is 1.67×10^{-27} kg.

Set Up

We'll first calculate the kinetic energy of an average neutron using Equation 14-13, which we learned in Section 14-3. From this we can find the speed and magnitude of momentum of an average neutron. Equation 26-10 will then allow us to calculate the de Broglie wavelength of such a neutron.

de Broglie wavelength:

$$\lambda = \frac{h}{p} \qquad (26\text{-}10)$$

Temperature and average translational kinetic energy of an ideal gas particle:

$$K_{\text{translational, average}} = \frac{1}{2}m(v^2)_{\text{average}} = \frac{3}{2}kT \quad (14\text{-}13)$$

Solve

Calculate the translational kinetic energy, speed, and momentum of an average neutron at $T = 293$ K.

From Equation 14-13,

$$K_{\text{translational, average}} = \frac{3}{2}kT = \frac{3}{2}(1.381 \times 10^{-23}\,\text{J/K})(293\,\text{K})$$
$$= 6.07 \times 10^{-21}\,\text{J}$$

Calculate the speed of a neutron with this kinetic energy:

$$K_{\text{translational, average}} = \frac{1}{2}mv^2 \text{ so}$$

$$v = \sqrt{\frac{2K_{\text{translational, average}}}{m}} = \sqrt{\frac{2(6.07 \times 10^{-21}\,\text{J})}{1.67 \times 10^{-27}\,\text{kg}}}$$
$$= 2.70 \times 10^3\,\text{m/s}$$

The magnitude of momentum of a neutron with this speed is

$$p = mv = (1.67 \times 10^{-27}\,\text{kg})(2.70 \times 10^3\,\text{m/s})$$
$$= 4.50 \times 10^{-24}\,\text{kg}\cdot\text{m/s}$$

Calculate the de Broglie wavelength of such a neutron.

From Equation 26-10,

$$\lambda = \frac{h}{p} = \frac{6.63 \times 10^{-34}\,\text{J}\cdot\text{s}}{4.50 \times 10^{-24}\,\text{kg}\cdot\text{m/s}} = 1.47 \times 10^{-10}\,\text{m} = 0.147\,\text{nm}$$

(Recall that $1\,\text{J} = 1\,\text{kg}\cdot\text{m}^2/\text{s}^2$ and $1\,\text{nm} = 10^{-9}\,\text{m}$.)

Reflect

The de Broglie wavelength of a thermal neutron is about 0.147 nm, a distance that is typical of the size of atoms and of the spacing between atoms within a molecule. For this reason, thermal neutrons are useful for studying the structure of complex molecules such as proteins: When the neutrons scatter from a protein molecule, they diffract and produce a diffraction pattern that is characteristic of the particular arrangement of atoms in the molecule. X rays can have the same wavelength, but they interact with the charges within atoms and so scatter only weakly from the relatively small atoms (with a small amount of internal charge) such as hydrogen, carbon, nitrogen, and oxygen found in proteins. Neutrons, by contrast, are electrically neutral and actually scatter more strongly from smaller atoms than from larger ones. This makes neutrons superior to X rays for studies of protein structure.

? Got the Concept? 26-3   Ranking de Broglie Wavelengths

Rank the following objects in order of their de Broglie wavelength, from longest to shortest. (a) A proton moving at 2.00×10^3 m/s; (b) a proton moving at 4.00×10^3 m/s; (c) an electron moving at 2.00×10^3 m/s; (d) an electron moving at 4.00×10^3 m/s.

Take-Home Message for Section 26-4

✔ Particles can exhibit wave properties such as diffraction.

✔ The wavelength of a particle is inversely proportional to its momentum. Hence wave effects are only noticeable for very small particles such as electrons, for which the momentum is very small.

26-5 The spectra of light emitted and absorbed by atoms show that atomic energies are quantized

We have seen that the late nineteenth and early twentieth centuries were years of tremendous change in physics. Studying the photoelectric effect and blackbody radiation led to the revolutionary concept that light has particle aspects, and de Broglie introduced the no less revolutionary idea that matter has wave aspects. During this same time, a key set of experiments radically transformed our understanding of the nature of atoms.

The Nuclear Atom

The early Greeks introduced the idea of the atom, a unit of matter so small that it could not be subdivided. (The word *atom* is derived from the Greek term for indivisible.) In 1897 the British physicist J. J. Thomson discovered that atoms are not in fact indivisible, but have an internal structure: All atoms contain negatively charged particles (electrons) that can be removed from the atom. It was known that atoms are electrically neutral, so there must also be positively charged material inside an atom. But what form does this positive charge take?

Thomson proposed that most of the mass of the atom is in the form of electrons and that the positively charged material is a low-density sort of jelly in which the electrons are embedded. This model is sometimes called the "plum pudding model," since Thomson envisioned electrons scattered throughout the positive charge much like raisins in the traditional English dessert. (If you're not familiar with plum pudding, think of electrons as pieces of fruit embedded in a gelatin dessert or salad.) A crucial experiment that tested this model was carried out in 1909 at the University of Manchester in England by the New Zealand–born British chemist and physicist Ernest Rutherford with his colleagues Hans Geiger and Ernest Marsden. They fired subatomic particles called *alpha particles* (which were known to be positive helium ions, thousands of times more massive than an electron) at a very thin gold foil. Rutherford expected that if the gold atoms had the structure described in Thomson's model, the alpha particles would be only slightly deflected from their initial direction as a result of passing through the diffuse, positive "pudding" of the atoms. Instead, Rutherford was startled to discover that alpha particles were sometimes scattered at large angles with respect to the initial direction, occasionally leaving the gold foil directly *backward*. (In reflecting on this experiment, Rutherford later said, "It was quite the most incredible event that ever happened to me in my life. It was as incredible as if you fired a 15-in. shell [a large projectile fired from a military weapon] at a piece of tissue paper and it came back and hit you. On consideration, I realized that this scattering backwards must be the result of a single collision, and when I made calculations I saw that it was impossible to get anything of that order of magnitude unless you took a system in which the mass of the atom was concentrated in a minute nucleus."

To account for this, Rutherford proposed a model of the atom in which negatively charged electrons orbit a small, positively charged *nucleus* that contains nearly all of the atom's mass. In Rutherford's model, most of the volume of each atom is empty, so most alpha particles fired at the gold foil would experience only slight deflections as they passed through. But once in a while, about one time out of every 10,000, an alpha particle would approach a gold nucleus almost head on and be scattered at a large angle, sometimes directly backward.

Unlike planets orbiting the Sun, electrons orbiting an atomic nucleus fit into a well-defined organizational structure. We'll explore this structure from three perspectives and in historical order: first, the early clues that hinted at the structure, then the development of mathematical models that describe the structure, and finally (in the following section) the theory that explains it.

Figure 26-11 **The absorption spectrum of the Sun** The dark lines in the spectrum of sunlight indicate that certain wavelengths are absorbed when light from the solar interior passes through the Sun's atmosphere.

The Discovery of Atomic Spectra

In the early part of the nineteenth century, the English scientist William Hyde Wollaston noted the appearance of dark lines in the spectrum of visible light coming from the Sun (Figure 26-11). While conducting experiments on glass-making, a German optician-turned-physicist, Joseph von Fraunhofer, independently discovered these lines. These lines are always in the same locations within the spectrum. Some years later Gustav Kirchhoff, the same physicist we encountered in our study of electric circuits, was able to reproduce these same dark lines in the laboratory. He passed light from a lamp (made to simulate sunlight) through vapors created by heating sodium. The light from the lamp itself had a continuous spectrum like that of a blackbody, but the spectrum of light that had passed through the sodium vapor had two dark lines. Kirchhoff concluded that the dark lines result from certain specific colors of light being absorbed by the sodium vapor. What is more, these lines were at the same position as the closely spaced pair of lines in the yellow-orange region of the Sun's spectrum. (You can easily find these lines in Figure 26-11.) The same mechanism must therefore be happening with sunlight: The light coming from the solar interior has a continuous, blackbody-like spectrum, but certain wavelengths of that light are absorbed by atoms in the Sun's atmosphere. The Sun's atmosphere must contain sodium atoms identical to those in Kirchhoff's laboratory. In light of Kirchhoff's discovery, a spectrum like that shown in Figure 26-11 is called an **absorption spectrum**, and the dark lines are called **absorption lines**.

Scientists soon discovered that each element produces its own characteristic absorption lines when light passes through a vapor containing atoms of that element. Thus an absorption spectrum acts as a "fingerprint" of the chemical composition of the vapor that produced the absorption lines. (This is how we know the chemical composition of the Sun's atmosphere. It's also how we determine the chemical composition of the atmospheres of distant stars like those in the photograph that opens this chapter, and how we know that all stars have basically the same chemical makeup.) What was not understood was *why* atoms should selectively absorb only light of certain wavelengths, and why the absorbed wavelengths should be different for atoms of different elements.

An important clue about the mystery of absorption spectra came from studying the light *emitted* by atoms. Physicists of the nineteenth century discovered that light created by heating a vapor gives rise to an **emission spectrum**, a spectrum that consists only of specific emitted wavelengths. What is more, if the vapor contains atoms of a certain element, the wavelengths in the emission spectrum from those atoms (the **emission lines**) are precisely the same as the wavelengths in the absorption spectrum of that same element (Figure 26-12). To explore this, we'll concentrate on the absorption and

(a) The absorption spectrum produced by passing white light through a gas of hydrogen atoms

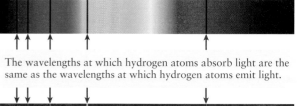

The wavelengths at which hydrogen atoms absorb light are the same as the wavelengths at which hydrogen atoms emit light.

(b) The emission spectrum produced by a heated gas of hydrogen atoms

Figure 26-12 **Absorption and emission spectra of atomic hydrogen** (a) When light passes through a gas, light of specific wavelengths is absorbed, forming dark lines. (b) When a gas is made to glow by passing an electric current through it, it emits only specific wavelengths of light.

- Atoms of each element absorb and emit light at wavelengths that are characteristic of that element.
- The characteristic wavelengths differ from one element to another.
- Hydrogen has the simplest arrangement of characteristic wavelengths of any element.

emission spectra of hydrogen, which has the simplest set of absorption and emission lines of any element. But the underlying physics applies to all elements.

Johann Balmer, a Swiss mathematician, made an analysis of the lines in the absorption and emission spectra of hydrogen. He devised a formula that both reproduced the wavelengths of lines that had been reported and correctly predicted the wavelengths of spectral lines of which Balmer was unaware. This was later extended by the Swedish physicist Johannes Rydberg. The Rydberg formula for the hydrogen spectral lines is

Wavelength of an absorption or emission line in the spectrum of atomic hydrogen

$$\frac{1}{\lambda} = R_H \left(\frac{1}{n^2} - \frac{1}{m^2} \right)$$

Rydberg constant = 1.09737×10^7 m^{-1}

n and m are integers: n can be 1, 2, 3, 4, ... and m can be any integer greater than n.

Rydberg formula for the spectral lines of hydrogen
(26-14)

The value of the constant R_H in Equation 26-14, called the *Rydberg constant*, is chosen to match the experimental data. To four significant figures, $R_H = 1.097 \times 10^7$ m^{-1}. As an example, the hydrogen absorption and emission lines shown in Figure 26-12 all correspond to $n = 2$ in Equation 26-14. The series of wavelengths for which $n = 2$ are called the *Balmer series*. For example, to get the wavelength of the red spectral line in Figure 26-12, set $n = 2$ and $m = 3$ in Equation 26-14:

$$\frac{1}{\lambda} = R_H \left(\frac{1}{2^2} - \frac{1}{3^2} \right) = (1.097 \times 10^7 \text{ m}^{-1}) \left(\frac{1}{4} - \frac{1}{9} \right) = 1.524 \times 10^6 \text{ m}^{-1}$$

Then take the reciprocal:

$$\lambda = \frac{1}{1.524 \times 10^6 \text{ m}^{-1}} = 6.563 \times 10^{-7} \text{ m} = 656.3 \text{ nm}$$

The spectral line to the left of this one in Figure 26-12 (in the blue-green part of the spectrum) corresponds to $n = 2$ and $m = 4$; you can show that for this spectral line, $\lambda = 486.2$ nm. The wavelengths with $n = 1$ are all in the ultraviolet and are called the *Lyman series*; the wavelengths with $n = 3$ are all in the infrared and are called the *Paschen series*.

What Balmer and Rydberg did not know was *why* the spectral hydrogen lines were given by this relatively simple formula. It was left to the Danish physicist Niels Bohr to provide the explanation.

Energy Quantization

Bohr realized that to fully understand the structure of the hydrogen atom, he had to be able to derive Balmer's formula using the laws of physics. He first made the rather wild assumption that the electron in a hydrogen atom can orbit the nucleus only in certain specific orbits. (This was a significant break with the ideas of Newton, in whose mechanics any orbit should be possible.) Figure 26-13 shows the four smallest of these **Bohr orbits**, labeled by the numbers $n = 1$, $n = 2$, $n = 3$, and so on.

Although confined to one of these allowed orbits while circling the nucleus, an electron can jump from one Bohr orbit to another. For an electron to do this, the hydrogen atom must gain or lose a specific amount of energy. The atom must absorb energy for the electron to go from an inner to an outer orbit; the atom must release energy for the electron to go from an outer to an inner orbit. As an example, Figure 26-14 shows an electron jumping between the $n = 2$ and $n = 3$ orbits of a hydrogen atom as the atom absorbs or emits a photon.

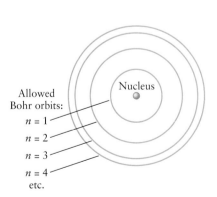

Allowed Bohr orbits:
$n = 1$
$n = 2$
$n = 3$
$n = 4$
etc.

Nucleus

Figure 26-13 Bohr orbits in the hydrogen atom In the model devised by Niels Bohr, electrons in the hydrogen atom are allowed to be in certain orbits only. (The radii of the orbits are not shown to scale.)

Figure 26-14 **The Bohr model explains absorption and emission spectra** When a photon is (a) absorbed or (b) emitted by a hydrogen atom, the electron makes a transition or jump between two allowed orbits. The photon energy equals the difference in energy between the upper and lower electron orbits.

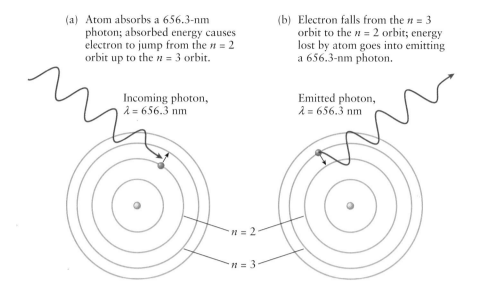

(a) Atom absorbs a 656.3-nm photon; absorbed energy causes electron to jump from the $n = 2$ orbit up to the $n = 3$ orbit.

Incoming photon, $\lambda = 656.3$ nm

(b) Electron falls from the $n = 3$ orbit to the $n = 2$ orbit; energy lost by atom goes into emitting a 656.3-nm photon.

Emitted photon, $\lambda = 656.3$ nm

$n = 2$

$n = 3$

When the electron jumps from one orbit to another, the energy of the photon that is emitted or absorbed equals the difference in energy between these two orbits. This energy difference, and hence the photon energy, is the same whether the jump is from a low orbit to a high orbit (Figure 26-14a) or from the high orbit back to the low one (Figure 26-14b). According to Einstein, if two photons have the same energy E, the relationship $E = hf$ (Equation 22-26) tells us that they must also have the same frequency f and hence the same wavelength $\lambda = c/f$. It follows that if an atom can emit photons of a given energy and wavelength, it can also absorb photons of precisely the same energy and wavelength. Thus, Bohr's picture explains Kirchhoff's observation that atoms emit and absorb the same wavelengths of light.

The Bohr picture also helps us visualize what happens to produce an emission spectrum. When a gas is heated, its atoms move around rapidly and can collide forcefully with each other. These energetic collisions excite the atoms' electrons into high orbits. The electrons then cascade back down to the innermost possible orbit, emitting photons whose energies are equal to the energy differences between different Bohr orbits. In this fashion, a hot gas produces an emission line spectrum with a variety of different wavelengths.

To produce an absorption spectrum, begin with a relatively cool gas, so that the electrons in most of the atoms are in inner, low-energy orbits. If a beam of light with a continuous spectrum is shone through the gas, most wavelengths will pass through undisturbed. Only those photons will be absorbed whose energies are just right to excite an electron to an allowed outer orbit. Hence, only certain wavelengths will be absorbed, and dark lines will appear in the spectrum at those wavelengths.

As in Figure 26-14, the energy of the photon that is absorbed or emitted in a jump between orbits must be equal to the *difference* between the energy of the atom with the electron in the larger-radius, higher-energy orbit and the energy of the atom with the electron in the smaller-radius, lower-energy orbit. Bohr concluded that the numbers n and m in the Rydberg formula correspond to the numbers of the orbits between which an electron jumps as it absorbs or emits a photon. The value of n is the number of the lower orbit, and the value of m is the number of the upper orbit. For example, the jump shown in Figure 26-14 corresponds to $n = 2$ and $m = 3$.

We can better understand Bohr's idea by combining the Rydberg formula, Equation 26-14, with the expressions $E = hf$ and $f = c/\lambda$ for a photon. Together these latter two expressions say that the energy of a photon of wavelength λ is $E = hc/\lambda$. So if we multiply Equation 26-14 by hc, we get an expression for the energy of a photon absorbed or emitted by a hydrogen atom:

(26-15)
$$E_{\text{photon}} = \frac{hc}{\lambda} = hcR_{\text{H}}\left(\frac{1}{n^2} - \frac{1}{m^2}\right) = \frac{hcR_{\text{H}}}{n^2} - \frac{hcR_{\text{H}}}{m^2} = \left(-\frac{hcR_{\text{H}}}{m^2}\right) - \left(-\frac{hcR_{\text{H}}}{n^2}\right)$$

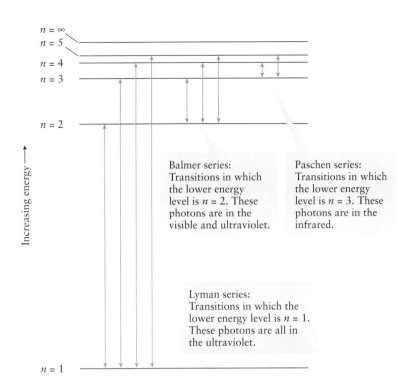

Figure 26-15 **Hydrogen energy levels** This figure shows some of the lower-lying energy levels of the hydrogen atom and possible transitions between those levels. Note that the energy difference is greatest between the $n = 1$ and $n = 2$ levels, less between the $n = 2$ and $n = 3$ levels, and even less between the $n = 3$ and $n = 4$ levels.

If we say that a hydrogen atom has energy $E_{\text{atom},n} = -hcR_{\text{H}}/n^2$ when the electron is in the nth orbit and has energy $E_{\text{atom},m} = -hcR_{\text{H}}/m^2$ when the electron is in the mth orbit, where m is greater than n, then we can rewrite Equation 26-15 as

$$E_{\text{photon}} = E_{\text{atom},m} - E_{\text{atom},n}$$
(26-16)

Equation 26-16 uses the idea that the energy of a hydrogen atom is **quantized**: That is, the energy can only have certain values. These energies are given by

$$E_{\text{atom},n} = -\frac{hcR_{\text{H}}}{n^2} \quad \text{where } n = 1, 2, 3, 4, \dots$$
(26-17)

Note that the energy of the atom is negative, and greater values of n correspond to energies that are less negative (that is, closer to zero). This agrees with the idea that the larger the orbit and the greater the value of n for that orbit, the higher the energy.

Each quantized value of the energy is called an **energy level**. Figure 26-15 shows several of the energy levels of the hydrogen atom, along with vertical arrows that show the energy of the photon that must be absorbed or emitted in a jump or transition between levels. The transitions that correspond to the Lyman series involve a photon with a very large amount of energy, so these photons have a high frequency and short wavelength: They are all in the ultraviolet part of the spectrum. By contrast, the transitions that correspond to the Paschen series involve a photon with a very small amount of energy, which is why these photons have a low frequency and long wavelength and are in the infrared part of the spectrum. The Balmer series is intermediate between these two; the wavelengths are either in the visible range (380 to 750 nm) or the ultraviolet range.

Elements other than hydrogen also absorb and emit light at specific wavelengths, although those wavelengths do not follow a simple mathematical pattern like the characteristic wavelengths of hydrogen given by Equation 26-14. The conclusion is that there are quantized energy levels for atoms of other elements, but the arrangement of energy levels is more complex than for hydrogen. In the following section we'll see how Niels Bohr justified the quantization of energy for the relatively simple case of the hydrogen atom, and how he was able to reproduce Equation 26-17 for the energy levels. We'll then use these ideas to gain insight into the structure of other atoms.

! Watch Out! Energy quantization is not just an obscure effect in atomic physics.

Discrete energy levels play an important role in modern technology. A *laser* is a device that emits an intense beam of light of a very specific wavelength. This is possible because the laser contains a material with two distinct energy levels. Light of the laser's characteristic wavelength is emitted when transitions take place from the upper level to the lower one. (What makes the laser unique is that these transitions occur coherently rather than at random. Energy is added to the material to pump its molecules into excited states. The excited molecules naturally want to transition to a lower state, but if left on their own, would do so at random times. However, when a photon of energy equal to the difference between two states is sent into the material, it can *stimulate* this transition and cause a second photon of that same energy to be released. This photon can stimulate further emission, and so the number of emitted photons increases. The emitted photons are all in phase and all travel in the same general direction. The net result is an intense beam.) If materials did not have quantized energy levels, the lasers found in Blu-ray and DVD players (which scan the video information encoded on the disc) could not exist. Fluorescent light bulbs also depend on energy quantization. The material inside the bulb emits ultraviolet photons when an electric current passes through it. These photons are absorbed by the white coating on the inner surface of the bulb. Since ultraviolet photons are very energetic, this excites the material of the coating to very high energy levels. The coating then drops down to its initial energy level in a series of small steps. (It's like taking a big leap to the top of a staircase, then coming carefully down the staircase one step at a time.) Each small step between closely spaced energy levels emits a low-energy, visible-light photon. There are so many such energy levels, with a variety of spacing between them, that the net result is that a mixture of almost all visible colors—that is, white light—is emitted from the bulb. These are just two of the many applications of energy quantization.

Example 26-4 Photon Possibilities

A collection of hydrogen atoms is excited to the $n = 3$ energy level. What are the possible wavelengths that these atoms could emit as they return to the lowest-energy ($n = 1$) level?

Set Up

There are two routes that an atom can take from the $n = 3$ level to the $n = 1$ level. One, it could drop down to the $n = 1$ level in a single step by emitting a single photon whose energy is equal to the difference between the energies of the $n = 3$ and $n = 1$ levels. The wavelength of this photon is given by Equation 26-14 with $n = 1$ and $m = 3$. Two, the atom could first drop to the $n = 2$ level by emitting a photon (with a wavelength given by Equation 26-14 with $n = 2$ and $m = 3$), then emit a second photon as it drops from the $n = 2$ level to the $n = 1$ level (with a wavelength given by Equation 26-14 with $n = 1$ and $m = 2$). So three different wavelengths can be emitted by the excited atoms. We'll calculate each of these in turn.

Rydberg formula for the spectral lines of hydrogen:

$$\frac{1}{\lambda} = R_H\left(\frac{1}{n^2} - \frac{1}{m^2}\right) \qquad (26\text{-}14)$$

Solve

Use Equation 26-14 to calculate each of the three possible wavelengths.

For the $n = 3$ to $n = 1$ transition,

$$\frac{1}{\lambda} = R_H\left(\frac{1}{1^2} - \frac{1}{3^2}\right) = (1.097 \times 10^7\,\text{m}^{-1})\left(1 - \frac{1}{9}\right)$$

$$= 9.751 \times 10^6\,\text{m}^{-1}$$

$$\lambda = \frac{1}{9.751 \times 10^6\,\text{m}^{-1}} = 1.026 \times 10^{-7}\,\text{m} = 102.6\,\text{nm}$$

This is an ultraviolet wavelength.

For the $n = 3$ to $n = 2$ transition,

$$\frac{1}{\lambda} = R_H\left(\frac{1}{2^2} - \frac{1}{3^2}\right) = (1.097 \times 10^7 \text{ m}^{-1})\left(\frac{1}{4} - \frac{1}{9}\right)$$

$$= 1.524 \times 10^6 \text{ m}^{-1}$$

$$\lambda = \frac{1}{1.524 \times 10^6 \text{ m}^{-1}} = 6.563 \times 10^{-7} \text{ m} = 656.3 \text{ nm}$$

This is a visible wavelength (in the red part of the spectrum).

For the $n = 2$ to $n = 1$ transition,

$$\frac{1}{\lambda} = R_H\left(\frac{1}{1^2} - \frac{1}{2^2}\right) = (1.097 \times 10^7 \text{ m}^{-1})\left(1 - \frac{1}{4}\right)$$

$$= 8.228 \times 10^6 \text{ m}^{-1}$$

$$\lambda = \frac{1}{8.228 \times 10^6 \text{ m}^{-1}} = 1.215 \times 10^{-7} \text{ m} = 121.5 \text{ nm}$$

This is another ultraviolet wavelength.

Reflect

Comparing with Figure 26-15 shows that the 656.3-nm wavelength represents an emission line of the Balmer series, while the 102.6-nm and 121.5-nm wavelengths represent emission lines of the Lyman series.

? Got the Concept? 26-4 Ranking Hydrogen Transitions

Rank the following transitions between hydrogen energy levels in terms of the energy of the photon involved, from highest to lowest energy. (a) An atom drops from the $n = 4$ level to the $n = 2$ level. (b) An atom rises from the $n = 3$ level to the $n = 5$ level. (c) An atom drops from the $n = 3$ level to the $n = 1$ level. (d) An atom drops from the $n = 4$ to the $n = 3$ level.

Take-Home Message for Section 26-5

✔ Atoms are composed of electrons orbiting a positively charged nucleus that has most of the mass of the atom.

✔ In the Bohr model, an electron's orbits around the nucleus can have only certain well-defined energies. Thus the energy of the atom is quantized and can have only certain values. The atom cannot have energies intermediate between those values.

✔ Electrons that make a transition from one allowed orbit to another either radiate or absorb a photon of a well-defined energy. This gives rise to the lines in emission and absorption spectra.

26-6 Models by Bohr and Schrödinger give insight into the intriguing structure of the atom

The force that pulls the Moon toward Earth is directed toward Earth's center, yet the Moon does not fall into Earth. In the same way, the negatively charged electron in an atom experiences a Coulomb force that points directly toward the positively charged protons at the center of the atom. Why does the electron orbit rather than fall in to the proton? You likely have an intuitive answer that both the Moon and the electron *orbit* rather than fall in. (In a real sense each *is* falling in, but it is always missing the target!) Niels Bohr based his model of the hydrogen atom, for which he received the Nobel Prize in Physics in 1922, on the physics of atomic orbits. The **Bohr model** provides a theoretical foundation for the physics of atomic spectra that we encountered in the previous section.

The Bohr Model

Bohr made two fundamental assumptions in describing the orbit of an electron around a positive atomic nucleus. First, he modeled the orbit as uniform circular motion, that is, as an electron moving at constant speed in a circular path. This is not exactly correct, but more than satisfactory to provide a broad and correct understanding of the atom. Bohr's second assumption was that only specific values are allowed for the angular momentum associated with each orbiting electron. We will return to this second assumption shortly.

Bohr considered a single electron orbiting a nucleus of charge $+Ze$, where Z, the *atomic number of an atom*, is the number of protons in the nucleus. The orbiting electron experiences only one force, the Coulomb attraction between it and the protons in the atomic nucleus. The magnitude of the Coulomb force on an electron orbiting in a circle of radius r is, from Equation 16-1,

$$F = \frac{k(Ze)(e)}{r^2} = \frac{kZe^2}{r^2}$$

Here $k = 8.99 \times 10^9 \, \text{N} \cdot \text{m}^2/\text{C}^2$ is the Coulomb constant. Newton's second law (Equation 4-2) requires that this force equal the mass of the electron m_e multiplied by the acceleration it experiences. For an object in uniform circular motion at speed v, the magnitude of the acceleration is, from Equation 3-17,

$$a = \frac{v^2}{r}$$

Combining the above two equations into Newton's second law gives

$$\frac{kZe^2}{r^2} = \frac{m_e v^2}{r}$$

If we multiply both sides of this equation by r, we get

(26-18)
$$\frac{kZe^2}{r} = m_e v^2$$

Let's set Equation 26-18 aside for a moment and examine Bohr's second assumption—a requirement that only specific values are allowed for the angular momentum associated with the electron. Bohr recognized that the dimensions of Planck's constant h are those of angular momentum, so he constrained the electron's angular momentum to be only multiples of h. Specifically, this requirement is

(26-19)
$$L = n\left(\frac{h}{2\pi}\right) = n\hbar$$

(orbital angular momentum in the Bohr model)

where L is the electron's orbital angular momentum and n is any integer starting from 1. The constant $\hbar$ (pronounced "h bar") is defined to be h divided by 2π.

To relate angular momentum to Equation 26-18, we express L in terms of the mass of the electron m_e, its speed v, and its distance from the center of the atom r using Equation 8-23:

Magnitude of the angular momentum of a particle

Component of the particle's linear momentum perpendicular to the vector $\vec{r}$ from rotation axis to particle

Magnitude of the angular momentum of a particle
(8-23)

$$L = rp_\perp = rp \sin\phi$$

Distance from the rotation axis to the particle

Since the electron moves in a circle, the vector from the rotation axis to the particle always has the same radius r and is always perpendicular to the momentum vector. So $\phi = 90°$, $\sin\phi = 1$, and

(26-20)
$$L = rm_e v$$

We can now rewrite Equation 26-18 in terms of the angular momentum L by multiplying the right-hand side by 1 in the form of $(r^2/r^2)(m_e/m_e)$:

$$\frac{kZe^2}{r} = \frac{r^2 m_e^2 v^2}{m_e r^2} = \frac{L^2}{m_e r^2}$$

Bohr's requirement that the electron's angular momentum is an integer multiple of $\hbar$ then gives

$$\frac{kZe^2}{r} = \frac{(n\hbar)^2}{m_e r^2}$$

or

$$r_n = \frac{n^2 \hbar^2}{m_e k Z e^2} \quad \text{where } n = 1, 2, 3, \ldots \tag{26-21}$$

(orbital radii in the Bohr model)

We add the subscript "n" to the variable r to indicate that the radius can take on only specific values and that the allowed values of radius depend on n. Because the values of r_n are proportional to the square of an integer, the orbital radii of electrons in an atom are quantized.

Notice that both n and Z in Equation 26-21 are dimensionless. Because r_n is a distance, all of the other terms on the right-hand side, taken as they appear in the equation, must have dimensions of distance as well. This distance, usually written as a_0 and called the Bohr radius, is

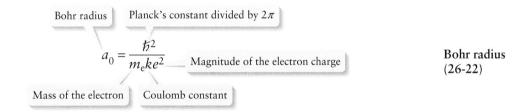

Bohr radius
(26-22)

Using the best measured values for $\hbar$, m_e, k, and e, we find that the value of the Bohr radius a_0 is approximately

$$a_0 = 0.529 \times 10^{-10} \text{ m} = 0.0529 \text{ nm}$$

In terms of a_0, the quantized radii of the electron orbits (Equation 26-21) are

Orbital radii in the Bohr model
(26-23)

The integer n identifies the orbit, where the $n = 1$ orbit is the closest to the nucleus. Because every element is distinguished by the number of protons it carries, the atomic number Z specifies a particular element. So, for example, setting Z equal to 1 gives the radii of the electron orbits in a hydrogen atom, and setting Z equal to 1 and n equal to 1 gives the radius of the first electron orbit in hydrogen. This is the normal state, usually referred to as the ground state, of hydrogen. Moreover, notice that the radius of the ground state orbit of hydrogen is the Bohr radius. In other words, the radius of a typical hydrogen atom is about 0.05 nm. We can also conclude from Equation 26-23 together with the value of the Bohr radius that, in general, atoms are no more than a few nm in radius. The Bohr model sets the scale for atomic sizes.

What is the energy of an electron orbiting an atomic nucleus according to the Bohr model? This is the sum of its kinetic energy and its electric potential energy. For two point charges q_1 and q_2, $U_{electric} = kq_1q_2/r$ (Equation 17-3), so

(26-24)
$$E = \frac{1}{2}m_e v^2 + \frac{k(-e)(Ze)}{r_n}$$

We can write the kinetic energy term in terms of angular momentum in a way that is similar to our approach in developing the relationship for r_n. Using Equation 26-20 shows that the kinetic energy term becomes

$$\frac{1}{2}m_e v^2 = \frac{1}{2}\frac{L^2}{m_e r_n^2} = \frac{n^2\hbar^2}{2m_e r_n^2}$$

so the total electron energy from Equation 26-24 is

$$E = \frac{n^2\hbar^2}{2m_e r_n^2} - \frac{kZe^2}{r_n}$$

Substituting Equation 26-23 for r_n and Equation 26-22 for a_0 gives the energy in terms of only the physical constants and the counting integer n:

$$E = \frac{n^2\hbar^2}{2m_e}\left(\frac{m_e kZe^2}{n^2\hbar^2}\right)^2 - kZe^2\frac{m_e kZe^2}{n^2\hbar^2}$$

Simplifying this is straightforward when you notice that the numerator of both terms is $m_e(kZe^2)^2$ and that both terms have $n^2\hbar^2$ in the denominator:

$$E = \frac{m_e(kZe^2)^2}{2n^2\hbar^2} - \frac{m_e(kZe^2)^2}{n^2\hbar^2}$$

or

(26-25)
$$E_n = -\frac{m_e(kZe^2)^2}{2n^2\hbar^2}, \quad n = 1, 2, 3, \ldots$$

(electron energies in the Bohr model)

We add the subscript n to the variable E to indicate that the orbital energy of the electron can take on only specific values, and that the allowed values of energy depend on n. Because the values of E_n are proportional to $1/n^2$, the orbital energy of electrons in an atom is quantized. This is in agreement with our conclusion in the previous section that for atomic spectral lines to occur only at specific, and always the same, wavelengths, electrons must orbit the hydrogen nucleus with specific, well-defined energies. Note also that the energy is equal to a negative constant divided by n^2, exactly in accordance with Equation 26-17. (We'll see below that the numerical value of the constant is the same in Equation 26-25 as in Equation 26-17.)

The value of energy of an electron in orbit around an atomic nucleus is negative, but closer and closer to zero for increasing values of n. In other words, the lowest energy orbit is the one closest to the nucleus, as we would expect. You might imagine that the negative values of energy suggest that the electron is at the bottom of a hole, and its orbital energy is insufficient to allow it to escape. That the electron energy is taken as negative emphasizes that the electron is bound to the nucleus, and that energy must be supplied in order to either move the electron to a higher orbit or to break the electron free from the nucleus altogether.

Because E_n is an energy, and because both n and Z are dimensionless, the combination of other quantities on the right-hand side of Equation 26-25 must have dimensions of energy as well. This energy, usually written as E_0, is called the Rydberg energy:

Coulomb constant   Magnitude of the electron charge

Mass of the electron

$$E_0 = \frac{m_e(ke^2)^2}{2\hbar^2} = 2.18 \times 10^{-18} \text{ J} = 13.6 \text{ eV}$$

Rydberg energy
(26-26)

Rydberg energy

Planck's constant divided by 2π

In Equation 26-26 we've given the value of the Rydberg energy in joules and in electron volts (eV). The energy that an electron acquires when it moves through a potential difference of 1 V is 1 eV, or approximately 1.602×10^{-19} J. To give you an idea of the amount of energy 1 eV represents, a photon of visible light carries between about 1.5 and 3 eV.

Using Equation 26-26, we can write Equation 26-25 for the quantized electron orbital energy in terms of the Rydberg energy E_0:

Electron energy for the nth allowed orbit in a single-electron atom   Atomic number of the atom

$$E_n = -\frac{Z^2}{n^2} E_0$$

Electron energies in the Bohr model
(26-27)

$n = 1, 2, 3, ...$ Rydberg energy = 13.6 eV

Again, the integer n identifies the orbit, and the atomic number Z specifies a particular element. Setting Z equal to 1 and n equal to 1 therefore tells us that the energy of the ground state of hydrogen is -13.6 eV. We can also conclude from Equation 26-27 together with the value of the Rydberg energy that, in general, the energy of electrons in orbit around an atomic nucleus is between about -10 eV and, for the largest elements (for which Z is about 100), -10^5 eV. The Bohr model sets the scale for atomic electron energies.

What does the Bohr model predict for the atomic spectral lines of hydrogen? Every line results from a transition between two electron orbits; that is, the energy of the emitted photon is the energy difference ΔE between two allowed orbits (see Equation 26-16). Since the energy of a photon is hf and its frequency $f = c/\lambda$, the wavelength λ of the photon is then

$$\lambda = \frac{hc}{\Delta E}$$

Recalling that the Rydberg formula is written in terms of the reciprocal of the photon wavelength, we write

$$\frac{1}{\lambda} = \frac{\Delta E}{hc}$$

(26-28)

Let's consider the transition of an electron from a higher orbit m down to a lower orbit n. We can determine the energy difference between these two orbits by applying Equation 26-27. Atomic number Z equals 1 for hydrogen, so

$$\Delta E = -\frac{1}{m^2}E_0 - \left(-\frac{1}{n^2}\right)E_0$$

or

$$\Delta E = \left(\frac{1}{n^2} - \frac{1}{m^2}\right)E_0$$

Substituting this into Equation 26-28 yields

$$\frac{1}{\lambda} = \frac{E_0}{hc}\left(\frac{1}{n^2} - \frac{1}{m^2}\right)$$

Compare this to the Rydberg formula, Equation 26-14. It has exactly the same form!
   How does the value of E_0/hc compare to the value of R_H, which equals approximately 1.10×10^7 m^{-1}? In SI units

$$\frac{E_0}{hc} = \frac{2.18 \times 10^{-18}\,\text{J}}{(6.63 \times 10^{-34}\,\text{J}\cdot\text{s})(3.00 \times 10^8\,\text{m/s})} = 1.10 \times 10^7\,\text{m}^{-1}$$

The ratio E_0/hc equals R_H, so $E_0 = hcR_H$. That's just what we expect if we compare Equation 26-17 (in which we deduced the energies from the Rydberg equation) and Equation 26-27 (in which we derived the energies from the Bohr assumptions). We conclude that the Rydberg formula for the spectral lines of hydrogen is entirely consistent with the Bohr model.

Example 26-5 Lowest Energy Level of a Lithium Ion

Find the (a) radius, (b) energy, and (c) speed of an electron in the lowest energy level of the doubly ionized lithium ion. An atom of lithium $(Z = 3)$ has three electrons, so this ion has just one electron.

Set Up

For this ion, $Z = 3$; for the lowest energy level, $n = 1$. We'll use Equation 26-23 to calculate the radius of this orbit, Equation 26-27 to calculate the energy, and Equation 26-19 for angular momentum to determine the electron speed.

Orbital radii in the Bohr model:

$$r_n = \frac{n^2 a_0}{Z}\quad\text{where } n = 1, 2, 3, \ldots$$
$$(26\text{-}23)$$

Electron energies in the Bohr model:

$$E_n = -\frac{Z^2}{n^2}E_0\quad\text{where } n = 1, 2, 3, \ldots$$
$$(26\text{-}27)$$

Orbital angular momentum in the Bohr model:

$$L = n\left(\frac{h}{2\pi}\right) = n\hbar\qquad(26\text{-}19)$$

Solve

(a) Calculate the radius of the orbit.

From Equation 26-23 with $Z = 3$ and $n = 1$,

$$r_1 = \frac{1^2 a_0}{3} = \frac{a_0}{3} = \frac{0.0529\,\text{nm}}{3} = 0.0176\,\text{nm}$$

(b) Calculate the energy.

From Equation 26-27 with $Z = 3$ and $n = 1$,

$$E_n = -\frac{3^2}{1^2}E_0 = -9E_0 = -9(13.6\,\text{eV}) = -122\,\text{eV}$$

(c) To find an expression for the electron speed, combine Equation 26-19 with the equation for the angular momentum of a particle moving in a circular orbit.

Equation 26-20 tells us that the angular momentum of an electron of mass m_e moving in an orbit of radius r at speed v is $L = rm_e v$. Set this equal to the expression for L in Equation 26-19:

$$rm_e v = n\hbar\qquad\text{so}$$

$$v = \frac{n\hbar}{rm_e}$$

We are interested in the case of $n = 1$, and calculated r_1 in part (a). Substituting the values for $\hbar = h/(2\pi)$ and the mass of the electron gives:

$$v = \frac{1(6.63 \times 10^{-34}\ \text{J} \cdot \text{s})}{2\pi\ (1.76 \times 10^{-11}\ \text{m})(9.11 \times 10^{-31}\ \text{kg})}$$

$$= 6.58 \times 10^6\ \text{m/s}$$

Reflect

The radius of an electron in the lowest level of the doubly ionized lithium ion ($Z = 3$) is one-third the radius of an electron in the ground state of hydrogen. Because there are three protons in the lithium nucleus, compared to one for hydrogen, the Coulomb force on the electron is greater, so it is reasonable that the orbital radius is smaller. Also, notice that the speed of the electron is about 2% of the speed of light in a vacuum.

Why did we insist that the problem be about a doubly ionized lithium ion, rather than a lithium atom? The difference is that when more than one electron is present, each electron is affected not only by the electric force from the nucleus but also by the electric force from the other electrons. We haven't taken the interaction between electrons into account in any of our calculations: The Bohr model is for *single-electron* atoms and ions only.

Although the Bohr model successfully predicts atomic spectra and other phenomena associated with hydrogen atoms, it nevertheless leaves us with an outstanding question. Why should the electron orbits be quantized in multiples of $\hbar$? For the answer we turn back to Louis de Broglie. Recall that de Broglie postulated that particles, such as electrons, have a wavelike nature. As a result, the electron's wave interferes with itself as the electron makes multiple revolutions around the nucleus. The only allowed orbits are those for which the electron waves constructively interfere and form standing waves.

For a nonrelativistic electron of mass m_e moving at speed v, the momentum has magnitude $p = m_e v$. From the de Broglie relation, Equation 26-10, the wavelength of such an electron is

$$\lambda = \frac{h}{m_e v}$$

For an electron orbiting an atomic nucleus we put this in terms of angular momentum by making use of Equation 26-20:

$$\lambda = \frac{hr}{m_e v r} = \frac{hr}{L}$$

To connect this more directly with the circular path of the orbit, we introduce a factor of 2π:

$$\lambda = \frac{hr}{L} = \frac{2\pi\hbar r}{L}$$

Now let L be an integer multiple of $\hbar$, as Bohr required. Then

$$\lambda = \frac{2\pi\hbar r}{n\hbar}$$

or

$$n\lambda = 2\pi r$$

We recognize $2\pi r$ as the circumference of the orbital path of the electron. An integer number of full electron waves fit into the circumference of the orbit only when Bohr's requirement is met. In a real sense, it is the wavelike nature of particles that results in the quantization of the energy of atomic electrons.

Confirming Energy Quantization

Bohr's model of the atom works relatively well for hydrogen, for a singly ionized helium ion (an atom with Z equal to 2 but only one electron), and for a doubly ionized lithium ion (an atom with Z equal to 3 but only one electron, as in Example 26-5). For atoms with more than one electron, it isn't possible to make calculations of energy levels using Bohr's physics. However, the general picture of the atom it provides, with electrons in quantized energy states, applies to all atoms. This was confirmed experimentally in 1914 by the German physicists James Franck and Gustav Hertz. (Franck and Hertz were awarded the 1925 Nobel Prize in Physics for their work. Gustav Hertz was a nephew of Heinrich Hertz, whom we encountered earlier.)

In what is now known as the *Franck-Hertz experiment*, Franck and Hertz used a device similar to the one shown schematically in Figure 26-16 to measure the effect of bombarding atoms in a gas with electrons. The cathode is heated in order to give off electrons, which are accelerated by a variable voltage toward the mesh grid. Some electrons pass through the grid and arrive at the anode, where the electron current is detected by the ammeter. Notice that a voltage is also applied between the grid and the anode, which acts against the electrons; only electrons that carry sufficient energy as they pass the grid will make it to the anode. These electrical components sit inside a tube filled with low-pressure mercury vapor, so collisions between an electron and a mercury (Hg) atom can occur. Franck and Hertz used mercury vapor because the spectral lines of low-pressure mercury gas were well studied; one prominent ultraviolet spectral line has a wavelength $\lambda = 253.7$ nm. The energy of this photon is

$$E = hf = \frac{hc}{\lambda} = \frac{(6.63 \times 10^{-34}\,\text{J}\cdot\text{s})(3.00 \times 10^8\,\text{m/s})}{253.7 \times 10^{-9}\,\text{m}}$$

$$= (7.84 \times 10^{-19}\,\text{J})\left(\frac{1\,\text{eV}}{1.60 \times 10^{-19}\,\text{J}}\right) = 4.90\,\text{eV}$$

This is the energy emitted when an atomic electron falls from an excited energy level state down to a lower energy level. To see the significance of this number, consider what happens in the collision of an electron and a mercury atom in the Franck-Hertz tube. In general, the more kinetic energy an electron carries as it approaches the grid, the more likely it will make it through. So anode current should grow as the voltage between the cathode and the grid is increased. However, if the energy levels of electrons in Hg (and all) atoms are quantized, when the kinetic energy of the incident electron equals the energy difference between two Hg energy levels, the Hg atom can absorb the electron's energy. When this happens it is less likely that the electron will reach the anode, so the anode current should decrease. Figure 26-17 shows a typical curve of current versus voltage from the Franck-Hertz experiment. The general trend, as we would expect, is that the anode current increases as the voltage is increased. But the current drops dramatically at certain values of accelerating voltage, in this case, at 4.9 V, and at 9.8 V and 14.7 V, which are multiples of 4.9 V. This is the very same energy of the mercury spectral line!

When the cathode-grid voltage is set to 4.9 V, the kinetic energy of electrons that leave the cathode reaches 4.9 eV just as they approach the mesh grid. (Recall that the unit of energy eV is the amount of energy gained by an electron when it experiences a potential difference of 1 V.) For this reason, electrons that collide with a mercury atom in close proximity to the grid give up their energy in the process of causing an electron in the atom to jump to a higher energy level. But because the collision occurs close to the grid, there is no opportunity for the electron to undergo another acceleration; in other words, there is no opportunity for the electron to acquire enough energy to reach the anode. For this reason, when the cathode-grid voltage is set to 4.9 V, so that the collisions

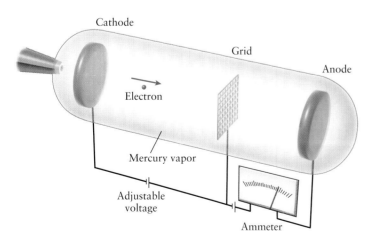

Figure 26-16 The Franck-Hertz experiment In this experiment, electrons are accelerated from a cathode toward a mesh grid in a tube filled with mercury vapor. Some electrons pass through the grid, and the current at the anode is measured by the ammeter.

between electrons and Hg atoms occur near the grid with electron kinetic energy equal to 4.9 eV, the number of electrons reaching the anode decreases. In addition, when the voltage is set to 9.8 V, accelerated electrons that collide with an Hg atom halfway between the cathode and the grid lose their energy, and then accelerate up to a kinetic energy of 4.9 eV again by the time they reach the grid. A collision there once again results in the electron transferring its energy to an Hg atom and being unable to reach the anode. A similar phenomenon occurs when the voltage is set to any integer multiple of 4.9 V. This rise and fall in anode current versus cathode-grid voltage is shown dramatically in Figure 26-17. The underlying explanation of this phenomenon is the quantization of electron energy levels in the atom, as predicted by Bohr.

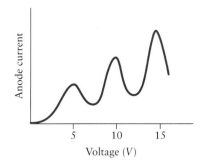

Figure 26-17 Evidence of energy quantization As the voltage between the cathode and the mesh grid in the Franck-Hertz experiment increases, the anode current increases. When the kinetic energy of the accelerated electrons equals the energy required to excite an electron in a mercury atom to a higher energy level, accelerated electrons lose energy. Not as many electrons reach the anode at the corresponding cathode-grid voltages, so the anode current decreases.

Beyond the Bohr Model: Quantum Mechanics

As powerful as the Bohr model is at providing an understanding of the atom and atomic spectra, it does not tell the whole story. The Bohr atom treats the electrons that orbit atomic nuclei as particles. Yet as the scientists of the early twentieth century were beginning to learn, electrons exhibit the properties of waves. No theoretical description of the atom can be complete unless it accounts for the wave nature of the electrons. The theoretical underpinning of our understanding of atoms that includes these wave properties is found in an equation developed by Austrian physicist Erwin Schrödinger. The *Schrödinger equation* relies on matter waves to describe the state of a system as a function of time, much like Newton's laws do while treating physical systems as particles. The Schrödinger equation is thus the fundamental equation of **quantum mechanics**, in which matter is treated as intrinsically wavelike in nature. Schrödinger was awarded the 1933 Nobel Prize in Physics for his work. (The Schrödinger equation itself is too mathematically ornate for the purposes of this book.)

Perhaps the most notable difference between Schrödinger's quantum-mechanical description and Newton's classical description of physics is seen in the ability to specify the position and velocity of objects. In Newton's description, at any instant of time we can identify a specific position in space and a specific velocity vector for any particle in a system, for example, an electron orbiting an atomic nucleus. A wave, however, is not localized in space, even after applying advanced mathematical notions that squeeze it into a relatively well-defined region of space. The result is that while Newton's laws predict the position and velocity of an object, the Schrödinger equation predicts the *probability* of finding a certain value of position or velocity. For this reason, electrons in the quantum model of the atom are described not as tiny marbles orbiting the nucleus at fixed radii, but rather as a charge distribution. This distribution, sometimes called a *probability cloud*, gives the probability of finding the electron at any given position; the denser the cloud in some region, the more likely it is that the electron will be found there.

Figure 26-18a shows the probability cloud associated with the ground state of hydrogen. The more dense the color in any region in this figure, the more likely it is that the electron will be found in that region. This probability distribution is spherically symmetric; that is, it only varies as a function of radius from the center of the nucleus. For that reason we can also express the same information in a curve of probability versus radius, as in Figure 26-18b. Notice that the most probable radius of an electron in the ground state of a hydrogen atom is a_0, the Bohr radius.

The Bohr model employs a single integer, or **quantum number**, to describe electron states. For the Bohr atom this integer is n, which determines the energy level of the electron. In the fully quantum-mechanical view of the atom, *four* quantum numbers are required. These are n, the principal quantum number; ℓ, the angular momentum (or orbital) quantum number; m_ℓ, the magnetic quantum number; and m_s, the electron spin quantum number. The specific values of each of these four quantum numbers completely describes the state of an electron in an atom.

The principal quantum number n plays a role in the quantum atom similar to that which n plays in the classical Bohr atom; in particular, it specifies the energy level or electron shell. The lowest energy state, or ground state, corresponds to n equal to 1.

The angular momentum quantum number ℓ is a measure of the angular momentum the electron carries. For any value of n, ℓ varies in integer steps from 0 to $n - 1$; each value of ℓ specifies a subshell, or electron orbital, within the energy level specified by n. The shape of each orbital is different. By convention, we refer to the orbitals with

Figure 26-18 The Schrödinger picture of the hydrogen atom (a) In the state of lowest energy, an electron in a hydrogen atom can be found anywhere relative to the nucleus. The darker (denser) the color in a region of the probability cloud, the more likely it is that the electron will be found in that region. (b) The probability cloud in part (a) is spherically symmetric, so we can represent the probability of finding the electron as a function of radius from the center of the nucleus. The probability peaks at a distance from the nucleus equal to a_0, the Bohr radius.

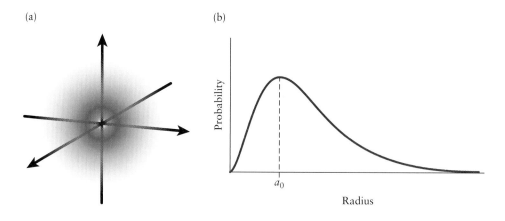

(a)

(b)

letters rather than integer numbers; the values of ℓ equal to 0, 1, 2, 3, 4, and 5 correspond to the orbitals s, p, d, f, g, and h. It is also standard to refer to an electron subshell by giving both n and this orbital letter code together. For example, an electron in the p subshell of the $n = 2$ energy level is said to be in the $2p$ subshell. The electron's energy is slightly dependent on the value of ℓ, an effect called *fine structure* not found in the Bohr model.

The magnetic quantum number m_ℓ specifies an orientation of an electron's subshell. It is so called because its value determines the (very small) energy associated with the interaction of a moving charge—the electron—with the magnetic field of the nucleus. The larger the value of ℓ, the more orientations are allowed; m_ℓ can take integer values between $-\ell$ and $+\ell$, including 0. Figure 26-19 shows the shapes of a number of electron orbitals.

The fourth quantum number m_s involves a new feature of the electron that was not discovered until the 1920s. Electrons have an intrinsic characteristic called **spin,** which is akin to the angular momentum of a rotating sphere. Even electrons that do not orbit an atomic nucleus possess spin, which can take on one of two values, often called spin "up" and spin "down." To fully describe an atomic electron, then, we must also specify its spin state. This is described by the electron spin quantum number m_s, which for an electron can be equal to either $+1/2$ or $-1/2$.

These quantum numbers play an important role in multi-electron atoms. Each electron in a multi-electron atom has a specific value of n, ℓ, m_ℓ, and m_s. (The details of the

Figure 26-19 Probability clouds for different quantum numbers The probability cloud for each atomic orbital, specified by a specific value of the quantum numbers n, ℓ, and m_ℓ, has a different shape.

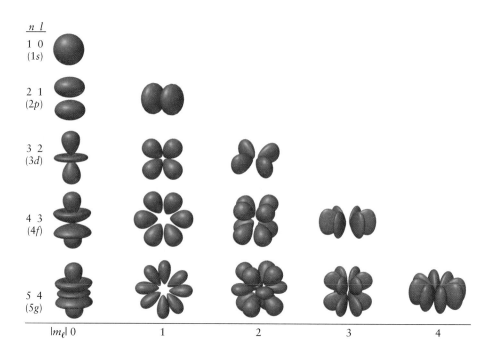

orbitals are affected by the presence of other electrons, but the same four quantum numbers still apply.) What is more, it turns out that there can be only *one* electron with a specific combination of these four quantum numbers. This fundamental restriction on electrons is called the **Pauli exclusion principle**, after the Swiss physicist Wolfgang Pauli who deduced this principle in 1925. (Pauli received the 1945 Nobel Prize in Physics for his work.) Let's see what the Pauli exclusion principle tells us about the structure of multi-electron atoms.

For the $n = 1$ shell, only one value of ℓ (equal to 0, which is the s orbital) and therefore only one value of m_ℓ, is allowed. Two values of m_s are always possible, so the maximum number of electrons that can occupy the $n = 1$ shell is $1 \times 2 = 2$, the product of the number of possible m_ℓ values and the number of possible m_s values. That is, two electrons can occupy the 1s orbital. For $n = 2$, ℓ is allowed to be either 0 or 1. When $\ell = 0$, the only possible value of the magnetic quantum number is $m_\ell = 0$. So including the factor of 2 for the electron spin quantum number, one possible value of m_ℓ and two possible values of m_s means two electrons can occupy the 2s orbital. However, when ℓ equals 1 (the p orbital), m_ℓ can be -1, 0, or $+1$. Including the factor of 2 for the electron spin quantum number, that means that the number of electrons that can occupy the 2p orbital is 3×2, or 6. Because two electrons can occupy the 2s orbital and six electrons can occupy the 2p orbital, an atom can have at most eight electrons in the $n = 2$ energy level.

The pattern above repeats for $n = 3$, up through ℓ equals 2, the d orbital. For this orbital, five values $(-2, -1, 0, +1, +2)$ are possible for m_ℓ, so the maximum number of electrons in this orbital is $5 \times 2 = 10$. Thus two electrons can occupy 3s, six can occupy 3p, and ten can occupy 3d. The maximum number of electrons in the $n = 3$ energy shell of an atom is therefore 18.

The arrangement of electrons, according to how many occupy which orbitals, is directly correlated to the chemical properties of the elements. Consider the 18 lightest elements and their electron configurations, listed in Table 26-1. (In the table, the number of electrons that occupy a particular orbital is given as a superscript, for example, $2p^4$ indicates that four electrons occupy the 2p orbital.) In hydrogen, lithium, and sodium (as well as potassium, rubidium, cesium, and francium), the outermost, or valence, electron is a single electron in an s orbital. These *alkali metals* share common chemical properties and occupy a single column in the periodic table.

Three of the *noble gases* are listed in Table 26-1. These are helium, neon, and argon; in each, the outermost subshell is completely full. As a result, all electrons are relatively tightly bound, so these elements do not easily gain, lose, or share electrons. For that reason, the noble gases are relatively inert. Helium, neon, and argon, as well as the other noble gases occupy a single column in the periodic table.

Halogens are elements that are highly reactive; that is, they easily form bonds with certain other elements, especially the alkali metals, to form molecules. Two halogens, fluorine and chlorine, are listed in Table 26-1. The outermost shell in both is one electron short of being full, which means that an atom of one of these elements can readily share an electron with another atom. This is particularly true for an atom of an alkali metal, which has one valence electron that is easily shared. So bring a sodium atom near a chlorine atom, and they will readily bond to form NaCl (table salt).

Watch Out! **As atomic number increases, electron orbitals do not fill in a continuous fashion.**

The configuration of the 18 electrons of argon, the heaviest element listed in Table 26-1, is $1s^2 2s^2 2p^6 3s^2 3p^6$. The next heaviest element is potassium, for which the configuration of the 19 electrons is $1s^2 2s^2 2p^6 3s^2 3p^6 4s^1$. Notice that the additional electron does not occupy the 3d orbital, although d follows p in our ordering (s, p, d, f, and so on). Orbitals fill according to the increase in energy required, and for that reason, do not fill according to counting up linearly in ℓ (orbital letters) and m_ℓ. Instead, orbitals fill in this order: 1s, 2s, 2p, 3s, 3p, 4s, 3d, 4p, 5s, 4d, 5p, 6s, 4f, 5d, 6p, 7s, 5f, 6d, 7p. Notice that 4s and not 3d follows the 3p orbital in this sequence, which is why the valence electron in potassium is in a 4s orbital.

The ideas of quantum mechanics find applications on scales even smaller than that of the atom. In the final two chapters of this book we will see how quantum-mechanical ideas help us understand the nature of the atomic nucleus and of the fundamental particles that are the essential building blocks of all ordinary matter.

 Go to Interactive Exercise 26-2 for more practice dealing with atomic energy

Table 26-1 Electron Configurations of Light Elements

Atomic Number	Element	Electron Configuration
1	Hydrogen (H)	$1s^1$
2	Helium (He)	$1s^2$
3	Lithium (Li)	$1s^2 2s^1$
4	Beryllium (Be)	$1s^2 2s^2$
5	Boron (B)	$1s^2 2s^2 2p^1$
6	Carbon (C)	$1s^2 2s^2 2p^2$
7	Nitrogen (N)	$1s^2 2s^2 2p^3$
8	Oxygen (O)	$1s^2 2s^2 2p^4$
9	Fluorine (F)	$1s^2 2s^2 2p^5$
10	Neon (Ne)	$1s^2 2s^2 2p^6$
11	Sodium (Na)	$1s^2 2s^2 2p^6 3s^1$
12	Magnesium (Mg)	$1s^2 2s^2 2p^6 3s^2$
13	Aluminum (Al)	$1s^2 2s^2 2p^6 3s^2 3p^1$
14	Silicon (Si)	$1s^2 2s^2 2p^6 3s^2 3p^2$
15	Phosphorus (P)	$1s^2 2s^2 2p^6 3s^2 3p^3$
16	Sulfur (S)	$1s^2 2s^2 2p^6 3s^2 3p^4$
17	Chlorine (Cl)	$1s^2 2s^2 2p^6 3s^2 3p^5$
18	Argon (Ar)	$1s^2 2s^2 2p^6 3s^2 3p^6$

? Got the Concept? 26-5 Ionization Energy

The *ionization energy* of an atom is the energy required to remove an electron from the atom. In terms of the Bohr model, it is equal to the energy difference between the $n = 1$ energy level (the level of lowest energy, in which the electron is closest to the nucleus) and the $n = \infty$ energy level (which has an infinite radius, so the electron has moved infinitely far away). Compared to the ionization energy of hydrogen, the energy required to remove the last electron from doubly ionized lithium is (a) 3 times greater; (b) 9 times greater; (c) 27 times greater; (d) 81 times greater; (e) 243 times greater.

Take-Home Message for Section 26-6

✔ In the Bohr model of the atom, electrons follow Newtonian orbits around the nucleus but with quantized values of orbital angular momentum. As a result only certain orbital radii and orbital energies are allowed.

✔ The Franck-Hertz experiment confirmed that atomic energies are quantized.

✔ The Schrödinger equation explains the hydrogen atom by describing the electron as a wave. It predicts the probability of finding the electron at a particular location within the atom.

✔ The Pauli exclusion principle allows us to understand the structure of multi-electron atoms.

Key Terms

absorption lines	de Broglie wavelength	photon
absorption spectrum	emission lines	quantized
blackbody	emission spectrum	quantum mechanics
blackbody radiation	energy level	quantum number
Bohr model	Pauli exclusion principle	spin
Bohr orbit	photoelectric effect	wave-particle duality
Compton scattering	photoelectrons	work function

Chapter Summary

Topic	Equation or Figure

The photoelectric effect: When light shines on a surface, the surface can emit electrons. However, no electrons are emitted if the frequency of the light is below a certain critical value. Einstein showed that this could be explained if light is absorbed in the form of photons whose energy is proportional to their frequency: $E = hf$, where h is Planck's constant.

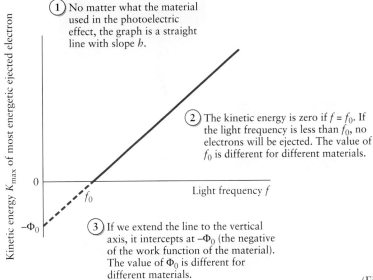

(1) No matter what the material used in the photoelectric effect, the graph is a straight line with slope h.

(2) The kinetic energy is zero if $f = f_0$. If the light frequency is less than f_0, no electrons will be ejected. The value of f_0 is different for different materials.

(3) If we extend the line to the vertical axis, it intercepts at $-\Phi_0$ (the negative of the work function of the material). The value of Φ_0 is different for different materials.

(Figure 26-3)

Blackbody radiation: Objects emit light due to their temperature. An ideal blackbody (one that does a perfect job of absorbing light) is also a perfect emitter of light. The details of the spectrum of a blackbody can be understood only if a blackbody emits light in the form of photons.

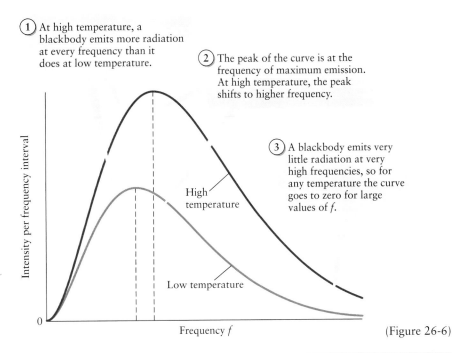

(1) At high temperature, a blackbody emits more radiation at every frequency than it does at low temperature.

(2) The peak of the curve is at the frequency of maximum emission. At high temperature, the peak shifts to higher frequency.

(3) A blackbody emits very little radiation at very high frequencies, so for any temperature the curve goes to zero for large values of f.

(Figure 26-6)

Compton scattering: Photons have momentum in inverse proportion to their wavelength. This is demonstrated by Compton scattering, in which a photon undergoes an increase in wavelength when it scatters from an electron.

Magnitude of the momentum of a photon   Energy of the photon

$$p = \frac{E}{c} = \frac{h}{\lambda}$$ Planck's constant   (26-7)

Speed of light in a vacuum   Wavelength of the photon

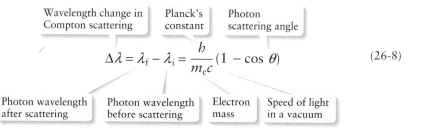

$$\Delta\lambda = \lambda_{\mathrm{f}} - \lambda_{\mathrm{i}} = \frac{h}{m_{e}c}(1 - \cos\theta)$$
(26-8)

Wavelength change in Compton scattering · Planck's constant · Photon scattering angle

Photon wavelength after scattering · Photon wavelength before scattering · Electron mass · Speed of light in a vacuum

---

Wave-particle duality: Just as photons have particle aspects, matter has wave aspects. The de Broglie wavelength of a particle is inversely proportional to the momentum (the same relationship between wavelength and momentum as for a photon).

A particle has a de Broglie wavelength... ...equal to Planck's constant...

$$\lambda = \frac{h}{p}$$
(26-10)

...divided by the momentum of the particle. The greater the momentum, the shorter the de Broglie wavelength.

Atomic structure and atomic spectra: Alpha particle scattering shows that the atom is made up of a small positive nucleus surrounded by electrons. The energy of an atom is quantized: It can have only certain definite values. The evidence for this comes from the spectra of atoms, which show that atoms absorb only specific wavelengths of light and that they emit the same wavelengths that they absorb.

(a) The absorption spectrum produced by passing white light through a gas of hydrogen atoms

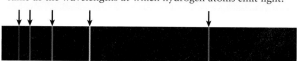

The wavelengths at which hydrogen atoms absorb light are the same as the wavelengths at which hydrogen atoms emit light.

(b) The emission spectrum produced by a heated gas of hydrogen atoms

- Atoms of each element absorb and emit light at wavelengths that are characteristic of that element.
- The characteristic wavelengths differ from one element to another.
- Hydrogen has the simplest arrangement of characteristic wavelengths of any element.

(Figure 26-12)

The Bohr model of the atom: In the Bohr model of hydrogen, a single electron orbits the nucleus much like a satellite orbiting Earth. The difference is that the orbit can have only certain values of angular momentum, radius, and energy. Transitions between the allowed orbits are the cause of absorption and emission spectra.

(a) Atom absorbs a 656.3-nm photon; absorbed energy causes electron to jump from the $n = 2$ orbit up to the $n = 3$ orbit.

(b) Electron falls from the $n = 3$ orbit to the $n = 2$ orbit; energy lost by atom goes into emitting a 656.3-nm photon.

Incoming photon, $\lambda = 656.3$ nm

Emitted photon, $\lambda = 656.3$ nm

$n = 2$

$n = 3$

(Figure 26-14)

The Schrödinger equation and the Pauli exclusion principle: In the more complete Schrödinger description of the atom, the electron is described by a wave. Four quantum numbers describe the state of the electron, and a probability cloud describes the probability of finding an electron at different positions within the atom. The Pauli exclusion principle states that there can be only one electron per state; this helps explain the properties of multi-electron atoms.

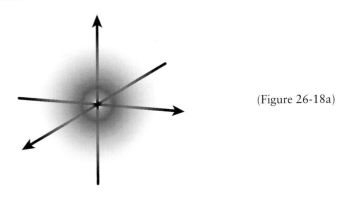

(Figure 26-18a)

Answer to What do you think? Question

(a)   The spectrum of light from a star is very similar to the spectrum of an ideal blackbody (Section 26-2). Blue light has a higher frequency than red light, so the blue stars have a frequency of maximum emission (at which they emit most strongly) that is greater than that of the red stars. The frequency of maximum emission of a blackbody increases with increasing temperature, so the blue stars must be hotter than the red stars. To determine the chemical composition of the stars, the color of stars isn't enough information: We need to look at the absorption spectrum of the stars (Section 26-5). This can't be inferred from a photograph like the one that opens this chapter. In fact, it turns out that all of the stars in this image have almost the same chemical composition: predominantly hydrogen and helium, with trace amounts of other elements.

Answers to Got the Concept? Questions

26-1   (e) Figure 26-6 shows that the higher the temperature, the more light a blackbody of a given size emits at *all* frequencies. Note that both the 3000-K blackbody and the 12,000-K blackbody emit light at visible frequencies, not just the invisible frequencies at which they emit most strongly. That's why you can see radiation from these objects. The 3000-K blackbody emits more red light than any other visible color, so it will appear red; the 12,000-K blackbody emits more blue light than any other visible color, so it will appear blue.

26-2   (d) If $\theta = 90°$, then $\cos \theta = \cos 90° = 0$ and the wavelength shift in Compton scattering is $\Delta\lambda = (h/m_ec)(1 - \cos\theta) = (h/m_ec)(1 - 0) = h/m_ec = \lambda_C$. The initial wavelength is λ_C, so the final wavelength is $\lambda_f = \lambda_i + \Delta\lambda = \lambda_C + \lambda_C = 2\lambda_C$. That is, the final photon has twice the wavelength and so half the energy of the initial photon. (The lost energy is transferred to the electron from which the photon scattered.)

26-3   (c), (d), (a), (b) Equation 26-10 tells us that the de Broglie wavelength λ is inversely proportional to the momentum p: $\lambda = h/p$. So a ranking from longest to shortest de Broglie wavelength is the same as a ranking from smallest to largest momentum. A proton has mass 1.67×10^{-27} kg and an electron has mass 9.11×10^{-31} kg, so the momentum in each of the four cases is (a) $(1.67 \times 10^{-27} \text{ kg})(2.00 \times 10^3 \text{ m/s}) = 3.34 \times 10^{-24}$ kg·m/s; (b) $(1.67 \times 10^{-27} \text{ kg})(4.00 \times 10^3 \text{ m/s}) = 6.68 \times 10^{-24}$ kg·m/s; (c) $(9.11 \times 10^{-31} \text{ kg})(2.00 \times 10^3 \text{ m/s}) = 1.82 \times 10^{-27}$ kg·m/s; (d) $(9.11 \times 10^{-31} \text{ kg})(4.00 \times 10^3 \text{ m/s}) =$

3.64×10^{-27} kg·m/s. So the ranking is (c) slow-moving electron, (d) fast-moving electron, (a) slow-moving proton, (b) fast-moving proton.

26-4   (c), (a), (b), (d) The photon energy is equal to the difference between the energies of the upper energy level of the atom and the lower energy level of the atom. To determine the ranking, just look at Figure 26-15 to see the spacing between the energy levels of the atom. The spacing is (c) greatest between the $n = 3$ and $n = 1$ levels, (a) second greatest between the $n = 4$ and $n = 2$ levels, (b) third greatest between the $n = 5$ and $n = 3$ levels, and (d) least between the $n = 4$ and $n = 3$ levels.

26-5   (b) The ionization energy is equal to $E_{\text{ionization}} = E_\infty - E_1$. From Equation 26-27,

$$E_{\text{ionization}} = E_\infty - E_1 = \left(-\frac{Z^2}{\infty^2} E_0 \right) - \left(-\frac{Z^2}{1^2} E_0 \right)$$
$$= 0 - (-Z^2 E_0) = Z^2 E_0$$

(The reciprocal of infinity is zero.) A doubly ionized lithium ion is like a hydrogen atom—both have a single electron—but with $Z = 3$ instead of $Z = 1$ for hydrogen. Since the ionization energy is proportional to Z^2, the value for doubly ionized lithium is $3^2 = 9$ times greater than for hydrogen. You can see that for hydrogen $E_{\text{ionization}} = E_0 = 13.6$ eV, while for doubly ionized lithium $E_{\text{ionization}} = 9E_0 = 9(13.6 \text{ eV}) = 122$ eV.

Questions and Problems

In a few problems, you are given more data than you actually need; in a few other problems, you are required to supply data from your general knowledge, outside sources, or informed estimate. Interpret as significant all digits in numerical values that have trailing zeros and no decimal points.

For all problems, use $g = 9.80 \text{ m/s}^2$ for the free-fall acceleration due to gravity. Neglect friction and air resistance unless instructed to do otherwise.

• Basic, single-concept problem

•• Intermediate-level problem, may require synthesis of concepts and multiple steps

••• Challenging problem

SSM *Solution is in Student Solutions Manual*

Conceptual Questions

1. •Does the de Broglie wavelength of a particle increase or decrease as its kinetic energy increases?

2. •How does the intensity of light from a blackbody change when its temperature is increased? What changes occur in the body's radiation spectrum?

3. •According to classical electromagnetic theory, an accelerated charge emits electromagnetic radiation. What would this mean for the electron in the Bohr atom? What would happen to its orbit?

4. •Are there quantities in classical physics that are quantized?

5. •What is the shortest wavelength of electromagnetic radiation that can be emitted by a hydrogen atom? SSM

6. •Prior to Einstein's description of the photoelectric effect, light was thought to act like a wave. Explain why the existence of a frequency below which photoelectrons are not emitted favors a description of light as a particle instead.

7. •Describe how the number of photoelectrons emitted from a metal plate in the photoelectric effect would change if (a) the intensity of the incident radiation were increased, (b) the wavelength of the incident radiation were increased, and (c) the work function of the metal were increased.

8. •Is it possible to observe photoelectrons emitted from a metal plate with relativistic speeds?

9. •Consider the photoelectric emission of electrons induced by monochromatic incident light. The incoming photons all have the same energy, but the emitted electrons have a range of kinetic energies. Why?

10. •A markedly nonclassical feature of the photoelectric effect is the fact that the energy of the emitted electrons doesn't increase as you increase the intensity of the light striking the metal surface. What change does occur as the intensity is increased?

11. •The Compton effect is practically unobservable for visible light. Why?

12. •Which of the two Compton scattering experiments more clearly demonstrates the particle nature of electromagnetic radiation: a collision of the photon with an electron or a collision with a proton? Explain your answer.

13. •An electron and a proton have the same kinetic energy. Which has the longer wavelength? SSM

14. •Is the wavelength of an electron the same as the wavelength of a photon if both particles have the same total energy?

15. •Why do you think Bohr's model was originally designed for the element hydrogen?

16. •Why do you think that Bohr's model of the hydrogen atom is still taught in undergraduate physics classes?

17. •Why do we never observe the wave nature of particles for everyday objects such as birds or bumblebees, for example? SSM

Multiple-Choice Questions

18. •An element emits a spectrum that
   A. is the same as all other elements.
   B. is evenly spaced.
   C. is unique to that element.
   D. is evenly spaced and unique to that element.
   E. is indistinguishable from most other elements.

19. •An ideal blackbody is an object that
   A. absorbs most of the energy that strikes it and emits a little of the energy it generates.
   B. absorbs a little of the energy that strikes it and emits most of the energy it generates.
   C. absorbs half of the energy that strikes it and emits half of the energy it generates.
   D. absorbs all the energy that strikes it and emits all the energy it generates.
   E. neither absorbs nor emits energy except at ultraviolet ("black light") wavelengths.

20. •The color of light emitted by a hot object depends on
   A. the size of the object.
   B. the shape of the object.
   C. the material from which the object is made.
   D. the temperature of the object.
   E. the color of the object.

21. •Which photon has more energy?
   A. a photon of ultraviolet radiation.
   B. a photon of green light.
   C. a photon of yellow light.
   D. a photon of red light.
   E. a photon of infrared radiation. SSM

22. •Light that has a wavelength of 600 nm strikes a metal surface, and a stream of electrons is ejected from the surface. If light of wavelength 500 nm strikes the surface, the maximum kinetic energy of the electrons emitted from the surface will
   A. be greater.
   B. be smaller.
   C. be the same.
   D. be 5/6 smaller.
   E. be unmeasurable.

23. •In the Compton effect experiment, the change in a photon's wavelength depends on
   A. the scattering angle.
   B. the initial wavelength.
   C. the final wavelength.

D. the density of the scattering material.

E. the atomic number of the scattering material.

24. •The maximum change in a photon's energy when it undergoes Compton scattering occurs when its scatter angle is at

A. 0°.

B. 45°.

C. 90°.

D. 135°.

E. 180°.

25. •As the scattering angle in the Compton effect increases, the energy of the scattered photon

A. increases.

B. stays the same.

C. decreases.

D. decreases by $\sin \theta$.

E. increases by $\sin \theta$. SSM

26. •The de Broglie wavelength depends only on

A. the particle's mass.

B. the particle's speed.

C. the particle's energy.

D. the particle's momentum.

E. the particle's charge.

Estimation/Numerical Analysis

27. •Estimate the order of magnitude of atomic ionization energies. Compare this with the order of magnitude for the nuclear binding energy, which is in the megaelectron volt (MeV) range.

28. •What is the approximate size of an atom?

29. •Estimate the amount of blackbody energy radiated from your body. SSM

30. •On average, an electron will exist in any given state in the hydrogen atom for about 10^{-8} s before jumping to a lower level. Estimate the number of revolutions about the nucleus that an electron makes in 10^{-8} s.

31. •Estimate the size of the electrostatic forces that hold the atom together compared to the size of the force due to gravity.

32. •Estimate the ratio of the mass of the proton to the mass of the electron.

33. •Estimate the speed of an electron at which the de Broglie wavelength approximates the size of an atom. SSM

34. •In the early twentieth century, Max Planck used an empirical mathematical equation to describe the data that was collected in measuring blackbody radiation. Here's the formula he conceived:

$$\frac{\Delta I}{\Delta \lambda} = \frac{2\pi hc^2}{\lambda^5}\left(\frac{1}{e^{hc/(\lambda k_B T)} - 1}\right)$$

where I is the intensity of light emitted, λ is the wavelength of light emitted, h is Planck's constant (6.63×10^{-34} J · s), c is the speed of light (3.00×10^8 m/s), k is the Boltzmann constant (1.38×10^{-23} J/K), and T is the absolute temperature of the radiating cavity in kelvin. Create a spreadsheet that generates at least 15 values of $\Delta I/\Delta \lambda$ (choose a range of 15+ values of wavelength and calculate the value of the above function for $\Delta I/\Delta \lambda$ for some constant temperature). Once you have the set of data, plot it in a graph and apply an appropriate curve fit to depict the unique shape of the curve. Now, change the temperature several times to see how the curve changes.

35. •Using Einstein's explanation of the photoelectric effect ($K = hf - \Phi_0$), derive a numerical value for Planck's constant (h) by plotting a graph of the following data.

Wavelength of light (nm)	Stopping potential (V)
400	1.60×10^{-19} J
450	1.06×10^{-19} J
500	6.09×10^{-20} J
550	2.00×10^{-20} J

Problems

26-1 Experiments that probe the nature of light and matter reveal the limits of classical physics

26-2 The photoelectric effect and blackbody radiation show that light is absorbed and emitted in the form of photons

36. ••Suppose a blackbody at 400 K radiates just enough heat in 15 min to boil water for a cup of tea. How long will it take to boil the same water if the temperature of the radiator is 500 K?

37. •Wien's displacement law states that the wavelength at which a blackbody emits the maximum power is given by the following formula:

$$\lambda_{max} = \frac{0.290 \text{ K} \cdot \text{cm}}{T}$$

At what wavelength (in nanometers) will a blackbody emit its highest value when its temperature is 400°C?

38. ••MIG (metal inert gas) welders can be approximated as blackbodies. At what wavelength will the peak radiation from such a welder be emitted if it operates at a temperature of 4000 K? Compare your answer to the wavelength for peak radiation from a TIG (tungsten inert gas) welder that operates at 6000 K. The peak wavelength is given by Wien's displacement law, shown in problem 37.

39. •An object that approximates a blackbody at a temperature of 300 K radiates heat into its immediate surroundings. Calculate the wavelength at which the maximum intensity per unit wavelength is emitted using Wien's displacement law (see problem 37). SSM

40. •Using Wien's displacement law (see problem 37), calculate the peak wavelength for blackbody radiation from (a) the polar ice cap at −40°C, (b) burning ethyl alcohol at 365 °C, and (c) deep space at 2.7 K.

41. •The Sun emits its maximum power at a wavelength of 475 nm. Using Wien's displacement law (see problem 37), determine the corresponding temperature of the outer layer of the Sun, assuming it is a blackbody.

42. •(a) If the human body acts like a blackbody, use Wien's displacement law (see problem 37) to calculate the wavelength of light for which the maximum amount of radiation is emitted. Assume an average skin temperature of 34°C. (b) In what part of the electromagnetic spectrum is the light?

43. •Calculate the range of photon frequencies and energies in the visible spectrum of light (approximately 380–750 nm). SSM

44. •What is the energy of a low-frequency 2000-Hz photon?

45. •What is the energy of a 0.200-nm photon?

46. •What are the wavelength and frequency of a 3.97×10^{-19}-J photon?

47. ••**Biology** Under most conditions, the human eye will respond to a flash of light if 100 photons hit photoreceptors at the back of the eye. Determine the total energy of such a flash if the wavelength is 550 nm (green light).

48. ••The threshold wavelength for the photoelectric effect for silver is 262 nm. (a) Determine the work function for silver. (b) What is the maximum kinetic energy of an emitted electron if the incident light has a wavelength of 222 nm?

49. ••Light that has a 195-nm wavelength strikes a metal surface, and photoelectrons are produced moving as fast as $0.004c$. (a) What is the work function of the metal? (b) What is the threshold wavelength for the metal above which no photoelectrons will be emitted? SSM

50. •What is the minimum frequency of light required to eject electrons from a metal with a work function of 6.53×10^{-19} J?

51. ••The work functions of aluminum, calcium, potassium, and cesium are 6.54×10^{-19} J, 4.65×10^{-19} J, 3.57×10^{-19} J, and 3.36×10^{-19} J, respectively. For which of the metals will photoelectrons be emitted when irradiated with visible light?

26-3 As a result of its photon character, light changes wavelength when it is scattered

52. ••The quantity $\lambda_C = h/(mc)$ is called the Compton wavelength. Calculate the numerical value of this quantity for (a) an electron, (b) a proton, and (c) a pi meson (which has a mass of 2.50×10^{-28} kg).

53. ••What is the momentum of a photon if the wavelength is (a) 550 nm and (b) 0.0711 nm? SSM

54. ••X rays that have wavelengths of 0.125 nm are scattered off free electrons at an angle of 30.0°. (a) Calculate the wavelength of the scattered electromagnetic radiation. (b) Calculate the *fractional wavelength change* $(\Delta\lambda/\lambda_i = (\lambda_f - \lambda_i)/\lambda_i)$ for the scattered X rays.

55. ••If a photon undergoes Compton scattering and experiences a fractional wavelength change of +7.25%, calculate the angle at which the scattered photons are directed if the original photons have a wavelength of 0.00335 nm.

56. ••Photons that have wavelengths of 0.00225 nm are Compton scattered at 45.0°. What is the energy of the scattered photons?

57. ••X ray photons that have wavelengths of 0.140 nm are scattered off carbon atoms (which possess essentially stationary electrons in their valence shells). What are the wavelengths of the Compton-scattered photons and the kinetic energies of the scattered electrons at angles of (a) 0.00°, (b) 30.0°, (c) 45.0°, (d) 60.0°, (e) 90.0°, and (f) 180°? SSM

58. ••A 0.0750-nm photon Compton scatters off a stationary electron. Determine the maximum speed of the scattered electron.

59. ••A photon Compton scatters off a stationary electron at an angle of 60.0°. The electron moves away with 1.28×10^{-17} J of kinetic energy. Determine the initial wavelength of the photon.

60. ••An X ray source is incident on a collection of stationary electrons. The electrons are scattered with a speed of 4.50×10^5 m/s, and the photon scatters at an angle of 60.0° from the incident direction of the photons. Determine the wavelength of the X ray source.

61. ••Arthur Holly Compton scattered photons that had wavelengths of 0.0711 nm off a block of carbon during his famous experiment of 1923 at Washington University in St. Louis, Missouri. (a) Calculate the frequency and energy of the photons. (b) What is the wavelength of the photons that are scattered at 90.0°? (c) What is the energy of the photons that are scattered at 90.0°? (d) What is the energy of the electrons that recoil from the Compton scattering with $\theta = 90.0°$?

26-4 Matter, like light, has aspects of both waves and particles

62. •Calculate the de Broglie wavelength of a 0.150-kg ball moving at 40 m/s. Comment on the significance of the result.

63. •Calculate the de Broglie wavelength of an electron that has a speed of $0.00730c$. SSM

64. •What is the de Broglie wavelength of a proton ($m = 1.67 \times 10^{-27}$ kg) moving at 400,000 m/s?

65. ••Restate de Broglie's formula for particle waves in the case that the speeds are relativistic.

66. •A proton ($m = 1.67 \times 10^{-27}$ kg) has a de Broglie wavelength of 1.20×10^{-15} m. Calculate the speed of the proton. Be careful of any relativistic corrections that might apply!

67. ••What is the de Broglie wavelength of an electron that has a kinetic energy of (a) 1.60×10^{-19} J, (b) 1.60×10^{-18} J, (c) 1.60×10^{-17} J, (d) 1.60×10^{-16} J, (e) 1.60×10^{-13} J, and (f) 1.60×10^{-10} J? SSM

68. ••Calculate the de Broglie wavelength of an alpha particle ($m_\alpha = 6.64 \times 10^{-27}$ kg) that has a kinetic energy of (a) 1.60×10^{-13} J, (b) 8.00×10^{-13} J, and (c) 1.60×10^{-12} J.

69. •Calculate the de Broglie wavelength of a thermal neutron that has a kinetic energy of about 6.41×10^{-21} J.

70. •A relativistic electron has a de Broglie wavelength of 346 fm (1 fm = 10^{-15} m). Determine its velocity.

71. ••Write an expression that relates the Newtonian kinetic energy ($K = (1/2)mv^2$) and mass of a nonrelativistic particle to its de Broglie wavelength. (Complete the expression $\lambda(K, m) = ?$)

26-5 The spectra of light emitted and absorbed by atoms show that atomic energies are quantized

72. ••The Balmer formula can be written as follows:

$$\lambda = (364.56 \text{ nm})\left(\frac{m^2}{m^2 - 4}\right)$$

where m is equal to any integer larger than 2. This represents the wavelengths of visible colors that are emitted from the hydrogen atom. Calculate the first four colors (wavelengths) that are observed in the spectrum of hydrogen.

73. ••**Astronomy** The 21-cm line emitted by hydrogen in interstellar gas clouds is the spectral line that is produced when an electron in the ground state of hydrogen switches spin states. Determine the energy difference between the two spin states in this *hyperfine* transition. SSM

74. ••Prove that the Balmer formula is a special case of the Rydberg formula with n set equal to 2.

$$\text{Rydberg formula: } \frac{1}{\lambda} = R_{\text{H}}\left(\frac{1}{n^2} - \frac{1}{m^2}\right)$$

$$R_{\text{H}} = 1.09737 \times 10^7\,\text{m}^{-1}$$

$$\text{Balmer formula: } \lambda = b\left(\frac{m^2}{m^2 - 4}\right) \quad b = 364.56\,\text{nm}$$

75. ••A hypothetical atom has four unequally spaced energy levels in which a single electron can be found. Suppose a collection of the atoms is excited to the highest of the four levels. (a) What is the maximum number of unique spectral lines that could be measured as the atoms relax and return to the lowest, ground state? (b) Suppose the previous hypothetical atom has 10 energy levels. Now what is the maximum number of unique spectral lines that could be measured in the emission spectrum of the atom?

76. •The Lyman series results from transitions of the electron in hydrogen in which the electron ends at the $n = 1$ energy level. Using the Rydberg formula for the Lyman series, calculate the wavelengths of the photons emitted in the transitions from the energy states that correspond to n equal to 2 through 6, and indicate the initial and final state of the transition corresponding to each wavelength.

77. •The Balmer series results from transitions of the electron in hydrogen in which the electron ends at the $n = 2$ energy level. Using the Rydberg formula for the Balmer series, calculate the wavelengths of the photons emitted in the transitions from the energy states that correspond to n equal to 3 through 6, and indicate the initial and final state of the transition corresponding to each wavelength. SSM

78. •The Paschen series results from transitions of the electron in hydrogen in which the electron ends at the $n = 3$ energy level. Using the Rydberg formula for the Paschen series, calculate the wavelengths of the photons emitted in the transitions from the energy states that correspond to n equal to 4 through 6, and indicate the initial and final state of the transition corresponding to each wavelength.

79. ••Calculate the shortest wavelength (and the highest energy) associated with emitted photons in the (a) Lyman, (b) Balmer, and (c) Paschen series.

80. ••Express the Balmer formula (See problem 74) in terms of the *frequency* of the photons that are emitted (rather than the wavelength). Extend all numerical values out to five significant figures.

81. ••Express the Rydberg formula in terms of the *frequency* of the photons that are emitted (rather than the wavelength). Extend all numerical values out to five significant figures. SSM

26-6 Models by Bohr and Schrödinger give insight into the intriguing structure of the atom

82. •For an electron in the nth state of the hydrogen atom, write expressions for (a) the angular momentum of the electron, (b) the radius of the electron's orbit, (c) the kinetic energy of the electron, (d) the total energy of the electron, and (e) the speed of the electron.

83. ••Set up a chart for the five quantities listed in problem 82 and calculate the values for $n = 1, 2, 3, 4,$ and 5 (4 significant figures, SI units). See the following table.

n	L_n	r_n	K_n	E_n	v_n
1					
2					
3					
4					
5					

84. ••Devise a straightforward method that allows you to calculate (a) the speed and (b) the angular momentum of an electron in the nth Bohr orbit. Your expressions should contain the number of the orbit, n, and constants.

85. ••Using the formula that you devised in problem 84, calculate the speed of an electron and the angular momentum of an electron in the tenth Bohr orbit. SSM

86. ••**Astronomy** Radio astronomers use radio frequency waves to identify the elements in distant stars. One of the standard lines that is often studied is designated the 272α line. This spectral line refers to the transition in hydrogen from $n_i = 273$ to $n_f = 272$. Calculate the wavelength and frequency of the electromagnetic radiation that is emitted for the 272α transition.

87. ••**Astronomy** For carbon ($Z = 6$), the frequencies of spectral lines resulting from a single electron transition are increased over those for hydrogen by the ratio of the Rydberg constants:

$$\frac{R_{\text{C}}}{R_{\text{H}}} = \frac{1 - m_e/m_{\text{C}}}{1 - m_e/m_{\text{H}}}$$

The mass of a hydrogen atom is 1837 times greater than the mass of the electron, and the mass of the carbon atom is 12 times greater than the mass of the hydrogen atom. Find the shift in frequency for the carbon 272α transition compared to the hydrogen 272α transition (see problem 86). Express your answer in megahertz (MHz).

88. ••The Lyman series ($n_f = 1$), the Balmer series ($n_f = 2$), and the Paschen series ($n_f = 3$) are commonly studied in basic chemistry and physics classes. The Brackett series ($n_f = 4$) and the Pfund series ($n_f = 5$) are not so well known. (a) Calculate the shortest and longest wavelengths for the spectral lines that are part of the Brackett series. (b) Calculate the shortest and longest wavelengths for the spectral lines that are part of the Pfund series.

89. •A hydrogen atom that has an electron in the $n = 2$ state absorbs a photon. (a) What wavelength must the photon possess to send the electron to the $n = 4$ state? (b) What possible wavelengths would be detected in the spectral lines that result from the deexcitation of the atom as it returns from $n = 4$ to the ground state? SSM

90. •How much energy is needed to ionize a hydrogen atom that starts in the Bohr orbit represented by $n = 3$? If an atom is ionized, its outer electron is no longer bound to the atom.

91. ••Derive the first 10 energy levels (sketch and label an energy-level diagram) for singly-ionized helium, He^+.

General Problems

92. ••**Biology** Vitamin D is produced in the skin when 7-dehydrocholesterol reacts with UVB rays (ultraviolet B) having wavelengths between 270 nm and 300 nm. What is the energy range of the UVB photons?

93. ••A typical HeNe laboratory laser produces light of wavelength 632.8 nm. The laser beam carries a power of 0.50 mW and strikes a target perpendicular to the beam. (a) How many photons per second strike the target? (b) At what rate does the laser beam deliver linear momentum to the target if the photons are all absorbed by the target?

94. ••**Astronomy** In 2009 astronomers detected gamma ray photons having energy ranging from 700 GeV to around 5 TeV coming from supernovae (exploding giant stars) in the galaxy M82. (a) What is the range of wavelengths of the gamma ray photons detected from M82? (b) Calculate the ratio of the energy of the 5 TeV photons to the energy of visible light having a wavelength of 500 nm.

95. •••Derive the Compton scattering formula for the change in wavelength between the scattered photon and the incident photon when they are exposed to a free electron, initially at rest.

96. ••A photon of frequency 4.81×10^{19} Hz scatters off a free stationary electron. Careful measurements reveal that the photon goes off at an angle of 125° with respect to its original direction. (a) How much energy does the electron gain during the collision? (b) What percent of its original energy does the photon lose during the collision?

97. ••**Chemistry** A laboratory oven that contains hydrogen molecules H_2 and oxygen molecules O_2 is maintained at a constant temperature. Each oxygen molecule is 16 times as massive as a hydrogen molecule. Find the ratio of the de Broglie wavelength of the hydrogen molecule to that of the oxygen molecule.

98. ••A hydrogen atom makes a transition from the $n = 5$ state to the ground state and emits a single photon of light in the process. The photon then strikes a piece of silicon, which has a photoelectric work function of 4.8 eV. Is it possible that a photoelectron will be emitted from the silicon? If not, why not? If so, find the maximum possible kinetic energy of the photoelectron.

99. ••Suppose the electron in the hydrogen atom were bound to the proton by gravitational forces (rather than electrostatic forces). Find (a) the radius and (b) the energy of the first orbit.

100. ••**Biology** The *E. coli* bacterium is about 2.0 μm long. Suppose you want to study it using photons of that wavelength or electrons having that de Broglie wavelength. (a) What is the energy of the photon and the energy of the electron? (b) Which one would be better to use, the photon or the electron? Explain why.

101. •••Use the de Broglie wave concept to fit circular standing waves into the orbits of the Bohr model of the hydrogen atom to prove Bohr's hypothesis of the quantization of angular momentum. Assume that in the first Bohr orbit, exactly one de Broglie wavelength matches up with the circumference, in the second orbit, two waves match up, in the third orbit, three waves match up, and so on (Figure 26-20).

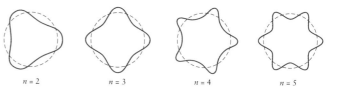

$n = 2$ $n = 3$ $n = 4$ $n = 5$

Figure 26-20 Problem 101

27

Nuclear Physics

In this chapter, your goals are to:

- (27-1) Explain why quantum ideas play an important role in nuclear physics.
- (27-2) Describe how we know that the force that holds the nucleus together is both strong and of short range.
- (27-3) Explain how and why the binding energy per nucleon in a nucleus depends on the size of the nucleus.
- (27-4) Calculate the energy released in nuclear fission.
- (27-5) Describe why very high temperatures are needed for nuclear fusion reactions.
- (27-6) Calculate what happens in the decay of a radioactive substance.

To master this chapter, you should review:

- (3-8) Uniform circular motion.
- (11-2) How to calculate the density of an object.
- (22-4) The energy of a photon.
- (25-7) The equivalence of mass and energy.
- (26-5) How an atom in an excited energy level emits a photon when the atom decays to a lower energy level.
- (26-6) The spin of an electron.

What do you think?

This remarkable painting on a wall in the Chauvet-Pont-d'Arc Cave in southern France is known to be 30,000 to 33,000 years old. This age is determined from the radioactive decay of carbon-14 (a type of carbon with 6 protons and 8 neutrons in its nucleus), which has a half-life of 5730 years. Of the carbon-14 that was present when this painting was made, the fraction that remains today is somewhat less than (a) half; (b) 1/4; (c) 1/8; (d) 1/16; (e) 1/32.

27-1 The quantum concepts that help explain atoms are essential for understanding the nucleus

We learned in Chapter 26 that a handful of radical ideas—the notion that energy is quantized, that electromagnetic waves come in packets called photons, and that the laws of quantum mechanics can specify only the probabilities that particles will behave in certain ways—are essential for understanding the nature and behavior of the atom. These same ideas apply on the even smaller scale of the atomic nucleus.

In this chapter we'll learn that unlike the atom, in which negatively charged electrons are bound to the positively charged nucleus by the attractive electric force, nuclei are bound by a *strong nuclear force* that keeps the protons and neutrons bound together. We'll also see that like atoms, nuclei have an intrinsic angular momentum or *spin*; magnetic resonance imaging, or MRI, makes use of how nuclear spins respond to an external magnetic field (**Figure 27-1a**). The interplay between the strong nuclear force and the electric force (which makes all of the protons in a nucleus try to repel each other) means

Figure 27-1 **Applications of nuclear physics** The properties of atomic nuclei explain (a) how magnetic resonance imaging (MRI) works, (b) how the Sun provides energy for life on Earth, and (c) how Earth sustains its internal energy.

When placed in a strong magnetic field, the nuclei of hydrogen atoms will orient their spins with the field. This effect is at the heart of the diagnostic technique called magnetic resonance imaging.

Photosynthesis in plants depends on energy from sunlight. This energy is released by nuclear reactions that take place in the core of the Sun.

More than 50% of the energy that powers our planet's geological activity, including volcanic eruptions, comes from the radioactive decay of unstable nuclei in Earth's interior.

(a)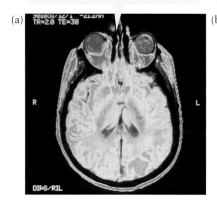
(b)
(c)

that nuclei of different sizes are more tightly or loosely bound. We'll see that as a result, the largest nuclei are prone to break into smaller fragments through a process called *fission*. We'll also learn that the smallest nuclei can release energy by *fusion*, in which two nuclei join together to form a larger one. Fusion reactions make the Sun shine, and the sunlight that they produce makes life on Earth possible (Figure 27-1b).

Understanding the binding energies of nuclei will also help us understand the three main types of radioactive decay: alpha, beta, and gamma. The energy released by alpha and beta decays in our planet's interior helps power geologic activity such as volcanic eruptions (Figure 27-1c). We'll find that the concept of *half-life* will help us understand the nature of radioactive decays of all kinds.

Take-Home Message for Section 27-1

✔ Many of the same concepts that help us understand atomic physics are also important in nuclear physics.

✔ The balance between the strong nuclear force and the electric force determines the stability of nuclei.

27-2 The strong nuclear force holds nuclei together

As we learned in Chapter 26, an atom has a small, positively charged nucleus at its center. There are two ways you can see that the nucleus must be positively charged. First, because atoms are neutral, positive charge is required to balance the negative charge of the electrons. Second, the Coulomb attraction between the nucleus and the negatively charged electrons provides the force that holds the atom together.

The *repulsive* Coulomb forces between the protons, however, must be large because they are close together inside the nucleus. To see just how large these forces are, note that a typical radius of a nucleus is about 5 fm (1 fm = 1 femtometer = 10^{-15} m). The electric force between two protons (each of charge $+e = 1.60 \times 10^{-19}$ C) separated by a distance $r = 5.00$ fm is

$$F = \frac{k(+e)(+e)}{r^2} = \frac{(8.99 \times 10^9 \, \text{N} \cdot \text{m}^2/\text{C}^2)(1.60 \times 10^{-19}\text{C})^2}{(5.00 \times 10^{-15}\text{m})^2}$$
$$= 9.21 \text{N}$$

This may not seem like a lot of force, but remember that this force is applied to a proton, which has a mass of only 1.67×10^{-27} kg. If the electric force were the only force acting on the protons in a nucleus, the nucleus would simply fly apart.

What's more, nuclei also have neutrons, the neutrally charged particle we mentioned briefly in Chapter 16. Neutrons and protons share many similar properties, including similar masses: The masses of the neutron and proton are 1.6749×10^{-27} kg and 1.6726×10^{-27} kg, respectively. Collectively, we refer to protons and neutrons as **nucleons**. Neutrons do not feel the electric force at all, so this force can't be responsible for keeping neutrons within the nucleus.

We conclude that there must be an additional *attractive* force that acts on all nucleons (both protons and neutrons) and that binds them together in the nucleus. This attractive force must be stronger than the repulsive electric force between protons, so we call it the **strong nuclear force**. To explain the properties of the nucleus, it must be that over short distances the strong nuclear force is hundreds of times stronger than the electrostatic force.

If the nuclear force is so strong compared to the electrostatic force, and if protons attract other protons by this force, why are neutrons necessary to help overcome the Coulomb repulsion between protons? The answer lies in the *range* of the strong nuclear force, which is the distance beyond which one nucleon no longer experiences a force due to another. Experiments show that the strong nuclear force diminishes rapidly as the separation between two nucleons increases, and has a range of about 2.0 fm. The radius of a proton or neutron is about 0.85 fm, so two nucleons must almost be touching to experience the strong nuclear force. By contrast, the electrostatic repulsion between protons separated by a distance r is proportional to $1/r^2$, and so is present even if the distance r is very large. We say that the strong nuclear force is a *short-range* force, whereas the electric force is a *long-range* force. As a very rough analogy, you can think of protons and neutrons as tiny spheres coated with very strong Velcro, which makes the nucleons stick together if they are brought close enough to each other.

Because the range of the nuclear force is smaller than the diameter of most nuclei (a few to perhaps 15 fm), each nucleon exerts an attractive nuclear force only on its nearest neighbors. Each proton in a nucleus, however, exerts a repulsive force on *every other* proton. As a result, the nuclear force between neighboring protons cannot overcome the Coulomb repulsion between all of the protons. To prevent a nucleus from spontaneously breaking apart (that is, for it to be *stable*), the nucleus must have about as many neutrons as protons or more (for larger nuclei). An unstable nucleus will eventually undergo a spontaneous transformation, termed a *decay*, in which it either splits apart or gives off energy in some other way. In addition, atoms that have more than about 20 protons require more neutrons than protons to be stable. Figure 27-2 shows the number of neutrons versus the number of protons in known, stable atomic nuclei. Notice that as the number of protons increases, more additional neutrons are required for stability.

Nuclides, Isotopes, and Nuclear Sizes

Atoms of each element have a unique number of protons and the same number of electrons: Hydrogen has one, helium two, lithium three, and so on. Many properties of atoms are related to this number, usually designated as the **atomic number** Z. In a similar way, many properties of nuclei arise from the **neutron number** N, that is, the number of neutrons in the nucleus. Although the value of Z of an elemental species is fixed (changing the number of protons changes the element), the value of N is not fixed for each element. Each combination of N and Z specifies a **nuclide**. For each element, the most common configuration of N and Z corresponds to the most stable nuclide; nuclei of that elemental species with a different number of neutrons are termed **isotopes**. For example, potassium has 19 protons (and 19 electrons). The most stable and most common nuclide of potassium has 20 neutrons, $N = 20$. We denote this nuclide with the symbol ^{39}K. The number of protons ($Z = 19$) is understood from the symbol K for potassium, and the total number of protons and neutrons, termed the **mass number** A, is given in the pre-superscript, 39. This potassium nuclide can also be referred to as potassium-39. Note that the mass number of any nucleus equals the atomic number plus the neutron number: $A = Z + N$.

More than 93% of all potassium atoms have a ^{39}K nucleus. About 7% of potassium atoms have a ^{41}K nucleus, however. These atoms contain 19 electrons orbiting a nucleus with 19 protons and 22 neutrons. We say that ^{39}K and ^{41}K are two isotopes of potassium.

Figure 27-2 **Neutrons versus protons in nuclei** The interplay between the strong nuclear force and the electric force explains the relationship between the numbers of protons and neutrons in nuclei.

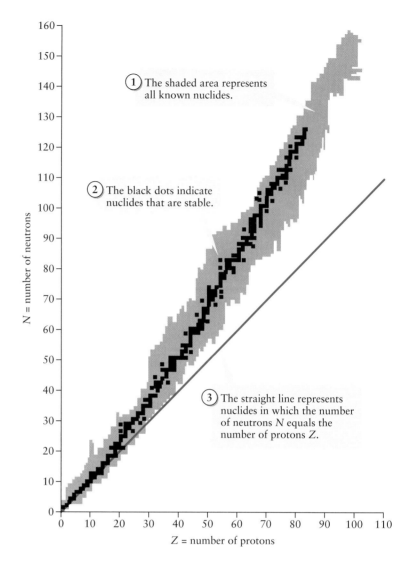

1. The shaded area represents all known nuclides.

2. The black dots indicate nuclides that are stable.

3. The straight line represents nuclides in which the number of neutrons N equals the number of protons Z.

N = number of neutrons

Z = number of protons

For most nuclides, more neutrons than protons must be present (N must be greater than Z) to prevent the protons from flying apart due to electrostatic repulsion.

The *size* of a nucleus (its radius and volume) is related to the mass number A. Experiments show that all nuclei are approximately spherical and have radii proportional to the cube root of A, or $A^{1/3}$:

Radius of a nucleus
(27-1)

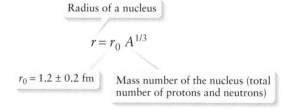

Radius of a nucleus

$$r = r_0\, A^{1/3}$$

$r_0 = 1.2 \pm 0.2$ fm

Mass number of the nucleus (total number of protons and neutrons)

This equation states that if we increase the number of nucleons in a nucleus by a factor of 10, say from $A = 20$ to $A = 200$, the radius of the nucleus will increase by a factor of $10^{1/3} = 2.2$ (Figure 27-3).

The volume of a sphere is proportional to r^3 and therefore proportional to $(A^{1/3})^3$, so the volume of a nucleus is directly proportional to the mass number A. In other words, the volume of a nucleus is proportional to the number of nucleons within that nucleus. This is consistent with our model of nucleons as spheres covered with Velcro. In this model, increasing the number of nucleons by a factor of 10 would result in a ball of nucleons (that is, a nucleus) with 10 times the volume. Because the attractive force is short-range, there is no tendency for the nucleons to be compressed together as the size of the nucleus increases. This is very different from planets, which are held together by the long-range gravitational force of attraction between all of the parts of the planet.

As an example, compare the planets Jupiter and Saturn (which have the same chemical composition): Jupiter has 3.3 times more mass than Saturn, but Jupiter has only 1.7 times greater volume because it is more highly compressed and has a greater density. Nuclei of different sizes do not behave like planets of different sizes: All nuclei have basically the same density (see Example 27-2 below). This is further evidence that the strong nuclear force is short-range rather than long-range.

Can we simply add more protons and more neutrons to make larger and larger nuclei? The answer is "no," and for the same reason that neutrons are required for nuclear stability. We have seen that as we work our way up the periodic table to atoms that have more and more protons, more *additional* neutrons are required for stability. Each additional proton exerts a repulsive force on all the others, but neutrons and protons can only attract their nearest neighbors. Those additional neutrons cause the size of the nucleus to grow, so that eventually, too many neutrons are near the surface of the nucleus and therefore not completely surrounded by neighbors. At that point, the nuclear forces holding the nucleus together are not large enough to overcome the Coulomb repulsion between the protons and the nucleus cannot be stable. The largest stable nuclide is lead-208 (^{208}Pb), which has 82 protons and 126 neutrons.

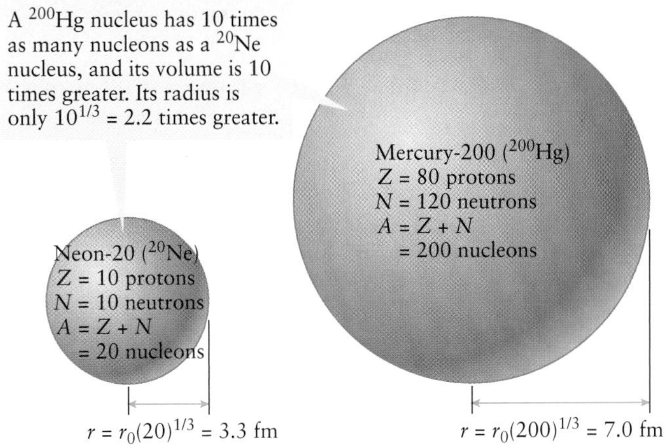

A ^{200}Hg nucleus has 10 times as many nucleons as a ^{20}Ne nucleus, and its volume is 10 times greater. Its radius is only $10^{1/3} = 2.2$ times greater.

Mercury-200 (^{200}Hg)
$Z = 80$ protons
$N = 120$ neutrons
$A = Z + N$
$= 200$ nucleons

Neon-20 (^{20}Ne)
$Z = 10$ protons
$N = 10$ neutrons
$A = Z + N$
$= 20$ nucleons

$r = r_0(20)^{1/3} = 3.3$ fm

$r = r_0(200)^{1/3} = 7.0$ fm

Figure 27-3 **Nuclear sizes** The volume of an atomic nucleus is proportional to the number of nucleons it contains (the mass number A). The nuclear radius is proportional to the cube root of A.

Example 27-1 Nuclear Radii

Estimate the radius of the nucleus of ^{12}C, a relatively small nucleus; ^{118}Sn, a nucleus of medium size; and ^{236}U, a relatively large nucleus. The nuclides ^{12}C and ^{118}Sn are stable; ^{236}U, like all other isotopes of uranium, is unstable.

Set Up

In each case, we use Equation 27-1 to calculate the radius of the nucleus. The value of A for each nucleus is given by the pre-superscript: $A = 12$ for the carbon (C) nucleus, $A = 118$ for the tin (Sn) nucleus, and $A = 236$ for the uranium (U) nucleus.

Radius of a nucleus:

$$r = r_0 A^{1/3} \qquad (27\text{-}1)$$

Solve

Apply Equation 27-1 to each nuclide.

We'll use $r_0 = 1.2$ fm in our calculations.
For ^{12}C, which has 6 protons and 6 neutrons,

$$r(^{12}\text{C}) = (1.2 \text{ fm})(12)^{1/3} = 2.7 \text{ fm}$$

For ^{118}Sn, which has 50 protons and 68 neutrons,

$$r(^{118}\text{Sn}) = (1.2 \text{ fm})(118)^{1/3} = 5.9 \text{ fm}$$

For ^{236}U, which has 92 protons and 144 neutrons,

$$r(^{236}\text{U}) = (1.2 \text{ fm})(236)^{1/3} = 7.4 \text{ fm}$$

Reflect

This calculation shows that typical nuclei have radii of just a few femtometers. Although ^{236}U has $236/12 = 19.7$ times as many nucleons as ^{12}C, it is only larger in radius by a factor of $(7.4 \text{ fm})/(2.7 \text{ fm}) = 2.7$. That's a consequence of the $A^{1/3}$ factor in Equation 27-1: Note that $(19.7)^{1/3} = 2.7$.

Notice that in carbon, the smallest of the three nuclei, the number of neutrons equals the number of protons. In tin, which has almost 10 times the mass number of carbon, the number of neutrons required for nuclear stability is 36% higher than the number of protons. And even with nearly 60% more neutrons than protons, the relatively large ^{236}U nucleus is not stable. So we see direct evidence that more and more additional neutrons, compared to the number of protons, are required for nuclear stability, and that at some size a nucleus is too large for the strong nuclear force to overcome the Coulomb repulsion between the protons.

Example 27-2 Nuclear Density

Estimate the density (in kg/m³) of a nucleus that has mass number A.

Set Up

The density of an object is its mass m divided by its volume V. We'll find the volume of a nucleus from Equation 27-1 and the formula for the volume of a sphere. To estimate the mass of a nucleus, we'll multiply the mass number A (the number of nucleons) by the average mass of a nucleon.

Radius of a nucleus:

$$r = r_0 A^{1/3} \qquad (27\text{-}1)$$

Definition of density:

$$\rho = \frac{m}{V} \qquad (11\text{-}1)$$

Volume of a sphere of radius r:

$$V = \frac{4}{3}\pi r^3$$

mass $m = A \times$ (average mass of a nucleon)
radius $r = r_0 A^{1/3}$
volume $V = (4/3)\pi r^3$
density $\rho = m/V$

Solve

Find the volume of a nucleus of mass number A.

Substitute Equation 27-1 into the expression for the volume of a sphere of radius r:

$$V = \frac{4}{3}\pi r^3 = \frac{4}{3}\pi (r_0 A^{1/3})^3 = \frac{4}{3}\pi r_0^3 (A^{1/3})^3$$

$$= \frac{4}{3}\pi r_0^3 A$$

Use $r_0 = 1.2 \text{ fm} = 1.2 \times 10^{-15}$ m as in Example 27-1:

$$V = \frac{4}{3}\pi (1.2 \times 10^{-15}\text{m})^3 A$$

$$= (7.2 \times 10^{-45}\text{m}^3)A$$

Take the average mass of a nucleon to be the average of the proton mass m_p and the neutron mass m_n. Use this to write an expression for the mass of a nucleus of mass number A.

The average mass of a nucleon is

$$m_\text{avg} = \frac{m_\text{p} + m_\text{n}}{2} = \frac{(1.6726 \times 10^{-27}\text{ kg}) + (1.6749 \times 10^{-27}\text{ kg})}{2}$$

$$= 1.6738 \times 10^{-27}\text{ kg}$$

A nucleus with A nucleons then has mass

$$m = m_\text{avg}A = (1.6738 \times 10^{-27}\text{ kg})A$$

Calculate the density of the nucleus.

The density of the nucleus is

$$\rho = \frac{m}{V} = \frac{(1.6738 \times 10^{-27}\text{ kg})A}{(7.2 \times 10^{-45}\text{m}^3)A} = \frac{1.6738 \times 10^{-27}\text{ kg}}{7.2 \times 10^{-45}\text{m}^3}$$

$$= 2.3 \times 10^{17}\text{ kg/m}^3$$

Reflect

Our final expression for the density ρ does not depend on A, the mass number of the nucleus. So the density of atoms of *all* nuclei is about the same. This agrees with our statements about the short-range character of the strong nuclear force. Note also that iridium, the densest of all stable elements, has a density of 22,650 kg/m³ (22.65 times the density of water). Our calculation shows that nuclei are 10^{13} times denser than iridium. Nuclei are *extremely* dense! This makes sense: Most of the mass of an atom is concentrated in its nucleus, which has a far smaller volume than the atom as a whole. So nuclear density (mass divided by volume) must be far greater than what we think of as the "ordinary" density of matter such as water.

Nuclear Spin and Magnetic Resonance Imaging

We learned in Section 26-6 that electrons have a type of intrinsic angular momentum called *spin*. (This is something of a misnomer because this angular momentum does not correspond directly to a spinning motion of the electrons.) Protons, and neutrons, too, have spin, and like an electron the spin of a proton or neutron can take on one of two values, often referred to as "spin up" and "spin down." In nuclei with an even number of nucleons, typically there are as many "spin up" nucleons as there are "spin down," and most such nuclei have zero net spin. (The orbital angular momentum of nucleons moving inside the nucleus can also contribute to the net spin of the nucleus.) But a nucleus with an odd number of nucleons must have a nonzero net spin. In particular, the nucleus of hydrogen—a single proton—has a net spin.

Measurements that make use of the spin of hydrogen nuclei enable us to localize hydrogen in an object or body, as well as to get information about the material in which those hydrogen atoms are embedded. That's the principle of **magnetic resonance imaging** (MRI). In an MRI scanner, spin information is used to form a three-dimensional map of the density of hydrogen atoms in a body. Because living organisms are composed largely of water—our bodies, for example, are 60 to 70% water—and because each water molecule contains two hydrogen atoms, living organisms are packed full of protons. As such, MRI is an ideal way to probe the internal structures in the body. The MRI scan in Figure 27-4 shows the leg bones rubbing against each other in the knee of a person with osteoarthritis. Figure 27-1a shows an MRI scan of a patient's head.

Protons and neutrons also have associated magnetic fields. This field is related to the spin, so because spin is either up or down, the field is a dipole in nature. Like a bar magnet, this magnetic field has a direction, and that direction is parallel to the direction of the angular momentum vector. As a result, when a nucleus that has an odd number of nucleons A is placed in a uniform external magnetic field, the magnetic force tends to align the spin direction either generally parallel to or antiparallel to the external field. To be precise, the spin direction aligns so that its component along the direction of the external field has a fixed positive or negative value. As such, the spin direction can rotate, or *precess*, around an axis defined by the field direction. Figure 27-5a shows this precession for a nucleus with spin aligned with the external field, and Figure 27-5b shows this precession for a nucleus with spin anti-aligned with (that is, opposed to) the external field.

Consider a large number of hydrogen atoms placed in a uniform magnetic field. About half of the protons will end up with spin aligned with the field and half with spin anti-aligned. Now, the energy of the spin-aligned orientation of a nucleus in an external magnetic field (Figure 27-5a) is higher than the energy of the anti-aligned orientation (Figure 27-5b) by an amount ΔE that depends on the magnetic field

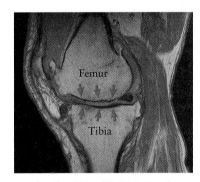

Figure 27-4 A magnetic resonance image The red arrows in this MRI image indicate where bones rub together in the knee of a patient suffering from osteoarthritis.

(a) A nucleus with its spin aligned with an external magnetic field

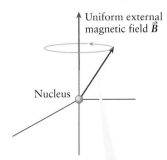

The spin angular momentum vector of the nucleus traces out a cone, but is generally aligned in the same direction as $\vec{B}$.

(b) A nucleus with its spin opposed to an external magnetic field

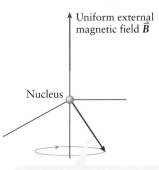

The spin angular momentum vector of the nucleus traces out a cone, but is generally aligned in the direction opposite to $\vec{B}$.

Figure 27-5 Nuclear spin When a nucleus is placed in an external magnetic field, the component of the spin along the direction of the external field is either (a) aligned with the external field or (b) anti-aligned with the external field.

strength. Suppose we now bathe the atoms in an additional alternating magnetic field with frequency f. The alternating field is made up of photons of frequency f and energy hf, where h is Planck's constant (see Section 26-2). If we choose the frequency f so that hf is equal to ΔE, that means the photon energy is just equal to the energy difference between the two spin states. As a result, protons that are initially in the lower energy state, with their spin anti-aligned with the field, can absorb a photon of energy $\Delta E = hf$ and flip their spin to align with the external field. What is more, protons that are initially in the higher energy state, with their spin aligned with the field, can be stimulated by the alternating field to emit a photon of energy $\Delta E = hf$ and so flip their spin to the lower energy, anti-aligned state. (We discussed this process of *stimulated emission* in Section 26-5.)

The difference in energy between the two spin states of a hydrogen atom depends on the strength of the external field. With the high field strength required to be able to clearly observe the spin-flipping phenomenon, typically around 3 T, the energy difference ΔE is about 5×10^{-7} eV. From Equation 22-26, the frequency of a photon of this energy is

$$f = \frac{\Delta E}{h} = \frac{5 \times 10^{-7} \text{ eV}}{4.14 \times 10^{-15} \text{ eV} \cdot \text{s}} = 1.2 \times 10^8 \text{ s}^{-1} = 120 \text{ MHz}$$

(We used the value of h in eV $\cdot$ s.) So the frequency of the oscillating field used to induce the hydrogen atoms to flip their spins in a strong magnetic field is around 100 MHz, which is in the radio-frequency part of the electromagnetic spectrum.

How do we "see" this spin flipping and so form an image of the location of hydrogen? Initially, there will always be more atoms in the lower energy state than the higher energy state (lower energy is more likely for a system than higher energy), so when the hydrogen atoms in an external magnetic field are exposed to radio-frequency waves, there is a net absorption of the electromagnetic energy. When the radio-frequency signal is turned off, the spins begin to return to their initial (equilibrium) state, and in so doing they generate a radio-frequency field. The MRI device detects that radio-frequency field and uses it to form a map of the density of hydrogen atoms in the body.

As we have mentioned, the difference in energy ΔE between the two hydrogen spin states depends on the strength of the external magnetic field. In an MRI device, the magnetic field is made to vary over a body's volume, so that the energy absorbed and then re-emitted in the spin-flip process also varies in different parts of the body. The exact frequencies of the radio energy detected by the MRI device therefore provide the information necessary to create images of high spatial resolution. In addition, the time it takes for the spins of the hydrogen nuclei to return to their equilibrium state depends on the particular molecules in the tissue. So, timing information in an MRI device provides the means to differentiate one type of tissue from another.

❓ Got the Concept? 27-1   Nuclear Radius and Density

A ^{20}Ne nucleus has 10 protons and 10 neutrons for a total of 20 nucleons, and a ^{160}Dy nucleus has 66 protons and 94 neutrons for a total of 160 nucleons. Compared to a ^{20}Ne nucleus, a ^{160}Dy nucleus has (a) double the radius and a greater density; (b) 8 times the radius and a greater density; (c) double the radius and the same density; (d) 8 times the radius and the same density; (e) double the radius and a lower density.

Take-Home Message for Section 27-2

✔ The strong nuclear force binds nucleons (protons and neutrons) together in the nucleus of an atom.

✔ The volume of a nucleus is proportional to its mass number (the total number of nucleons in the nucleus).

✔ To be stable, a light nucleus must have about as many neutrons as protons. More massive nuclei with more than about 20 protons require more neutrons than protons to be stable. Very large nuclei with mass number greater than 208 are always unstable.

✔ Protons and neutrons have spin. Magnetic resonance imaging uses the difference in energy between a state in which a nuclear spin is aligned with an external magnetic field and the state in which the spin is anti-aligned with the field.

27-3 Some nuclei are more tightly bound and more stable than others

Release a ball at the top of a hill and it rolls down. Pull an object attached to the free end of a spring away from its equilibrium position and it tends to return to that position. In both cases, the systems are finding their way to a more stable configuration. All physical systems do the same; if a more stable configuration exists for a system, it will eventually find itself in that configuration as long as nature provides a mechanism for the transition to take place.

In this context, consider the nucleus of a helium atom, which consists of two protons and two neutrons. The four nucleons remain bound together in a configuration that must be more stable than when they are separate, otherwise the helium nucleus would end up broken apart. This stability results because the attraction of the strong nuclear force between the four nucleons overwhelms the electrostatic repulsive force between the two protons. But let's look at stability from another perspective.

The total mass M_{tot} of the two protons and two neutrons in the nucleus of ^{4}He equals the sum of two proton masses (m_p) and two neutron masses (m_n):

$$M_{tot} = 2m_p + 2m_n$$
$$= 2(1.6726 \times 10^{-27} \text{ kg}) + 2(1.6749 \times 10^{-27} \text{ kg}) = 6.695 \times 10^{-27} \text{ kg}$$

But the actual mass of a helium nucleus is 6.645×10^{-27} kg, which is *less* than the total mass of the two protons and two neutrons. How is this possible? The answer is the key to nuclear stability; the energy equivalent of the difference in mass is tied up in binding the nucleons together. Recall from Section 25-7 that energy and mass are equivalent: An object of mass m has a rest energy $E_0 = mc^2$ (Equation 25-21). Since a ^{4}He nucleus has a smaller mass than its constituent nucleons, it also has a smaller rest energy. The difference between the rest energy of the ^{4}He nucleus and the rest energy of its constituent nucleons is the **binding energy** E_B. You can think of this as the energy that is released when the four nucleons come together to form a ^{4}He nucleus. Alternatively, you can think of the binding energy as the energy that would be required to separate a ^{4}He nucleus into two protons and two neutrons. The greater the binding energy per nucleon in a nucleus, the more tightly the nucleus is bound and therefore the more stable it is.

Figure 27-6 is a graph of the binding energy per nucleon (E_B/A) as a function of mass number. The energy values on the vertical axis are given in MeV, where 1 MeV = 10^6 eV. As we have learned before, one eV, or one electron volt, is the amount of energy acquired by an electron when it experiences a potential difference of one volt: 1 eV = 1.60×10^{-19} J. Figure 27-6 shows that the value of E_B/A for all nuclides is in the range from 1 to 9 MeV, which is why these units are convenient. (By comparison, the binding energy of a hydrogen *atom*—the energy required to separate the single electron in a hydrogen atom from the proton—is 13.6 eV, about 10^{-5} as great as the binding energy of even the most weakly bound nucleus. This indicates how small the forces are on electrons within atoms compared to the forces within the nucleus.)

Figure 27-6 shows that the nuclide with the smallest binding energy per nucleon is ^{2}H, an isotope of hydrogen with one proton and one neutron. (Just 0.0115% of hydrogen atoms are of this isotope.) As A increases, E_B/A increases rapidly because more nucleons are surrounded by other nucleons to which they are attracted and so are more tightly bound. The value of E_B/A peaks at about 8.8 MeV for A in the range of 56 to 62, and then decreases slowly for higher and higher values of A. The decrease happens because the number of protons Z also increases as A increases, and the electric repulsion between protons destabilizes the nucleus. The most stable nuclei are ^{56}Fe and ^{58}Fe (two isotopes of iron) and ^{62}Ni; the binding energy per nucleon of these nuclei places them near the peak of the E_B/A curve in Figure 27-6.

Whenever possible, nuclei will rearrange themselves to maximize their stability and so maximize their binding energy per nucleon. Figure 27-6 shows that nuclei of relatively large values of mass number A can become more stable by decreasing the number of nucleons. One process of this kind is **nuclear fission**, in which nuclei split into smaller pieces. For example, atoms of curium-244 can spontaneously fission into xenon-135 and molybdenum-109. Curium has 96 protons, so ^{244}Cm has 148 neutrons. There are 54 protons in xenon and 42 in molybdenum or 96 total. The total number of

Figure 27-6 The curve of binding energy The binding energy per nucleon in a nucleus depends on the mass number *A*.

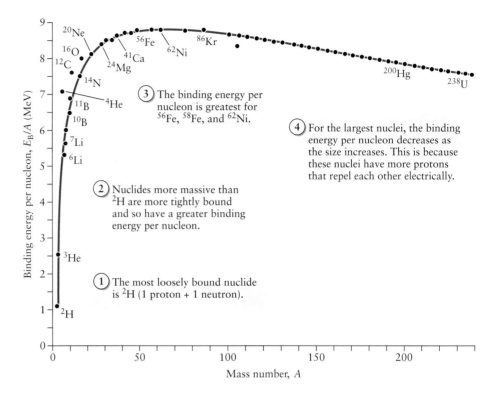

③ The binding energy per nucleon is greatest for ^{56}Fe, ^{58}Fe, and ^{62}Ni.

④ For the largest nuclei, the binding energy per nucleon decreases as the size increases. This is because these nuclei have more protons that repel each other electrically.

② Nuclides more massive than ^{2}H are more tightly bound and so have a greater binding energy per nucleon.

① The most loosely bound nuclide is ^{2}H (1 proton + 1 neutron).

nucleons in ^{135}Xe and ^{109}Mo, 244, equals the number of nucleons in ^{244}Cm. In other words, ^{135}Xe and ^{109}Mo contain the same 96 protons and 148 neutrons as in the original ^{244}Cm; the curium atom has split into two fragments. We'll discuss fission in more detail in Section 27-4.

Figure 27-6 also shows that nuclei with relatively small values of *A* can become more stable by increasing the number of nucleons. One way to do this is by **nuclear fusion**, in which two small nuclei join together to make a larger one. For example, the 12 protons and 12 neutrons in two carbon-12 nuclei can fuse to form a magnesium-24 nucleus. More than one atom can also be formed by fusion processes. For example, when a helium-3 nucleus fuses with a lithium-6 nucleus, the reaction forms two helium-4 nuclei and one hydrogen-1 nucleus. (Count the nucleons: ^{3}He has two protons and one neutron, and ^{6}Li has three protons and three neutrons, for a total of five protons and four neutrons. The two ^{4}He nuclei have two protons and two neutrons each, leaving one more proton to form a ^{1}H atom.) In Section 27-5 we'll discuss nuclear fusion more carefully.

Let's see how to calculate the binding energy per nucleon of a nucleus such as ^{4}He. We can do this by finding the total mass of the protons and neutrons that make up the nucleus, and comparing it to the mass of the actual nucleus. We then convert that difference to the equivalent energy. Instead of measuring mass in kilograms, it's most convenient to use units that take advantage of the equivalence between mass and energy. Since rest energy equals mass multiplied by c^2, we'll measure masses in units of MeV/c^2. Note that $1\ MeV/c^2 = 1.7827 \times 10^{-30}$ kg.

In these units, the mass of a proton is $938.27\ MeV/c^2$, the mass of a neutron is $939.57\ MeV/c^2$, and the mass of a ^{4}He nucleus is $3727.4\ MeV/c^2$. The difference Δ between (i) the mass of the two protons and two neutrons separately and (ii) the mass of the helium nucleus is

$$\Delta = 2m_p + 2m_n - m_{He}$$
$$= 2(938.27\ MeV/c^2) + 2(939.57\ MeV/c^2) - 3727.4\ MeV/c^2$$
$$= 28.3\ MeV/c^2$$

The energy equivalent of any mass is obtained by multiplying it by c^2, so the energy equivalent of this difference is 28.3 MeV. (Notice how straightforward it is to find the energy equivalent of mass when we write mass in units of MeV/c^2.) In other words, the binding energy of ^{4}He is 28.3 MeV. There are four nucleons in the helium nucleus, so E_B/A is about 7.1 MeV. You can verify this result from the curve in Figure 27-6.

The binding energy per nucleon is much higher for ^{4}He than for other light nuclei. (Figure 27-6 shows that E_B/A is about 2.5 MeV for ^{3}He and about 5.3 MeV for ^{6}Li.) Thanks to its relatively high binding energy per nucleon, ^{4}He is far more stable than other light nuclei. For this reason, when large nuclei break apart to transform to a more stable, more energetically favorable state, in many cases, they do so by emitting two protons and two neutrons bound together. In such processes the four bound protons and neutrons are referred to as an *alpha particle* (α particle), and we say that the alpha is emitted as nuclear radiation. Nuclear radiation is the emission by a nucleus of either energy or one of a small number of particles. We'll discuss nuclear radiation in more detail in Section 27-6.

To find the binding energy of ^{4}He, we subtracted the mass of the nucleus from the mass of the two protons and two neutrons separately, and then multiplied by c^2 to find the equivalent energy. In general, for a nucleus consisting of N neutrons and Z protons, E_B is

$$E_B = (Nm_n + Zm_p - m_{nucleus})c^2$$

where m_n is the mass of a neutron, m_p is the mass of a proton, and $m_{nucleus}$ is the mass of the nucleus. In practice, it's easier to measure the masses of neutral *atoms* (including their electrons) than the masses of isolated atomic nuclei. In terms of these masses, we can write the binding energy of a nucleus as

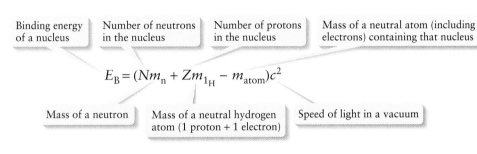

| Binding energy of a nucleus | Number of neutrons in the nucleus | Number of protons in the nucleus | Mass of a neutral atom (including electrons) containing that nucleus |

$$E_B = (Nm_n + Zm_{1_H} - m_{atom})c^2$$

Mass of a neutron — Mass of a neutral hydrogen atom (1 proton + 1 electron) — Speed of light in a vacuum

Binding energy of a nucleus (27-2)

The terms Zm_{1_H} and m_{atom} each include the mass of Z electrons, so the electron masses cancel. The masses of neutral atoms are given in Appendix C. Note that the values in that appendix are given in units of atomic mass units (u or amu), where

$$1\,u = 931.494\,\text{MeV}/c^2$$

? **Got the Concept? 27-2**
Stability

Rank these nuclides in order from most stable to least stable: (a) ^{11}B, (b) ^{20}Ne, (c) ^{86}Kr, (d) ^{200}Hg.

Example 27-3 The Binding Energy of ^{4}He

Previously, we determined the binding energy per nucleon in the ^{4}He nucleus using the mass of the nucleus. Use the values from Appendix C to determine the binding energy per nucleon (in MeV/c^2) in the ^{4}He nucleus using the atomic mass of ^{4}He.

Set Up

We'll use Equation 27-2 and the values given in Appendix C for the neutron mass, the atomic mass of ^{1}H, and the atomic mass of ^{4}He.

Binding energy of a nucleus:

$$E_B = (Nm_n + Zm_{1_H} - m_{atom})c^2$$

(27-2)

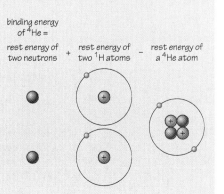

binding energy of ^{4}He =

rest energy of two neutrons + rest energy of two ^{1}H atoms − rest energy of a ^{4}He atom

Solve

Calculate the binding energy of the ^{4}He nucleus, which has two neutrons ($N = 2$) and two protons ($Z = 2$).

From Appendix C,

neutron mass $= m_n = 1.008665$ u

atomic mass of ^{1}H $= m_{1_H} = 1.007825$ u

atomic mass of ^{4}He $= m_{4_{He}} = 4.002602$ u

Substitute these into Equation 27-2, with $m_{atom} = m_{4_{He}}$:

$$E_B = (2(1.008665 \text{ u}) + 2(1.007825 \text{ u}) - 4.002602 \text{ u})c^2$$
$$= 0.030378 \text{ u}c^2$$

Since 1 u $= 931.494$ MeV/c^2,

$$E_B = 0.030378 \text{ u}c^2\left(\frac{931.494 \text{ MeV}/c^2}{1 \text{ u}}\right)$$

$$= 28.297 \text{ MeV}$$

The binding energy per nucleon equals the binding energy of the nucleus divided by the number of nucleons in the nucleus.

The ^{4}He nucleus has four nucleons (two neutrons and two protons), so the binding energy per nucleon is

$$\frac{E_B}{A} = \frac{28.297 \text{ MeV}}{4} = 7.0742 \text{ MeV}$$

Reflect

Previously, we calculated $E_B/A = 7.1$ MeV to two significant figures using the mass of the ^{4}He nucleus; our new calculation is consistent with this.

Why would we do this kind of calculation using atomic masses rather than nuclear masses? The reason is that in general, the masses of neutral atoms have been well measured, but precise measurements of the masses of atomic nuclei in isolation are difficult to obtain.

Take-Home Message for Section 27-3

✔ The binding energy of a nucleus is the energy that would be required to separate it into its individual nucleons.

✔ The greater the binding energy per nucleon in a nucleus, the more tightly the nucleus is bound and therefore the more stable it is.

✔ The most stable nuclides are ^{56}Fe, ^{58}Fe, and ^{62}Ni. Smaller and larger nuclei have a lower binding energy per nucleon.

27-4 The largest nuclei can release energy by undergoing fission and splitting apart

As we saw in the previous section, nuclei with higher values of binding energy per nucleon (E_B/A) are more stable than those with lower values. Figure 27-6, a plot of the binding energy E_B per nucleon in nuclei versus mass number A, shows that E_B/A decreases as A increases beyond 60 or so. In other words, large nuclei are less stable than smaller ones for A greater than about 60. As a consequence of this instability, these large nuclei can undergo processes that result in the fragmentation or *fission* of the nucleus into smaller nuclei. Fragmentation of a large nucleus can happen spontaneously, or it can be induced by imparting energy to the nucleus through a collision. In either case, the smaller fragments of a large nucleus have a higher value of E_B/A and are therefore more stable.

Let's take a look at one of the most important processes of this kind, called **neutron-induced fission**. As an example, the collision of a neutron with a ^{235}U nucleus will cause it to fission. Figure 27-7 shows one possible result. For a brief time the neutron and ^{235}U nucleus remain stuck together as ^{236}U*. (The asterisk indicates that this is an excited and short-lived state of ^{236}U.) This excited nucleus quickly fissions into fragments. In this particular reaction the fragments are an isotope of tellurium, an isotope of zirconium, and three neutrons. Because these fragments are all more stable than the

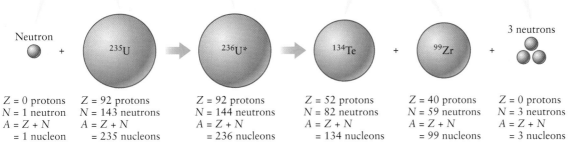

① A uranium nucleus (^{235}U) absorbs a neutron.

② The result is a uranium nucleus (^{236}U) in an excited state.

③ The excited uranium nucleus fissions into two smaller, more tightly bound nuclei...

④ ...as well as a few neutrons. These can trigger the fission of other ^{235}U nuclei.

Neutron

^{235}U

^{236}U*

^{134}Te

^{99}Zr

3 neutrons

$Z = 0$ protons
$N = 1$ neutron
$A = Z + N$
  $= 1$ nucleon

$Z = 92$ protons
$N = 143$ neutrons
$A = Z + N$
  $= 235$ nucleons

$Z = 92$ protons
$N = 144$ neutrons
$A = Z + N$
  $= 236$ nucleons

$Z = 52$ protons
$N = 82$ neutrons
$A = Z + N$
  $= 134$ nucleons

$Z = 40$ protons
$N = 59$ neutrons
$A = Z + N$
  $= 99$ nucleons

$Z = 0$ protons
$N = 3$ neutrons
$A = Z + N$
  $= 3$ nucleons

 Energy is released in this fission reaction: The total kinetic energy of the fission fragments is much greater than the total kinetic energy of the initial neutron and ^{235}U nucleus.

Figure 27-7 **Neutron-induced fission** When one of the largest nuclei absorbs a slow-moving neutron, it can fission into smaller, more stable fragments.

original nucleus, energy is released. The process described by Figure 27-7 occurs even when the colliding neutron is moving very slowly and so has essentially zero kinetic energy. So the released energy is almost entirely due to the change in binding energy between the initial ^{235}U nucleus and the fission products.

We can estimate the energy released in the process shown in Figure 27-7 by comparing the binding energy of the ^{235}U nucleus to the binding energies of the ^{134}Te and ^{99}Zr fragments. (There is no binding energy associated with the initial or final neutrons; they are not bound to any other particle.) If you make measurements on the graph in Figure 27-6, you'll see that the binding energy per nucleon E_B/A is about 7.6 MeV for $A = 235$, about 8.4 MeV for $A = 134$, and about 8.7 MeV for $A = 99$. The binding energy of each nucleus (E_B) equals the binding energy per nucleon (E_B/A) multiplied by the number of nucleons (A). The energy released in the fission reaction equals the difference between the total binding energy of the fragments and the binding energy of the initial ^{235}U nucleus:

(energy released)

= (binding energy of ^{134}Te) + (binding energy of ^{99}Zr)

    − (binding energy of ^{235}U)

= (134)(8.4 MeV) + (99)(8.7 MeV) − (235)(7.6 MeV)

= 200 MeV

We've given our result to just one significant figure because the values of E_B/A that we measured from Figure 27-6 are just estimates. The actual amount of energy released during this process is about 185 MeV, which is quite close to our estimate. This illustrates the tremendous amount of energy released in fission. By contrast, combustion (a chemical process that involves the electrons in molecules, not the nuclei of atoms) yields only a few eV for every molecule of fuel consumed. The energy release in fission is greater by a factor of several million!

When a heavy nucleus like ^{235}U undergoes fission, a wide variety of fragments can result. Figure 27-7 shows one possible result. Two others are

$$n + {}^{235}U \rightarrow {}^{236}U^* \rightarrow {}^{143}Ba + {}^{90}Kr + 3n$$

$$n + {}^{235}U \rightarrow {}^{236}U^* \rightarrow {}^{140}Xe + {}^{92}Sr + 4n$$

In each case the total number of protons and the total number of neutrons both remain the same. The ^{235}U nucleus has 92 protons and $235 − 92 = 143$ neutrons. In the first of these two processes, ^{143}Ba has 56 protons and 87 neutrons, and ^{90}Kr has 36 protons and 54 neutrons. The total number of protons after the fission has occurred is then $56 + 36 = 92$. The total number of neutrons before the fission is $143 + 1 = 144$ (including the neutron that starts the process). After the fission, the number of neutrons is $87 + 54 + 3 = 144$ (including the three neutrons released in the process). You can easily verify that the number of protons and the number of neutrons remain the same in the second process above.

Uranium-235 can also undergo *spontaneous* fission, in which the nucleus fragments without undergoing a collision with a neutron. The following example shows how to calculate the energy released in this process.

Example 27-4 Spontaneous Uranium Fission

Determine the energy released when a ^{235}U nucleus spontaneously undergoes fission to ^{140}Xe, ^{92}Sr, and three neutrons. The binding energy per nucleon in the nuclei of ^{235}U, ^{140}Xe, and ^{92}Sr are 7.59 MeV, 8.29 MeV, and 8.65 MeV, respectively.

Set Up

The energy released during the fission process is the difference between the binding energy of ^{140}Xe and ^{92}Sr nuclei and the binding energy of the ^{235}U nucleus. (There is no binding energy associated with the three neutrons.) The binding energy for each nucleus equals the binding energy per nucleon for that nucleus multiplied by the number of nucleons A.

(energy released)

= (total binding energy of fragments)
 − (binding energy of original ^{235}U nucleus)

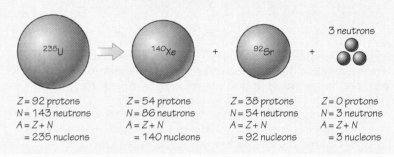

^{235}U	^{140}Xe	^{92}Sr	3 neutrons
$Z = 92$ protons	$Z = 54$ protons	$Z = 38$ protons	$Z = 0$ protons
$N = 143$ neutrons	$N = 86$ neutrons	$N = 54$ neutrons	$N = 3$ neutrons
$A = Z + N$	$A = Z + N$	$A = Z + N$	$A = Z + N$
$= 235$ nucleons	$= 140$ nucleons	$= 92$ nucleons	$= 3$ nucleons

Solve

Calculate the energy released.

The binding energies of the individual nuclei are

For ^{235}U: $(235)(7.59 \text{ MeV}) = 1784 \text{ MeV}$

For ^{140}Xe: $(140)(8.29 \text{ MeV}) = 1161 \text{ MeV}$

For ^{92}Sr: $(92)(8.65 \text{ MeV}) = 796 \text{ MeV}$

The energy released in the spontaneous fission is then

(energy released) = (binding energy of ^{140}Xe) + (binding energy of ^{92}Sr)
 − (binding energy of ^{235}U)

= 1161 MeV + 796 MeV − 1784 MeV

= 173 MeV

Reflect

This result is consistent with our earlier claim that a typical amount of energy released in the fission of ^{235}U is around 200 MeV.

Spontaneous fission of ^{235}U is a *very* unlikely process: A given ^{235}U nucleus has a 50% chance of decaying in a period of 7.04×10^8 years, and the probability that it will decay by spontaneous fission is only 7.0×10^{-11}. (Here 7.04×10^8 y is the *half-life* for spontaneous fission. We'll discuss this concept more carefully in Section 27-6. In that section we'll see that ^{235}U normally decays by a different process called *alpha emission*.) By contrast, once a ^{235}U nucleus merges with a neutron to form an excited ^{236}U* nucleus as in Figure 27.7, it typically undergoes fission within a fraction of a second.

All of the examples of fission that we've described result in the release of neutrons. Imagine what can happen if a large number of ^{235}U atoms are close to each other. Should one nucleus be struck by a neutron and fission as shown in Figure 27-7, there would then be three neutrons moving through the uranium. Should each of these neutrons strike a ^{235}U nucleus and start a fission process, there would be nine neutrons, so possibly nine more fissions. With a sufficient number of ^{235}U atoms present, this *chain reaction* quickly grows, with an accompanying rapid increase in energy released.

Isotopes that are capable of sustaining a fission chain reaction are used as nuclear fuels. Such isotopes are termed *fissile*; the most common fissile nuclear fuels are ^{233}U, ^{235}U, ^{239}Pu, and ^{241}Pu. The fission reactions they undergo are characterized by the production of, typically, two or three neutrons in addition to larger fragments. In addition, a chain reaction in a fissile material can be induced by a neutron carrying essentially zero kinetic energy. Indeed, in fuels such as ^{235}U, slower, less energetic neutrons are more efficiently absorbed by the fissile nuclei.

The process that produces electricity in most nuclear reactors uses the energy released in a fission chain reaction to heat water and produce steam to drive an electric generator (see Section 20-4). There are several challenges to producing energy in this way. First, a minimum amount, or *critical mass*, of fissile material must be present to sustain a chain reaction. As it happens, only a small percentage, about 0.7%, of the naturally occurring uranium in the world is ^{235}U, and the other three commonly used fissile isotopes do not occur naturally. Most naturally occurring uranium is ^{238}U. To use uranium as a nuclear fuel, then, it is necessary to separate the ^{235}U atoms from the ^{238}U, a costly and difficult process known as *enrichment*. In addition, the ^{238}U atoms that inevitably remain tend to absorb free neutrons and thereby inhibit a chain reaction. Once the critical mass of ^{235}U has been assembled, controlling the chain reaction is another challenge. If too many of the neutrons produced in the fissions result in a second fission, the energy released increases so rapidly that the fuel and whatever vessel is used to contain it can be damaged or even melt. To control a fission chain reaction in a nuclear reactor, control rods made of a substance that is a good absorber of neutrons are inserted between pieces of fuel.

Operating a fission nuclear reactor safely is perhaps the most significant challenge posed by nuclear power generation. First, the fission fragments are radioactive. Many of these fragments, or the fragments produced when they decay, are long-lived and tend to produce dangerous radiation for years, centuries, or even longer. In addition, reactors commonly use water as a *moderator*, a material that tends to slow the free neutrons (in order to make them more easily absorbed by a ^{235}U nucleus). Should the containment vessel rupture, this hot water can be released into the atmosphere in the form of steam carrying radioactive particles. Perhaps the most significant nuclear accident occurred at the Chernobyl nuclear power plant in Ukraine in 1986, in which an uncontrolled chain reaction caused a catastrophic power increase, leading to a series of explosions and the release of large quantities of radioactive steam, fuel, and smoke into the environment.

? Got the Concept? 27-3 Fission

Consider the spontaneous fission process ^{20}Ne $\rightarrow$ ^{10}B + ^{10}B. This process does not occur in nature. Why not? (a) The number of protons does not remain constant; (b) the number of neutrons does not remain constant; (c) both (a) and (b); (d) this process would absorb energy, not release it; (e) this process would neither release not absorb energy.

Take-Home Message for Section 27-4

✔ The binding energy per nucleon of nuclides with more than about 60 neutrons and protons decreases with increasing values of A. For this reason, the fission process, in which a nucleus breaks up into smaller fragments, leads to more stable configurations of the nucleons in bigger nuclei.

✔ Fission can be triggered by allowing a slow-moving neutron to merge with a fissile nucleus.

27-5 The smallest nuclei can release energy if they are forced to fuse together

In the previous section, we explored nuclear fission, processes by which a large nucleus splits into smaller fragments to move to a more stable configuration of the neutrons and protons it contains. Smaller is not always better, however. The larger the binding energy per nucleon (E_B/A) in a nucleus the more stable it is, and on the plot of E_B/A versus mass number (Figure 27-6), the very largest nuclei have lower values of E_B/A than smaller ones when A is greater than about 60. This is not the case, however, for small nuclei. With a few exceptions, as we move up toward a mass number A of around 60, the binding energy per nucleon increases. In other words, for small nuclei, moving to a more stable configuration requires processes that make the nucleus larger. These are *fusion* processes.

Figure 27-8 shows the process by which two ^{3}He nuclei (each with two protons and one neutron) fuse together to form a ^{4}He nucleus (with two protons and two neutrons). Two protons are left over, and without more neutrons there is no way for the protons to be bound together in a single nucleus. As a result, these protons fly off separately. Energy is released during this process because the final configuration of the protons and neutrons is more stable than the initial configuration. A photon carries away the energy released.

How much energy is released in the process shown in Figure 27-8? As in fission, the energy released in fusion is the difference between the total binding energy of the final nuclei and the total binding energy of the original nuclei. From Figure 27-6, the binding energy per nucleon for ^{3}He is approximately 2.5 MeV. Each ^{3}He has three nucleons, so the total binding energy of ^{3}He is 3(2.5 MeV) = 7.5 MeV and the two ^{3}He nuclei together have a combined binding energy equal to 7.5 MeV + 7.5 MeV = 15.0 MeV. In Section 27-3 we found that a ^{4}He nucleus has a binding energy of 28.3 MeV. The two protons are single particles, so they make no contribution to the binding energy. The energy released in the fusion process is therefore

$$\text{(energy released)} = \text{(binding energy of } ^4\text{He nucleus)} - \text{(binding energy of two } ^3\text{He nuclei)}$$
$$= (28.3 \text{ MeV}) - (15.0 \text{ MeV}) = 13.3 \text{ MeV}$$

The actual value, found using more accurate values of the binding energies, is closer to 12.86 MeV.

The process shown in Figure 27-8 is the final step in the *proton–proton cycle*, the fusion process that is the source of the Sun's energy. The cycle begins with the fusing of two protons (the nuclei of ^{1}H) to form ^{2}H. This nucleus fuses with another proton, forming ^{3}He, and finally, two ^{3}He nuclei fuse to form ^{4}He. We can summarize these three steps as

$$\text{Step 1: } ^1\text{H} + ^1\text{H} \rightarrow ^2\text{H} + \text{e}^+ + \nu_e$$
$$\text{Step 2: } ^2\text{H} + ^1\text{H} \rightarrow ^3\text{He} + \gamma$$
$$\text{Step 3: } ^3\text{He} + ^3\text{He} \rightarrow ^4\text{He} + ^1\text{H} + ^1\text{H} + \gamma$$

In Step 1, the e$^+$ particle is a **positron**, a particle with the same mass as an electron but with positive charge $+e$. The particle named ν_e (the Greek letter nu with a subscript e)

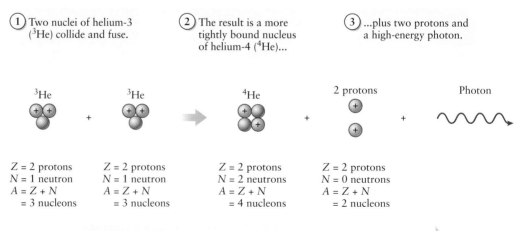

① Two nuclei of helium-3 (^{3}He) collide and fuse.

② The result is a more tightly bound nucleus of helium-4 (^{4}He)...

③ ...plus two protons and a high-energy photon.

| ^{3}He | ^{3}He | ^{4}He | 2 protons | Photon |

Z = 2 protons
N = 1 neutron
$A = Z + N$
= 3 nucleons

Z = 2 protons
N = 1 neutron
$A = Z + N$
= 3 nucleons

Z = 2 protons
N = 2 neutrons
$A = Z + N$
= 4 nucleons

Z = 2 protons
N = 0 neutrons
$A = Z + N$
= 2 nucleons

Figure 27-8 Nuclear fusion
The fusion of two ^{3}He nuclei to make a ^{4}He nucleus is one of the energy-releasing reactions that takes place in the core of the Sun.

Energy is released in this fusion reaction: The total kinetic energy of the ^{4}He nucleus and protons, plus the energy of the photon, is much greater than the total kinetic energy of the initial ^{3}He nuclei.

is a **neutrino**, a nearly massless, neutral particle. Also note that this step involves a proton being converted into a neutron. We'll discuss this conversion, called beta-plus decay, in Section 27-6. Step 1, and the subsequent interaction of the positron with an electron in the Sun (in which the two particles annihilate each other and convert into photons), release 1.44 MeV of energy. The energy released in Step 2 is 5.49 MeV. Both of these steps must occur twice before Step 3 can occur (because Step 3 requires two ^{3}He nuclei). So six protons are used in Steps 1 and 2, and in Step 3 two of those protons are returned along with a ^{4}He nucleus. The net result is therefore that four ^{1}H nuclei disappear and are replaced by one ^{4}He nucleus. The net energy release is 2(1.44 MeV) from the Step 1 happening twice, plus 2(5.49 MeV) from Step 2 happening twice, plus 12.86 MeV from the fusion of two ^{3}He nuclei in Step 3. The sum of these is a net energy release of 26.7 MeV as four ^{1}H nuclei are transformed into a ^{4}He nucleus.

Fission processes release more energy than the proton–proton fusion cycle, around 200 MeV compared to 26.7 MeV. Therefore, it might seem that fission is a more effective way to convert fuel to energy. Consider, however, that while 235 nucleons in ^{235}U are spent in order to release 200 MeV, in the fusion process described above, the 26.7 MeV released come at the expense of only four nucleons. Comparing energy per nucleon (think miles per gallon), fission provides less than 1 MeV per nucleon, while proton–proton fusion gives 26.7 MeV divided by 4, or nearly 7 MeV per nucleon. Fusion processes are more efficient at releasing the energy held in binding nuclei together.

Hydrogen makes up about 75% of the Sun's mass of approximately 2×10^{30} kg. If the proton–proton cycle leads to more stability, why doesn't all of that hydrogen quickly fuse to form ^{4}He, leaving the Sun a gigantic (and cool) ball of helium gas? The answer lies in the same forces at play within nuclei: the Coulomb force that repels protons from each other and the short-ranged strong nuclear force that draws them together. In order for two protons to fuse to form ^{2}H, they must come within a few femtometers of each other, at which point the strong attraction is able to overcome the Coulomb repulsion. This requires the protons to have considerable kinetic energy, which can come from being at high temperature. A temperature of more than 4×10^6 K is required for the proton–proton cycle to start. The temperature at the core of the Sun is around 15×10^6 K, so the proton–proton cycle can and does occur there. Even at that temperature, however, the probability that two nearby protons will fuse is small. This means that only a small fraction—about 4×10^{-19}—of the hydrogen in the Sun is undergoing fusion at any one time. The Sun won't burn out for a long time.

As a star ages, its core temperature increases and additional fusion reactions become possible. For example, three ^{4}He nuclei can fuse to form a ^{12}C nucleus. This requires a temperature of about 10^8 K because the ^{4}He nuclei are more massive than protons and repel each other more strongly due to their greater charge. At even higher temperatures, a ^{4}He nucleus can fuse with a ^{12}C nucleus to form a ^{16}O nucleus, a ^{4}He nucleus can fuse with a ^{16}O nucleus to form a ^{20}Ne nucleus, and so on. So as stars age, they manufacture heavier and heavier chemical elements. In the most massive stars, which have the highest core temperatures, so much kinetic energy is available at the very end of the star's evolution that fusion processes can produce even the heaviest nuclei up to uranium. Making these massive nuclei by fusion absorbs rather than releases energy, which is why it can happen only in very special circumstances.

These fusion reactions in stars make life on Earth possible. Here's why: When the universe first originated some 13.8 billion years ago, almost all ordinary matter was in the form of hydrogen or helium. (We'll discuss the origin of the universe in Chapter 28.) All heavier elements had to be manufactured by fusion within stars. After an aging star produces elements heavier than helium and goes through its final stages of evolution, it disperses much of its material into interstellar space (**Figure 27-9**). This material, which is enriched in heavy elements, can then be incorporated into a later generation of stars. We now understand that our Sun is such a "second-generation" star, with an elevated abundance of elements heavier than helium. As part of the process by which the Sun formed, some of these elements went into forming Earth and the other planets that orbit the Sun. This means that all of the nuclei of carbon in the organic compounds that make up your body, all of the nuclei in the oxygen that you breathe, and all of the nuclei in the silicon in the sand on Earth's beaches were manufactured in stars that died billions of years ago. This is one of the great lessons of nuclear physics: You are made of star-stuff.

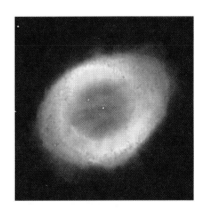

Figure 27-9 Seeding space with fusion products The Ring Nebula is a cloud of gas emitted by an aging star. The cloud includes nitrogen (shown in red) and oxygen (shown in green) produced by fusion reactions within the star.

? Got the Concept? 27-4 Fusion

Ordinary hydrogen gas is in the form of diatomic hydrogen (H_2), and more than 99.9% of the atoms in hydrogen gas have a nucleus that is a single proton (^1H). Why don't the two hydrogen nuclei in an H_2 molecule spontaneously undergo fusion as in Step 1 of the proton–proton cycle: $^1H + {}^1H \rightarrow {}^2H + e^+ + \nu_e$? (a) The nuclei are too far apart; (b) the nuclei are moving too slowly relative to each other; (c) the nuclei in an atom are different from those outside an atom; (d) both (a) and (b); (e) all of (a), (b), and (c).

Take-Home Message for Section 27-5

✔ The binding energy per nucleon of nuclides that have fewer than about 60 neutrons and protons is larger for increasing values of mass number A. For this reason, fusion of two small nuclei to form a larger one leads to more stable configuration of the nucleons.

✔ Fusion can only take place if the fusing nuclei come very close to each other. Therefore very high temperatures are required so the nuclei can overcome their mutual electric repulsion.

27-6 Unstable nuclei may emit alpha, beta, or gamma radiation

For many people, phrases such as radioactivity and nuclear radiation have come to be synonymous with danger. Not all nuclear radiation is dangerous, however. In addition, nuclear radiation in our environment is not limited to that released during the explosion of a bomb or an incident at a power plant. In this section we'll explore a number of aspects of nuclear radiation.

All naturally occurring nuclear processes take place because the final state is more stable, more energetically favorable, than the initial state. This is true of fission and fusion processes, and it is true of radiation processes, too. A relatively few nuclides—266 out of over 3000—are stable. All the rest are **radioactive**; that is, they decay into another nuclide by radiating away one or more particles. It's also possible for a nucleus in an excited state to radiate energy in the form of a photon as the nucleus transitions to a less excited state. Depending on the process, this radiation can carry away a small or a large amount of energy.

The three most common modes by which radiation occurs are *alpha*, *beta*, and *gamma* radiation. The terms were coined by Ernest Rutherford, who in his research between 1899 and 1903 classified radiation according to the depth that a radiation particle was able to penetrate other objects. Alpha particles penetrated the least, beta particles more, and gamma particles the most. (Rutherford received the 1908 Nobel Prize in Chemistry for this research, which was the first to show that one element can change into another through radioactive processes.) We now know that alpha particles are the nuclei of ^4He; beta particles are electrons; and gamma particles are photons of very high energy that can be millions of times greater than the energies of visible-light photons.

Before we consider the properties of these specific types of radioactive processes, let's look at a concept that is common to all of them—the idea of *radioactive half-life*.

Radioactive Decay and Half-Life

Nuclear radiation of all kinds—alpha, beta, or gamma—involves physics on the very small scale of the nucleus, and so is governed by quantum mechanics. As we learned in Section 26-6, quantum mechanics cannot tell us the position or velocity of a particular object at any time. It can, however, tell us the *probability* that an object will be at a particular place or have a particular velocity at any given instant. In the same way, quantum mechanics cannot tell us when a given radioactive nucleus will decay, but it can predict the probability that this nucleus will decay within a given time interval. Because all we can state is probabilities, it follows that radioactive decay is a statistical process.

As an example, consider the emission of a beta particle in the decay of ^{137}Cs (cesium), in which a neutron is converted into a proton, which proceeds as

$$^{137}\text{Cs} \rightarrow {}^{137}\text{Ba} + \text{e}^- + \bar{\nu}_e$$

The beta particle is the electron (e$^-$), and $\bar{\nu}_e$ is an **antineutrino**, related to the light, neutral neutrino particle we discussed in the context of fusion in the previous section. Every ^{137}Cs nucleus can, and eventually will, decay radioactively to a barium nucleus (^{137}Ba). In addition, we know that if a sample of ^{137}Cs contains, say, 10,000 atoms, after 30.17 years, only about 5000 will be left. However, if we select any one individual ^{137}Cs atom of those 10,000 atoms, we have no idea when its nucleus will decay. Perhaps it will decay in the next second, or perhaps not for 1000 years or more.

We can't make a definitive claim about when any particular atomic nucleus will decay, but we can quantify the *rate* at which a group of radioactive atoms decays, that is, the number of decays per second. Because radioactive decay is a statistical process, the probability λ that any one nucleus of a given type will decay in the next second is the same for all such nuclei. The quantity λ is called the **decay constant**. It has units of s^{-1}, since it refers to a probability per second. The value of λ is different for different radioactive nuclides: It is greater for nuclides that decay rapidly, and smaller for nuclides that decay slowly.

If we have a sample of N such nuclei, the total number of decays that take place in the next second—that is, the *decay rate*—will be equal to the product of the decay constant λ (the probability that any one nucleus decays in the next second) and N (the number of radioactive nuclei present). So the decay rate of a radioactive sample is greater if the sample is larger (the number of nuclei N in the sample is greater) or the nuclei have a higher decay constant λ. The SI unit of decay rate is the **becquerel** (Bq), after the nineteenth-century French physicist Antoine Henri Becquerel, who along with Marie Skłodowska-Curie and Pierre Curie won the 1903 Nobel Prize in Physics for their discovery of radioactivity. The becquerel is equivalent to one radioactive decay per second. Physicists also commonly use units of curies (Ci) for decay rate; 1 Ci $= 3.7 \times 10^{10}$ Bq $= 3.7 \times 10^{10}$ decays/s.

After an elapsed time of Δt seconds, the total number of decays will be equal to $\lambda N \Delta t$. This means that in a time Δt, the number of radioactive nuclei decreases by $\lambda N \Delta t$. So the *change* in the number of nuclei present is

$$\Delta N = -\lambda N \Delta t \tag{27-3}$$

The minus sign in Equation 27-3 indicates that the number of nuclei decreases as a result of the decays. If we divide both sides of Equation 27-3 by the elapsed time Δt, we get an expression for the rate of change $\Delta N / \Delta t$ of the number of nuclei:

In a sample of radioactive material, the rate of change of the number of radioactive nuclei...

... is negative because the number of nuclei decreases due to decay...

$$\frac{\Delta N}{\Delta t} = -\lambda N$$

Radioactive decay equation
(27-4)

...is proportional to the decay constant (the probability that a given nucleus decays in a one-second interval)...

...and is proportional to the number of radioactive nuclei that remain.

Equation 27-4 tells us that as time goes by, the decay rate will decrease because the number of radioactive nuclei will decrease. Using the tools of calculus, we can solve Equation 27-4 to find the number of nuclei present as a function of time, $N(t)$. The result is

$$N(t) = N_0 e^{-\lambda t} \tag{27-5}$$

In Equation 27-5 N_0 is the number of nuclei present at a specific time that we choose to call $t = 0$. This equation tells us that the number of nuclei present decreases

Figure 27-10 Nuclear decay and half-life This exponential curve shows the evolution of a sample that originally contains 10,000 cesium-137 (^{137}Cs) nuclei, which decay to barium-137 (^{137}Ba).

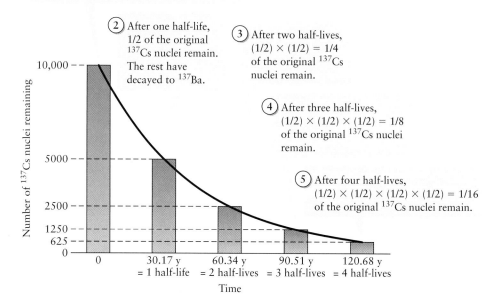

(1) Initially we have 10,000 ^{137}Cs nuclei. These nuclei are unstable, and decay to ^{137}Ba.

(2) After one half-life, 1/2 of the original ^{137}Cs nuclei remain. The rest have decayed to ^{137}Ba.

(3) After two half-lives, (1/2) × (1/2) = 1/4 of the original ^{137}Cs nuclei remain.

(4) After three half-lives, (1/2) × (1/2) × (1/2) = 1/8 of the original ^{137}Cs nuclei remain.

(5) After four half-lives, (1/2) × (1/2) × (1/2) × (1/2) = 1/16 of the original ^{137}Cs nuclei remain.

- The half-life of a nuclear decay is the time required for one-half of the nuclei present initially to decay.
- Radioactive decay is a statistical process: It's impossible to predict when any one individual unstable nucleus will decay.

exponentially, as Figure 27-10 shows. The number of decays per second at time t is equal to the decay constant λ (the decay probability per second per nucleus), multiplied by $N(t)$, the number of nuclei remaining at time t. We can write this as

$$R(t) = \lambda N(t) = \lambda N_0 e^{-\lambda t}$$

In this equation λN_0 is equal to the decay rate at $t = 0$, which we call R_0. So the decay rate as a function of time is

(27-6)
$$R(t) = R_0 e^{-\lambda t}$$

Equation 27-6 tells us that the decay rate, too, will decrease exponentially as time goes by. With the passage of time, a radioactive sample will undergo fewer and fewer decays.

Figure 27-10 shows that for the beta decay of ^{137}Cs, the number of ^{137}Cs nuclei remaining decreases by one-half every 30.17 years. Because the decay rate is proportional to the number of ^{137}Cs nuclei remaining, the decay rate also decreases by one-half every 30.17 years. This time is called the **half-life** of the radioactive decay, to which we give the symbol $\tau_{1/2}$ (the Greek letter tau). If there are N_0 radioactive nuclei present at $t = 0$, there will be $N_0/2$ present at $t = \tau_{1/2}$. If we substitute this into Equation 27-5, we get

$$\frac{N_0}{2} = N_0 e^{-\lambda \tau_{1/2}}$$

Divide both sides of this equation by N_0, then take the natural logarithm of both sides:

(27-7)
$$\ln\left(\frac{1}{2}\right) = \ln\left(e^{-\lambda \tau_{1/2}}\right)$$

Why do we do this? The reason is that the natural logarithm "undoes" the exponential function: For any x, $\ln(e^x) = x$. Furthermore, the natural logarithm has the property that for any x, $\ln(1/x) = -\ln x$. If we apply these to Equation 27-7, we get

$$-\ln 2 = -\lambda \tau_{1/2}$$

or

$$\tau_{1/2} = \frac{\ln 2}{\lambda} \qquad (27\text{-}8)$$

Equation 27-8 says that the half-life is inversely proportional to the decay constant λ, the probability that a given nucleus of a certain type will decay in a one-second interval. The greater the decay constant, the shorter the half-life and the more rapidly a sample of that nucleus will decay.

The half-life of radioactive sources varies widely, from far less than one second to billions of years or more. Commonly used radioactive sources include ^{32}P, a beta emitter used in DNA research that has a half-life of 14.3 days, ^{241}Am, an alpha emitter often found in household smoke detectors that has a half-life of 432.2 y, and ^{238}U, an alpha emitter with a half-life of 4.47×10^9 y. Notice that none of these have a half-life on the order of a second or less; radioactive isotopes with half-lives that short aren't terribly useful because they don't stay around long enough.

Example 27-5 Technetium

A form of an isotope of technetium, ^{99m}Tc, undergoes gamma decay with a half-life of 6.01 h. (The "m" stands for "metastable." Technetium-99 is widely used as a radioactive tracer for medical purposes, because its gamma radiation is easily detected and because technetium doesn't stay in the body for long. Hence the total radiation delivered to the patient is low.) (a) What is the decay constant λ, the probability that a nucleus of ^{99m}Tc will decay per second, in this sample? Does λ change as time goes on? (b) What fraction of the initial number of ^{99m}Tc nuclei will be left after 1.00 day? (c) What fraction will be left after 4.00 days have elapsed?

Set Up

We'll use Equation 27-8 to relate the decay constant λ to the half-life. Equation 27-5 will tell us the number of ^{99m}Tc nuclei remaining after a time t in terms of the initial number of nuclei N_0.

Half-life of a radioactive substance:

$$\tau_{1/2} = \frac{\ln 2}{\lambda} \qquad (27\text{-}8)$$

Number of radioactive nuclei present at time t:

$$N(t) = N_0 e^{-\lambda t} \qquad (27\text{-}5)$$

Solve

(a) Determine the decay constant for ^{99m}Tc.

We can rewrite Equation 27-8 as

$$\lambda = \frac{\ln 2}{\tau_{1/2}}$$

Substitute $\tau_{1/2} = 6.01$ h:

$$\lambda = \frac{\ln 2}{(6.01\ \text{h})}\left(\frac{1\ \text{h}}{60\ \text{min}}\right)\left(\frac{1\ \text{min}}{60\ \text{s}}\right) = 3.20 \times 10^{-5}\ \text{s}^{-1}$$

The probability that a given ^{99m}Tc nucleus will decay in a one-second interval is 3.20×10^{-5}, corresponding to odds of 1 in $(1/3.20 \times 10^{-5}) = 31{,}200$. This value depends only on the half-life, so it does not vary with time.

(b) To find the fraction of nuclei remaining after $t = 1.00$ d, substitute this value of t into Equation 27-5.

The fraction of ^{99m}Tc nuclei remaining after a time t equals the number remaining $N(t)$ divided by the number N_0 present initially:

$$\frac{N(t)}{N_0} = \frac{N_0 e^{-\lambda t}}{N_0} = e^{-\lambda t}$$

From part (a) we know the value of λ in s^{-1}, so we need to express t in seconds:

$$t = (1.00\ \text{d})\left(\frac{24\ \text{h}}{1\ \text{d}}\right)\left(\frac{60\ \text{min}}{1\ \text{h}}\right)\left(\frac{60\ \text{s}}{1\ \text{min}}\right) = 8.64 \times 10^4\ \text{s}$$

The fraction of nuclei remaining is then

$$\frac{N(1.00 \text{ d})}{N_0} = e^{-(3.20 \times 10^{-5} \text{ s}^{-1})(8.64 \times 10^4 \text{ s})}$$

$$= e^{-2.77} = 0.0628$$

(c) Repeat part (b) with $t = 4.00$ d.

The elapsed time is now

$$t = (4.00 \text{ d})\left(\frac{8.64 \times 10^4 \text{ s}}{1 \text{ d}}\right) = 3.46 \times 10^5 \text{ s}$$

and the fraction of nuclei remaining is

$$\frac{N(4.00 \text{ d})}{N_0} = e^{-(3.20 \times 10^{-5} \text{ s}^{-1})(3.46 \times 10^5 \text{ s})}$$

$$= e^{-11.1} = 1.55 \times 10^{-5}$$

Reflect

We can check our results by noting that $t = 1.00$ d is almost exactly four times the 6.01-h half-life of ^{99m}Tc, and $t = 4.00$ d is almost exactly 16 times the half-life. This check agrees with our calculations, as it should.

After each half-life, the number of ^{99m}Tc nuclei remaining decreases by one-half. So after four half-lives, the fraction of ^{99m}Tc nuclei remaining will be

$$\frac{1}{2} \times \frac{1}{2} \times \frac{1}{2} \times \frac{1}{2} = \frac{1}{2^4} = \frac{1}{16} = 0.0625$$

The fraction actually remaining at $t = 1.00$ d (slightly less than four 6.01-h half-lives) is 0.0628, very close to our estimate.
After 16 half-lives, the fraction of ^{99m}Tc remaining will be

$$\frac{1}{2^{16}} = \frac{1}{65,536} = 1.53 \times 10^{-5}$$

The fraction actually remaining at $t = 4.00$ d (slightly less than 16 times the 6.01-h half-life) is 1.55×10^{-5}, which again is very close to our estimate.

Alpha Radiation

When a large nucleus breaks into smaller fragments during a nuclear fission process, the new configuration of those protons and neutrons is more stable because the total binding energy increases. There are many ways, of course, for a large nucleus to split into smaller pieces, but the more likely decay products would be those that are more stable, that is, those with larger binding energy per nucleon. It is also probable, however, for smaller fragments to be ejected from a large nucleus. The radioactive emission of a ^{4}He nucleus is the most likely decay process for a large nucleus. As Figure 27-6 shows, ^{4}He has a greater binding energy per nucleon than any other nucleus with a small value of A. So the nucleus of the ^{4}He atom is far more stable than other small nuclei and therefore a far more probable decay product of large nuclei. Another name for a ^{4}He nucleus is an **alpha particle** (symbol α), and emission of an alpha particle is called **alpha decay**.

The alpha radiation process reduces the number of protons Z of the initial, or parent, nucleus by two, and reduces the number of neutrons of the parent by two. The result is a daughter nucleus that has an atomic number $Z-2$ and a mass number $A-4$, accompanied by an alpha particle (**Figure 27-11**).

The daughter nucleus and the alpha particle both carry away the energy released in alpha decay. However, the kinetic energy of the alpha particle is far greater than the kinetic energy of the daughter. We can confirm this by looking at the ratio of the kinetic energy K_α of the alpha particle to the kinetic energy K_D of the daughter. The alpha particle has mass m_α and is emitted with speed v_α, and the daughter nucleus has

③ ...plus an alpha particle (a ^{4}He nucleus).

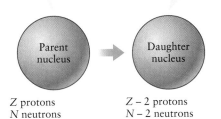

① This parent nucleus is large and unstable.

② The parent decays into a daughter nucleus with two fewer protons and two fewer neutrons...

Alpha (α) particle

Parent nucleus → Daughter nucleus +

Z protons
N neutrons
$Z + N = A$ nucleons

$Z - 2$ protons
$N - 2$ neutrons
$(Z - 2) + (N - 2)$
$= A - 4$ nucleons

$Z = 2$ protons
$N = 2$ neutrons
$Z + N = 4$ nucleons

Figure 27-11 Alpha decay Large nuclei can increase their stability by emitting an alpha particle and becoming a smaller, more stable daughter nucleus.

Energy is released in this decay: The daughter nucleus is more tightly bound (has a greater binding energy) than the parent.

mass m_D and is emitted with speed v_D. The kinetic energies of the alpha particle and the daughter are then

$$K_\alpha = \frac{1}{2}m_\alpha v_\alpha^2$$

$$K_D = \frac{1}{2}m_D v_D^2$$

The ratio of these kinetic energies is

$$\frac{K_\alpha}{K_D} = \frac{(1/2)m_\alpha v_\alpha^2}{(1/2)m_D v_D^2} = \frac{m_\alpha}{m_D}\left(\frac{v_\alpha}{v_D}\right)^2 \tag{27-9}$$

We can relate the speeds v_α and v_D by noting that momentum must be conserved in the alpha decay (because no external forces act on the parent nucleus as it decays). If the parent nucleus is at rest, the total momentum is zero before the decay and so must be zero after the decay. The alpha particle and daughter nucleus must therefore fly off in opposite directions, and each must have the same magnitude of momentum:

$$m_\alpha v_\alpha = m_D v_D \quad \text{so} \quad \frac{v_\alpha}{v_D} = \frac{m_D}{m_\alpha}$$

If we substitute this into Equation 27-9, we find that the ratio of the alpha particle's kinetic energy to that of the daughter nucleus is

$$\frac{K_\alpha}{K_D} = \frac{m_\alpha}{m_D}\left(\frac{v_\alpha}{v_D}\right)^2 = \frac{m_\alpha}{m_D}\left(\frac{m_D}{m_\alpha}\right)^2 = \frac{m_D}{m_\alpha}$$

The mass of the daughter nucleus is much larger than the mass of the alpha particle, so the fraction m_D/m_α is much greater than one and K_α is large compared to K_D.

All nuclei with more than 82 protons are unstable and have some probability of alpha decay. As an example, the element thorium ($Z = 90$ protons) undergoes alpha decay to radium which contains two fewer protons ($Z = 88$). The α decay of ^{228}Th, for example, is

$$^{228}\text{Th} \rightarrow {}^{224}\text{Ra} + \alpha$$

Just as ^{228}Th decays to ^{224}Ra, ^{224}Ra undergoes alpha decay to ^{220}Rn. This process continues until the daughter nucleus has $Z = 82$ or less. In this case the final alpha decay is to lead, with $Z = 82$:

$$^{228}\text{Th} \rightarrow {}^{224}\text{Ra} + \alpha \quad (\tau_{1/2\,\text{Th}} = 1.91\text{ y})$$
$$\quad \hookrightarrow {}^{220}\text{Rn} + \alpha \quad (\tau_{1/2\,\text{Ra}} = 3.63\text{ d})$$
$$\quad \hookrightarrow {}^{216}\text{Po} + \alpha \quad (\tau_{1/2\,\text{Rn}} = 55.6\text{ s})$$
$$\quad \hookrightarrow {}^{212}\text{Pb} + \alpha \quad (\tau_{1/2\,\text{Po}} = 0.145\text{ s})$$

Note that in each alpha decay the daughter nucleus has two fewer protons and two fewer neutrons than its parent.

Another nucleus that decays by alpha radiation is ^{235}U, which has a half-life $\tau_{1/2} = 7.04 \times 10^8$ y. (In Example 27-4 in Section 27-4 we looked at the spontaneous fission of ^{235}U. This is a very rare decay mode; ^{235}U undergoes alpha decay rather than spontaneous fission almost 100% of the time.) It is now understood that substantial amounts of ^{235}U, ^{238}U ($\tau_{1/2} = 4.47 \times 10^9$ y), and ^{232}Th ($\tau_{1/2} = 1.40 \times 10^{10}$ y) are found in Earth's core, and that the energy released by the alpha decay of these isotopes helps to sustain our planet's high internal temperatures and so keep much of the interior in a fluid state. All of Earth's geologic activity, including earthquakes, volcanic eruptions (Figure 27-1c), and the drifting of continents, is powered by the motions of our planet's fluid interior. So alpha decay plays an important role in Earth's dynamic geology.

Example 27-6 Alpha Decay of ^{238}U

The uranium isotope ^{238}U undergoes alpha decay to ^{234}Th. The binding energy per nucleon is 7.570 MeV in ^{238}U and 7.597 MeV in ^{234}Th. Find the energy released in the process ^{238}U $\rightarrow$ ^{234}Th $+ \alpha$.

Set Up

We'll use the same principle that we used in Example 27-4 (Section 27-4) to find the energy released in fission: The released energy equals the total binding energy of the nuclei present after the decay, minus the binding energy of the original (parent) nucleus. As in that example, we'll find the binding energy for each nucleus by multiplying the binding energy per nucleon times the number of nucleons A.

(energy released)

= (total binding energy of daughter plus alpha particle)
  − (binding energy of original ^{238}U nucleus)

Solve

From Example 27-3 (Section 27-3), the binding energy of a ^{4}He nucleus (an alpha particle) is

$E_B(^4\text{He}) = 28.297$ MeV

The binding energy of a ^{234}Th nucleus ($A = 234$) is

$E_B(^{234}\text{Th}) = (234)(7.597 \text{ MeV}) = 1777.7$ MeV

and the binding energy of a ^{238}U nucleus ($A = 238$) is

$E_B(^{238}\text{U}) = (238)(7.570 \text{ MeV}) = 1801.7$ MeV

The released energy is then

$E_{\text{released}} = E_B(^{234}\text{Th}) + E_B(^4\text{He}) - E_B(^{238}\text{U})$
$= 1777.7 \text{ MeV} + 28.297 \text{ MeV} - 1801.7 \text{ MeV}$
$= 4.3$ MeV

Reflect

The alpha particles emitted by radioactive isotopes with long half-lives, such as ^{238}U, tend to have kinetic energies in the 4 to 5 MeV range.

Beta Radiation

For most possible mass numbers A, there exist a number of nuclides with that same value of A. For example, molybdenum, technetium, ruthenium, and rhodium each have an isotope with 99 nucleons: ^{99}Mo has 42 protons and 57 neutrons, ^{99}Tc has 43 protons and 56 neutrons, ^{99}Ru has 44 protons and 55 neutrons, and ^{99}Rh has 45 protons and 54 neutrons. (It is also possible to create isotopes of other elements with A equal to 99.) The binding energy per nucleon in each is slightly different, however, so that only one is the most stable: For $A = 99$, the most stable isotope is ^{99}Ru. If there were a process whereby ^{99}Tc could convert one of its neutrons into a proton, or ^{99}Rh could convert one of its protons into a neutron, either of these nuclei could transform into the more stable ^{99}Ru.

The process that makes this possible is called **beta decay**. There are actually two varieties of beta decay. In **beta-minus decay** a neutron (charge zero) changes into a proton (charge $+e$). The net charge cannot change, so an electron (charge $-e$), also known as a beta-minus (β^-) particle, is also produced and escapes from the nucleus. To account for other conservation requirements a third particle, the neutral and nearly massless antineutrino ($\bar{\nu}_e$), is also created in this process. The full process, then, is

$$n \rightarrow p + e^- + \bar{\nu}_e \quad \text{(beta-minus decay)}$$

In **beta-plus decay** a proton ($+e$) changes into a neutron (charge zero). To conserve charge, a positively charged electron or *positron*, also called a beta-plus (β^+) particle, is also produced and escapes from the nucleus, along with a neutral and nearly massless neutrino (which, for our purposes, is essentially the same particle as an antineutrino). This process is

$$p \rightarrow n + e^+ + \nu_e \quad \text{(beta-plus decay)}$$

Figure 27-12a depicts the beta-minus decay of a nucleus with too many neutrons such as ^{99}Tc. We can write this process as

$$^{99}\text{Tc} \rightarrow {}^{99}\text{Ru} + e^- + \bar{\nu}_e$$

The number of nucleons ($A = 99$) is the same before and after the decay, but the number of protons Z has increased by 1 (from 43 to 44) and the number of

(a) Nuclei with too many neutrons can undergo beta-minus (β^-) decay.

①The daughter nucleus has one more proton and one fewer neutron than the parent.

②The nucleus also emits an electron (so electric charge is conserved) and an antineutrino.

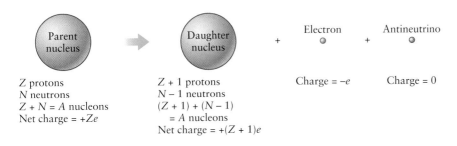

(b) Nuclei with too many protons can undergo beta-plus (β^+) decay.

①The daughter nucleus has one fewer proton and one more neutron than the parent.

②The nucleus also emits a positron (so electric charge is conserved) and a neutrino.

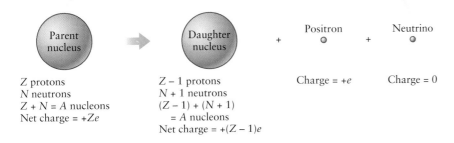

Figure 27-12 Beta decay Nuclei can increase their stability by (a) converting one neutron into a proton or (b) converting one proton into a neutron.

neutrons N has decreased by 1 (from 56 to 55). The decay of ^{137}Cs to ^{137}Ba, which we discussed at the beginning of this section, is another example of beta-minus decay. In this case Z increases from 55 to 56 and N decreases from 82 to 81.

One particularly important example of beta-plus decay is the decay of potassium-40 (^{40}K) to argon-40 (^{40}Ar). This process has a very long half-life of 1.25×10^9 y. Since potassium is abundant in Earth's interior, the energy released by this decay makes a substantial contribution to keeping our planet's interior in a fluid state and powering its geologic activity.

Figure 27-12b depicts the beta-plus decay of a nucleus with too many protons such as ^{99}Rh. We can write this process as

$$^{99}\text{Rh} \rightarrow {}^{99}\text{Ru} + e^+ + \nu_e$$

As for beta-minus decay, the number of nucleons remains the same (in this case, $A = 99$). But now the number of protons Z decreases by 1 (from 45 to 44), and the number of neutrons increases by 1 (from 54 to 55).

An important application of beta decay is in carbon-14 dating, a technique used to measure the age of objects that are composed, or partially composed, of organic matter. Almost all carbon atoms in Earth's atmosphere—for example, the carbon in carbon dioxide, CO_2—has a nucleus with a stable isotope of carbon, either ^{12}C (98.9%) or ^{13}C (1.1%). But about 1 in 10^{12} of those carbon atoms has a ^{14}C nucleus. This radioactive isotope of carbon is a β^- emitter with a half-life of 5730 y:

$$^{14}\text{C} \rightarrow {}^{14}\text{N} + e^- + \bar{\nu}_e$$

Carbon-14 is constantly produced in the atmosphere by cosmic rays slamming into ^{14}N nuclei. As a result, even though ^{14}C radioactively decays, the ratio of ^{14}C to ^{12}C in the atmosphere has remained relatively constant for at least tens of thousands of years. The ^{14}C/^{12}C ratio is the same in living organisms, for example, plants that breathe in CO_2, as it is in the atmosphere. However, once an organism dies, it no longer replenishes its supply of carbon, so the ^{14}C/^{12}C ratio decreases as the ^{14}C decays. As an example, a measurement of the ^{14}C/^{12}C ratio in, say, the smoke stains in the Chauvet-Pont-d'Arc Cave in southern France (see the figure that opens this chapter), which contains the earliest known cave paintings, allows a determination of time since the firewood that created the smoke was part of a living tree. Carbon-14 dating tells us that the paintings were made between 30,000 and 33,000 years ago.

Example 27-7 Ötzi the Iceman

In 1991, two German hikers discovered a human corpse in the Ötztal Alps on the border between Austria and Italy. The remains were not those of the victim of a climbing accident, but rather a well-preserved natural mummy of a man who lived during the last Ice Age. The rate of radioactive decay of ^{14}C in the mummy of "Ötzi the Iceman" was measured to be 0.121 Bq per gram. In a living organism, the rate of radioactive decay of ^{14}C is 0.231 Bq per gram. How long ago did Ötzi the Iceman live?

Set Up

Once Ötzi died, his body stopped taking in ^{14}C. After that time the number $N(t)$ of ^{14}C nuclei in his body decreased due to beta-minus decay. The ^{14}C decay rate $R(t)$, which is proportional to $N(t)$, decreased in the same manner. We are given $R(t) = 0.121$ Bq/g for the present-day decay rate, and $R_0 = 0.231$ Bq/g (the decay rate for a living organism and hence the decay rate at time $t = 0$, the last date on which Ötzi was still alive). We'll solve Equation 27-6 for the present time t (the elapsed time since Ötzi died), using Equation 27-8 to find the decay constant λ from the known half-life $\tau_{1/2} = 5730$ y of ^{14}C.

Decay rate as a function of time:

$$R(t) = R_0 e^{-\lambda t} \tag{27-6}$$

Half-life of a radioactive substance:

$$\tau_{1/2} = \frac{\ln 2}{\lambda} \tag{27-8}$$

Solve

Rearrange Equations 27-6 and 27-8 to find an expression for the time t since Ötzi died.

We know the present-day decay rate $R(t)$ and the initial decay rate in a living organism R_0. We want to find the time t since Ötzi died, so we rearrange Equation 27-6. Divide both sides by R_0:

$$\frac{R(t)}{R_0} = e^{-\lambda t}$$

Take the natural logarithm of both sides and recall that $\ln e^x = x$:

$$\ln\left(\frac{R(t)}{R_0}\right) = \ln e^{-\lambda t} = -\lambda t$$

Divide both sides by $-\lambda$:

$$t = -\frac{1}{\lambda}\ln\left(\frac{R(t)}{R_0}\right)$$

To get an expression for $1/\lambda$, divide both sides of Equation 27-8 by $\ln 2$:

$$\frac{1}{\lambda} = \frac{\tau_{1/2}}{\ln 2}$$

Putting everything together, the time t since Ötzi died is

$$t = -\frac{\tau_{1/2}}{\ln 2}\ln\left(\frac{R(t)}{R_0}\right)$$

Substitute the given values into the expression for t.

We are given $\tau_{1/2} = 5730$ y for ^{14}C, $R(t) = 0.121$ Bq/g, and $R_0 = 0.231$ Bq/g:

$$t = -\frac{(5730\text{ y})}{\ln 2}\ln\left(\frac{0.121\text{ Bq/g}}{0.231\text{ Bq/g}}\right) = -\frac{(5730\text{ y})}{0.693}\ln 0.524$$

$$= -\frac{(5730\text{ y})}{0.693}(-0.647) = 5350\text{ y}$$

Reflect

Our result shows that Ötzi died 5350 y ago. His well-preserved mummy thus gives us a unique look into life in prehistoric Europe.

Carbon-14 dating can only be used on objects less than about 50,000 years old, or about 8 to 10 half-lives of ^{14}C. For older objects the decay rate of ^{14}C has decreased to such a small value that it is difficult to measure accurately, and so any determination of age becomes difficult with this technique. For much older objects such as rocks, a similar approach is used but with isotopes with much longer half-lives. For example, the age of meteorites that fall to Earth is determined by looking at the ratio of uranium to lead (the endpoint of a series of alpha decays that starts with uranium); the oldest of these is more than 4.5×10^9 years old.

Gamma Radiation

A nucleus in an excited state radiates energy in a way analogous to the emission of a photon when an electron in an excited atomic state falls to a state of lower energy (see Section 26-5). Just like atoms, nuclei have excited states of definite energy, and when they transition from an excited state to a less excited one, they will emit a photon of energy equal to the difference in energy between the initial and final states.

The most common way that a nucleus can become excited is following an alpha decay or a beta decay. Although these decay processes result in a more stable

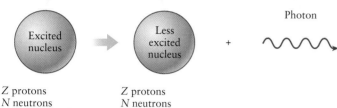

① In gamma (γ) decay, a nucleus in an excited state drops into a less excited state.

② The energy lost by the nucleus goes into a high-energy photon (gamma ray).

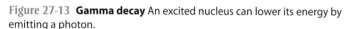

Photon

Excited nucleus → Less excited nucleus + 〰️〰️〰️〰️

Z protons
N neutrons
Z + N = A nucleons

Z protons
N neutrons
Z + N = A nucleons

In gamma decay, there is no change in the number of protons or the number of neutrons in the nucleus.

Figure 27-13 Gamma decay An excited nucleus can lower its energy by emitting a photon.

configuration of the nucleons, the nucleons that remain in the daughter nucleus may not be, initially, in the most stable arrangement for that particular nuclide. This excited daughter nucleus decays to a more stable configuration, giving off energy in the form of a gamma (γ) ray (Figure 27-13). This process is called **gamma decay**.

The energy carried away when a nucleus in an excited state decays is on the order of 1 MeV. From Equation 22-26 in Section 22-4, $E = hf$, the frequency and wavelength of a photon of energy $E = 1.00$ MeV $= 1.00 \times 10^6$ eV are

$$f = \frac{E}{h} = \frac{1.00 \times 10^6 \text{ eV}}{4.14 \times 10^{-15} \text{ eV} \cdot \text{s}} = 2.42 \times 10^{20} \text{ Hz}$$

$$\lambda = \frac{c}{f} = \frac{3.00 \times 10^8 \text{ m/s}}{2.42 \times 10^{20} \text{ Hz}}$$

$$= 1.24 \times 10^{-12} \text{ m} = 1.24 \times 10^{-3} \text{ nm}$$

Recall that the wavelength of visible light photons is in the range from 380 to 750 nm; a photon emitted by an excited nucleus is in the gamma radiation range, far from the visible part of the spectrum.

Gamma radiation does not change the atomic number Z or the neutron number N of a nucleus; that is, the number of protons and the number of neutrons remain the same after a γ ray is emitted. As an example, earlier we considered the beta-minus decay of ^{137}Cs to ^{137}Ba. In 95% of those decays, the ^{137}Ba nucleus is formed in an excited state that we denote as ^{137}Ba*:

$$^{137}\text{Cs} \rightarrow {}^{137}\text{Ba*} + \text{e}^- + \bar{\nu}_e$$

The excited barium nucleus then decays to its ground state by emission of a photon of energy 0.662 MeV:

$$^{137}\text{Ba*} \rightarrow {}^{137}\text{Ba} + \gamma$$

The values of $Z = 56$ and $N = 81$ for the barium nucleus do not change in this second step of the radiation process.

? Got the Concept? 27-5 Half-Lives

A certain radioactive isotope has a half-life of 5 days. You are given a sample containing a number of nuclei of this isotope. About how long would you have to wait before about 1/1000 of the initial number of nuclei remained? (a) 20 days; (b) 40 days; (c) 50 days; (d) 100 days; (e) 1000 days.

Take-Home Message for Section 27-6

✔ Radioactive decay is a statistical process. The number of nuclei and the decay rate both decrease exponentially with time, and both decrease by one-half in a time equal to one half-life.

✔ The three most common modes by which radiation occurs are alpha, beta, and gamma radiation.

✔ Alpha particles are ^{4}He nuclei and are emitted by large nuclei with $Z > 82$. In alpha emission the proton number and neutron number each decrease by 2.

✔ Beta particles are either negatively-charged electrons or positively-charged positrons. In beta-minus emission a neutron in the nucleus changes into a proton; in beta-plus emission a proton in the nucleus changes into a neutron.

✔ Gamma particles are high-energy photons, typically with energies of about 1 MeV. They are emitted when a nucleus decays from an excited state to a less excited one.

Key Terms

alpha decay
alpha particle
antineutrino
atomic number
becquerel
beta decay
beta-minus decay
beta-plus decay
binding energy

decay constant
gamma decay
half-life
isotope
magnetic resonance imaging
mass number
neutrino
neutron-induced fission
neutron number

nuclear fission
nuclear fusion
nucleon
nuclide
positron
radioactive
strong nuclear force

Chapter Summary

Topic	Equation or Figure
The strong nuclear force and nuclear sizes: The strong nuclear force is a short-range attractive force that acts between nucleons (protons or neutrons). Because the force is short-ranged, the volume of a nucleus is proportional to the mass number (number of nucleons). In larger nuclei, the number of neutrons must exceed the number of protons in order to counterbalance the electric repulsion between protons.	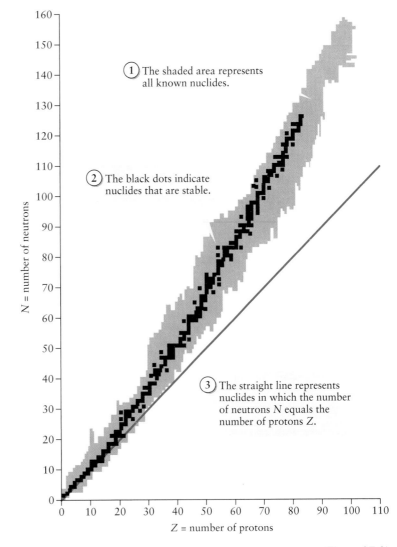

 For most nuclides, more neutrons than protons must be present (N must be greater than Z) to prevent the protons from flying apart due to electrostatic repulsion.

(Figure 27-2)

Nuclear binding energy: The binding energy of a nucleus is the energy required to separate it into its constituent nucleons. The binding energy per nucleon is greatest for nuclei with around 60 nucleons; for larger nuclei, the electric repulsion between protons makes nuclei less stable. The binding energy of a particular isotope can be calculated from the mass of a neutral atom containing that isotope.

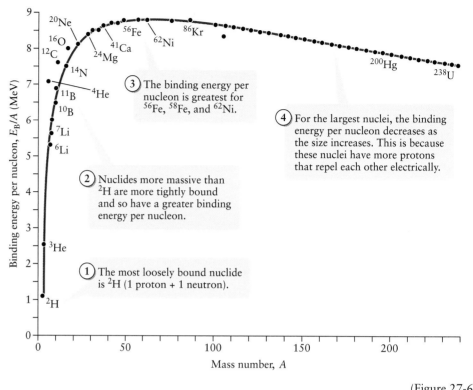

(3) The binding energy per nucleon is greatest for ^{56}Fe, ^{58}Fe, and ^{62}Ni.

(4) For the largest nuclei, the binding energy per nucleon decreases as the size increases. This is because these nuclei have more protons that repel each other electrically.

(2) Nuclides more massive than ^{2}H are more tightly bound and so have a greater binding energy per nucleon.

(1) The most loosely bound nuclide is ^{2}H (1 proton + 1 neutron).

(Figure 27-6)

Binding energy of a nucleus | Number of neutrons in the nucleus | Number of protons in the nucleus | Mass of a neutral atom (including electrons) containing that nucleus

$$E_B = (Nm_n + Zm_{1_H} - m_{atom})c^2 \qquad (27\text{-}2)$$

Mass of a neutron | Mass of a neutral hydrogen atom (1 proton + 1 electron) | Speed of light in a vacuum

Nuclear fission: When the largest nuclei absorb a neutron, they fragment (fission) into two smaller nuclei plus a few neutrons. The fragments are more tightly bound than the original nucleus, so energy is released in this process. In a sustained fission reaction, the released neutrons trigger other, nearby nuclei to also undergo fission.

(1) A uranium nucleus (^{235}U) absorbs a neutron.

(2) The result is a uranium nucleus (^{236}U) in an excited state.

(3) The excited uranium nucleus fissions into two smaller, more tightly bound nuclei...

(4) ...as well as a few neutrons. These can trigger the fission of other ^{235}U nuclei.

Neutron + ^{235}U → ^{236}U* → ^{134}Te + ^{99}Zr + 3 neutrons

Neutron	^{235}U	^{236}U*	^{134}Te	^{99}Zr	3 neutrons
$Z = 0$ protons	$Z = 92$ protons	$Z = 92$ protons	$Z = 52$ protons	$Z = 40$ protons	$Z = 0$ protons
$N = 1$ neutron	$N = 143$ neutrons	$N = 144$ neutrons	$N = 82$ neutrons	$N = 59$ neutrons	$N = 3$ neutrons
$A = Z + N$ = 1 nucleon	$A = Z + N$ = 235 nucleons	$A = Z + N$ = 236 nucleons	$A = Z + N$ = 134 nucleons	$A = Z + N$ = 99 nucleons	$A = Z + N$ = 3 nucleons

 Energy is released in this fission reaction: The total kinetic energy of the fission fragments is much greater than the total kinetic energy of the initial neutron and ^{235}U nucleus.

(Figure 27-7)

Nuclear fusion: The smallest nuclei can merge together to form a larger nucleus, releasing energy in the process. These fusion processes require very high temperatures so that the fusing nuclei have enough kinetic energy to overcome their mutual electric repulsion.

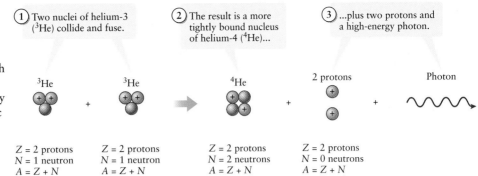

1. Two nuclei of helium-3 (^{3}He) collide and fuse.

2. The result is a more tightly bound nucleus of helium-4 (^{4}He)...

3. ...plus two protons and a high-energy photon.

^{3}He

^{3}He

^{4}He

2 protons

Photon

Z = 2 protons
N = 1 neutron
$A = Z + N$
= 3 nucleons

Z = 2 protons
N = 1 neutron
$A = Z + N$
= 3 nucleons

Z = 2 protons
N = 2 neutrons
$A = Z + N$
= 4 nucleons

Z = 2 protons
N = 0 neutrons
$A = Z + N$
= 2 nucleons

 Energy is released in this fusion reaction: The total kinetic energy of the ^{4}He nucleus and protons, plus the energy of the photon, is much greater than the total kinetic energy of the initial ^{3}He nuclei.

(Figure 27-8)

Nuclear decay and half-life: The decay of unstable nuclei is a statistical process. This means that the rate of decay is proportional to the number of unstable nuclei present. As a result, the number of nuclei and the decay rate both decline in an exponential manner. The time for the number of nuclei and the decay rate to decrease by one-half is called the half-life.

- The half-life of a nuclear decay is the time required for one-half of the nuclei present initially to decay.
- Radioactive decay is a statistical process: It's impossible to predict when any one individual unstable nucleus will decay.

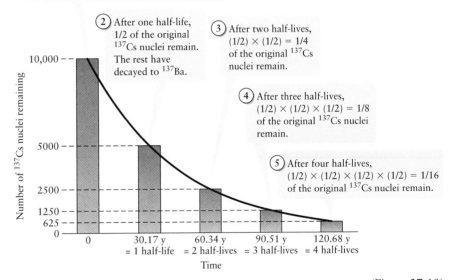

1. Initially we have 10,000 ^{137}Cs nuclei. These nuclei are unstable, and decay to ^{137}Ba.

2. After one half-life, 1/2 of the original ^{137}Cs nuclei remain. The rest have decayed to ^{137}Ba.

3. After two half-lives, (1/2) × (1/2) = 1/4 of the original ^{137}Cs nuclei remain.

4. After three half-lives, (1/2) × (1/2) × (1/2) = 1/8 of the original ^{137}Cs nuclei remain.

5. After four half-lives, (1/2) × (1/2) × (1/2) × (1/2) = 1/16 of the original ^{137}Cs nuclei remain.

(Figure 27-10)

Alpha, beta, and gamma decays: Large nuclei release energy and become more stable by emitting an alpha particle (a ^{4}He nucleus). Other nuclei with too many neutrons undergo beta-minus decay, in which one neutron changes into a proton; those with too many protons undergo beta-plus decay, in which one proton changes into a neutron. In gamma decay, an excited state of a nucleus (indicated by an asterisk) transitions to a less excited state and emits a gamma-ray photon (γ).

An alpha decay:

$$^{228}\text{Th} \rightarrow {}^{224}\text{Ra} + \alpha$$

A beta-minus decay:

$$^{99}\text{Tc} \rightarrow {}^{99}\text{Ru} + e^- + \bar{\nu}_e$$

A beta-plus decay:

$$^{99}\text{Rh} \rightarrow {}^{99}\text{Ru} + e^+ + \nu_e$$

A gamma decay:

$$^{137}\text{Ba}^* \rightarrow {}^{137}\text{Ba} + \gamma$$

Answer to What do you think? Question

(e) In a radioactive material, the amount that remains decreases by 1/2 after one half-life. After two half-lives, the amount that remains is $(1/2) \times (1/2) = (1/2)^2 = 1/4$; after three half-lives, $(1/2) \times (1/2) \times (1/2) = (1/2)^3 = 1/8$; and so on. The age of the cave painting is between $(30,000 \text{ y})/(5730 \text{ y}) = 5.2$ half-lives and $(33,000 \text{ y})/(5730 \text{ y}) = 5.8$ half-lives, so more than 5 half-lives have elapsed since the painting was made. Therefore the amount of carbon-14 remaining is less than $(1/2) \times (1/2) \times (1/2) \times (1/2) \times (1/2) = (1/2)^5 = 1/32$ of the original.

Answers to Got the Concept? Questions

27-1 (c) Equation 27-1 tells us that the radius of a nucleus is proportional to $A^{1/3}$, the cube root of the number of nucleons (the mass number). Compared to a ^{20}Ne nucleus, a ^{160}Dy nucleus has $160/20 = 8$ times as many nucleons, so its radius is larger by a factor of $8^{1/3} = 2$. As we discussed in Example 27-2, nuclei with any value of A have the same density.

27-2 (c), (b), (d), (a) A ranking in order of stability is a ranking in order of binding energy per nucleon. Figure 27-6 shows that this is greatest for ^{86}Kr (more than 8.5 MeV), second greatest for ^{20}Ne (slightly more than 8 MeV), third greatest for ^{200}Hg (slightly less than 8 MeV), and least for ^{11}B (about 6.9 MeV).

27-3 (d) The number of protons and the number of neutrons both remain the same: A ^{20}Ne nucleus has 10 protons and 10 neutrons, and each ^{10}B nucleus has 5 protons and 5 neutrons. But the binding energy per nucleon for ^{10}B is *less* than for ^{20}Ne (see Figure 27-6). For fission to occur and energy to be released, the fragments must have a greater binding energy per nucleon than the initial nucleus. The process ^{20}Ne $\rightarrow$ ^{10}B + ^{10}B would require a substantial amount of energy to be spontaneously added to the ^{20}Ne nucleus, and there's no way that can happen.

27-4 (d) In order for two ^{1}H nuclei to fuse, they must come to within a few femtometers (1 fm = 10^{-15} m) of each other. But in a typical molecule, including H_2, the separation between nuclei in adjacent atoms is on the order of 10^{-10} m. So the nuclei are normally too far apart to fuse. If they were moving rapidly with respect to each other and so had a large kinetic energy, the nuclei could overcome their mutual Coulomb repulsion and come close enough to fuse. But atomic nuclei move hardly at all within molecules, and so fusion cannot happen this way. Option (c) is incorrect: The properties of a nucleus are affected not at all by the presence or absence of electrons orbiting the nucleus.

27-5 (c) After each 5-day half-life, the number of nuclei decreases by a factor of 1/2. For the number to decrease by a factor of 1/1000 requires about 10 half-lives, since $(1/2)^{10} = 1/2^{10} = 1/1024$.

Questions and Problems

In a few problems, you are given more data than you actually need; in a few other problems, you are required to supply data from your general knowledge, outside sources, or informed estimate. Interpret as significant all digits in numerical values that have trailing zeros and no decimal points.

For all problems, use $g = 9.80 \text{ m/s}^2$ for the free-fall acceleration due to gravity. Neglect friction and air resistance unless instructed to do otherwise.
* Basic, single-concept problem
** Intermediate-level problem, may require synthesis of concepts and multiple steps
*** Challenging problem
SSM *Solution is in Student Solutions Manual*

Conceptual Questions

1. •What is an isotope?

2. •What is the difference between atomic number and mass number?

3. •Describe two characteristics of the binding energy that are comparable to the work function (from the photoelectric effect) and two characteristics that are dissimilar to the concept of the work function.

4. •(a) Describe what is meant by the phrase "larger nuclei are neutron rich." (b) Why do most nuclei contain at least as many neutrons as protons?

5. •A simple idea of nuclear physics can be stated as follows: "The whole nucleus weighs less than the sum of its parts." Explain why.

6. •Describe the basic characteristics of the nuclear force that exists between nucleons. What other competing force is present in the nucleus?

7. •Some historians would claim that without Einstein's special theory of relativity, nuclear physics would never have developed. Explain why. SSM

8. •What is the difference between fission and fusion?

9. •(a) Which elements in the periodic table are more likely to undergo nuclear fission? (b) Which are more likely to undergo nuclear fusion?

10. •**Astronomy** (a) Describe the nuclear reactions that occur in our Sun. (b) Discuss how the equilibrium state of the Sun is not permanent and discuss the eventual future of our solar system.

11. •Explain how conservation of energy and momentum would be violated if a neutrino were not emitted in beta decay. SSM

12. •To date, the decay constant of a radioactive nucleus is just that, *constant*. It does not depend on the size of the nuclear sample or the temperature or any external fields (such as gravity, electricity, or magnetism). Define the decay constant and

comment on how nuclear radioactivity would change if the quantity were dependent on temperature.

13. •At any given instant, a sample of radioactive uranium contains many, many different isotopes of atoms that are *not* uranium. Explain why.

14. •(a) Explain how radioactive ^{14}C is used to determine the age of ancient artifacts. (b) Which types of artifacts can have their age determined in this way and which types cannot?

15. •If atomic masses are used, explain why the mass of a beta particle is *not* accounted for in the basic beta decay

$$n \rightarrow p + e^- + \bar{\nu}_e$$

Assume that the mass of the antineutrino ($\bar{\nu}_e$) is very small and can be neglected. SSM

16. •**Medical** Describe, in broad terms, the health risks associated with the three major forms of radioactivity: alpha, beta, and gamma. Focus on the dangers due to inherent health risks and the ability of each to penetrate shielding material.

Multiple-Choice Questions

17. •In an atomic nucleus, the nuclear force binds _____ together.
   A. electrons
   B. neutrons
   C. protons
   D. neutrons and protons
   E. neutrons, protons, and electrons

18. •The mass of a nucleus is _____ the sum of the masses of its nucleons.
   A. always less than
   B. sometimes less than
   C. always more than
   D. always equal to
   E. sometimes equal to

19. •Which of the following statements is true?
   A. Fusion absorbs energy and fission releases energy.
   B. Fusion releases energy and fission absorbs energy.
   C. Both fusion and fission absorb energy.
   D. Both fusion and fission release energy.
   E. Both fusion and fission can release or absorb energy. SSM

20. •In fission processes, which of the following statements is true?
   A. Only the total number of nuclei remains the same.
   B. Only the total number of protons remains the same.
   C. Only the total number of neutrons remains the same.
   D. The total number of protons and the total number of nuclei both remain the same.
   E. The total number of protons and the total number of neutrons both remain the same.

21. •In a spontaneous fission reaction, the total mass of the products is _____ the mass of the original elements.
   A. greater than
   B. less than
   C. the same as
   D. double
   E. one-half

22. • What is the source of the Sun's energy?
   A. chemical reactions
   B. fission reactions
   C. fusion reactions
   D. gravitational collapse
   E. both fusion reactions and fission reactions

23. •In a spontaneous fusion reaction, the total mass of the products is _____ the mass of the original elements.
   A. greater than
   B. less than
   C. the same as
   D. double
   E. one-half SSM

24. • The decay constant λ depends only on
   A. the number of atoms at the initial time.
   B. the initial decay rate.
   C. the half-life.
   D. the binding energy per nucleon.
   E. whether the decay is alpha, beta, or gamma.

25. • The number of radioactive atoms in a radioactive sample
   A. decreases linearly with time.
   B. increases linearly with time.
   C. decreases exponentially with time.
   D. increases exponentially with time.
   E. remains constant.

26. • The decay rate for any isotope
   A. decreases linearly with time.
   B. increases linearly with time.
   C. decreases exponentially with time.
   D. increases exponentially with time.
   E. remains constant.

Estimation/Numerical Analysis

27. •Estimate the relative size of the nuclear force compared to the electrostatic force between two protons in the nucleus.

28. •Estimate the size of the nucleus compared to the size of the atom.

29. •Estimate the density of an atomic nucleus compared to an atom. SSM

30. •Estimate the energy of a typical nuclear reaction compared to the energy of a typical chemical reaction.

31. •Estimate the ratio of the nuclear force between two nucleons when they are separated by a distance of 0.5 fm compared to a separation distance of 2.0 fm.

32. •Estimate the number of half-lives that must go by before 10% of an isotope remains. What about 1%?

33. •Estimate the mass of a basketball that has the density of nuclear material.

34. •Estimate the number of nuclei that are in a 50-kg human body.

35. •If the half-life of a radioactive isotope is 1 day, (a) estimate how long it takes before the sample is reduced to 62.5% of its original amount and (b) how long before it is reduced to 6.25% of its original amount. SSM

36. •(a) Using a spreadsheet or programmable calculator, calculate the binding energy per nucleon for the following isotopes of the first five elements. Masses given are atomic masses.

hydrogen-1: 1.007825 u	lithium-8: 8.022486 u	boron-8: 8.024605 u
hydrogen-2: 2.014102 u	lithium-9: 9.026789 u	boron-10: 10.012936 u
hydrogen-3: 3.016049 u	lithium-11: 11.043897 u	boron-11: 11.009305 u
helium-3: 3.016029 u	beryllium-7: 7.016928 u	boron-12: 12.014352 u
helium-4: 4.002602 u	beryllium-9: 9.012174 u	boron-13: 13.017780 u
helium-6: 6.018886 u	beryllium-10: 10.013534 u	boron-14: 14.025404 u
helium-8: 8.033922 u	beryllium-11: 11.021657 u	boron-15: 15.031100 u
lithium-6: 6.015121 u	beryllium-12: 12.026921 u	
lithium-7: 7.016003 u	beryllium-14: 14.024866 u	

(b) Now calculate the binding energy per nucleon for the following isotopes of the last five naturally occurring elements. The masses provided are atomic masses.

radium-221: 221.01391 u	thorium-228: 228.028716 u	uranium-231: 231.036264 u
radium-223: 223.018499 u	thorium-229: 229.031757 u	uranium-232: 232.037131 u
radium-224: 224.020187 u	thorium-230: 230.033127 u	uranium-233: 233.039630 u
radium-226: 226.025402 u	thorium-231: 231.036299 u	uranium-234: 234.040946 u
radium-228: 228.031064 u	thorium-232: 232.038051 u	uranium-235: 235.043924 u
actinium-227: 227.027749 u	thorium-234: 234.043593 u	uranium-236: 236.045562 u
actinium-228: 228.031015 u	protactinium-231: 231.035880 u	uranium-238: 238.050784 u
thorium-227: 227.027701 u	protactinium-234: 234.043300 u	uranium-239: 239.054290 u

(c) Compare your results and comment on any patterns or trends that are obvious.

Problems

Note: In all cases, unless otherwise stated, binding energy, energy released, and the like, are to be expressed in MeV. Also, all atomic masses should be calculated to the nearest 10^{-6}.

27-1 The quantum concepts that help explain atoms are essential for understanding the nucleus

27-2 The strong nuclear force holds nuclei together

37. •Provide the elemental abbreviation, and give the number of protons, the number of neutrons, and the mass number for each of the following isotopes:

   A. hydrogen-3   C. aluminum-26
   B. beryllium-8   D. gold-197

E. technetium-100   G. osmium-190
F. tungsten-184   H. plutonium-239

38. ••Calculate the radius of each of the nuclei in problem 37.

39. •Name the element, and give the number of protons, the number of neutrons, and the mass number for each of the following nuclei:

   A. ^{2}H   D. ^{12}C   G. ^{131}I
   B. ^{4}He   E. ^{56}Fe   H. ^{235}U
   C. ^{6}Li   F. ^{90}Sr

40. ••If our Sun (mass $= 1.99 \times 10^{30}$ kg, radius $= 6.96 \times 10^8$ m) were to collapse into a neutron star (an object composed of tightly packed neutrons with roughly the same density as neutrons within a nucleus), what would the new radius of our "neutron-sun" be?

41. ••Given that a nucleus is approximately spherical and has a radius $r = r_0 A^{1/3}$ (where r_0 is about 1.2 fm), determine its approximate mass density. Express your answer in SI units and convert to tons per cubic inch, units that might be used in a news report. SSM

27-3 Some nuclei are more tightly bound and more stable than others

42. ••Calculate the atomic mass of each of the isotopes listed below. Give your answer in atomic mass units (u) and in grams (g). The values will include the mass of Z electrons.

   A. ^{1}H   D. ^{12}C   G. ^{131}I
   B. ^{4}He   E. ^{56}Fe   H. ^{238}U
   C. ^{9}Be   F. ^{90}Sr

43. •What is the binding energy of carbon-12? Give your answer in MeV. SSM

44. ••What is the binding energy per nucleon for the following isotopes?

   A. ^{2}H   D. ^{12}C   G. ^{129}I
   B. ^{4}He   E. ^{56}Fe   H. ^{235}U
   C. ^{6}Li   F. ^{90}Sr

45. •What minimum energy is needed to remove a neutron from ^{40}Ca and convert it to ^{39}Ca? The atomic masses of the two isotopes are 39.96259098 u and 38.97071972 u, respectively.

46. •What is the binding energy of the last neutron of carbon-13? The atomic mass of carbon-13 is 13.003355 u.

47. ••**Medical** Iodine-131 is a radioactive isotope that is used in the treatment of cancer of the thyroid. The natural tendency of the thyroid to take up iodine creates a pathway for which radiation (β- and γ) that is emitted from this unstable nucleus can be directed onto the cancerous tumor with very little collateral damage to surrounding healthy tissue. Another advantage of the isotope is its relatively short half-life (8 days). Calculate the binding energy of iodine-131 and the binding energy per nucleon. The mass of iodine-131 is 130.906124 u.

27-4 The largest nuclei can release energy by undergoing fission and splitting apart

48. •Calculate the energy released in the following nuclear fission reaction:

$$^{239}\text{Pu} + \text{n} \rightarrow {}^{98}\text{Tc} + {}^{138}\text{Sb} + 4\text{n}$$

Recall that the atomic masses are ^{239}Pu $= 239.052157$ u, ^{98}Tc $= 97.907215$ u, and ^{138}Sb $= 137.940793$ u.

49. •Complete the following nuclear fission reaction of thorium-232 and calculate the energy released in the reaction: ^{232}Th + n → ^{99}Kr + ^{124}Xe + __? The atomic masses are ^{232}Th = 232.038051 u, ^{99}Kr = 98.957606 u, and ^{124}Xe = 123.905894 u. SSM

50. •Complete the following fission reactions:
   A. ^{235}U + n → ^{128}Sb + ^{101}Nb + __?__
   B. ^{235}U + n → __?__ + ^{116}Pd + 4n
   C. ^{238}U + n → ^{99}Kr + __?__ +11n
   D. __?__ + n → ^{101}Rb + ^{130}Cs + 8n

51. •Complete the following fission reactions:
   A. ^{242}Am + __?__ → ^{90}Sr + ^{149}La + 4n
   B. ^{244}Pa + n → __?__ + ^{131}Sb + 12n
   C. __?__ + n → ^{92}Se + ^{153}Sm + 6n
   D. ^{262}Fm + n → ^{112}Rh + __?__ + 9n

52. •Calculate the energy (in MeV) released in the following nuclear fission reaction:

$$^{242}\text{Am} + \underline{\quad?\quad} \rightarrow {}^{90}\text{Sr} + {}^{149}\text{La} + 4n$$

Start by completing the reaction and use the following nuclear masses: ^{242}Am = 242.059549 u, ^{90}Sr = 89.9077387 u, and ^{149}La = 148.934733 u.

53. ••Assuming that in a fission reactor a neutron loses half its energy in each collision with an atom of the moderator, determine how many collisions are required to slow a 200-MeV neutron to an energy of 0.04 eV.

54. ••Knowing that the binding energy per nucleon for uranium-235 is about 7.6 MeV/nucleon and the binding energy per nucleon for typical fission fragments is about 8.5 MeV/nucleon, find an average energy release per uranium-235 fission reaction in MeV.

55. ••How many kilograms of uranium-235 must completely fission to produce 1000 MW of power continuously for one year? SSM

56. ••Repeat problem 55 in the more realistic case where the fission reactions are about 30% efficient in producing 1000 MW of power over 1 year of continuous operation.

57. ••Calculate the number of fission reactions per second that take place in a 1000-MW reactor. Assume that there are 200 MeV/reaction released.

27-5 The smallest nuclei can release energy if they are forced to fuse together

58. •Complete the following fusion reactions:
   A. ^{2}H + ^{3}H → ^{4}He + __?__
   B. ^{4}He + ^{4}He → ^{7}Be + __?__
   C. ^{2}H + ^{2}H → ^{3}He + __?__
   D. ^{2}H + ^{1}H → γ + __?__
   E. ^{2}H + ^{2}H → ^{3}H + __?__

59. ••Calculate the energy released in each of the fusion reactions in problem 58. Give your answers in MeV.

60. ••Astronomy Consider the proton–proton cycle that occurs in most stars (including our own Sun):

$$\text{Step 1: } {}^{1}\text{H} + {}^{1}\text{H} \rightarrow {}^{2}\text{H} + e^{+} + \nu_{e}$$
$$\text{Step 2: } {}^{2}\text{H} + {}^{1}\text{H} \rightarrow {}^{3}\text{He} + \gamma$$
$$\text{Step 3: } {}^{3}\text{He} + {}^{3}\text{He} \rightarrow {}^{4}\text{He} + 2\,{}^{1}\text{H} + \gamma$$

Calculate the net energy released from the three steps. Do *not* ignore the mass of the positron in Step 1. (You may ignore the mass of the neutrino.)

61. ••Each D–T fusion reaction releases about 20 MeV. How much tritium (T or ^{3}H) is needed to create 1014 J of energy, assuming that an endless supply of deuterium (D or ^{2}H) is available? SSM

62. ••How many fusion reactions per second must be sustained to operate a deuterium–tritium fusion power plant that outputs 1000 MW, operating at 33% efficiency?

27-6 Unstable nuclei may emit alpha, beta, or gamma radiation

63. •Complete the following conversions (cpm is counts per minute):
   A. 100 μCi = _____ Bq
   B. 1500 cpm = _____ Bq
   C. 16,500 Bq = _____ Ci
   D. 7.55×10^{10} Bq = _____ cpm

64. •The curie unit is defined as 1 Ci = 3.7×10^{10} Bq, which is about the rate at which radiation is emitted by 1.00 g of radium. (a) Calculate the half-life of radium from the definition. (b) What does your calculation tell you about the radiation emission rate of radium?

65. •A certain radioactive isotope has a decay constant of 0.00334 s^{-1}. Find the half-life in seconds and days. SSM

66. ••A radioactive sample is monitored with a radiation detector to produce 5640 counts per minute. Twelve hours later, the detector reads 1410 counts per minute. Calculate the decay constant and the half-life of the sample.

67. •What fraction of a sample of ^{32}P will be left after 4 months? Its half-life is 14.3 days.

68. •What fraction of a radioactive sample will be left after 6 half-lives? What about 7.5 half-lives?

69. ••Medical A patient is injected with 7.88 μCi of radioactive iodine-131 that has a half-life of 8.02 days. Assuming that 90% of the iodine ultimately finds its way to the thyroid, what decay rate do you expect to find in the thyroid after 30 days? SSM

70. ••The ratio of carbon-14 to carbon-12 in living wood is 1.3×10^{-12}. How many decays per second are there in 550 g of wood?

71. ••You take a course in archaeology that includes field work. An ancient wooden totem pole is excavated from your archaeological dig. The beta activity is measured at 150 cpm. If the totem pole contains 225 g of carbon and the ratio of carbon-14 to carbon-12 in living trees is 1.3×10^{-12}, what is the age of the pole?

72. ••Twelve centuries after a tree limb is cut, determine the decay rate for 500 g of carbon from the tree.

73. ••How many nuclei of radon-222 are present in a sample with an activity of 485 cpm? SSM

74. ••The ages of rocks that contain fossils can be determined using the isotope ^{87}Rb. This isotope of rubidium undergoes beta decay with a half-life of 4.75×10^{10} y. Ancient samples contain a ratio of ^{87}Sr to ^{87}Rb of 0.0225. Given that ^{87}Sr is a stable product of the beta decay of ^{87}Rb, and there was originally no ^{87}Sr present in the rocks, calculate the age of the rock sample.

Assume that the decay rate is constant over the relatively short lifetime of the rock compared to the half-life of ^{87}Rb.

75. •Complete the following alpha decays:
    A. $^{238}U \rightarrow \alpha + \_\_?\_\_$
    B. $^{234}Th \rightarrow \_\_?\_\_ + ^{-?}Ra$
    C. $\_\_?\_\_ \rightarrow \alpha + ^{236}U$
    D. $^{214}Bi \rightarrow \alpha + \_\_?\_\_$

76. •Complete the following beta decays:
    A. $^{14}C \rightarrow e^- + \bar{\nu}_e + \_\_?\_\_$
    B. $^{239}Np \rightarrow e^- + \bar{\nu}_e + \_\_?\_\_$
    C. $\_\_?\_\_ \rightarrow e^- + \bar{\nu}_e + ^{60}Ni$
    D. $^3H \rightarrow e^- + \bar{\nu}_e + \_\_?\_\_$
    E. $^{13}N \rightarrow e^+ + \_\_?\_\_ + \_\_?\_\_$

77. •Complete the following gamma decays:
    A. $^{131}I^* \rightarrow \gamma + \_\_?\_\_$
    B. $^{145}Pm^* \rightarrow ^{145}Pm + \_\_?\_\_$
    C. $\_\_?\_\_ \rightarrow \gamma + ^{24}Na$

78. ••Nickel-64 has an excited state 1.34 MeV above the ground state. The atomic mass of the ground state of this isotope of nickel is 63.927967 u. (a) What is the mass of the atom when the nucleus is in this excited state? (b) What is the wavelength of the gamma ray that is emitted when the nucleus decays to the ground state?

General Problems

79. ••(a) What is the approximate radius of the ^{238}U nucleus? (b) What electric force do two protons on opposite ends of the ^{238}U nucleus exert on each other? (c) If the electric force in part (b) were the only force acting on the protons, what would be their acceleration just as they left the nucleus? (d) Why do the protons in part (b) not accelerate apart? SSM

80. ••The semi-empirical binding energy formula is given as follows:

$$E_B = (15.8\,\text{MeV})A - (17.8\,\text{MeV})A^{2/3}$$
$$- (0.71\,\text{MeV})\frac{Z(Z-1)}{A^{1/3}} - (23.7\,\text{MeV})\frac{(N-Z)^2}{A}$$

where A is the mass number, N is the number of neutrons, and Z is the number of protons. Using the formula, calculate the binding energy per nucleon for fermium-252. Compare your answer with the standard common expression for the binding energy: $E_B = (Nm_n + Zm_p - m_{whole})c^2$.

81. ••The *fissionability parameter* is defined as the atomic number squared divided by the mass number for any given nucleus (Z^2/A). It can be shown that when this parameter is less than 44, a nucleus will be stable against small deformation; essentially, the nucleus will be stable against spontaneous fission. Calculate the value of this parameter for (a) ^{235}U, (b) ^{238}U, (c) ^{239}Pu, (d) ^{240}Pu, (e) ^{246}Cf, and (f) ^{254}Cf.

82. •The stable isotope of sodium is ^{23}Na. What kind of radioactivity would be expected from (a) ^{22}Na and (b) ^{24}Na?

83. ••In 2010, physicists for the first time created the heavy element number 117 by colliding calcium-48 and berkelium-249 at high energy. The result was two isotopes of the new element, one of which had a half-life of 14 ms and contained 176 neutrons. (a) What is the radius of the nucleus of the new element 117? (b) What percent of the newly created isotope was left 1.0 s after its creation? SSM

84. ••Natural uranium is made up of two isotopes: ^{235}U and ^{238}U. The half-life of ^{235}U is 7.04×10^8 y, and the half-life of ^{238}U is 4.47×10^9 y. Assuming that all uranium isotopes were created simultaneously and in equal amounts at the same time that Earth was formed, estimate the age of Earth. The current percent abundance of ^{235}U is 0.72% and for ^{238}U it is 99.28%.

85. ••**Astronomy** The atom technetium (Tc) has no stable isotopes, yet its spectral lines have been detected in red giant stars (stars at the end of their lifetimes). Tc can be produced artificially on Earth. Its longest-lived isotope, ^{98}Tc, has a half-life of 4.2 million years. (a) If any ^{98}Tc was present when Earth formed 4.5×10^9 y ago, what percentage of it is still present? (Careful! You cannot do this calculation with your calculator. You must use logarithms to express the answer in scientific notation.) (b) What percent of the original ^{98}Tc would be present in a red giant that is 10 billion years old? (Careful again! You'll need to use logarithms.) (c) Explain why the detection of technetium in old stars is strong evidence that stars manufacture the atoms in the universe.

86. ••An old wooden bowl unearthed in an archeological dig is found to have one-fourth of the amount of carbon-14 present in a similar sample of fresh wood. Determine the age of the bowl.

87. ••In an attempt to determine the age of the cave paintings in Chauvet-Pont-d'Arc Cave in France, scientists used carbon-14 dating to measure the age of bones of bears found in the cave. The bears are depicted in the paintings, so presumably the bones are approximately the same age as the paintings. The results showed that the level of ^{14}C was reduced to 2.35% of its present-day level. How old were the bones (and presumably the paintings)? SSM

88. ••In one common type of household smoke detector, the radioactive isotope americium-241 decays by alpha emission. The alpha particles produce a small electrical current because they are charged. If smoke enters the detector, it blocks the alpha particles, which reduces the current and causes the alarm to go off. The half-life of ^{241}Am is 433 y, and its atomic weight is 241 g/mol. Typical decay rates in smoke detectors are 690 Bq. (a) Write the decay reaction of ^{241}Am and identify the daughter nucleus. (b) By how much does the alpha particle current decrease in 1.0 y due to the decay of the americium? How much in 50 y? (c) How many grams of ^{241}Am are there in a typical smoke detector?

89. ••In March 2011, a giant tsunami struck the Fukushima nuclear reactor in Japan, resulting in very large radiation leaks, including cesium-137. The isotope has a 30-y half-life and is a beta-minus emitter. (a) What daughter nucleus is left after cesium-137 decays? (b) How long after the release will it take for the decay rate of the cesium-137 to be reduced by 99%?

90. ••Three isotopes of aluminum are given in the following table:

Isotope	Atomic mass (u)	E_B/nucleon	Decay process
^{26}Al	25.986892		
^{27}Al	26.981538		Stable
^{28}Al	27.981910		

Calculate the binding energy per nucleon for each isotope and make a prediction of the decay processes for the unstable isotopes aluminum-26 and aluminum-28.

91. ••**Biology** In February 2010, the discovery of leaks due to aging pipes of the carcinogen tritium (^{3}H) at 27 U.S. nuclear reactors was announced. In one well in Vermont, contaminated water registered 70,500 pCi/L, whereas the federal safety limit was 20,000 pCi/L. Tritium is a β^- emitter with a half-life of 12.3 y. (a) How many protons and neutrons does the tritium nucleus contain? (b) Write out the decay equation for tritium and identify the daughter nucleus. (c) If the leak at the Vermont site is stopped, how long will it take for the water in the contaminated well to reach the federal safety level?

92. ••**Medical** Iodine-125 is used to treat, among other things, brain tumors and prostate cancer. It decays by gamma decay with a half-life of 59.4 days. Patients who fly soon after receiving ^{125}I implants are given medical statements from the hospital verifying such treatment because their radiation could set off radiation detectors at airports. If the initial decay rate was 525 μCi, (a) what will the rate be at the end of the first year, and (b) how many months after the treatment will the decay rate be reduced by 90%?

93. ••**Medical** Ruthenium-106 is used to treat melanoma in the eye. This isotope decays by β^- emission with a half-life of 373.59 days. One source of the isotope is reprocessed nuclear reactor fuel. (a) How many protons and neutrons does the ^{106}Ru nucleus contain? (b) Could we expect to find significant amounts of ^{106}Ru in ore mined from the ground? Why or why not? (c) Write the decay equation for ^{106}Ru and identify the daughter nucleus. (d) How many years after ^{106}Ru is implanted in the eye does it take for its decay rate to be reduced by 75%?

94. ••**Medical** You are asked to prepare a sample of ruthenium-106 for a radiation treatment. Its half-life is 373.59 days, it is a beta emitter, its atomic weight is 106/g/mol, and its density at room temperature is 12.45/g/cm^3. (a) How many grams will you need to prepare a sample having an

activity rate of 125 μCi? (b) If the sample in part (a) is a spherical droplet, what will be its radius?

95. ••Electron capture by a proton is *not* allowed in nature. Explain why. Specifically, describe why the following nuclear reaction does *not* occur (and for good reason!): SSM

$$e^- + p \nrightarrow n + \nu_e$$

96. ••Taking into account the recoil (kinetic energy) of the daughter nucleus, calculate the energy of the alpha particle in the following decay:

$$^{235}U \rightarrow \alpha + {}^{231}Th$$

97. ••Several radioactive decay series are observed in nature. Four of the most well known are as shown here. The neptunium decay series actually is extinct. The first three all end in different, stable isotopes of lead.

   A. Thorium Decay Series: $A = 4n$
      Starting isotope: ^{232}Th Ending isotope: ^{208}Pb
   B. Radium or Uranium Decay Series: $A = 4n + 2$
      Starting isotope: ^{238}U Ending isotope: ^{206}Pb
   C. Actinium Decay Series: $A = 4n + 3$
      Starting isotope: ^{235}U Ending isotope: ^{207}Pb
   D. Neptunium Decay Series: $A = 4n + 1$
      Starting isotope: ^{237}Np Ending isotope: ^{209}Bi

Trace out the pathway (keeping count of the total number of alpha and beta emissions) that terminates with a stable nucleus for each of the radioactive decay series above.

98. ••A friend suggests that the world's energy problems could be solved if only physicists were to pursue the fusion of *heavy* nuclei rather than the fusion of *light* nuclei. To prove his point, he suggests that the following fusion reaction should be considered: ^{157}Nd + ^{80}Ge → ^{235}U + 2n.

Using the insights that you have acquired in this chapter, show that his argument is flawed. The atomic mass of neodymium-157 is 156.939032 u, and the mass of germanium-80 is 79.925373 u.

28

Particle Physics and Beyond

What do you think?

The collision of an electron and its antimatter equivalent, a positron, produces two sprays of particles at the center of an experiment at CERN, the European particle physics laboratory. The blue and purple lines show the outline of the experimental apparatus, and the other lines show the paths of the particles. Among the particles produced by such a collision, you are likely to find ones made of (a) matter; (b) antimatter; (c) a mixture of matter and antimatter; (d) both (a) and (b); (e) all of (a), (b), and (c).

In this chapter, your goals are to:

- (28-1) Explain why physicists examine both the smallest and largest objects in the universe.
- (28-2) Describe the difference between hadrons and leptons, and explain what hadrons are made of.
- (28-3) Explain how fundamental particles interact with each other, and the differences among the four fundamental forces.
- (28-4) Calculate the distance to a remote galaxy from its recessional velocity.

To master this chapter, you should review:

- (10-4) How the speed of an object in a gravitational orbit depends on the strength of the gravitational force (distance from the massive central object).
- (13-10) How the Doppler effect describes the shift in frequency of a sound wave coming from a moving object.
- (15-3) How an ideal gas cools when it undergoes an adiabatic expansion.
- (22-4) The energy of a photon.
- (25-7) The equivalence of mass and energy.
- (26-2) How the frequency spectrum of blackbody radiation depends on temperature.
- (26-5) Atomic spectra and Rutherford's discovery of the nucleus.
- (27-6) The process of beta decay.

28-1 Studying the ultimate constituents of matter helps reveal the nature of the universe

"What is the world made of?" "Where did I come from?" You probably asked questions like these when you were a small child. Physicists and astronomers ask these questions throughout their professional lives, in search of ever more sophisticated answers about the nature and origin of the physical universe. In this final chapter we'll take a brief look at our present understanding of **particle physics**, the branch of physics that concerns the fundamental constituents of matter and how they interact. We'll also see evidence that our universe began in a state of tremendously high temperature and immense density some 13.8 billion years ago.

We'll begin by looking at the fundamental particles that make up all of the ordinary matter that you see around you, including the matter that makes up your body.

Oranges are a good source of potassium. A small fraction of the potassium is radioactive ^{40}K, which decays by emitting a positron—a bit of antimatter. This decay also involves a quark, a neutrino, and an exchange particle called a W$^+$.

(a)

When a television is disconnected from the source, about 1% of the "static" on the screen is from cosmic background radiation—the afterglow of the Big Bang at the beginning of time.

(b)

Figure 28-1 **Particle physics and cosmology** Ordinary objects and common technology connect us to (a) particle physics, the study of fundamental particles and their interactions, and (b) cosmology, the study of the nature and evolution of the universe.

These include leptons, of which the electron is the best-known example, and quarks, which have the curious property that their charges are a fraction of the fundamental charge e. We'll see that these fundamental particles interact with each other by exchanging particles back and forth, a process that actually involves violating the law of conservation of energy (but in a way that's nonetheless compatible with the laws of physics). These interactions help explain the strong forces that hold the nucleus together, the electric forces that keep electrons in the atom, and the weak forces that cause radioactive beta decay (Figure 28-1a).

We'll conclude the chapter by redirecting our attention from the smallest particles to some of the largest structures in the universe, the galaxies. We'll learn that our universe is expanding and is filled with electromagnetic radiation that is left over from the first few hundred thousand years of the history of the universe (Figure 28-1b). We'll also find that most of the matter in the universe is in a form whose nature is almost a complete mystery, and that most of the energy in the universe is even more mysterious in its character. There is much we know about the physical universe, but there is a great deal more yet to be learned!

> ## Take-Home Message for Section 28-1
>
> ✔ Ordinary matter such as atoms is composed of a small variety of different fundamental particles.
>
> ✔ Studying distant galaxies tells us about the nature and evolution of our universe.

28-2 Most forms of matter can be explained by just a handful of fundamental particles

By the late nineteenth century, scientists had come to the conclusion that all matter was composed of atoms and that atoms could not be subdivided into more elementary particles. The notion was that a hydrogen atom was fundamentally different from a helium atom, which in turn was fundamentally different from a carbon atom, and so on. As we learned in Section 26-5, this conclusion was incorrect. In 1897 J. J. Thomson discovered the electron, a particle of charge $-e = -1.602 \times 10^{-19}$ C, which turned out to be a constituent of the atoms of every element. In 1909 Ernest Rutherford discovered the atomic nucleus, and in 1917 he found evidence that all nuclei contain a positively charged particle of charge $+e$ that is identical with a hydrogen nucleus—that is, what we now call a proton. In 1932 the English physicist James Chadwick discovered the neutron, which has zero charge. As we learned in Chapter 27, all nuclei are composed of protons and neutrons (referred to collectively as nucleons) and all atoms are made of nuclei plus electrons. So by 1932 it seemed that these three particles—electron, proton, and neutron—were the truly fundamental building blocks of which all matter is made. That conclusion, too, turned out to be wildly incorrect.

Since 1932, literally hundreds of other subatomic particles have been discovered. All of these are unstable and decay to other particles with a radioactive half-life of a

(a)

Electrons begin their
acceleration here...

...and strike the
target here, 3.2 km away.

(b)

When high-energy electrons
scatter from protons, a
substantial number scatter
backwards.

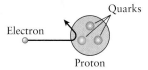

Electron

Quarks

Proton

This is evidence that there
are small charged objects
(quarks) inside the proton.

Figure 28-2 Discovering quarks
(a) In a seminal experiment at the
Stanford Linear Accelerator Center,
electrons were accelerated to a kinetic
energy of 20 GeV $= 2 \times 10^4$ MeV and
fired into a target containing protons.
(b) Measuring how the electrons
scattered from the protons provided
evidence of the existence of quarks.

fraction of a second. (In this aspect they resemble the neutron: A free neutron that is not incorporated into a nucleus undergoes beta decay with a half-life of about 15 minutes.) But none of these additional particles can be regarded as simple combinations of protons, neutrons, and electrons. They include the neutrino, which has no charge, interacts hardly at all with other particles, and has a mass so close to zero that it has yet to be accurately measured; the muon, which resembles an electron in almost every way except that it is 207 times more massive; the pion, which like the proton and neutron experiences the strong nuclear force but has only about one-seventh the mass of a proton; and the delta, which resembles a proton or neutron but is about 30% more massive and comes in four varieties, with charges $+2e$, $+e$, 0, and $-e$. The discovery of these additional particles forced physicists to once again ask the question: What *are* the fundamental building blocks of matter?

Hadrons and Quarks

To answer this question, it's useful to distinguish between particles that experience the strong nuclear force, including the proton and neutron, and those that do not, such as the electron. Since protons and neutrons are relatively heavy and electrons are relatively light, we use the term **hadrons** (from the Greek word for stout or thick) for particles that experience the strong force and the term **leptons** (from the Greek word for small or delicate) for those that do not. (The photon is considered to be in a special category of its own, to which we will return later.) The vast majority of new particles discovered since 1932 are hadrons, so we'll look at these first.

In 1964 the American physicists Murray Gell-Mann and George Zweig independently proposed that all hadrons are made of more fundamental entities that Gell-Mann whimsically named **quarks**. The first evidence that quarks really exist came from experiments carried out at the Stanford Linear Accelerator Center, or SLAC, in 1967 (**Figure 28-2a**). These experiments were the same in principle as Rutherford's 1909 experiment that led to the discovery of the atomic nucleus. As we saw in Section 26-5, Rutherford aimed a beam of alpha particles at a target of gold. Had the charge inside the atom been distributed more or less uniformly, the alpha particles would have undergone only gentle deflections as they passed through the gold atoms. Instead, Rutherford found that some alpha particles were scattered by very large angles. This was evidence that charge was highly concentrated into a very small object within the atom—namely, the nucleus. The experiment at SLAC used electrons instead of alpha particles and aimed these electrons at a target of protons (the nuclei of hydrogen atoms inside a tank of liquid hydrogen). The 3.2-km-long accelerator gave each electron a tremendous kinetic energy—some 20,000 MeV, compared to the 7 MeV of Rutherford's alpha particles—and a correspondingly large momentum. This was done so that the electron would have a de Broglie wavelength much smaller than the size of the proton and so would be sensitive to fine details of the proton's internal structure as it passed through the proton.

Much as in the Rutherford experiment six decades before, many physicists expected that the electrons would undergo only small-angle deflections because they thought the charge inside the proton was distributed uniformly over its volume. And just as in the Rutherford experiment, what they found was that a substantial number of electrons were scattered by very large angles (**Figure 28-2b**). The conclusion was that the proton's charge is carried by smaller entities inside the proton, which are the quarks.

These experiments and a host of others confirm that Gell-Mann and Zweig were correct and that quarks are the fundamental building blocks of all hadrons. To explain all of the hundreds of hadron varieties currently known, we need six varieties or *flavors* of quarks. These are known as the up (u), down (d), charm (c), strange (s), top (t), and bottom (b) quarks. **Table 28-1** lists the six quarks, along with their masses and charges. The quarks are divided into three groups, or *generations*, of two quarks: u and d in the first generation, c and s in the second generation, and t and b in the third generation. As seen in the table, the quarks in each generation are more massive than those in the previous generation. Physicists usually write the quark generations as

$$\binom{u}{d}\binom{c}{s}\binom{t}{b}$$

The quarks along the top row (u, c, and t) all have the same charge, as do the quarks along the bottom row (d, s, and b). However, all six quarks differ not only in mass but also in other subtle properties. (These properties are beyond our scope in this brief introduction.) All attempts to find evidence of a fourth-generation quark have failed; as best we know, there are only three generations of quarks.

Note that each flavor of quark has a charge that is a *fraction* of e, either $+2e/3$ or $-e/3$. The explanation is that the proton, neutron, and other related particles are actually combinations of three quarks. Figure 28-3 shows four examples of such combinations. Any hadron that can be made up of three quarks is called a **baryon**. Baryons always have a net charge that is an integer multiple of e. Two examples are the least massive baryons, the proton (Figure 28-3a) of net charge $+e$ and the neutron (Figure 28-3b) with net charge zero. This picture explains why the neutron, which has zero net charge, nonetheless produces a magnetic field of its own: The u quark and two d quarks within the neutron are each charged, and the motions of these charged particles within the neutron generates a magnetic field.

Note that in Table 28-1 we list only an *approximate* mass for each quark. We know quite precise values of the masses of other particles; for example, the mass of the proton is known to seven significant figures. By contrast, we know only rough values for the masses of the quarks. That's because, unlike protons, neutrons, or electrons, quarks are never found as solitary particles. Instead, they are found only in combinations like those shown in Figure 28-3. This situation is quite unlike that in atoms, in which electrons can be removed by ionization, or in nuclei, in which protons or neutrons can be removed from a nucleus in a sufficiently energetic collision. By contrast, it seems to be impossible to remove an individual quark from a baryon. This is called **quark confinement**: Quarks within baryons are confined there and cannot be removed and isolated.

Note also that the mass of a baryon is generally much greater than the sum of the masses of its constituent quarks. For example, a proton is composed of two u quarks and a d quark, which from Table 28-1 have a combined mass of approximately $2(2 \text{ MeV}/c^2) + 5 \text{ MeV}/c^2 = 9 \text{ MeV}/c^2$. Yet the mass of the proton is $938.1 \text{ MeV}/c^2$. The difference is associated with the forces that bind the quarks together and keep them confined within the proton. We'll explore these forces in the following section.

Other hadrons are made up of a quark and an **antiquark**, the **antimatter** version of a quark. Every matter particle has an antimatter partner that is identical to it in every way except that it is oppositely charged. We've already encountered one example of antimatter, the positron (e^+) produced in β^+ decay (Section 27-5), which has the same mass as the electron. In fact, the positron and electron are identical except for the sign of their charges ($-e$ for the electron, $+e$ for the positron); the electron and the positron are a matter–antimatter pair. Each of the six quarks has an antiquark associated with it. Antiquarks are signified by adding "bar" to the name or placing a bar over the quark's symbol. For example, the antiquark associated with the up quark (u) is the up-bar or $\bar{\text{u}}$. The $\bar{\text{u}}$ antiquark has charge $-2e/3$, the opposite of the $+2e/3$ charge of the u quark. In the same way, the $\bar{\text{d}}$ antiquark carries charge $+e/3$, the opposite of the $-e/3$ charge of the d quark.

Figure 28-4 shows three examples of hadrons made up of a quark and antiquark. Such quark-antiquark combinations are called **mesons**. The charge of a meson is always an integer multiple of e. Pions are the least massive mesons; the π^+ (Figure 28-4a) and the π^- (Figure 28-4b) both have a mass of $139.6 \text{ MeV}/c^2$. There is also a neutral pion (π^0), which is made up of a combination of $\text{u}\bar{\text{u}}$ and $\text{d}\bar{\text{d}}$ and has a slightly lower mass of $135.0 \text{ MeV}/c^2$. Pions, like all other mesons, are unstable; they decay with a half-life that is a small fraction of a second.

When a particle and its antimatter partner meet, their total mass can be converted to its equivalent energy. For example, the collision of an electron and a positron results in two or more photons that carry the total energy of the two particles, a process known as *annihilation*. A quark and antiquark can also annihilate each other; we'll see examples of these processes in the next section.

Table 28-1 The Six Quarks

Quark	Symbol	Charge	Approximate Mass
up	u	$+\frac{2}{3}e$	$2 \text{ MeV}/c^2$
down	d	$-\frac{1}{3}e$	$5 \text{ MeV}/c^2$
charm	c	$+\frac{2}{3}e$	$1.3 \text{ GeV}/c^2$
strange	s	$-\frac{1}{3}e$	$0.1 \text{ GeV}/c^2$
top	t	$+\frac{2}{3}e$	$173 \text{ GeV}/c^2$
bottom	b	$-\frac{1}{3}e$	$5 \text{ GeV}/c^2$

Note: $1 \text{ GeV}/c^2 = 10^3 \text{ MeV}/c^2 = 10^9 \text{ eV}/c^2$.

(a) Proton (p)

Net charge $= q_u + q_u + q_d$

$= \left(+\frac{2}{3}e\right) + \left(+\frac{2}{3}e\right) + \left(-\frac{1}{3}e\right) = +e$

(b) Neutron (n)

Net charge $= q_d + q_d + q_u$

$= \left(-\frac{1}{3}e\right) + \left(-\frac{1}{3}e\right) + \left(+\frac{2}{3}e\right) = 0$

(c) Delta-plus-plus (Δ^{++})

Net charge $= q_u + q_u + q_u$

$= \left(+\frac{2}{3}e\right) + \left(+\frac{2}{3}e\right) + \left(+\frac{2}{3}e\right) = +2e$

(d) Sigma-minus (Σ^-)

Net charge $= q_s + q_d + q_d$

$= \left(-\frac{1}{3}e\right) + \left(-\frac{1}{3}e\right) + \left(-\frac{1}{3}e\right) = -e$

Figure 28-3 **Baryons** All baryons, including (a) the proton and (b) the neutron, are made of combinations of three quarks. The proton is the only stable baryon; the neutron and (c), (d), all others decay into simpler particles.

(a) Positive pion (π^+)

Net charge $= q_u + q_{\bar{d}}$

$= \left(+\frac{2}{3}e\right) + \left(+\frac{1}{3}e\right) = +e$

(b) Negative pion (π^-)

Net charge $= q_d + q_{\bar{u}}$

$= \left(-\frac{1}{3}e\right) + \left(-\frac{2}{3}e\right) = -e$

(c) Strange D-plus (D_s^+)

Net charge $= q_c + q_{\bar{s}}$

$= \left(+\frac{2}{3}e\right) + \left(+\frac{1}{3}e\right) = +e$

Figure 28-4 **Mesons** All mesons are made of a quark and an antiquark.

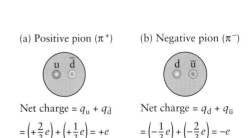

Table 28-2 The Six Leptons

Lepton	Symbol	Charge	Mass
electron	e^-	$-e$	$0.5110 \text{ MeV}/c^2$
electron neutrino	ν_e	0	$<2 \text{ eV}/c^2$
muon	μ^-	$-e$	$105.7 \text{ MeV}/c^2$
muon neutrino	ν_μ	0	$<0.18 \text{ MeV}/c^2$
tau	τ^-	$-e$	$1777 \text{ MeV}/c^2$
tau neutrino	ν_τ	0	$<18.2 \text{ MeV}/c^2$

Just as three quarks make up a baryon, three antiquarks make up an *antibaryon*. For each variety of baryon, there is a corresponding variety of antibaryon. For example, the proton has quark content uud and charge $2e/3 + 2e/3 + (-e/3) = e$; the corresponding antibaryon is the antiproton, which has quark content $\overline{u}\,\overline{u}\,\overline{d}$ and charge $(-2e/3) + (-2e/3) + e/3 = -e$.

Leptons

While all hadrons are composed of quarks or antiquarks, there are other particles that are not composed of quarks at all. The *leptons* are another category of particles that are as fundamental as the quarks; to the best of our knowledge, leptons are not made up of smaller constituents. The electron is a lepton, as is the muon that we encountered in Section 25-4 (see Example 25-3). The electron neutrino that is created in hydrogen fusion (Section 27-5) and in beta decay (Section 27-6) is also a lepton. There are six leptons, listed in Table 28-2, as well as their antimatter partners.

Notice that in Table 28-2 we have listed the leptons in three groups of two. As is the case for the quarks, leptons form three generations, which we usually write as

$$\begin{pmatrix} e^- \\ \nu_e \end{pmatrix} \begin{pmatrix} \mu^- \\ \nu_\mu \end{pmatrix} \begin{pmatrix} \tau^- \\ \nu_\tau \end{pmatrix}$$

As for the quarks listed in Table 28-1, the leptons in each successive generation are more massive than their counterparts in the preceding generation. In many ways, the muon and tau are more massive versions of the electron, so they share many properties and interact with other particles in similar ways. One way in which electrons, muons, and tau particles *are* significantly different (in addition to the differences in mass) is that while the electron is stable and does not decay, the other two are unstable: The muon has a half-life of 1.56 μs (see Example 25-3) and the tau has a half-life of about 2.0×10^{-13} s.

Each lepton also has an antimatter particle associated with it. We have already encountered the electron and positron as a matter–antimatter pair. The antimuon (μ^+) and antitau (τ^+) are identical to the muon and tau in every way except that μ^+ and τ^+ carry positive rather than negative charge.

Unlike quarks, which appear in groups of three to form baryons or in quark-antiquark combinations to form mesons, leptons do not group together to form other particles. In the following section we'll see the reason for this. We'll also discover an essential third class of particles in addition to hadrons and leptons, the *exchange particles*.

Conservation Laws for Hadrons and Leptons

In earlier chapters we encountered a number of important conservation laws, including the conservation of energy and the conservation of momentum. There are several additional conservation laws that govern the behavior of hadrons and leptons.

- *Baryon number is conserved.* Every baryon (composed of three quarks) is assigned *baryon number B* equal to $+1$, and every antibaryon is assigned B equal to -1. Mesons are assigned $B = 0$, as are leptons. Experiments show that in every process that involves baryons, the sum of the values of B for all particles present before the process equals the sum of the values of B for all particles present after the process. Consider, for example, β^- decay of a neutron:

(28-1)
$$\begin{array}{cccc} n \rightarrow & p & + e^- & + \overline{\nu}_e \\ B = 1 & 1 & 0 & 0 \end{array}$$

The neutron and proton are both baryons, so each has $B = +1$. The electron and the antineutrino are leptons, each with $B = 0$. The total baryon number equals 1 before the decay and equals $1 + 0 + 0 = 1$ after the decay, so baryon number is conserved. Quarks have a fractional baryon number: Every quark has $B = +1/3$ and every antiquark has

$B = -1/3$. That's consistent with a baryon with three quarks having $B = 3(+1/3) = +1$ and a meson with a quark and antiquark having $B = (+1/3) + (-1/3) = 0$.

- *Lepton number is conserved.* Every electron (e^-) and electron neutrino (ν_e) is assigned *electron-lepton number* $L_e = +1$, and every positron (e^+) and electron antineutrino ($\bar{\nu}_e$) is assigned L_e equal to -1. Muons, muon neutrinos, tau particles, and tau neutrinos each have a similarly defined *muon-lepton number* L_μ or a *tau-lepton number* L_τ. A particle that is not a lepton is assigned L_e, L_μ, and L_τ equal to zero. Experiment shows that each of these lepton numbers is separately conserved.

As an example, consider the β^- decay of a neutron from Equation 28-1:

$$\text{n} \rightarrow \text{p} + \text{e}^- + \bar{\nu}_e$$
$$L_e = 0 \quad\quad 0 \quad\quad 1 \quad\quad -1 \tag{28-2}$$

Both the neutron and the proton are baryons, so each has $L_e = 0$; the electron has $L_e = +1$ and the electron antineutrino has $L_e = -1$. The total value of L_e before the decay is zero, and afterward it is $0 + 1 + (-1) = 0$. The electron-lepton number is conserved in β^- decay. A second example is the decay of a tau into an electron, an electron antineutrino, and a tau neutrino:

$$\tau^- \rightarrow \text{e}^- + \bar{\nu}_e + \nu_\tau$$
$$L_e = 0 \quad\quad 1 \quad\quad -1 \quad\quad 0 \tag{28-3}$$
$$L_\tau = 1 \quad\quad 0 \quad\quad 0 \quad\quad 1$$

The tau (τ^-) and tau neutrino (ν_τ) each have $L_e = 0$ and $L_\tau = 1$, the electron (e^-) has $L_e = 1$ and $L_\tau = 0$, and the electron antineutrino ($\bar{\nu}_e$) has $L_e = -1$ and $L_\tau = 0$. You can see that the total electron-lepton number equals 0 before and after the decay, and the tau-lepton number equals 1 before and after the decay. Both of these lepton numbers are separately conserved in the β^- decay.

Physicists have searched for processes in which either baryon number or one of the lepton numbers is not conserved. No such process has ever been observed. It appears that these four laws—conservation of baryon number, electron-lepton number, muon-lepton number, and tau-lepton number—are as universal and fundamental as the law of conservation of electric charge (which says that the net electric charge has the same value before and after any process).

 Got the Concept? 28-1 **Particles That May or May Not Exist**

Which of the following particles could possibly exist? (a) A meson with charge $-2e$; (b) a baryon with charge $-2e$; (c) an antibaryon with charge $-2e$; (d) more than one of these; (e) none of these.

 Got the Concept? 28-2 **Processes That May or May Not Happen**

Electric charge is conserved in each of these processes, but some never occur. Which are possible? (a) $\text{n} + \text{p} \rightarrow \text{n} + \text{p} + \text{p} + \bar{\text{p}}$ ($\bar{\text{p}}$ is an antiproton); (b) $\text{n} + \text{p} \rightarrow \text{n} + \text{n} + \text{p}$; (c) $\pi^- \rightarrow \mu^- + \bar{\nu}_e$; (d) more than one of these; (e) none of these.

Take-Home Message for Section 28-2

✔ Quarks are the constituents of hadrons, the particles that experience the strong nuclear force. There are six varieties of quarks, each of which has a charge that is a fraction of e.

✔ Baryons such as the proton are made up of three quarks. Mesons are made up of a quark and an antiquark.

✔ Leptons, which include electrons and neutrinos, have no constituent particles. They do not experience the strong force.

✔ In the interactions of subatomic particles, baryon number and the three lepton numbers are conserved.

28-3 Four fundamental forces describe all interactions between material objects

Since early in our study of physics we have seen the importance of *forces*, the pushes and pulls that one object exerts on another. We've encountered three fundamental kinds of forces: gravity, the electromagnetic force, and the strong nuclear force. Gravity is an attractive force that draws objects that have mass closer together. The gravitational force attracts you to Earth and keeps Earth in orbit around the Sun. Electric and magnetic forces, two manifestations of the electromagnetic force, cause charged objects to accelerate. As a result, electrons can be bound to atomic nuclei, and atoms can bond together. Any contact force, such as the normal force between two objects that touch or the force of friction are electromagnetic in nature; they arise from the electromagnetic interactions between the atoms of the two surfaces in contact. As we explored in Chapter 27, the strong force binds protons and neutrons together to form atomic nuclei. To this list, we will add a force that we have not yet named but the effects of which we encountered in Section 27-6; this **weak force** is at the heart of the interaction that governs beta decay.

Now that we are exploring the fundamental constituents of matter, we are in a position to ask a central question about force: How do objects exert forces on each other? And what is fundamentally different about the four different kinds of forces: gravitational, electromagnetic, strong, and weak? If we can understand how fundamental entities such as quarks and leptons exert forces of different kinds on each other, we will be closer to answering these questions.

As we will see, the manner in which fundamental particles exert forces on each other—that is, how they *interact*—comes from a remarkable aspect of nature: It is possible to violate the law of conservation of energy, provided we do it for a sufficiently short time.

The Heisenberg Uncertainty Principle

To see how it's possible to violate energy conservation, let's consider the photon. We learned in Section 22-4 that the energy of a photon is proportional to its frequency:

Energy of a photon
(22-26)

Energy of a photon   Wave frequency

$$E = hf$$

Planck's constant $= 6.62606957 \times 10^{-34}$ J • s

For a photon to have a definite energy, it must therefore have a definite frequency. However, a wave ψ with a definite frequency also has an infinite duration (Figure 28-5a). This means that wave has always been present and will always be present. A more realistic description of a wave is one that has a finite duration: for example, the wave produced when you turn a source of waves (like a laser pointer) on and then off again. Mathematically, a wave of a finite duration Δt can be expressed as a sum of waves of infinite duration like the one shown in Figure 28-5a, but with a range of frequencies of breadth Δf (Figure 28-5b). To make a shorter-duration wave, we have to add together infinite-duration waves from a broader range of frequencies (Figure 28-5c). We can express the relationship between Δt (the duration of a wave) and Δf (the breadth of frequencies that go into that wave) as

(28-4)

$$\Delta f \Delta t \geq \frac{1}{4\pi}$$

Equation 28-4 says that the product of the wave duration Δt and the frequency breadth Δf cannot be less than $1/4\pi$. (This number arises from the specific way in which Δt and Δf are defined mathematically.) It says that to minimize the duration of a wave necessarily means increasing the range of frequencies that make up the wave. So the shorter the duration of a wave, the less precisely we can answer the question "What is the frequency of the wave?" In other words, a wave of finite duration does not have a single definite frequency.

(a)

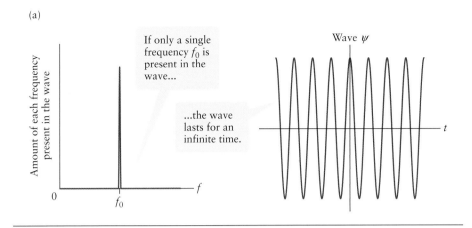

(b)

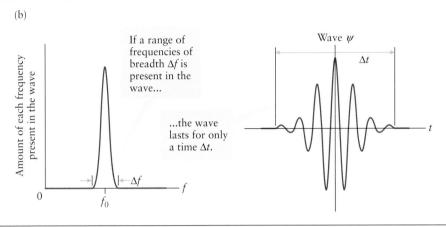

(c)

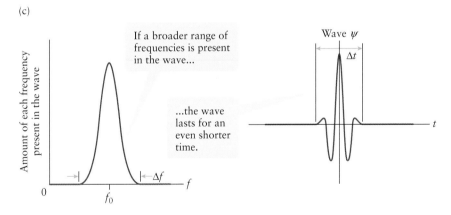

Figure 28-5 **Wave frequency and wave duration** (a) A wave of a single, discrete frequency has an infinite duration. (b), (c) Combining waves over a breadth of frequencies yields a total wave that has a finite duration.

Equation 28-4 is true for waves of all kinds, from ocean waves to sound waves to seismic waves. If we apply it to electromagnetic waves and multiply both sides of the equation by Planck's constant h, we get

$$h\Delta f \Delta t \geq \frac{h}{4\pi} \quad \text{or} \quad \Delta(hf)\Delta t \geq \frac{\hbar}{2} \tag{28-5}$$

Here $\Delta(hf)$ is the breadth of values of the quantity hf that must be included in the wave, and $\hbar = h/2\pi$. (We introduced $\hbar = h/2\pi$, or "h bar," in Section 26-6.) But Equation 22-26 tells us that $E = hf$ is the energy of a photon associated with the wave. So $\Delta(hf) = \Delta E$, the uncertainty in photon energy, and Equation 28-5 becomes

$$\Delta E \Delta t \geq \frac{\hbar}{2} \text{ for a photon} \tag{28-6}$$

This means that just as a wave of finite duration does not have a single definite frequency, a photon of finite duration does not have a single definite energy, but includes energy values that extend over a range of breadth ΔE. We can think of ΔE as the *uncertainty* in the energy of the photon. Equation 28-6 says that if a photon has a duration Δt, the product of Δt and uncertainty in energy ΔE of the photon cannot be less than $\hbar/2$, and so ΔE cannot be less than $\hbar/(2\Delta t)$. This energy uncertainty is *not* a result of the limitations of an experimental apparatus that we might use to measure the energy of a photon. Rather, it is intrinsic to photons because of their wave nature.

We have used the equation $E = hf$ to apply to photons only. But this relationship applies to particles of *all* kinds: We can think of anything with an energy E as having an associated frequency f given by $E = hf$. Then Equation 28-6 applies to phenomena in general, not just to photons. This implies that any physical phenomenon that has a finite duration Δt will necessarily have an uncertainty ΔE in energy given by Equation 28-6. The minimum value of this uncertainty is found by replacing the $\geq$ sign (greater than or equal to) in Equation 28-6 with an equals sign. The shorter the duration Δt of the phenomenon, the greater the minimum value of the energy uncertainty ΔE and the more uncertain the energy of the phenomenon. The German physicist Werner Heisenberg first expressed this idea in 1927, which is why it is known as the **Heisenberg uncertainty principle**:

Heisenberg uncertainty principle
(28-7)

The energy of a phenomenon is necessarily uncertain by an amount ΔE.

$$\Delta E \Delta t \geq \frac{\hbar}{2}$$

The shorter the duration Δt of the phenomenon, the greater the energy uncertainty ΔE.

$\hbar = \dfrac{h}{2\pi}$
= Planck's constant divided by 2π

Another way to interpret Equation 28-7 is to say that it places limits on how precisely the law of conservation of energy must be obeyed. Suppose a system undergoes some kind of process that lasts for a time Δt. Equation 28-7 says that the energy of the system is necessarily uncertain during that process, and the minimum energy uncertainty is given by $\Delta E \Delta t = \hbar/2$ or $\Delta E = \hbar/(2\Delta t)$. So it's fundamentally impossible to measure the energy of the system with an uncertainty less than $\hbar/(2\Delta t)$. This means that during the process, the energy of the system could actually vary by as much as $\hbar/(2\Delta t)$, and there would be no way that we could tell that the energy had changed value—that is, that the energy was not conserved. The shorter the duration Δt of the process, the greater the amount $\hbar/(2\Delta t)$ by which energy conservation can be (temporarily) violated during that time Δt. Stated another way, it's acceptable to violate the law of conservation of energy by an amount ΔE, provided the duration of time during which the law is violated is no more than $\Delta t = \hbar/(2\Delta E)$.

Exchange Particles: The Electromagnetic Force

The Heisenberg uncertainty principle helps us understand the following bold statement: *All forces result from the exchange of particles.* To see what we mean by this statement, first consider the electromagnetic force. Up to this point we've used the idea that charged particles exert electromagnetic forces on each other even at a distance, with no physical contact required. But a more sophisticated way to look at this force is to envision that when two charged particles exert an electric or magnetic force on each other, they do so by exchanging a photon: One of the charged particles emits the photon and the other absorbs it. The exchanged photon has energy, and it violates the law of conservation of energy for this photon to spontaneously appear and be emitted by one of the charged particles. But as we have seen, it's perfectly acceptable to violate the law of conservation of energy by an amount ΔE, provided we do so for a time no longer than $\Delta t = \hbar/(2\Delta E)$. So the uncertainty principle proves that what we've described can take place provided the exchanged photon is absorbed by the second particle (and so disappears) within a time $\hbar/(2\Delta E)$ after the photon was emitted by the first particle.

In this picture we say that the electromagnetic force is *mediated* by the exchange of a photon and that the exchanged photon is the *mediator* of the force. The particles that mediate forces are called **exchange particles**. You should try to envision a continuous stream of photons going back and forth between charged particles, so that the two particles continuously exert an electromagnetic force on each other. Any charged particle can and does emit and absorb photons exchanged in this way. These photons are not the same as the photons emitted by a light bulb or a laser, however; they can exist for only a finite time before they must disappear. We call them **virtual particles**.

We can represent the exchange process by a diagram such as the one shown in Figure 28-6. In this slightly simplified version of a *Feynman diagram*, invented by American physicist Richard Feynman to visualize and analyze processes that involve fundamental particles, time runs from left to right. The lines associated with each particle do not represent actual paths that the particles take through space, but only indicate which particles interact with which other particles. In Figure 28-6, two electrons exchange a photon, and in so doing each exerts an electromagnetic force on the other. A particle such as an electron is drawn as a solid, straight line in Feynman diagrams. An exchange particle is drawn either as a wavy line (as for the photon) or a spiral line.

This picture helps us understand Coulomb's law, which tells us that the electric force between two charged particles separated by a distance r decreases in proportion to $1/r^2$. In other words, the electric force goes to zero only when the charges are infinitely far apart. This is possible because the photon has zero mass, so its minimum energy $E = hf$ is zero. (If the photon did have a mass m, its minimum energy would be its rest energy mc^2.) If one particle violates conservation of energy by creating and emitting a photon of energy ΔE, the uncertainty principle says that this photon can exist no longer than $\Delta t = \hbar/(2\Delta E)$. Even traveling at the speed of light c, this photon can travel no farther than $c\Delta t = c\hbar/(2\Delta E)$, so that distance is the maximum separation at which two charged particles can interact by exchanging photons of energy ΔE. But because the lower limit on the energy of a photon is zero, the amount ΔE by which the particle violates conservation of energy can be as small as we like. So the distance $c\Delta t = c\hbar/(2\Delta E)$ can be arbitrarily large, and two charged particles can interact via the electromagnetic force at any distance out to infinity. But because only these low-energy photons can be exchanged between charged particles separated by great distances r, the force is quite weak if r is large—just as in Coulomb's law.

Exchange Particles: The Strong Force

We use a similar picture to explain the strong interaction between quarks. The Feynman diagram in Figure 28-7 shows two quarks that exert forces on each other by exchanging a particle called the **gluon**. This rather whimsical name expresses the idea that the exchange of gluons provides the force that confines, or glues, quarks inside baryons. Gluon exchange is also how the quark and antiquark inside a meson interact with each other and how the antiquarks inside an antibaryon interact.

Like the photon, the gluon has zero mass, so the lower limit on the energy of a gluon is zero. Using the same argument we made above for photons, it follows that quarks or antiquarks separated by any distance, no matter how large, can interact by exchanging gluons. But unlike photons, the kinds of particles that emit and absorb gluons are not simply those with electric charge: They are particles that have a different attribute called *color*. (This is yet another of the light-hearted names associated with quark physics. It has nothing to do with the wavelength of light or the perception of color by the human eye.) Quarks and antiquarks carry color, as do gluons themselves. So unlike photons, gluons can interact with each other by exchanging other gluons. (By contrast, photons have no electric charge and cannot interact directly with other photons.) As a consequence of this

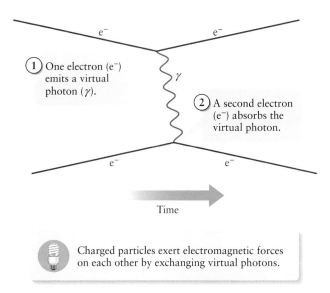

Figure 28-6 **Photon exchange and the electromagnetic interaction** All electric and magnetic interactions between charged particles involve the exchange of virtual photons.

1. One electron (e^-) emits a virtual photon (γ).

2. A second electron (e^-) absorbs the virtual photon.

Time

Charged particles exert electromagnetic forces on each other by exchanging virtual photons.

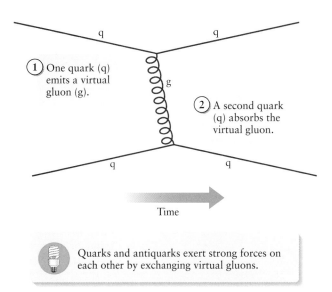

1. One quark (q) emits a virtual gluon (g).

2. A second quark (q) absorbs the virtual gluon.

Time

Quarks and antiquarks exert strong forces on each other by exchanging virtual gluons.

Figure 28-7 **Gluon exchange and the strong interaction between quarks** The forces that bind quarks together inside baryons and mesons are the result of the exchange of virtual gluons.

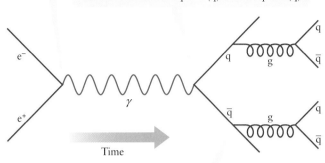

(1) In electron-positron annihilation, an electron (e⁻) and a positron (e⁺) collide and are transformed into a virtual photon.

(2) If the e⁻ and e⁺ had sufficient energy, the photon can transform into a quark (q) and antiquark (q̄).

(3) The quark and antiquark can emit gluons, which transform into quark-antiquark pairs. The shower of quarks and antiquarks coalesces into hadrons, including baryons (three quarks), antibaryons (three antiquarks), and/or mesons (quark-antiquark pairs).

Figure 28-8 Electron-positron annihilation and hadron production When an electron and positron collide, the result can be a shower of hadrons produced via virtual photons and gluons.

curious aspect of gluons, the force that gluons mediate between quarks gets stronger, not weaker, as the quarks move farther apart. This helps to explain why this force makes it impossible to remove an isolated quark from a hadron.

Since quarks are charged, they also interact electromagnetically by exchanging photons. But these interactions have a relatively small effect compared to the dominant strong interaction mediated by gluons.

As we mentioned in the previous section, when an electron and a positron (matter and antimatter) collide, they annihilate. The Feynman diagram in Figure 28-8 shows one possible outcome of such an annihilation. The electron of charge $-e$ and the positron of charge $+e$ disappear and are replaced by a single photon. This *virtual* photon does not have the proper relationship between energy and momentum that a real photon must have, and so it violates the conservation laws that govern the electron-positron collision. Like an exchange photon, this virtual photon can exist for only a finite time before it must disappear. In some cases the virtual photon will transform back into an electron-positron pair. But as Figure 28-8 shows, it's also possible for the virtual photon to become a quark and an antiquark of the same flavor (for instance, a u quark and a ū antiquark, or an s quark and an s̄ antiquark). The quark and antiquark can themselves emit virtual gluons, which can in turn transform into quark-antiquark pairs. Depending on how much energy is available in the original collision between electron and positron, many such quark-antiquark pairs can be produced. These will sort themselves into combinations of quarks (baryons), combinations of antiquarks (antibaryons), and quark-antiquark pairs (mesons). The result will be a shower of hadrons emanating from the site of the electron-positron collision. The image that opens this chapter shows two "jets" of hadrons emerging from the site of just such a collision. Many varieties of hadrons were first identified in electron-positron collision experiments of this sort.

The notion of gluon exchange also helps us understand the strong force in the context in which we first encountered it in Section 27-2: a force that attracts nucleons (protons or neutrons) to each other. Figure 28-9 shows how this force between nucleons arises. The meson that is produced in this way acts as an exchange particle between the nucleons, and this exchange gives rise to the attractive force between nucleons. The meson exchange particle has a substantial mass $m_{exchange}$, and so the minimum energy that an exchanged meson can have is its rest energy $m_{exchange}c^2$. Producing such a meson means violating the conservation of energy by an amount of at least $\Delta E = m_{exchange}c^2$, which means that the meson can exist for a time no longer than $\Delta t = \hbar/(2\Delta E) = \hbar/2m_{exchange}c^2$. Even moving at the speed of light, the maximum distance that this exchange meson can travel during a time Δt is

$$(28\text{-}8) \qquad c\Delta t = \frac{c\hbar}{2m_{exchange}c^2} = \frac{\hbar}{2m_{exchange}c}$$

(range of a force mediated by an exchange particle of mass $m_{exchange}$)

Equation 28-8 tells us the *range* of the force mediated by the exchange meson. There are many types of mesons, but the type whose exchange will have the longest range is the one with the smallest mass. (That's because the range in Equation 28-8 is inversely proportional to the mass of the exchange particle.) The mass of the lightest meson, the neutral pion (π^0), is $135.0 \text{ MeV}/c^2 = 135.0 \times 10^6 \text{ eV}/c^2$, so the maximum range of the strong force between nucleons should be

$$\frac{\hbar}{2m_{neutral\ pion}c} = \frac{\hbar c}{2m_{neutral\ pion}c^2} = \frac{hc}{4\pi m_{neutral\ pion}c^2}$$

$$= \frac{(4.136 \times 10^{-15} \text{ eV} \cdot \text{s})(3.00 \times 10^8 \text{ m/s})}{4\pi(135.0 \times 10^6 \text{ eV}/c^2)c^2}$$

$$= 7.31 \times 10^{-16} \text{ m} = 0.731 \text{ fm}$$

① One of the quarks in the proton emits a virtual gluon.

② The virtual gluon transforms into a virtual quark (q) and a virtual antiquark ($\bar{q}$), which together constitute a virtual meson.

Figure 28-9 **The strong force between nucleons** The force that binds protons and neutrons together has its origin in the interaction between quarks and gluons.

③ The virtual quark and antiquark transform back into a virtual gluon, which is absorbed by one of the quarks in the neutron.

- Nucleons (protons and neutrons) exert the strong force on each other by exchanging virtual mesons.
- This exchange is actually due to interactions between quarks and gluons.

Because this range is so small, the strong force between nucleons is a short-range force, just as we discussed in Section 27-2. Indeed, the rough value of 0.731 fm that we calculated here is of the same order of magnitude as the value we gave in Section 27-2 for the range of the strong force between nucleons.

Exchange Particles: The Weak Force and the Gravitational Force

The weak force is mediated by *three* different exchange particles. These are the neutral Z^0 (charge zero) and the positively and negatively charged W particles, W^+ (charge $+e$) and W^- (charge $-e$). (These particles are really three manifestations of the same particle, but have different names for historical reasons.) These particles are not massless: The Z^0 has mass 91.2 GeV/c^2, about 100 times the mass of a proton, and the W^+ and W^- both have mass 80.4 GeV/c^2. The calculation above shows that the range of the force mediated by an exchange particle of mass m_{exchange} is inversely proportional to the mass. So since the Z^0, W^+, and W^- have roughly 600 times the mass of the neutral pion (135.0 MeV/c^2), the range of the weak force is roughly 1/600 that of the strong force between nucleons, or about 10^{-18} m $= 10^{-3}$ fm. The weak force is not only weak, it is of *extremely* short range!

It turns out that any two particles can exert a weak force on each other. For example, Figure 28-10 shows two examples of the weak interaction between a neutrino and an electron. Note that even though the weak interaction does not directly involve electric charge, charge is still conserved. In Figure 28-10a the neutrino remains neutral, and the electron retains its charge of $-e$. In Figure 28-10b the neutrino emits a W^+ of charge $+e$ and becomes an electron of charge $-e$, so charge is conserved. The W^+ combines with an electron of charge $-e$ and becomes a neutrino of charge zero, and again charge is conserved.

Figure 28-11 shows how the weak force gives rise to the β^- decay of a neutron (quark content udd) to a proton (quark content uud):

$$n \rightarrow p + e^- + \bar{\nu}_e$$

Compared to a u quark, a d quark is substantially more massive (see Table 28-1) and so has a substantially greater rest

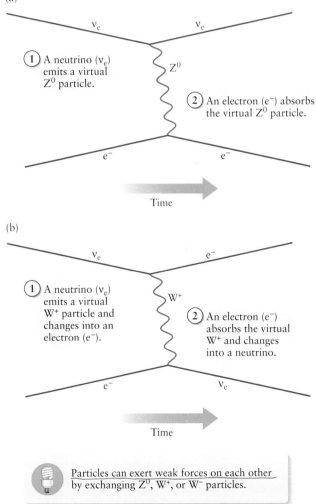

(a)

① A neutrino (ν_e) emits a virtual Z^0 particle.

② An electron (e^-) absorbs the virtual Z^0 particle.

Time

(b)

① A neutrino (ν_e) emits a virtual W^+ particle and changes into an electron (e^-).

② An electron (e^-) absorbs the virtual W^+ and changes into a neutrino.

Time

Particles can exert weak forces on each other by exchanging Z^0, W^+, or W^- particles.

Figure 28-10 **The weak interaction** The weak force between particles involves the exchange of massive virtual particles. These come in both (a) neutral and (b) charged varieties.

Figure 28-11 Beta decay of a neutron The weak interaction describes how a free neutron decays into a proton, an electron, and an antineutrino.

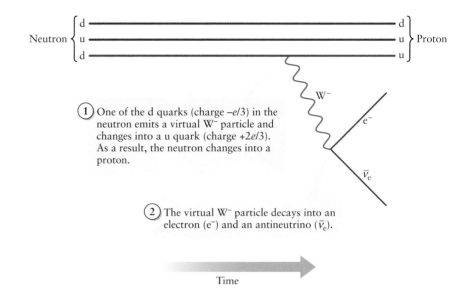

(1) One of the d quarks (charge $-e/3$) in the neutron emits a virtual W^- particle and changes into a u quark (charge $+2e/3$). As a result, the neutron changes into a proton.

(2) The virtual W^- particle decays into an electron (e^-) and an antineutrino ($\bar{\nu}_e$).

Time

energy. It's therefore favorable for the d quark to lower its energy by becoming a u quark. It does this by emitting a W^- particle as shown. This is a very short-lived virtual particle, since the rest energy of the W^- ($80.4\ \text{GeV}/c^2$) is far greater than the rest energy of the original d quark (about $5\ \text{MeV}/c^2 = 5 \times 10^{-3}\ \text{GeV}/c^2$), and so emitting a W^- means that we've violated energy conservation by quite a bit. The W^- of charge $-e$ quickly decays into an electron of charge $-e$ and a neutral antineutrino. (It has to be an *anti*neutrino to conserve electron-lepton number L_e: The W^- is not a lepton and therefore has $L_e = 0$, so the particles into which the W^- decays must also have a net electron-lepton number of zero. The electron, with $L_e = +1$, must therefore be accompanied by a neutral particle with $L_e = -1$, which means an antineutrino rather than a neutrino.)

As we discussed in Section 27-6, in some nuclei it's energetically favorable for β^+ decay to take place, in which a proton changes to a neutron. This is similar to the process shown in Figure 28-11, except that one of the u quarks in a proton changes to a d quark by emitting a W^+ rather than a W^-. The W^+ then decays into a positron (e^+) and a neutrino (ν_e). One nucleus in which this happens is the potassium isotope ^{40}K, which occurs in nature and is found in foods such as oranges (see Figure 28-1a). The bit of antimatter produced in this way—the positron—quickly encounters an atomic electron and annihilates. Happily, this happens very infrequently in your food (and releases very little energy when it does happen). So you need not fear the harmful effects of biting into a bit of antimatter when you eat fruit!

A particle named the *graviton* is thought to mediate the gravitational force. So far, no experimental evidence has confirmed the existence of this particle, although no experimental evidence has excluded it either. Because the gravitational force between two masses separated by a distance r is proportional to $1/r^2$, just like the electric force between two charged particles, the graviton is thought to have the same mass as the photon: zero. The graviton also carries no charge.

Table 28-3 summarizes the four fundamental forces and the exchange particles that mediate these forces. The picture that this table summarizes—in which two classes of fundamental particles, the six quarks and the six leptons, interact by means of six exchange particles, the graviton, the Z^0, the W^+, the W^-, the photon, and the gluon—is called the **Standard Model**. What this very simple table does not reflect is the decades of experimental and theoretical effort (and several Nobel Prizes in Physics) that have gone into constructing this model. A number of questions about the Standard Model remain unanswered and are the subject of active research by physicists around the globe.

Table 28-3 The Forces

Force	Range	Mediator(s)	Strength Relative to the Strong Force
gravity	infinite	graviton	10^{-40}
weak	$\sim 10^{-3}$ fm	Z^0, W^+, W^-	10^{-6}
electromagnetic	infinite	photon (γ)	10^{-2}
strong	~ 1 fm	gluon (g)	1

? Got the Concept? 28-3
A Quark Decay

A c quark can decay into an s quark via the weak force. In this process, what does the c quark emit? (a) A W^+, which decays into a positron and a neutrino; (b) a W^+, which decays into a positron and an antineutrino; (c) a W^-, which decays into an electron and a neutrino; (d) a W^-, which decays into an electron and an antineutrino.

Take-Home Message for Section 28-3

✔ Particles exert forces on each other through the exchange of virtual particles that are emitted by one particle and absorbed by another.

✔ The photon is the exchange particle that mediates the electromagnetic force, the gluon mediates the strong force, and the Z^0, W^+, and W^- particles mediate the weak force. The graviton, which has not yet been detected, is thought to mediate the gravitational force.

28-4 We live in an expanding universe, and the nature of most of its contents is a mystery

The Standard Model gives us a description of all the normal matter in the world around us. All of the atoms in our bodies and our surroundings are made of leptons and hadrons, the hadrons are all made of quarks, and all of these particles interact by means of exchange particles. Why, you may ask, do we specify that this description applies to "normal" matter only? What other sort of matter could there be, and how much of the contents of the physical universe is so-called normal matter? To answer these questions, we must turn our attention from the smallest fundamental particles to galaxies, some of the largest structures in the universe, and to the nature of the universe itself—the subject of the science called **cosmology**.

The Hubble Law and the Expansion of the Universe

Figure 28-12 is an image made with the Hubble Space Telescope of a small portion of the night sky. At first glance it may appear that the bright dots in this image are individual stars. But closer inspection shows that each is actually a **galaxy**, a collection of a tremendous number of stars (typically 10^{10} to 10^{12}). Our Sun is one of approximately 2×10^{11} stars in the Milky Way, the local galaxy of which we are part.

In the 1920s the American astronomer Edwin Hubble was the first to show that galaxies are very distant. (He did this by identifying within other galaxies certain types of stars that are found in our own galaxy and whose light output is known. These stars appear very dim when seen in other galaxies, so they must be very far away.) In collaboration with the astronomer Milton Humason, he also measured the spectra of many galaxies. These are really the spectra of the combined light from all of the stars that make up each galaxy, and so they have absorption lines (see Section 26-5). Hubble and Humason found that most galaxies show a **redshift** in their spectrum: That is, the absorption lines in the spectra of galaxies are all shifted to longer wavelengths and lower frequencies than in the spectra of nearby stars in our own galaxy. (We use the term *redshift* because in the visible part of the electromagnetic spectrum, red light has the longest wavelength and lowest frequency.) What's more, the greater the distance to a galaxy, the greater the wavelength shift. What did this discovery mean?

To answer this question, recall what we learned in Section 13-10 about the *Doppler effect*: If a source of waves is moving away from us, the wave that we receive from that source is shifted to a lower frequency and a longer wavelength. In Section 13-10 we introduced this idea in the context of sound waves, but the same effect also applies to electromagnetic waves. Hubble and Humason concluded that other galaxies are moving away from us, and so the light that we receive from those galaxies is at a longer wavelength. Their results showed that the speed v at which a distant galaxy is receding from us is directly proportional to the distance d to that galaxy (Figure 28-13):

Watch Out!

A galaxy is not a solar system, nor is it the entire universe.

In everyday language, many people use the terms *galaxy*, *solar system*, and *universe* interchangeably. In fact, a galaxy is a large collection of stars and other matter, while a solar system is a single star and its retinue of planets. Our galaxy contains some 200 billion stars, many of which have planets, and the observable universe contains literally billions of galaxies.

Figure 28-12 **A universe of galaxies** Each blob of light in this image from the Hubble Space Telescope is a galaxy, a grouping of billions of stars and other matter. The area shown is a very small patch of sky, only about 1/12 the width of the full Moon.

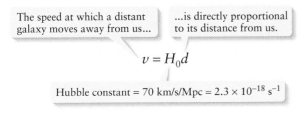

The speed at which a distant galaxy moves away from us...

...is directly proportional to its distance from us.

$$v = H_0 d$$

Hubble constant = 70 km/s/Mpc = 2.3×10^{-18} s^{-1}

The Hubble law
(28-9)

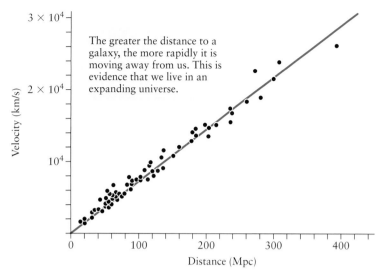

The greater the distance to a galaxy, the more rapidly it is moving away from us. This is evidence that we live in an expanding universe.

Figure 28-13 The Hubble law Each black dot in this graph represents the speed at which an individual galaxy is moving away from us, as well as the distance to that galaxy. The straight line is the best fit to that data, corresponding to a value of $H_0 = 70$ km/s/Mpc in Equation 28-9.

This direct proportionality is called the **Hubble law**, and the constant H_0 in Equation 28-9 is called the **Hubble constant**. Astronomers typically measure distances in the cosmos in parsecs (pc), a unit equal to 3.26 light-years. One light-year (ly) is the distance that light travels in one year, equal to 3.09×10^{16} m. The current best value of H_0 is about 70 km/s/Mpc, where 1 Mpc = 1 megaparsec = 10^6 pc = 3.26×10^6 ly.

What can we deduce from the observations that distant galaxies are moving away from us and that the speed at which they move away is proportional to the distance? The interpretation is that because remote galaxies are getting farther and farther apart as time goes on, the universe is expanding. A good analogy is a loaf of raisin bread being baked in an oven. As the loaf expands during baking, the amount of space between the raisins gets larger and larger. In the same way, as the universe expands, the amount of space between widely separated galaxies increases. The expansion of the universe is actually the expansion of space.

Watch Out! The universe is expanding, but the galaxies (and you) are not.

It's important to realize that the expansion of the universe occurs primarily in the vast spaces that separate clusters of galaxies. Just as the raisins in a loaf of raisin bread don't expand as the loaf expands in the oven, galaxies themselves do not expand. Einstein and others have established that an object that is held together by its own gravity, such as a galaxy or a cluster of galaxies, is always contained within a patch of nonexpanding space. A galaxy's gravitational field produces this nonexpanding region. Thus, Earth and your body, for example, are not getting any bigger. Only the distance between widely separated galaxies increases with time.

Although distant galaxies are moving away from us, that does *not* mean that we are at the center of the universe. In an expanding universe, every galaxy moves away from every other galaxy, so an alien astronomer in a distant galaxy would see the same relationship between the speeds and distances of galaxies—that is, the same Hubble law—as does an Earth astronomer. Since every point in the universe appears to be at the center of the expansion, it follows that our universe has no center at all.

Watch Out! "If the universe is expanding, what is it expanding into?"

This commonly asked question arises only if we take too literally our raisin bread analogy, in which the loaf (representing the universe) expands in three-dimensional space into the surrounding air. But the actual universe includes *all* space; there is nothing "beyond" it, because there is no "beyond." Asking "What lies beyond the universe?" is as meaningless as asking "Where on Earth is north of the North Pole?"

The ongoing expansion of space explains why the light from remote galaxies is redshifted. Imagine a photon coming toward us from a distant galaxy. As the photon travels through space, the space is expanding, so the photon's wavelength becomes stretched. When the photon reaches our eyes, we see an increased wavelength: The photon has been redshifted. The greater the distance the photon has had to travel and so the longer the amount of time it has had to travel to reach us, the more its wavelength will have been stretched. Thus, photons from distant galaxies have larger redshifts than those of photons from nearby galaxies, as expressed by the Hubble law.

A redshift caused by the expansion of the universe is properly called a **cosmological redshift**. It is *not* the same as a Doppler shift. Doppler shifts are caused by an object's motion through space, whereas a cosmological redshift is caused by the expansion of space.

Example 28-1 Measuring the Distance to a Galaxy from Its Redshift

Find the distances to the following galaxies: (a) NGC 4889, whose redshifted spectrum shows that it is moving away from us at 6410 km/s (2.14% of the speed of light); (b) 1255-0, which has a larger redshift that shows it to be moving away from us at 0.822c. Express the distances in megaparsecs and in light-years.

Set Up

For each galaxy we know the speed v at which it is moving away from us due to the expansion of the universe. So we can use the Hubble law to determine the distance d to that galaxy.

The Hubble law:

$$v = H_0 d \qquad (28\text{-}9)$$

Solve

(a) Use Equation 28-9 and the value $H_0 = 70$ km/s/Mpc to calculate the distance to NGC 4889.

Rewrite Equation 28-9 to solve for the distance d:

$$d = \frac{v}{H_0}$$

We are given $v = 6410$ km/s, so

$$d = \frac{6410 \text{ km/s}}{70 \text{ km/s/Mpc}} = 92 \text{ Mpc}$$

$$= (92 \text{ Mpc})\left(\frac{3.26 \times 10^6 \text{ ly}}{1 \text{ Mpc}}\right) = 3.0 \times 10^8 \text{ ly}$$

(b) Repeat the calculations for the galaxy 1255-0, which is moving away at 0.822c.

For this galaxy,

$$v = 0.822c = 0.822(3.00 \times 10^5 \text{ km/s})$$
$$= 2.47 \times 10^5 \text{ km/s}$$

Use this to calculate the distance to the galaxy as in part (a):

$$d = \frac{2.47 \times 10^5 \text{ km/s}}{70/\text{km/s/Mpc}} = 3.5 \times 10^3 \text{ Mpc}$$

$$= (3.5 \times 10^3 \text{ Mpc})\left(\frac{3.26 \times 10^6 \text{ ly}}{1 \text{ Mpc}}\right) = 1.1 \times 10^{10} \text{ ly}$$

Reflect

NGC 4889 is some 300 million light-years away, so the light we see from NGC 4889 left that galaxy 300 million years ago, before even the first dinosaurs appeared on Earth. Galaxy 1255-0 is far more distant, some 11 *billion* light-years away. When the light we receive from 1255-0 left that galaxy some 11 billion years ago, our Earth—which is a mere 4.56×10^9 years old—had not yet formed. When we use telescopes to look at distant astronomical objects, we are not only looking out into space; we are also looking back in time.

The Big Bang, Cosmic Background Radiation, and the Origin of Matter

If the universe is expanding, it must be that in the past the matter in the universe must have been closer together and therefore denser than it is today. If we look far enough into the very distant past, there must have been a time when the density of matter was almost inconceivably high. This leads us to conclude that some sort of tremendous event caused ultradense matter to begin the expansion that continues to the present day. This event, called the **Big Bang**, marks the beginning of the universe. If we use the observed expansion rate of the universe and work backwards, we find that the age of the universe is approximately 13.8 billion (13.8×10^9) years.

We learned in Section 15-3 that the temperature of an ideal gas decreases when it undergoes an adiabatic expansion (one in which there is no heat transfer to its surroundings). The expansion of the universe is much like an adiabatic expansion, so it follows that the average temperature of the universe has decreased over the past 13.8 billion years. (There are places in the universe that are at very high temperature, such as the interiors of stars. But the average temperature of space is very cold.) If we work backward in time, it follows that the early universe must have been at very high temperatures indeed. The hot early universe must therefore have been filled with many

! Watch Out!

It's not correct to think of the Big Bang as an explosion.

When a bomb explodes, pieces of debris fly off into space from a central location. If you could trace all the pieces back to their origin, you could find out exactly where the bomb had been. This process is not possible with the universe, however, because the universe itself always has and always will consist of all space. The present-day universe is infinite, and was infinite when it first originated in the Big Bang.

high-energy photons of high frequency and short wavelength. The properties of this radiation field depended on its temperature, as described by Planck's blackbody law (Section 26-2).

The universe has expanded so much since those ancient times that all those short-wavelength photons have had their wavelengths stretched by a tremendous factor. As a result, they have become low-energy, long-wavelength photons. The temperature of this cosmic radiation field is now only a few degrees above absolute zero, and the blackbody spectrum of this radiation has its peak intensity at low frequencies in the microwave part of the electromagnetic spectrum (wavelengths of approximately 1 mm). Hence, this radiation field, which fills all of space, is called the **cosmic microwave background** or **cosmic background radiation**. It represents the "afterglow" of the very high-temperature conditions that prevailed when the universe was young.

It's actually possible to detect the cosmic background radiation using an ordinary television (Figure 28-1b). Using much more sensitive detectors, scientists have found that the cosmic background radiation does indeed have a blackbody spectrum with a temperature of 2.725 K, which we can regard as the average temperature of the present-day universe. However, we observe different cosmic background radiation coming from different parts of the sky. Even when the effects of Earth's motion are accounted for, there remain variations in the temperature of the radiation field of about 200 μK (200 microkelvins, or 2×10^{-4} K) above or below the average 2.725 K temperature (Figure 28-14). These tiny temperature variations indicate that matter and radiation were not distributed in a totally uniform way in the early universe. Regions that were slightly denser than average were also slightly cooler than average; less dense regions were slightly warmer. Over time, the denser regions evolved to form the first galaxies, and within them the first stars. By studying these nonuniformities, we are really studying our origins.

The map of the sky shown in Figure 28-14 shows the universe as it was some 380,000 years after the Big Bang, when the universe was less than 0.003% of its present age. Prior to that time, the universe was so dense as to be opaque. (Think of a hot, dense, luminous fog.) Events prior to that date are forever hidden from our direct view. But by using the laws of particle physics that we described in Sections 28-2 and 28-3, we can infer a great deal about what conditions must have been like in the very early universe.

Our best understanding is that the universe began at extremely high temperatures and that the immense energy gave rise to a sea of quarks, antiquarks, leptons, and antileptons. As the universe expanded and cooled, by 10^{-6} s after the Big Bang quarks and antiquarks had coalesced into baryons, antibaryons, and mesons. But now we have a dilemma. If there had been perfect symmetry between particles and antiparticles, then for every proton there should have been an antiproton. For every electron, there should likewise have been a positron. By the time the universe was 1 second old, every particle would have been annihilated by an antiparticle, leaving no matter at all in the universe! Obviously, this did not happen.

The resolution to this dilemma is that the laws of physics are very slightly asymmetrical between matter and antimatter: For every 10^9 antimatter particles that were created out of the energy of the Big Bang, 10^9 plus one matter particles were created. When those 10^9 antimatter particles encountered 10^9 matter particles, annihilations resulted in no more particles and lots of energy in the form of photons. (These are the photons that gave rise to the cosmic background radiation that we see today.) That one extra matter particle avoided annihilation. All the normal matter that we see today is a result of the very slight imbalance between matter and antimatter in the early universe. Experiments on particle interactions are consistent with this very slight asymmetry between matter and antimatter.

By 15 minutes after the Big Bang, temperatures had dropped enough that protons and neutrons could coalesce into the first atomic nuclei. Only the four lightest elements (hydrogen, helium, lithium, and beryllium) were present in appreciable numbers. The heavier elements would be formed only much later, once stars had formed and nuclear reactions within those stars could manufacture carbon, nitrogen, oxygen, and all the other elements.

Figure 28-14 Temperature variations in the cosmic microwave background This map shows small variations in the temperature of the cosmic background radiation across the entire sky. Lower-temperature regions (shown in blue) show where the early universe was slightly denser than average.

Keep in mind that only *nuclei* formed in the first 15 minutes of the history of the universe. It would be another 380,000 years before temperatures became low enough for these nuclei to combine with electrons to form atoms. When this happened, the universe became transparent for the first time. The map of the cosmic background radiation in Figure 28-14 shows the universe at that moment in cosmic history.

Dark Matter and Dark Energy

At this point we might feel that we have understood, at least in broad outline, the nature and origin of matter. But have we in fact accounted for all of the matter in the universe? To answer this question we again look at galaxies, in particular *spiral galaxies* like the ones shown in Figure 28-15. These galaxies have a flattened disk shape, and all of the material in the galaxy rotates around the center. Each part of the galaxy is held in its orbit around the center by the gravitational attraction of the other parts of the galaxy; the greater that gravitational attraction on a given part, the faster that part must move to remain in a circular orbit (see Section 10-4). So if we can measure the speed at which parts of the galaxy orbit around the center, we can calculate the total amount of mass in the galaxy. Most of the atoms in the galaxy are in its stars, and we can estimate the combined mass of all the stars from the brightness of the galaxy. If the total amount of mass in the galaxy is a close match to the combined amount of mass in its stars, then we can conclude that ordinary matter accounts for most or all of the mass of the galaxy.

We can measure orbital speeds most directly for spiral galaxies that happen to be oriented edge-on to us, like the one shown in Figure 28-15b. We do this by looking at the spectrum of light from the two sides of the galaxy: The spectral lines will be Doppler shifted toward shorter wavelengths on the side where material is rotating toward us, and toward longer wavelengths on the side where material is rotating away from us. (These shifts are in addition to the cosmological redshift in the galaxy's spectrum as a whole, caused by the expansion of the universe.) The amount of Doppler shift tells us the speeds at which material in various parts of the galaxy are moving around the center.

These measurements have been made for a wide variety of spiral galaxies, and the results are always the same: The total mass of the galaxy deduced from the orbital speeds is about *five times greater* than the combined mass of all of the stars in the galaxy. So ordinary matter, which is concentrated in the galaxy's stars, is only a small fraction of the mass of the galaxy. The remaining mass is in a form that does not emit electromagnetic radiation of any kind, and may not even be composed of the fundamental particles that we described in Sections 28-2 and 28-3. This mysterious matter, which is the dominant form of matter in spiral galaxies and apparently in the universe as a whole, is called **dark matter**. As of this writing, its nature remains a mystery to science.

There is yet another complication to our understanding of the universe. As we look at increasingly distant galaxies, and so look farther back in the history of the universe, we find that as expected these distant galaxies are moving away from us due to the expansion of the universe. But the expansion rate has *not* been constant; observations show that for the past billion years, the expansion of the universe has been *speeding up!* To date, the only viable explanation for this increased speed is that, in addition to ordinary matter and dark matter, the universe is suffused with a curious form of energy that causes an accelerated expansion of the universe. We cannot detect this energy from its gravitational effects (the technique astronomers use to detect dark matter), and it does not emit detectable radiation of any kind. We refer to this curious energy as **dark energy**. The nature of dark energy is even more mysterious than that of dark matter.

Since matter and energy are equivalent through the Einstein relation $E = mc^2$, we can compare the relative importance of ordinary matter, dark matter, and dark energy by asking what fraction of the total energy in the universe each one represents. The graph in Figure 28-16 shows the results: Dark energy represents 68.3% of the total, dark matter another 26.8%, and "ordinary" normal matter just 4.9%. This is a sobering graph, for it tells us that more than 95% of the energy in the universe is in forms that we do not understand! The quest to understand both dark energy and dark matter is one of the most challenging pursuits in contemporary physics.

(a) We see spiral galaxy NGC 3982 face-on.

All of the matter in the galaxy rotates around the center.

(b) We see spiral galaxy NGC 4013 edge-on.

Stars at one edge rotate toward us...

...and stars at the other edge rotate away from us.

Figure 28-15 Measuring the mass of a spiral galaxy (a) The material in a spiral galaxy is held in orbit around the center by the gravitational attraction of all the other material in the galaxy. (b) For a galaxy that is edge-on to us, we can measure the orbital speeds, and hence the mass of the galaxy, using the Doppler effect.

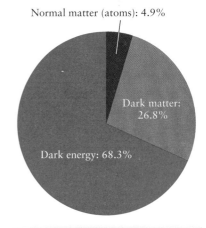

Normal matter (atoms): 4.9%

Dark matter: 26.8%

Dark energy: 68.3%

What we think of as "normal" matter makes up less than 5% of the contents of the universe.

Figure 28-16 Recipe for a universe This pie chart shows the three constituents of the universe as a whole. The numbers show what percentage of the total energy content of the universe each constituent represents.

? Got the Concept? 28-4 Dark Matter

In spiral galaxies like those shown in Figure 28-15, the ratio of dark matter to visible matter is about five to one. The percentage of the light from a spiral galaxy that is emitted by dark matter is closest to (a) 83%; (b) 80%; (c) 20%; (d) 17%; (e) zero.

Take-Home Message for Section 28-4

✔ The redshifts of galaxies reveal that we live in an expanding universe.

✔ The universe began with a Big Bang, with tremendously high temperatures present in the early universe. The cosmic background radiation is the "afterglow" of that early epoch in the history of the universe.

✔ The dominance of matter over antimatter in the universe is the result of a small but important asymmetry in the laws of particle physics.

✔ Most of the matter in the universe is in the form of dark matter, and most of the energy in the universe is in the form of dark energy. The nature of dark matter and the nature of dark energy are unsolved problems in physics.

Key Terms

antimatter
antiquark
baryon
Big Bang
cosmic background radiation
cosmic microwave background
cosmological redshift
cosmology
dark energy

dark matter
exchange particle
galaxy
gluon
hadron
Heisenberg uncertainty principle
Hubble constant
Hubble law
lepton

meson
particle physics
quark
quark confinement
redshift
Standard Model
virtual particle
weak force

Chapter Summary

Topic	Equation or Figure

Hadrons and leptons: Hadrons, including the proton and neutron, are particles that experience the strong force; leptons, including the electron and neutrino, are particles that do not. Hadrons are composed of more fundamental particles called quarks; these are confined to the interior of hadrons and cannot exist in isolation. Hadrons include baryons, which are made of three quarks, and mesons, which are made of a quark and antiquark. Leptons are themselves fundamental particles; they are not made of anything simpler.

Baryons

(a) Proton (p)

Net charge = $q_u + q_u + q_d$

$= \left(+\frac{2}{3}e\right) + \left(+\frac{2}{3}e\right) + \left(-\frac{1}{3}e\right) = +e$

(b) Neutron (n)

Net charge = $q_d + q_d + q_u$

$= \left(-\frac{1}{3}e\right) + \left(-\frac{1}{3}e\right) + \left(+\frac{2}{3}e\right) = 0$

(Figure 28-3)

Mesons

(a) Positive pion (π^+)

Net charge = $q_u + q_{\bar{d}}$

$= \left(+\frac{2}{3}e\right) + \left(+\frac{1}{3}e\right) = +e$

(b) Negative pion (π^-)

Net charge = $q_d + q_{\bar{u}}$

$= \left(-\frac{1}{3}e\right) + \left(-\frac{2}{3}e\right) = -e$

(Figure 28-4)

The four forces and exchange particles: All interactions between particles can be understood in terms of four basic forces. Each of these forces involves the exchange of virtual particles, which is permitted by the Heisenberg uncertainty principle. The strong force between quarks involves the exchange of gluons, and the electromagnetic force involves the exchange of photons; both gluons and photons have zero mass, so these are long-range forces. The weak force responsible for beta decay involves the exchange of massive particles, so this force has a very short range. The gravitational force involves the exchange of gravitons, which have not yet been detected experimentally.

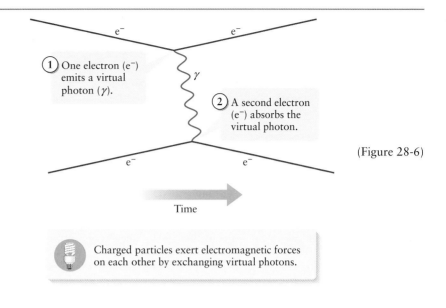

(1) One electron (e^-) emits a virtual photon (γ).

(2) A second electron (e^-) absorbs the virtual photon.

Time

(Figure 28-6)

Charged particles exert electromagnetic forces on each other by exchanging virtual photons.

The energy of a phenomenon is necessarily uncertain by an amount ΔE.

$$\Delta E \Delta t \geq \frac{\hbar}{2}$$

(28-7)

The shorter the duration Δt of the phenomenon, the greater the energy uncertainty ΔE.

$\hbar = \dfrac{h}{2\pi}$
= Planck's constant divided by 2π

Cosmology, dark matter, and dark energy: The Hubble law—the more distant a galaxy is from us, the faster it moves away from us—tells us that the universe is expanding and was once very highly compressed and at very high temperature. The cosmic background radiation is a relic of this ancient epoch. Processes in the early universe had a slight preference for matter over antimatter, which is why the present-day universe contains matter but almost no antimatter. Normal matter is actually just a small component of the universe: Mysterious dark matter is about five times as prevalent. Even more important and even more mysterious is dark energy, which causes the expansion of the universe to speed up.

The speed at which a distant galaxy moves away from us... ...is directly proportional to its distance from us.

$$v = H_0 d$$

(28-9)

Hubble constant = 70 km/s/Mpc = 2.3×10^{-18} s^{-1}

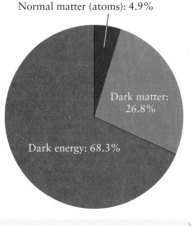

Normal matter (atoms): 4.9%

Dark matter: 26.8%

Dark energy: 68.3%

(Figure 28-16)

What we think of as "normal" matter makes up less than 5% of the contents of the universe.

Answer to What do you think? Question

(e) A collision such as this can produce other electrons (matter) and positrons (antimatter), but can also produce a shower of quarks and antiquarks. These can group into baryons (particles that are combinations of quarks, which are matter), antibaryons (particles that are combinations of antiquarks, which are antimatter), and mesons (particles that are a combination of a quark and an antiquark, and hence a mixture of matter and antimatter).

Answers to Got the Concept? Questions

28-1 (c) A quark has charge $+2e/3$ or $-e/3$, and an antiquark has charge $-2e/3$ or $+e/3$ (see Table 28-1). Mesons are made up of one quark and one antiquark, and no quark-antiquark combination has a net charge $-2e$. Baryons are made up of three quarks, and no combination of three quarks has a net charge $-2e$. Antibaryons are made up of three antiquarks; if each of the three has charge $-2e/3$, the total charge of the antibaryon will be $(-2e/3) + (-2e/3) + (-2e/3) = -2e$.

28-2 (a) The neutron (n) and proton (p) each have baryon number $B = 1$ and the antiproton ($\overline{p}$) has $B = -1$. Baryon number is conserved in process (a), since total B equals $1 + 1 = 2$ before the process and equals $1 + 1 + 1 + (-1) = 2$ after the process. This process can occur if the initial neutron and proton have sufficient kinetic energy so that, when they collide, this energy can be converted into the rest energy of the additional proton and antiproton. Baryon number is *not* conserved in process (b), since total B equals $1 + 1 = 2$ before the process and $1 + 1 + 1 = 3$ after the process. So (b) cannot occur. In process (c) the π^- is a meson with $L_e = 0$ and $L_\mu = 0$, the muon μ^- has $L_e = 0$ and $L_\mu = 1$, and the electron antineutrino $\overline{\nu}_e$ has $L_e = -1$ and $L_\mu = 0$. The total electron-lepton number is 0 before the process and $0 + -1 = -1$ after the process, and the total muon-lepton number is zero before the process and $1 + 0 = 1$ after the process. So neither electron-lepton number L_e nor muon-lepton number L_μ is conserved in process (c), and process (c) cannot occur.

28-3 (a) Table 28-1 shows that the c quark has charge $+2e/3$ and the s quark has a charge $-e/3$. The quark must therefore emit a W^+ with charge $+e$, so that the total charge of the W^+ and s quark equals the initial charge of the c quark: $(+e) + (-e/3) = +2e/3$. The W^+ must decay into a positron (charge $+e$) to conserve charge. Since the W^+ has electron-lepton number $L_e = 0$ (it's not a lepton) and the positron has $L_e = -1$, the positron must be accompanied by a neutral particle with $L_e = +1$. That must be a neutrino, not an antineutrino (which has $L_e = -1$).

28-4 (e) Dark matter is so called because it emits *no* electromagnetic radiation of any kind.

Questions and Problems

In a few problems, you are given more data than you actually need; in a few other problems, you are required to supply data from your general knowledge, outside sources, or informed estimate.

Interpret as significant all digits in numerical values that have trailing zeros and no decimal points.

For all problems, use $g = 9.80 \text{ m/s}^2$ for the free-fall acceleration due to gravity. Neglect friction and air resistance unless instructed to do otherwise.

* Basic, single-concept problem
** Intermediate-level problem, may require synthesis of concepts and multiple steps
*** Challenging problem

SSM *Solution is in Student Solutions Manual*

Conceptual Questions

1. •Define these terms: (a) baryon, (b) meson, (c) quark, (d) lepton, and (e) antiparticle.

2. •What do exchange particles do?

3. •Discuss the similarities and differences between a photon and a neutrino.

4. •Write the quark content for the antiparticle of each of the following: (a) e, (b) n, (c) p, and (d) π^+.

5. •Considering only the up (u), down (d), and strange (s) quarks, how many unique combinations of quarks resulting in a baryon can be produced (neglect the antimatter baryons) from just these three quarks? SSM

6. •A positron is stable; that is, it does not decay. Why, then, does a positron have only a short existence?

7. •When a positron and an electron annihilate at rest, why must more than one photon be created?

8. •A meson and a baryon come very close to one another. The particles could interact by which of the fundamental forces (gravitational, electromagnetic, weak, or strong)?

9. •Explain the evidence in favor of the existence of dark matter.

10. •Why are scientists convinced that dark energy must exist?

Multiple-Choice Questions

11. •A particle composed of a quark and an antiquark is classified as a
    A. baryon.
    B. meson.
    C. photon.
    D. pion.
    E. particon. SSM

12. •A particle composed of three quarks is classified as a
    A. baryon.
    B. meson.

C. photon.
D. pion.
E. particon.

13. •The quark composition of an antiproton is
   A. uud.
   B. uu$\overline{\text{d}}$.
   C. $\overline{\text{uu}}$d.
   D. $\overline{\text{uu}}\overline{\text{d}}$.
   E. uuu.

14. •The beta decay process is mediated by the
   A. gravitational force.
   B. electromagnetic force.
   C. strong force.
   D. weak force.
   E. dark force.

15. •Through which force does the electrically neutral pion interact with other particles?
   A. strong
   B. electromagnetic
   C. weak
   D. gravitational
   E. dark SSM

16. •Through which force does the photon interact with other particles?
   A. strong
   B. electromagnetic
   C. weak
   D. gravitational
   E. dark

17. •Particles that interact using the strong force are
   A. neutrinos.
   B. leptons.
   C. photons.
   D. quarks.
   E. gravitons.

18. •Recent observations indicate that the universe is dominated by
   A. matter.
   B. dark matter.
   C. dark energy.
   D. photons.
   E. none of the above.

19. •A mysterious energy that seems to cause the expansion of the universe to accelerate is
   A. gravitational potential energy.
   B. electric potential energy.
   C. thermal energy.
   D. dark energy.
   E. kinetic energy.

Estimation/Numerical Analysis

20. •One type of meson that can be exchanged between nucleons is the ρ (rho), with a mass of 775 MeV/c^2. Estimate the range of the interaction due to the exchange of ρ mesons. *Hint:* The range of the meson must be less than $R = c\hbar/(2mc^2)$, and the radius of the nucleus is about 1.5 fm.

21. •Estimate the distance to a galaxy that is moving away from us at one-tenth of the speed of light due to the expansion of the universe.

Problems

28-1 Studying the ultimate constituents of matter helps reveal the nature of the universe

28-2 Most forms of matter can be explained by just a handful of fundamental particles

22. ••Is the reaction $n \rightarrow \pi^+ + \pi^- + \mu^+ + \mu^-$ possible? Explain your answer.

23. ••Is the reaction $p \rightarrow e^+ + \gamma$ possible? Explain your answer. SSM

24. ••Is the reaction $e^- + p \rightarrow n + \overline{\nu}_e$ possible? Explain your answer.

25. ••(a) Describe the particle that is composed of the quark combination uds. (b) How would your answer change if the quark combination were instead uss? Explain your answer.

26. ••What is the quark structure of a π^- meson?

27. ••Two protons collide in a particle accelerator and generate new particles from their kinetic energy. Which of the following reactions is possible? Which is not possible? If a reaction is not possible, state which conservation law(s) is (are) violated.
   a. $p + p \rightarrow p + p + p + \overline{p}$
   b. $p + p \rightarrow p + p + n + \overline{n}$
   c. $p + p \rightarrow p + K^+$

28. ••Which of the following two possibilities for the weak decay of a sigma particle (a baryon) are possible? Why?
   a. $\Sigma^- \rightarrow \pi^- + p$
   b. $\Sigma^- \rightarrow \pi^- + n$

29. ••Which of the following reactions is possible? If a reaction is not possible, tell which conservation law(s) is (are) violated.
   a. $n \rightarrow p + e^- + \overline{\nu}_e$
   b. $\mu^- \rightarrow e^- + \overline{\nu}_e + \nu_\mu$
   c. $\pi^- \rightarrow \mu^- + \overline{\nu}_\mu$ SSM

30. ••One possible mode of decay of a photon is proton-antiproton pair production:

$$\gamma \rightarrow p + \overline{p}$$

What is the maximum wavelength of such a photon?

31. ••A neutral η (eta) meson at rest decays into two gamma rays according to

$$\eta \rightarrow \gamma + \gamma$$

Calculate the energy, momentum, and wavelength of each of the photons. The mass of the eta particle is 547 MeV/c^2.

32. ••Two protons collide in a particle accelerator, with the same speed, causing the reaction:

$$p + p \rightarrow p + p + \pi^0$$

Calculate the minimum kinetic energy of each of the incident protons.

33. ••A high-energy photon in the vicinity of a nucleus can create an electron-positron pair by pair production:

$$\gamma \rightarrow e^- + e^+$$

(a) What minimum energy photon is required? (b) Why is the nucleus needed? SSM

34. ••How much energy, in the form of gamma rays, would result due to the annihilation of a positron with kinetic energy 34 MeV and an electron with kinetic energy 16 MeV?

35. ••A proton-antiproton annihilation takes place, and the resulting photons have a total energy of 2.5 GeV. Find the kinetic energy of the proton and antiproton if the proton has (a) the same kinetic energy as the antiproton and (b) 1.25 times as much kinetic energy as the antiproton. SSM

36. ••The kinetic energy of a neutral pion (π^0) is 860 MeV. This pion decays to two photons, one of which has energy 640 MeV. Calculate the energy of the other photon.

28-3 Four fundamental forces describe all interactions between material objects

37. ••Draw the Feynman diagram for beta-plus decay wherein a proton changes into a neutron, a positron, and a neutrino.

38. ••Draw the Feynman diagram for the pair production of an electron and a positron from a gamma ray.

39. ••Draw the Feynman diagram for pair annihilation of a proton and an antiproton. SSM

40. ••Draw the Feynman diagram for the beta decay of the antineutron.

41. ••Suppose that a new fundamental force were discovered having a range about equal to the radius of a hydrogen atom. Estimate what you would expect to be the mass (in MeV/c^2) of the mediating particle of the new force. How does the mass compare with the mass of the electron?

28-4 We live in an expanding universe, and the nature of most of its contents is a mystery

42. •A galaxy is observed to recede from Earth with an approximate speed of $0.8c$. (a) Approximately how far from Earth is this galaxy? (b) How long ago was this light emitted by the galaxy?

43. •The galaxy NGC 3982 shown in Figure 28-15a is 6.8×10^7 ly from Earth. Approximately how fast is it receding from Earth?

General Problems

44. ••Suppose that a fifth fundamental force were discovered and that it was mediated by electrons and positrons. Estimate the range of the force.

45. ••(a) What stable nucleus would have a diameter equal to 10 times the approximate range of the strong force? (b) What would be the electric repulsion on a proton at the surface of the nucleus in part (a)? SSM

46. ••According to the U.S. Energy Information Administration, the United States used approximately 4.12×10^{12} kWh of electrical energy during 2010. (a) If we could generate all the energy using a matter–antimatter reactor, how many kilograms of fuel would the reactor need, assuming 100% efficiency in the annihilation? (b) Suppose the fuel consisted of

iron and anti-iron, each of density 7800 kg/m^3. If the iron and anti-iron were each stored as a cubical pile, what would be the dimensions of each cube?

47. •••Suppose that an electron neutrino and an electron antineutrino, both of which are just barely moving, encounter each other in space and completely annihilate to form two photons of equal energy. In view of the uncertainty about the mass of the electron neutrino (see Table 28-2), what is the shortest wavelength of light that could be emitted by the annihilation? Would the light be visible to the human eye?

48. ••Baryons are said to be "color neutral." This refers to the fact that each of the three quarks that comprise the baryon has a "color" quantum number (which is either red, green, or blue). (There is no actual color to see with quarks; color is only a quantum characteristic that can be likened to the electric charge of an electron, for example). Just as with visible light, when you mix red with green with blue you get white light, so it is with the color nature of quarks. A proton is not simply made up of uud; rather, it is $u_R u_G d_B$, for example. All the other quantum numbers that you have learned about (charge, baryon number, strangeness, etc.) still apply. The color quantum numbers "add up" to a neutral white ($R + G + B =$ white). Express all the possible color quantum states that might exist for an ordinary proton ($u_R u_G d_B$, $u_B u_R d_G$, etc.).

49. ••Like baryons, mesons are also color neutral. (See problem 48 for a discussion of color neutrality.) However, because mesons are made up of quark-antiquark pairs, it is not the combination of colors that cancels, but rather the subtraction of "anticolor" that creates the color neutrality. For example, a π^+ meson is made up of $u_R \overline{d}_R$. The antidown red quark cancels the up red quark much like the negative charge of an electron is canceled by the positive charge of a positron. Write the various possible color combinations for the three pions (π^+, π^0, π^-).

50. ••**Astronomy** The universe has evolved through several distinct *epochs* over the course of the last 13.8 billion years. The first epoch that occurred after the Big Bang, called the *Planck Epoch*, took place during the first 10^{-43} s after the Big Bang. This time interval is called the *Planck time*, t_P. (To understand what happened during the Planck Epoch, a theory that describes gravity in terms of quantum mechanics is required. No such theory yet exists.) Calculate t_P by first using dimensional analysis and finding the proper combination of the fundamental constants: G, c, and $\hbar$. Then use the current numerical values of each to calculate the precise number for t_P. (Recall that G is Newton's gravitational constant and is equal to 6.6738×10^{-11} N·m^2/kg^2, c is the speed of light and is equal to 2.9979×10^8 m/s, and $\hbar$ is Planck's constant divided by 2π or 1.0546×10^{-34} J·s). Start your calculation by asking what power would each of the three constants need to have in order to yield units of seconds: $[t_P] = [G]^x[c]^y[\hbar]^z$. Solve for x, y, and z to derive the formula for t_P.

51. •••In July 2011, physicists at the Tevatron at Fermilab near Chicago announced the discovery of a long-predicted but short-lived particle, the *xi-sub-b baryon*, Ξ_b. It is made up of a strange quark, an up quark, and a bottom quark, and is so short-lived that it travels less than a millimeter during its lifetime. (a) What is the charge of the Ξ_b baryon? (b) Estimate its lifetime if it travels about 0.5 mm before it decays.

APPENDIX A
SI Units and Conversion Factors

Base Units*

Length	The *meter* (m) is the distance traveled by light in a vacuum in 1/299,792,458 s.
Time	The *second* (s) is the duration of 9,192,631,770 periods of the radiation corresponding to the transition between the two hyperfine levels of the ground state of the ^{133}Cs atom.
Mass	The *kilogram* (kg) is the mass of the international standard body preserved at Sèvres, France.
Mole	The *mole* (mol) is the amount of substance of a system which contains as many elementary entities as there are atoms in 0.012 kg of carbon-12.
Current	The *ampere* (A) is that constant current which, if maintained in two straight parallel conductors of infinite length, of negligible circular cross section, and placed 1 m apart in vacuum, would produce between the conductors a force equal to 2×10^{-7} N/m of length.
Temperature	The *kelvin* (K) is 1/273.16 of the thermodynamic temperature of the triple point of water.
Luminous intensity	The *candela* (cd) is the luminous intensity in a given direction, of a source that emits monochromatic radiation of frequency 540×10^{12} Hz and that has a radiant intensity, in that direction of 1/683 W/steradian.

*These definitions are found on the Internet at http://physics.nist.gov/cuu/Units/current.html.

Derived Units

Force	newton (N)	$1 \text{ N} = 1 \text{ kg} \cdot \text{m/s}^2$
Work, energy	joule (J)	$1 \text{ J} = 1 \text{ N} \cdot \text{m}$
Power	watt (W)	$1 \text{ W} = 1 \text{ J/s}$
Frequency	hertz (Hz)	$1 \text{ Hz} = \text{cy/s}$
Charge	coulomb (C)	$1 \text{ C} = 1 \text{ A} \cdot \text{s}$
Potential	volt (V)	$1 \text{ V} = 1 \text{ J/C}$
Resistance	ohm (Ω)	$1 \Omega = 1 \text{ V/A}$
Capacitance	farad (F)	$1 \text{ F} = 1 \text{ C/V}$
Magnetic field	tesla (T)	$1 \text{ T} = 1 \text{ N/(A} \cdot \text{m)}$
Magnetic flux	weber (Wb)	$1 \text{ Wb} = 1 \text{ T} \cdot \text{m}^2$
Inductance	henry (H)	$1 \text{ H} = 1 \text{ J/A}^2$

Conversion Factors

Conversion factors are written as equations for simplicity; relations marked with an asterisk are exact.

Length

1 km = 0.6214 mi

1 mi = 1.609 km

1 m = 1.0936 yard = 3.281 ft = 39.37 in.

*1 in. = 2.54 cm

*1 ft = 12 in. = 30.48 cm

*1 yard = 3 ft = 91.44 cm

1 light-year = 1 $c \cdot y$ = 9.461 $\times$ 10^{15} m

*1 Å = 0.1 nm

Area

*1 m^2 = 10^4 cm^2

1 km^2 = 0.3861 mi^2 = 247.1 acres

*1 in.^{2} = 6.4516 cm^2

1 ft^2 = 9.29 $\times$ 10^{-2} m^2

1 m^2 = 10.76 ft^2

*1 acre = 43 560 ft^2

1 mi^2 = 640 acres = 2.590 km^2

Volume

*1 m^3 = 10^6 cm^3

*1 L = 1000 cm^3 = 10^{-3} m^3

1 gal = 3.785 L

1 gal = 4 qt = 8 pt = 128 oz = 231 in.^{3}

1 in.^{3} = 16.39 cm^3

1 ft^3 = 1728 in.^{3} = 28.32 L
    = 2.832 $\times$ 10^4 cm^3

Time

*1 h = 60 min = 3.6 ks

*1 d = 24 h = 1440 min = 86.4 ks

1 y = 365.25 day = 3.156 $\times$ 10^7 s

Speed

*1 m/s = 3.6 km/h

1 km/h = 0.2778 m/s = 0.6214 mi/h

1 mi/h = 0.4470 m/s = 1.609 km/h

1 mi/h = 1.467 ft/s

Angle and Angular Speed

*π rad = 180°

1 rad = 57.30°

1° = 1.745 $\times$ 10^{-2} rad

1 rev/min = 0.1047 rad/s

1 rad/s = 9.549 rev/min

Mass

*1 kg = 1000 g

*1 tonne = 1000 kg = 1 Mg

1 u = 1.6605 $\times$ 10^{-27} kg
    931.49 MeV/c^2

1 kg = 6.022 $\times$ 10^{26} u

1 slug = 14.59 kg

1 kg = 6.852 $\times$ 10^{-2} slug

Density

*1 g/cm^3 = 1000 kg/m^3 = 1 kg/L

(1 g/cm^3)g = 62.4 lb/ft^3

Force

1 N = 0.2248 lb = 10^5 dyn

*1 lb = 4.448222 N

(1 kg)g = 2.2046 lb

Pressure

*1 Pa = 1 N/m^2

*1 atm = 101.325 kPa = 1.01325 bar

1 atm = 14.7 lb/in.^{2} = 760 mmHg
    = 29.9 in.Hg = 33.9 ftH$_2$O

1 lb/in.^{2} = 6.895 kPa

1 torr = 1 mmHg = 133.32 Pa

1 bar = 100 kPa

Energy

*1 kW $\cdot$ h = 3.6 MJ

*1 cal = 4.186 J

1 ft $\cdot$ lb = 1.356 J = 1.286 $\times$ 10^{-3} BTU

*1 L $\cdot$ atm = 101.325 J

1 L $\cdot$ atm = 24.217 cal

1 BTU = 778 ft $\cdot$ lb = 252 cal = 1054.35 J

1 eV = 1.602 $\times$ 10^{-19} J

1 u $\cdot$ c^2 = 931.49 MeV

*1 erg = 10^{-7} J

Power

1 horsepower = 550 ft $\cdot$ lb/s = 745.7 W

1 BTU/h = 2.931 $\times$ 10^{-4} kW

1 W = 1.341 $\times$ 10^{-3} horsepower
    = 0.7376 ft $\cdot$ lb/s

Magnetic Field

*1 T = 10^4 G

Thermal Conductivity

1 W/(m $\cdot$ K) = 6.938 BTU $\cdot$ in./(h $\cdot$ ft^2 $\cdot$ °F)

1 BTU $\cdot$ in./(h $\cdot$ ft^2 $\cdot$ °F) = 0.1441 W/(m $\cdot$ K)

APPENDIX B
Numerical Data

Terrestrial Data

Free-fall acceleration g

    Standard value (at sea level at 45° latitude)* $9.806\ 65$ m/s^2; 32.1740 ft/s^2

    At equator* 9.7804 m/s^2

    At poles* 9.8322 m/s^2

Mass of Earth M_E 5.98×10^{24} kg

Radius of Earth R_E, mean 6.38×10^6 m; 3960 mi

Escape speed 1.12×10^4 m/s; 6.96 mi/s

Solar constant[†] 1.37 kW/m^2

Standard temperature and pressure (STP):

    Temperature 273.15 K

    Pressure 101.3 kPa (1.00 atm)

Molar mass of air 28.97 g/mol

Density of air (273.15 K, 101.3 kPa), ρ_{air} 1.29 kg/m^3

Speed of sound (273.15 K, 101.3 kPa) 331 m/s

Latent heat of fusion of H_2O (0°C, 1 atm) 334 kJ/kg

Latent heat of vaporization of H_2O (100°C, 1 atm) 2.26 MJ/kg

---

* Measured relative to Earth's surface.

[†] Average power incident normally on 1 m^2 outside Earth's atmosphere at the mean distance
  from Earth to the Sun.

Astronomical Data*

Earth

    Distance to the Moon, mean[†] 3.844×10^8 m; 2.389×10^5 mi

    Distance to the Sun, mean[†] 1.496×10^{11} m; 9.32×10^7 mi; 1.00 AU

    Orbital speed, mean 2.98×10^4 m/s

Moon

    Mass 7.35×10^{22} kg

    Radius 1.737×10^6 m

    Period 27.32 day

    Acceleration of gravity at surface 1.62 m/s^2

Sun

    Mass 1.99×10^{30} kg

    Radius 6.96×10^8 m

---

* Additional solar system data are available from NASA at http://nssdc.gsfc.nasa.gov/planetary/
planetfact.html.

[†] Center to center.

Physical Constants*

Universal constant of gravitation	G	$6.673\ 84(80) \times 10^{-11}$ N · m^2/kg^2
Speed of light	c	$2.997\ 924\ 58 \times 10^8$ m/s
Fundamental charge	e	$1.602\ 176\ 565(35) \times 10^{-19}$ C
Avogadro's constant	N_A	$6.022\ 141\ 29(27) \times 10^{23}$ particles/mol
Gas constant	R	$8.314\ 462\ 1(75)$ J/(mol · K)
		$1.987\ 206\ 5(36)$ cal/(mol · K)
		$8.205\ 746(15) \times 10^{-2}$ L · atm/(mol · K)
Boltzmann constant	$k = R/N_A$	$1.380\ 648\ 8(13) \times 10^{-23}$ J/K
		$8.617\ 332\ 4(81) \times 10^{-5}$ eV/K
Stefan-Boltzmann constant	$\sigma = (\pi^2/60)k^4/(\hbar^3c^2)$	$5.670\ 373(21) \times 10^{-8}$ W/(m^2 · K^4)
Atomic mass constant	$m_u = (1/12)m(^{12}\text{C})$	$1.660\ 538\ 921(73) \times 10^{-27}$ kg $= 1$ u
Permeability of free space	μ_0	$4\pi \times 10^{-7}$ N/A^2
		$1.256\ 637 \ldots \times 10^{-6}$ N/A^2
Permittivity of free space	$\epsilon_0 = 1/(\mu_0c^2)$	$8.854\ 187\ 817 \ldots \times 10^{-12}$ C^2/(N · m^2)
Coulomb constant	$k = 1/(4\pi\epsilon_0)$	$8.987\ 551\ 787 \ldots \times 10^9$ N · m^2/C^2
Planck's constant	h	$6.626\ 069\ 57(29) \times 10^{-34}$ J · s
		$4.135\ 667\ 516(91) \times 10^{-15}$ eV · s
	$\hbar = h/(2\pi)$	$1.054\ 571\ 726(47) \times 10^{-34}$ J · s
		$6.582\ 119\ 28(15) \times 10^{-16}$ eV · s
Mass of electron	m_e	$9.109\ 382\ 91(40) \times 10^{-31}$ kg
		$0.510\ 998\ 928(11)$ MeV/c^2
Mass of proton	m_p	$1.672\ 621\ 777(74) \times 10^{-27}$ kg
		$938.272\ 046(21)$ MeV/c^2
Mass of neutron	m_n	$1.674\ 927\ 351(74) \times 10^{-27}$ kg
		$939.565\ 379(21)$ MeV/c^2
Bohr magneton	$m_B = e\hbar/(2m_e)$	$9.274\ 009\ 68(20) \times 10^{-24}$ J/T
		$5.788\ 381\ 806\ 6(38) \times 10^{-5}$ eV/T
Nuclear magneton	$m_n = e\hbar/(2m_p)$	$5.050\ 783\ 53(11) \times 10^{-27}$ J/T
		$3.152\ 451\ 258(99) \times 10^{-8}$ eV/T
Magnetic flux quantum	$\phi_0 = h/(2e)$	$2.067\ 833\ 758(46) \times 10^{-15}$ T · m^2
Quantized Hall resistance	$R_K = h/e^2$	$2.581\ 280\ 744\ 34(84) \times 10^4$ Ω
Rydberg constant	R_H	$1.097\ 373\ 156\ 853\ 9(55) \times 10^7$ m^{-1}
Josephson frequency–voltage quotient	$K_J = 2e/h$	$4.835\ 978\ 70(11) \times 10^{14}$ Hz/V
Compton wavelength	$\lambda_C = h/(m_ec)$	$2.426\ 310\ 238\ 9(16) \times 10^{-12}$ m

* The values for these and other constants may be found on the Internet at http://physics.nist.gov/cuu/Constants/index.html. The numbers in parentheses represent the uncertainties in the last two digits. (For example, 2.044 43(13) stands for 2.044 43 ± 0.000 13.) Values without uncertainties are exact, including those values with ellipses (such as the value of pi is exactly 3.1415...).

1																		18
1 H	2												13	14	15	16	17	2 He
3 Li	4 Be												5 B	6 C	7 N	8 O	9 F	10 Ne
11 Na	12 Mg	3	4	5	6	7	8	9	10	11	12		13 Al	14 Si	15 P	16 S	17 Cl	18 Ar
19 K	20 Ca	21 Sc	22 Ti	23 V	24 Cr	25 Mn	26 Fe	27 Co	28 Ni	29 Cu	30 Zn		31 Ga	32 Ge	33 As	34 Se	35 Br	36 Kr
37 Rb	38 Sr	39 Y	40 Zr	41 Nb	42 Mo	43 Tc	44 Ru	45 Rh	46 Pd	47 Ag	48 Cd		49 In	50 Sn	51 Sb	52 Te	53 I	54 Xe
55 Cs	56 Ba	57–71 Lanthanoids	72 Hf	73 Ta	74 W	75 Re	76 Os	77 Ir	78 Pt	79 Au	80 Hg		81 Tl	82 Pb	83 Bi	84 Po	85 At	86 Rn
87 Fr	88 Ra	89–103 Actinoids	104 Rf	105 Db	106 Sg	107 Bh	108 Hs	109 Mt	110 Ds	111 Rg	112 Cn							

Lanthanoids	57 La	58 Ce	59 Pr	60 Nd	61 Pm	62 Sm	63 Eu	64 Gd	65 Tb	66 Dy	67 Ho	68 Er	69 Tm	70 Yb	71 Lu
Actinoids	89 Ac	90 Th	91 Pa	92 U	93 Np	94 Pu	95 Am	96 Cm	97 Bk	98 Cf	99 Es	100 Fm	101 Md	102 No	103 Lr

* From http://old.iupac.org/reports/periodic_table/IUPAC_Periodic_Table-21Jan11.pdf.

Atomic Numbers and Atomic Weights*

Atomic Number	Name	Symbol	Weight	Atomic Number	Name	Symbol	Weight
1	Hydrogen	H	[1.007 84; 1.008 11]	57	Lanthanum	La	138.90547(7)
2	Helium	He	4.002602(2)	58	Cerium	Ce	140.116(1)
3	Lithium	Li	[6.938; 6.997]	59	Praseodymium	Pr	140.90765(2)
4	Beryllium	Be	9.012182(3)	60	Neodymium	Nd	144.242(3)
5	Boron	B	[10.806; 10.821]	61	Promethium	Pm	
6	Carbon	C	[12.009 6; 12.011 6]	62	Samarium	Sm	150.36(2)
7	Nitrogen	N	[14.006 43; 14.007 28]	63	Europium	Eu	151.964(1)
8	Oxygen	O	[15.999 03; 15.999 77]	64	Gadolinium	Gd	157.25(3)
9	Fluorine	F	18.9984032(5)	65	Terbium	Tb	158.92535(2)
10	Neon	Ne	20.1797(6)	66	Dysprosium	Dy	162.500(1)
11	Sodium	Na	22.98976928(2)	67	Holmium	Ho	164.93032(2)
12	Magnesium	Mg	24.3050(6)	68	Erbium	Er	167.259(3)
13	Aluminum	Al	26.9815386(8)	69	Thulium	Tm	168.93421(2)
14	Silicon	Si	[28.084; 28.086]	70	Ytterbium	Yb	173.054(5)
15	Phosphorus	P	30.973762(2)	71	Lutetium	Lu	174.966 8(1)
16	Sulfur	S	[32.059; 32.076]	72	Hafnium	Hf	178.49(2)
17	Chlorine	Cl	[35.446; 35.457]	73	Tantalum	Ta	180.94788(2)
18	Argon	Ar	39.948(1)	74	Tungsten	W	183.84(1)
19	Potassium	K	39.0983(1)	75	Rhenium	Re	186.207(1)
20	Calcium	Ca	40.078(4)	76	Osmium	Os	190.23(3)
21	Scandium	Sc	44.955912(6)	77	Iridium	Ir	192.217(3)
22	Titanium	Ti	47.867(1)	78	Platinum	Pt	195.084(9)
23	Vanadium	V	50.9415(1)	79	Gold	Au	196.966569(4)
24	Chromium	Cr	51.9961(6)	80	Mercury	Hg	200.59(2)
25	Manganese	Mn	54.938045(5)	81	Thallium	Tl	[204.382; 204.385]
26	Iron	Fe	55.845(2)	82	Lead	Pb	207.2(1)
27	Cobalt	Co	58.933195(5)	83	Bismuth	Bi	208.98040(1)
28	Nickel	Ni	58.6934(2)	84	Polonium	Po	
29	Copper	Cu	63.546(3)	85	Astatine	At	
30	Zinc	Zn	65.38 (2)	86	Radon	Rn	
31	Gallium	Ga	69.723(1)	87	Francium	Fr	
32	Germanium	Ge	72.63 (1)	88	Radium	Ra	
33	Arsenic	As	74.92160(2)	89	Actinium	Ac	
34	Selenium	Se	78.96(3)	90	Thorium	Th	232.03806(2)
35	Bromine	Br	79.904(1)	91	Protactinium	Pa	231.03588(2)
36	Krypton	Kr	83.798(2)	92	Uranium	U	238.02891(3)
37	Rubidium	Rb	85.4678(3)	93	Neptunium	Np	
38	Strontium	Sr	87.62(1)	94	Plutonium	Pu	
39	Yttrium	Y	88.90585(2)	95	Americium	Am	
40	Zirconium	Zr	91.224(2)	96	Curium	Cm	
41	Niobium	Nb	92.90638(2)	97	Berkelium	Bk	
42	Molybdenum	Mo	95.96 (2)	98	Californium	Cf	
43	Technetium	Tc		99	Einsteinium	Es	
44	Ruthenium	Ru	101.07(2)	100	Fermiun	Fm	
45	Rhodium	Rh	102.90550(2)	101	Mendelevium	Md	
46	Palladium	Pd	106.42(1)	102	Nobelium	No	
47	Silver	Ag	107.8682(2)	103	Lawrencium	Lr	
48	Cadmium	Cd	112.411(8)	104	Rutherfordium	Rf	
49	Indium	In	114.818(3)	105	Dubnium	Db	
50	Tin	Sn	118.710(7)	106	Seaborgium	Sg	
51	Antimony	Sb	121.760(1)	107	Bohrium	Bh	
52	Tellurium	Te	127.60(3)	108	Hassium	Hs	
53	Iodine	I	126.90447(3)	109	Meitnerium	Mt	
54	Xenon	Xe	131.293(6)	110	Darmstadtium	Ds	
55	Cesium	Cs	132.9054519(2)	111	Roentgenium	Rg	
56	Barium	Ba	137.327(7)	112	Copernicium	Cn	

* Some weights are listed as intervals ([a; b]; a ≤ atomic weight ≤ b) because these weights are not constant but depend on the physical, chemical, and nuclear histories of the samples used. Atomic weights are not listed for some elements because these elements do not have stable isotopes. Exceptions are thorium, protactinium, and uranium. Elements 113 to 118 are not listed in this table although they have been reported (IUPAC has not named them). From *Atomic Weights of the Elements 2009 (IUPAC Technical Report)*, Pure Appl. Chem., Vol. 83(2), pp. 359–396, 2011.

Table of Atomic Masses

Element	Symbol	Mass number (*indicates radioactive)	Atomic mass	Percent abundance	Half-life and decay mode (if unstable)	
(Neutron)	*n*	1*	1.008665		10.4 m	β⁻
Hydrogen	H	1	1.007825	99.985		
Deuterium	D	2	2.014102	0.015		
Tritium	T	3*	3.016049		12.33 y	β⁻
Helium	He	3	3.016029	0.00014		
		4	4.002602	99.99986		
		6*	6.018886		0.81 s	β⁻
		8*	8.033922		0.12 s	β⁻
Lithium	Li	6	6.015121	7.5		
		7	7.016003	92.5		
		8*	8.022486		0.84 s	β⁻
		9*	9.026789		0.18 s	β⁻
		11*	11.043897		8.7 ms	β⁻
Beryllium	Be	7*	7.016928		53.3 d	ec
		9	9.012174	100		
		10*	10.013534		1.5×10^{6} y	β⁻
		11*	11.021657		13.8 s	β⁻
		12*	12.026921		23.6 ms	β⁻
		14*	14.042866		4.3 ms	β⁻
Boron	B	8*	8.024605		0.77 s	β⁺
		10	10.012936	19.9		
		11	11.009305	80.1		
		12*	12.014352		0.0202 s	β⁻
		13*	13.017780		17.4 ms	β⁻
		14*	14.025404		13.8 ms	β⁻
		15*	15.031100		10.3 ms	β⁻
Carbon	C	9*	9.031030		0.13 s	β⁺
		10*	10.016854		19.3 s	β⁺
		11*	11.011433		20.4 m	β⁺
		12	12.000000	98.90		
		13	13.003355	1.10		
		14*	14.003242		5730 y	β⁻
		15*	15.010599		2.45 s	β⁻
		16*	16.014701		0.75 s	β⁻
		17*	17.022582		0.20 s	β⁻

(Continued)

Element	Symbol	Mass number (*indicates radioactive)	Atomic mass	Percent abundance	Half-life and decay mode (if unstable)	
Nitrogen	N	12*	12.018613		0.0110 s	β^+
		13*	13.005738		9.96 m	β^+
		14	14.003074	99.63		
		15	15.000108	0.37		
		16*	16.006100		7.13 s	β^-
		17*	17.008450		4.17 s	β^-
		18*	18.014082		0.62 s	β^-
		19*	19.017038		0.24 s	β^-
Oxygen	O	13*	13.024813		8.6 ms	β^+
		14*	14.008595		70.6 s	β^+
		15*	15.003065		122 s	β^+
		16	15.994915	99.71		
		17	16.999132	0.039		
		18	17.999160	0.20		
		19*	19.003577		26.9 s	β^-
		20*	20.004076		13.6 s	β^-
		21*	21.008595		3.4 s	β^-
Fluorine	F	17*	17.002094		64.5 s	β^+
		18*	18.000937		109.8 m	β^+
		19	18.998404	100		
		20*	19.999982		11.0 s	β^-
		21*	20.999950		4.2 s	β^-
		22*	22.003036		4.2 s	β^-
		23*	23.003564		2.2 s	β^-
Neon	Ne	18*	18.005710		1.67 s	β^+
		19*	19.001880		17.2 s	β^+
		20	19.992435	90.48		
		21	20.993841	0.27		
		22	21.991383	9.25		
		23*	22.994465		37.2 s	β^-
		24*	23.993999		3.38 m	β^-
		25*	24.997789		0.60 s	β^-
Sodium	Na	21*	20.997650		22.5 s	β^+
		22*	21.994434		2.61 y	β^+
		23	22.989767	100		
		24*	23.990961		14.96 h	β^-
		25*	24.989951		59.1 s	β^-
		26*	25.992588		1.07 s	β^-
Magnesium	Mg	23*	22.994124		11.3 s	β^+
		24	23.985042	78.99		
		25	24.985838	10.00		
		26	25.982594	11.01		
		27*	26.984341		9.46 m	β^-
		28*	27.983876		20.9 h	β^-
		29*	28.375346		1.30 s	β^-

Element	Symbol	Mass number (*indicates radioactive)	Atomic mass	Percent abundance	Half-life and decay mode (if unstable)	
Aluminum	Al	25*	24.990429		7.18 s	β+
		26*	25.986892		7.4×10^5 y	β+
		27	26.981538	100		
		28*	27.981910		2.24 m	β−
		29*	28.980445		6.56 m	β−
		30*	29.982965		3.60 s	β−
Silicon	Si	27*	26.986704		4.16 s	β+
		28	27.976927	92.23		
		29	28.976495	4.67		
		30	28.973770	3.10		
		31*	30.975362		2.62 h	β−
		32*	31.974148		172 y	β−
		33*	32.977928		6.13 s	β−
Phosphorus	P	30*	29.978307		2.50 m	β+
		31	30.973762	100		
		32*	31.973762		14.26 d	β−
		33*	32.971725		25.3 d	β−
		34*	33.973636		12.43 s	β−
Sulfur	S	31*	30.979554		2.57 s	β+
		32	31.972071	95.02		
		33	32.971459	0.75		
		34	33.967867	4.21		
		35*	34.969033		87.5 d	β−
		36	35.967081	0.02		
Chlorine	Cl	34*	33.973763		32.2 m	β+
		35	34.968853	75.77		
		36*	35.968307		3.0×10^5 y	β−
		37	36.965903	24.23		
		38*	37.968010		37.3 m	β−
Argon	Ar	36	35.967547	0.337		
		37*	36.966776		35.04 d	ec
		38	37.962732	0.063		
		39*	38.964314		269 y	β−
		40	39.962384	99.600		
		42*	41.963049		33 y	β−
Potassium	K	39	38.963708	93.2581		
		40*	39.964000	0.0117	1.28×10^9 y	β+, ec, β−
		41	40.961827	6.7302		
		42*	41.962404		12.4 h	β−
		43*	42.960716		22.3 h	β−

(Continued)

Element	Symbol	Mass number (*indicates radioactive)	Atomic mass	Percent abundance	Half-life and decay mode (if unstable)	
Calcium	Ca	40	39.962591	96.941		
		41*	40.962279		1.0×10^5 y	ec
		42	41.958618	0.647		
		43	42.958767	0.135		
		44	43.955481	2.086		
		46	45.953687	0.004		
		48	47.952534	0.187		
Scandium	Sc	41*	40.969250		0.596 s	β^+
		43*	42.961151		3.89 h	β^+
		45	44.955911	100		
		46*	45.955170		83.8 d	β^-
Titanium	Ti	44*	43.959691		49 y	ec
		46	45.952630	8.0		
		47	46.951765	7.3		
		48	47.947947	73.8		
		49	48.947871	5.5		
		50	49.944792	5.4		
Vanadium	V	48*	47.952255			
		50*	49.947161	0.25	15.97 d	β^+
		51	50.943962	99.75	1.5×10^{17} y	β^+
Chromium	Cr	48*	47.954033		21.6 h	ec
		50	49.946047	4.345		
		52	51.940511	83.79		
		53	52.940652	9.50		
		54	53.938883	2.365		
Manganese	Mn	53*	52.941292		3.74×10^6 y	ec
		54*	53.940361		312.1 d	ec
		55	54.938048	100		
		56*	55.938908		2.58 h	β^-
Iron	Fe	54	53.939613	5.9		
		55*	54.938297		2.7 y	ec
		56	55.934940	91.72		
		57	56.935396	2.1		
		58	57.933278	0.28		
		60*	59.934078		1.5×10^6 y	β^-
Cobalt	Co	57*	56.936294		271.8 d	ec
		58*	57.935755		70.9 h	ec, β^+
		59	58.933198	100		
		60*	59.933820		5.27 y	β^-
		61*	60.932478		1.65 h	β^-
Nickel	Ni	58	57.935346	68.077		
		59*	58.934350		7.5×10^4 y	ec, β^+
		60	59.930789	26.223		
		61	60.931058	1.140		
		62	61.928346	3.634		
		63*	62.929670		100 y	β^-
		64	63.927967	0.926		

Element	Symbol	Mass number (*indicates radioactive)	Atomic mass	Percent abundance	Half-life and decay mode (if unstable)
Copper	Cu	63	62.929599	69.17	
		64*	63.929765		12.7 h ec
		65	64.927791	30.83	
		66*	65.928871		5.1 m β^-
Zinc	Zn	64	63.929144	48.6	
		66	65.926035	27.9	
		67	66.927129	4.1	
		68	67.924845	18.8	
		70	69.925323	0.6	
Gallium	Ga	69	68.925580	60.108	
		70*	69.926027		21.1 m β^-
		71	70.924703	39.892	
		72*	71.926367		14.1 h β^-
Germanium	Ge	69*	68.927969		39.1 h ec, β^+
		70	69.924250	21.23	
		72	71.922079	27.66	
		73	72.923462	7.73	
		74	73.921177	35.94	
		76	75.921402	7.44	
		77*	76.923547		11.3 h β^-
Arsenic	As	73*	72.923827		80.3 d ec
		74*	73.923928		17.8 d ec, β^+
		75	74.921594	100	
		76*	75.922393		1.1 d β^-
		77*	76.920645		38.8 h β^-
Selenium	Se	74	73.922474	0.89	
		76	75.919212	9.36	
		77	76.919913	7.63	
		78	77.917307	23.78	
		79*	78.918497		$\leq 6.5 \times 10^4$ y β^-
		80	79.916519	49.61	
		82*	81.916697	8.73	1.4×10^{20} y $2\beta^-$
Bromine	Br	79	78.918336	50.69	
		80*	79.918528		17.7 m β^+
		81	80.916287	49.31	
		82*	81.916802		35.3 h β^-
Krypton	Kr	78	77.920400	0.35	
		80	79.916377	2.25	
		81*	80.916589		2.11×10^5 y ec
		82	81.913481	11.6	
		83	82.914136	11.5	
		84	83.911508	57.0	
		85*	84.912531		10.76 y β^-
		86	85.910615	17.3	

(Continued)

Element	Symbol	Mass number (*indicates radioactive)	Atomic mass	Percent abundance	Half-life and decay mode (if unstable)
Rubidium	Rb	85	84.911793	72.17	
		86*	85.911171		18.6 d β^-
		87*	86.909186	27.83	4.75×10^{10} y β^-
		88*	87.911325		17.8 m β^-
Strontium	Sr	84	83.913428	0.56	
		86	85.909266	9.86	
		87	86.908883	7.00	
		88	87.905618	82.58	
		90*	89.907737		29.1 y β^-
Yttrium	Y	88*	87.909507		106.6 d ec, β^+
		89	88.905847	100	
		90*	89.914811		2.67 d β^-
Zirconium	Zr	90	89.904702	51.45	
		91	90.905643	11.22	
		92	91.905038	17.15	
		93*	92.906473		1.5×10^6 y β^-
		94	93.906314	17.38	
		96	95.908274	2.80	
Niobium	Nb	91*	90.906988		6.8×10^2 y ec
		92*	91.907191		3.5×10^7 y ec
		93	92.906376	100	
		94*	93.907280		2×10^4 y β^-
Molybdenum	Mo	92	91.906807	14.84	
		93*	92.906811		3.5×10^3 y ec
		94	93.905085	9.25	
		95	94.905841	15.92	
		96	95.904678	16.68	
		97	96.906020	9.55	
		98	97.905407	24.13	
		100	99.907476	9.63	
Technetium	Tc	97*	96.906363		2.6×10^6 y ec
		98*	97.907215		4.2×10^6 y β^-
		99*	98.906254		2.1×10^5 y β^-
Ruthenium	Ru	96	95.907597	5.54	
		98	97.905287	1.86	
		99	98.905939	12.7	
		100	99.904219	12.6	
		101	100.905558	17.1	
		102	101.904348	31.6	
		104	103.905428	18.6	
Rhodium	Rh	102*	101.906794		207 d ec
		103	102.905502	100	
		104*	103.906654		42 s β^-

Element	Symbol	Mass number (*indicates radioactive)	Atomic mass	Percent abundance	Half-life and decay mode (if unstable)
Palladium	Pd	102	101.905616	1.02	
		104	103.904033	11.14	
		105	104.905082	22.33	
		106	105.903481	27.33	
		107*	106.905126		6.5×10^6 y β^-
		108	107.903893	26.46	
		110	109.905158	11.72	
Silver	Ag	107	106.905091	51.84	
		108*	107.905953		2.39 m ec, β^+, β^-
		109	108.904754	48.16	
		110*	109.906110		24.6 s β^-
Cadmium	Cd	106	105.906457	1.25	
		108	107.904183	0.89	
		109*	108.904984		462 d ec
		110	109.903004	12.49	
		111	110.904182	12.80	
		112	111.902760	24.13	
		113*	112.904401	12.22	9.3×10^{15} y β^-
		114	113.903359	28.73	
		116	115.904755	7.49	
Indium	In	113	112.904060	4.3	
		114*	113.904916		1.2 m β^-
		115*	114.903876	95.7	4.4×10^{14} y β^-
		116*	115.905258		54.4 m β^-
Tin	Sn	112	111.904822	0.97	
		114	113.902780	0.65	
		115	114.903345	0.36	
		116	115.901743	14.53	
		117	116.902953	7.68	
		118	117.901605	24.22	
		119	118.903308	8.58	
		120	119.902197	32.59	
		121*	120.904237		55 y β^-
		122	121.903439	4.63	
		124	123.905274	5.79	
Antimony	Sb	121	120.903820	57.36	
		123	122.904215	42.64	
		125*	124.905251		2.7 y β^-
Tellurium	Te	120	119.904040	0.095	
		122	121.903052	2.59	
		123*	122.904271	0.905	1.3×10^{13} y ec
		124	123.902817	4.79	
		125	124.904429	7.12	
		126	125.903309	18.93	
		128*	127.904463	31.70	$>8 \times 10^{24}$ y $2\beta^-$
		130*	129.906228	33.87	1.2×10^{21} y $2\beta^-$

(Continued)

Element	Symbol	Mass number (*indicates radioactive)	Atomic mass	Percent abundance	Half-life and decay mode (if unstable)
Iodine	I	126*	125.905619		13 d ec, β^+, β^-
		127	126.904474	100	
		128*	127.905812		25 m β^-, ec, β^+ β^-
		129*	128.904984		1.6×10^7 y
Xenon	Xe	124	123.905894	0.10	
		126	125.904268	0.09	
		128	127.903531	1.91	
		129	128.904779	26.4	
		130	129.903509	4.1	
		131	130.905069	21.2	
		132	131.904141	26.9	
		134	133.905394	10.4	
		136	135.907215	8.9	
Cesium	Cs	133	132.905436	100	
		134*	133.906703		2.1 y β^-
		135*	134.905891		2×10^6 y β^-
		137*	136.907078		30 y β^-
Barium	Ba	130	129.906289	0.106	
		132	131.905048	0.101	
		133*	132.905990		10.5 y ec
		134	133.904492	2.42	
		135	134.905671	6.593	
		136	135.904559	7.85	
		137	136.905816	11.23	
		138	137.905236	71.70	
Lanthanum	La	137*	136.906462		6×10^4 y ec
		138*	137.907105	0.0902	1.05×10^{11} y ec, β^+
		139	138.906346	99.9098	
Cerium	Ce	136	135.907139	0.19	
		138	137.905986	0.25	
		140	139.905434	88.43	
		142	141.909241	11.13	
Praseodymium	Pr	140*	139.909071		3.39 m ec, β^+
		141	140.907647	100	
		142*	141.910040		25.0 m β^-
Neodymium	Nd	142	141.907718	27.13	
		143	142.909809	12.18	
		144*	143.910082	23.80	2.3×10^{15} y α
		145	144.912568	8.30	
		146	145.913113	17.19	
		148	147.916888	5.76	
		150	149.920887	5.64	
Promethium	Pm	143*	142.910928		265 d ec
		145*	144.912745		17.7 y ec
		146*	145.914698		5.5 y ec
		147*	146.915134		2.623 y β^-

Element	Symbol	Mass number (*indicates radioactive)	Atomic mass	Percent abundance	Half-life and decay mode (if unstable)
Samarium	Sm	144	143.911996	3.1	
		146*	145.913043		1.0×10^8 y α
		147*	146.914894	15.0	1.06×10^{11} y α
		148*	147.914819	11.3	7×10^{15} y α
		149	148.917180	13.8	
		150	149.917273	7.4	
		151*	150.919928		90 y β^-
		152	151.919728	26.7	
		154	153.922206	22.7	
Europium	Eu	151	150.919846	47.8	
		152*	151.921740		13.5 y ec, β^+
		153	152.921226	52.2	
		154*	153.922975		8.59 y β^-
		155*	154.922888		4.7 y β^-
Gadolinium	Gd	148*	147.918112		75 y α
		150*	149.918657		1.8×10^6 y α
		152*	151.919787	0.20	1.1×10^{14} y α
		154	153.920862	2.18	
		155	154.922618	14.80	
		156	155.922119	20.47	
		157	156.923957	15.65	
		158	157.924099	24.84	
		160	159.927050	21.86	
Terbium	Tb	158*	157.925411		180 y ec, β^+, β^-
		159	158.925345	100	
		160*	159.927551		72.3 d β^-
Dysprosium	Dy	156	155.924277	0.06	
		158	157.924403	0.10	
		160	159.925193	2.34	
		161	160.926930	18.9	
		162	161.926796	25.5	
		163	162.928729	24.9	
		164	163.929172	28.2	
Holmium	Ho	165	164.930316	100	
		166*	165.932282		1.2×10^3 y β^-
Erbium	Er	162	161.928775	0.14	
		164	163.929198	1.61	
		166	165.930292	33.6	
		167	166.932047	22.95	
		168	167.932369	27.8	
		170	169.935462	14.9	
Thulium	Tm	169	168.934213	100	
		171*	170.936428		1.92 y β^-

(Continued)

Element	Symbol	Mass number (*indicates radioactive)	Atomic mass	Percent abundance	Half-life and decay mode (if unstable)
Ytterbium	Yb	168	167.933897	0.13	
		170	169.934761	3.05	
		171	170.936324	14.3	
		172	171.936380	21.9	
		173	172.938209	16.12	
		174	173.938861	31.8	
		176	175.942564	12.7	
Lutetium	Lu	173*	172.938930		1.37 y ec
		175	174.940772	97.41	
		176*	175.942679	2.59	3.8×10^{10} y β^-
Hafnium	Hf	174*	173.940042	0.162	2.0×10^{15} y α
		176	175.941404	5.206	
		177	176.943218	18.606	
		178	177.943697	27.297	
		179	178.945813	13.629	
		180	179.946547	35.100	
Tantalum	Ta	180	179.947542	0.012	
		181	180.947993	99.988	
Tungsten (Wolfram)	W	180	179.946702	0.12	
		182	181.948202	26.3	
		183	182.950221	14.28	
		184	183.950929	30.7	
		186	185.954358	28.6	
Rhenium	Re	185	184.952951	37.40	
		187*	186.955746	62.60	4.4×10^{10} y β^-
Osmium	Os	184	183.952486	0.02	
		186*	185.953834	1.58	2.0×10^{15} y α
		187	186.955744	1.6	
		188	187.955744	13.3	
		189	188.958139	16.1	
		190	189.958439	26.4	
		192	191.961468	41.0	
		194*	193.965172		6.0 y β^-
Iridium	Ir	191	190.960585	37.3	
		193	192.962916	62.7	
Platinum	Pt	190*	189.959926	0.01	6.5×10^{11} y α
		192	191.961027	0.79	
		194	193.962655	32.9	
		195	194.964765	33.8	
		196	195.964926	25.3	
		198	197.967867	7.2	
Gold	Au	197	196.966543	100	
		198*	197.968217		2.70 d β^-
		199*	198.968740		3.14 d β^-

Element	Symbol	Mass number (*indicates radioactive)	Atomic mass	Percent abundance	Half-life and decay mode (if unstable)
Mercury	Hg	196	195.965806	0.15	
		198	197.966743	9.97	
		199	198.968253	16.87	
		200	199.968299	23.10	
		201	200.970276	13.10	
		202	201.970617	29.86	
		204	203.973466	6.87	
Thallium	Tl	203	202.972320	29.524	
		204*	203.973839		3.78 y β^-
		205	204.974400	70.476	
	(Ra E″)	206*	205.976084		4.2 m β^-
	(Ac C″)	207*	206.977403		4.77 m β^-
	(Th C″)	208*	207.981992		3.053 m β^-
	(Ra C″)	210*	209.990057		1.30 m β^-
Lead	Pb	202*	201.972134		5×10^4 y ec
		204	203.973020	1.4	
		205*	204.974457		1.5×10^7 y ec
		206	205.974440	24.1	
		207	206.975871	22.1	
		208	207.976627	52.4	
	(Ra D)	210*	209.984163		22.3 y β^-
	(Ac B)	211*	210.988734		36.1 m β^-
	(Th B)	212*	211.991872		10.64 h β^-
	(Ra B)	214*	213.999798		26.8 m β^-
Bismuth	Bi	207*	206.978444		32.2 y ec, β^+
		208*	207.979717		3.7×10^5 y ec
		209	208.980374	100	
	(Ra E)	210*	209.984096		5.01 d α, β^-
	(Th C)	211*	210.987254		2.14 m α
	(Ra C)	212*	211.991259		60.6 m α, β^-
		214*	213.998692		19.9 m β^-
		215*	215.001836		7.4 m β^-
Polonium	Po	209*	208.982405		102 y α
	(Ra F)	210*	209.982848		138.38 d α
	(Ac C′)	211*	210.986627		0.52 s α
	(Th C′)	212*	211.988842		0.30 μs α
	(Ra C′)	214*	213.995177		164 μs α
	(Ac A)	215*	214.999418		0.0018 s α
	(Th A)	216*	216.001889		0.145 s α
	(Ra A)	218*	218.008965		3.10 m α
Astatine	At	215*	214.998638		$\approx$100 μs α
		218*	218.008685		1.6 s α
		219*	219.011297		0.9 m α
Radon	Rn				
	(An)	219*	219.009477		3.96 s α
	(Tn)	220*	220.011369		55.6 s α
	(Rn)	222*	222.017571		3.823 d α

(Continued)

Element	Symbol	Mass number (*indicates radioactive)	Atomic mass	Percent abundance	Half-life and decay mode (if unstable)
Francium		221*	221.01425		4.18 m α
	Fr	222*	222.017585		14.2 m β^-
	(Ac K)	223*	223.019733		22 m β^-
Radium	Ra	221*	221.01391		29 s α
	(Ac X)	223*	223.018499		11.43 d α
	(Th X)	224*	224.020187		3.66 d α
		225*			14.9 d β^-
	(Ra)	226*	226.025402		1600 y α
	(MsTh$_1$)	228*	228.031064		5.75 y β^-
Actinium	Ac	225*			10 d α
	(Ms Th$_2$)	227*	227.027749		21.77 y β^-
		228*	228.031015		6.15 h β^-
		229*			1.04 h β^-
Thorium	Th				
	(Rd Ac)	227*	227.027701		18.72 d α
	(Rd Th)	228*	228.028716		1.913 y α
		229*	229.031757		7300 y α
	(Io)	230*	230.033127		75,000 y α, sf
	(UY)	231*	231.036299	100	25.52 h β^-
	(Th)	232*	232.038051		1.40×10^{10} y α
	(UX$_1$)	234*	234.043593		24.1 d β^-
Protactinium	Pa	231*	231.035880		32,760 y α
	(UZ)	234*	234.043300		6.7 h β^-
Uranium	U	231*	231.036264		4.2 d β^+
		232*	232.037131		69 y α
		233*	233.039630		1.59×10^5 y α
	(UII)	234*	234.040946	0.0055	2.45×10^5 y α
	(Ac U)	235*	235.043924	0.720	7.04×10^8 y α
	(UI)	236*	236.045562		2.34×10^7 y α
		238*	238.050784	99.2745	4.47×10^9 y α
		239*	239.054290		23.5 m β^-
Neptunium	Np	235*	235.044057		396 d α
		236*	236.046559		1.54×10^5 y ec
		237*	237.048168		2.14×10^6 y α
Plutonium	Pu	236*	236.046033		2.87 y α, sf
		238*	238.049555		87.7 y α, sf
		239*	239.052157		24,120 y α, sf
		240*	240.053808		6560 y α, sf
		241*	241.056846		14.4 y β^-
		242*	242.058737		3.7×10^5 y α, sf
		244*	244.064200		8.1×10^7 y α, sf
Americium	Am	240*	240.055285		2.12 d ec
		241*	241.056824		432 y α, sf
Curium	Cm	247*	247.070347		1.56×10^7 y α
		248*	248.072344		3.4×10^5 y α, sf

Element	Symbol	Mass number (*indicates radioactive)	Atomic mass	Percent abundance	Half-life and decay mode (if unstable)
Berkelium	Bk	247* 249*	247.070300 249.074979		1380 y α 327 d β⁻
Californium	Cm	250* 251*	250.076400 251.079580		13.1 y α, sf 898 y α
Einsteinium	Es	252* 253*	252.082974 253.084817		1.29 y α 2.02 d α, sf
Fermium	Fm	253* 254*	253.085173 254.086849		3.00 d ec 3.24 h α, sf
Mendelevium	Md	256* 258*	256.093988 258.098594		75.6 m ec, β⁺ 55 d α
Nobelium	No	257* 259*	257.096855 259.100932		25 s α 58 m α, sf
Lawrencium	Lr	259* 260*	259.102888 260.105346		6.14 s α, sf 3.0 m α, sf
Rutherfordium	Rf	260* 261*	260.160302 261.108588		24 ms sf 65 s α, sf
Dubnium	Db	261* 262*	261.111830 262.113763		1.8 s α 35 s α
Seaborgium	Sg	263*	263.118310		0.78 s α, sf
Bohrium	Bh	262*	262.123081		0.10 s α, sf
Hassium	Hs	265* 267*	265.129984 267.131770		1.8 ms α 60 ms α
Meitnerium	Mt	266* 268*	266.137789 268.138820		3.4 ms α, sf 70 ms α
Darmstadtium	Ds	269* 271* 273*	269.145140 271.146080 272.153480		0.17 ms α 1.1 ms α 8.6 ms α
Roentgenium	Rg	272*	272.153480		1.5 ms α
Copernicium	Cn	277*	?		0.2 ms α
Ununtrium	Unt	284*	?		? α
Ununquadium	Unq	289*	?		? α
Ununpentium	Unp	288*	?		? α
Ununhexium	Unh	292*	?		? α
Ununseptium	Uus				
Ununoctium	Uno	294*	?		? α

Math Tutorial

In this tutorial, we review some of the basic results of algebra, geometry, trigonometry, and calculus. In many cases, we merely state results without proof. Table M-1 lists some mathematical symbols.

M-1 Significant figures

Many numbers we work with in science are the result of measurement and are therefore known only within a degree of uncertainty. This uncertainty should be reflected in the number of digits used. For example, if you have a 1-meter-long rule with scale spacing of 1 cm, you know that you can measure the height of a box to within a fifth of a centimeter or so. Using this rule, you might find that the box height is 27.0 cm. If there is a scale with a spacing of 1 mm on your rule, you might perhaps measure the box height to be 27.03 cm. However, if there is a scale with a spacing of 1 mm on your rule, you might not be able to measure the height more accurately than 27.03 cm because the height might vary by 0.01 cm or so, depending on where you measure the height of the box. When you write down that the height of the box is 27.03 cm, you are stating that your best estimate of the height is 27.03 cm, but you are not claiming that it is exactly 27.030000 . . . cm high. The four digits in 27.03 cm are called **significant figures**. Your measured length, 27.03 cm, has four significant digits. Significant figures are also called significant digits.

The number of significant digits in an answer to a calculation will depend on the number of significant digits in the given data. When you work with numbers that have uncertainties, you should be careful not to include more digits than the certainty of measurement warrants. *Approximate* calculations (order-of-magnitude estimates) always result in answers that have only one significant digit or none. When you multiply, divide, add, or subtract numbers, you must consider the accuracy of the results. Listed below are some rules that will help you determine the number of significant digits of your results.

(1) When multiplying or dividing quantities, the number of significant digits in the final answer is no greater than that in the quantity with the fewest significant digits.

(2) When adding or subtracting quantities, the number of decimal places in the answer should match that of the term with the smallest number of decimal places.

(3) Exact values have an unlimited number of significant digits. For example, a value determined by counting, such as 2 tables, has no uncertainty and is an exact value. In addition, the conversion factor 0.0254000 . . . m/in. is an exact value because 1.000 . . . inches is exactly equal to 0.0254000 . . . meters. (The yard is, by definition, equal to exactly 0.9144 m, and 0.9144 divided by 36 is exactly equal to 0.0254.)

(4) Sometimes zeros are significant and sometimes they are not. If a zero is before a leading nonzero digit, then the zero is not significant. For example, the number 0.00890 has three significant digits. The first three zeroes are not significant digits but are merely markers to locate the decimal point. Note that the zero after the nine is significant.

Table M-1 Mathematical Symbols	
$=$	is equal to
$\neq$	is not equal to
$\approx$	is approximately equal to
$\sim$	is of the order of
$\propto$	is proportional to
$>$	is greater than
$\geq$	is greater than or equal to
$\gg$	is much greater than
$<$	is less than
$\leq$	is less than or equal to
$\ll$	is much less than
Δx	change in x
$|x|$	absolute value of x
$n!$	$n(n-1)(n-2)\ldots 1$
Σ	sum

(5) Zeros that are between nonzero digits are significant. For example, 5603 has four significant digits.

(6) The number of significant digits in numbers with trailing zeros and no decimal point is ambiguous. For example, 31,000 could have as many as five significant digits or as few as two significant digits. To prevent ambiguity, you should report numbers by using scientific notation or by using a decimal point.

Example M-1 Finding the Average of Three Numbers

Find the average of 19.90, −7.524, and −11.8179.

Set Up
You will be adding 3 numbers and then dividing the result by 3. The first number has four significant digits, the second number has four, and the third number has six.

Solve
Sum the three numbers.

$$19.90 + (-7.524) + (-11.8179) = 0.5581$$

If the problem only asked for the sum of the three numbers, we would round the answer to the least number of decimal places among all the numbers being added—the answer would be 0.56 (0.5581 rounds up to 0.56 to two significant digits). However, we must divide this intermediate result by 3, so we use the intermediate answer with the two extra digits (italicized and red).

$$\frac{0.5581}{3} = 0.1860333\ldots$$

Only two of the digits in the intermediate answer, 0.5581..., are significant digits, so we must round the final number to get our final answer. The number 3 in the denominator is a whole number and has an unlimited number of significant digits. Thus, the final answer has the same number of significant digits as the numerator, which is 2.

The final answer is 0.19.

Reflect
The sum in step 1 has two significant digits following the decimal point, the same as the number being summed with the least number of significant digits after the decimal point.

M-2 Equations

An **equation** is a statement written using numbers and symbols to indicate that two quantities, written on either side of an equal sign (=), are equal. The quantity on either side of the equal sign may consist of a single term, or of a sum or difference of two or more **terms**. For example, the equation $x = 1 - (ay + b)/(cx - d)$ contains three terms, x, 1, and $(ay + b)/(cx - d)$.

You can perform the following operations on equations:

(1) The same quantity can be added to or subtracted from each side of an equation.

(2) Each side of an equation can be multiplied or divided by the same quantity.

(3) Each side of an equation can be raised to the same power.

These operations are meant to be applied to each *side* of the equation rather than each term in the equation. (Because multiplication is distributive over addition, operation 2— and only operation 2—of the preceding operations also applies term by term.)

Caution: Division by zero is forbidden at any *stage in solving an equation; results (if any) would be invalid.*

Adding or Subtracting Equal Amounts
To find x when $x - 3 = 7$, add 3 to both sides of the equation: $(x - 3) + 3 = 7 + 3$; thus, $x = 10$.

Multiplying or Dividing by Equal Amounts
If $3x = 17$, solve for x by dividing both sides of the equation by 3; thus, $x = \frac{17}{3}$, or 5.7.

Example M-2 Simplifying Reciprocals in an Equation

Solve the following equation for x:

$$\frac{1}{x} + \frac{1}{4} = \frac{1}{3}$$

Equations containing reciprocals of unknowns occur in many circumstances in physics. Two instances of this are geometric optics and electric circuit analysis.

Set Up

In this equation, the term containing x is on the same side of the equation as a term not containing x. Furthermore, x is found in the denominator of a fraction. We'll start by isolating the $1/x$ term, find common denominators, and then multiply both sides of the equation by appropriate quantities.

Solve

Subtract $\dfrac{1}{4}$ from each side.

$$\frac{1}{x} = \frac{1}{3} - \frac{1}{4}$$

Simplify the right side of the equation by using the lowest common denominator.

Begin by multiplying both terms on the right-hand side by appropriate forms of 1.

$$\frac{1}{x} = \frac{1}{3}\frac{4}{4} - \frac{1}{4}\frac{3}{3} = \frac{4}{12} - \frac{3}{12}$$

$$= \frac{4-3}{12} = \frac{1}{12} \quad \text{so} \quad \frac{1}{x} = \frac{1}{12}$$

Multiply both sides of the equation by $12x$ to determine the value of x.

$$12x\frac{1}{x} = 12x\frac{1}{12}$$

$$12 = x$$

Reflect

To check our answer, substitute 12 for x in the left side of original equation.

$$\frac{1}{x} + \frac{1}{4} = \frac{1}{12} + \frac{3}{12} = \frac{4}{12} = \frac{1}{3}$$

M-3 Direct and inverse proportions

When we say variable quantities x and y are **directly proportional**, we mean that as x and y change, the ratio x/y is constant. To say that two quantities are proportional is to say that they are directly proportional. When we say variable quantities x and y are **inversely proportional**, we mean that as x and y change, the ratio xy is constant.

Relationships of direct and inverse proportion are common in physics. Objects moving at the same velocity have momenta directly proportional to their masses. The ideal gas law ($PV = nRT$) states that pressure P is directly proportional to (absolute) temperature T, when volume V remains constant, and is inversely proportional to volume, when temperature remains constant. Ohm's law ($V = IR$) states that the voltage V across a resistor is directly proportional to the electric current in the resistor when the resistance remains constant.

Constant of Proportionality

When two quantities are directly proportional, the two quantities are related by a *constant of proportionality*. If you are paid for working at a regular rate R in dollars per day, for example, the money m you earn is directly proportional to the time t you work;

the rate R is the constant of proportionality that relates the money earned in dollars to the time worked t in days:

$$\frac{m}{t} = R \quad \text{or} \quad m = Rt$$

If you earn \$400 in 5 days, the value of R is \$400/(5 days) = \$80/day. To find the amount you earn in 8 days, you could perform the calculation

$$m = (\$80/\text{day})(8 \text{ days}) = \$640$$

Sometimes the constant of proportionality can be ignored in proportion problems. Because the amount you earn in 8 days is $\frac{8}{5}$ times what you earn in 5 days, this amount is

$$m_{8 \text{ days}} = 8_{\text{days}} \frac{\$400}{5_{\text{days}}} = \$640$$

Example M-3 Painting Cubes

You need 15.4 mL of paint to cover one side of a cube. The area of one side of the cube is 426 cm². What is the relation between the volume of paint needed and the area to be covered? How much paint do you need to paint one side of a cube on which the one side has an area of 503 cm²?

Set Up
To determine the amount of paint for the side whose area is 503 cm² we will set up a proportion.

Solve
The volume V of paint needed increases in proportion to the area A to be covered.

V and A are directly proportional.

That is, $\dfrac{V}{A} = k$ or $V = kA$

where k is the proportionality constant

Determine the value of the proportionality constant using the given values $V_1 = 15.4$ mL and $A_1 = 426$ cm².

$$k = \frac{V_1}{A_1} = \frac{15.4 \text{ mL}}{426 \text{ cm}^2} = 0.0362 \text{ mL/cm}^2$$

Determine the volume of paint needed to paint a side of a cube whose area is 503 cm² using the proportionality constant in step 1.

$$V_2 = kA_2 = (0.0362 \text{ mL/cm}^2)(503 \text{ cm}^2)$$
$$= 18.2 \text{ mL}$$

Reflect
Our value for V_2 is greater than the value for V_1, as expected. The amount of paint needed to cover an area equal to 503 cm² should be greater than the amount of paint needed to cover an area of 426 cm² because 503 cm² is larger than 426 cm².

M-4 Linear equations

A **linear equation** is an equation of the form $x + 2y - 4z = 3$. That is, an equation is linear if each term either is constant or is the product of a constant and a variable raised to the first power. Such equations are said to be linear because the plots of these equations form straight lines or planes. The equations of direct proportion between two variables are linear equations.

Graph of a Straight Line

A linear equation relating y and x can always be put into the standard form

(M-1) $y = mx + b$

where m and b are constants that may be either positive or negative. Figure M-1 shows a graph of the values of x and y that satisfy Equation M-1. The constant b, called the **y intercept,** is the value of y at $x = 0$. The constant m is the **slope** of the line, which equals the ratio of the change in y to the corresponding change in x. In the figure, we

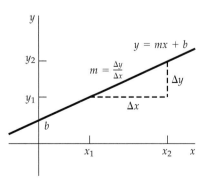

Figure M-1 Graph of the linear equation $y = mx + b$, where b is the y intercept and $m = \Delta y/\Delta x$ is the slope.

have indicated two points on the line, (x_1, y_1) and (x_2, y_2), and the changes $\Delta x = x_2 - x_1$ and $\Delta y = y_2 - y_1$. The slope m is then

$$m = \frac{y_2 - y_1}{x_2 - x_1} = \frac{\Delta y}{\Delta x}$$

If x and y are both unknown in the equation $y = mx + b$, there are no unique values of x and y that are solutions to the equation. Any pair of values (x_1, y_1) on the line in Figure M-1 will satisfy the equation. If we have two equations, each with the same two unknowns x and y, the equations can be solved simultaneously for the unknowns. Example M-4 shows two methods for simultaneously solving two linear equations.

Example M-4 Using Two Equations to Solve for Two Unknowns

Find any and all values of x and y that simultaneously satisfy

$$3x - 2y = 8 \qquad \textbf{(M-2)}$$

and

$$y - x = 2 \qquad \textbf{(M-3)}$$

Set Up
Graph the two equations. At the point where the lines intersect, the values of x and y satisfy both equations.

We can solve two simultaneous equations by first solving either equation for one variable in terms of the other variable and then substituting the result into the second equation.

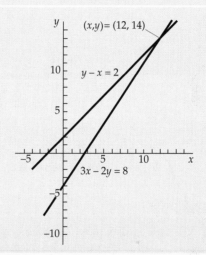

Figure M-2 Graph of Equations M-2 and M-3. At the point where the lines intersect, the values of x and y satisfy both equations.

Solve

Solve Equation M-3 for y.	$y = x + 2$
Substitute this value for y into Equation M-2.	$3x - 2(x + 2) = 8$
Simplify the equation and solve for x.	$3x - 2x - 4 = 8$ $x - 4 = 8$ $x = 12$
Use your solution for x and one of the given equations to find the value of y.	Return to Equation M-3 and substitute $x = 12$. $y - x = 2$, where $x = 12$ $y - 12 = 2$ $y = 2 + 12 = 14$

Reflect

An alternative method is to multiply one equation by a constant such that one of the unknown terms is eliminated when the equations are added or subtracted.	We can multiply through Equation M-3 by 2 $2(y - x) = 2(2)$ $2y - 2x = 4$
Add the result to Equation M-2 and solve for x:	$\begin{aligned} 2y - 2x &= 4 \\ 3x - 2y &= 8 \end{aligned}$ $3x - 2x = 12 \Rightarrow x = 12$
Substitute into Equation M-3 and solve for y:	$y - 12 = 2 \Rightarrow y = 14$

M-5 Quadratic equations and factoring

A **quadratic equation** is an equation of the form $ax^2 + bxy + cy^2 + ex + fy + g = 0$, where x and y are variables and and a, b, c, e, f, and g are constants. In each term of the equation the powers of the variables are integers that sum to 2, 1, or 0. The designation *quadratic equation* usually applies to a much simpler equation of one variable that can be written in the standard form

(M-4)
$$ax^2 + bx + c = 0$$

where a, b, and c are constants. The quadratic equation has two solutions or **roots**—values of x for which the equation is true.

Factoring

We can solve some quadratic equations by **factoring**. Very often terms of an equation can be grouped or organized into other terms. When we factor terms, we look for multipliers and multiplicands—which we now call **factors**—that will yield two or more new terms as a product. For example, we can find the roots of the quadratic equation $x^2 - 3x + 2 = 0$ by factoring the left side to get $(x - 2)(x - 1) = 0$. The roots are $x = 2$ and $x = 1$.

Factoring is useful for simplifying equations and for understanding the relationships between quantities. You should be familiar with the multiplication of the factors $(ax + by)(cx + dy) = acx^2 + (ad + bc)xy + bdy^2$.

You should readily recognize some typical factorable combinations:

1. Common factor: $2ax + 3ay = a(2x + 3y)$
2. Perfect square: $x^2 - 2xy + y^2 = (x-y)^2$ (If the expression on the left side of a quadratic equation in standard form is a perfect square, the two roots will be equal.)
3. Difference of squares: $x^2 - y^2 = (x + y)(x - y)$

Also, look for factors that are prime numbers (2, 5, 7, etc.) because these factors can help you simplify terms quickly. For example, the equation $98x^2 - 140 = 0$ can be simplified because 98 and 140 share the common factor 2. That is, $98x^2 - 140 = 0$ becomes $2(49x^2 - 70) = 0$, so we have $49x^2 - 70 = 0$.

This result can be further simplified because 49 and 70 share the common factor 7. Thus, $49x^2 - 70 = 0$ becomes $7(7x^2 - 10) = 0$, so we have $7x^2 - 10 = 0$.

The Quadratic Formula

Not all quadratic equations can be solved by factoring. However, *any* quadratic equation in the standard form $ax^2 + bx + c = 0$ can be solved by the **quadratic formula,**

(M-5)
$$x = \frac{-b \pm \sqrt{b^2 - 4ac}}{2a} = -\frac{b}{2a} \pm \frac{1}{2a}\sqrt{b^2 - 4ac}$$

When b^2 is greater than $4ac$, there are two solutions corresponding to the + and − signs, respectively. **Figure M-3** shows a graph of y versus x where $y = ax^2 + bx + c$. The curve, a **parabola**, crosses the x axis twice. (The simplest representation of a parabola in (x, y) coordinates is an equation of the form $y = ax^2 + bx + c$.) The two roots of this equation are the values for which $y = 0$; that is, they are the x *intercepts*.

When b^2 is less than $4ac$, the graph of y versus x does not intersect the x axis, as is shown in **Figure M-4**; there are still two roots, but they are not real numbers. When $b^2 = 4ac$, the graph of y versus x is tangent to the x axis at the point $x = -b/2a$; the two roots are each equal to $-b/2a$.

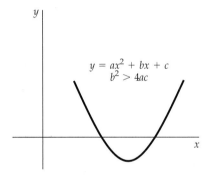

Figure M-3 Graph of y versus x when $y = ax^2 + bx + c$ for the case $b^2 > 4ac$. The two values of x for which $y = 0$ satisfy the quadratic equation (Equation M-4).

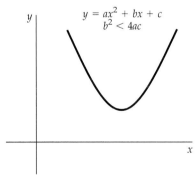

Figure M-4 Graph of y versus x when $y = ax^2 + bx + c$ for the case $b^2 < 4ac$. In this case, there are no real values of x for which $y = 0$.

Example M-5 Factoring a Second-Degree Polynomial

Factor the expression $6x^2 + 19xy + 10y^2$.

Set Up
We examine the coefficients of the terms to see whether the expression can be factored without resorting to more advanced methods. Remember that the multiplication $(ax + by)(cx + dy) = acx^2 + (ad + bc)xy + bdy^2$.

Solve

The coefficient of x^2 is 6, which can be factored two ways.

$ac = 6$
$3 \cdot 2 = 6 \quad$ or $\quad 6 \cdot 1 = 6$

The coefficient of y^2 is 10, which can also be factored two ways.

$bd = 10$
$5 \cdot 2 = 10 \quad$ or $\quad 10 \cdot 1 = 10$

List the possibilities for a, b, c, and d in a table. Include a column for $ad + bc$.
  If $a = 3$, then $c = 2$, and vice versa. In addition, if $a = 6$, then $c = 1$, and vice versa. For each value of a there are four values for b.

a	b	c	d	$ad + bc$
3	5	2	2	16
3	2	2	5	19
3	10	2	1	23
3	1	2	10	32
2	5	3	2	19
2	2	3	5	16
2	10	3	1	32
2	1	3	10	23
6	5	1	2	17
6	2	1	5	32
6	10	1	1	16
6	1	1	10	61
1	5	6	2	32
1	2	6	5	17
1	10	6	1	61
1	1	6	10	16

Find a combination such that $ad + bc = 19$. As you can see from the table there are two such combinations.

$ad + bc = 19$
$3 \cdot 5 + 2 \cdot 2 = 19$ and
$2 \cdot 2 + 5 \cdot 3 = 19$

It doesn't matter which combination we choose. To finish this problem we will use the combination in the second row of the table to factor the expression in question:

$6x^2 + 19xy + 10y^2 = (3x + 2y)(2x + 5y)$

Reflect
As a check, expand $(3x + 2y)(2x + 5y)$ to see if we return to the original equation.

$$(3x + 2y)(2x + 5y) = 6x^2 + 15xy + 4xy + 10y^2$$
$$= 6x^2 + 19xy + 10y^2$$

You should be able to show that the combination in the fifth row is also an acceptable factoring.

M-6 Exponents and logarithms

Exponents

The notation x^n stands for the quantity obtained by multiplying x by itself n times. For example, $x^2 = x \cdot x$ and $x^3 = x \cdot x \cdot x$. The quantity n is called the **power,** or the **exponent,** of x (the **base**). Listed below are some rules that will help you simplify terms that have exponents.

(1) When two powers of x are multiplied, the exponents are added:

$$(x^m)(x^n) = x^{m+n} \tag{M-6}$$

Example: $x^2 x^3 = x^{2+3} = (x \cdot x)(x \cdot x \cdot x) = x^5$.

(2) Any number (except 0) raised to the 0 power is defined to be 1:

$$x^0 = 1$$

(M-7)

(3) Based on rule 2,

$$x^n x^{-n} = x^0 = 1$$

(M-8)

$$x^{-n} = \frac{1}{x^n}$$

(4) When two powers are divided, the exponents are subtracted:

(M-9)

$$\frac{x^n}{x^m} = x^n x^{-m} = x^{n-m}$$

(5) When a power is raised to another power, the exponents are multiplied:

(M-10)

$$(x^n)^m = x^{nm}$$

(6) When exponents are written as fractions, they represent the roots of the base. For example,

$$x^{1/2} \cdot x^{1/2} = x$$

so

$$x^{1/2} = \sqrt{x} \quad (x > 0)$$

Example M-6 Simplifying a Quantity That Has Exponents

Simplify $\dfrac{x^4 x^7}{x^8}$.

Set Up

According to rule 1, when two powers of x are multiplied, the exponents are added.

$$(x^m)(x^n) = x^{m+n} \qquad \text{(M-6)}$$

Rule 4 states that when two powers are divided, the exponents are subtracted.

$$\frac{x^n}{x^m} = x^n x^{-m} = x^{n-m} \qquad \text{(M-9)}$$

Solve

Simplify the numerator $x^4 x^7$ using rule 1.

$$x^4 x^7 = x^{4+7} = x^{11}$$

Simplify $\dfrac{x^{11}}{x^8}$ using rule 4.

$$\frac{x^{11}}{x^8} = x^{11} x^{-8} = x^{11-8} = x^3$$

Reflect

Use the value $x = 2$ to test our answer.

$$\frac{2^4 2^7}{2^8} = 2^3 = 8$$

$$\frac{2^4 2^7}{2^8} = \frac{(16)(128)}{256} = \frac{2048}{256} = 8$$

Logarithms

Any positive number can be expressed as some power of any other positive number except one. If y is related to x by $y = a^x$, then the number x is said to be the **logarithm** of y to the **base** a, and the relation is written

$$x = \log_a y$$

Thus, logarithms are *exponents*, and the rules for working with logarithms correspond to similar laws for exponents. Listed below are some rules that will help you simplify terms that have logarithms.

(1) If $y_1 = a^n$ and $y_2 = a^m$, then

$$y_1 y_2 = a^n a^m = a^{n+m}$$

Correspondingly,

(M-11)

$$\log_a y_1 y_2 = \log_a a^{n+m} = n + m = \log_a a^n + \log_a a^m = \log_a y_1 + \log_a y_2$$

It then follows that

$$\log_a y^n = n \log_a y \tag{M-12}$$

(2) Because $a^1 = a$ and $a^0 = 1$,

$$\log_a a = 1 \tag{M-13}$$

and

$$\log_a 1 = 0 \tag{M-14}$$

There are two bases in common use: logarithms to base 10 are called **common logarithms**, and logarithms to base e (where $e = 2.718\ldots$) are called **natural logarithms.**

In this text, the symbol ln is used for natural logarithms and the symbol log, without a subscript, is used for common logarithms. Thus,

$$\log_e x = \ln x \quad \text{and} \quad \log_{10} x = \log x \tag{M-15}$$

and $y = \ln x$ implies

$$x = e^y \tag{M-16}$$

Logarithms can be changed from one base to another. Suppose that

$$z = \log x \tag{M-17}$$

Then

$$10^z = 10^{\log x} = x \tag{M-18}$$

Taking the natural logarithm of both sides of Equation M-18, we obtain

$$z \ln 10 = \ln x$$

Substituting $\log x$ for z (see Equation M-17) gives

$$\ln x = (\ln 10) \log x \tag{M-19}$$

Example M-7 Converting between Common Logarithms and Natural Logarithms

The steps leading to Equation M-19 show that, in general, $\log_b x = (\log_b a)\log_a x$, and thus that conversion of logarithms from one base to another requires only multiplication by a constant. Describe the mathematical relation between the constant for converting common logarithms to natural logarithms and the constant for converting natural logarithms to common logarithms.

Set Up
We have a general mathematical formula for converting logarithms from one base to another. We look for the mathematical relation by exchanging a for b and vice versa in the formula.

Solve
We have a formula for converting logarithms from base a to base b.

$$\log_b x = (\log_b a)\log_a x$$

To convert from base b to base a, exchange all a for b and vice versa.

$$\log_a x = (\log_a b)\log_b x$$

Divide both sides of the equation in step 1 by $\log_a x$.

$$\frac{\log_b x}{\log_a x} = \log_b a$$

Divide both sides of the equation in step 2 by $(\log_a b)\log_a x$.

$$\frac{1}{\log_a b} = \frac{\log_b x}{\log_a x}$$

The results show that the conversion factors $\log_b a$ and $\log_a b$ are reciprocals of one another.

$$\frac{1}{\log_a b} = \log_b a$$

Reflect
For the value of $\log_{10} e$, your calculator will give 0.43429. For ln 10, your calculator will give 2.3026. Multiply 0.43429 by 2.3026; you will get 1.0000.

M-7 Geometry

The properties of the most common **geometric figures**—bounded shapes in two or three dimensions whose lengths, areas, or volumes are governed by specific ratios—are a basic analytical tool in physics. For example, the characteristic ratios within triangles give us the laws of *trigonometry* (see Section M-8), which in turn give us the theory of vectors, essential in analyzing motion in two or more dimensions. Circles and spheres are essential for understanding, among other concepts, angular momentum and the probability densities of quantum mechanics.

Basic Formulas in Geometry

Circle The ratio of the circumference of a circle to its diameter is a number π, which has the approximate value

$$\pi = 3.141\ 592$$

The circumference C of a circle is thus related to its diameter d and its radius r by

(M-20) $C = \pi d = 2\pi r$ circumference of circle

The area of a circle is (Figure M-5)

(M-21) $A = \pi r^2$ area of circle

Parallelogram The area of a parallelogram is the base b multiplied by the height h (Figure M-6):

$$A = bh$$

Triangle The area of a triangle is one-half the base multiplied by the height (Figure M-7):

$$A = \frac{1}{2}bh$$

Sphere A sphere of radius r (Figure M-8) has a surface area given by

(M-22) $A = 4\pi r^2$ surface area of sphere

and a volume given by

(M-23) $V = \frac{4}{3}\pi r^3$ volume of sphere

Cylinder A cylinder of radius r and length L (Figure M-9) has a surface area (not including the end faces) of

(M-24) $A = 2\pi r L$ surface of cylinder

and volume of

(M-24) $V = \pi r^2 L$ volume of cylinder

Area of a circle $A = \pi r^2$

Figure M-5 Area of a circle.

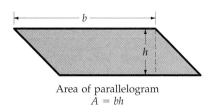

Area of parallelogram
$A = bh$

Figure M-6 Area of a parallelogram.

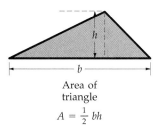

Area of
triangle
$A = \frac{1}{2}bh$

Figure M-7 Area of a triangle.

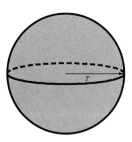

Spherical surface area
$A = 4\pi r^2$
Spherical volume
$V = \frac{4}{3}\pi r^3$

Figure M-8 Surface area and volume of a sphere.

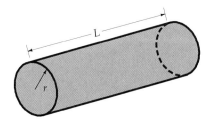

Cylindrical surface area
$A = 2\pi r L$
Cylindrical volume
$V = \pi r^2 L$

Figure M-9 Surface area (not including the end faces) and the volume of a cylinder.

Example M-8 Calculating the Volume of a Spherical Shell

An aluminum spherical shell has an outer diameter of 40.0 cm and an inner diameter of 38.0 cm. What is the volume of the aluminum in this shell?

Set Up	
The volume of the aluminum in the spherical shell is the volume that remains when we subtract the volume of the inner sphere having $d_i = 2r_i = 38.0$ cm from the volume of the outer sphere having $d_o = 2r_o = 40.0$ cm.	Spherical Volume: $$V = \frac{4}{3}\pi r^3 \qquad \text{(M-23)}$$

Solve	
Subtract the volume of the sphere of radius r_i from the volume of the sphere of radius r_o.	$$V = V_o - V_i = \frac{4}{3}\pi r_o^3 - \frac{4}{3}\pi r_i^3 = \frac{4}{3}\pi\left(r_o^3 - r_i^3\right)$$
Substitute 20.0 cm for r_o and 19.0 cm for r_i.	$$V = \frac{4}{3}\pi\left[(20.0\,\text{cm})^3 - (19.0\,\text{cm})^3\right]$$ $$= 4.78 \times 10^3\,\text{cm}^3$$

Reflect	
The volume calculated is less than the volume of the outer sphere.	$$V_o = \frac{4}{3}\pi r_o^3 = \frac{4}{3}\pi(20.0\,\text{cm})^3$$ $$= 3.35 \times 10^4\,\text{cm}^3$$

M-8 Trigonometry

Trigonometry, which gets its name from Greek roots meaning "triangle" and "measure," is the study of some important mathematical functions, called **trigonometric functions**. These functions are most simply defined as ratios of the sides of right triangles. However, these right-triangle definitions are of limited use because they are valid only for angles between zero and 90°. However, the validity of the right-triangle definitions can be extended by defining the trigonometric functions in terms of the ratio of the coordinates of points on a circle of unit radius drawn centered at the origin of the xy plane.

In physics, we first encounter trigonometric functions when we use vectors to analyze motion in two dimensions. Trigonometric functions are also essential in the analysis of any kind of periodic behavior, such as circular motion, oscillatory motion, and wave mechanics.

Angles and Their Measure: Degrees and Radians

The size of an angle formed by two intersecting straight lines is known as its **measure**. The standard way of finding the measure of an angle is to place the angle so that its **vertex**, or point of intersection of the two lines that form the angle, is at the center of a circle located at the origin of a graph that has Cartesian coordinates and one of the lines extends rightward on the positive x axis. The distance traveled *counterclockwise* on the circumference from the positive x axis to reach the intersection of the circumference with the other line defines the measure of the angle. (Traveling clockwise to the second line would simply give us a negative measure; to illustrate basic concepts, we position the angle so that the smaller rotation will be in the counterclockwise direction.)

One of the most familiar units for expressing the measure of an angle is the **degree,** which equals $1/360$ of the full distance around the circumference of the circle. For greater precision, or for smaller angles, we either show degrees plus minutes (′) and seconds (″), with $1' = 1°/60$ and $1'' = 1'/60 = 1°/3600$; or show degrees as an ordinary decimal number.

For scientific work, a more useful measure of an angle is the **radian** (rad). Again, place the angle with its vertex at the center of a circle and measure counterclockwise rotation around the circumference. The measure of the angle in radians is then defined as the length of the circular arc from one line to the other divided by the radius of the

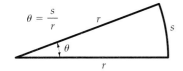

Figure M-10 The angle θ in radians is defined to be the ratio s/r, where s is the arc length intercepted on a circle of radius r.

circle (Figure M-10). If s is the arc length and r is the radius of the circle, the angle θ measured in radians is

(M-26)
$$\theta = \frac{s}{r}$$

Because the angle measured in radians is the ratio of two lengths, it is dimensionless. The relation between radians and degrees is

$$360° = 2\pi \text{ rad}$$

or

$$1 \text{ rad} = \frac{360°}{2\pi} = 57.3°$$

Figure M-11 shows some useful relations for angles.

The Trigonometric Functions

Figure M-12 shows a right triangle formed by drawing the line segment BC perpendicular to AC. The lengths of the sides are labeled a, b, and c. The right-triangle definitions of the trigonometric functions $\sin \theta$ (the **sine**), $\cos \theta$ (the **cosine**), and $\tan \theta$ (the **tangent**) for an acute angle θ are

(M-27)
$$\sin \theta = \frac{a}{c} = \frac{\text{opposite side}}{\text{hypotenuse}}$$

(M-28)
$$\cos \theta = \frac{b}{c} = \frac{\text{adjacent side}}{\text{hypotenuse}}$$

(M-29)
$$\tan \theta = \frac{a}{b} = \frac{\text{opposite side}}{\text{adjacent side}} = \frac{\sin \theta}{\cos \theta}$$

(**Acute angles** are angles whose positive rotation around the circumference of a circle measures less than 90° or $\pi/2$.) Three other trigonometric functions—the **secant** (sec), the **cosecant** (csc), and the **cotangent** (cot), defined as the reciprocals of these functions—are

(M-30)
$$\csc \theta = \frac{c}{a} = \frac{1}{\sin \theta}$$

(M-31)
$$\sec \theta = \frac{c}{b} = \frac{1}{\cos \theta}$$

(M-32)
$$\cot \theta = \frac{b}{a} = \frac{1}{\tan \theta} = \frac{\cos \theta}{\sin \theta}$$

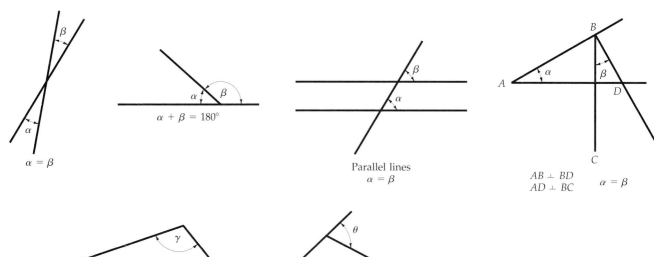

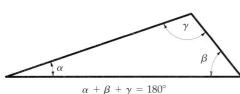

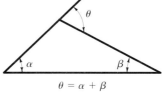

Figure M-11 Some useful relations for angles.

The angle θ, whose sine is x, is called the arcsine of x, and is written $\sin^{-1} x$. That is, if

$$\sin \theta = x$$

then

$$\theta = \arcsin x = \sin^{-1} x \tag{M-33}$$

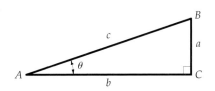

Figure M-12 A right triangle with sides of length a and b and a hypotenuse of length c.

The arcsine is the inverse of the sine. The inverse of the cosine and tangent are defined similarly. The angle whose cosine is y is the arccosine of y. That is, if

$$\cos \theta = y$$

then

$$\theta = \arccos y = \cos^{-1} y \tag{M-34}$$

The angle whose tangent is z is the arctangent of z. That is, if

$$\tan \theta = z$$

then

$$\theta = \arctan z = \tan^{-1} z \tag{M-35}$$

Trigonometric Identities

We can derive several useful formulas, called **trigonometric identities**, by examining relationships between the trigonometric functions. Equations M-30 through M-32 list three of the most obvious identities, formulas expressing some trigonometric functions as reciprocals of others. Almost as easy to discern are identities derived from the **Pythagorean theorem**,

$$a^2 + b^2 = c^2 \tag{M-36}$$

Simple algebraic manipulation of Equation M-36 gives us three more identities. First, if we divide each term in Equation M-36 by c^2, we obtain

$$\frac{a^2}{c^2} + \frac{b^2}{c^2} = 1$$

or, from the definitions of $\sin \theta$ (which is a/c) and $\cos \theta$ (which is b/c),

$$\sin^2 \theta + \cos^2 \theta = 1 \tag{M-37}$$

Similarly, we can divide each term in Equation M-36 by a^2 or b^2 and obtain

$$1 + \cot^2 \theta = \csc^2 \theta \tag{M-38}$$

and

$$1 + \tan^2 \theta = \sec^2 \theta \tag{M-39}$$

Table M-2 lists these last three and many more trigonometric identities. Notice that they fall into four categories: functions of sums or differences of angles, sums or differences of squared functions, functions of double angles (2θ), and functions of half angles ($\frac{1}{2}\theta$). Notice that some of the formulas contain paired alternatives, expressed with the signs $\pm$ and $\mp$; in such formulas, remember to always apply the formula with either all the upper or all the lower alternatives.

Table M-2   Trigonometric Identities

$$\sin(A \pm B) = \sin A \cos B \pm \cos A \sin B$$

$$\cos(A \pm B) = \cos A \cos B \mp \sin A \sin B$$

$$\tan(A \pm B) = \frac{\tan A \pm \tan B}{1 \mp \tan A \tan B}$$

$$\sin A \pm \sin B = 2 \sin\left[\frac{1}{2}(A \pm B)\right]\cos\left[\frac{1}{2}(A \mp B)\right]$$

$$\cos A + \cos B = 2 \cos\left[\frac{1}{2}(A + B)\right]\cos\left[\frac{1}{2}(A - B)\right]$$

$$\cos A - \cos B = 2 \sin\left[\frac{1}{2}(A + B)\right]\sin\left[\frac{1}{2}(B - A)\right]$$

$$\tan A \pm \tan B = \frac{\sin(A \pm B)}{\cos A \cos B}$$

$$\sin^2 \theta + \cos^2 \theta = 1; \sec^2 \theta - \tan^2 \theta = 1; \csc^2 \theta - \cot^2 \theta = 1$$

$$\sin 2\theta = 2 \sin \theta \cos \theta$$

$$\cos 2\theta = \cos^2 \theta - \sin^2 \theta = 2 \cos^2 \theta - 1 = 1 - 2 \sin^2 \theta$$

$$\tan 2\theta = \frac{2 \tan \theta}{1 - \tan^2 \theta}$$

$$\sin \frac{1}{2}\theta = \pm \sqrt{\frac{1 - \cos \theta}{2}}; \cos \frac{1}{2}\theta = \pm \sqrt{\frac{1 + \cos \theta}{2}};$$

$$\tan \frac{1}{2}\theta = \pm \sqrt{\frac{1 - \cos \theta}{1 + \cos \theta}}$$

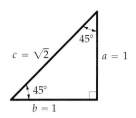

Figure M-13 An isosceles right triangle.

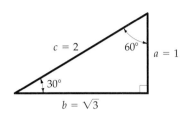

Figure M-14 A 30°–60°–90° right triangle.

Some Important Values of the Functions

Figure M-13 is a diagram of an *isosceles* right triangle (an isosceles triangle is a triangle with two equal sides), from which we can find the sine, cosine, and tangent of 45°. The two acute angles of this triangle are equal. Because the sum of the three angles in a triangle must equal 180° and the right angle is 90°, each acute angle must be 45°. For convenience, let us assume that the equal sides each have a length of 1 unit. The Pythagorean theorem gives us a value for the hypotenuse of

$$c = \sqrt{a^2 + b^2} = \sqrt{1^2 + 1^2} = \sqrt{2} \text{ units}$$

We calculate the values of the functions as follows:

$$\sin 45° = \frac{a}{c} = \frac{1}{\sqrt{2}} = 0.707 \quad \cos 45° = \frac{b}{c} = \frac{1}{\sqrt{2}} = 0.707 \quad \tan 45° = \frac{a}{b} = \frac{1}{1} = 1$$

Another common triangle, a 30°–60°–90° right triangle, is shown in Figure M-14. Because this particular right triangle is in effect half of an *equilateral triangle* (a 60°–60°–60° triangle or a triangle having three equal sides and three equal angles), we can see that the sine of 30° must be exactly 0.5 (Figure M-15). The equilateral triangle must have all sides equal to c, the hypotenuse of the 30°–60°–90° right triangle. Thus, side a is one-half the length of the hypotenuse, and so

$$\sin 30° = \frac{1}{2}$$

To find the other ratios within the 30°–60°–90° right triangle, let us assign a value of 1 to the side opposite the 30° angle. Then

$$c = \frac{1}{0.5} = 2 \qquad\qquad b = \sqrt{c^2 - a^2} = \sqrt{2^2 - 1^2} = \sqrt{3}$$

$$\cos 30° = \frac{b}{c} = \frac{\sqrt{3}}{2} = 0.866 \qquad \tan 30° = \frac{a}{b} = \frac{1}{\sqrt{3}} = 0.577$$

$$\sin 60° = \frac{b}{c} = \cos 30° = 0.866 \qquad \cos 60° = \frac{a}{c} = \sin 30° = \frac{1}{2}$$

$$\tan 60° = \frac{b}{a} = \frac{\sqrt{3}}{1} = 1.732$$

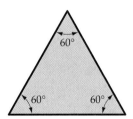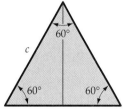

Figure M-15 (a) An equilateral triangle. (b) An equilateral triangle that has been bisected to form two 30°–60°–90° right triangles.

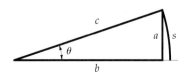

Figure M-16 For small angles, $\sin \theta = a/c$, $\tan \theta = a/b$, and the angle $\theta = s/c$ are all approximately equal.

Small-Angle Approximation

For small angles, the length a is nearly equal to the arc length s, as can be seen in Figure M-16. The angle $\theta = s/c$ is therefore nearly equal to $\sin \theta = a/c$:

(M-40) $\qquad\qquad \sin \theta \approx \theta \quad$ for small values of θ

Similarly, the lengths c and b are nearly equal, so $\tan \theta = a/b$ is nearly equal to both θ and $\sin \theta$ for small values of θ:

(M-41) $\qquad\qquad \tan \theta \approx \sin \theta \approx \theta \quad$ for small values of θ

Equations M-40 and M-41 hold only if θ is measured in radians. Because $\cos \theta = b/c$, and because these lengths are nearly equal for small values of θ, we have

(M-42) $\qquad\qquad \cos \theta \approx 1 \quad$ for small values of θ

Figure M-17 shows graphs of θ, $\sin \theta$, and $\tan \theta$ versus θ for small values of θ. If accuracy of a few percent is needed, small-angle approximations can be used only for angles of about a quarter of a radian (or about 15°) or less. Below this value, as the angle becomes smaller, the approximation $\theta \approx \sin \theta \approx \tan \theta$ is even more accurate.

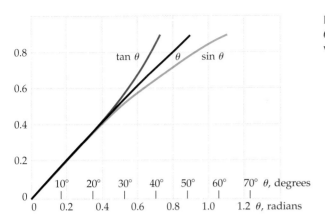

Figure M-17 Graphs of tan θ, θ, and sin θ versus θ for small values of θ.

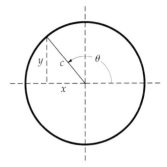

Trigonometric Functions as Functions of Real Numbers

So far we have illustrated the trigonometric functions as properties of angles. Figure M-18 shows an *obtuse* angle with its vertex at the origin and one side along the x axis. The trigonometric functions for a "general" angle such as this are defined by

Figure M-18 Diagram for defining the trigonometric functions for an obtuse angle.

$$\sin \theta = \frac{y}{c} \qquad \text{(M-43)}$$

$$\cos \theta = \frac{x}{c} \qquad \text{(M-44)}$$

$$\tan \theta = \frac{y}{x} \qquad \text{(M-45)}$$

It is important to remember that values of x to the left of the vertical axis and values of y below the horizontal axis are negative; c in the figure is always regarded as positive. Figure M-19 shows plots of the general sine, cosine, and tangent functions versus θ. The sine and cosine functions have a period of 2π rad. Thus, for any value of θ, $\sin(\theta + 2\pi) = \sin \theta$, and so forth. That is, when an angle changes by 2π rad, the function returns to its original value. The tangent function has a period of π rad. Thus, $\tan(\theta + \pi) = \tan \theta$, and so forth. Some other useful relations are

$$\sin(\pi - \theta) = \sin \theta \qquad \text{(M-46)}$$

$$\cos(\pi - \theta) = -\cos \theta \qquad \text{(M-47)}$$

$$\sin\left(\frac{1}{2}\pi - \theta\right) = \cos \theta \qquad \text{(M-48)}$$

$$\cos\left(\frac{1}{2}\pi - \theta\right) = \sin \theta \qquad \text{(M-49)}$$

Because the radian is dimensionless, it is not hard to see from the plots in Figure M-21 that the trigonometric functions are functions of all real numbers.

(a)

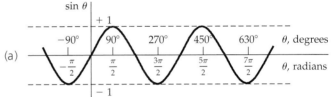

(b)

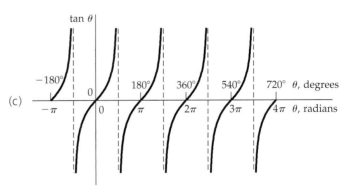

(c)

Figure M-19 The trigonometric functions sin θ, cos θ, and tan θ versus θ.

Example M-9 Cosine of a Sum

Using the suitable trigonometric identity from Table M-2, find $\cos(135° + 22°)$. Give your answer with four significant figures.

Set Up
As long as all angles are given in degrees, there is no need to convert to radians, because all operations are numerical values of the functions. Be sure, however, that your calculator is in degree mode. The suitable identity is $\cos(A \pm B) = \cos A \cos B \mp \sin A \sin B$, where the upper signs are appropriate.

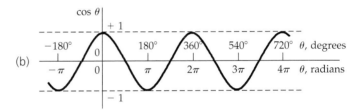

Solve

Write the trigonometric identity for the cosine of a sum, with $A = 135°$ and $B = 22°$:

$$\cos(135° + 22°) = (\cos 135°)(\cos 22°) - (\sin 135°)(\sin 22°)$$

Using a calculator, find $\cos 135°$, $\sin 135°$, $\cos 22°$, and $\sin 22°$:

$$\cos 135° = -0.7071$$
$$\cos 22° = 0.9272$$
$$\sin 135° = 0.7071$$
$$\sin 22° = 0.3746$$

Enter the values in the formula and calculate the answer:

$$\cos(135° + 22°) = (-0.7071)(0.9272) - (0.7071)(0.3746)$$
$$= -0.9205$$

Reflect

The calculator shows that $\cos(135° + 22°) = \cos(157°) = -0.9205$.

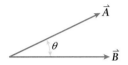

Figure M-20 Two vectors separated by an angle θ.

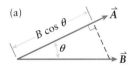
(a)

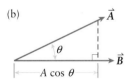
(b)

Figure M-21 The dot product is a measure of how parallel two vectors are. (a) $B \cos \theta$ is the component of $\vec{B}$ that is parallel to $\vec{A}$. (b) $A \cos \theta$ is the component of $\vec{A}$ that is parallel to $\vec{B}$.

M-9 The Dot Product

For two vectors $\vec{A}$ and $\vec{B}$ separated by angle θ, as shown in Figure M-20, their dot product C is defined as

$$\text{(M-50)} \qquad C = \vec{A} \cdot \vec{B} = AB \cos \theta$$

which you can read as, "C equals A dot B." In Equation M-50, A and B are the magnitudes of vectors $\vec{A}$ and $\vec{B}$ respectively. As a result, the dot product of two vectors is a scalar quantity. This is why $\vec{A} \cdot \vec{B}$ is also called the **scalar product** of $\vec{A}$ and $\vec{B}$.

Physically, the dot product $\vec{A} \cdot \vec{B}$ is a measure of how parallel the two vectors are. We can think of it as the magnitude of vector $\vec{A}$ multiplied by the component of vector $\vec{B}$ that is parallel to $\vec{A}$. Referring to Figure M-21a, we see that $B \cos \theta$ is the component of $\vec{B}$ that is parallel to $\vec{A}$. That is, $B \cos \theta$ tells us how much of $\vec{B}$ points in the direction of $\vec{A}$. Alternatively, the dot product $\vec{A} \cdot \vec{B}$ can be thought of as the magnitude of vector $\vec{B}$ multiplied by the component of vector $\vec{A}$ parallel to $\vec{B}$ (Figure M-21b).

The dot product is commutative; the order of the vectors in a dot product does not affect the result:

$$\vec{A} \cdot \vec{B} = \vec{B} \cdot \vec{A}$$

The dot product is also distributive, which means

$$\vec{A} \cdot (\vec{B} + \vec{C}) = \vec{A} \cdot \vec{B} + \vec{A} \cdot \vec{C}$$

Three special cases of the dot product are particularly important in physics. First, the dot product of two vectors $\vec{A}$ and $\vec{B}$ that point in the same direction (so $\theta = 0$ and $\cos \theta = \cos 0 = 1$) equals the product of their magnitudes:

$$\text{(M-51)} \qquad \vec{A} \cdot \vec{B} = AB \cos 0 = AB$$
(if $\vec{A}$ and $\vec{B}$ point in the same direction)

(As an example, the dot product of a vector $\vec{A}$ with itself is equal to the square of its magnitude: $\vec{A} \cdot \vec{A} = AA \cos 0 = A^2$.)

Second, the dot product of two perpendicular vectors $\vec{A}$ and $\vec{B}$ (so $\theta = 90°$ and $\cos \theta = \cos 90° = 0$) is zero:

$$\text{(M-52)} \qquad \vec{A} \cdot \vec{B} = AB \cos 90° = 0$$
(if $\vec{A}$ and $\vec{B}$ are perpendicular)

Third, if two vectors $\vec{A}$ and $\vec{B}$ point in opposite directions (so $\theta = 180°$ and $\cos \theta = \cos 180° = -1$), their dot product equals the *negative* of the product of their magnitudes:

$$\text{(M-53)} \qquad \vec{A} \cdot \vec{B} \cos 180° = -AB$$
(if $\vec{A}$ and $\vec{B}$ point in opposite directions)

Finally, it's useful to know how to calculate the dot product of two vectors $\vec{A}$ and $\vec{B}$ that are expressed in terms of their components A_x, A_y, A_z, and B_x, B_y, and B_z:

$$\text{(M-54)} \qquad \vec{A} \cdot \vec{B} = A_x B_x + A_y B_y + A_z B_z$$

You can verify that Equation M-54 is correct by thinking of $\vec{A}$ as the sum of three vectors: $\vec{A}_1$, which has only an x-component A_x; $\vec{A}_2$, which has only a y-component A_y; and $\vec{A}_3$, which has only a z-component A_z. From the definition of the dot product, $\vec{A}_1 \cdot \vec{B}$ is equal to A_x multiplied by the component of $\vec{B}$ in the direction of $\vec{A}_1$, or $\vec{A}_1 \cdot \vec{B} = A_x B_x$. Similarly, $\vec{A}_2 \cdot \vec{B} = A_y B_y$ and $\vec{A}_3 \cdot \vec{B} = A_z B_z$. Since $\vec{A} = \vec{A}_1 + \vec{A}_2 + \vec{A}_3$ and the dot product is distributive, it follows that

$$\vec{A} \cdot \vec{B} = (\vec{A}_1 + \vec{A}_2 + \vec{A}_3) \cdot \vec{B} = \vec{A}_1 \cdot \vec{B} + \vec{A}_2 \cdot \vec{B} + \vec{A}_3 \cdot \vec{B} = A_x B_x + A_y B_y + A_z B_z$$

That's the same as Equation M-54. If the vectors have only x- and y-components, Equation M-54 simplifies to $\vec{A} \cdot \vec{B} = A_x B_x + A_y B_y$.

Example M-10 The Dot Product

(a) Calculate the dot product of vector $\vec{A}$ with magnitude 5.00 pointed in a horizontal direction 36.9° north of east and vector $\vec{B}$ of magnitude 1.50 pointed in a horizontal direction 53.1° south of west. (b) What is the dot product of vector $\vec{C}$ with components $C_x = 4.00$, $C_y = 3.00$ and vector $\vec{D}$ with components $D_x = -0.900$, $D_y = -1.20$?

Set Up

In part (a) we know the magnitude and direction of the vectors, so we'll use Equation M-50. In part (b) the vectors are given in terms of components, so we'll evaluate the dot product using Equation M-54.

$$\vec{A} \cdot \vec{B} = AB \cos \theta \qquad \text{(M-50)}$$

Dot product of two vectors in terms of components:

$$\vec{A} \cdot \vec{B} = A_x B_x + A_y B_y + A_z B_z \qquad \text{(M-54)}$$

Solve

(a) The drawing shows that the angle between $\vec{A}$ and $\vec{B}$ is $\theta = 163.8°$. We use this in Equation M-50 to evaluate the dot product.

$$\begin{aligned}
\vec{A} \cdot \vec{B} &= AB \cos \theta \\
&= (5.00)(1.50) \cos 163.8° \\
&= (5.00)(1.50)(-0.960) \\
&= -7.20
\end{aligned}$$

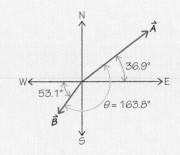

(b) Both $\vec{C}$ and $\vec{D}$ are in the x-y plane and have no z components, so we just need the first two terms in Equation M-54 to calculate their dot product.

$$\begin{aligned}
\vec{C} \cdot \vec{D} &= C_x D_x + C_y D_y \\
&= (4.00)(-0.900) + (3.00)(-1.20) \\
&= -7.20
\end{aligned}$$

Reflect

It's not a coincidence that we got the same result in part (b) as in part (a): Vectors $\vec{A}$ and $\vec{C}$ are the same, as are vectors $\vec{B}$ and $\vec{D}$. (You can verify this by using the techniques from Chapter 3 to calculate the components of the vectors $\vec{A}$ and $\vec{B}$ in part (a). You'll find that the components are the same as those of $\vec{C}$ and $\vec{D}$ in part (b).) This should give you confidence that the method of calculating the dot product using components gives you the same result as the method that involves the magnitudes and directions of the vectors.

Notice that the angle between vectors $\vec{A}$ and $\vec{B}$ is between 90° and 180°, and the dot product is negative.

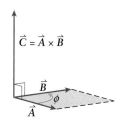

Figure M-22 The cross product is a vector $\vec{C}$ that is perpendicular to both $\vec{A}$ and $\vec{B}$, and has a magnitude $AB \sin \phi$, which equals the area of the parallelogram shown.

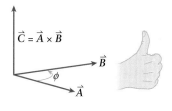

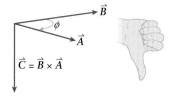

Figure M-23 (a) To find the direction of , point the fingers of your right hand in the direction of vector $\vec{A}$, then curl them toward vector $\vec{B}$. Your thumb points in the direction of the cross product. (b) The direction of $\vec{B} \times \vec{A}$ points in the opposite direction of.

M-10 The Cross Product

The dot product, described in section M-9, is only one way to multiply two vectors. We can also multiply two vectors $\vec{A}$ and $\vec{B}$ using the **cross product**

$$\text{(M-55)} \qquad \vec{C} = \vec{A} \times \vec{B}$$

The symbol "$\times$" represents the mathematical operation known as the cross product. As you can see from Equation M-55, the result of taking the cross product of two vectors is also a vector. The magnitude of the resulting vector is the product of the magnitudes of the two vectors and the sine of the angle between them. That is, the magnitude of the cross product of $\vec{A}$ and $\vec{B}$ is

$$\text{(M-56)} \qquad C = |\vec{A} \times \vec{B}| = AB \sin \phi$$

where according to convention ϕ is defined as the angle that goes from $\vec{A}$ to $\vec{B}$. $\vec{C}$ points in the direction perpendicular to both $\vec{A}$ and $\vec{B}$ as shown in **Figure M-22**.

The magnitude of the cross product $\vec{A} \times \vec{B}$ can be interpreted as the magnitude of vector $\vec{A}$ multiplied by the component of vector $\vec{B}$ perpendicular to $\vec{A}$, or the magnitude of vector $\vec{B}$ multiplied by the component of vector $\vec{A}$ perpendicular to $\vec{B}$.

Note that the order of the two vectors in a cross product makes a difference. The cross product of $\vec{B}$ and $\vec{A}$ is the negative of the cross product of $\vec{A}$ and $\vec{B}$ or

$$\text{(M-57)} \qquad \vec{A} \times \vec{B} = -\vec{B} \times \vec{A}$$

This results from the definition of the angle ϕ in Equation M-56. Since ϕ is directed from the first vector to the second vector, if you travel the angle from the second vector to the first—in reverse direction-ϕ becomes negative. And the sine of a negative angle is also negative.

In addition, the cross product obeys the distributive law under addition:

$$\text{(M-58)} \qquad \vec{A} \times (\vec{B} + \vec{C}) = \vec{A} \times \vec{B} + \vec{A} \times \vec{C}$$

To determine the direction of the cross product $\vec{C} = \vec{A} \times \vec{B}$, you can use the right-hand rule. To apply this rule, point the fingers of your right hand in the direction of the first vector of the cross product (in this case $\vec{A}$). Then curl your fingers toward the second vector, $\vec{B}$. If you stick your thumb straight out, it points in the direction of the cross product, vector $\vec{C}$ (**Figure M-23a**). If you instead want to find the direction of the cross product $\vec{B} \times \vec{A}$, begin by pointing the fingers of your right hand in the direction of vector $\vec{B}$. Then curl them toward vector $\vec{A}$. Your thumb again points in the direction of the cross product (**Figure M-23b**). Note that because you must curl your fingers in the opposite direction as for $\vec{C} = \vec{A} \times \vec{B}$, the cross product of $\vec{B} \times \vec{A}$ points in the opposite direction of $\vec{A} \times \vec{B}$, which is just what we stated in Equation M-58.

There are two special cases of the cross product that are worth pointing out. The first is the cross product for two perpendicular vectors, for which $\phi = 90°$, so $\sin \phi = 1$.

$$|\vec{A} \times \vec{B}| = AB \sin 90° = AB(1) = AB$$

(magnitude of the cross product of two perpendicular vectors)

The second special case is the cross product of two parallel vectors, for which $\phi = 0$, so $\sin \phi = 0$.

$$|\vec{A} \times \vec{B}| = AB \sin 0 = AB(0) = 0$$

(magnitude of the cross product for two parallel vectors)

One example of a cross product of two parallel vectors is the cross product of a vector with itself: $\vec{A} \times \vec{A} = 0$.

Example M-11 The Cross Product

Evaluate $\vec{A} \times \vec{B}$, in which the components of vector $\vec{A}$ are $A_x = 5$, $A_y = 0$, and the components of vector $\vec{B}$ are $B_x = 9$, $B_y = 7$.

Set Up

We will use the definition of the magnitude of the cross product, Equation M-56, to find the magnitude of the cross product, and the right-hand rule to determine the direction of the cross product.

We will have to use the components of vector $\vec{B}$ to determine its magnitude and the angle it makes with the x axis and vector $\vec{A}$.

$$C = |\vec{A} \times \vec{B}| = AB \sin \phi$$

$$(M-56)$$

Finding vector magnitude and direction from vector components:

$$A = \sqrt{A_x^2 + A_y^2}$$

$$\tan \theta = \frac{A_y}{A_x} \qquad (3\text{-}2)$$

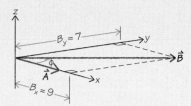

Solve

Begin by determining the magnitude and direction of vector $\vec{B}$ using its components and Equations 3-2.

Determine the magnitude of vector $\vec{B}$ from its components:

$$B = \sqrt{B_x^2 + B_y^2} = \sqrt{9^2 + 7^2} = 11.4$$

Determine the angle $\vec{B}$ makes with the x axis (and vector $\vec{A}$) from its components:

$$\tan \phi = \frac{B_y}{B_x} = \frac{7}{9} = 0.778, \text{ so}$$

$$\phi = \arctan 0.778 = 37.9°$$

Because vector $\vec{A}$ has only an x component, its magnitude is equal to its x component, and the angle it makes with the x axis is 0.

Determine the magnitude of vector $\vec{A}$ from its components:

$$A = \sqrt{A_x^2 + A_y^2} = \sqrt{5^2 + 0^2} = 5$$

Because $\vec{A}$ has only an x component, it makes an angle of zero degrees with the x axis.

Now that we know both the magnitude and direction of the vectors, we can use Equation M-56 to determine the magnitude of the cross product.

Apply Equation M-56 to the two vectors:

$$\begin{aligned}
|\vec{A} \times \vec{B}| &= AB \sin \phi \\
&= (5.00)(11.4)\sin 37.9° \\
&= (5.00)(11.4)(0.614) \\
&= 35.0
\end{aligned}$$

Use the right-hand rule to determine the direction of the cross product.

From the figure, if we first point the fingers of our right hand in the direction of $\vec{A}$ (along the x axis), and then curl them toward $\vec{B}$, we see that the thumb points in the positive z direction. So the cross product $\vec{A} \times \vec{B}$ has a magnitude of 35 in the $+z$ direction.

Reflect

The vectors $\vec{A}$ and $\vec{B}$ lie in the xy plane, so the cross product, which must be perpendicular to both vectors, should point along the z axis, which is just what we found.

Official MCAT® Prep Resources

MCAT | **AAMC**
Medical College
Admission Test

$30

Second Edition
The Official Guide
to the MCAT® Exam

Learn the Basics

www.aamc.org/mcat

- Free content outlines
- MCAT essentials
- *The Official Guide to the MCAT® Exam*

Find Out Where You Stand

www.e-mcat.com

- Get a free practice test
- Take timed test to simulate actual test experience
- Use estimated score as a baseline

Identify Your Strengths and Weaknesses

www.aamc.org/mcatsap

- Buy The Official MCAT® Self-Assessment Package
- Analyze your MCAT knowledge
- Use the data to customize your study plan

Get More Practice

www.e-mcat.com

- Choose from seven additional practice tests
- Use test results to monitor your progress

$104 for Package

The Official MCAT® Self-Assessment Package

$35

e-MCAT Practice Tests

MCAT® is a program of the
Association of American Medical Colleges

MCAT® Appendix

The section that follows includes material from previously administered MCAT® items and is reprinted with permission of the Association of American Medical Colleges (AAMC).

Passage 13 (81–85)

Tennis balls must pass a rebound test before they can be certified for tournament play. To qualify, balls dropped from a given height must rebound within a specified range of heights. Measuring rebound height can be difficult because the ball is at its maximum height for only a brief time. It is possible to perform a simpler indirect measurement to calculate the height of rebound by measuring how long it takes the ball to rebound and hit the floor again. The diagram below illustrates the experimental setup used to make the measurement.

The ball is dropped from a height of 2.0 m, and it hits the floor and then rebounds to a height h. A microphone detects the sound of the ball each time it hits the floor, and a timer connected to the microphone measures the time (t) between the two impacts. The height of the rebound is $h = gt^2/8$ where $g = 9.8$ m/s^2 is the acceleration due to gravity. Care must be taken so that the measured times do not contain systematic error. Both the speed of sound and the time of impact of the ball with the floor must be considered. Four balls were tested using the method. The results are listed in the table below.

Ball	Time (s)
A	1.01
B	1.05
C	0.97
D	1.09

81. If NO air resistance is present, which of the following quantities remains constant while the ball is in the air between the first and second impacts?

- A. Kinetic energy of the ball
- B. Potential energy of the ball
- C. Momentum of the ball
- D. Horizontal speed of the ball

82. What measurement is made by the timer?

- A. Duration of the impacts
- B. Time that the ball is at maximum height
- C. Time between the first and second bounces
- D. Decrease in time of successive bounces

83. Balls C and D failed the test, while balls A and B passed. A ball that had which of the following measured times would definitely pass the test?

- A. 0.95 s
- B. 0.99 s
- C. 1.03 s
- D. 1.07 s

84. What percentage of its original potential energy did ball C lose between the start of the experiment and its rebound to maximum height?

- A. 24%
- B. 42%
- C. 51%
- D. 58%

85. With what approximate vertical speed does a ball in the experiment strike the floor?

- A. 4 m/s
- B. 6 m/s
- C. 10 m/s
- D. 20 m/s

Passage X (Questions 134–138)

The study of the flight of projectiles has many practical applications. The main forces acting on a projectile are air resistance and gravity. The path of a projectile is often approximated by ignoring the effects of air resistance. Gravity is then the only force acting on the projectile. When air resistance is included in the analysis, another force F_R is introduced. F_R is proportional to the square of the velocity, v. The direction of the air resistance is exactly opposite the direction of motion. The equation for air resistance is $F_R = pv^2$, where p is a proportionality constant that depends on such factors as the density of the air and the shape of the projectile.

Air resistance was studied by launching a 0.5-kg projectile from a level surface. The projectile was launched with a speed of 30 m/s at a 40° angle to the surface. (Note: Assume air resistance is present unless otherwise specified. Acceleration due to gravity is $g = 9.8$ m/s^2; sin 40° = 0.64; cos 40° = 0.77.)

134. If a 0.5-kg projectile is launched straight up and is given an initial kinetic energy of 259 J, to what maximum height will the projectile rise? (Note: Assume that the effects of air resistance are negligible.)

- A. 26 m
- B. 51 m
- C. 102 m
- D. 204 m

135. Which of the following statements is (are) true about the kinetic energy of the projectile when it returns to the elevation from which it was launched?

- I. It is the same as when it was launched.
- II. It is dependent on the initial velocity.
- III. It is dependent on the value of p.
- A. II only
- B. I and II only
- C. I and III only
- D. II and III only

136. What is the approximate total kinetic energy of the projectile at the highest point in its path? (Note: Assume that the effects of air resistance are negligible.)
- A. 92 J
- B. 133 J
- C. 184 J
- D. 225 J

137. Which of the following graphs best illustrates the relationship between the total speed of the projectile (*v*) and its horizontal distance from the launch point (*x*)? (Note: Assume that the effects of air resistance are negligible and that the left axis represents the location of the launch point.)

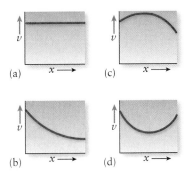

(a)   (c)

(b)   (d)

138. What is the magnitude of the horizontal component of air resistance on the projectile at any point during the flight? (Note: v_x = horizontal speed.)
- A. $(pv^2)\cos 40°$
- B. $pv^2/2$
- C. $(pv_x^2)\sin 40°$
- D. pv_x^2

Passage 12 (Questions 73–80)

A pressure difference between the ends of a horizontal, uniform pipe is required to maintain the steady flow of a viscous fluid through the pipe. The fluid pressure at the downstream end of the pipe (P_2) depends on the tube length (L), inside radius of the pipe (r), viscosity (η), volume flow rate (f), and the pressure at the beginning of the pipe (P_1). These quantities are related by Poiseuille's equation $\Delta P = P_2 - P_1 = \dfrac{(8\eta Lf)}{\pi r^4}$, where ΔP is the pressure differential.

Two experiments were conducted to investigate the flow of distilled water and glycerin through various pipes.

Experiment 1
Distilled water flowed through pipes that were 5.0 m long and had different radii. The pressure differential between the beginning and end of the pipes was measured while the flow rate was held constant. The results are listed in Table 2. (Note: At 20°C, distilled water has a viscosity of 1×10^{-3} kg/m·s and a density of 1×10^3 kg/m³.)

Trial	L(m)	r(m)	f(m³/s)	ΔP (N/m²)
1	5.0	0.05	0.003	6.0
2	5.0	0.04	0.003	15.0
3	5.0	0.03	0.003	47.0
4	5.0	0.02	0.003	240
5	5.0	0.01	0.003	3820

Experiment 2
Glycerin flowed through pipes that had different lengths but the same radius. The flow rates were measured while the pressure differential between the beginning and end of the pipes was held constant. The results are listed in Table 3. (Note: At 20°C, glycerin has a viscosity of 1.5×10^{-3} kg/m·s and a density of 1.3×10^3 kg/m³.)

Trial	L(m)	r(m)	ΔP (N/m²)	f(m³/s)
1	1.0	0.01	100	2.6×10^{-7}
2	1.5	0.01	100	1.7×10^{-7}
3	2.0	0.01	100	1.3×10^{-7}
4	2.5	0.01	100	1.0×10^{-7}
5	3.0	0.01	100	8.7×10^{-8}

73. When viscosity increases, an increase in pressure differential is required to maintain the same volume flow rate. This is because an increase in viscosity is related to an increase in which of the following factors?
- A. Pipe radius
- B. Friction
- C. Fluid density
- D. Turbulence

74. When fluid flow is compared to electricity, which of the following fluid characteristics is analogous to electrical current?
- A. Pressure differential
- B. Volume flow rate
- C. Viscosity
- D. Flow speed

75. In which trial of Experiment 2 did a unit length of glycerin have the greatest kinetic energy?
- A. Trial 1
- B. Trial 2
- C. Trial 4
- D. Trial 5

76. If an additional trial of Experiment 1 were conducted using a pipe with a radius of 0.06 m, what would be the expected pressure differential?
- A. 0.4 N/m²
- B. 1.0 N/m²
- C. 1.5 N/m²
- D. 2.9 N/m²

77. If water and glycerin flowed through identical pipes and had the same pressure differential, which fluid would have the greater flow rate?
- A. Water, because it has a lower viscosity
- B. Water, because it has a lower density
- C. Glycerin, because it has a higher viscosity
- D. Glycerin, because it has a higher density

78. If ΔP is held constant and r is reduced by 33%, by what percentage does f decrease?
- A. 33%
- B. 56%
- C. 80%
- D. 98%

79. If an additional trial of Experiment 2 were conducted using a pipe that was 4.0 m long, what would be the expected flow rate?
    A. 2.6×10^{-8} m³/s
    B. 4.3×10^{-8} m³/s
    C. 6.5×10^{-8} m³/s
    D. 7.8×10^{-8} m³/s

80. Which of the following graphs best shows how the volume flow rate of a fluid flowing out of a uniform pipe varies as the pressure differential between the beginning and end of the pipe is changed?

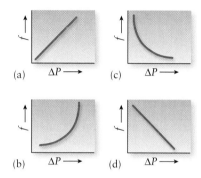

Passage 7 (Questions 41–48)

Three experiments were conducted to examine collisions between masses suspended at the ends of simple pendula. Each pendulum consisted of a mass, or bob, attached to the bottom of a rod of negligible mass. The top of the rod was anchored to a solid surface from which it could swing free of friction. For sufficiently small amplitudes, each pendulum executed simple harmonic motion with a period $T = 2\sqrt{\dfrac{l}{g}}$, where l is the length of the pendulum, and g is the acceleration due to gravity. (Note: Assume air resistance is negligible in the experiments.)

Experiment 1

Two identical pendula with steel balls for bobs were hung next to each other so that their bobs barely touched when they hung vertically. The left pendulum bob was moved to the side and then released. The bobs collided, and immediately after the collision, the left bob was at rest while the right bob moved with the same speed that the left bob had before the collision.

Experiment 2

The steel bobs were replaced with nonidentical balls made of an unknown substance. The left pendulum bob was moved to the same position as in Experiment 1 and then released. After the collision, the right bob moved to the right while the left bob rebounded to the left.

Experiment 3

The bobs were replaced with identical balls made of a deformable, sticky material. The left pendulum bob was moved to the same position as in Experiment 1 and then released. After the collision, the bobs stuck together, and the final speed of the combined bobs was less than the speed of the left bob before the collision.

41. Which of the following diagrams best describes the subsequent motion of a bob that detaches from a simple pendulum? When the bob moves through the lowest point of its arc?

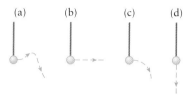

42. Which of the following quantities is NOT conserved in the collision in Experiment 3?
    A. Linear momentum
    B. Angular momentum
    C. Potential energy
    D. Kinetic energy

43. Which of the following would occur if the collision in Experiment 1 was perfectly elastic?
    A. The right bob would rise to the same height that the left bob started from.
    B. The support rod would stretch enough that the right bob would move in a straight line.
    C. Both bobs would move away from the impact with equal and opposite velocities.
    D. Both bobs would move to the right after the collision.

44. Which of the following accurately describes the difference between the pendulum bobs used in Experiment 2?
    A. The left bob was more elastic than the right bob.
    B. The left bob was less elastic than the right bob.
    C. The left bob weighed less than the right bob.
    D. The left hob weighed more than the right bob.

45. If the speed of the left bob in Experiment 3 was v immediately before the collision, what was its speed immediately after the collision?
    A. $v/4$
    B. $v/2$
    C. $v/\sqrt{2}$
    D. Approximately equal to but slightly less than v.

46. If the maximum height of the left bob in Experiment 1 was increased by a factor of 2, by what factor would the maximum height of the right bob increase?
    A. $\sqrt{2}$
    B. $\frac{3}{2}$
    C. 2
    D. 4

47. The speed with which the left bob rebounds in Experiment 2 is
    A. less than its speed before the collision.
    B. equal to its speed before the collision.
    C. greater than its speed before the collision.
    D. equal to the speed of the right bob before the collision.

48. A pendulum bob is replaced with one that weighs 2 times as much. How does the time it takes for the new bob to move from a small angle of displacement to the lowest point of its arc compare to the time it took the old bob?

A. It takes $\frac{1}{2}$ as long.
B. It takes the same amount of time.
C. It takes $\sqrt{2}$ times as long.
D. It takes 2 times as long.

Passage 1 (Questions 1–6)

Ultrasonic waves are used in many applications by analyzing their reflections and refractions from interfaces between different media. Different intensities of reflections of a scanned object can provide a picture or pattern that can be electronically constructed and displayed on a monitor. The ultrasonic wavelength used for investigation should be equal to or smaller than the size of the object being scanned.

Ultrasonic waves are used for diagnostic purposes in medicine. Waves are directed into a body, and reflections occur at interfaces between different tissues or fluids. A practical depth of penetration of this technique for diagnostic scans is about 200 times the wavelength of the incident ultrasonic waves.

Another use of ultrasonic waves is an application of the Doppler effect. Doctors can detect and monitor blood flow speed as blood moves directly toward the incident ultrasonic waves. Calculations of flow toward the incident waves are made by using the relationship $\frac{f_o}{f_s} = \frac{v}{v - 2v_s}$, where f_o is the observed frequency, f_s is the incident frequency, v is the speed of ultrasonic waves in the medium, and v_s is the speed of the moving blood.

Ultrasonic waves that are reflected from a moving object are shifted in frequency. If the incident and reflected waves are mixed, this change in frequency causes beats to occur at a frequency that is equal to the absolute value of the difference between the incident and the reflected frequency. This phenomenon can also occur when waves that have 2 separate frequencies and that have been emitted from 2 stationary sources are mixed.

1. The speed of sound in tissue for an ultrasonic scan is 1,500 m/s. From the information given in the passage, what maximum frequency could be used for a scan when a depth of investigation of 0.20 m is required?
   A. 1.2×10^3 Hz
   B. 6.0×10^3 Hz
   C. 1.2×10^6 Hz
   D. 1.5×10^6 Hz

2. When ultrasonic waves are detected by a transducer that feeds a signal into a display monitor, which of the following is the best description of the energy transformation that occurs at the transducer?
   A. mechanical to electrical
   B. kinetic to mechanical
   C. potential to kinetic
   D. potential to electrical

3. When a trumpet is tuned by comparison with a 512-Hz note from a piano, a beat frequency of 4 Hz is produced. The trumpet could have produced which of the following pairs of frequencies?
   A. 504 Hz and 508 Hz
   B. 508 Hz and 512 Hz

   C. 508 Hz and 516 Hz
   D. 512 Hz and 516 Hz

4. An ultrasonic wave enters a body with a speed of 1,500 m/s, and a reflection is noted at the same position on the surface 4.0×10^{-5} s later. What is the distance between the surface and the reflecting interface?
   A. 1.3×10^{-8} m
   B. 2.7×10^{-8} m
   C. 3.0×10^{-2} m
   D. 6.0×10^{-2} m

5. According to the passage, what is the lowest wave frequency that should be used to provide an image of an object that is 10^{-3} m on each side? (Note: Assume that the wave travels at 1,500 m/s.)
   A. 1.5 Hz
   B. 1.5×10^6 Hz
   C. 6.7×10^6 Hz
   D. 6.7×10^7 Hz

6. Which of the following best explains why ultrasonic waves are reflected at boundaries within a human body?
   A. Reflections occur at the boundary between objects with different shapes.
   B. Reflections occur at the boundary between objects with different densities.
   C. Reflections occur when the frequency of the ultrasonic wave decreases.
   D. Reflections occur when the frequency of the ultrasonic wave increases.

Passage 2 (Questions 7-13)

A student performed 2 experiments to investigate the behavior of springs.

Experiment 1

The student investigated the stretching of a horizontal spring on a smooth table. One end of the spring was fastened to a vertical wall, and the spring was stretched by a string attached over a pulley to a hanging mass (M) as shown in Figure 1. The experimental results showing how the total spring length (L) changed with variations to M are given in Table 4.

Trial	M (kg)	L (m)
1	0	0.400
2	0.1	0.405
3	0.4	0.420
4	2.0	0.500

Experiment 2

The student slid a mass (m) with speed (v) along a horizontal tabletop with which it had negligible friction. The mass collided with a magnetic bumper plate that was attached to a horizontal spring fastened to a wall, as shown in Figure 2. When the spring was compressed a distance (x), the spring had a potential energy of $1/2kx^2$, where k is the spring constant. Maximum potential energy occurred at maximum recoil distance, x_{max}. Values of mass and speed used in Experiment 2 are given in Table 5.

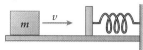

Trial	m (kg)	v (m/s)
1	1	5
2	2	4
3	3	3
4	4	2
5	5	1

7. What is the value of k for the spring used in Experiment 1?
A. 4 N/m
B. 25 N/m
C. 100 N/m
D. 200 N/m

8. For which of the following trials of Experiment 2 is the recoil distance the greatest?
A. Trial 1
B. Trial 2
C. Trial 3
D. Trial 4

9. In Experiment 2, the mass clings to the magnetic bumper plate and vibrates back and forth. Which of the following best compares f_1, the vibration frequency of Trial 1 to f_4, the vibration frequency of Trial 4?(Note: The period of vibration of this system is $T = 2\pi\sqrt{\dfrac{m}{k}}$.)

A. $f_1 = \dfrac{f_4}{4}$

B. $f_1 = \dfrac{f_4}{2}$

C. $f_1 = f_4$

D. $f_1 = 2f_4$

10. In Experiment 2, what minimum force must exist between the mass and the bumper plate so that the mass can vibrate back and forth?

A. $\dfrac{k(x_{max})^2}{4}$

B. $\dfrac{k(x_{max})^2}{2}$

C. kx_{max}

D. $k(x_{max})^2$

11. Which of the following graphs best describes the potential energy (PE) of the spring as it is compressed to x_{max} in Experiment 2?

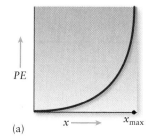
(a)

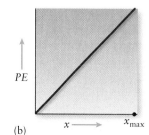
(b)

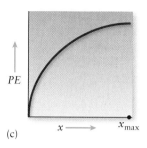
(c)

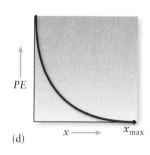
(d)

12. Two identical springs are hung side by side and are attached to the same mass. If each spring has a spring constant of 12 N/m, what is the spring constant of the 2-spring system?
A. 6 N/m
B. 12 N/m
C. 18 N/m
D. 24 N/m

13. After Trial 3 of Experiment 1, an additional mass of 3.0 kg was added to the 0.4-kg mass that was already attached to the system. How much farther did the mass descend when the 3.0-kg mass was added?
A. 0.15 m
B. 0.17 m
C. 0.55 m
D. 0.57 m

Passage II (Questions 74–80)

An electromagnetic railgun is a device that can fire projectiles by using electromagnetic energy instead of chemical energy. A schematic of a typical railgun is shown below.

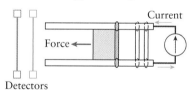

The operation of the railgun is simple. Current flows from the current source into the top rail, through the movable conducting armature into the bottom rail, then back to the current source. The current in the two rails produces a magnetic field directly proportional to the amount of current. This field produces a force on the charges moving through the movable armature. The force pushes the armature and the projectile along the rails.

The force is proportional to the square of the current running through the railgun. For a given current, the force and the magnetic field will be constant along the entire length of the railgun. The detectors placed outside the railgun give off signal when the projectile passes them. This information can be used to determine the exit speed and kinetic energy of the projectile. Projectile mass, rail current, and exit speed are listed in the table below.

Projectile mass (kg)	Rail current (A)	Exit speed (km/s)
0.01	10.0	2.0
0.01	15.0	3.0
0.02	10.0	1.4
0.04	10.0	1.0

74. Which of the following graphs best represents the dependence of exit speed on projectile mass?

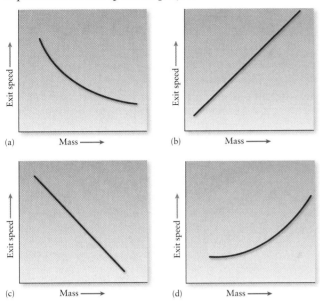

(a)   Mass ⟶

(b)   Mass ⟶

(c)   Mass ⟶

(d)   Mass ⟶

75. What change made to the railgun would reduce power consumption without lowering the exit speeds?
 A. lowering the rail current
 B. lowering the rail resistivity
 C. lowering the rail cross-sectional area
 D. reducing the magnetic field strength

76. Starting from a resting position at the right end of the railgun, the armature applies a constant force of 3.0 N to a projectile with mass of 0.06 kg. How long will it take for the projectile to move 1.0 m?
 A. 0.02 s
 B. 0.04 s
 C. 0.20 s
 D. 0.40 s

77. Lengthening the rails would increase the exit speed because of
 A. an increased rail resistance.
 B. a stronger magnetic field between the rails.
 C. a larger force on the armature.
 D. a longer distance over which the force is present.

78. For a given mass, if the current were decreased by a factor of 2, the exit speed would
 A. increase by a factor of 2.
 B. increase by a factor of $\sqrt{2}$.
 C. decrease by a factor of $\sqrt{2}$.
 D. decrease by a factor of 2.

79. If a projectile leaves a railgun near the surface of Earth with a speed of 2.0 km/s horizontally, how far will it fall in 0.01 s?
 A. 4.9×10^{-4} m
 B. 4.9×10^{-2} m
 C. 9.8×10^{-2} m
 D. 2.0×10^{1} m

80. If a projectile with a mass of 0.1 kg accelerates from a resting position to a speed of 10 m/s in 2 s, what is the average power supplied by the railgun to the projectile?
 A. 0.5 W
 B. 2.5 W
 C. 5.0 W
 D. 10.0 W

Passage 4 (Questions 19-26)

The Hubble telescope was carried into Earth's orbit and released by the space shuttle *Discovery* while orbiting at 7.81×10^{3} m/s. Before releasing the telescope, it was necessary for *Discovery* to achieve an orbit higher than any shuttle had reached in the past. This was because high levels of solar activity had heated Earth's upper atmosphere, causing it to expand outward. In order to avoid atmospheric drag which could cause the telescope to fall back to Earth, the telescope had to be placed as high as possible above the outer fringes of the atmosphere. The Hubble telescope was placed in orbit at an altitude of 6.11×10 m, a distance equivalent to approximately 4.8% of Earth's diameter. This altitude was as high as *Discovery's* chemically powered engines could lift the 1.1×10^{4}-kg telescope.

The Hubble telescope is the largest telescope ever placed into orbit. Its primary concave mirror has a diameter of 2.4 m and a focal length of approximately 13 m. In addition to having optical detectors, the telescope is equipped to detect ultraviolet light, which does not easily penetrate Earth's atmosphere. (Note: Use $g = 9.8$ m/s^2.)

19. When the Hubble telescope is focused on a very distant object, the image from its primary mirror is
 A. real and erect.
 B. real and inverted.
 C. virtual and erect.
 D. virtual and inverted.

20. Assuming that the Hubble telescope is traveling in a circular orbit, which of the following expressions is equal to its orbital speed? (Note: G is the gravitational constant, M_e is the mass of Earth, M_t is the mass of the telescope, and r is the distance of the telescope from the center of Earth.)
 A. $\dfrac{GM_E}{r}$

 B. $\sqrt{\dfrac{GM_E}{r}}$

 C. $\dfrac{GM_t}{r}$

 D. $\dfrac{GM_t}{r^2}$

21. If the acceleration due to gravity were nearly constant from Earth's surface to the altitude of the Hubble telescope, what would the telescope's approximate potential energy be relative to Earth's surface?
 A. 3.3×10^{9} J
 B. 6.6×10^{9} J
 C. 3.3×10^{10} J
 D. 6.6×10^{10} J

22. The magnification of a telescope is determined by dividing the focal length of the primary mirror by the focal length of the eyepiece. If an eyepiece with a focal length of 2.5×10^{-2} m could be used with the primary mirror of the Hubble telescope, it would produce an image magnified approximately how many times?
 A. 10
 B. 96
 C. 520
 D. 960

23. Which of the following is the best explanation of why ultraviolet light does not penetrate Earth's atmosphere as easily as visible light does?

A. Ultraviolet light has a shorter wavelength and is more readily absorbed by the atmosphere.
B. Ultraviolet light has a lower frequency and is more readily absorbed by the atmosphere.
C. Ultraviolet light contains less energy and cannot travel as far through the atmosphere.
D. Ultraviolet light is reflected away from Earth by the upper atmosphere.

24. As the Hubble telescope was transported into orbit aboard *Discovery*, which of the following best describes the main energy conversion(s) taking place?
A. kinetic to gravitational potential
B. gravitational potential to kinetic
C. gravitational potential to kinetic to chemical
D. chemical to kinetic to gravitational potential

25. What is the approximate kinetic energy of the Hubble telescope while in orbit?
A. 4.3×10^7 J
B. 3.3×10^{11} J
C. 6.6×10^{11} J
D. 4.3×10^{12} J

26. The image of a very distant object that is produced by a mirror such as that used in the Hubble telescope will be at what location in relationship to the mirror and focal point?
A. behind the mirror
B. between the mirror and the focal point
C. at the focal point
D. outside the focal point

Passage IV (Questions 94–99)

Retroreflecting arrays consist of spherical beads. Light is refracted as it enters a bead, then reflected off the back of the bead. Arrays of retroreflectors are attached to flexible sheets used as safety reflectors on clothing and bicycles. Ideal retroreflectors return a beam to its source regardless of the angle of incidence of the beam to it. Each light ray is returned on a path no farther than the diameter of a bead from the source ray. Thus, if a distortion that changes the path of light is placed in front of the retroreflecting array, the incident and reflected rays will pass through the same distortion. When this occurs, the ray perfectly retraces itself, thereby canceling the distortion.

An experiment is conducted with the setup illustrated in Figure 3. A light source covered by a screen with a pinhole in it provides a point source of light. A glass pane acts as a beam splitter. Some of the source light is reflected out of the experiment; the remainder of the light is incident upon the array. Some of the light returning from the array is reflected at a right angle to a viewing screen by the beam splitter. The wave front reflected from a retroreflecting array will be irregular. However, a pinhole image will still appear on the screen because the human eye cannot perceive the separation of the wave fronts.

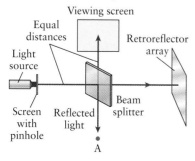

94. The beam splitter in Figure 3 is set at what angle to the incident beam?
A. 15°
B. 30°
C. 45°
D. 60°

95. Which of the following figures shows a possible path for a ray of light passing through a retroreflecting bead? (Note: Assume that the index of refraction of the bead is greater than that of air.)

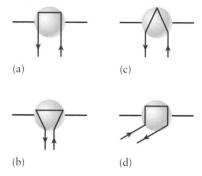

(a)   (c)

(b)   (d)

96. The glass that is used as a beam splitter is replaced with glass that is identical except that it has a 10% higher index of refraction. Which of the following changes will occur to the pinhole?
A. It will move.
B. It will become larger.
C. It will become smaller.
D. It will become clearer.

97. What is the approximate number of wavelengths of light that can travel in one direction within a retroreflecting bead that has a diameter of 5×10^{-5} m? (Note: The speed of light $= 3 \times 10^8$ m/s and its frequency is approximately 10^{15} Hz.)
A. 0.6
B. 1.7×10^2
C. 1.5×10^4
D. 3.3×10^6

98. If a single-wavelength laser is used in place of the light source and pinhole, how will the image on the viewing screen change?
A. It will be larger.
B. It will be less bright.
C. It will be shifted.
D. It will show stronger patterns of interference.

99. With the screen removed, an observer at Point A sees a virtual image of the pinhole. Which of the following indicates that the observer is viewing a virtual image?
A. It moves as the observer changes the angle of view.
B. It is larger than the real image.
C. It is dimmer than the real image.
D. It is fuzzier than the real image.

Answers to Odd Problems

Chapter 16

1. Similarities: The force varies as $1/r^2$; the force is directly proportional to the product of the masses or charges; the force is directed along the line that connects the two particles.
   Differences: Newton's law includes a minus sign, whereas Coulomb's law does not. Like charges repel; like masses always attract. The gravitational constant G is many orders of magnitude smaller than the Coulomb constant k, the gravitational force is many orders of magnitude weaker than the Coulomb force.

3. There would be no difference. The charges were originally arbitrarily defined as positive and negative.

5. The mass decreases because electrons are removed from the object to make it positively charged.

7. Our current understanding is that like charges repel and opposite charges attract. If a charged insulating object either repels or attracts both charged glass (positive) and charged rubber (negative), then you may have discovered a new kind of charge. However, you may also just be observing polarization (to be discussed in later chapters).

9. When the comb is run through your hair, electrons are transferred to the comb. The paper is polarized by the charged comb. The paper is then attracted to the comb. When they touch, a small amount of charge is transferred to the paper so now the paper and comb are similarly charged and repel each other.

11. (a) No, the object could be attracted after being polarized. (b) Yes, to be repelled, the suspended object should be charged positively.

13. (a) If the charges creating the field move, the fact that the field propagates at the speed of light also allows us to understand the changes in the field, and hence force on the other charged objects. With the electric field, we see that charged particles take some time to experience the effects of other charges moving near them. (b) In electrostatics the field is just a computational device, and using it is merely a matter of convenience. However, in electrodynamics the field is necessary for energy and momentum to be conserved, so it is more than just convenience that leads us to the electric field.

15. The electric forces are equal in magnitude but opposite in direction, and the force on the proton is downward while the force on the electron is upward.

17. The electric field is the total electric field due to all charges both inside and outside the Gaussian surface. However, the charges outside the surface create a net zero electric flux through the surface.

19. B

21. C

23. A

25. C

27. B

29. Around 10^{-11} or 10^{-12} C

31. Around 1–10 μC.

33. About 0.05 N·m²/C.

35. 4.6×10^{-18} C

37. 2.7×10^7 C

39. 2.7×10^{13} electrons

41. (a) 2.33×10^5 C/m³; (b) Insulating. If it were conducting, none of the charge would be in the volume of the sphere. Charge would only be on the surface.

43. (a) $-1.25\ \mu$C; (b) $7.8 \times 10^{12}\ e^-$; (c) No more than three steps

45. 5.08 m

47. 0.0110 N to the right

49. -27.3 m

51. $F_A = 36.3$ N at angle $\theta = 24.4°$ below the negative x axis
    $F_B = 43.5$ N at angle $\theta = 10.9°$ below the positive x axis
    $F_C = 25.2$ N at angle $\theta = 67.4°$ above the negative x axis

53. (a) q_1 and q_2 have opposite charges (one is positive and the other is negative). (b) The charge of q_2 is larger in magnitude to counter the effect of being a larger distance away.

55. (a) 6.88×10^7 N/C in the $+x$ direction; (b) -24.1 cm

57. (a) 5.09×10^5 N/C pointing 25.9° above the negative x axis; (b) 8.15×10^{-14} N pointing 25.9° below the positive x axis

59. 2.18×10^6 m/s

61. 0.075 N·m²/C

63. 6.64×10^{-5} C/m

65. $E = \sigma/\varepsilon_0$ pointing radially outward.

67. (a) We're not told the direction of the electric field, so the charge could be positive or negative. We'll assume positive. $Q = 1.78 \times 10^{-11}$ C; (b) 3.54×10^{-9} C/m²

69. $+3.20\ \mu$C on the inner shell, $-9.00\ \mu$C on the outer shell. The field points radially inward with magnitude 6.61×10^7 N/C

71. (a) With 94 protons confined to a small space, they are likely to be concentrated into an approximately uniform sphere of charge, which is equivalent to a point charge for points outside it. (b) 1.1×10^{29} m/s²

73. (a) 1.6×10^7 electrons; (b) $m_{el}/m_{cell} = 1.6 \times 10^{-10}$. So the extra mass is not significant.
    (c) $\sigma = -0.014$ C/m² $= 9.0 \times 10^{16}$ electrons/m²

75. 7.9×10^{-9} C

77. (a) 1.9×10^{-16} C; (b) 1200 electrons

79. 2.16×10^{10} N/C at an angle of 60° with respect to the $+x$ axis

81. 14.2 N/C in the direction of travel of the electron

83. (a) 1.8×10^{21} N/C, outward from the center of the nucleus. (b) 1.5×10^{13} N/C, outward from the center of the nucleus. (c) 2.6×10^{24} m/s², outward from the center of the nucleus.

85. (a) The electric field is pointing downward. (b) Gravity pulls downward on the drop, so the electric force on it must be upward. Since the drop is negative, the electric field must point downward for the force to point upward. (c) 9070 N/C; (d) 7

87. (a) 0; (b) $\dfrac{\sigma_i R_i^2}{\epsilon_0 r^2}$ pointing radially outward;

(c) $\dfrac{\sigma_i R_i^2 - \sigma_o R_o^2}{\epsilon_0 r^2}$ pointing radially outward

Chapter 17

1. The electric potential is the electric potential energy per unit charge and is a scalar. The electric field is the electric force per unit charge and is a vector. The electric potential depends on both the electric field and the region over which the field extends.

3. Both electric field and potential result from the existence of a charge. If a charge exists, it will create both of these physical quantities. In order to have a nonzero potential energy, a force must be able to do work on an object as it is moved. If only one charge exists, then there is no electric force acting on the charge. If there is no force, then that force can do no work, so there is no potential energy.

5. This statement makes sense only if the zero point of the electric potential has been previously defined.

7. A topographical map details the lines surrounding a mountain (for example) where the elevation is the same. These plots let a hiker know how the terrain changes in elevation as one moves from point A to point B. If the lines of elevation are very close together, that will indicate that the mountain is very steep. Walking along these lines of constant elevation is much easier (no change in gravitational potential energy). With an equipotential line, the quantity that stays the same is the voltage. With no change in voltage, there is no change in electric potential energy (hence, no work is done when an electric charge moves along an equipotential line). These lines will never cross (they would be multivalued and that is not possible), and the closer together they are, the greater the electric field is in that region.

9. (a) Yes, a region of constant potential must have zero electric field. (b) No, if the electric field is zero, the potential need only be constant.

11. Increase plate area, decrease separation, and increase the dielectric constant.

13. Connecting capacitors in series allows the total potential difference to be split among the individual capacitors. Every capacitor has a maximum allowable voltage that cannot be exceeded. If the maximum voltage across a capacitor is exceeded, the resulting electric field will cause dielectric breakdown, destroying the capacitor. By connecting the capacitors in series, each capacitor is kept below its maximum voltage.

15. The energy stored in the capacitor increases.

17. For a given potential across them, the capacitors in parallel have a greater total area and so can store a greater amount of charge, thus having a larger capacitance.

19. The larger dielectric strength increases the maximum voltage that can be applied across the electrodes. The larger dielectric constant decreases the voltage required to store a given amount of charge, thus increasing the capacitance. The dielectric material can also be used to maintain the separation of the electrodes.

21. D

23. C

25. C

27. B

29. A

31. 10^6 m

33. 25 V

35. (a) -3.20×10^{-6} J; (b) 3.20×10^{-6} J; (c) The potential difference between the two points would remain the same. However, if a negative charge were placed at rest at point a, then it would never reach point b unless an external force acted upon it. The charge would accelerate in the $-x$ direction, away from point b.

37. 242 V

39. 18 J

41. 28.8 V

43. -2.78×10^{-8} C

45. 7.14 cm and 16.7 cm

47. 2.13×10^6 V

49. (a) 1 V/m to the left; (b) 1 V/m to the left

51.

(a)

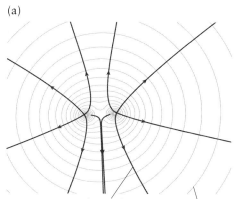

Electric field line   Equipotential line

(b)

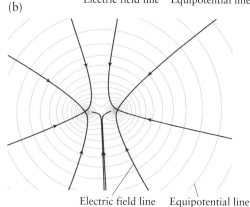

Electric field line   Equipotential line

53.

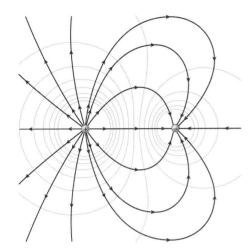

55. 24 μC
57. 8.85×10^{-9} F
59. 1.00 mm
61. 6.37×10^{-7} J
63. 2000 V
65. 1000 J
67. (a) 55.0 μF; (b) 5.00 μF
69. 4.00 μF and 4.00 μF
71. 11.0 μC on the 0.0500 μF capacitor and 22.0 μF on the 0.100 μF capacitor
73. 4.0 μF
75. (a) 7.41 μF; (b) 88.9 μC on each capacitor; (c) $V_{10} = 8.89$ V, $V_{40} = 2.22$, $V_{100} = 0.889$ V
77. 8020
79. 2.15×10^{-7} C
81. (a) $C_{eq} = \dfrac{\epsilon_0 A}{2d}\left(\dfrac{3\kappa_2 \kappa_3}{\kappa_2 + \kappa_3} + \dfrac{\kappa_1}{2}\right)$

(b) $\kappa_3 = 1$ so $C_{eq} = \dfrac{\epsilon_0 A}{2d}\left(\dfrac{3\kappa_2}{\kappa_2 + 1} + \dfrac{\kappa_1}{2}\right)$
83. 1.45×10^7 m/s
85. (a) $U = \dfrac{3kq^2}{a}$

(b) $U = \dfrac{2kq^2}{a} + \dfrac{kq^2}{b}$

(c) $U = \dfrac{5kq^2}{a} + \dfrac{kq^2}{b}$
87. (a) 231 N; (b) Not large enough to move you, but large enough to feel. (c) 2.31×10^{-13} J; (d) 5.04×10^8 m/s; (e) It is faster than the speed of light!
89. 4.4 pF
91. 1.15 pF
93. 7.2 μC
95. 720 μJ
97. (a) C increases by a factor of 4; (b) Q increases by a factor of 4; (c) V remains constant; (d) The stored potential energy increases by a factor of 4; (e) The capacitance will still increase by a factor of 4, the charge will remain constant, V is reduced by a factor of 4, and the energy is reduced by a factor of 4.

99. (a) $C_{\text{before}} = \dfrac{\epsilon_0 A}{d}$, $C_{\text{after}} = \dfrac{\epsilon_0 A}{d - d'}$

(b) The effect is independent of the location of the slab.
101. (a) $Q = 75$ μC, $U = 1.1 \times 10^{-4}$ J
(b) $Q_{20} = 33.3$ μC, $Q_{25} = 41.7$ μC, $U_{20} = 27.8$ μJ, $U_{25} = 34.7$ μJ
103. (a) 2.00×10^6 V/m; (b) $\sigma = 1.77 \times 10^{-5}$ C/m^2; (c) Since the charge remains constant, the surface charge density also remains constant. This means that the electric field also remains constant. The voltage between the plates, however, is reduced because the quantity d is reduced.

Chapter 18

1. Current is the amount of charge passing a point in the circuit per second. In this sense, it is more like a flux than a vector. It has a direction and magnitude, but is not really a vector.
3. There is no contradiction. If a conductor is in electrostatic equilibrium, the electric field within it must be zero. However, any conductor that is carrying a current is definitely not in electrostatic equilibrium.
5. This energy is dissipated as heat in the wire. The amount of energy dissipated is related to the amount of current running through the wire and the resistance of the wire.
7. A small resistance will dissipate more power and generate more heat.
9. The bird will not be electrocuted because there is no potential difference between the bird's feet. The bird grabs only one high-voltage wire, and it is not completing a circuit.
11. The voltage drop across each bulb in the old series string was about 1/50 of 110 V, or 2.2 V. The modern parallel connection puts the full 110 V across each bulb. Placing 110 V across one of the old bulbs, designed to operate at 2.2 V, would result in excessive current in the filament, which would burn out the bulb immediately, perhaps in a spectacular manner.
13. It cannot be changed instantaneously because the resistor in the circuit limits the current (the rate of flow of charge). Because the current is finite, it requires time for the charge to flow on and off of the capacitor plates.
15. A
17. E
19. D
21. C
23. B
25. A laptop computer draws about 2 A of current.
A hair dryer draws about 10 A of current.
A compact fluorescent light bulb draws about 0.2 A of current.

27. About 10 Ω
29. About 100 mA
31. 1 kΩ, 100 μF
33. 1.12×10^{22} electrons
35. 1.09×10^{22} electrons
37. (a) 4.82×10^{3} A; (b) Potassium ions are positive, so the current is in the same direction that the ions flow.
39. 0.0220 Ω
41. 16.0 Ω
43. 0.109 Ω
45. The resistor that carries 112 pA is 10 times larger in diameter than the other.
47. (a) 0.77; (b) 1.3
49. 12,000 K^+ ions
51. 12 Ω
53. 4.0 V
55. 24.0 Ω
57. (a) 0.750 A in each resistor; (b) $V_9 = 6.75$ V, $V_3 = 2.25$ V
59. (a) 0.60 A in each of the 6-Ω resistors and 0.30 A in the 12-Ω resistor. (b) 1.5 A
61. 29 Ω
63. $R_A = 6.0$ Ω, $R_B = 3.0$ Ω
65. 12.5 A
67. 2.7 kW
69. 58 MW
71. $1.32
73. 12.0 s
75. 0.500 A
77. (a) (i) 0.00216 s, (ii) 0.00108 s; (b) (i) 0.0667 A, (ii) 0.211 A
79. 5 Ω: 45 W; 8 Ω: 28 W; 10 Ω: 90 W; 16 Ω: 56 W
81. (a) 30 Ω; (b) 3 hours and 45 minutes longer
83. (a) 16 Ω; (b) 20 Ω: 0.12 A, 12 Ω: 0.20 A, 8 Ω: 0.32 A; (c) 20 Ω: 0.29 W, 12 Ω: 0.49 W, 8 Ω: 0.84 W; (d) It is 18% of the total power dissipation.
85. (a) 4.8 Ω·m; (b) 24 W
87. 39 kΩ
89. (a) 36 μC; (b) 0.060 A; (c) 6.0×10^{-4} s after closing the switch
91. (a) Parallel because both branches (the circuit element and the probe) span the same potential difference; (b) (i) Very large resistance because this will cause the total resistance of the two components in parallel to have approximately the same resistance as the original element, (ii) Very small capacitance so that the time constant is small. The measurements will quickly reach a steady state because many time constants elapse over a short time.
93. (a) 1000 F; (b) 1100 J, or 10^6 times more energy; (c) 17% of the energy of a AAA battery; (d) 0.060 Ω

Chapter 19

1. Use both ends of one iron rod to approach the other iron rods. If both ends of the rod you are holding attract both ends of the other two rods, then the one you are holding is not magnetized iron.

3. The wire is aligned with the magnetic field.
5. (a) The electric field points into the page. (b) The beam is deflected into the page. (c) The electron beam is not deflected.
7. Yes, because the width of the field determines the length of the wire in the magnetic field. The larger the length of wire in the magnetic field, the larger the force exerted on the wire will be.
9. Since the directions of the two currents are opposite, the magnetic field from one wire is opposite to the other. This means the magnetic fields will cancel each other to some extent.
11. There will be no net force on the wires, but there will be a torque. When two wires are perpendicular, each wire will experience a force on one end and another force, equal in magnitude but opposite in direction, on the other end. This results in zero net force. However, the forces will produce a nonzero torque that will make the wires want to align with each other such that the current in each wire is traveling in the same direction as the current in the neighboring wire.
13. C
15. D
17. A
19. A
21. E
23. about 4×10^{-6} T
25. 6 μs
27.

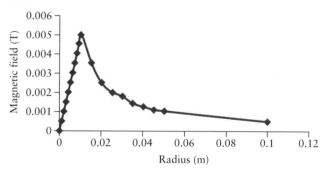

For $r < 0.01$ m: $B(r) = (0.501\ \text{T/m})r$
For $r > 0.01$ m: $B(r) = (0.000050\ \text{T} \cdot \text{m})/r$
29. (a) out of the page; (b) down; (c) down; (d) to the left; (e) out of the page; (f) right
31. 0.7 N
33. 5.8×10^{-18} N to the left
35. 3.4×10^{-12} N in the $-z$ direction
37. 5×10^{-3} T in the $+x$ direction
39. 1.5 μm
41. 0.511 T
43. 0 N
45. 2.5 T
47. 0.664 N·m at 30°
0.755 N·m at 10°
0.493 N·m at 50°
49. B_O and B_P point in the $+z$ direction (out of the page). B_Q and B_R point in the $-z$ direction (into the page).
51. $B = \dfrac{\mu_0 I}{4}\left(\dfrac{R - r}{rR}\right)$ into the page

53. $B = 4 \times 10^{-6}$ T $= 0.08\, B_{Earth}$
55. 120 m
57. 1.03×10^{-8} N away from wire 2
59. 0.313 N
61. 0.0109 T
63. 9.12×10^{-31} kg
65. (a) $B = 1.0$ T at $r = 1.0$ m, $B = 0.001$ T at $r = 1.0$ km. Compared to Earth's magnetic field, the lightning bolt has a magnetic field 20,000 times as large at 1.0 m and 20 times as large at 1.0 km. (b) $B = 2.0 \times 10^{-6}$ T at $r = 1.0$ m, $B = 2.0 \times 10^{-9}$ T at $r = 1.0$ km. In each case, the field due to the lightning is 500,000 times greater than the household current. (c) $r = 2.0$ mm
67. 0.40 N
69. (a) 298 A; (b) 0.62 T; (c) The easiest way to achieve the same field with less current is to increase the number of windings, N.
71. (a) 15.9 mA; (b) The loop will initially rotate to point its dipole moment in the direction of the magnetic field. At this instant, there is no net torque on the loop, but because of its rotational momentum it will continue to rotate past this point, slowing down. After it momentarily comes to rest, it will rotate back toward its initial position. If the system is frictionless, this harmonic oscillation will continue indefinitely.
73. 650 V
75. $B = \dfrac{\mu_0 i}{2\sqrt{2}\pi d}$ pointing down and to the right so that the vector is 45° below a horizontal line pointing to the right.
77. $B = 1.5 \times 10^{-6}$ T $= 0.031\, B_{Earth}$. Since the field is so much smaller than the Earth's magnetic field, there should be little or no cause for concern.
79. (a) $B = 0.7155 \dfrac{\mu_0 Ni}{R}$ in the $+x$ direction.

    (b) $B = 0.221 \dfrac{\mu_0 Ni}{R}$ in the $+x$ direction.
81. (a) 2.64×10^{-6} T pointing westward; (b) 2.64×10^{-7} T pointing vertically upward; (c) 2.64×10^{-6} T pointing southward; (d) In (a) and (c), B is about 5% of B_{Earth}. This is fairly small, but it could possibly be enough to interfere with navigation. In part (b), B is about 0.5% of B_{Earth}, which seems too small to cause much of a problem.

Chapter 20

1. The falling bar magnet induces a current in the wall of the pipe. In accordance with Lenz's law, the direction of this induced current is such that it produces a magnetic field that exerts a force on the magnet opposing its motion. The speed of the magnet therefore increases more slowly than in free-fall until it reaches a terminal speed at which the magnetic and gravitational forces are equal and opposite. After this point, the magnet continues to fall, but at a constant speed.

3. (a) The magnetic coil produces a rapidly varying magnetic field that flows through the plate. From Lenz's law, this induces a current in the plate such that the magnetic field of the plate opposes the field of the magnetic coil, repelling the plate and causing it to levitate. (b) The large induced current in the plate dissipates a large amount of power in the plate, heating it up. (c) If the plate were an insulating material, this trick would not work because a current could not be induced.
5. A current is induced in loop b while the current in loop a is changing. If the current is in the direction shown and is increasing, the flux of its magnetic field through loop b is upward and increasing. In accordance with Lenz's law, the direction of the induced current in loop b is such that the flux of its magnetic field through loop b is downward, opposing the change in flux that produced it. This means that the current in loop b is in the opposite direction as the current in loop a, and the two loops repel. After the current in loop a stops changing, then the current in loop b becomes 0 and there is no force between the loops.
7. D
9. C
11. A
13. About 16,500 A
15. About 5 V/km or 15,000 V total. This is such a large variation that it would cause the cable to exceed the maximum current allowed and cease to function.

17.

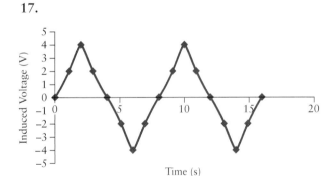

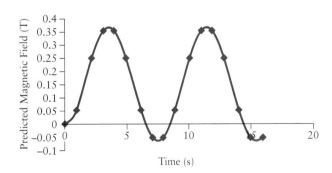

19. (a) 0.204 Wb; (b) 0.220 V
21. (a) counterclockwise; (b) clockwise; (c) clockwise; (d) counterclockwise; (e) clockwise; (f) no induced current

23.

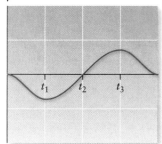

25. (a) For part a and part c, $F = 0$ on all segments.
For part b, using x to represent the length of the
given segment still in the magnetic field, $F = 0$ on the
right-hand segment, $F = (0.0790 \text{ N/m})x$ in the upward
direction for the top segment, $F = (0.0790 \text{ N/m})x$ in
the downward direction for the bottom segment, and
$F = 0.0790 \text{ N}$ to the left for the left-hand segment.
(b) The current will be counterclockwise as viewed.
The force on the left-hand segment will be zero.
The force on the top and bottom segments will be
$F = (0.0790 \text{ N/m})x$ downward and upward,
respectively. The force on the right-hand segment will
be $F = 0.0790 \text{ N}$ to the left.

27. 380 V

29. 1.9×10^6 rev/s. This is nearly 2 million rev/s, which
does not seem feasible.

31. (a) 0.12 V; b) 0.0047 A; (c) counterclockwise

33. 2.1 V

35. (a) 0.01 A; (b) up; (c) 0.001 N

Chapter 21

1. It means that the current peaks one-fourth of a period
after the voltage drop peaks. One-fourth of a period
corresponds to a phase difference of 90°.

3. Ohm's law ($V = iR$) is not precisely valid for AC circuits.
But if you use the rms voltage and the rms current, it
can still be used to describe a resistor in a circuit.

5. For example:

Country	V_{rms}	Frequency (Hz)
Angola	220	50
Botswana	231	50
Ecuador	120–127	60
Mexico	127	60
Slovenia	220	50

7. No. These devices usually work over a range of
voltages and will not function properly if a transformer
is used. They contain "smart" circuitry that senses if the
voltage is 120 V or 240 V. (Many people needlessly buy
transformers for their laptops. What may be needed,
depending on destination, is an adapter plug.)

9. PCBs were widely used for many applications,
especially as dielectric fluids in transformers,
capacitors, and coolants. PCBs have low water
solubility ($0.0027 - 0.42$ ng/L for Aroclors) and low
vapor pressures at room temperature, but they have
high solubility in most organic solvents, oils, and fats.

They have high dielectric constants, very high thermal
conductivity, and high flash points (from 170 to 380°C)
and are chemically fairly inert, being extremely resistant
to oxidation, reduction, addition, elimination, and
electrophilic substitution.

11. The electrical energy stored in the capacitor is
$U_E = \dfrac{q^2}{2C}$, which varies as $\cos^2 \omega t$, and the magnetic
energy stored in the inductor is $U_B = \dfrac{Li^2}{2}$, which varies
as $\sin^2 \omega t$. Therefore, when a maximum amount of
energy is stored in the capacitor, no energy is stored in
the inductor, and vice versa.

13. The capacitor has energy $U_E = \dfrac{q^2}{2C}$ and the inductor
has energy $U_B = \dfrac{Li^2}{2}$. We see that both energies are
proportional to the square of a quantity related to
charge (either i or q). However, the magnetic energy
stored is proportional to the inductance, but the electric
energy stored is inversely proportional to the
capacitance.

15. (a) Self-inductance drops by a factor of 2;
(b) self-inductance increases by a factor of 2

17. Impedance describes the overall opposition to the flow
of charge in a circuit driven by a time-varying voltage.
The SI units of impedance are ohms (Ω).

19. D

21. A

23. C

25. C

27. Most household appliances draw between 1 and
20 amps. Most households have a main circuit breaker
that "trips" at 150 A or so.

29. The voltage is graphed in red, oscillating between +12
V and −12 V. The current is shown in black, and
oscillates between +0.001 A and −0.001 A.

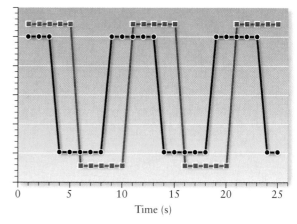

Time (s)

31. $V(t) = 120\sqrt{2}\text{V} \sin{(120\pi t)}$

33. 11.6 V

35. 849 V

37. $V_{max} = \sqrt{2} i_{rms} R$
(a) 0.354 A; (b) 707 V; (c) 156 Ω

39. 22 turns

41. 2 A rms

43. (a) 21 turns; (b) 0.788 A

45. (a) 42; (b) $i_s = 0.0150 A$, $i_p = 0.625 A$; (c) 192 Ω

47. 0.19 mH

49. (a) 9.65 mH; (b) The inductance doubles, assuming the number of turns and the length are held constant because the inductance is proportional to A: $\left(L = \dfrac{\mu N^2 A}{\ell} \right)$.

51. 41 Hz

53. 286 H/F

55. 1/4

57. 5.07 nF

59. 3.2 A

61. (a) 4.50 μF; (b) The rms current will increase. As ω is increased, X_c is reduced, so i_{rms} will increase.

63. (a) $i = 0.566$ A, $P = 16.0$ W; (b) $i = 1.45$ A, $P = 0$ W; (c) $i = 1.78$ A, $P = 0$ W

65. $V_R = 1.00$ V, $V_C = 10.3$ V, $V_L = -1.25$ V

67. 590 Ω, 65°

69. (a) 690 Ω; (b) 0.15 A; (c) $I(t) = (0.15\ \text{A}) \sin (1000\ \pi t + 1.56)$

71. 104.5 MHz, $i = 0.009$ A

73. 5.07 nF

75. (a) $\kappa = 6.03 \times 10^7$; (b) Table 17-1 tells us that this is not a feasible dielectric constant since it is much greater than that of ordinary materials. However, the required capacitance ($C = 670\ \mu$F) is easily within the range of ultracapacitors.

77. For DC, $V = 60$ V. For AC, $V_{rms} = 15$ V and $V_0 = 21$ V

79. (a) $i_0 = 1.53$ A, $V_0 = 13.0$ V; (b) 10.0 W; (c) 3.60×10^4 J

81. (a) $V_R = 14.9$ V, $V_C = 53.9$ V, $V_L = 23.4$ V; (b) $V_R = 33.9$ V, $V_C = 80.9$ V, $V_L = 80.9$ V; (c) The sum of the voltages is larger than the applied voltage because the maximum voltages in each circuit element are not reached simultaneously. Since each element reaches its maximum voltage at a different time, the sum of the maximum voltages is not required to equal the maximum applied potential. (d) At resonance, the impedance of the capacitor is equal to the impedance of the inductor. Both circuit elements have impedance of $\sqrt{L/C}$. Therefore, the maximum voltage of each circuit element will be the same.

Chapter 22

1. (a) e, c, b, d, a; (b) Gamma rays have the highest energy per photon. Microwaves have the lowest energy per photon.

3. The electric field is perpendicular to the magnetic field and the velocity of the wave. That means, for example, that the electric fields will vibrate along the x axis, the magnetic field along the y axis, and the EM radiation will move along the z axis. These three vectors are mutually perpendicular to each other.

5. The speed of light is proportional to the frequency ($c = \lambda f$).

7. All EM waves move at the same speed ($c = 3.00 \times 10^8$ m/s). However, the wavelength times the frequency equals this constant ($c = \lambda f$). Therefore, the wavelength can vary (smaller/larger) in proportion to the frequency (larger/smaller).

9. For example, using visible light to see in your house, using wireless Internet transmitted as radio waves, and using microwaves to heat food in a microwave

11. C

13. D

15. E

17. 1.3 s

19. 2.25 eV

21. About 10 nm

23. People 300 km away receive the news first. $t_{radio} = 0.001$ s and $t_{sound} = 0.0088$ s.

25. (a) 7.25×10^{-8} m, UV; (b) 4.29×10^{-7} m, Purple Visible; (c) 3.75×10^{-9} m, X ray; (d) 1.00×10^{-5} m, IR; (e) 3.33×10^{-5} m, IR; (f) 8.72×10^{-10} m, X ray; (g) 3.65×10^{-8} m, UV; (h) 5.00×10^{-8} m, UV

27. (a) 4.29×10^{14} Hz; (b) 5.00×10^{14} Hz; (c) 6.00×10^{14} Hz; (d) 7.50×10^{14} Hz; (e) 3.00×10^{15} Hz; (f) 9.01×10^{18} Hz; (g) 6.00×10^{11} Hz; (h) 4.47×10^{18} Hz

29. 2.78 m to 3.41 m

31. 3 m

33. 1.28 s

35. (a) 2.09×10^7 m; (b) $\omega = 3.77 \times 10^{15}$ rad/s, $f = 6.00 \times 10^{14}$ Hz, $\lambda = 5.00 \times 10^{-7}$ m

37. (a) 1.26×10^{11} m^{-1}; (b) $12\pi \times 10^{18}$ rad/s; (c) 4.99×10^{-11} m; (d) 6.02×10^{18} Hz

39. 2.8 A

41. 1.0×10^{-5} T

43. (a) 6.67×10^{-7} T; (b) 53.1 W/m^2

45. (a) $\lambda = 853$ nm, $f = 3.52 \times 10^{14}$ Hz; (b) $\lambda = 442$ nm, $f = 6.79 \times 10^{14}$ Hz; (c) $\lambda = 621$ nm, $f = 4.83 \times 10^{14}$ Hz; (d) $\lambda = 232$ nm, $f = 1.29 \times 10^{15}$ Hz; (e) $\lambda = 19.6$ nm, $f = 1.53 \times 10^{16}$ Hz; (f) $\lambda = 142$ nm, $f = 2.11 \times 10^{15}$ Hz; (g) $\lambda = 627$ nm, $f = 4.78 \times 10^{14}$ Hz; (h) $\lambda = 273$ nm, $f = 1.10 \times 10^{15}$ Hz

47. (a) 62.1 nm to 414 nm; (b) It extends from violet visible light into the UV.

49. (a) 0.45 m; (b) 101,000 J/m^3

Chapter 23

1. Christiaan Huygens suggested that every point along the front of a wave be treated as many separate sources of tiny "wavelets" that themselves move at the speed of the wave. This is important to see how light moves from one medium to another, different, medium and allows you to predict the bending that occurs in Snell's law.

3. The light coming from the Sun is refracted as it passes through Earth's atmosphere. Because of the wavelength dependence of the index of refraction, red light bends less than orange than yellow than green than blue than violet. In addition, bluer light is scattered more than red light as it passes through the atmosphere. Because of these effects, more red/orange colored light can pass into the shadow of the Earth, making the Moon appear red.

5. The swimming pool seems shallower. If you follow a light ray that bounces off the bottom of the swimming pool and ends up entering your eyes, the light exits the water at an angle from the normal larger than the angle at which it arrived at the water–air boundary. This is because the index of refraction of water is greater than that of air. Your brain assumes that light travels in a straight line, so you interpret the origin of the ray on the floor of the swimming pool to be along the line of the light ray as it enters your eyes. The distance you measure to the origin of the light ray is fixed by your stereo vision, so the swimming pool appears to be shallower than it really is.

7. When light moves from a medium of larger index of refraction toward a medium with a smaller index of refraction, the angle of refraction will *increase*.
   If you make the incident angle larger and larger, the refracted angle will ultimately approach 90°. After the light moves past this critical value, it is reflected rather than refracted. In mathematical terms, $\sin\theta$ cannot be larger than 1. When Snell's law would require the sine of the angle of refraction to be greater than one, total internal reflection will occur.

9. When light enters a material with a negative index of refraction, the light refracts "back" away from the normal to the surface as seen in the picture below:

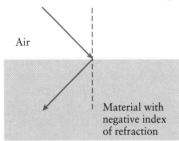

11. 1. Soap bubbles reflect different colors. 2. Thin film coatings on photographic lenses ("nonreflective coatings"). 3. Oil floating in a puddle of water will show different colors

13. Yes, diffraction occurs for all wavelike phenomena. Sound waves are diffracted through an open window, and water waves are diffracted around an obstacle in the water, for example.

15. Sunlight includes all of the colors of the rainbow, and one of the delightful aspects of a diamond is its strong dispersion. The diamond bends violet light much more than red light so that sunlight is separated into its different colors as it passes through the diamond surface.

17. D

19. B

21. B

23. A

25. A

27. air–diamond: $\sim(1.0$–$2.4)$

29. (a) 300,000,000 m; (b) 2×10^{10} m; (c) 9.5×10^{15} m

31. $n_2/n_1 = 1.56$

33. 1.00029

35. 763 km

37. (a) 13.2°; (b) 22.1°; (c) 30.9°; (d) 20.1°

39. (a) 41.8°; (b) 48.8°; (c) 58.5°; (d) no critical angle

41. 41.8°

43. 48.8°

45. 59.1°

47. (a) 59.1°; (b) 0.00793 cm

49. 0.0534 cm

51. 60° and 120°

53. 150 W/m²; the light is polarized 75° from the vertical

55. 40.9°

57. 36.9°

59. 452 nm

61. 648 nm

63. $t = 93.8$ nm; $t = 282$ nm; $t = 469$ nm; $t = 657$ nm

65. $\lambda = 631$ nm and $\lambda = 473$ nm are not seen in the film.

67. $\theta_1 = 36.4°$, additional fringes are not visible.

69. 583 nm

71. 633 nm

73. 17 mm

75. 7.71×10^{-6} degrees $= 1.34 \times 10^{-7}$ radians

77. 3.46×10^{-5} m

79.

Optical material with percentage of time required for light to pass through compared to an equal length of vacuum	Speed of light	Index of refraction
100%	3.00×10^8 m/s	1.00
125%	2.40×10^8 m/s	1.25
150%	2.00×10^8 m/s	1.50
200%	1.50×10^8 m/s	2.00
500%	6.00×10^7 m/s	5.00
1000%	3.00×10^7 m/s	10.00

81. $n_1 \sin(\theta_1) = n_n \sin(\theta_n)$ is true

83. 40.5°

85. (a) 0.969 m; (b) You would have to change the index of refraction of water to be smaller than the index of refraction of air. Under this condition, total internal reflection cannot occur.

87. 9.14 m

89. (a) 9.38 W/m²; (b) 45°

91. 902 nm

93. 188 μm

95. (a) 1.77; (b) 310 nm, 465 nm, and 620 nm

97. (a) 500 mi/hr; (b) 1.2 hr; (c) The formula $w\sin\theta = m\lambda$ applies only if the wavelength is smaller than the slit width. In this case, $w = 100$ mi and $\lambda = 500$ mi, so the formula would not apply.

99. 9.35×10^{-4} rad. This is larger than the pupil, which makes sense because the pupil is larger in diameter than the hole.

101. (a) Ring 1 occurs at 3.05×10^{-4} rad $= 1.75 \times 10^{-2}$ degrees, ring 2 occurs at 5.58×10^{-4} rad $= 3.19 \times 10^{-2}$ degrees, ring 3 occurs at 8.10×10^{-4} rad $= 4.64 \times 10^{-2}$ degrees. (b) $x_1 = 7.6 \times 10^{-3}$ mm, $x_2 = 1.4 \times 10^{-2}$ mm, $x_3 = 2.0 \times 10^{-2}$ mm. (c) The dark rings, and hence the bright rings, are so close together that the bright rings are essentially right next to each other and mask the dark rings, thus eliminating the diffraction effect and giving us only bright light.

103. (a) about 55–60 cm; (b) Telescopes are built larger than the diffraction-limited diameter for greater light-collecting power (it allows you to see dimmer objects).

105. (a) 13.1 cm; (b) 96.9 cm. Since the resolution with atmospheric turbulence is less than the diffraction

limited resolution, the atmospheric resolution is the limiting factor, so the telescope can resolve only objects separated by about 1 m.

Chapter 24

1. Actually, a plane mirror does neither, but instead inverts objects back to front. If the mirror inverted right and left, then the object's right hand that points east would appear on the image as a right hand pointing toward the west. Because the image is inverted back to front, the object facing north is transformed into an image that faces south. Also, the object's right hand is transformed into a left hand in the image.

3. A real image forms where light rays come together but no light rays actually meet where a virtual image forms.

5. This phrase refers to the fact that the curved mirrors produce images that are not located at the same point as an image in a plane mirror. In fact, the mirrors are designed so that if you see a vehicle in the rearview mirror, you should not change lanes because they are too close.

7. Additional information is needed. In accordance with the lensmaker's equation, if the radius of curvature of the front surface is the larger, it is a diverging lens; if the radius of curvature of the front surface is the smaller, it is a converging lens.

9. The full image formed, but it is dimmer than before because it is formed with half as much light.

11. Since the brain can be "trained" to interpret all the nerve inputs that it receives, it is feasible that upright and inverted could be "redefined." It must be very disconcerting, however, for those several days when "up is down and down is up!"

13. The condition known as presbyopia (farsightedness) is associated with a weakening of the muscles around the eye and inflexibility in the crystalline lens system as a person ages. This loss in adaptive amplitude leads to the inability to focus on objects that are close up. Myopia, on the other hand, is a disorder that affects as many as 50% of the world's population, but it is not brought on by the aging process.

15. B

17. A

19. B

21. B

23. B

25. 3 m

27. For example, a shiny spoon (f about 5 cm), a shiny salad bowl (f about 15 cm), and a reflector in a flashlight (f about 2 cm).

29. About 22.86 cm

31. 60°

33. 0.900 m

35. The first five images to the left are 3 m, 17 m, 23 m, 37 m, and 43 m to the left of the left mirror. The first five images to the right are 7 m, 13 m, 27 m, 33 m, and 47 m to the right of the right mirror.

37.
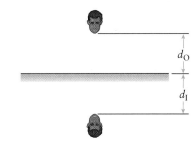

39. The camera will focus the lens on the surface of the mirror (2 m away from the virtual image). It will be out of focus a little bit.

41. As long as $d_O > f$, the image will be real in a concave, spherical mirror. In this case, the focal point is located between the mirror and the object.

43. (a)
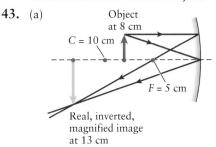

(b) $d_I = 13$ cm, $m = -1.7$. The image is real and inverted.

45. (a)
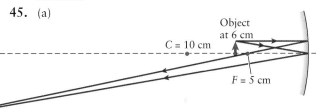

(b) $d_I = 30$ cm, $h = -5.0$ cm

47. (a) $d_I = 30$ cm, $h = 60$ cm, the image is real and inverted; (b) $d_I = 12$ cm, $h = 12$ cm, the image is real and inverted; (c) $d_I = 8.11$ cm, $h = 1.62$ cm, the image is real and inverted

49.
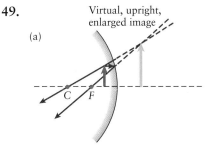

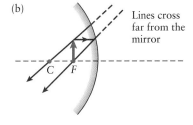

(c)

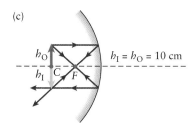

51. In both cases, the images are virtual. The image of an "up close" object is larger than the image of the same object "far away." In both cases the image is smaller than the object.

53. A convex mirror is the opposite of a concave mirror. Use a phrase such as "The cave goes in" to remember the shape of a concave mirror.

55. (a) $d_I = -6.67$ cm, $h = 3.33$ cm. The image is virtual and upright.

    (b) $d_I = -8.33$ cm, $h = 1.67$ cm. The image is virtual and upright.

    (c) $d_I = -9.09$ cm, $h = 0.909$ cm. The image is virtual and upright.

57. $d_I = -4.29$ m, $m = 0.429$. The image is virtual and upright.

59. $d_I = -0.962$ cm, length = 0.926 cm. The image is virtual and upright.

63. $d_I = -6.32$ cm, $h = 0.947$. The image is virtual and upright.

65. No, a real image in a converging lens occurs on the opposite side of the lens from the object.

67. (a) Nearsightedness is corrected by a diverging lens.
    (b) Farsightedness is corrected by a converging lens.

69. (a) $d_I = -6.7$ cm, $h_I = 13$ cm. The image is virtual and upright.

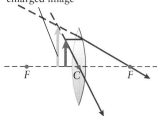

Upright, virtual, enlarged image

    (b) $d_I = -20$ cm, $h_I = 20$ cm. The image is virtual and upright.

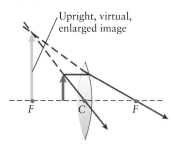

Upright, virtual, enlarged image

(c) No image is formed.

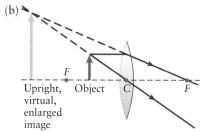

Parallel lines do not cross. No image formed.

(d) $d_I = 33$ cm, $h_I = 6.7$ cm. The image is real and inverted.

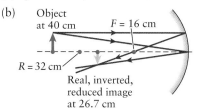

Inverted, real, reduced image

71. (a) $d_I = -45$ cm, $h_I = 5.0$ cm. The image is virtual and upright.
    (b)

Upright, virtual, enlarged image

Object

73. (a) $R = 0.191$ m; (b) $f = 31.8$ cm
75. 360
77. (a) $d_I = 26.7$ cm, $m = -0.668$
    (b)

Object at 40 cm $F = 16$ cm

$R = 32$ cm

Real, inverted, reduced image at 26.7 cm

79. (a) at minimum power: $f_{min} = 8.72$ mm, $P_{min} = 115$ D; at maximum power: $f_{max} = 6.73$ mm, $P_{max} = 150$ D; (b) 6.92 mm; (c) This image would not fall on the retina of the human eye.

81. 1.67 cm

83. 2.16

85. (a) The image is 750 cm from the mirror on the same side as the object. (b) The image is real and inverted. (c) The image is now 15.3 cm past the mirror on the opposite side from the lens. The image is real and upright.

87. (a) The final image is 43.75 cm behind the mirror. The image is virtual and inverted. $h = 10.5$ cm. (b) The final image is 116.1 cm behind the mirror. The image is virtual and erect. $h = 3.00$ cm.

89. (a) 3.1 mm in front of the retina. (b) Reflection is not affected by the index of refraction, so the answer would be the same as in part (a).

91. 2.67 D

93. (a) Both the near point and far point are too close, so the patient needs bifocals to correct both problems. (b) $f_{near} = 37.5$ cm, $f_{distant} = 275$ cm; (c) $P_{near} = -2.67$ D, $P_{distant} = -0.364$ D

95. (a) 203 mm; (b) 0.51 cm; (c) 27.5 mm

97. (a) -790; (b) The negative sign tells us that the image is inverted when viewed in the eyepiece.

Chapter 25

1. There is always a great deal of resistance to change, in any context, but especially in academia. Physics in this era (c. 1900) was predicated on Newtonian theories regarding motion, gravity, and cosmology. In addition, Maxwell's theory of electromagnetism was firmly in place. These two scientists were held in the highest esteem and were "above reproach." It was heretical for anyone, much less a young, upstart physicist who was not even part of a university, to challenge these time-tested, universally accepted underpinnings of physics. On a much more pragmatic level, it is always a "tough sell" when the new theory is complicated and not intuitive.

3. A frame of reference or reference frame is a coordinate system with respect to which we will make observations or measurements. An inertial frame is one that moves at constant speed relative to another; that is, we refer to a frame of reference attached to a nonaccelerating object as an inertial frame.

5. It means that there is no preferential frame of reference that is "fixed in place." Even the velocity of a car relative to the Earth can be rephrased as the velocity of the Earth relative to the car!

7. If the object is moving relative to you, you measure its length by finding the difference between the coordinates of its endpoints at the same time.

9. Yes, it is possible. See Equation 25-15. If two light bulbs at opposite ends of a meter stick flash simultaneously as measured in a frame at rest relative to the meter stick (one bulb at $x = 0$ and the other at $x = 1$ m), then the order of events can change depending on the circumstances of the observer. An observer moving in the $-x$ direction would measure the bulb at 1 m flashing after the bulb at 0 m, while an observer moving in the $+x$ direction would measure the bulb at 0 m flashing after the bulb at 1 m.

11. Writers like to use travel through higher dimensions, which could presumably shorten the distance traveled, making it appear that the traveler has moved faster than the speed of light. They also often employ wormholes, or connections between two points in space–time, created when the space–time "surfaces" become deformed enough to "touch" each other and become physically connected. These sorts of deformations are possible (at least in theory) where high-mass objects like black holes exist.

13. One important prediction of general relativity is that light bends as it passes through a gravitational field.

15. D

17. D

19. A

21. 0.417c

23. Yes. If $v/c = 0.6$, then $\gamma = 1.25$. If $v/c = 0.8$, then $\gamma = 1.67$.

25. About 5% of the world population

27.

v/c	γ	v/c	γ
0	1.0000	0.7	1.4003
0.1	1.0050	0.8	1.6667
0.2	1.0206	0.9	2.2942
0.3	1.0483	0.99	7.0888
0.4	1.0911	0.999	22.3663
0.5	1.1547	0.9999	70.7124
0.6	1.2500		

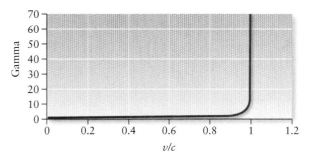

(a) 0.1; (b) $v/c = 0.1$, slope $= 0.102$; $v/c = 0.5$, slope $= 0.770$; $v/c = 0.9$, slope $= 10.9$; $v/c = 0.999$, slope $= 11,200$

29. (a) $x' = x - 4t$, $y' = y$, $z' = z$; (b) $x' = -14$ m, $y' = 1$ m, $z' = 0$ m

31. (a) $x' = x - 15t$, $y' = y$, $z' = z$; (b) $x'' = x - 4t$, $y'' = y$, $z'' = z$; (c) $x' = x'' - 11t$, $y' = y''$, $z' = z''$

33. (a) $x = 12$ m, $y = 4$ m, $z = 6$ m, $t = 4 \times 10^{-3}$ s; (b) $x = 8$ m, $y = 4$ m, $z = 6$ m, $t = 4 \times 10^{-3}$ s

35. 39.1 m/s in the direction of 39.8° north of the east direction

37. The massive slab of marble was used to keep any vibrations from disrupting the light as it moved through the interferometer. The slab needed to be rotated at different points in the experiment, so the mercury made it easier to move.

39. (a) 6.61×10^{-7} μs; (b) 0.00667 μs; (c) 0.675 μs; (d) 23.8 μs; (e) 396 μs

41. 1.67

43. 2.19 μs

45. 0.9999999995c

47. 1.5 ns

49. 1.92 m

51. 0.980 m

53. $v = 2.99 \times 10^8$ m/s, $v/c = 0.997$

55. The velocity of the truck relative to the car is $-0.110c$. The velocity of the car relative to the truck is $0.110c$.

57. 0.548c

59. 0.99999999999989c

61. (a) 1.35×10^{-22} kg·m/s; (b) 9.51×10^{-15} J; (c) 8.20×10^{-14} J; (d) 9.15×10^{-14} J

63. 0.866c

65. 0.209c

67. 8.2 m/s² in the upward direction
69. 3.29 μs
71. (a) 0.946c; (b) 0.946c; (c) Consider part a. If the nucleus is "Earth," then the laboratory goes by at a speed of 0.8c, which corresponds to rocket B in part b. The nucleus emits an electron in the forward direction, at a speed of 0.6c, and this electron corresponds to rocket A in part b. In part a, we determined the speed of the electron relative to the lab frame, which is then equivalent to the speed of rocket A relative to rocket B in this problem. Since two objects must always have the same speed relative to each other, our calculation of the speed of rocket B relative to rocket A better give the same answer.
73. (a) The muon does not make it to Earth's surface. (b) 0.977c
75. 6×10^{15} m
77. 0.986c
79. 0.252 mg
81. (a) The proper time is the time measured by the particle because the two events occur at the same location in that reference frame. The proper length is in the Earth reference frame because the atmosphere and Earth are at rest in that frame. (b) 40.4 μs. (c) 92.6 μs
83. (a) 46.3 beats/min; (b) 70 beats/min
85. $K_{Newton} = 3.645 \times 10^{19}$ J, $K_{Einstein} = 1.647 \times 10^{20}$ J, % error = 69%. The Newtonian expression underestimates the kinetic energy.
87. (a) 0.77c; (b) 96.7 years

Chapter 26

1. The de Broglie wavelength decreases as the kinetic energy increases.
3. The electron in the Bohr atom would be constantly accelerating, and so it would be constantly radiating. As the electron radiates, it would lose energy and slowly spiral into the nucleus.
5. The shortest wavelength is the wavelength that is emitted when a free electron is captured into the lowest energy state. This wavelength is 91.2 nm.
7. (a) As long as the radiation has a frequency above the cutoff value, the greater the intensity the larger the flux of photoelectrons that flow off of the metal plate. If the frequency of the light is below the cutoff, it does not matter how much the intensity is increased. No photoelectrons will be emitted. (b) If the wavelength is increased, the energy of the radiation will decrease. If it slips below the work function of the metal plate, the photoelectric effect will cease. (c) If the work function of the metal is increased, it will eventually grow past the energy of the incident photons and the effect will cease.
9. The exact amount of energy required to free an electron from a surface depends on a variety of factors, for example, the energy level of tht electron and the electron's motion at the moment it is struck by a photon. The kinetic energy of any electron is the difference between the energy of a photon and this varying energy required to liberate the electron.

11. Visible light can Compton scatter. However, because the Compton wavelength is so small compared to the wavelength of visible light (0.0024 nm compared to 400 to 750 nm), the change in the wavelength of a visible photon would be negligibly small.
13. The electron has the longer wavelength.
15. Hydrogen is the first, most elemental, simplest element in the periodic table. Any cogent theory of atoms must describe hydrogen first and then work its way up to the larger, more complicated elements. In addition, because hydrogen only has one electron, it presents no complications to the orbit of an electron due to the interaction of other electrons.
17. Because the rest mass of a macroscopic object is so large, the de Broglie wavelength is too small to observe. The wavelength of everyday objects is many orders of magnitude less than the radius of an atom. This is far too small for us to observe.
19. D
21. A
23. A
25. C
27. 5–25 eV; this is 10^5 times smaller than the nuclear binding energy.
29. Between 500 and 1000 W
31. The electrostatic force is 10^{39} times greater than the gravitational force.
33. 7.3×10^6 m/s
35.

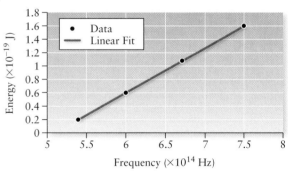

Using this data, we calculate $h = 6.8 \times 10^{-34}$ J·s = 4.12×10^{-15} eV·s
37. 4310 nm
39. 9670 nm
41. 6110 K
43. $f_{violet} = 7.85 \times 10^{14}$ Hz, $f_{red} = 3.85 \times 10^{14}$ Hz; $E_{violet} = 3.27$ eV, $E_{red} = 1.59$ eV
45. $E = 6200$ eV = 9.94×10^{-16} J
47. $E = 225$ eV = 3.61×10^{-17} J
49. (a) $\phi = 2.27$ eV; (b) 546 nm
51. Ca, K, and Cs
53. (a) 1.21×10^{-27} kg·m/s = 2.25 eV/c; (b) 9.32×10^{-24} kg·m/s = 17400 eV/c
55. 25.8°
57. (a) λ = 0.140 nm, K = 0 eV
    (b) λ = 0.14033 nm, K = 20.8 eV
    (c) λ = 0.14071 nm, K = 44.7 eV
    (d) λ = 0.1412 nm, K = 75.3 eV
    (e) λ = 0.14243 nm, K = 151 eV
    (f) λ = 0.14486 nm, K = 297 eV
59. 0.137 nm

61. (a) $f = 4.22 \times 10^{18}$ Hz, $E = 17440$ eV;
(b) 0.0735 nm; (c) 16,860 eV; (d) 576 eV

63. 3.32×10^{-10} m

65. $\lambda = \dfrac{h}{\gamma m v}$

67. (a) 1.23×10^{-9} m
(b) 3.88×10^{-10} m
(c) 1.23×10^{-10} m
(d) 3.88×10^{-11} m
(e) 8.80×10^{-13} m
(f) 1.24×10^{-15} m

69. 1.43×10^{-10} m

71. $\lambda = \dfrac{h}{\sqrt{2Km}}$

73. 5.91×10^{-6} eV

75. (a) 6 lines; (b) 45 lines

77. $n_i = 6 \rightarrow n_f = 2$, $\lambda = 411$ nm
$n_i = 5 \rightarrow n_f = 2$, $\lambda = 434$ nm
$n_i = 4 \rightarrow n_f = 2$, $\lambda = 486$ nm
$n_i = 3 \rightarrow n_f = 2$, $\lambda = 656$ nm

79. $\lambda_{\text{Lyman}} = 91.1$ nm
$\lambda_{\text{Balmer}} = 365$ nm
$\lambda_{\text{Paschen}} = 820$ nm

81. $f = (3.2898 \times 10^{15} \text{ Hz})\left(\dfrac{1}{n^2} - \dfrac{1}{m^2}\right)$

83.

n	L_n (J·s)	r_n (m)	K_n (J)
1	1.06×10^{-34}	5.29×10^{-11}	2.18×10^{-18}
2	2.11×10^{-34}	2.11×10^{-10}	5.44×10^{-19}
3	3.16×10^{-34}	4.76×10^{-10}	2.42×10^{-19}
4	4.22×10^{-34}	8.46×10^{-10}	1.36×10^{-19}
5	5.27×10^{-34}	1.32×10^{-9}	8.70×10^{-20}

n	E_n (J)	v_n (m/s)
1	-2.18×10^{-18}	2.19×10^{6}
2	-5.44×10^{-19}	1.09×10^{6}
3	-2.42×10^{-19}	7.29×10^{5}
4	-1.36×10^{-19}	5.47×10^{5}
5	-8.70×10^{-20}	4.38×10^{5}

85. $v = 2.19 \times 10^5$ m/s, $L = 1.055 \times 10^{-33}$ J·s

87. $\Delta f = 0.2$ MHz

89. (a) 486 nm; (b) 1879 nm, 656 nm, 486 nm, 122 nm, 103 nm, and 97.3 nm

91. $E_{10} = -0.0340$ eV
$E_9 = -0.0420$ eV
$E_8 = -0.0531$ eV
$E_7 = -0.0694$ eV
$E_6 = -0.0944$ eV
$E_5 = -0.136$ eV
$E_4 = -0.213$ eV
$E_3 = -0.378$ eV
$E_2 = -0.85$ eV
$E_1 = -3.40$ eV

93. (a) 1.59×10^{15} photons/s; (b) 1.67×10^{-12} kg·m/s

95. $\lambda_f - \lambda_i = \dfrac{h}{m_e c}(1 - \cos\theta)$

97. $\dfrac{\lambda_H}{\lambda_O} = 4$

99. (a) 1.20×10^{29} m; (b) $E = -4.21 \times 10^{-97}$ J

Chapter 27

1. Isotopes are nuclides with the same atomic number, Z, but a different number of neutrons, N. So, they have different mass numbers, $A = Z + N$.

3. Binding energy is a measure of how difficult it is to break the nucleus into its constituent "pieces," similar to the way that the work function measures the difficulty in removing an electron from the surface of a conductor. Also, the binding energy is different for different isotopes, similar to the way that the work function is tabulated for different conductors studied in the photoelectric effect.
The work function is usually a few eV (maybe ~ 10 eV at the maximum), while the binding energy in the nucleus is on the order of MeV. There is really no definitive means of predicting the work function. It is measured empirically, whereas for the binding energy we can use Einstein's equation to calculate the expected values ($E = \Delta mc^2$).

5. The whole nucleus weighs less than the sum of its parts because some of that mass is converted to energy when the parts are fused. In order to break apart the nucleus, you must add energy equivalent to the binding energy, which becomes mass after fission.

7. Certainly, the central ideas of binding energy and "Q values" of nuclear reactions would not have been possible to develop without the $E = mc^2$ concept. The Manhattan project in World War II, where nuclear weapons were designed and successfully tested, would not have been possible had it not been for Einstein's theory of the equivalence of mass and energy (as well as his letter to President Roosevelt endorsing this project in the early 1940s).

9. (a) Elements with higher atomic number than iron are more likely to fission. (b) Elements with lower atomic number than iron are more likely to undergo fusion.

11. Consider the typical beta decay, such as the decay of a neutron: n $\rightarrow$ p $+ \beta^- + \nu_e$. If the electron (β^-) were the only decay product, application of conservation of energy and momentum to the two-body decay would require that the β particle be ejected with a single unique energy. Instead, we observe experimentally that β particles are produced with energies that range from zero to a maximum value. Further, because the original neutron had spin $1/2$, conservation of angular momentum would be violated if the final decay products consisted of only the two particles p and β, each with spin $1/2$.

13. As uranium begins to decay, the daughter nuclei are themselves often radioactive, decaying by various paths (which all lead down to a stable isotope in the periodic table). So, at any instant there are isotopes from all the many possible decay modes of all the radioactive progeny of the parent nuclei.

15. Because the atomic masses include Z electron masses in their values, it is already taken into account. There is 1

electron mass included in the atomic mass of the proton. There is no need to add in another one to account for the beta particle.

17. D

19. E

21. B

23. B

25. C

27. The nuclear force is about 100 times larger than the electrostatic force.

29. 5×10^{17}

31. About 40 times greater. In some models, the force is actually repulsive for small separation distances (< 0.7 fm) and attractive for larger separation distances.

33. 2×10^{16} kg

35. (a) 0.678 days or 16.3 hours; (b) 4 days

37. (a) H, 1 proton and 2 neutrons, mass number 3
(b) Be, 4 protons and 4 neutrons, mass number 8
(c) Al, 13 protons and 13 neutrons, mass number 26
(d) Au, 79 protons and 118 neutrons, mass number 197
(e) Tc, 43 protons and 57 neutrons, mass number 100
(f) W, 74 protons and 110 neutrons, mass number 184
(g) Os, 76 protons and 114 neutrons, mass number 190
(h) Pu, 94 protons and 145 neutrons, mass number 239

39. (a) Hydrogen, 1 proton and 1 neutron, mass number 2
(b) Helium, 2 protons and 2 neutrons, mass number 4
(c) Lithium, 3 protons and 3 neutrons, mass number 6
(d) Carbon, 6 protons and 6 neutrons, mass number 12
(e) Iron, 26 protons and 30 neutrons, mass number 56
(f) Strontium, 38 protons and 52 neutrons, mass number 90
(g) Iodine, 53 protons and 78 neutrons, mass number 131
(h) Uranium, 92 protons and 143 neutrons, mass number 235

41. 12.7 km

43. 92.2 MeV

45. 15.643 MeV

47. 1100 MeV binding energy, 8.42 MeV/nucleon

49. 10n, 89.95 MeV is released

51. (a) n; (b) ^{102}Zr; (c) ^{250}Cm; (d) ^{142}Cs

53. About 33 collisions

55. 366 kg

57. 3.1×10^{19} reactions/second

59. (a) 17.6 MeV; (b) 19.0 MeV; (c) 3.27 MeV; (d) 5.49 MeV; (e) 4.03 MeV

61. 1.56×10^{-9} g

63. (a) 3.70×10^6 Bq; (b) 25 Bq; (c) 4.46×10^{-7} Ci; (c) 4.53×10^{12} cpm

65. 208 s = 0.00240 days

67. 0.298% of the original sample

69. 0.530 μCi

71. 25,740 years

73. 3.85×10^6 nuclei

75. (a) ^{238}U $\rightarrow$ ^{234}Th + α
(b) ^{234}Th $\rightarrow$ ^{230}Ra + α
(c) ^{240}Pu $\rightarrow$ ^{236}U + α
(d) ^{214}Bi $\rightarrow$ ^{210}Tl + α

77. (a) ^{131}I* $\rightarrow$ ^{131}I + γ
(b) ^{145}Pm* $\rightarrow$ ^{145}Pm + γ
(c) ^{24}Na* $\rightarrow$ ^{24}Na + γ

79. (a) 7.4 fm; (b) 1.0 N; (c) 6.2×10^{26} m/s^2; (d) The strong nuclear force holds them together.

81. (a) 36; (b) 36; (c) 37; (d) 37; (e) 39; (f) 38

83. (a) 8.0 fm; (b) 3.1×10^{-20}%

85. (a) 10^{-320}%; (b) $10^{-714.7}$%; (c) Since essentially none of the original Tc would be left at the end of the star's lifetime, it must be produced recently in the star. This means that the star is manufacturing the atom.

87. 31,000 years

89. (a) barium-137; (b) 200 years

91. (a) 1 proton and 2 neutrons; (b) ^{3}H $\rightarrow$ ^{3}He + e$^-$. The daughter is helium-3. (c) 22.4 years

93. (a) 44 protons and 62 neutrons; (b) We would not find any in mined ores. Such ores are many millions of years old (at the least), so if any ruthenium-106 were present initially, it would have decayed to essentially zero. (c) ^{106}Ru $\rightarrow$ ^{106}Rh + e$^-$. The daughter is rhodium-106. (d) 2.05 years

95. The Q value for this forbidden reaction is -0.274 MeV. Since $Q < 0$, this reaction cannot occur. If it did, all low-energy electrons could react with protons and turn into neutrons. This would disrupt all chemical reactions, making life impossible.

97. (a) 6 α and 4 e$^-$ particles are emitted; (b) 18 α and 6 e$^-$ particles are emitted; (c) 9 α and 4 e$^-$ particles are emitted; (d) 8 α and 4 e$^-$ particles are emitted

Chapter 28

1. (a) Baryons are hadrons made up of three quarks. (b) Mesons are hadrons made up of quark–antiquark pairs. (c) Quarks are fractionally charged particles that make up hadronic material (baryons and/or mesons). (d) Leptons are particles that are driven by the electroweak force and mediated with photons and W/Z bosons. e) Antiparticles are the negative-energy counterpart to every particle. All particles possess antiparticles that are equal in mass but opposite in charge.

3. The photon and the neutrino are both neutral in charge and possess no baryon quantum number. There are some early models in which the electron neutrino is massless; in that case, it would travel at the speed of light. These two particles are driven by the electroweak force.

5. uud, udd, ddd, uuu, uus, dds, uds, uss, dss, and sss.

7. If only one photon were created, linear momentum could not be conserved because a zero momentum photon is not possible.

9. Measurements of the orbital speed of spiral galaxies show that the galaxies rotate faster than should be allowed given the amount of visible matter in the galaxy. The mass needed to hold the rotating galaxy together is about six times the total mass of the visible matter in the galaxy. This additional necessary matter is dark matter.

11. B

13. D

15. A

17. D

19. D

21. 430 Mpc

23. No. It is impossible because neither baryon number nor lepton number is conserved.

25. (a) Λ or Σ^0 because these particles have the correct charge and quark configuration.
(b) Ξ_0 because this particle has the correct charge and quark configuration.

27. (a) and (b) are possible, (c) is not possible because baryon number is not conserved.

29. All are allowed.

31. $E_\gamma = 273.5$ MeV, $\lambda = 4.53 \times 10^{-6}$ nm, and $p = 1.46 \times 10^{-19}$ kg$\cdot$m/s for each photon.

33. (a) 1.02 MeV; (b) The nucleus is required to absorb the momentum so that conservation of momentum is not violated.

35. (a) $K_p = 0.312$ GeV $= K_{p^-}$; (b) $K_p = 0.347$ GeV and $K_{p^-} = 0.277$ GeV

37.

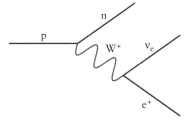

39.

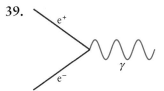

41. $m = 1.87 \times 10^{-3}$ MeV/c^2, so $m_e/m = 270$

43. 1500 km/s

45. (a) Ge; (b) 290 N

47. 620 nm. These photons will appear reddish-orange in color to the human eye.

49. $\pi^+ = (u_R \overline{d_R}, u_B \overline{d_B}, u_G \overline{d_G})$
$\pi^0 = (u_R \overline{u_R}, u_B \overline{u_B}, u_G \overline{u_G})$
$\pi^- = (d_R \overline{u_R}, d_B \overline{u_B}, d_G \overline{u_G})$

51. (a) It is neutral ($Q = 0$). (b) About 2 ps

MCAT® Question Answers

The section that follows includes material from previously administered MCAT® items and is reprinted with permission of the Association of American Medical Colleges (AAMC).

Passage 13 (B)

81. D
82. C
83. C
84. B
85. B

Passage X (G)

134. B
135. D
136. B
137. D
138. D

Passage 12 (A)

73. B
74. B
75. A
76. D
77. A
78. C
79. C
80. A

Passage 7

41. C
42. D
43. A
44. C
45. B
46. C
47. A
48. B

Passage 1

1. D
2. A
3. C
4. C
5. B
6. B

Passage 2

7. D
8. B
9. D
10. C
11. A
12. D
13. A

Passage II (E)

74. A
75. B
76. C
77. D
78. D
79. A
80. B

Passage 4 (C)

19. B
20. B
21. D
22. C
23. A
24. D
25. B
26. C

Passage IV (F)

94. C
95. C
96. A
97. B
98. D
99. A

Credits

Photo Credits

Cover: Kim Taylor/Hotspot Media   **p. 673:** Exactostock/
SuperStock   **p. 674 (left):** Biophoto Associates/Science Source
p. 674 (right): KingWu/iStockphoto   **p. 676:** Image Source/
Getty Images   **p. 681:** Kung-Ming Jan and Shu Chien, 1973
Rockefeller University Press. Originally published in The
Journal of General Physiology. 61:638–654.   **p. 689:** NOAA
p. 694: Reprinted by permission from Macmillan Publishers
Ltd., Shinoda, T., Ogawa, H., Cornelius, F., Toyoshima, C.,
Research Collaboratory for Structural Bioinformatics, Crystal
structure of the sodium-potassium pump at 2.4 A resolution.
Nature (2009) 459: 446-50, copyright 2009.   **p. 712:**
muratseyit/iStockphoto   **p. 713 (left):** iStockphoto/Thinkstock
p. 713 (right): NHPA/SuperStock   **p. 734:** iStockphoto/
Thinkstock   **p. 735:** NASA   **p. 758:** Charles D. Winters/
Science Source   **p. 759 (left):** iStockphoto/Thinkstock
p. 759 (center): Stockbyte/Thinkstock   **p. 759 (right):** Asia
Images Group Pte Ltd/Alamy   **p. 770:** David Tauck   **p. 794:**
Biophoto Associates/Science Source   **p. 804:** iStockphoto/
Thinkstock   **p. 805 (left):** GIPHOTOSTOCK/Science Photo
Library   **p. 805 (center):** Todd Ruskell   **p. 805 (right):**
Cordelia Molloy/Science Photo Library   **p. 805 (bottom):**
Alchemy/Alamy   **p. 806:** Eli Sidman, Technical Services
Group, MIT.   **p. 815 (left):** Jupiterimages/Thinkstock
p. 815 (right): Christina Micek   **p. 819 (left):** iStockphoto/
Thinkstock   **p. 819 (right):** iStockphoto/Thinkstock   **p. 823:**
Elekta   **p. 825:** Arno Massee/Science Photo Library   **p. 828
(top):** Paul Silverman/Fundamental Photographs   **p. 828
(bottom):** David Tauck   **p. 846:** iStockphoto/Thinkstock
p. 847 (left): Digital Vision/Thinkstock   **p. 847 (right):** "Keck,
et. al. ""Transcranial magnetic stimulation as a therapeutic
tool in psychiatry: what do we know about the neurobiological
mechanisms?" Journal of Psychiatric Research 35 (2001)
193–215   **p. 867:** iStockphoto/Thinkstock   **p. 868 (left):**
iStockphoto/Thinkstock   **p. 868 (right):** iStockphoto/
Thinkstock   **p. 872:** Peter & Georgina Bowater/Aurora
Photos   **p. 880:** Todd Ruskell   **p. 905:** iStockphoto/
Thinkstock   **p. 906 (left):** S.Kafka and K.Honeycutt, Indiana
University/WIYN/NOAO/NSF   **p. 906 (right):** Will & Deni
McIntyre/Science Photo Library/Science Source   **p. 907:**
Siebeck, et. al. "A Species of Reef Fish that Uses Ultraviolet
Patterns for Covert Face Recognition" Current Biology 20
(March 9, 2010) 407–410.   **p. 935:** Medioimages/Photodisc/
Thinkstock   **p. 936 (left):** David Tauck   **p. 936 (center):**
iStockphoto/Thinkstock   **p. 936 (right):** Carol and Mike
Werner/Phototake, Inc.   **p. 936 (bottom):** George Resch/
Fundamental Photographs   **p. 937:** iStockphoto/Thinkstock
p. 942: David Tauck   **p. 943:** GIPhotoStock/Science Source
p. 944: GIPhotoStock/Photo Researchers/Getty Images
p. 945: Kent Wood/Science Source   **p. 946:** David Tauck
p. 947: GIPhotoStock/Science Source   **p. 952:** David Tauck
p. 955: Michael Gray/Dreamstime.com   **p. 962 (left):** George
Resch/Fundamental Photographs   **p. 962 (right):** George
Resch/Fundamental Photographs   **p. 962 (bottom):** Edward
Kinsman/Science Photo Library/Science Source   **p. 963:**
Edward Kinsman/Science Photo Library/Science Source
p. 966 (left): Richard Megna/Fundamental Photographs
p. 966 (center): Richard Megna/Fundamental Photographs
p. 966 (right): Richard Megna/Fundamental Photographs
p. 966 (bottom): sciencephotos/Alamy   **p. 969:** NASA, ESA,
and STScI   **p. 980:** Stockbyte/Thinkstock   **p. 981 (right):**
Syracuse Newspapers/Blume/The Image Works   **p. 981 (left):**
BananaStock/Thinkstock   **p. 985:** David Tauck   **p. 985
(top):** David Tauck   **p. 985 (bottom):** David Tauck   **p. 987
(top):** David Tauck   **p. 987 (bottom):** David Tauck   **p. 988
(top):** David Tauck   **p. 988 (middle):** David Tauck   **p. 988
(bottom):** David Tauck   **p. 995:** Melissa Gaskell/Alamy
p. 1002 (left): Monika Graff/The Image Works   **p. 1002
(right):** Rachael Lynn Beaton/University of Virginia   **p. 1005:**
Derrick Alderman/Alamy   **p. 1024:** Pixtal/SuperStock
p. 1025 (left): iStockphoto/Thinkstock   **p. 1025 (right):**
Comstock/Thinkstock   **p. 1025 (bottom):** © Eitan Simanor/
Alamy   **p. 1026:** OJO Images/SuperStock   **p. 1036:** Sir John
Sulston   **p. 1053:** © 2011 CERN, for the benefit of the CMS
Collaboration   **p. 1057 (left):** NASA   **p. 1057 (right):** NASA,
ESA, A. Bolton (Harvard-Smithsonian CfA) and the SLACS
Team   **p. 1066:** T.A. Rector (University of Alaska Anchorage),
Richard Cool (University of Arizona) and WIYN/NOAO
p. 1067 (left): U.S. Air Force /Staff Sgt. Jacob Bragg/Alamy
p. 1067 (right): iStockphoto/Thinkstock   **p. 1067 (bottom):**
Dr. Adam P. Hitchcock, McMaster University   **p. 1071 (left):**
Richard Megna/Fundamental Photographs   **p. 1071 (center):**
Richard Megna/Fundamental Photographs   **p. 1071 (right):**
Richard Megna/Fundamental Photographs   **p. 1072:**
iStockphoto/Thinkstock   **p. 1081:** Frederick Murphy/CDC
p. 1109: Stephane Marc/Maxppp/Zuma Press/Newscom
p. 1110 (left): Dr. Leon Kaufman. University Of California, San
Francisco   **p. 1110 (center):** Brand X Pictures/Thinkstock
p. 1110 (right): iStockphoto/Thinkstock   **p. 1115:** Science
Source   **p. 1125:** C. F. Claver/WIYN/NOAO/NSF   **p. 1146:**
CERN   **p. 1147 (left):** iStockphoto/Thinkstock   **p. 1147
(right):** Hemera/Thinkstock   **p. 1148:** Stanford Linear
Accelerator Center/Science Source   **p. 1159:** NASA, ESA, R.
Ellis (Caltech), and the UDF 2012 Team   **p. 1162:** ESA and
the Planck Collaboration   **p. 1163 (top):** NASA, ESA, and the
Hubble Heritage Team (STScI/AURA)   **p. 1163 (bottom):**
C. Howk (JHU), B. Savage (U. Wisconsin), N.A.Sharp
(NOAO)/WIYN/NOAO/NSF

Illustration and Table Credits

Figure 23-19: Adapted from OpenStax College, Polarization,
Connexions, May 25, 2012, Fig. 11, http://cnx.org/content/
m42522/1.2/.

INDEX

and Maxwell's equation, 911–912
and water/electric flux, 694–695
Gauss's law for the magnetic field, 912–913
Geckos, wall-climbing ability of, 159–160
Geiger, Hans, 1083
Gell-Mann, Murray, 1148
General theory of relativity, 1033, 1055–1058
Generations of quarks, 1148
Generators, 847, 858–860, 868
Genetic fingerprinting, 688
Geometrical center of system, 275
Geometrical optics, 980–1017
  concave mirrors, 984–995
  convex lenses, 1002–1015
  convex mirrors, 995–1007
  defined, 980
  images in mirrors/lenses, 980–981
  plane mirrors, 981–984
Geometry, M10–M11
Geostationary satellites, 404–405
Germanium, A5, A6, A11
Germer, Lester, 1081
Glare reduction, 952–953, 959
Glass, curvature in mirrors vs., 1004
Global Positioning System (GPS), 401, 1025
Global warming, 614
Gluons, 1155–1157
Gold, A5, A6, A16
GPS (Global Positioning System), 401, 1025
Gram (unit), 4
Gravitation, 381–415
  and apparent weightlessness, 411–413
  importance of, 381–382
  and orbit of Moon, 382–392
  and orbits of planets/satellites, 400–411
  universal, 381–392, 400–411
Gravitational constant, 385, 387–388, 681
Gravitational field, 686
Gravitational force, 118
  as conservative force, 392–393
  and distance from Earth's surface, 398
  in free-body diagrams, 129–131
  as fundamental force, 1158
  and weight, 125, 127
  work done by, 201, 713–714
Gravitational lensing, 1057
Gravitational potential energy, 219–220, 392–400
  attributes of, 394–395
  and conservation of total mechanical energy, 395–397
  and distance from Earth's surface, 392–394
  and escape speed, 397–400
  for satellite in orbit, 403
  and work done by gravitational force, 713–714
Gravitons, 1158
Gravity. *See also* Acceleration due to gravity
  deflection of light by, 1056–1058
  and general theory of relativity, 1055–1058
Greenhouse effect, 613–614
Greenhouse gases, 613–614
Ground terminal, 765
Guitar, 547–549, 556

H

Hadrons, 1148, 1150–1151, 1156
Hafnium, A5, A6, A16
Hagen-Poiseuille equation, 465–466
Hair dryers, traveling with, 871

Hair growth, 8–9
Half-life, 1126–1130
Halogens, 1099
Harmonic property, 489
Harmonics, 547–548, 552, 554
Hassium, A5, A6, A19
Hawks
  hunting by, 244
  inelastic collision of pigeon and, 263–264
  two-dimensional motion of, 81–83
Headlights, in cars, 985, 988
Hearing, 536, 559
Heart
  angle of, 79–80
  and blood pressure, 435–436
  magnetic fields in, 808, 809
Heat, 599–604, 612–618
  and adiabatic processes, 637–638
  conduction, 612, 616–618
  convection, 612, 615–616
  in first law of thermodynamics, 629–632
  in heat engine, 649
  internal, 600
  latent, 605–610
  molar specific, 642–645
  and phase changes, 600, 605–606
  radiation, 612–614
  in reversible processes, 652
  specific, 600–601, 641–647
  and temperature, 599–604
  and temperature change in ideal gases, 641–647
  temperature vs., 600, 638
Heat death of universe, 665
Heat engines
  efficiency of, 628–629, 649–651
  perfect, 649
  reservoir for, 649
  and second law of thermodynamics, 649–652
Heated objects, radiation from, 1071
Heating, isochoric, 638
Height, gravitational potential energy and, 221
Heisenberg, Werner, 1154
Heisenberg uncertainty principle, 1152–1154
Helium, 1117, A5–A7
Helium-4 (^{4}He)
  binding energy of, 1118–1119
  formation of, by fusion, 1124–1125
Henry, 877, 878
Hertz (unit), 483
Hertz, Gustav, 1096
Holes, semiconductor, 896
Holes, thermal expansion of, 598
Hollow cylinder, moment of inertia for, 305
Holmium, A5, A6, A15
Honeybees, navigation by, 73, 950, 951
Hooke's law, 215–219
  for compressed spring, 217–219
  for elastic materials, 371
  limitations of, 359
  for pendulum, 503
  and restoring force, 486–496
  for simple harmonic motion, 489, 492
  for stretched spring, 215–217
  for tensile/compressive stress, 355–357
  and uniform circular motion, 486–488
  for volume stress, 361–363
Horizontal motion, graphing, 26
Horsepower (unit), 557

Physical Constants*

atomic mass constant	$m_u = \frac{1}{12}m(^{12}C)$	$1\ u = 1.660\ 538\ 921(73) \times 10^{-27}\ kg$
avogadro's constant	N_A	$6.022\ 141\ 29(27) \times 10^{23}$ particles/mol
Boltzmann constant	$k = R/N_A$	$1.380\ 648\ 8(13) \times 10^{-23}\ J/K = 8.617\ 332\ 4(81) \times 10^{-5}\ eV/K$
Bohr magneton	$m_B = e\hbar/(2m_e)$	$9.274\ 009\ 68(20) \times 10^{-24}\ J/T = 5.788\ 381\ 806\ 6(38) \times 10^{-5}\ eV/T$
Coulomb constant	$k = 1/(4\pi\epsilon_0)$	$8.987\ 551\ 787\ldots \times 10^9\ N \cdot m^2/C^2$
Compton wavelength	$\lambda_C = h/(m_e c)$	$2.426\ 310\ 238\ 9(16) \times 10^{-12}\ m$
fundamental charge	e	$1.602\ 176\ 565(35) \times 10^{-19}\ C$
gas constant	R	$8.314\ 462\ 1(75)\ J/(mol \cdot K) = 1.987\ 206\ 5(36)\ cal/(mol \cdot K)$ $= 8.205\ 746(15) \times 10^{-2}\ L \cdot atm/(mol \cdot K)$
universal constant of gravitation	G	$6.673\ 84(80) \times 10^{-11}\ N \cdot m^2/kg^2$
mass of electron	m_e	$9.109\ 382\ 91(40) \times 10^{-31}\ kg = 0.510\ 998\ 928(11)\ MeV/c^2$
mass of proton	m_p	$1.672\ 621\ 777(74) \times 10^{-27}\ kg = 938.272\ 046(21)\ MeV/c^2$
mass of neutron	m_n	$1.674\ 927\ 351(74) \times 10^{-27}\ kg = 939.565\ 379(21)\ MeV/c^2$
permeability of free space	μ_0	$4\pi \times 10^{-7}\ N/A^2$
permittivity of free space	$\epsilon_0 = 1/(\mu_0 c^2)$	$8.854\ 187\ 817\ldots \times 10^{-12}\ C^2/(N \cdot m^2)$
Planck's constant	h	$6.626\ 069\ 57(29) \times 10^{-34}\ J \cdot s = 4.135\ 667\ 516(91) \times 10^{-15}\ eV \cdot s$
	$\hbar = h/(2\pi)$	$1.054\ 571\ 726(47) \times 10^{-34}\ J \cdot s = 6.582\ 119\ 28(15) \times 10^{-16}\ eV \cdot s$
speed of light	c	$2.997\ 924\ 58 \times 10^8\ m/s$
Stefan–boltzmann constant	σ	$5.670\ 373(21) \times 10^{-8}\ W/(m^2 \cdot K^4)$

* The values for these and other constants can be found in Appendix B as well as on the Internet at http://physics.nist.gov/cuu/Constants/index.html where you may find updated values for the constants. The numbers in parentheses represent the uncertainties in the last two digits. (For example, 2.044 43(13) stands for 2.044 43 ± 0.000 13.) Values without uncertainties are exact. Values with ellipses are exact (like the number $\pi = 3.1415\ldots$), but are not completely specified.

Vector Products

$$\vec{A} \cdot \vec{B} = AB \cos\theta \qquad |\vec{A} \times \vec{B}| = AB \sin\theta$$

Quadratic Formula

If $ax^2 + bx + c = 0$, then $x = \dfrac{-b \pm \sqrt{b^2 - 4ac}}{2a}$

Geometry and Trigonometry

$C = \pi d = 2\pi r$	definition of π
$A = \pi r^2$	area of circle
$V = \frac{4}{3}\pi r^3$	spherical volume
$A = 4\pi r^2$	spherical surface area
$V = A_{base}L = \pi r^2 L$	cylindrical volume
$A = 2\pi rL$	cylindrical surface area

$o = h\sin\theta$
$a = h\cos\theta$

$\sin^2\theta + \cos^2\theta = 1$
$\sin(A \pm B) = \sin A \cos B \pm \cos A \sin B$
$\cos(A \pm B) = \cos A \cos B \mp \sin A \sin B$
$\sin A \pm \sin B = 2\sin[\frac{1}{2}(A \pm B)]\cos[\frac{1}{2}(A \mp B)]$

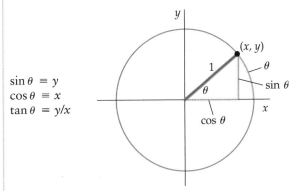

$\sin\theta \equiv y$
$\cos\theta \equiv x$
$\tan\theta \equiv y/x$

If $|\theta| \ll 1$, then $\cos\theta \approx 1$ and $\tan\theta \approx \sin\theta \approx \theta$ (θ in radians)